Photophysics and Photochemistry in the Vacuum Ultraviolet

NATO ASI Series

Advanced Science Institutes Series

A series presenting the results of activities sponsored by the NATO Science Committee, which aims at the dissemination of advanced scientific and technological knowledge, with a view to strengthening links between scientific communities.

The series is published by an international board of publishers in conjunction with the NATO Scientific Affairs Division

A	Life Sciences	Plenum Publishing Corporation
B	Physics	London and New York
C	Mathematical and Physical Sciences	D. Reidel Publishing Company Dordrecht, Boston and Lancaster
D	Behavioural and Social Sciences	Martinus Nijhoff Publishers
E	Engineering and Materials Sciences	The Hague, Boston and Lancaster
F	Computer and Systems Sciences	Springer-Verlag
G	Ecological Sciences	Berlin, Heidelberg, New York and Tokyo

Series C: Mathematical and Physical Sciences Vol. 142

Photophysics and Photochemistry in the Vacuum Ultraviolet

edited by

S. P. McGlynn

Department of Chemistry, Louisiana State University,
Baton Rouge, Louisiana, U.S.A.

G. L. Findley

Department of Chemistry,
New York University, New York, NY, U.S.A.

and

R. H. Huebner

Argonne National Laboratory,
Argonne, Illinois, U.S.A.

D. Reidel Publishing Company

Dordrecht / Boston / Lancaster

Published in cooperation with NATO Scientific Affairs Division

Proceedings of the NATO Advanced Study Institute on
Photophysics and Photochemistry in the Vacuum Ultraviolet
Lake Geneva, Wisconsin, U.S.A.
August 15-28, 1982

Library of Congress Cataloging in Publication Data

NATO Advanced Study Institute on Photophysics and Photochemistry in the Vacuum Ultraviolet (1982 : Lake Geneva, Wis.)
Photophysics and photochemistry in the vacuum ultraviolet.

(NATO ASI series. Series C, Mathematical and physical sciences; vol. 142)
"Proceedings of the NATO Advanced Study Institute on Photophysics and Photochemistry in the Vacuum Ultraviolet, Lake Geneva, Wisconsin, U.S.A., August, 15-28, 1982"—T.p. verso.
"Published in cooperation with NATO Scientific Affairs Division."
Bibliography: p.
Includes index.
1. Vacuum ultraviolet spectroscopy—Congresses. 2. Atomic spectroscopy—Congresses. 3. Molecular spectroscopy—Congresses. 4. Photochemistry—Congresses. 5. Multiphoton processes—Congresses. 6. Chemistry, Physical and theoretical—Congresses. I. McGlynn, S. P., 1931- . II. Findley, G. L., 1952- . III. Huebner, R. H., 1931- . IV. North Atlantic Treaty Organization. Scientific Affairs Division. V. Title. VI. Series: NATO ASI series. Series C, Mathematical and physical sciences; no. 142.
QC459.N38 1982 535.8'44 84-18238

ISBN-13: 978-94-010-8827-5 e-ISBN-13: 978-94-009-5269-0
DOI: 10.1007/978-94-009-5269-0

Published by D. Reidel Publishing Company
P.O. Box 17, 3300 AA Dordrecht, Holland

Sold and distributed in the U.S.A. and Canada
by Kluwer Academic Publishers,
190 Old Derby Street, Hingham, MA 02043, U.S.A.

In all other countries, sold and distributed
by Kluwer Academic Publishers Group,
P.O. Box 322, 3300 AH Dordrecht, Holland

D. Reidel Publishing Company is a member of the Kluwer Academic Publishers Group

TABLE OF CONTENTS

PREFACE xv

PHOTOPHYSICS OF HIGHLY-EXCITED STATES 1

G. L. Findley, J. A. Wilder, P. Hochmann and S. P. McGlynn

1. Introduction 1
2. Quantum Defect Theory 8
3. Core Effects 14
4. Intermolecular Rydberg Correlations 22
5. Field Effects 27
6. Rydberg States in Condensed Media 33
7. Concluding Remarks 37

METHODS FOR STUDYING HIGHER EXCITED STATES IN MOLECULES AND MOLECULAR CRYSTALS BY MEANS OF SYNCHROTRON RADIATION 41

Ernst-Eckhard Koch and Peter Gürtler

1. Introduction 41
2. Properties of Synchrotron Radiation Sources: A Short Overview 46
3. Some General Aspects of Synchrotron Radiation Instrumentation 54
4. Absorption and Reflection Spectroscopy 57
5. Time-Resolved and Energy-Resolved VUV Luminescence Spectroscopy 67
6. Photoemission 72
7. Perspectives 89
8. Appendix 92

NONLINEAR OPTICS AND LASER SPECTROSCOPY IN THE VACUUM ULTRAVIOLET 105

Stephen C. Wallace

1. Introduction 105
2. Vacuum Ultraviolet Lasers 106
3. Nonlinear Optical Methods for Producing Tunable VUV Radiation 110
4. Applications to Chemical Physics 121
5. Future Goals 127

MULTIPHOTON IONIZATION AND THIRD-HARMONIC GENERATION IN ATOMS AND MOLECULES 133

John C. Miller and R. N. Compton

1. Introduction 133
2. Resonantly-Enhanced Multiphoton Ionization 134
3. Multiphoton Ionization and Third-Harmonic Generation in Rare Gases 145

ELECTRON-IMPACT SPECTROSCOPY OF MOLECULES 163

H. H. Brongersma, H. A. van Sprang and G. J. Verhaart

1. Introduction 163
2. General Theory 164
3. Experimental 170
4. Experimental Results 173

ELEMENTS OF QUANTUM DEFECT THEORY. I. INTRODUCTION AND FORMALISM 191

A. R. P. Rau

1. Introduction to Quantum Defect Theory 191
2. Multichannel Quantum Defect Theory in Atoms 209

ELEMENTS OF QUANTUM DEFECT THEORY. II. A UNIFIED THEORY OF RYDBERG AND AUTOIONIZING STATES 217

K. T. Lu

1. Introduction 217
2. Eigenchannel Quantum Defect Theory 221
3. *Ab initio* Calculations of Quantum Defect Parameters 234
4. Applications 235

ELEMENTS OF QUANTUM DEFECT THEORY. III. DIATOMIC MOLECULES 245

Chris H. Greene

1. Introduction to Molecular QDT 245
2. Comparison with Experiment 251
3. Inclusion of Molecular Dissociation 253

NEGATIVE-ION STATES 261

R. N. Compton

1. Introduction 261
2. Negative-Ion Properties 262
3. Unimolecular Electron Attachment 265
4. Multiply-Charged Negative Ions 270
5. Photoabsorption by Negative Ions 271
6. Recent Studies of Long-Lived He Negative Ions 274
7. Production of Negative Ions by Photon Reactions 279
8. Collision of High-Rydberg Atoms with Electron-Attaching Molecules and the Possible Influence of Dipole Bound States 281
9. Negative Ions of Molecular Clusters 286

SUPEREXCITED AND IONIC STATE RELAXATION PROCESSES IN VACUUM ULTRAVIOLET EXCITED POLYATOMIC MOLECULES 297

Sydney Leach

1. Introduction 297

2. Fundamental Processes 298
3. Competing Decay Channels of Superexcited States 305
4. Methods for Studying Superexcited and Ionic States and Their Relaxation Processes 310
5. Further Remarks on Molecular Autoionization 317
6. Decay of Ion States 321

PHOTOIONIZATION DYNAMICS OF SMALL MOLECULES 341

J. L. Dehmer, Dan Dill and A. C. Parr

1. Introduction 341
2. Shape Resonances in Molecular Photoionization 343
3. Molecular Autoionization 377
4. Concluding Remarks 396

PHOTODISSOCIATION DYNAMICS OF GAS-PHASE SMALL MOLECULES 409

M. N. R. Ashfold

1. Introduction 409
2. ICN Photodissociation in the Near Ultraviolet 413
3. VUV Photodissociation of HCN 426
4. VUV Photodissociation of the Cyanogen Halides 443
5. Conclusions 457

VUV SPECTROSCOPY OF RARE-GAS VAN DER WAALS DIMERS 467

P. M. Dehmer and S. T. Pratt

1. Introduction 467
2. General Background 468
3. Experimental Apparatus 471
4. Dependence of the Cluster Ion Signals on Nozzle Stagnation Pressure 474
5. Appearance Potentials of the Dimer Ions and Dissociation Energies of the Dimer-Ion Ground States 475

6. Molecular Rydberg Structure Between the Atomic Fine-Structure Thresholds: Dissociation Energies of $X_2^+(C\ ^2\Pi_{1/2u})$ and $XY^+(B\ ^2\Pi_{1/2}$ and $D\ ^2\Sigma^+_{1/2})$ 482
7. Molecular Rydberg Structure in the Spectra of NeY Near the Ne 3s Atomic Resonance 493
8. Molecular Rydberg Structure in the Spectra of ArKr and ArXe Near the Ar 5s and 3d Atomic Resonances 500
9. Molecular Rydberg Structure in the Spectra of X_2 and XY Near the High-Lying Atomic Rydberg States 502
10. Summary and Conclusions 510

EXCITONS AND ENERGY TRANSFER IN INSULATORS 515

J. Jortner, E. E. Koch and N. Schwentner

1. Prologue 515
2. Frenkel Excitons 517
3. Wannier Excitons 525
4. Deep Wannier Atomic Impurity States 534
5. Excitons and Impurity States in Rare-Gas Fluids 540
6. Molecular Wannier Impurity States in Rare-Gas Solids 545
7. Excited-State Dynamics in Rare-Gas Solids 551
8. Self-Trapping of Excitons in Rare-Gas Solids 552
9. Lattice Relaxation Around Impurity States 558
10. Electronic Relaxation of Impurity States 564
11. Electronic Energy Transfer in Rare-Gas Solids 567
12. Perspectives 573

REGULAR AND IRREGULAR MOTION IN CLASSICAL AND QUANTUM SYSTEMS 579

Marko Robnik

1. Introduction: Order and Chaos 579
2. Integrable Systems 581
3. Perturbation of Integrable Systems and the KAM-Theorem 587
4. Area-Preserving Mappings 591
5. Chaotic Systems 597

6. A Generic System is Neither Integrable Nor Ergodic, But Shows a Stochastic Transition 601
7. The Birkhoff-Gustavson Normal Form and Invariant Tori Below the Critical Energy 604
8. Do Quantum Systems Exhibit Chaotic Behavior? 606
9. Integrable Systems and Tori Quantization 608
10. Quantization of Chaotic Systems 612
11. Statistics of Energy Levels 618
12. Quantizing a Generic System 623
13. Conclusions and Discussion 625

NEW TRENDS IN ATOMIC DIAMAGNETISM 631

J. C. Gay

1. Introduction 631
2. The Coulomb Spectrum 633
3. The Landau Spectrum 638
4. The Atom in a Magnetic Field 650
5. Classical Mechanics of Diamagnetism 654
6. Quantum Mechanical Approach 670
7. Semi-Classical Methods 686
8. Experimental Studies in Atomic Physics 691
9. Prospects and Conclusions 700

EVIDENCE FOR MOTIONAL ELECTRIC FIELD EFFECTS IN THE ABSORPTION SPECTRUM OF LITHIUM VAPOR IN A MAGNETIC FIELD 707

H. M. Crosswhite, F. S. Tomkins, H. Crosswhite and W. R. S. Garton

1. Introduction 707
2. Results and Discussion 708

ON THE SPECTRUM OF $V(\rho) = -\frac{qt}{\rho} + \frac{rt}{\rho^2} + st\rho + t\rho^2$. PHYSICAL IMPLICATIONS FOR A VARIETY OF PROBLEMS 721

J. A. C. Gallas and R. F. O'Connell

1. Introduction 721
2. Solution of the Eigenvalue Problem 722
3. Applications 724

4. Summary 727

PERTURBATION SPECTROSCOPY 729

John D. Scott

1. Introduction 729
2. Perturbation Theory and MCD Spectroscopy 736
3. Electric Field-Effect Spectroscopy 756
4. Pressure-Effect Spectroscopy 760

CIRCULAR DICHROISM AND MAGNETIC CIRCULAR DICHROISM STUDIES IN THE VACUUM ULTRAVIOLET 765

A. Gedanken

1. Introduction 765
2. Instrumentation 769
3. Results 770

ELECTRIC FIELD STUDIES IN THE VACUUM ULTRAVIOLET 783

Daniel D. Altenloh and B. R. Russell

1. Introduction 783
2. Experimental 784
3. General Results 786
4. Excited-State Parameters 791
5. Arc-Suppressor "Solvent" Effects 807
6. Conclusions 816

VALENCE-SHELL AND RYDBERG TRANSITIONS IN LARGE MOLECULES 819

C. Sandorfy and L. S. Lussier

1. Introduction 819
2. The Rydberg Concept 820
3. From Spectra to Photochemistry 826
4. Rydberg States of Transition Metal Complexes 832

MULTIPHOTON SPECTROSCOPY AND PHOTOCHEMISTRY 841

M. B. Robin

1. Introduction 841
2. Nonlinear Photochemistry 843
3. Two Examples 845
4. Other Systems 848

EXCESS ENERGY DEPENDENCE OF VIBRATIONAL RELAXATION AND PHOTOPHYSICAL BRANCHING RATIOS IN ISOLATED AROMATIC MOLECULES: RELEVANCE TO VACUUM ULTRAVIOLET PHOTOCHEMISTRY 855

E. C. Lim

1. Introduction 855
2. Spectral Characteristics of Gas-Phase Fluorescence 856
3. Excess-Energy Dependence of Radiationless Transitions: Theory 859
4. Excess-Energy Dependence of Radiationless Transitions: Experiment 863
5. Intramolecular Vibrational Relaxation 867
6. Photochemical Implications of the Energy Dependence of Electronic Relaxation Processes 870

THEORETICAL STUDIES OF THE ELECTRONIC STRUCTURE AND SPECTRA OF NH_3^+ 875

Angelo R. Rossi and Phaedon Avouris

1. Introduction 875
2. Franck-Condon Analysis of the $^2A_1(NH_3^+) \leftarrow {}^1A_1(NH_3)$ Spectrum 878
3. *Ab initio* Calculations on Low-Lying States of NH_3^+ 881

THEORETICAL CORRELATIONS OF ORGANIC PHOTOCHEMICAL REACTIONS IN THE VUV 895
Angelo R. Rossi and R. Srinivasan

1. Introduction 895
2. 1,4-Cyclohexadiene 897
3. Methylene Cyclopropane 903

PHOTOCHEMISTRY OF SATURATED ALCOHOLS AND OPEN-CHAIN ETHERS AT 185 nm IN THE LIQUID PHASE 913

Clemens von Sonntag

1. Introduction 913
2. Discussion 913

PHOTOLYSIS OF CYCLIC ETHERS AND ACETALS AT 185 nm IN THE LIQUID PHASE 927

Clemens von Sonntag

1. Discussion 927

LIST OF PARTICIPANTS 945

SUBJECT INDEX 951

PREFACE

This volume contains the texts of the Invited Lectures presented at the NATO Advanced Study Institute PHOTOPHYSICS AND PHOTOCHEMISTRY IN THE VACUUM ULTRAVIOLET, which was held at the Interlaken Lodge in Lake Geneva, Wisconsin USA, August 15-28, 1982. This ASI was the third in a series of NATO Institutes devoted to a study of the physical and chemical transformations undergone by atoms and molecules after absorption of high-energy (VUV $\gtrsim$ 6 eV) radiation. [The previous Institutes were CHEMICAL SPECTROSCOPY AND PHOTOCHEMISTRY IN THE VACUUM ULTRAVIOLET (Val Morin, Canada, 1973) and HIGH-ENERGY CHEMICAL SPECTROSCOPY AND PHOTOCHEMISTRY (Breukelen, The Netherlands, 1976).]

The purpose of this ASI was to provide a forum for the exchange of attitudes and viewpoints between physicists and chemists working in the VUV area. This area represents one of the most fertile regions for chemistry/physics interfacing, yet the chemists remain largely unaware of the fact that the physicists have implicitly solved many of the chemists' problems (and vice versa); and the physicists remain largely unaware of the nature of the chemists' problems (and vice versa). Consequently, a major aim of the Institute was to construct a proper Chemistry/Physics Interface, to familiarize each group with the techniques, both experimental and theoretical, of the other group, and to bring the major successes and failures of each group, within the VUV area, to light.

The above stipulations concerning the Institute also serve as desiderata for the present volume. The individual authors, as well as the editors, have attempted to construct a book which is pedagogically useful, both to chemists and to physicists, but which still serves as an introduction to state-of-the-art research on highly-excited states of atoms and molecules. Only the reader, however, can gauge our success.

* * * * * * * *

It remains to us the pleasant task of thanking those organizations and individuals who made the Institute, and this volume, possible.

<u>Organizations</u>: NATO, of course, provided the basic grant for the Institute, and to the NATO Scientific Committee in general, but to Dr. Mario di Lullo in particular, we extend our thanks.

Generous additional travel support for the American participants was provided by the United States Department of Energy.

Dr. Frank P. Hudson arranged for this support and to him we express our gratitude.

Our respective institutions -- Louisiana State University, New York University and Argonne National Laboratory -- provided organizational and clerical support, as well as travel support for us and for some of the participants. To the various Deans, Chairmen, Directors, etc. who were involved, we extend our thanks.

Finally, we wish to thank the following companies, each of which contributed financially to the Institute as a Corporate Sponsor:

Schoeffel/McPherson Instrument Co.
Lambda Physik
Instruments SA (Jobin-Yvon)
Spectra Gases, Inc.
Oriel Corporation
EMR Photoelectric
Quanta-Ray
Lumonics Inc.
New England Nuclear

Individuals: Ms. Miriam L. Holden (Director, ANL Conference Planning and Management) arranged for the meeting site and provided much needed expertise concerning the practical aspects of hosting a conference; Ms. Sue Scott (LSU) served as the Institute secretary. Both of these persons contributed greatly to the smooth running of the Institute.

Dr. W. S. Felps (LSU) and Dr. D. Kumar (LSU) generously consented to manage the audio-visual equipment for all of the Lectures.

Ms. Marie M. Wittig (Interlaken Lodge) was very gracious in sorting out who owed what to whom after the Institute.

Ms. Pat Smith (NYU), Ms. Lee France (NYU) and Ms. Lisa Carswell (NYU) contributed to the physical production of the book either by typing or by helping to prepare the index.

Finally, we reserve special thanks for Dr. A. M. Findley (NYU), who participated in every stage of the editorial process; indeed, without this participation the present volume could not have been produced.

S. P. McGlynn
Louisiana State University
Baton Rouge, Louisiana

G. L. Findley
New York University
New York, New York

R. H. Huebner
Argonne National Laboratory
Argonne, Illinois

PHOTOPHYSICS OF HIGHLY-EXCITED STATES

G. L. Findley[a], J. A. Wilder[a], P. Hochmann[b] and S. P. McGlynn[c]

[a]Department of Chemistry
New York University
New York, NY 10003, USA

[b]Department of Chemistry
University of Texas at San Antonio
San Antonio, TX 78285, USA

[c]Department of Chemistry
Louisiana State University
Baton Rouge, LA 70803, USA

1. INTRODUCTION

Photophysics and photochemistry in the vacuum ultraviolet (VUV; energies greater than ∿6 eV) is dominated, to one extent or another, by the presence of highly-excited (Rydberg, or *R*) states. Indeed, a large majority of questions in modern atomic and molecular physics impinge on the structure of Rydberg states and their continuum counterparts, questions as diverse as, for example, the nature of electron transfer in gaseous and condensed phases; the effects of electric and magnetic fields on highly-excited states; field-ionization of *R* states as a means of constructing weak-field detectors; the use of highly-excited states as tunable microwave sensors and in the construction of tunable microwave lasers; etc. This listing could obviously be extended; and, in fact, it is extended -- both in breadth and in depth -- in the following chapters in this volume.

The purpose of this introductory chapter is two-fold:

--- To provide a general overview of the photophysics and,

S. P. McGlynn et al. (eds.), Photophysics and Photochemistry in the Vacuum Ultraviolet, 1–40.

to a more limited extent, the photochemistry of highly-excited states. This overview is, of necessity, a personal one; it stresses those topics with which we have been engrossed for the past ten or so years.

--- To orient the reader with respect to the following chapters in this volume. (Reference to these chapters will be made by placing the authors' names in square brackets.)

1.1 Qualitative Motivation: What is a Rydberg?

Notwithstanding the above comments concerning the importance of highly-excited states, it is only recently (within the last 15-20 years, say) that these states have come to be studied extensively. There are at least two reasons for this tardiness.

--- Experimental. The generation of sufficiently intense VUV radiation has always been problematic using continuum sources, especially given the high resolution necessary for detailed Rydberg spectral studies. However, these difficulties have been largely overcome by the proliferation of synchrotron radiation sources; by the development (either currently available or soon-to-be available) of VUV and XUV lasers; and by the increased resolution which can be achieved today in electron-impact spectroscopy. These new experimental techniques, as well as some of their application, are described in [E. E. Koch and P. Gürtler], synchrotron radiation sources; in [S. C. Wallace] and [J. C. Miller and R. N. Compton], VUV lasers and third-harmonic generation of VUV radiation; and in [H. H. Brongersma, H. A. van Sprang and G. J. Verhaart], electron-impact spectroscopy. Indeed, the recent explosion of research in the area of highly-excited states has been spurred almost totally by the exchange of "new lamps for old" -- which seems a remarkably apt phrase, given that the newly commissioned electron storage ring at the University of Wisconsin is named "Aladdin"!

--- Theoretical. There have been, traditionally, two approaches to the theoretical description of highly-excited states. The first approach, that of the chemists, has its genesis in the LCAO-MO-SCF treatment of many-electron states. From this viewpoint, valence states are of primary importance, with the incorporation of Rydberg character into an excited state being vested in the inclusion of "diffuse" orbitals in the LCAO basis. (For a description of this treatment, and its attendant difficulties, see [C. Sandorfy and L. S. Lussier].) The second approach, that of the physicists, gives primary importance to scattering states, with the highly-excited (bound) states being considered as simply the discrete portion of a scattering "channel". (An introduction to this approach is given in [A. R. P. Rau].) Both the chemical and physical descriptions of highly-excited states are convenient within certain limits and each approach, when appropriately circumscribed, has its own advantages and disadvantages. The greatest drawback in the existence of two distinct theories of highly-excited states, however, lies in the inability of the practitioners of one approach to translate their terminology into the language of the other approach. And, this "language barrier" has led to two disparate lines of research in the study of Rydberg states: Molecular Rydberg states are usually discussed from a valence viewpoint, while atomic Rydberg states are treated within a scattering framework. As a result, the structural similarities between atomic and molecular highly-excited states are often obscured, or overlooked altogether. Recently, however, with the work of the Seaton and Fano schools, a third approach to highly-excited states has been developed, namely, the *quantum defect method* (QDM). The QDM is intermediate between the two limiting approaches described above and, indeed, provides a smooth joining of the valence viewpoint with the scattering viewpoint. A comprehensive introduction to the QDM, for both atoms and molecules, is given in [A. R. P. Rau], [K. T. Lu] and [C. H. Greene]; however, quantum defect terminology has become so pervasive that it permeates almost every chapter in this volume.

For our purposes, however, we must give at least a first-order answer to the question "What is a Rydberg?", and we provide this answer in the form of three alternatives.

Alternative 1. In atomic physics, transition to an *R* state occurs whenever an electron is promoted from a ground-state orbital to an excited-state orbital possessing an *aufbau* quantum number greater than that of the valence shell. In molecular physics, this intuitive clarity is lost in the inability to assign a "principal" quantum number to the valence shell, an inability only partially alleviated by heuristic and ambiguous "united atom" arguments.

Alternative 2. An *R* transition is any member of a series which obeys the Rydberg equation

$$E_{j\alpha} = I - 1/\nu_{j\alpha}^2 \quad , \qquad (1)$$

where $E_{j\alpha}$ is the energy of that spectral feature thought to be the jth member of the αth *R* series converging on the ionization limit I, and where $\nu_{j\alpha}$ is the effective quantum number. This characterization is empirically limited by the need for spectra which exhibit high resolution over large spectral regions; furthermore, it precludes interactions between *R* series.

Alternative 3. This last alternative is operational: In this case, we say quite simply that an *R* state is any state defined with respect to an asymptotically hydrogenic Hamiltonian. Several conclusions immediately obtain:

(i) Energy spacings of *R* states decrease as $1/\nu_{j\alpha}^2$, where $\nu_{j\alpha} = n_j - \mu_\alpha$ (n_j is an integer and μ_α is the quantum defect).

(ii) The average radius $\langle r \rangle$ of an *R* orbital is proportional to $\nu_{j\alpha}^2$. Hence, highly-excited states are spatially enormous, which has led to their being termed "giant" molecules [1] or atoms.

(iii) For a fixed core, the spin-orbit coupling parameter remains approximately constant, whereas the exchange integral K decreases at least as fast as $1/\nu_{j\alpha}^2$. Thus, the coupling regime index K/ζ is proportional to $\nu_{j\alpha}^{-k}$, $k > 2$, which indicates that highly-excited molecular states are in a pure (Ω,ω) regime.

For our discussion of highly-excited states, we shall adopt *alternative 3* and then investigate the scope and limitations of the concomitant operational definitions (i)-(iii). In order to do so, however, we must first clarify what we mean by an "asymptotically hydrogenic" Hamiltonian.

1.2 Rydberg Hamiltonians [1-4]

Consider an (N+1)-electron molecule (or atom) M and, in correspondence with it, an N-electron cation M^+. We assume, for the moment, that M and M^+ have identical nuclear distributions given by the set $\{\underset{\sim}{R}_\omega\}$, where $\omega = 1,\ldots,K$ and where K is the number of nuclei. The electronic Hamiltonian of the (N+1)-electron species can then be written as

$$H(M) = H(1) + H(M^+) + G(1;2,\ldots,N+1) \quad . \tag{2}$$

In Eq. (2), H(1) describes a single electron (i.e., the optical electron) in the electrostatic field of all K nuclei,

$$H(1) = -(\hbar^2/2m)\,\nabla^2(1) - e^2 \sum_{\omega=1}^{K} Z_\omega/|\underset{\sim}{r}_1 - \underset{\sim}{R}_\omega| \quad , \tag{3}$$

where Z_ω is the atomic number of the ωth nucleus and $\underset{\sim}{r}_1$ is the position vector of the optical electron; $H(M^+)$ is the electronic Hamiltonian of the N-electron cationic species; and, finally, $G(1;2,\ldots,N+1)$, which represents the interactions between the optical electron and the N-electron subsystem (including spin-orbit coupling effects), is given by

$$G(1;2,\ldots,N+1) = \sum_{i=2}^{N+1} g(1,i) \quad . \tag{4}$$

We now assume that those electronic wavefunctions of M which correspond to R states may be written as an antisymmetrized product of a 1-electron function $|M;1\rangle$ and an (antisymmetrized) N-electron function $|M^+;2,\ldots,N+1\rangle$; that is,

$$|M;1,\ldots,N+1\rangle = A_{1,N}|M;1\rangle|M^+;2,\ldots,N+1\rangle \quad , \tag{5}$$

where $A_{1,N}$ is a residual antisymmetrizer defined, in terms of the identity permutation $\hat{e}$ and the transpositions $\hat{t}_{1i}$, by

$$A_{1,N} = (N+1)^{-1}\left[\hat{e} - \sum_{i=2}^{N+1} \hat{t}_{1i}\right] \quad . \tag{6}$$

We also impose the following orthogonality restriction (i.e., the so-called "strong orthogonality" condition [5]):

$$\langle M;j|M^+; \ldots j \ldots \rangle_{(j)} = 0, \quad j = 1,\ldots,N+1 \quad . \tag{7}$$

The expectation value of the Hamiltonian of Eq. (2) and the norm of the function of Eq. (5) follow immediately from Eqs. (2-7):

$$\langle M;1,\ldots,N+1|H(M)|M;1,\ldots,N+1\rangle = (N+1)^{-1} \langle M;1|[H(1)$$

$$+ \langle M^+;2,\ldots,N+1|\sum_{i=2}^{N+1} g(1,i)[\hat{e} - \hat{t}_{1i}]|M^+;2,\ldots,N+1\rangle$$

$$+ \langle M^+;2,\ldots,N+1|H(M^+)|M^+;2,\ldots,N+1\rangle]|M;1\rangle \tag{8}$$

$$\langle M;1,\ldots,N+1|M;1,\ldots,N+1\rangle =$$

$$(N+1)^{-1} \langle M;1|\langle M^+;2,\ldots,N+1|M^+;2,\ldots,N+1\rangle|M;1\rangle \quad . \tag{9}$$

If we now choose $|M^+;2,\ldots,N+1\rangle$ to be the μth eigenstate of $H(M^+)$, with eigenvalue E_μ, we find that minimization of Eq. (8) [constrained by Eq. (9)] with respect to $|M;1\rangle$ yields the following eigenequation:

$$[H(1) + \langle M^+,\mu;2,\ldots,N+1|\sum_{i=2}^{N+1} g(1,i)[\hat{e} - \hat{t}_{1i}]|M^+,\mu;2,\ldots,N+1\rangle$$

$$- \varepsilon I]|M;1\rangle = 0 \quad , \tag{10}$$

where I is the identity operator. The eigenvalue ε is related to E_μ and to the expectation value $\langle M|H(M)|M\rangle \equiv E$ via

$$\varepsilon = E - E_\mu \quad . \tag{11}$$

If, further, we denote the energy of the electronic ground state of M by E_{GS}, we can write

$$-\varepsilon = (E_\mu - E_{GS}) - (E - E_{GS})$$

or, since $E_\mu - E_{GS} \equiv I_\mu$ (the μth ionization potential) and $E - E_{GS} \equiv E_{ex}$ (the excitation energy),

$$E_{ex} = I_\mu + \varepsilon \quad . \tag{12}$$

Thus, we see that $-\varepsilon$ represents the term value of a Rydberg state and $|M;1\rangle$ represents a Rydberg spin orbital.

If we introduce the first-order density matrix [5]

$$\Gamma_\mu(2'|2) = \langle M^+,\mu;2',3,\ldots,N+1|M^+,\mu;2,3,\ldots,N+1\rangle \tag{13}$$

into Eq. (10), we find

$$[(-\hbar^2/2m)\,\nabla^2(1) - e^2 \sum_{\omega=1}^{K} Z_\omega/|\underset{\sim}{r}_1 - \underset{\sim}{R}_\omega|$$

$$+ N \int d\tau_2 g(1,2)[\hat{e} - \hat{t}_{12}]\Gamma_\mu(2'|2) - \varepsilon I]|M;1\rangle = 0 \tag{14}$$

where τ_2 represents both space and spin variables of electron 2.

At this point, we make the following observations.

Observation 1. If we perform a normal coordinate analysis of M, with the normal coordinates being denoted $\underset{\sim}{Q} = (Q_1,\ldots Q_{K'})$, $K' < K$, the Coulomb term of Eq. (14) [i.e., the second term] can be denoted $V_C(\underset{\sim}{r},\underset{\sim}{Q})$, where we have dropped the index "1" on the spatial coordinates of the optical electron for notational convenience. If we now relax the constraint that the geometries of M and M^+ be identical (i.e., if we relax our "rigid core" restriction) and assume, instead, only that these geometries are adiabatically related (i.e., if we assume only the Born-Oppenheimer approximation), we may expand $V_C(\underset{\sim}{r},\underset{\sim}{Q})$ about the equilibrium internuclear configuration $\underset{\sim}{Q} = 0$. Thus,

$$V_C(\underset{\sim}{r},\underset{\sim}{Q}) = -\frac{Ze^2}{r} + \left.\frac{\partial V}{\partial \underset{\sim}{Q}}\right|_{\underset{\sim}{Q}=0} \cdot \underset{\sim}{Q} + \ldots \tag{15}$$

where $Z = \sum_{\omega=1}^{K} Z_\omega$. Writing $Z = Z_c + N$ ($Z_c \equiv$ core charge; $Z_c = 1$ for M neutral) and subsuming all $\underset{\sim}{Q}$-dependent terms in $V_v(\underset{\sim}{r},\underset{\sim}{Q})$, Eq. (15) becomes

$$V_C(\underset{\sim}{r},\underset{\sim}{Q}) = -\frac{Z_c e^2}{r} - \frac{Ne^2}{r} + V_v(\underset{\sim}{r},\underset{\sim}{Q}) \tag{16}$$

where $V_v(\underset{\sim}{r},\underset{\sim}{Q})$ represents a vibronic interaction potential.

Observation 2. For a highly-excited state, the third term of Eq. (14), when coupled with $- Ne^2/r$ of Eq. (16), becomes [1]

$$-\frac{Ne^2}{r} + N\int d\tau_2 g(1,2)[\hat{e} - \hat{t}_{12}]\,\Gamma_\mu(2'|2)$$

$$\rightarrow V_{ra}(r) + V_{rm}(\underset{\sim}{r}) + V_{so}(\underset{\sim}{r},\sigma) + V_{sv}(\underset{\sim}{r},\sigma,\underset{\sim}{Q}) \quad ,$$

where V_{ra} and V_{rm} are residual atomic and molecular potentials, respectively [6]; both of these potentials are short-ranged:

$$\lim_{r\to\infty} r\,V_{ra} = \lim_{r\to\infty} r\,V_{rm} = 0 \quad .$$

$V_{so}(\underset{\sim}{r},\sigma)$, where $(\underset{\sim}{r},\sigma)$ are the space-spin coordinates of the optical electron, represents the orbital-orbital and spin-orbital interactions between the optical electron and the electrons of M^+. And, finally, $V_{sv}(\underset{\sim}{r},\sigma,\underset{\sim}{Q})$, where the $\underset{\sim}{Q}$ functionality follows from the $\underset{\sim}{Q}$-dependence of Γ_μ (per *Observation 1*), represents the effects of spin-vibronic interactions.

Observation 3. In many cases, we are interested in the effects of external fields -- either electric, magnetic or molecular (i.e., condensed phase) -- on highly-excited states. To account for these effects, we must include a field term F in Eq. (14).

With the aid of *Observations 1-3*, the (radial) Hamiltonian of Eq. (14) becomes

$$H = \frac{\hbar^2}{2m}\left[-\frac{d^2}{dr^2} + \frac{\ell(\ell+1)}{r^2}\right] - \frac{Z_c e^2}{r} + V_{ra}(r) + V_{rm}(\underset{\sim}{r})$$

$$+ V_{so}(\underset{\sim}{r},\sigma) + V_v(\underset{\sim}{r},\underset{\sim}{Q}) + V_{sv}(\underset{\sim}{r},\sigma,\underset{\sim}{Q}) + F \quad , \tag{17}$$

where $\ell(\ell + 1)/r^2$ is the centrifugal barrier which results from a transformation to spherical-polar coordinates, followed by a separation of radial and angular variables. Eq. (17) vindicates our use of the term "asymptotically hydrogenic" in our description of a Rydberg Hamiltonian: Indeed, at large r the dominant (molecular, nonrelativistic) potential is the Coulomb term, $- Z_c e^2/r$.

The rest of this chapter is concerned with the justification of Eq. (17) for Rydberg states. In Section 2, we first present an approximate treatment of V_{ra} and V_{rm} from the viewpoint of single-channel quantum defect theory (SQDT), and then refine this treatment and extend it to include part of V_{so} by the introduction of multichannel quantum defect theory (MQDT) interactions. In Section 3, core effects (i.e., those perturbations arising from V_v, V_{sv} and anomalous exchange and spin-orbit coupling interactions) are discussed. In addition, the importance and prevalence of Rydberg-valence mixing is investigated. In Section 4, interatomic and intermolecular correlations of Rydberg term values and ionization energies are introduced as an alternative method of assessing the nature of the short-range core potentials. Finally, electric and magnetic field effects are treated in Section 5, while condensed media effects are discussed in Section 6.

2. QUANTUM DEFECT THEORY [7]

(This topic is developed in detail in [A. R. P. Rau], [K. T. Lu] and [C. H. Greene].)

2.1 Single-Channel Quantum Defect Theory

One-electron Coulomb problem:

We begin with the hydrogenic atom. The radial Schrödinger equation (in Rydberg units) is

$$\left[\frac{d^2}{dr^2} - \frac{\ell(\ell+1)}{r^2} + \frac{2Z_c}{r} + 2\,\varepsilon_j\right] f_j(r) = 0 \quad , \tag{18}$$

where

$$\varepsilon = - Z_c^{\,2}/k^2 \tag{19}$$

and

$$k = \begin{cases} \nu_j & \text{for } \varepsilon < 0 \\ i\gamma & \text{for } \varepsilon > 0 \end{cases} \quad . \tag{20}$$

The bound-state ($\varepsilon < 0$) wavefunctions are given by the regular Coulomb function f, which asymptotically is

$$f \sim u(\nu,r)\,\sin\pi\nu - v\,\cos\pi\nu$$

where u is a rising exponential and v is a falling exponential in r. Since the bound-state wavefunctions must vanish at $r \to \infty$, we find

$$\sin \pi\nu = 0 \tag{21}$$

Thus, $\nu = n$, where n is an integer, and

$$\varepsilon = - Z_c^2/n^2 \quad . \tag{22}$$

Introduction of the residual atomic potential:

Upon inclusion of the residual atomic potential, Eq. (18) becomes

$$\left[\frac{d^2}{dr^2} - \frac{\ell(\ell + 1)}{r^2} + \frac{2Z_c}{r} - 2\, V_{ra}(r) + 2\, \varepsilon_{j\ell}\right] F_{j\ell}(r) = 0 \quad . \tag{23}$$

Since $V_{ra}(r)$ is short-ranged, Eq. (23) asymptotically reverts to Eq. (18) at distances greater than r_0, the cut-off distance for V_{ra}. In order to ensure smooth joining at $r = r_0$, we find

$$F_{j\ell}(r) = f_j(r) \cos \pi\mu_\ell - g_j(r) \sin \pi\mu_\ell \quad . \tag{24}$$

Using the appropriate asymptotic forms for f and g, Eq. (24) becomes

$$F_{j\ell} \sim u_j(r) \sin \pi(\nu + \mu_\ell) - v_j \cos \pi(\nu + \mu_\ell) \; , \tag{25}$$

from which it follows that

$$\sin \pi(\nu + \mu_\ell) = 0 \tag{26}$$

or

$$\nu_{j\ell} = n_j - \mu_\ell \quad . \tag{27}$$

From Eqs (19,20) then,

$$\varepsilon_{j\ell} = - Z_c^2/(n_j - \mu_\ell)^2 \quad . \tag{28}$$

Thus, the Rydberg equation [i.e., Eq. (1)] is established.

Introduction of the residual molecular potential:

The above procedure for V_{ra} can be extended to molecules if V_{rm} is indeed short-ranged. This, in turn, implies a significant

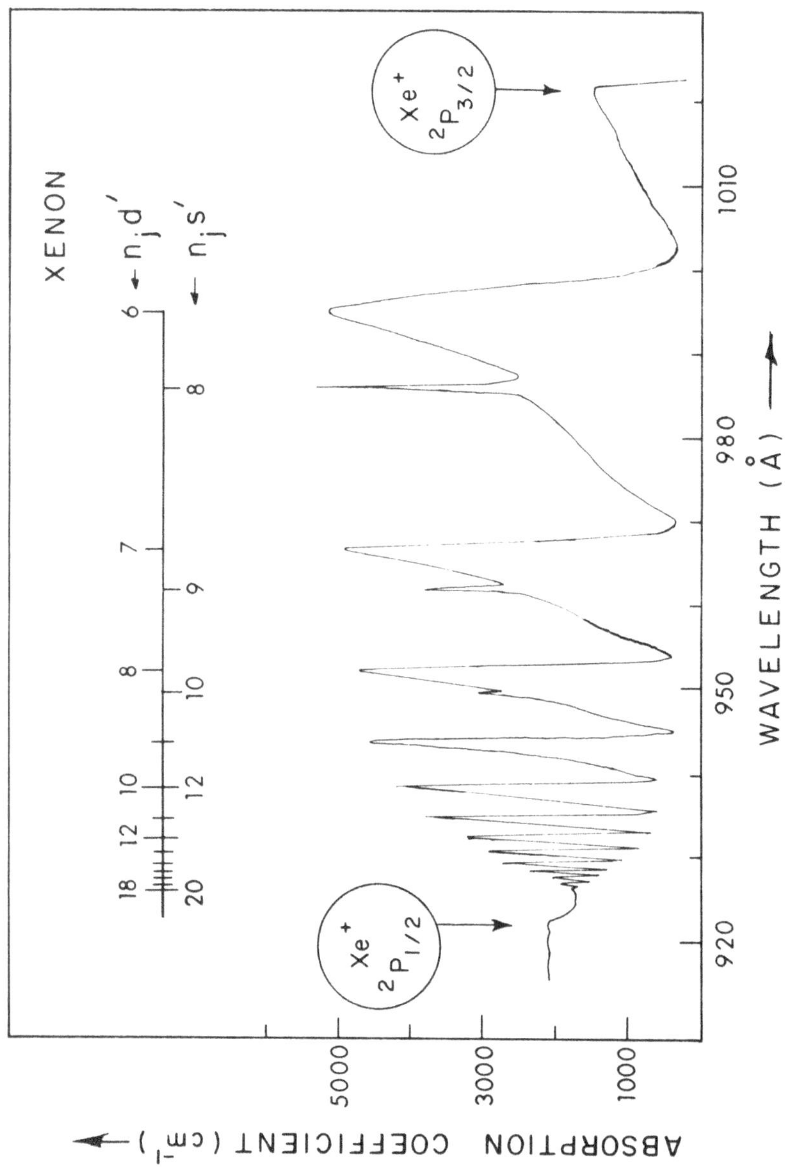

Fig. 1. Figure caption follows Fig. 2.

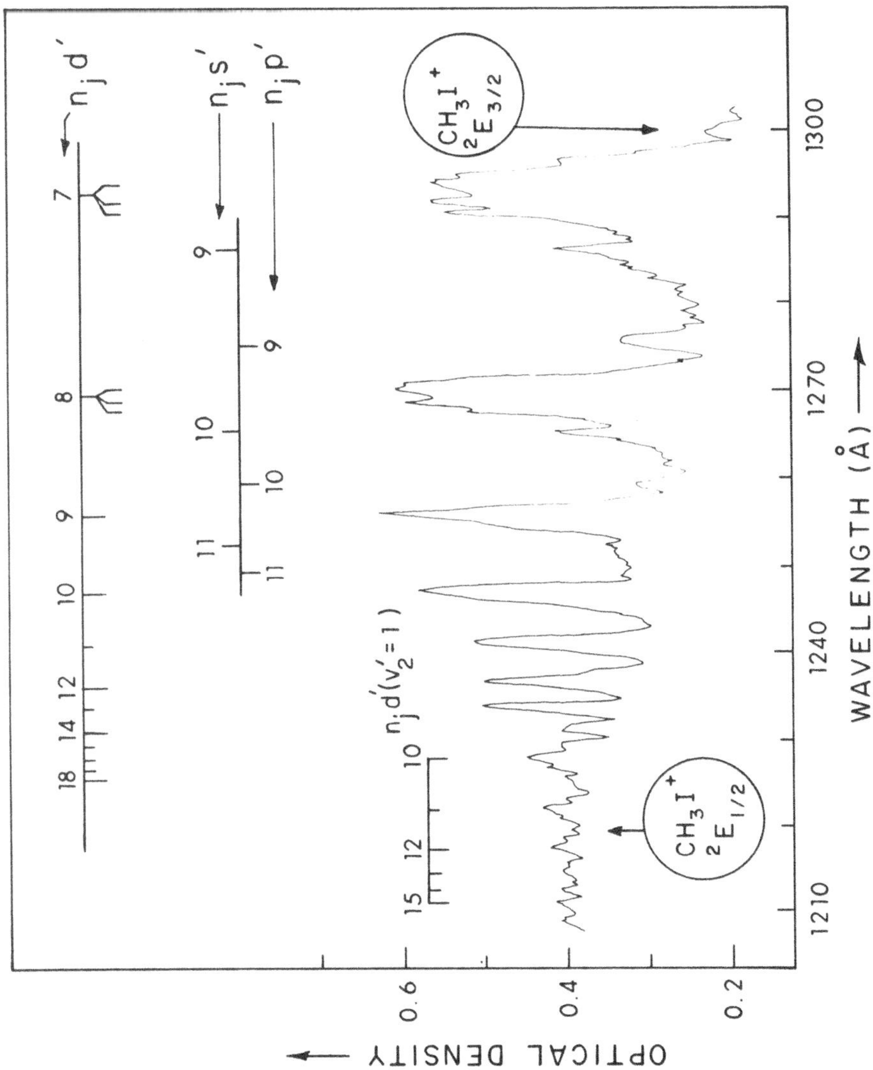

Fig. 2. Figure caption follows on next page.

Fig. 1. The autoionization spectrum of xenon between the $^2P_{3/2}$ and $^2P_{1/2}$ ionization limits [adapted from R. G. Huffman, Y. Tanaka and J. C. Larrabee, J. Chem. Phys. 39, 902 (1963)]. The primed symbol $n_j\ell'$ distinguishes series converging on $^2P_{1/2}$ from those converging on $^2P_{3/2}$. (Taken from [6].)

Fig. 2. The absorption spectrum of methyliodide between the $^2E_{3/2}$ and $^2E_{1/2}$ ionization limits. The primed symbol $n_j\ell'$ distinguishes series converging on $^2E_{1/2}$ from those converging on $^2E_{3/2}$. (Taken from [6].)

similarity between an atomic *R* spectrum and a molecular *R* spectrum if, in both instances, the *total* residual potential is very nearly the same. That this assumption is reasonable in some cases is established [6] for Xe/CH_3I by a comparison of Figs. 1 and 2.

Further information concerning the residual potentials may be gained through introduction of the phase-amplitude method [8]. This *ansatz* transforms the Schrödinger equation into two coupled first-order differential equations, thereby effecting the separation

$$F_{j\ell}(r) = a_{j\ell}(r)[f_j(r)\cos\pi\,\mu_{j\ell}(r) - g_j(r)\sin\pi\,\mu_{j\ell}(r)] \quad , \qquad (29)$$

where $a_{j\ell}(r)$ is an amplitude function while $\mu_{j\ell}(r)$ is a quantum defect function. Within the phase-amplitude method, it is possible to show that [8]

$$\pi\mu_{j\ell}(r) = 2W^{-1}\int_0^r [V_{ra}(r') + V_{rm}(r')]\,[f_j(r')\cos\pi\mu_{j\ell}(r') - g_j(r')\sin\pi\mu_{j\ell}(r')]^2\,dr' \quad , \qquad (30)$$

where W is the Wronskian of the f and g functions. (In Eq. (30), we have assumed that the dominant part of V_{rm} is spherically symmetric.)

If the residual potentials are negative and monotonically increasing, it follows from Eq. (30), and from the detailed structure of the f and g functions, that [6]

$$\nu_{j\ell} - \nu_{(j-1)\ell} \geq 1 \quad . \tag{31}$$

However, if the residual potentials are positive and monotonically decreasing,

$$\nu_{j\ell} - \nu_{(j-1)\ell} \leq 1 \quad . \tag{32}$$

This is illustrated for the rare gases (V_{rm} = 0) in Fig. 3. Thus, we see that V_{ra} is attractive for each of the rare gases. Application of this procedure to CH_3I [6] shows that V_{rm} is repulsive for j < 2, but that it becomes negligible at higher j because of the dominance of V_{ra} at larger r. The effect is even more pronounced for ℓ-waves with ℓ ≥ 2 since, in this case, the centrifugal barrier acts to impose effective spherical symmetry even on

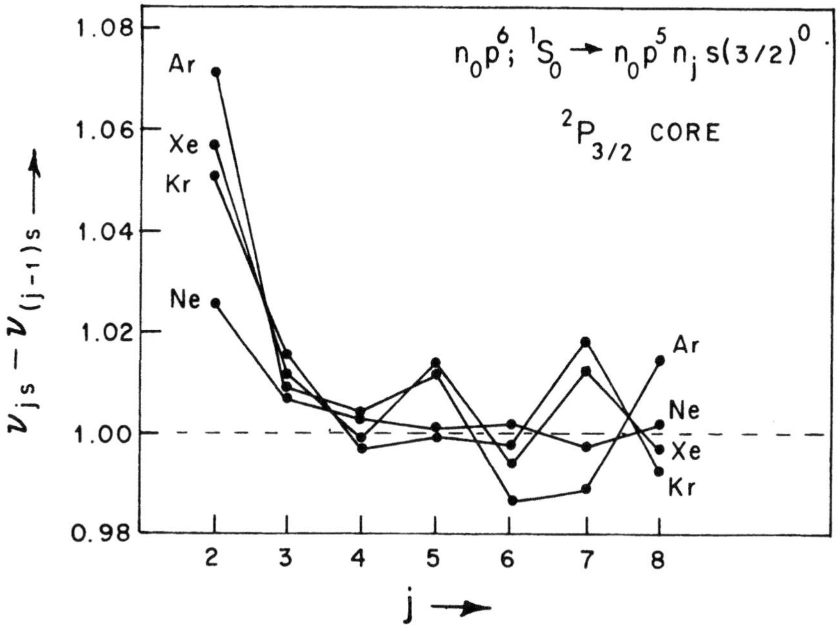

Fig. 3. A plot of $\nu_{js} - \nu_{(j-1)s}$ for the $n_0p^6\ ^1S_0 \rightarrow n_0p^5n_j\ s^0_{3/2}$, J = 1 series ($^2P_{3/2}$ core) of the rare gases [data from C. E. Moore, Natl. Bur. Stand. Circ. 467 (U. S. Gov. Print. Off., Washington, 1958)]. (Taken from [6].)

the molecular system [6]. (A comprehensive discussion of centrifugal-barrier effects in molecules, including the induction of shape-resonance phenomena, is given in [J. L. Dehmer, D. Dill and A. C. Parr].)

2.2 Multichannel Quantum Defect Theory

If two or more ionization limits are adjacent, two *R* series (one converging on each limit) may interact, presuming, of course, that each series has the same total angular momentum [9]. Of even more importance, however, a discrete state of one series may interact with the continuum of the other series, thereby preionizing (autoionizing). A "channel" is the union of the discrete members of an *R* series with the continuum of the same series, and interactions such as those described above are known as multichannel interactions. That such interactions occur in Xe [10] and CH_3I [6,11] is shown in Figs. 1 and 2: We see the characteristic Beutler-Fano [12-14] autoionization profiles in both systems. Within the MQDT, however, the similarities between Xe and CH_3I can be further strengthened [15], as will now be shown.

In the spectral region of interest in Xe and CH_3I, there are two ionization limits: $I_1 = I(^2P_{3/2}$ or $^2E_{3/2})$ and $I_2 = I(^2P_{1/2}$ or $^2E_{1/2})$. Thus, each level must be referenced to both ionization limits:

$$E = I_1 - \frac{1}{\nu_1^2} = I_2 - \frac{1}{\nu_2^2} \quad . \tag{33}$$

Boundary conditions on the bound-state wavefunctions impose a consistency constraint which, when coupled with Eq. (33), allows us to fit experimental data to a ν_1 (modulo 1) versus ν_2 (modulo 1) plot and, thus, to determine quantum defects [10,14,16]. This procedure is illustrated in Fig. 4. The similarities between the two plots are striking and further indicate the close relationship between the residual potential terms for Xe and CH_3I.

Finally, we note that manifestations of autoionization in polyatomic systems are known [17] which are more esoteric (e.g., J dependence and q reversal) than that discussed above. In addition, rotational and vibrational preionization and molecular predissociation have been treated in an MQDT format [14,18-20]. (Concerning this last topic, see [C. H. Greene] and [J. L. Dehmer, D. Dill and A. C. Parr].)

3. CORE EFFECTS

Most, if not all, of the core effects considered here should be included within the channel framework of Section 2.2. Unfor-

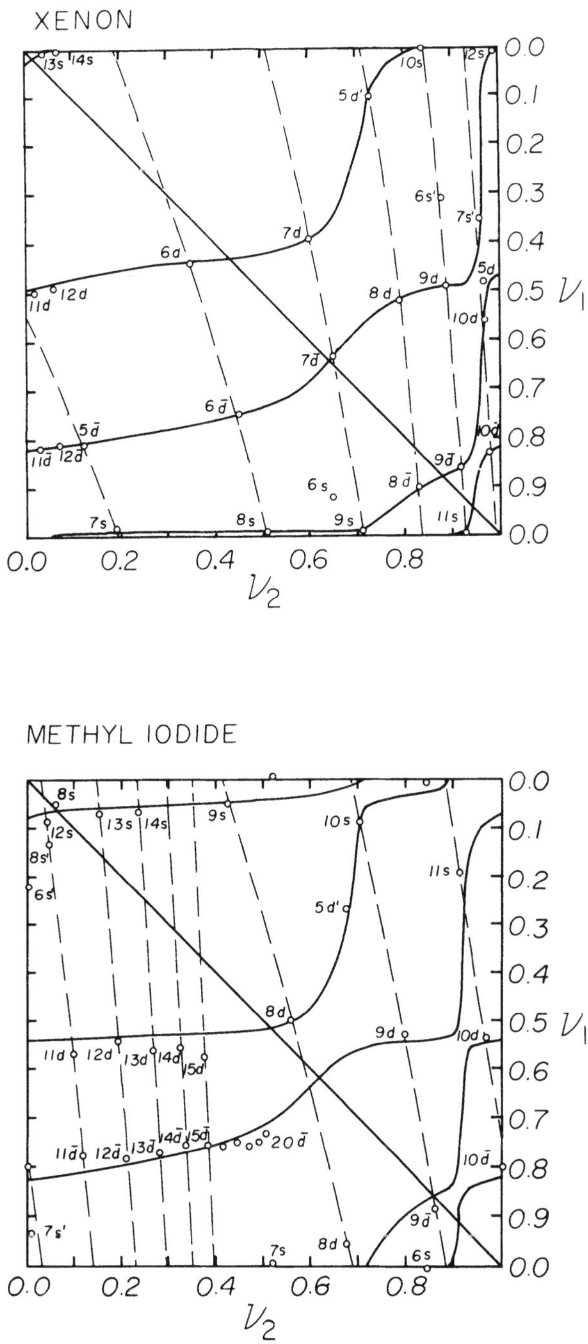

Fig. 4. Figure caption follows on next page.

Fig. 4. Lu-Fano [10,14,16] plots for the discrete states of xenon and methyliodide. The xenon plot is adapted from [10]. ν_1 and ν_2 are effective quantum numbers (modulo 1) defined with respect to the first and second ionization potentials, respectively (see text). The full diagonal lines are given by $\nu_1 = \nu_2$; the broken cruves are plots of $\nu_2(\nu_1)$, the functionality being defined by Eq. (33) of the text. The full curves are the loci of the consistency constraint mentioned in the text and discussed more fully in [10,15] and in [K. T. Lu]. The data point notation is as follows: $d = d_{5/2}\ (I_1)$; $\bar{d} = d_{3/2}\ (I_1)$; $s = s_{1/2}\ (I_1)$; $s' = s_{1/2}\ (I_2)$; $d' = d_{3/2}\ (I_2)$. (Taken from [15].)

tunately, the theoretical basis for such an inclusion is not yet available for polyatomics and is even incomplete for diatomics [14,18-20]. Indeed, even if the tactical approach were developed fully, considerations of the multichannel nature would require high-resolution data over broad spectral ranges. Such data, unfortunately, are not often available.

The general type of high-resolution data covers small spectroscopic ranges (e.g., the first s complex in CH_3I [21]) and spectral effects may be evident which, because of the narrow observational range, require discussion in terms of interactions within that range. Even within the MQDT, sudden perturbations may occur which, because of artificial restrictions on the number of interacting channels, simply cannot be treated within the channel framework.

As a result, we now feel free to discuss a variety of core effects, of which at least one (i.e., spin-orbit coupling) has already been considered within the MQDT.

3.1 Spin-Orbit Coupling

We use the $5p \rightarrow 6s$ *R* excitation of HI as an example. It is found that the linear C-I bond of alkyl, alkene and alkyne iodides dominates spin-orbit coupling in these moieties and, therefore, that the HI considerations have a wider range of applicability than one might at first think. Indeed, because the spin-orbit coupling constants for chlorine, bromine and iodine are all much larger than that for carbon, these same considerations also apply to chlorides and bromides.

Those states which arise from $5p \rightarrow 6s$ configurational excitation are shown in Fig. 5 in both the (Λ,S) and (Ω,ω) limits. If these states are labeled 1,2,3 and 4 in order of increasing energy, then the intermediate coupling regime yields the following set of energy differences and intensity ratios [22]:

$$E(4) - E(2) = 2(K^2 + \zeta^2)^{1/2}$$
$$E(3) - E(1) = 2\zeta$$
$$E(2) - E(1) = K - (K^2 + \zeta^2)^{1/2} + \zeta$$
$$E(4) - E(3) = K + (K^2 + \zeta^2)^{1/2} - \zeta$$

$$I(^2\Pi_{3/2}) - I(^2\Pi_{1/2}) = 2$$

$$f(4)/f(2) = \left[\frac{K}{\zeta} + \left(1 + \frac{K^2}{\zeta^2}\right)^{1/2}\right]^2 .$$

Here, K is the exchange integral, ζ is the spin-orbit coupling constant ($K \equiv \langle 5p6s|1/r_{12}|6s5p\rangle$ and $\zeta \equiv 1/2 \langle 5p|A|5p\rangle$) and $f(\beta)$

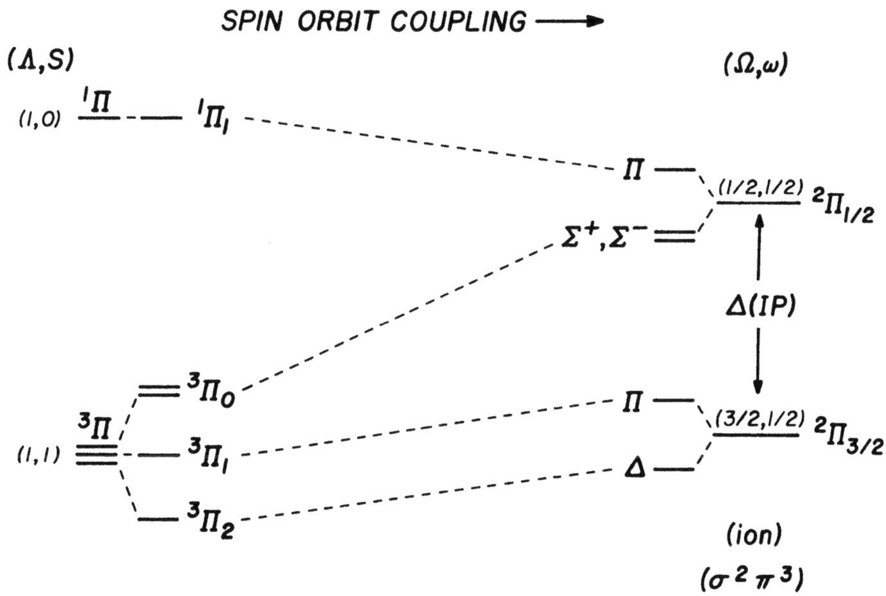

Fig. 5. Spin-orbit coupling in the $\ldots\sigma^2\pi^3 6s$ configuration of HI. (Taken from [1].)

is the oscillator strength of the transition $\tilde{X} \to \beta$. The capability of this model in fitting details of intermediate coupling in the simple halides is extensive [23-25], as may be seen in Fig. 6.

3.2 Vibronic and Spin-Vibronic Coupling

If $\underset{\sim}{q}$, $\underset{\sim}{Q}$ and $\underset{\sim}{\sigma}$ represent, respectively, sets of electron spatial coordinates, atom spatial coordinates (or, more appropriately, normal coordinates) and electron spin coordinates, the molecular Hamiltonian, including spin-vibronic effects, may be designated $H(\underset{\sim}{q}, \underset{\sim}{\sigma}, \underset{\sim}{Q})$. If we assume that spin-orbit coupling is small, a Herzberg-Teller expansion in the normal coordinates $\underset{\sim}{Q}$ about the equilibrium configuration $\underset{\sim}{Q}_o$ of a given electronic state yields

$$H(\underset{\sim}{q},\underset{\sim}{\sigma},\underset{\sim}{Q}) = H(\underset{\sim}{q};\underset{\sim}{Q}_o) + H(\underset{\sim}{q},\underset{\sim}{\sigma};\underset{\sim}{Q}_o) + \left.\frac{\partial H(\underset{\sim}{q},\underset{\sim}{Q})}{\partial \underset{\sim}{Q}}\right|_{\underset{\sim}{Q}_o} \cdot (\underset{\sim}{Q}-\underset{\sim}{Q}_o) + \left.\frac{\partial H(\underset{\sim}{q},\underset{\sim}{\sigma},\underset{\sim}{Q})}{\partial \underset{\sim}{Q}}\right|_{Q_o} \cdot (\underset{\sim}{Q}-\underset{\sim}{Q}_o) + \cdots \tag{34}$$

which, upon truncation, may be written

$$H(\underset{\sim}{q}, \underset{\sim}{\sigma}, \underset{\sim}{Q}) = H_e + H_{so} + H_v + H_{sv} \quad , \tag{35}$$

where H_e is the Born-Oppenheimer electronic Hamiltonian at configuration $\underset{\sim}{Q}_o$; H_{so} is the spin-orbit Hamiltonian, also at $\underset{\sim}{Q}_o$; H_v and H_{sv}, in obvious notations, are the vibronic and spin-vibronic Hamiltonians; and all coordinate dependences to the right of a semicolon are parametric. If the H_e/H_{so} separation is valid, then $H_{so} < H_e$ and, hence, H_v and H_{sv} are, respectively, first and second order in nature. If H_{so} is large, it is advisable to rewrite Eq. (34) as

$$H(\underset{\sim}{q},\underset{\sim}{\sigma},\underset{\sim}{Q}) = H(\underset{\sim}{q},\underset{\sim}{\sigma};\underset{\sim}{Q}_o) + \left.\frac{\partial H(\underset{\sim}{q},\underset{\sim}{\sigma},\underset{\sim}{Q})}{\partial \underset{\sim}{Q}}\right|_{\underset{\sim}{Q}_o} \cdot (\underset{\sim}{Q} - \underset{\sim}{Q}_o) + \cdots \quad , \tag{36}$$

in which case H_{sv}, now the second term, is formally first order.

In the MQDT, as presented in Section 2.2, the zeroth-order electronic and spin-orbit interactions are included in the channel formalism and they need not be considered here. However, the vibronic and spin-vibronic interactions are not a part of the MQDT, at least as we have formulated it. Hence, these effects can give rise to spectral perturbations which are not only *not* accounted for within the MQDT but which can confuse the entire MQDT deperturbational tactic. Thus, it is very important to recognize vibronic and spin-vibronic intrusions, particularly be-

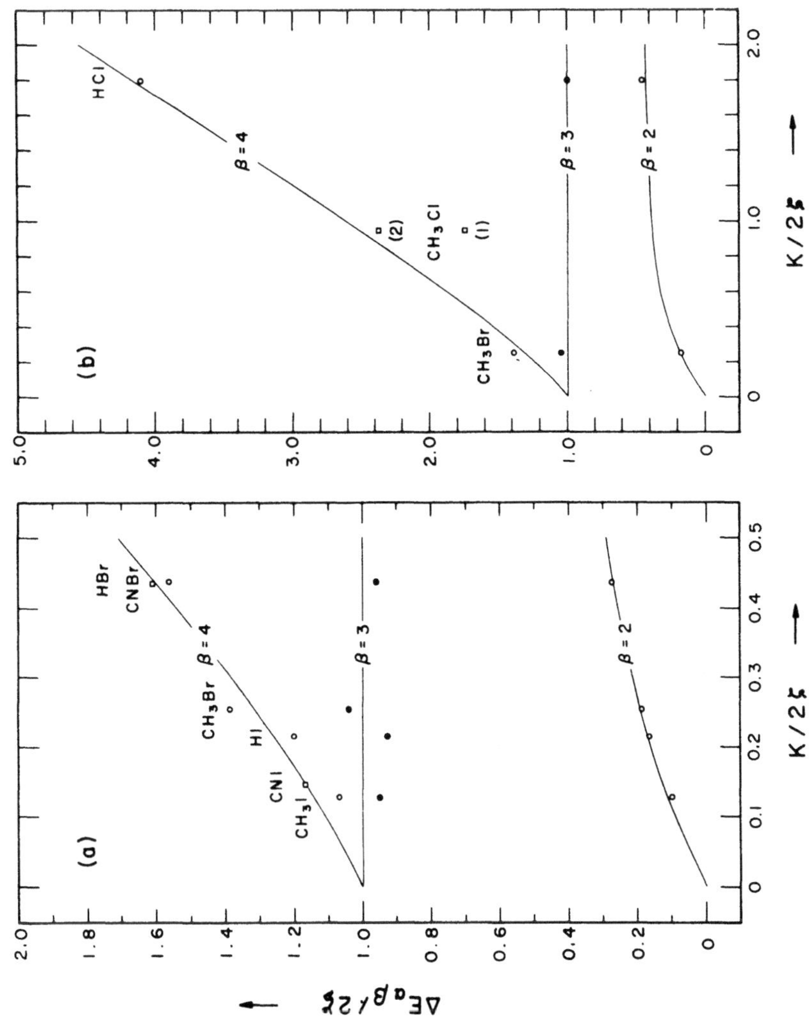

Fig. 6. Figure caption follows on next page.

Fig. 6. A plot of $\Delta E_{\alpha\beta}$, $\alpha = 1$ and $\beta = 2,3,4$, versus the exchange:spin-orbit ratio for various halides. The solid lines are theoretical curves which follow from the intermediate-coupling model of Section 3.1; the points are experimental data. (Taken from [11].)

cause these intrusions are both common and large.

We again adopt the linear HI molecule as an exemplar of vibronic and spin-vibronic coupling. This molecule possesses only one normal mode, which forms a basis for the Σ^+ representation of $C_{\infty v}$. The extension to CH_3I is accomplished here, for simplicity, by "permitting" HI to possess non-totally symmetric Π normal modes. Thus, in the point group $C_{\infty v}$, the CH_3I normal modes transform as Q_1, Q_2, $Q_3 \subset \Sigma^+$ and Q_4, Q_5, $Q_6 \subset \Pi$. Consequently, we are now in a position to tabulate the representations of $C_{\infty v}$ for which the various vector collections form bases (cf. Table 1).

Now, in the (Λ,S) regime (Table 1, second column), only the $\tilde{X} \to 4$ transition, namely $^1\Sigma^+ \to {}^1\Pi$, is electric-dipole allowed. In the (Ω,ω) regime (Table 1, third column), spin-orbit mixing of states 2 and 4 confers allowedness on the $\tilde{X} \to 2$ transition. The transitions $\tilde{X} \to 1$ and $\tilde{X} \to 3$, however, remain forbidden and retain this forbiddenness until vibronically coupled to either state 2 or state 4 by Π normal modes (Table 1, fifth column). We thus have the interesting result that states 2 and 4 may be coupled in a nominally first-order way by totally symmetric modes, whereas states 1 and 3 can couple with either of states 2 and 4 in ways which are nominally second order, with this coupling being mediated only by non-totally symmetric modes. An example [26] of 3-4 state coupling in CD_3I, mediated by Q_6, is shown in Fig. 7. Exam-

Table 1. $C_{\infty v}$ Representations of the States of Fig. 5.

State	$\tilde{q}$	$\tilde{q}$, $\tilde{\sigma}$	$\tilde{q}$, $\tilde{\sigma}$, $Q_{1,2,3}$	$\tilde{q}$, $\tilde{\sigma}$, $Q_{4,5,6}$
1	Π	Δ	Δ	Φ, Π
2	Π	Π	Π	Π, Σ^+, Σ^-
3	Π	Σ^+, Σ^-	Σ^+, Σ^-	Π
4	Π	Π	Π	Π, Σ^+, Σ^-

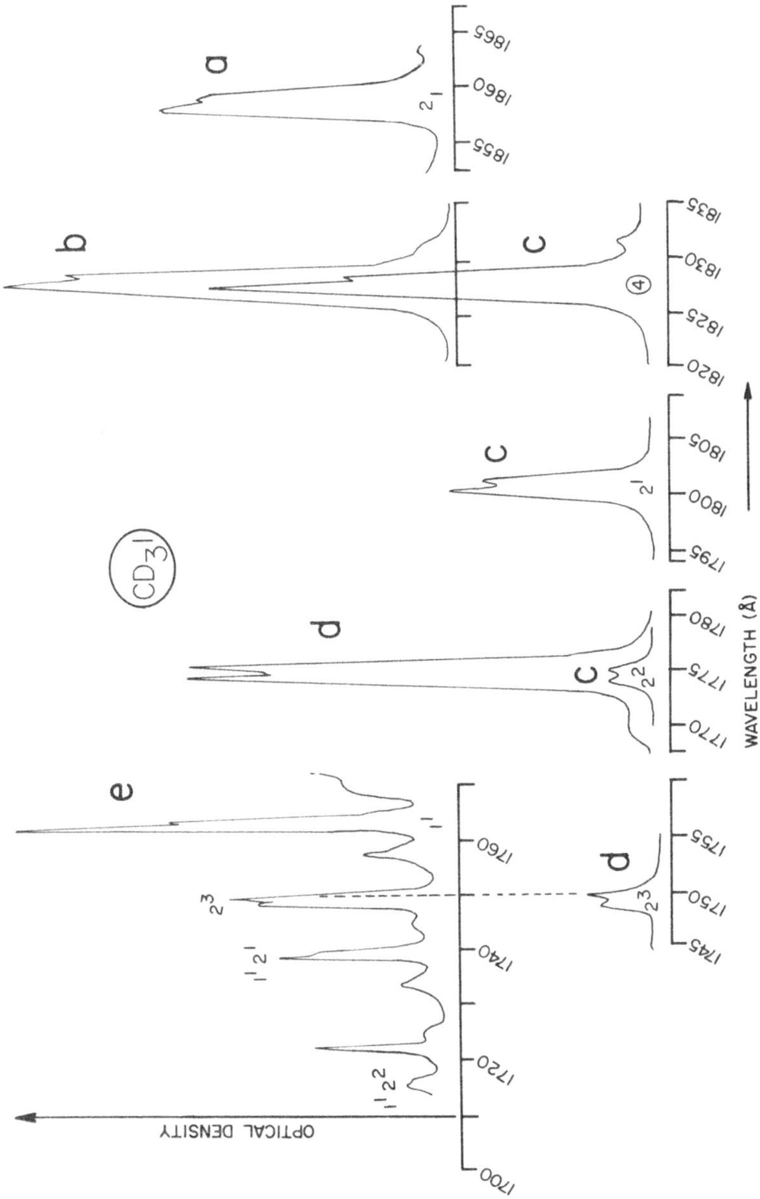

Fig. 7. Figure caption follows on next page.

Fig. 7. Spin-vibronic doubling in CD_3I mediated by the non-totally symmetric (Π) mode Q_6. The interactions in this case occur between states 3 (Q_6) and 4 and are carried throughout the v_2 progression of Q_2 built on the transition 0 → 4. The alternation in intensity is associated with a slight difference in the vibrational frequency ν_2 in states 3 and 4. The doubling is absent in CH_3I because the increase in the Q_6 frequency in the fully protonated derivative destroys the near degeneracy. Conditions for the various absorption spectra were as follows: (a) P = 1.4 torr, T = 295 K; (b) 0.009 torr, 295 K; (c) 0.008 torr, 234 K; (d) 0.1 torr, 234 K; (e) 0.325 torr, 234 K. (Taken from [26].)

ples [27] of 2-4 mixing abound. The matrix elements of the nominally first-order and second-order effects are as follows: C_2H_5Br [27] (2-4 mixing), 145 cm^{-1}; CNCℓ [28] (4-intravalence mixing), 180 cm^{-1}; CH_3I [26] (3-4 mixing), 15 cm^{-1}.

4. INTERMOLECULAR RYDBERG CORRELATIONS

An alternative approach to the assessment of core effects for highly-excited states is vested in the provision of correlation algorithms for ionization energies and Rydberg term values. Such algorithms have become increasingly common [2-4,29-36] and are often presented in the form

$$T(M;\mu,\alpha,j) = a + bI(M;\mu) \quad , \tag{37}$$

where $T(M;\mu,\alpha,j)$ is the term value, for molecule M, of the jth R state (of α symmetry) convergent upon the μth cationic state, and where $I(M;\mu)$ represents the ionization energy required to excite the ground state of M to the μth state of M^+.

Eq. (37) is applied in an *inter*molecular fashion: That is, (μ, α, j) is fixed and Eq. (37) is plotted as a function of M, within a group of "related" molecules. Examples are given in Fig. 8, for atoms [32], and in Fig. 9, for molecules [33]. "Related", in the sense which we intend, is a model-dependent con-

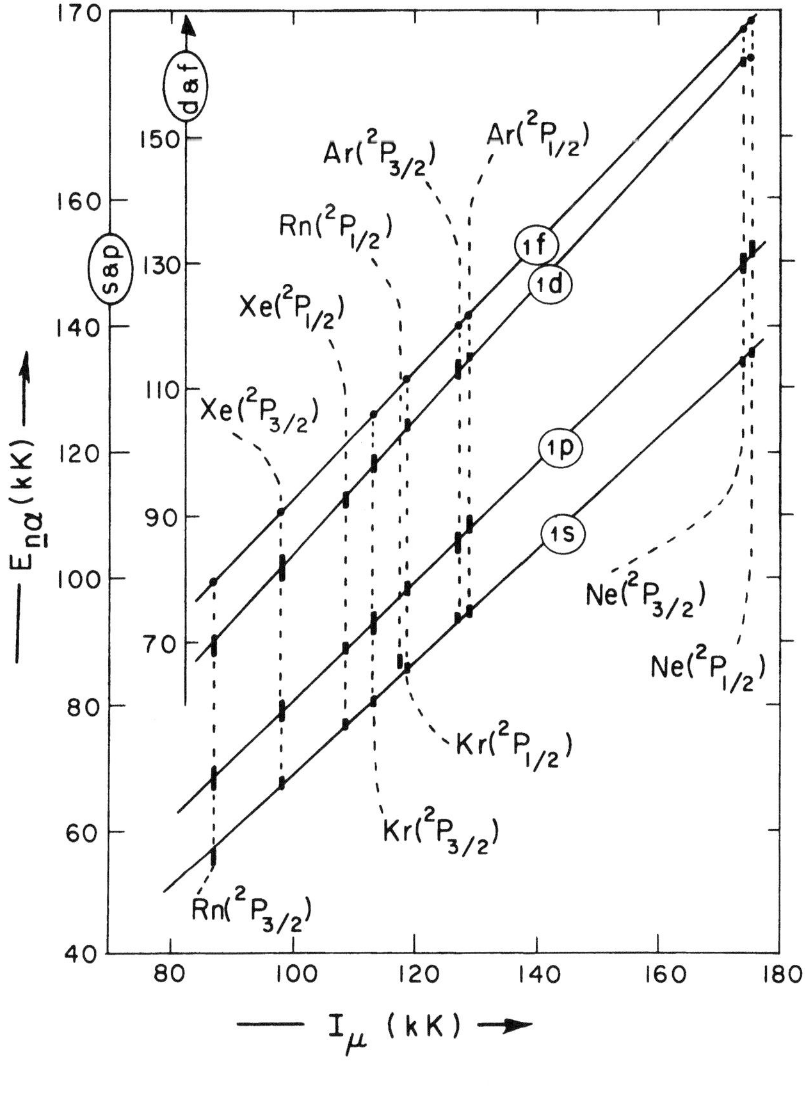

Fig. 8. Plots of $E_{n\alpha}$, $n \equiv n_j$, for j = 1 versus I_μ for the rare gases. Energies are in the units 1kK $\equiv$ 10^3 cm^{-1}. (Taken from [32].)

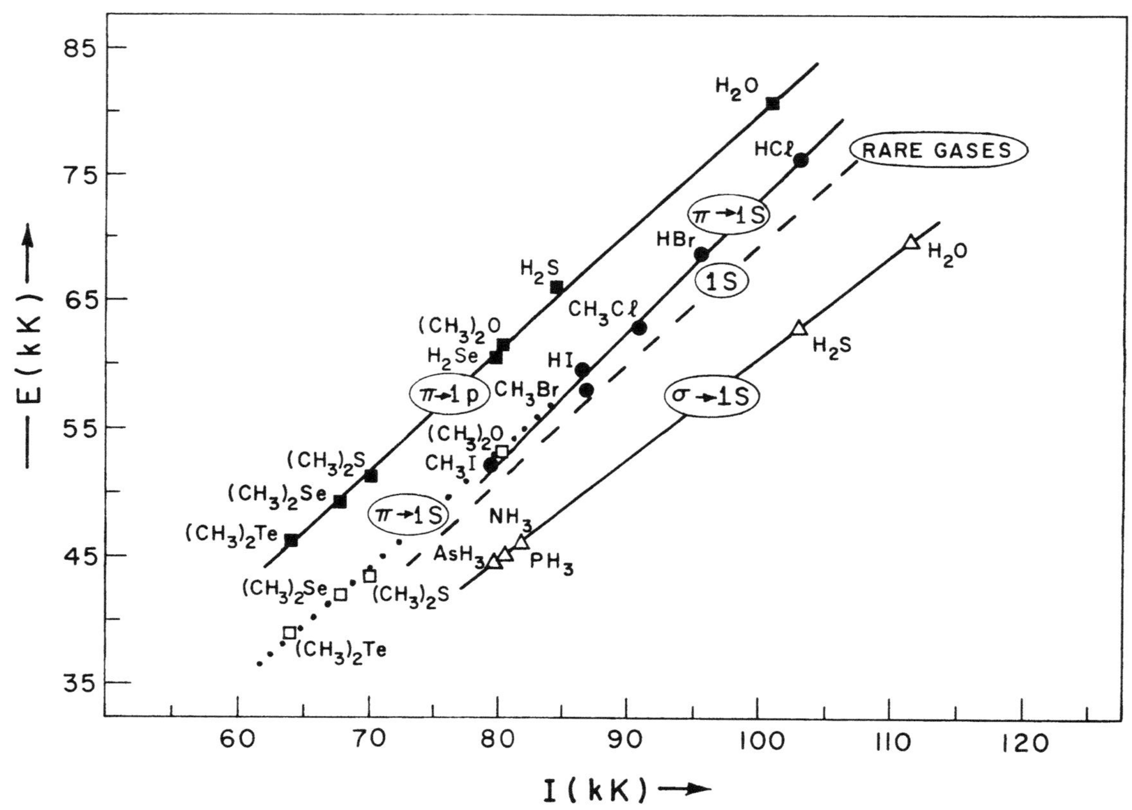

Fig. 9. Plots of Eq. (37) for a variety of molecular Rydberg states. (Taken from [33].)

cept and, in the usual case, represents a requirement that the atoms or molecules under study be valence isoelectronic [2].

Eq. (37) is a restricted version of a more general algorithm [2]

$$T(M;\mu,\alpha,j) - a' + b'T(M;\nu,\beta,k) \qquad (38)$$

which correlates two different (but fixed) term values within a group of molecules. (Indeed, Eq. (37) results from Eq. (38) when one sets $\mu = \nu$ and considers the ionization energy of a molecule to be the deepest term value [2,4].) In addition, similar relationships have been found to apply to Feshbach resonances [36] and to ionization energies expressed as functions of structural parameters for molecular fragments [3,4].

Clearly, the above correlations must be capable of specification within the MQDT. To date, however, this extension has not been made. As a result, we feel free to return to the viewpoint of Section 1, and to provide a model, starting from the Hamiltonian of Eq. (14), which rationalizes the existence of Rydberg correlations.

Recall the one-electron Hamiltonian of Eq. (14), namely,

$$H(M,\mu;1) = -(\hbar^2/2m)\,\nabla^2(1) - e^2\sum_{\omega=1}^{K} Z_\omega/|\underset{\sim}{r}_1 - \underset{\sim}{R}_\omega|$$
$$+ N\int d\tau_2 g(1,2)[\hat{e} - \hat{t}_{12}]\,\Gamma_\mu(2'|2) \quad .$$

In addition to the above Hamiltonian, we make the following two assumptions:

A1. We restrict our attention to molecules of the form XS, where X (the *chromophore*) dominates the affairs of the optical electron, while S (the *substituent*) serves as a perturbation.

A2. Given a chromophore/substituent partitioning, we presume that the X and S charge distributions are expandable in ways which reflect the major variations of the core potential caused by changes in the structure of X and/or S.

This general model, when processed in sufficient detail, predicts the existence of Rydberg correlation algorithms and permits a theoretical evaluation of the algorithmic constants (a,a',etc.) which appear in Eqs. (37,38). We provide a brief outline below.

4.1 Chromophore/Substituent Partitioning

For convenience, we rewrite Eq. (14) as

$$\hat{H}(M,\mu;1) = -(\hbar^2/2m)\,\nabla^2(1) - (N + Z_c)\int d\underset{\sim}{r}_2\; h(1,\underset{\sim}{r}_2)\;\Gamma^{(n)}(\underset{\sim}{r}_2)$$

$$+ N\int d\tau_2\; g(1,2)\;[\hat{e} - \hat{t}_{12}]\;\Gamma_\mu(2'|2) \quad , \tag{39}$$

where $h(1, \underset{\sim}{r}_2) \equiv e^2/|\underset{\sim}{r}_1 - \underset{\sim}{r}_2|$ and where $\Gamma^{(n)}(\underset{\sim}{r}_2)$ is the (singular, normalized) nuclear charge density defined by

$$\Gamma^{(n)}(\underset{\sim}{r}_2) = (N + Z_c)^{-1}\sum_{\omega=1}^{K} Z_\omega\;\delta(\underset{\sim}{r}_2 - \underset{\sim}{R}_\omega) \quad . \tag{40}$$

In line with the determinative role of X in the affairs of the optical electron (i.e., in line with the model constraint *A1*), we can in general partition $H(M,\mu;1)$ as [2-4]

$$H(XS,\mu;1) = H(X/S,\mu;1) + V(S/X,\mu;1) \quad , \tag{41}$$

where the first letter of the symbol X/S or S/X implies the pre-emptive functionality, and where V(S/X), in contradistinction to H(X/S), contains no zeroth-order terms. Eq. (41) results from a partitioning of core densities [2-4]:

$$\Gamma(XS) = N^{-1}[N_X\;\Gamma(X/S) + N_S\;\Gamma(S/X)] \tag{42}$$

and

$$\Gamma^{(n)}(XS) = (N{+}Z_c)^{-1}[(N_X{+}Z_c)\Gamma^{(n)}(X/S) + (N_S{+}Z_c)\Gamma^{(n)}(S/X)] \quad , \tag{43}$$

where N_X and N_S are the numbers of core electrons on X and S, respectively. Thus,

$$H(X/S,\mu;1) = -(\hbar^2/2m)\,\nabla^2(1) - (N_X{+}Z_c)\int d\underset{\sim}{r}_2 h(1,\underset{\sim}{r}_2)\Gamma^{(n)}(X/S;\underset{\sim}{r}_2)$$

$$+ N_X\int d\tau_2\; g(1,2)[\hat{e} - \hat{t}_{12}]\;\Gamma_\mu(X/S;2'|2) \tag{44}$$

and

$$V(S/X,\mu;1) = -(N_S + Z_c)\int d\underset{\sim}{r}_2\; h(1,\underset{\sim}{r}_2)\;\Gamma^{(n)}(S/X;\underset{\sim}{r}_2)$$

$$+ N_S\int d\tau_2\; g(1,2)[\hat{e} - \hat{t}_{12}]\;\Gamma_\mu(S/X;2'|2) \quad . \tag{45}$$

Finally, we note that a given partitioning of the core nuclei into chromophoric and substituent parts [i.e., Eq. (43)] does not determine uniquely the electron partitioning $N = N_X + N_S$ [i.e., Eq. (42)]. This ambiguity is removed by the implicit requirement of model constraint *A1* that the positive hole (monopole Z_C) be entirely located on the chromophoric part of the XS^+ core. This means, for example, that X must have a much lower ionization potential than S. (In this context, see the *R*-state analysis of homonuclear and heteronuclear rare-gas dimers given in [P. M. Dehmer and S. T. Pratt].)

4.2 Correlation Algorithms

The derivation of Rydberg correlation algorithms from the Hamiltonian of Eqs. (41,44,45) devolves on model constraint *A2*. For a group of molecules XS, there are three basic types of algorithm:

(i) S is held constant while X is varied [2]. This is illustrated in Fig. 8 for atoms (i.e., S absent) and in Fig. 9 for molecules (e.g., the sequence STe, SSe, SS, SO, where $S = (CH_3)_2$).

(ii) X is held constant while S is varied [3]. This is illustrated in Fig. 10 for the alkylmonoiodides, the alkylmonobromides and the olefins.

(iii) Both X and S are varied simultaneously [4]. This is illustrated by the correlation lines of Fig. 9.

Type (i) results from a (truncated) Taylor series expansion of the X/S charge density; type (ii) results from a (truncated) multipole expansion of the S/X charge density; and, type (iii) incorporates both expansions of types (i) and (ii). All of types (i)-(iii) ultimately provide correlation algorithms of the general form of Eqs. (37,38), within first-order perturbation theory. And, explicit expressions for the algorithmic constants, as functions of X/S and S/X potentials, have been given in every case [2-4].

5. FIELD EFFECTS

(The problem of *R* states in electric and magnetic fields is treated at great length, both experimentally and theoretically, elsewhere in this volume. Some of the relevant chapters include: atoms, [M. Robnik], [J. C. Gay], [H. M. Crosswhite, F. S. Tomkins, H. Crosswhite and W. R. S. Garton] and [J. A. C. Gallas and R. F. O'Connell]; molecules, [J. D. Scott], [A. Gedanken] and [D. D. Altenloh and B. R. Russell].

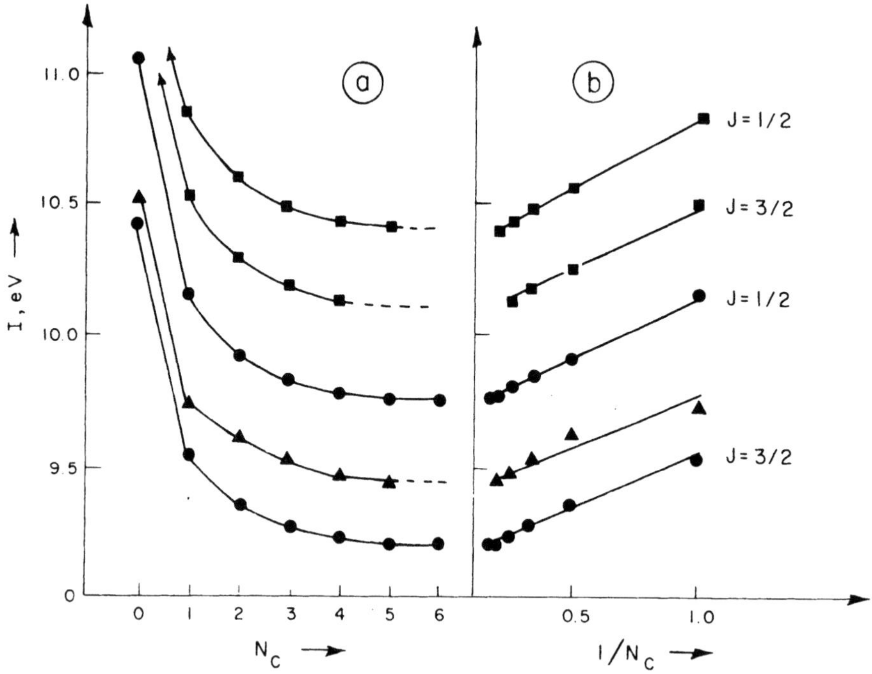

Fig. 10. (a) Plot of lowest ionization energy I of monosubstituted alkanes with unbranched alkyl chains versus the number of alkyl carbon atoms N_c. Data point symbols are as follows: circle, alkylmonoiodides; square, alkylmonobromides; triangle, olefins. The curves labeled 3/2 and 1/2 connect data that correspond to the lower and higher doublet components, respectively. The data for $N_c = 0$ are those for HI and $CH_2{=}CH_2$; the points for HBr are not included, but their positions are indicated by arrows. (b) As in (a), except that the abscissa is $1/N_c$. (Taken from [3].)

5.1 Magnetic Field Effects

The application of a magnetic field $\underset{\sim}{B}$ introduces the Zeeman term $H_B = \underset{\sim}{\mu} \cdot \underset{\sim}{B}$ (where $\underset{\sim}{\mu}$ is the magnetic moment) into the Hamiltonian. Given the normal bandwidths for polyatomic molecules, it is improbable that line splittings due to the Zeeman effect will be observed in such systems, at least for $B \lesssim 20$ T. The observation of Zeeman effects then devolves on the measurement of band-shape differences, for example, the measurement of magnetic circular dichroism (MCD) [37], which is the difference in molecular absorptivity of left and right circularly-polarized light in the presence of a magnetic field. The primary aim of such studies is to make state assignments. Unfortunately, difficulties intrude. For example, μ is different in the various molecular spin-orbit limits of which, in contrast to the atom, there are more than two [i.e., (Λ,S) and (Ω,ω)]. Indeed, the only well-defined quantum number is a resultant of molecular rotational and electronic motions. The situation, then, is very complicated in a diatomic molecule; in a polyatomic molecule, because of large coriolis effects, it can become even worse.

It has been shown [23,26,38], however, that MCD can provide rather precise information. We exemplify, again using the first s complex of CH_3I in a primitive $C_{\infty v}$ point group (cf. Section 3). In this instance, we find the spectra shown in Figs. 11 and 12 [26]. The MCD signal is [37]

$$\Delta A^H(\tilde{\nu}_0) \equiv A_-^H(\tilde{\nu}_0) - A_+^H(\tilde{\nu}_0) = aB(dA^0(\tilde{\nu})/d\tilde{\nu})_{\tilde{\nu}_0} + bBA^0(\tilde{\nu}_0) \quad , \tag{46}$$

where A denotes absorptivity, the superscripts H and O denote the presence or absence of the field $\underset{\sim}{B}$, respectively, and the subscripts + and - denote right circularly-polarized and left circularly-polarized light, respectively. Since the ground state $^1\Sigma^+$ possesses no angular momentum and, therefore, a is proportional to μ_β of the excited state β, the derivative component of Eq. (46) will be large when μ_β is large and absent when μ_β is zero. The second term in Eq. (46), attributable to field-induced mixing, is usually small and will only be important when a is zero. The relevant factors are that a derivative signal (+ or - phase) denotes an excited-state magnetic moment the size of which, namely $\mu_\beta = -(\Lambda_c + 2\Sigma_c + \lambda + 2\sigma)\hbar^{-1}\mu_{Bohr}$ in (Ω,ω) coupling, is proportional to signal height, and that an absorptive-type signal (+ or - phase) denotes an excited state possessing zero magnetic moment.

The experimental magnetic moments of states 1,2,3 and 4, as evaluated [38] from the data of Figs. 11 and 12, are 1, 0.33, 0 and 0.27, respectively, whereas those obtained from the μ_β expression given above are 1, 0.33, 0 and 0.33, in excellent agreement with experiment. A great deal of other information is contained in the phasing and is discussed in [38,39] and in [J. D. Scott].

The quadratic field effect, attributable to diamagnetism (see [J. C. Gay]), is usually very small. However, since the diamagnetic interactions are proportional to the Rydberg cross-sectional area, they increase as ν^4, whereas the electrostatic bind-

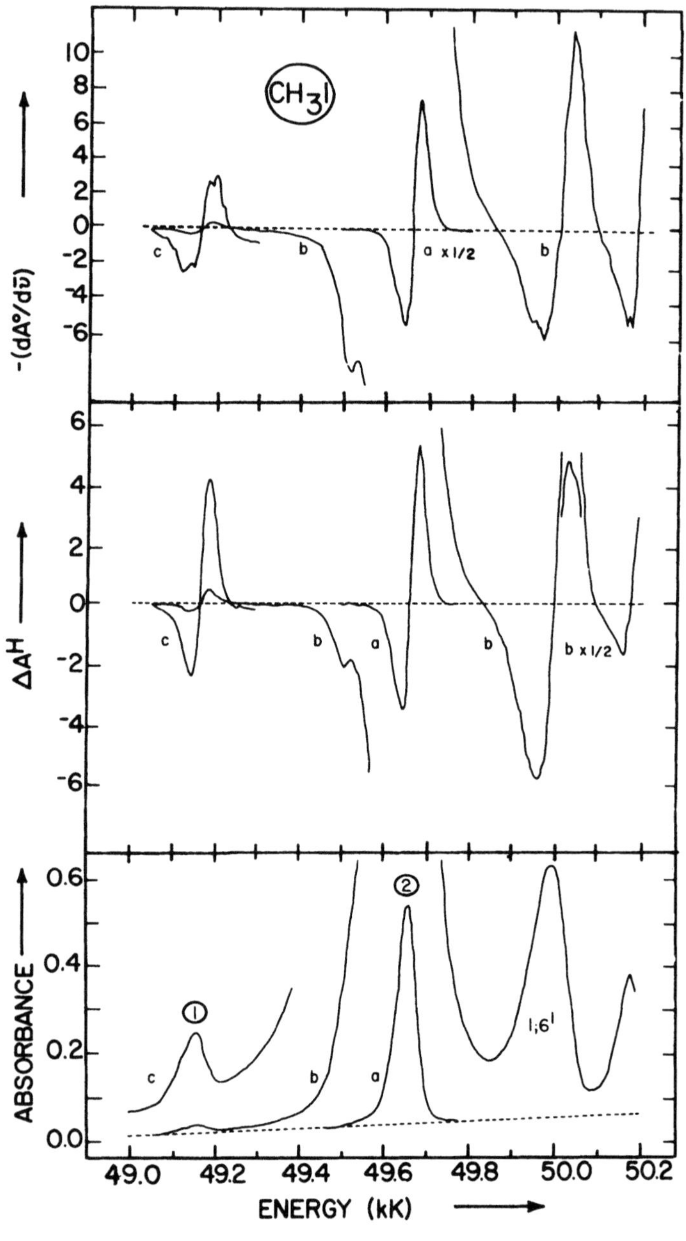

Fig. 11. Absorption spectra A^{o}, MCD spectra ΔA^{H} and derivative spectra $dA^{o}/d\bar{\nu}$ of CH_3I in the 6s Rydberg complex. a = 0.01, b = 1 and c = 10 torr. The energy units are 1 kK $\equiv 10^{3}$ cm^{-1}. (Taken from [26].)

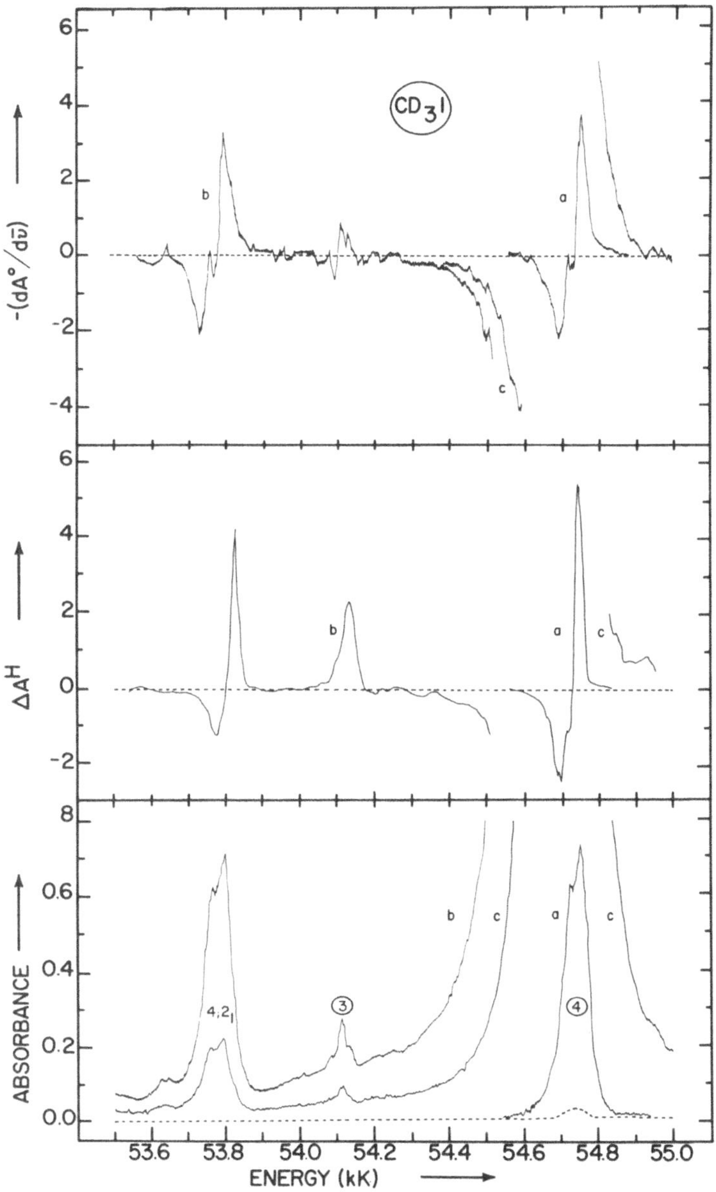

Fig. 12. Absorption spectra A^o, MCD spectra ΔA^H and derivative spectra $dA^o/d\bar{\nu}$ of CD_3I in the 6s Rydberg complex. a = 0.0085, b = 2.26, c = 0.665 torr. The energy units are 1 kK ≡ 10^3 cm^{-1}. (From [26].)

ing energies decrease as ν^{-2}. Thus, the ratio of magnetic energy to electrostatic binding energy increases as ν^6 and at n = 30, for example, is about 10^9 times larger than for n = 1. In this case, the quadratic field effect not only dominates the Zeeman effect but also overwhelms the electrostatic binding energy and converts the system into a "magnetic system". Quadratic field effects have been observed in atoms [40,41] and a striking example is given in [H. M. Crosswhite, F. S. Tomkins, H. Crosswhite and W. R. S. Garton]. We [42] have searched CH_3I and C_6H_6 for quadratic field effects in the vicinities of both I_1 and I_2; and, although distinct field perturbations were evident, no behavior reminiscent of a magnetic system was observed.

5.2 Electric Field Effects

The study of molecules in strong external electric fields, while burgeoning [43], is still in its infancy. One goad for research is provided by the development of focused, pulsed laser systems, with power densities in the GW/cm^2 range, which are strong enough to produce massive distortions of molecular electronic charge distributions. The Rydberg regime provides a facile means of performing such "high field" investigations. Since the mean Coulomb field of a Rydberg molecule is $Z_c/\langle r^2\rangle \propto \nu^{-4}$, it follows that even a very weak external electric field can induce ionization at large n values. Furthermore, since level densities increase as ν^5, it follows that weak-field-induced mixings and, hence, massive nonlinearities can occur in high-n R states. Unfortunately, few or no investigations of high-n molecular states are known. Indeed, the little VUV work available [37,44,45] is concerned mainly with state identifications. (In this context, see [D. D. Altenloh and B. R. Russell].)

The electron of a hydrogenic system in a static electric field finds itself in the potential

$$V(r) = -\frac{Z_c}{r} + \underset{\sim}{E} \cdot \underset{\sim}{r} \quad . \tag{47}$$

The effect of the term $H_E = \underset{\sim}{E}\cdot\underset{\sim}{r}$ is to couple the discrete states of the field-free atom, producing changes in the atomic structure (i.e., Stark effects); however, it can also produce discrete-state/continuum coupling, leading to field ionization and to a real alteration in the ionization threshold. The situation in a molecule is very different from that in the atom: The molecule possesses dipole moments, different in both ground (unprimed) and excited (primed) states, which the atom does not. Thus, if we define $\Delta\underset{\sim}{\mu} = \underset{\sim}{\mu} - \underset{\sim}{\mu}'$, asserting that both dipole moments are either parallel or antiparallel to the principal molecular axis (which itself lies at an angle θ to $\underset{\sim}{E}$), and if we define $\Delta\alpha = \alpha - \alpha'$, where α is a mean isotropic polarizability, we can write [44]

$$H_E(\theta) = (hc)^{-1}\, \Delta\underset{\sim}{\mu} \cdot \underset{\sim}{E} - \frac{1}{2}\, \Delta\alpha E^2 \quad . \tag{48}$$

The total effects are considerably more complex than that implied above. Yet, for the purposes of this discussion, Eq. (48) contains the essence of our interests. Thus, the Stark measurements (low n or low field) provide a means of measuring μ', and the effect of measurements at high n, quadratic because α varies as ν^6, will yield α'.

As in the case of MCD, because of bandwidth problems the measurement of molecular Stark effects usually devolves on the detection of bandshape alterations. An applicable method has been discussed [37] and a few examples [44,45] of its use are available. Briefly, the experimental format known as electric linear dichroism (ELD) is preferable. If the electric vector of the incident light is plane-polarized and at an angle γ to the external electric field $\underset{\sim}{E}$ (which in turn is perpendicular to the direction of light incidence), the field-on/field-off absorbance difference (i.e., the ELD signal) is [37]

$$\Delta A^E_\gamma(\tilde{\nu}_o) \equiv A^E_\gamma(\tilde{\nu}_o) - A^o(\tilde{\nu}_o) = E^2 \Big\{ C^\gamma_1 A^o(\tilde{\nu}_o) + C^\gamma_2 \Big[(dA^o(\tilde{\nu})/d\tilde{\nu})_{\tilde{\nu}_o} - \frac{A^o(\tilde{\nu}_o)}{\tilde{\nu}_o} \Big] + C^\gamma_3 \Big[(d^2A^o(\tilde{\nu})/d\tilde{\nu})_{\tilde{\nu}_o} - \frac{2}{\tilde{\nu}_o}\, (dA^o(\tilde{\nu})/d\tilde{\nu})_{\tilde{\nu}_o} + 2\, \frac{A^o(\tilde{\nu}_o)}{\tilde{\nu}_o^{\,2}} \Big] \Big\} \quad . \tag{49}$$

This expression is obviously more complicated than that in the magnetic field case, primarily because of the presence of the second derivative. Expressions for the coefficients C_1, C_2 and C_3 are available [37] in terms of dipole moments, polarizability tensor components and field-induced mixing parameters, and experimental means of extracting them are known (see [J. D. Scott]).

6. RYDBERG STATES IN CONDENSED MEDIA

It is impossible to discuss molecular *R* states in condensed media without an understanding of them in the gas phase. Unfortunately, our understanding of gas-phase behavior is not very good and, for that reason, we feel compelled to initiate our discussion of the condensed phase by a retreat to the gas.

6.1 Gas Phase

In the vapor, molecules may exhibit the following characteristics:

C1. A well-developed *R* series, almost hydrogen-like in that the intensity decreases as $1/\nu^3$.

C2. A well-developed *R* series, but with maximum intensity at intermediate values of n.

C3. A collapsed *R* series, often with only the lowest member observable (e.g., the lowest *R*(s) transition in H_2O, amides, saturated hydrocarbons, etc., characterized usually by being anomalously broad, intense and quantum-defect deviant [46]).

C4. A series such as *C1* or *C2*, but with an anomalous quantum defect for the first few members [47] (see, for example, our discussion of SQDT in Section 2.1).

Rydberg series may, of course, exhibit many perturbations, the recognition and identification of which has been the content of Sections 2 and 3. We now concern ourselves primarily with the series characteristics *C1*-*C4*. It seems to us that a particularly simple model capable of rationalizing items *C1*-*C4* is a double-well potential (cf. Fig. 13).

For atomic *R* states, V(r) is a monotonically-increasing negative-definite function of r. However, the effective potential

$$V_{eff}(r) = V(r) + \frac{\ell(\ell + 1)}{r^2} \tag{50}$$

is monotonic only for s-waves. For $\ell \neq 0$ ℓ-waves, $V_{eff}(r)$ is generally double-welled and exhibits a centrifugal barrier (CB) such that the bound states may be approximate eigenstates of either the inner well (IW) or the outer well (OW). The CB controls the finite number of states, some perhaps resonant, which can be sustained in the IW. The OW potential is asymptotically hydrogenic and it sustains an infinite number of *R* states.

For molecular *R* states, since $V_{rm}(\underset{\sim}{r})$ is short-ranged, the effective potential is essentially spherically symmetric at large r. At small r, however, the effects of $V_{rm}(\underset{\sim}{r})$ are drastic. From an atomic point of view, the imposition of $V_{rm}(\underset{\sim}{r})$ on an atomic potential effects a symmetry breaking of the sphere group. Retaining $V_{eff}(\underset{\sim}{r}) \simeq V_{eff}(r)$, the result is the induction of ℓ-wave mixing in a spherically symmetric potential. Thus, even for a parental s-wave, the effective potential for a molecule will be of the double-well form but the barrier may be negative-definite: That is, $V_{eff}(r)$ may be a double-well potential for molecular *R* states regardless of the parental ℓ value. As for atoms, the nature of the barrier determines the number of bound levels in

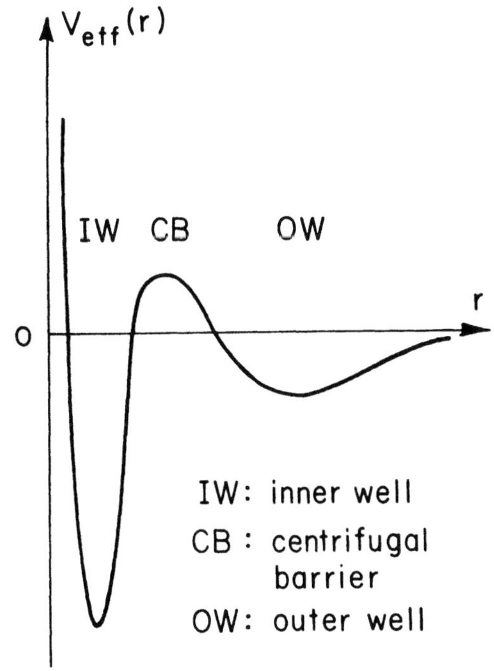

Fig. 13. A model double-well potential: IW, inner well; CB, centrifugal barrier; OW, outer well. (Taken from [1].)

the IW and, more importantly, the localization of wave amplitude in either the IW or the OW region.

We now attempt to rationalize the gas-phase characteristics. In order to do so, we note that the initial orbital in the absorptive event is largely confined in the IW and, consequently, that only those terminal *R* orbitals with some IW amplitude will couple optically with the ground state. In this view, then, explanation proceeds as follows:

E1. The CB is either very weak or absent.

E2. The CB is weak and narrow and, while the IW cannot sustain even one bound level, all levels are bimodal (i.e., possess both IW and OW amplitude), with this bimodal component maximizing at intermediate n.

E3. One or more levels can be sustained in the IW, the CB being such that all eigenfunctions are either wholly IW or wholly OW, with little bimodality (i.e., mixing) occurring.

E4. There is always a bimodal component in the first s *R* member.

6.2 Condensed Phase

Consider now a molecule-doped insulator, either liquid or solid. The dopant molecules exhibit the following behavior upon formation of the condensed phase:

C5. Collapsed *R* series (cf. *C3*, above) remain largely unaffected [46].

C6. Total quenching of *R* series is common [47-50].

C7. A Wannier exciton behavior results and a few members of the exciton *R* series are observed [51]. (The subject of Wannier excitons in doped rare-gas solids is treated comprehensively in [J. Jortner, E. E. Koch and N. Schwentner].)

C8. Anions which do not support any bound states in the gas phase may do so in the condensed phase; moreover, some of these states may be resonant and may lead to electron transfer (i.e., charge-transfer-to-solvent (CTTS) [52]) phenomena in aqueous solution and single crystal hosts [46]. (In this context, see the discussion of negative-ion states given in [R. N. Compton].)

The alterations in $V_{eff}(r)$ caused by the condensation of the guest into a condensed insulator host are as follows. To first order, $V_{eff}(r)$ remains unaffected by the crystal field of the host at small r; at large r, the conduction band of the host will dominate $V_{eff}(r)$ so that

$$\lim_{r\to\infty} I\ (\text{dopant}) = E\ (\text{conduction band})\ .$$

At intermediate r, further substructure (i.e., negative-definite barriers) may or may not appear in the OW region. If the dopant is a charged entity, such as an anion, the resulting host-guest interactions become so drastic that even the IW region is grossly altered [53]. Indeed, in this case the deepening of the IW may support bound states and, if substructure appears in the intermediate r region, some of these states may become resonant (i.e., may autoionize to the host conduction band). The behavioral characteristics *C5*–*C8*, then, are explained as follows:

E5. These states, already localized in the IW, remained largely unaffected by the environment, except possibly for a rigid shift in energy.

E6. The host conduction band must lie near or below the first *R* series member and V_{eff} substructure in the OW region must be minimal.

E7. The host conduction band must lie above the first few *R* series members and/or a large substructure of V_{eff} must occur in the OW region.

E8. The events already discussed for a monopolar dopant must occur [53].

CONCLUDING REMARKS

We have tried to assess broadly the Rydberg area of molecular electronic structure and interactions in the frame of modern atomic physics, building in the appropriate molecular extensions where necessary. In the course of this effort, it became totally obvious that the chemical literature is not only *not* keeping pace with the rapidly advancing area of atomic physics, but that it is also losing ground in molecular physics. We believe this to be most unfortunate because it puts the chemist at a distance from the cutting edge of his subject.

The intent of the present volume (and, indeed, of the NATO Advanced Study Institute from which it is derived) is to provide an effective chemistry/physics interface for the study of highly-excited states. The need for such an interface is manifest: The growing awareness within the physics and chemistry communities that they are, after all, discussing the same problem requires that each community learn the other's language. We have attempted in the present chapter -- and, as editors, in the collation of the following chapters -- to construct an initial lexicon.

We are convinced that high-energy chemistry is tied, in large measure, to our understanding of molecular *R* states (see, for example, [M. N. R. Ashfold], [M. B. Robin], [A. R. Rossi and R. Srinivasan] and [C. von Sonntag]). Perhaps one simple example will suffice. Consider a prototypical light-induced charge-exchange reaction,

$$D + A + h\nu \rightarrow D^+ + A^- \quad ,$$

where D and A, which may be either atoms or molecules, are an electron donor and acceptor, respectively. From a chemical point of view, a natural transition state for this reaction would be

$$D + A + h\nu \rightarrow (D^+ \ldots e^- \ldots A) \rightarrow D^+ + A^- \quad .$$

Obviously, this transition state can be viewed as a highly-excited state D* which is interacting with an external field, namely that provided by A; and, depending upon the nature of the potential generated by A, A^- may be either stable (i.e., bound) or metastable (e.g., shape-resonant). Moreover, the efficiency of this reaction will be determined largely by the de-excitation channels available to D* (see, for example, [S. Leach] and [E. C. Lim]).

The above example is of more than academic interest: Such reactions are common in the upper atmosphere. Indeed, even at lower energies than the VUV, if D has a low enough ionization potential that its Rydberg states overlap the valence region (e.g., D = chlorophyll), D* may be a Rydberg/valence mixed state (cf. [C. Sandorfy and L. S. Lussier]). The transition state, then, may well be dominated by field ionization of the D* Rydberg component, even at low energies. And, since many biological reactions are of the charge-exchange type (e.g., the primary photophysical event in photosynthesis), an intrusion of *R*-state concepts into this area will assuredly provide new insights.

ACKNOWLEDGMENT

This work was supported by the United States Department of Energy.

REFERENCES

1. S. P. McGlynn and G. L. Findley, J. Photochem. 17, 461 (1981).
2. P. Hochmann, G. L. Findley and S. P. McGlynn, Z. Naturforsch. 35a, 595 (1980); Errata, Z. Naturforsch. 35a, 1431 (1980).
3. P. Hochmann, S. P. McGlynn and G. L. Findley, Int. J. Quantum Chem., Quantum Biol. Symp. 9, 173 (1982).
4. A. M. Langsjoen, P. Hochmann, P. Brint, S. P. McGlynn and G. L. Findley, Phys. Rev. A, submitted.
5. See, for example, R. McWeeny and B. T. Sutcliffe, *Methods of Molecular Quantum Mechanics* (Academic Press, New York, 1969).
6. H. -t. Wang, W. S. Felps, G. L. Findley, A. R. P. Rau and S. P. McGlynn, J. Chem. Phys. 67, 3940 (1977).
7. For a recent, masterful summary, see: M. J. Seaton, Rep. Prog. Phys. 46, 167 (1983).
8. J. L. Dehmer and U. Fano, Phys. Rev. A2, 304 (1970).
9. M. J. Seaton, Proc. Phys. Soc. 88, 801,815 (1966).
10. K. T. Lu, Phys. Rev. A4, 579 (1971).
11. M. A. Baig, J. P. Connerade, J. Dagata and S. P. McGlynn, J. Phys. B 14, L25 (1981).
12. H. Beutler, Z. Phys. 93, 177 (1935).
13. U. Fano, Phys. Rev. 124, 1866 (1961).
14. U. Fano, Phys. Rev. A2, 353 (1970).

15. J. A. Dagata, G. L. Findley, S. P. McGlynn, J. P. Connerade and M. A. Baig, Phys. Rev. A24, 2485 (1981).
16. K. T. Lu and U. Fano, Phys. Rev. A2, 81 (1970).
17. J. P. Connerade, M. A. Baig and S. P. McGlynn, J. Phys. B 14, L67 (1981).
18. O. Atabek and C. Jungen, in: *Electron and Photon Interactions with Atoms*, H. Kleinpoppen and M. R. C. McDowell, eds. (Plenum, New York, 1976).
19. D. Dill and Ch. Jungen, J. Phys. Chem. 84, 2116 (1980).
20. Ch. Jungen and D. Dill, J. Chem. Phys. 73, 3338 (1980).
21. S. Felps, P. Hochmann, P. Brint and S. P. McGlynn, J. Mol. Spectrosc. 59, 355 (1976).
22. R. S. Mulliken, Phys. Rev. 57, 500 (1940).
23. S. P. McGlynn, J. D. Scott, W. S. Felps and G. L. Findley, J. Chem. Phys. 72, 421 (1980).
24. S. P. McGlynn, W. S. Felps and G. L. Findley, Chem. Phys. Lett. 78, 89 (1981).
25. W. S. Felps, G. L. Findley and S. P. McGlynn, Chem. Phys. Lett. 81, 490 (1981).
26. S. P. McGlynn, W. S. Felps, J. D. Scott and G. L. Findley, J. Chem. Phys. 73, 4925 (1980).
27. W. S. Felps, J. D. Scott, G. L. Findley and S. P. McGlynn, J. Chem. Phys. 74, 4832 (1981).
28. W. S. Felps, S. P. McGlynn and G. L. Findley, J. Mol. Spectrosc. 86, 71 (1981).
29. H. J. Maria, J. L. Meeks, P. Hochmann, J. F. Arnett and S. P. McGlynn, Chem. Phys. Lett. 19, 309 (1973).
30. S. Foster, P. Hochmann, W. S. Felps, H. -t. Wang and S. P. McGlynn, in: *Chemical Spectroscopy and Photochemistry in the Vacuum Ultraviolet*, C. Sandorfy, P. J. Ausloos and M. B. Robin, eds. (D. Reidel, Dordrecht, 1974).
31. P. Hochmann and S. P. McGlynn, in: *Vacuum Ultraviolet Radiation Physics*, E. E. Koch, R. Haensel and C. Kunz, eds. (Pergamon, Braunschweig, 1975).
32. S. Chattopadhyay, P. Hochmann and S. P. McGlynn, J. Chem. Phys. 65, 3341 (1976).
33. S. P. McGlynn, S. Chattopadhyay, P. Hochmann and H. -t. Wang, J. Chem. Phys. 68, 4738 (1978).
34. J. H. Fock, P. Gürtler and E. E. Koch, Chem. Phys. 47, 87 (1980).
35. G. L. Findley, P. Hochmann and S. P. McGlynn, Ann. Israel Phys. Soc. 6, 161, 164 (1983).
36. D. Spence, J. Chem. Phys. 66, 669 (1977); Phys. Rev. A15, 883 (1977).
37. J. D. Scott, W. S. Felps and S. P. McGlynn, Nuc. Instrum. Meth. 152, 231 (1978).
38. J. D. Scott, W. S. Felps, G. L. Findley and S. P. McGlynn, J. Chem. Phys. 68, 4678 (1978).

39. J. D. Scott, W. S. Felps and S. P. McGlynn, in: *Proceedings of the Colloque International CNRS*(Aussois, France, 1982), J. de Physique (France), Colloque C2, Supplement 11 (1982).
40. W. R. S. Garton and F. S. Tomkins, Astrophys. J. 158, 839 (1969).
41. C. D. Harper and M. D. Levinson, Opt. Commun. 20, 107 (1977).
42. J. P. Connerade, W. R. S. Garton, M. A. Baig, J. Hormes and S. P. McGlynn, unpublished.
43. J. E. Bayfield, Phys. Rep. 51, 317 (1979).
44. J. D. Scott, and B. R. Russell, J. Chem. Phys. 63, 3243 (1975).
45. G. C. Causley, J. D. Scott and B. R. Russell, Rev. Sci. Instrum. 48, 264 (1977).
46. S. P. McGlynn and G. L. Findley, in: *Proc. Workshop on Far VUV Spectroscopy and Photochemistry* (Mülheim a.d. Ruhr, 1981), pp. 99-102.
47. M. B. Robin, *Higher Excited States of Polyatomic Molecules* (Academic Press, New York, 1975).
48. M. B. Robin and N. A. Kuebler, J. Mol. Spectrosc. 33, 274 (1970).
49. I. Messing, B. Raz and J. Jortner, Chem. Phys. 23, 351 (1977).
50. J. W. Lewis, R. V. Nauman, D. B. Bouler and S. P. McGlynn, J. Phys. Chem. 87, 3611 (1983).
51. J. Jortner, in: *Vacuum Ultraviolet Radiation Physics*, E. E. Koch, R. Haensel and C. Kunz, eds. (Pergamon, Braunschweig, 1974), p. 263.
52. M. Fox, in: *Concepts of Inorganic Photochemistry*, A. W. Adamson and P. D. Fleischer, eds. (Wiley-Interscience, New York, 1975), p. 333.
53. G. L. Findley and S. P. McGlynn, Rev. Mod. Phys., to be published.

METHODS FOR STUDYING HIGHER EXCITED STATES IN MOLECULES AND MOLECULAR CRYSTALS BY MEANS OF SYNCHROTRON RADIATION

Ernst-Eckhard Koch and Peter Gürtler

Hamburger Synchrotronstrahlungslabor HASYLAB
Deutsches Elektronen-Synchrotron DESY
2000 Hamburg 52, West Germany

1. INTRODUCTION

Progress in spectroscopy, historically, has been intimately linked to the availability of radiation sources. Synchrotron Radiation (SR) sources have only recently been developed for broad application and, at present, there is a rapidly expanding worldwide effort in the use of these sources [1-4].

Synchrotron radiation is produced when a fast, charged particle, with energy $\gg m_0c^2$, is deflected in a strong magnetic field. Circular electron and positron accelerators and storage rings, wherein the particles move with relativistic energies, are the "natural" sources of SR, which exhibits a well-known and unique combination of the following characteristics:

--- wide spectral continuum from the infrared out to the x-ray region.

--- high intensity (brightness).

--- high degree of collimation ($\sim$1 mrad).

--- well-defined linear or elliptical polarization properties.

--- subnanosecond time structure; pulse duration as short as 100 ps.

--- accurately calculable spectral distribution.

S. P. McGlynn et al. (eds.), Photophysics and Photochemistry in the Vacuum Ultraviolet, 41–103.

--- high temporal and spatial stability (storage rings) and clean environment (10^{-9} Torr).

Thus, SR bridges the gap between the far-UV and the x-ray range. Since 1953, when synchrotron radiation was used for the first time for optical experiments in the VUV by Hartman and Tomboulian [5], spectroscopic work employing this excellent VUV light source has matured in many ways. SR, which is considered by the accelerator builder as the fundamental limit in his ability to raise the energy of circular accelerators and storage rings indefinitely, has emerged within the last 15 years as a very powerful tool. The main stages of this emergence were brought about by the outstanding characteristics of this light source for spectroscopy and structural research and may be characterized as follows:

--- In the 1940's, prediction [6] observation [7,8] and theoretical treatment [9,10] of synchrotron radiation [11].

--- In 1953, first spectroscopic application of SR by Hartman and Tomboulian [5].

--- Since about 1963, broad application of SR in programs devoted to the investigation of the electronic structure of atoms, molecules and solids by absorption and reflection experiments.

--- Since about 1970, application of SR in studies of secondary processes, such as luminescence, fragmentation and photoemission spectroscopy, in several laboratories.

--- Since about 1970, soft x-ray diffraction experiments.

--- Beginning of the 1970's, use of storage rings with greatly improved currents and stability as VUV and x-ray sources.

--- Late 1970's and early 1980's, commissioning of large synchrotron radiation centers (e.g., "Synchrotron Radiation Source" SRS in Great Britain, "Photon Factory" in Japan, "National SR Light Source" in the U.S.).

--- Late 1970's and early 1980's, use of wigglers and undulators [12].

--- Early 1980's, construction of free-electron lasers [13].

As a result of the excellent research opportunities which became evident to a larger group of scientists from different disciplines in the 1970's, the demand for SR sources significantly outran the supply. In response, a number of SR facilities have recently increased their capacities, and a number of new dedicated storage rings have been built or are nearing completion.

Even without wigglers and undulators, SR has made important contributions to spectroscopy. Interactions of photons with matter can be classified according to three different processes, namely, absorption, inelastic scattering and elastic scattering. The two first processes are essentially used to obtain information about electronic structures of atoms, molecules and condensed materials; the third process provides structural information, especially in the x-ray region. SR has been mainly used to study electronic excitations by absorption and reflection spectroscopy or by following excited systems via fluorescence, photoelectrons or photoions. X-ray scattering experiments with SR are more recent, are also extremely promising, and have become widely practiced. The ability to perform these experiments using energy-variable SR, which can be selected from the continuum by various types of monochromators, is a major advantage. Furthermore, there are a number of experiments which can very elegantly make use of the white continuum of SR.

Among the areas of current research, where the use of SR as a light source has enabled important contributions, are:

--- Atomic and molecular spectroscopy of highly-excited states, including absorption, photoelectron and mass spectroscopy as well as time-resolved spectroscopy.

--- Investigation of the energy band structure of solids and surfaces by means of angle-resolved photoelectron spectroscopy and of excitons; deep level spectroscopy.

--- X-ray absorption and x-ray scattering experiments for the investigation of the geometrical structure of crystalline and amorphous solids, dilute systems and biological materials.

--- Use of SR for applied research in soft x-ray optics, soft x-ray microscopy and x-ray lithography.

All of the above developments have become possible as a result of remarkable breakthroughs in accelerator and storage-ring technology, and in instrumentation. In the present paper we discuss a few of these newly developed methods. In doing so, we concentrate rather narrowly on those experimental methods used to study higher excited electronic states in molecules and molecular crystals by means of SR, and on the interplay of these methods with current problems in VUV spectroscopy. Necessarily, a brief paper of this kind can give only a small selection from the many interesting and ingenious developments and cannot be exhaustive. Instead, new developments are emphasized, with particular reference to work done in our laboratory and, therefore, more familiar to us. This is, of necessity, an arbitrary selection, which largely reflects our own personal interests. The reader who is interested in more general reviews is referred to recent books and conference proceedings [2-4,14].

The experimental programs currently on-going at SR laboratories cover many areas and, even restricting our discussion to the spectroscopy of molecules and molecular crystals, there are many diversified experiments using different techniques, such as absorption, reflection, photoelectron, fluorescence and mass spectroscopy. Fig. 1 and Table 1 may give an idea of the wide variety of experiments and spectroscopic techniques. Often, these experiments are extensions and improvements of existing techniques employing conventional sources. But, as we shall see below, the use of SR has added, in most cases, new dimensions or has enabled completely new classes of experiments.

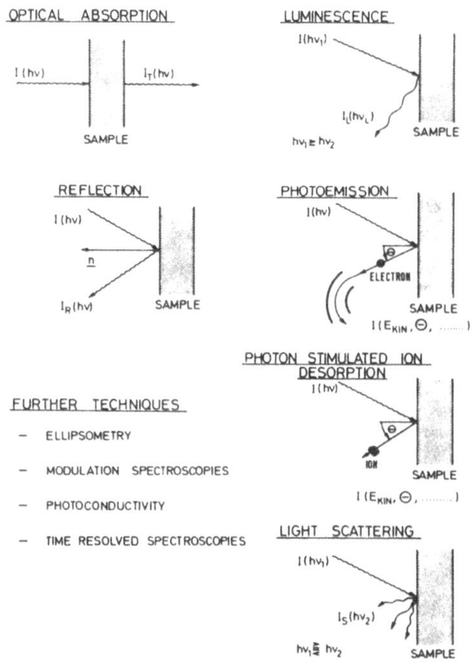

Fig. 1. Schematic illustration of various spectroscopic techniques using SR.

Table 1. Experiments in Molecular Spectroscopy Using Synchrotron Radiation (for general reference see [25])

Experiment/Technique	Information
<u>Absorption/Reflection</u>	
Excitation of valence and inner-shell electrons Various classes of molecules VUV matrix-isolation spectroscopy Molecules on surfaces	total cross-sections excitation energies ionization potentials electronic, vibrational and rotational energy levels potential energy curves Franck-Condon factors molecular structure (EXAFS) correlation effects
<u>Photoelectron Emission</u>	
Angle-dependent hν-dependent Adsorbed species Resonant photoemission	total and partial cross-sections ionization potentials electronic and vibrational energy levels orbital symmetry, FC factors β-parameters relaxation effects
<u>Fluorescence</u>	
From excited primary molecule From excited dissociative product Time-resolved	fluorescence yield radiative and radiationless decay processes lifetimes, branching ratios partial density of states
<u>Mass Spectrometry</u>	fragmentation cross-sections relaxation phenomena photochemical reactions dissociation, photodesorption
<u>Coincidence Experiments</u>	detailed information on the final state prepared by photoabsorption or photon scattering

2. PROPERTIES OF SYNCHROTRON RADIATION SOURCES: A SHORT OVERVIEW

Before turning to a discussion of experimental methods, we first briefly recall the basic properties of SR which make it such a versatile tool. SR properties can be derived by applying the methods of classical electrodynamics to the motion of relativistic electrons (and positrons) in circular orbits. Major properties of SR can be easily understood from the picture of an oscillating dipole (a classical antenna) subjected to a Lorentz transformation. In particular, because of the relativistic speed of the electrons, the angular distribution of emitted intensity will be strongly distorted into a narrow cone in the instantaneous direction of motion of the electron (cf. Fig. 2) and the emitted spectrum will be strongly Doppler-shifted. The generation of a continuous spectrum extending from the microwave region to the hard x-ray region can be explained by considering the harmonics of the frequency of revolution, which is in the microwave range. The relativistic transformation of the radiation pattern from the rest frame of the electron into the laboratory frame leads to a distribution of intensity into higher harmonics of the revolution frequency. The higher the e^- energy, the higher will be the har-

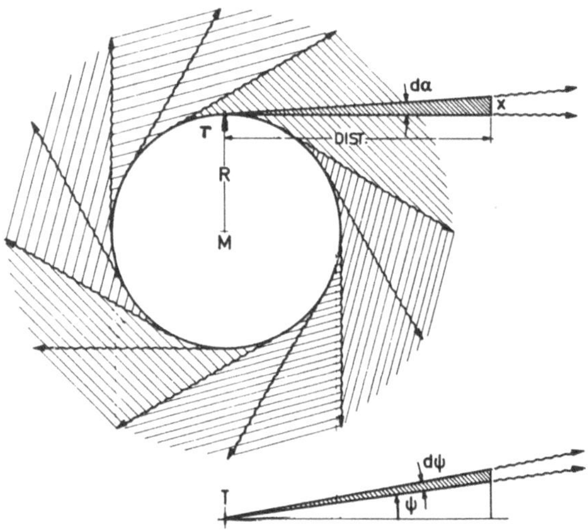

Fig. 2. Geometry of synchrotron radiation emission. The plane of the accelerator or storage ring is filled with radiation (top), while the intensity in the off-plane direction drops rapidly with the increasing angle (bottom).

monics content; and, at high revolution frequencies, the density of lines merges into a white continuum. Further, since there is a certain spread in the revolution frequency, the spectrum can be considered white even at short wavelengths. The strong polarization of the dipole radiation remains.

2.1 Basic Equations

The theory of synchrotron radiation, based on the work of Ivanenko and Pomeranchuk [6] and Schwinger [10], has been described in detail during recent years, both in convenient summaries (e.g., [1-4]) as well as in basic textbooks [15]. For a quantitative description of the properties of synchrotron radiation, a convenient starting point is Schwinger's formula for the total power radiated by a monoenergetic relativistic electron on a circular orbit (cgs units):

$$P = \iint I(\lambda,\psi)d\lambda d\psi = (2/3)(e^2c/R^2)(E/mc^2)^4 \tag{1}$$

where

$$I(\lambda,\psi) = (27/32\pi^3)(e^2c/R^3)(\lambda_c/\lambda)^4 \gamma^8[1 + (\gamma\psi)^2]^2$$

$$\times \left\{K^2_{2/3}(\xi) + \frac{(\gamma\psi)^2}{1 + (\gamma\psi)^2} K^2_{1/3}(\xi)\right\} . \tag{2}$$

λ_c is the "cut-off" wavelength given by

$$\lambda_c = (4\pi R/3)\gamma^{-3}; \qquad \lambda_c(\text{Å}) = 5.59\ R(\text{meter})\cdot\{E(\text{GeV})\}^{-3} \tag{3}$$

and $\xi \equiv (\lambda_c/2\lambda)[1+(\gamma\psi)^2]^{3/2}$. In the above expressions, $\gamma = E/m_oc^2$ is the electron rest-mass energy (0.511 MeV); c is the speed of light; λ is the wavelength of the emitted radiation; R is the bending radius of the electron orbit; ψ is the azimuthal angle and $K_{1/2}(\xi)$, $K_{2/3}(\xi)$ are modified Bessel functions of the second kind (see, for instance, Watson [16]).

The strong dependence of the total radiated power on E^4 [Eq. (1)] explains why machine builders are obliged to increase rapidly the radius of the electron storage rings when they want to reach higher energies, in order to keep the radiated power at a reasonable value. For example, the PETRA e^+e^--storage ring in Hamburg, with a maximum energy of 20 GeV, has a circumference of 2.3 km; the LEP machine at CERN, with a maximum energy of 130 GeV, is planned to have a circumference of 30 km.

Some basic relationships of synchrotron radiation produced

by bending magnets, relationships which are of prime importance for spectroscopy, x-ray scattering experiments and other applications, are summarized in Table 2. The spectral distribution, angular distribution and pulse structure of the radiation from a storage ring are illustrated in Fig. 3, where we have taken storage ring DORIS as an example (see also Fig. 5 for the spectral

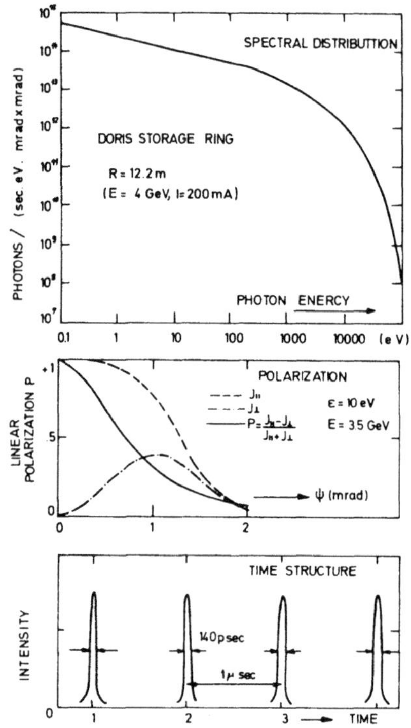

Fig. 3. Spectral distribution of SR into a solid angle of 1 mrad x 1 mrad for the storage ring DORIS (upper panel). Angular distribution of intensity components with electrical vector parallel ($I_{||}$) and normal ($I_{\perp}$) to the plane of the synchrotron, and linear polarization at $h\nu$ = 10 eV for DORIS at E = 3.5 GeV; Ψ is the elevation angle perpendicular to the orbital plane (middle panel). Time structure of DORIS operating in the single-bunch mode with pulses of 140 psec duration and separated by 1 μsec (lower panel).

Table 2. Basic Synchrotron Radiation Relationships

ELECTRON ENERGY:	E (GeV)
RADIUS OF CURVATURE:	R (meters)
ELECTRON CURRENT:	I (Amperes)
MAGNETIC FIELD:	B (Gauss)

Total power radiated by a relativistic electron

$$P = \int I(\lambda,\psi)\,d\lambda\,d\psi = \frac{2}{3}\frac{e^2c}{R^2}\frac{E^4}{(m_oc^2)^4}$$

Energy loss per turn, per electron

$$\delta E\ (\text{keV}) = 88.5\ \frac{E^4\ (\text{GeV})}{R\ (\text{m})}$$

Total power radiated

$$P_{tot}\ (\text{kW}) = 2.66 \cdot E^3\ (\text{GeV}) \cdot B\ (\text{K}\Gamma) \cdot I\ (\text{A})$$

Critical wavelength

$$\lambda_c\ (\text{Å}) = 5.59\ \frac{R\ (\text{m})}{E^3\ (\text{GeV})} = \frac{186.4}{B\ (\text{K}\Gamma) \cdot E^2\ (\text{GeV})}$$

Characteristic energy

$$\varepsilon_c\ (\text{eV}) = 2218 \cdot \frac{E^3\ (\text{GeV})}{R\ (\text{m})} = 2.96 \times 10^{-7}\ \frac{\gamma^3}{R\ (\text{m})}$$

Emission angle $\simeq 1/\gamma = m_oc^2/E$

with $\gamma = E/m_oc^2 = 1957 \cdot E\ (\text{GeV})$

Energy of a photon

$$\varepsilon\ (\text{eV}) = 12.40/\lambda\ (\text{Å})$$

References:
D. Ivanenko, *et al.*, Phys. Rev. 65, 343 (1944).
J. Schwinger, Phys. Rev. 70, 798 (1946); 75, 1912 (1949).
G. K. Green, "Spectra and Optics of Synchrotron Radiation," BNL Report 50522 (Brookhaven Natl. Lab., 1976).

distribution).

The wide spectral range and intense continuum, excellent degree of collimation, well-defined polarization, fast time structure, clean environment and high stability, as well as the quantitatively known characteristics, are the major properties which make SR so attractive. These properties must be compared to those of other alternative conventional sources of radiation in the VUV and x-ray region. Such comparisons, each with a different emphasis, have been given in the past (see, e.g., [2-4]) and will not be repeated here.

2.2 Wigglers and Undulators

The majority of today's SR experiments make use of the SR originating from the circular trajectories of electrons in bending magnets of storage rings. It is possible, however, to design other magnetic structures which may be inserted into electron storage rings and which have significantly higher brightness and/or better spectral properties than those possible from storage-ring bending magnets. The principles of the most frequently discussed devices, namely wigglers and undulators, are sketched in Fig. 4. We only briefly describe these systems here, and refer the reader to recent review papers available on the subject (e.g., [12,18]) for more details. Both wiggler and undulator magnets may be inserted into straight sections of a storage ring and the function of the bending magnets (i.e., to maintain a circulating beam of electrons in a closed orbit) remains the same. Within the wiggler or undulator, the electron beam is periodically deflected in an alternating-polarity transverse magnetic field. However, no net deflection or displacement of the electron beam occurs.

A *wiggler* is a magnetic structure which forces the electron beam to execute a trajectory having a smaller local radius of curvature than that executed in the dipole bending magnets, by using a larger local magnetic field. Thus, the effect of the wiggler on the emitted spectrum is to increase the critical energy and to shift the overall spectrum to higher energies. Typically, $N \simeq 30$ wiggles can be achieved by using $2N + 1$ poles. Furthermore, by matching the angular deflection of the electron beam, and the angular emission of the radiation, from each pole to the angular acceptance of the SR beam line and experiment, enormous enhancements of the brightness, as compared to that possible from ring bending magnets, may be obtained and the radiation from many poles can be collected in a small angular cone for one experiment. The effect on the emitted spectrum is illustrated in Fig. 5.

Similar to the N-pole wiggler, an *undulator* is a periodic electromagnetic structure which forces electrons to oscillate

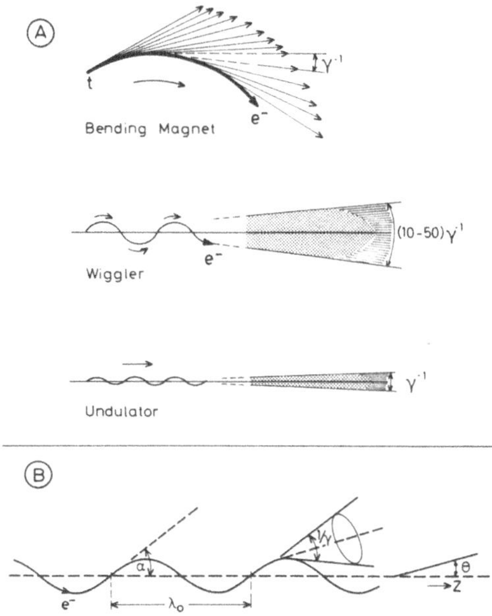

Fig. 4. (A), schematic drawing of the emission characteristic from a bending magnet, a wiggler and an undulator. (B), schematic drawing of a planar multipole wiggler/undulator. A horizontal periodic magnetic field of period λ_0 forces electrons to oscillate in the vertical plane with the same period. Instantaneously, the light is emitted in a $1/\gamma$ cone centered on the tangent of the trajectory. θ is the angle of observation. If α is the maximum deflection angle of this trajectory with the z-axis, the nonrelativistic transverse case ($K < 1$) corresponds to $\alpha < 1/\gamma$, and the relativistic case ($K > 1$) to $\alpha > 1/\gamma$.

transversally in a given plane (planar undulator), or to describe a helix along the mean path of the structure (helical undulator). Undulator design aims for the production of quasi-monochromatic SR. The angular deflection of the electron beam is now kept smaller than or equal to the natural emission angle of SR, given by $\langle\psi\rangle = 1/\gamma$. Interference effects in undulator radiation result in peaks at one or a few wavelengths.

Although the general treatment of the radiation from undula-

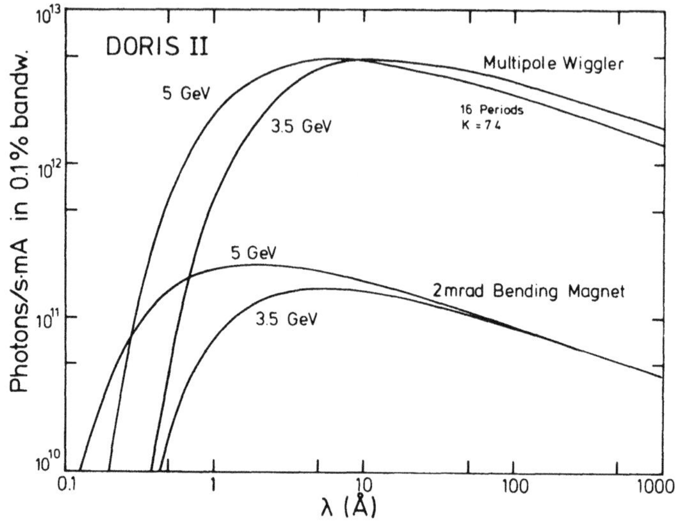

Fig. 5. Comparison of the intensity of SR from a bending magnet and from a multipole wiggler with λ_0 = 120 mm for the DORIS II storage ring at electron energies of 3.5 GeV and 5 GeV.

tors is quite complicated, we attempt to illustrate qualitatively its properties for the case of the so-called "weak-field" undulator. The oscillation of the electron in the undulator with period λ_0 is depicted in Fig. 4. Instantaneously emitted photons are still focused in a very small cone with angular width $1/\gamma$, but now the axis of this cone oscillates around the z-axis with a maximum angle α. For one electron, the principal interference effects can be easily derived by comparing the path and time differences betweeen the electron travelling between two poles and the emitted photons. In this description the undulator is very similar to an optical grating and, for a given angle of observation θ (cf. Fig. 4), the emission is quasi-monochromatic; its wavelength is given by

$$\lambda = \frac{\lambda_0}{2\gamma^2}\left(1 + \frac{K^2}{2} + \gamma^2\theta^2\right) \tag{4}$$

Where $K = (eB_0\lambda_0)/(2\pi m_0 c) = \alpha\gamma = 0.0934\ B_0\lambda_0$ (K is the so-called "deflection parameter" and is independent of the electron energy); B_0 is the peak field in Tesla; and λ_0 is the magnet period in mm. The spectral width of the emission at $\theta = 0$ results from the interferences and $\Delta\lambda/\lambda$ is of the order of $1/N$, where N is the number of periods of the undulator. The qualitative aspects of the emitted spectrum, for different values of the parameters $K = \alpha\gamma$ and σ' (σ' being the angular spread of the electron beam), are sketched in Fig. 6. At least three different modes of operation can be proposed [19]:

Mode 1. Curve (a) of Fig. 6: With a small deflection and a small angular spread, only one harmonic will be emitted through a small pinhole on the axis of the undulator. This harmonic will be almost independent of B and will have a low power as compared to the case where $K \gg 1$.

Mode 2. Curve (c) of Fig. 6: With a large deflection and a small angular spread, high power is emitted through the same pinhole in many bands, which can be tuned with the magnetic field B. For a well-collimated electron beam ($\sigma'\gamma \ll 1/\sqrt{N}$), the increase

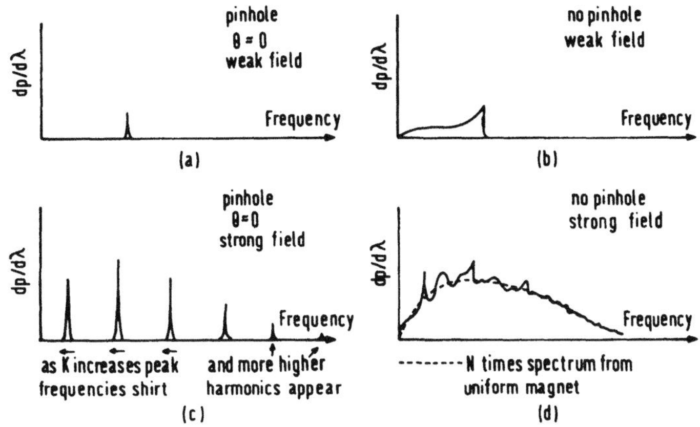

Fig. 6. Expected spectral distribution of the light emitted by an undulator in different modes of operation. (a), Low-field case ($K \ll 1$); observation through a small pinhole with a very parallel electron beam. (b), Low-field case ($K \ll 1$); angular integrated with a very parallel electron beam (or through a small pinhole, but with a divergent electron beam). (c), High-field case ($K > 1$); observation through a small pinhole with a very parallel electron beam. (d), High-field case ($K > 1$); angular integrated with a very parallel electron beam (or through a small pinhole, but with a divergent electron beam). (From [19].)

in spectral brilliance is proportional to N^2.

Mode 3. Curve (d) of Fig. 6: With a large deflection and no pinhole, the spectrum is similar to the usual SR, but with a power multiplied by N ("multiple wiggler").

The usefulness of these devices, and their unique properties, are currently being actively exploited at a number of places and they are increasingly in demand as preferred radiation sources [12,18]. The possibility of increasing the spectral brightness by factors of up to N^2 ($N \simeq 50$) is very exciting, particularly for the most difficult experiments requiring high resolution in energy, time or space.

3. SOME GENERAL ASPECTS OF SYNCHROTRON RADIATION INSTRUMENTATION

A systematic discussion of methods for studying higher excited states in molecules and molecular crystals by means of SR would have to cover all aspects summarized in the block diagram of Fig. 7, namely, the source, beamline, various types of monochromators, experimental setups and sample stages, as well as detectors and data handling. In the preceding section we gave a short overview on the properties of SR sources. We have omitted interesting topics, such as the comparison of SR sources to alternative light sources (see, e.g., [17]), or the detailed discussion of optical engineering for beamline and monochromator design. Before we turn to specific techniques, experiments and examples in the following sections, we first discuss a few general aspects

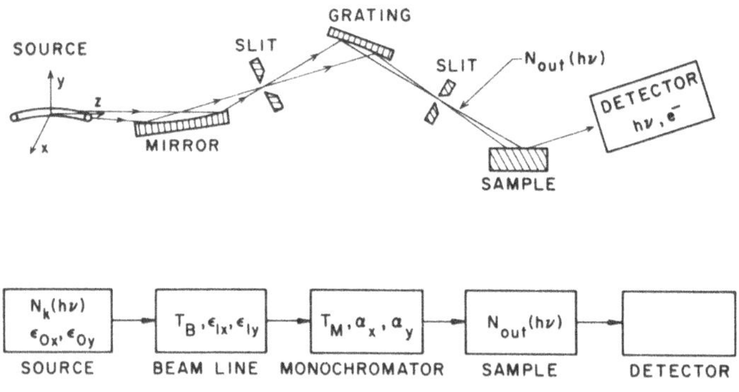

Fig. 7. Schematic of source, beam-line optics, monochromator, sample and detector configuration. (From Nat. Acad. Science Report on SR, Washington, 1976.)

of SR instrumentation.

Once the merits of the light source, for a given experiment, have been considered, the matching of the source to the experiment has to be given attention. In this, probably the most important aspect at a SR source is the fact that, unlike conventional spectroscopy, the source, beamline with optical components, monochromator and experiment have to be treated as a whole. Most frequently, the most important parameter is the flux per unit bandwidth at a given photon energy in the range of interest. Also, the flux density (i.e., the focused spot size on the sample) is of interest. The main factors characterizing a beam line and experiment, with respect to these parameters, are (cf. Fig. 7): (i), the source, with a flux $N(\lambda)$ per horizontal angle and per Å, and with horizontal and vertical emittances ε_x, ε_y; (ii), the beam line and beamline optics, with emittances ε_x', ε_y' at the monochromator and with a transmission coefficient T_B; (iii), the monochromator, with acceptances α_x, α_y and with a transmission coefficient T_M; (iv), the sample, with an effective area; and (v), the detector, with its efficiency. The monochromatized flux falling onto the sample is then given by

$$N_{out}(\lambda) = N(\lambda)\ \theta_x\ T_B\ T_M \frac{\alpha_x}{\varepsilon_x'} \frac{\alpha_y}{\varepsilon_y'} \qquad (5)$$

where θ_x is the horizontal beamline acceptance angle and where $\alpha_i/\varepsilon_i' \leq 1$. Thus, in the design of an experiment, questions relating to optical matching of the beamline and to the transmission coefficient of the beamline optics, or to the matching and efficiency of monochromators in order to select the desired frequency, become important.

It is interesting to note that, for example, the fabrication of ultrasmooth aspherical mirrors with complex surfaces, as used in beamline optics, turned out to be a considerable technological problem, particularly since these mirrors have to withstand intense x-ray radiation. This latter fact has led to a search for low expansion mirror materials. Other solutions, such as coated metallic mirrors or chemical vapor deposited SiC mirrors, are being studied presently [20].

Improvements in reflecting coatings for mirror surfaces constitute another area of current research. Experiments with SR have provided us with useful quantitative information concerning the optical properties of these coatings in various spectral ranges. A systematic compilation of these data is now available [21]. In Fig. 8, we reproduce the reflectance of gold, which is frequently used as a coating material for mirrors. A few points are worth noting: (i) In the $h\nu \gtrsim 10$ eV range of the VUV, the

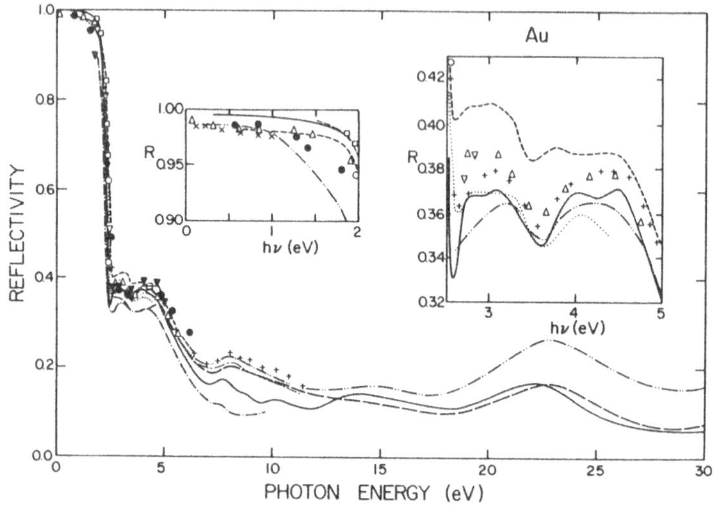

Fig. 8. Reflectivity of gold from the infrared to the VUV. The figure illustrates that there still exists some scatter in the published optical properties in the VUV range. The same is true for the soft x-ray region. (From [21].)

reflectance is below 20%. For higher photon energies, the normal incidence reflectance decreases rapidly. Thus, for example, at 100 eV the reflectance is below 1% and, consequently, grazing-incidence optics must be used. (ii) In the visible and near-UV, optical constants are known fairly accurately; however, there is still quite some spread in the data points at higher photon energies. Certainly, the accuracy of optical measurements and the factors influencing changes (e.g., adsorbate layers of contaminating materials) deserve further studies. (iii) The precise knowledge of optical data in the VUV and soft x-ray regions enable us to extend optical engineering into these ranges. As an example, we mention the enhancement of reflectance, in certain spectral ranges, by multilayer coatings [20].

Monochromators

The methods used for monochromatizing synchrotron radiation depend on the energy range of interest (cf. Fig. 9). Normal-incidence monochromators are employed in the VUV region from ∿5 to 40 eV (2000 - 300 Å), while grazing-incidence grating monochromators cover the energy region from ∿30 to ∿800 eV (400 - 20 Å) [22]. In the hard x-ray energy region above 3 keV (= 4 Å), crystal monochromators are used; these are operated under a hel-

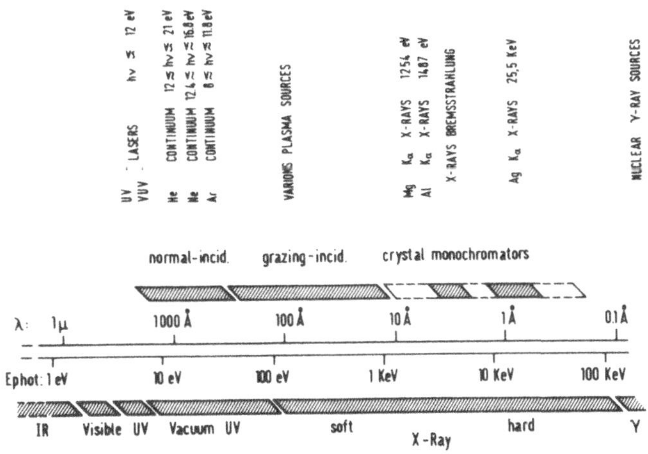

Fig. 9. Spectral ranges covered by SR and alternative light sources. The ranges for operating different types of monochromators are also indicated.

ium atmosphere separated from the storage ring vacuum by beryllium windows and, therefore, the transmitted radiation is limited to energies above ∿3 keV. While a number of well-tested monochromators are available for these regions (see, e.g., [23]), there is a lack of suitable monochromators for the remaining gap in the soft x-ray region (800 eV - 3000 eV). This gap could be covered by either grating or crystal monochromators. Because the K-shell absorption edges of several light elements (e.g., elements which have important biological applications, as well as intrinsically interesting spectra) fall into this range, much effort is presently being devoted to the development of crystal monochromators with focusing optics for this energy region [24]. We also mention interesting developments in the area of transmission-grating monochromators, or zone-plate monochromators [22], which might, if successful and routinely available, help to fill this gap.

4. ABSORPTION AND REFLECTION SPECTROSCOPY

4.1 Studies of Gas-Phase Molecules

In the study of electronic excitations of molecules, absorption spectroscopy and reflection spectroscopy, over extended energy regions, constitute a prime source of basic information [25]. In the dipole approximation, the absorption coefficient μ is given by [26]

$$\mu(\omega) = \frac{2\pi^2 e^2}{mc} N f(\omega,o) \tag{6}$$

where $f(\omega,o)$ is the optical oscillator strength; N is the density of absorbing molecules; and ω is the frequency of the absorbed photon. In optical experiments, the $\underset{\sim}{k}$ dependence of the generalized oscillator stregth $f(\omega,\underset{\sim}{k})$ can be neglected. This $\underset{\sim}{k}$ dependence, however, plays an eminent role in electron energy-loss experiments, as discussed, for example, in [27]. The optical absorption coefficient is related to the total photoabsorption cross-section $\sigma(\omega)$ via

$$\mu(\omega) = \sigma(\omega)\ N\ . \tag{7}$$

$\sigma(\omega)$ contains the microscopic information about the energy levels of the system under investigation [26], since $f(\omega,o)$ is given by

$$f(\omega,o) = \frac{2m\omega}{\hbar} \left|\langle\psi_f(\omega)\left|\sum_i \underset{\sim}{r}_i\cdot\hat{\underset{\sim}{E}}\right|\psi_i\rangle\right|^2 \tag{8}$$

where $\hat{E}(\equiv \underset{\sim}{E}/\|\underset{\sim}{E}\|)$ is the direction of the dielectric field vector; $\underset{\sim}{r}_i$ denotes the position of the ith electron; and $|\psi_i\rangle$ and $|\psi_f(\omega)\rangle$ are the ground and excited states of the molecule, respectively. For a molecule or molecular crystal, the excited-state energy E consists of vibrational and rotational energy, in addition to electronic energy, and the eigenstates of the molecule reflect a combination of all three modes.

In order to resolve the associated spectral features of higher excited states, a resolution on the order of a few meV, at 10-40 eV, is necessary. In Fig. 10, an experimental set-up capable of high-resolution absorption and reflection spectroscopy is depicted [28]. This 3 m instrument, installed at the DORIS storage ring, has a vertical dispersion plane; it provides a bandwidth of 0.03 Å with 10 μm slits and with a 1200 lines/mm grating. The wavelength range covered extends from 3000 Å to 300 Å (4 eV to 40 eV). A toroidal mirror with a reflecting surface of 500 x 60 mm^2 focuses (with a deflection angle of 15°) SR onto the entrance slit. A cone of 1 mrad x 1 mrad is accepted in an s-polarizing geometry.

The capabilities of this instrument are demonstrated by the absorption spectrum of Ar gas, which is shown in Fig. 11 [29]. The spectrum is recorded photoelectrically and Rydberg states up to very high quantum numbers are visible. Another example is shown in Fig. 12, where the partially-resolved rotational finestructure of one vibrational band of the $b\,^1\Pi$ excitation of N_2 is compared with a numerical fit to the data [30]. It is clear from

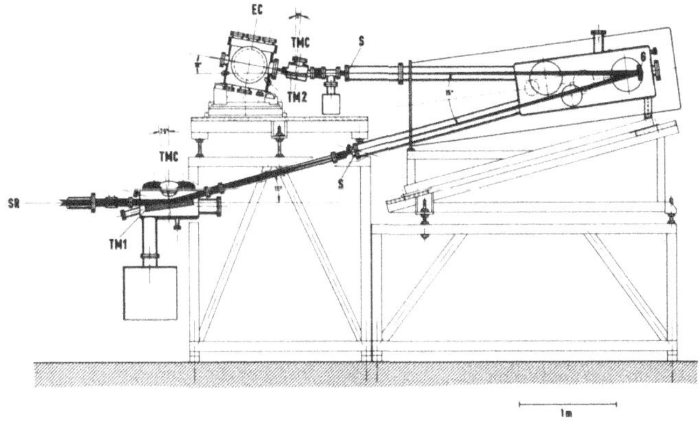

Fig. 10. Schematic drawing of the beam line and the 3 m normal-incidence monochromator, cut perpendicular to the storage ring plane. SR enters from the left, is deflected 15° out of its direction by a plane mirror (not shown) and reflected upwards 15° and focused by the toroidal mirror TM 1 onto the entrance slit (S) of the monchromator. SR hits the grating (G). A monochromatic beam emerges from the exit slit (S) and is focused by a second toroidal mirror TM 2 into the experimental chamber (EC). [From J. H. Fock, Diplomarbeit (Univ. of Hamburg, 1979).]

these examples that, aside from high resolution, the ability to obtain quantitative (photoelectrically recorded) cross-section data is of great advantage and importance for spectral analyses.

Over the last 5 years, a number of remarkable experiments in solid-state spectroscopy, and molecular and atomic spectroscopy, have been performed with this instrument with great success [25]. We mention, in particular, the quantitative analysis of the spectra of N_2 [31], CO [32], NO [33], H_2O [34], and C_6H_6 and related compounds [35]. In addition to the measurement of VUV valence spectra of gas-phase molecules, work on condensed gases (see Section 4.2) has profited immensely from the high resolution and stability of the instrument.

As an example of high-resolution gas-phase work, we show, in Fig. 13, the absorption cross-section of CO [32] in the photon energy range from 10.1 eV to 14.3 eV, which covers the region of Rydberg transitions converging to the first ionic state $X^2\Sigma^+$ of

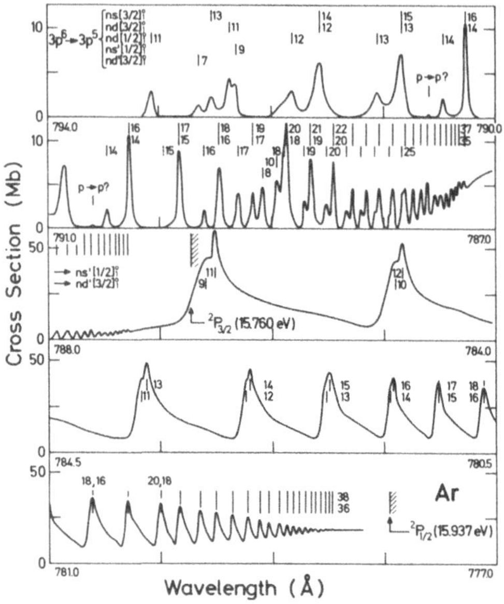

Fig. 11. Absorption spectrum of Ar near the first two ionization limits. Between the ionization limits, Fano-Beutler profiles are observed. Note that the whole spectrum covers only 450 meV. (From [29].)

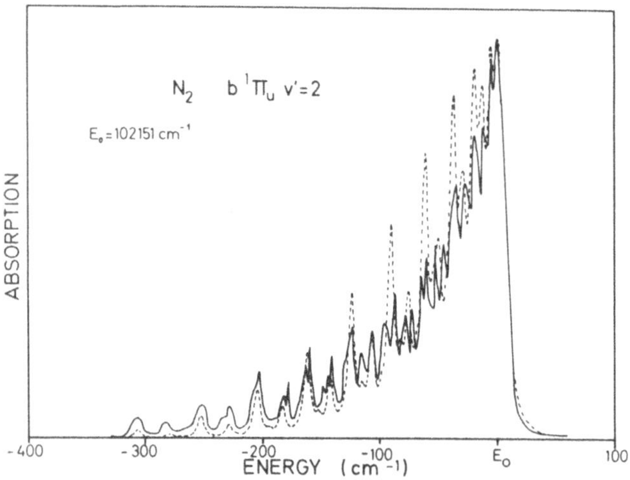

Fig. 12. Measured absorption spectrum (solid line) and calculated spectrum using the model of a nonrigid rotator (dashed line) for the v'=2 band in the $b^1\Pi_u$ progression of N_2. (From P. Gürtler, unpublished.)

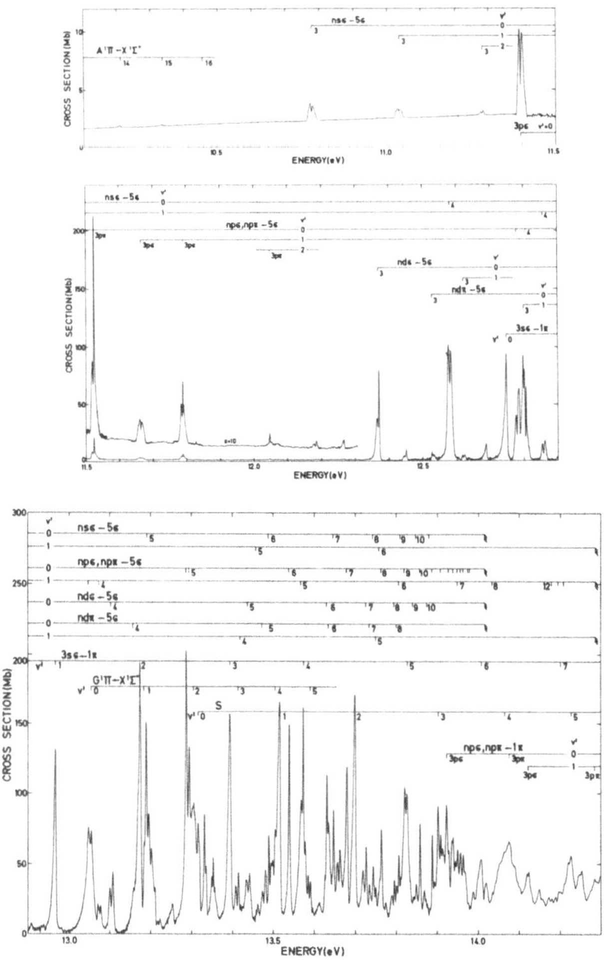

Fig. 13. Absorption spectrum of CO in the photon energy range from 10.1 eV to 14.3 eV, covering the region of Rydberg transitions converging on the first ionic state $X^2\Sigma^+$ of CO^+. The Rydberg transitions $5\sigma \rightarrow np\sigma$ and $5\sigma \rightarrow np\pi$ are given on top of the spectrum. In the photon energy range below 12.0 eV, the apparent background absorption is due to higher-order intensity from the monochromator. (From [32].)

CO^+. The interpretation and correct assignment of such VUV absorption spectra is tedious because of the large number of occupied valence orbitals, and because of the many accessible states which usually overlap heavily in the absorption spectrum. Fortunately, quite frequently a large number of absorption bands can be arranged in hydrogen-like Rydberg series according to the well-known Rydberg formula

$$E_{n\alpha} = IP^k - R/(n - \delta_{n\alpha})^2 \tag{9}$$

where $E_{n\alpha}$ is the energy of the nth member of the αth Rydberg series converging to the kth ionization potential IP^k; R denotes the Rydberg constant (13.6055 eV for CO); and $\delta_{n\alpha}$ is a quantum defect which is characteristic for the symmetry of the Rydberg series [36].

A certain routine has been developed, in the past few years, for making assignments with more confidence; but, with more and more experimental data becoming available, it is apparent that in many cases much remains to be done to arrive at satisfactory assignments. In an ideal situation, long non-overlapping series are observed, yielding unequivocal assignments [36,37]. For many molecules, comparison of the absorption spectrum with a highly-resolved photoelectron spectrum, and with a good quantum chemical calculation for the ground state, has provided key information about the orbitals from which the transitions originate, as well as a guideline for the interpretation of vibrational progressions associated with the Rydberg transitions (see, e.g., [26,36]).

The next important step is the ordering of the observed bands into ns, np, nd,... type Rydberg series [36,38]. A powerful tool for the further analysis is the comparison of equal members of corresponding Rydberg series within a group of related molecules, as demonstrated convincingly by Robin [36]. This method has been given a quantitative formulation by Maria, *et al.* [39], who found a linear relationship between the excitation energies (or term values) of the Rydberg transitions and the corresponding ionization potentials. This relationship can be written in the form

$$T_{n\alpha} = a_{n\alpha}\, IP^k + b_{n\alpha} \tag{10}$$

where $a_{n\alpha}$ and $b_{n\alpha}$ are constants for a certain Rydberg series and for a certain member n within a set of related molecules, and where $T_{n\alpha}$ is a term value

$$T_{n\alpha} = IP^k - E_{n\alpha} . \tag{11}$$

This linear term value/ionization energy correlation has been successfully tested for atoms and for several groups of molecules by Hochmann, McGlynn and their co-workers [37,39]. Furthermore, it is interesting to note that, as demonstrated by Spence [40], Feshbach resonances (which are closely related to Rydberg states) show similar linear behavior and, in turn, can be used for an equally accurate prediction and assignment of Rydberg states. An obvious advantage of the correlation method is that it can be used even if only a very few Rydberg states are known, and especially for series members of low n, for which the most significant deviations from a constant quantum defect are often observed. It is not clear *a priori*, however, what kind of "relatedness" is necessary in the group of molecules one chooses for a comparison.

In order to investigate the above concept of "relatedness", Fock, *et al.* [32] have tested the correlation method for simple molecules and have checked the limits of the method's applicability by using high-resolution absorption spectra and by applying Eq. (10) to the Rydberg series of the sequence of molecules BF, CO and N_2. The "relatedness" of these molecules is that they are isoelectronic; however, each molecule has a quite different geometry of the total electron density since, for both constituent atoms, the nuclear charge is changed by one at each step in the sequence. Thus, for these simple molecules the influence of the molecular geometry can be investigated while keeping the number of valence electrons constant. The test was based on data from the absorption spectrum of CO (cf. Fig. 13), the absorption spectrum of the BF molecule, previously determined by Caton and Douglas [41], and measurements of the N_2 absorption spectrum by Gürtler, *et al.* [31].

Fig. 13 displays those parts of the CO spectrum which contain the $5\sigma \rightarrow np\sigma(0,0)$ and $5\sigma \rightarrow np\pi(0,0)$ Rydberg transitions. Our assignment closely follows the analysis reported by Ogawa and Ogawa [42]. For $n \geq 6$, both series coalesce and no separate maxima are observed. The detailed analysis and intercomparison of the data by Fock, *et al.* [32] yields a remarkable result: In Fig. 14, a plot of T_n versus the ionization energy IP for the $np\sigma$ and $np\pi$ Rydberg transitions, n = 3-9, is shown for the three molecules. The linear correlation predicted by Eq. (10) is obvious. Such a correlation is especially remarkable for the lower members of both the $np\sigma$ and $np\pi$ Rydberg series; for these members, the data points follow the linear behavior to within 10 meV. Fig. 14 clearly shows the applicability of Eq. (10) to a group of isoelectronic molecules: That is, it clearly demonstates the existence of a linear relationship between the Rydberg term values and the corresponding ionization limits. The slope of the correlation lines, especially for n = 3, reflects the influence of the decreasing asymmetry, from BF to N_2, of the total electron density. This asymmetry also causes the large $p\sigma$ and $p\pi$ splitting, which

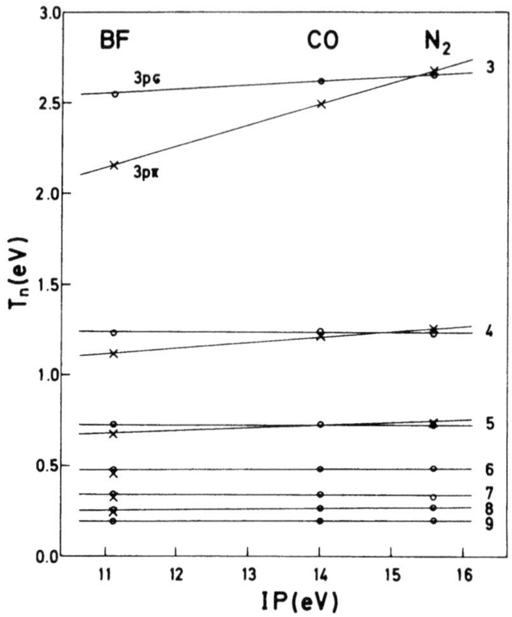

Fig. 14. Term value versus ionization energy correlation lines for corresponding Rydberg states in the isoelectronic molecules BF, CO and N_2. Circles represent data for the npσ(0,0) states, while crosses stand for those of npπ(0,0). (From [32].)

is almost negligible for N_2. Furthermore, these results provide a strong argument in favor of the correctness of the individual assignments in all three molecules. We can anticipate that similar detailed studies will soon become available, with the commissioning of additional high-resolution VUV monochromators which take advantage of the intense SR continuum.

4.2 Studies of Condensed Gases, Matrices and Molecular Crystals

Optical studies of condensed media are described by the frequency-dependent complex dielectric function $\tilde{\varepsilon}(\omega) = \varepsilon_1 + i\varepsilon_2$ or, equivalently, by the complex conductivity $\tilde{\sigma}(\omega) = \sigma_1 + i\sigma_2$ or the complex refractive index $\tilde{n}(\omega) = n(\omega) + ik(\omega)$, which are used with Maxwell's equations but which are descriptions of the material properties being studied. These are fundamental quantities and can be calculated quantum mechanically from microscopic models of the solid [43]. These quantities, which are connected via

$$\tilde{\varepsilon}(\omega) = 1 + \frac{4\pi i\tilde{\sigma}(\omega)}{\omega} ; \qquad \tilde{\varepsilon}(\omega) = \tilde{n}^2(\omega) \tag{12}$$

represent the elementary excitation spectrum. For instance, ε_2 (interband) provides a measure of the interband absorption and can be expressed as

$$\varepsilon_2\omega^2 = \frac{e^2\hbar^2}{3\pi m^2} \sum_{i,f} \int_0^\infty d^3k |\langle\psi_f|\underset{\sim}{p}|\psi_i\rangle|^2 \; \delta(E_f(\underset{\sim}{k}) - E_i(\underset{\sim}{k}) - \hbar\omega) \tag{13}$$

where the electric-dipole approximation is used for the electron-photon interaction Hamiltonian; $|\psi_i\rangle$ and $|\psi_f\rangle$ are the initial (occupied) and final (empty) states; and $\underset{\sim}{k}$, the electron wave-vector, is conserved through direct transitions.

The most direct methods for determining optical properties are measurements of transmission and reflectance spectra. Above the fundamental absorption gap, which is at $\hbar\omega \lesssim 12$ eV for most solids and somewhat larger for rare-gas solids up to E_{gap}^{Neon} = 21.58 eV, the absorption starts to become so high that samples with thicknesses well below 1000 Å must be used. In many instances (e.g., with single crystals), transmission experiments in this spectral range are practically impossible. And, in such cases $\varepsilon_1(\omega)$ and $\varepsilon_2(\omega)$ must be determined from different modes of reflectivity measurements: (i), from a Kramers-Kronig analysis of reflectance spectra over extended spectral ranges [43]; (ii), from measurements of $R_s(\omega)$ and $R_p(\omega)$, the reflectance coefficients at non-normal incidence, and subsequent use of Fresnel's equation; or (iii), from ellipsometry (see, e.g., [21,43]).

In the present section, we concern ourselves only with absorption spectroscopy and reflection spectroscopy. An experimental set-up for the simultaneous determination of reflection and transmission is depicted in Fig. 15. This set-up has been developed for low-temperature experiments involving studies of excitons in large-gap insulators [29,30]. Commonly, for spectroscopic investigations these solids (including rare gases and solid N_2, O_2, CO, etc.) are condensed as thin films onto a cooled substrate. Alloys are usually prepared by mixing the constituents in the gas phase, with appropriate partial pressures, and by depositing them *in situ* on a cryostat in the form of a thin film. In this kind of preparation, care must be taken to work at extremely low base pressures in the gas handling system and in the UHV experimental vessel (in the range of 10^{-10} Torr or better). Only UHV bakeable stainless-steel systems satisfy these conditions. The concentration in the sample may differ from the gas phase composition as a result of demixing during the transfer from the storage vessel to the substrate, and becuase of different sticking coefficients for both components on the cold sub-

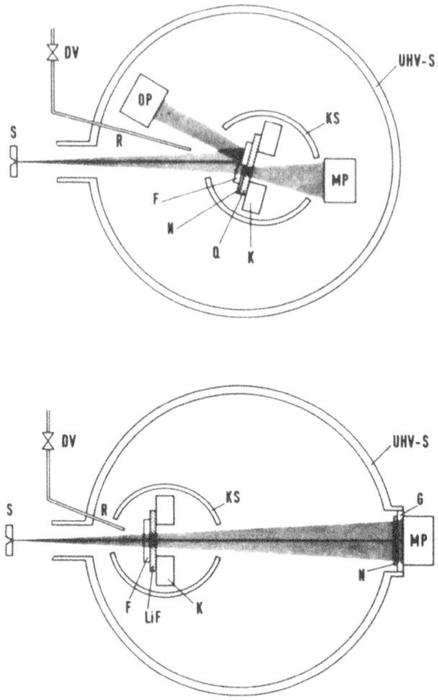

Fig. 15. Sketch of an experimental arrangement for low-temperature reflection and transmission experiments. For $\lambda \gtrsim 1040$ Å, a transparent LiF substrate can be used for transmission experiments; for $\lambda \lesssim 1040$ Å, the light transmitted through the sample is converted by a sodium salycilate film into longer wavelength photons which are transmitted by the quartz substrate. S, exit slit; DV, dosing valve; R, gas inlet; UHV-S, ultrahigh vacuum system; KS, cryoshield; K, cryostat; Q, quartz plate; LiF, window; F, sample film; N, sodium salycilate film; G, window; MP, closed multiplier; OP, open multiplier. (From [29].)

strate. The thickness of the film, typically some tenths of a nm up to some hundred μm, can be determined by monitoring the sequence of interference fringes observed for light reflected from the sample during evaporation. Either laser light or, in order to get higher contrast and higher accuracy for film thicknesses of ∿1 Å, VUV light (but still in the transparent region of the sample) is used.

The many interesting results for electronic excitations in condensed rare gases have been discussed recently in an extensive review [44] and a current state-of-the-art summary can be found elsewhere in this volume [45]. The interrelationship between molecular concepts and solid-state concepts is clearly evident for Frenkel type excitons, which constitute the lowest electronic excitations in simple molecular crystals (e.g., solid N_2 and CO).

Results of a quantitative high-resolution measurement in the range of the $w^1\Delta_u \leftarrow X^1\Sigma_g^+$ and $a^1\Pi_g \leftarrow X^1\Sigma_g^+$ derived exciton progressions in solid N_2 appear in Figs. 16 and 17 [46,47]. Both transitions are dipole forbidden in the gas phase. In solid N_2, long progressions with sharp bands are observed. The individual peaks of the fine-structure observed for the $w^1\Delta_u$ state, with halfwidths $\lesssim$ 1 meV (FWHM), are among the sharpest structures observed for excited states in solids. These data, including temperature-dependence effects, have been unambiguously explained within the framework of the strong exciton-phonon coupling model. The observed fine-structure reveals the zero-phonon line corresponding to the purely electronic transition without lattice contributions, whereas the phonon sideband reflects the phonon density of states of the N_2 crystal coupled to excited states (cf. Fig. 17). We note in passing that, for N_2 in a Ne matrix, we were able to determine the coupling of the N_2 electronic transitions to the acoustical phonons of the Ne host lattice [48]. Finally, in an appendix to this paper, we have summarized the spectroscopic data for gaseous, solid and matrix-isolated N_2 [30, 31,46-48].

5. TIME-RESOLVED AND ENERGY-RESOLVED VUV LUMINESCENCE SPECTROSCOPY

Absorption and photoemission measurements are, in general, insufficient to characterize the final state of the absorption process. Much information is contained in the luminescence emitted by excited molecules, molecular ions and fragments, or molecular crystals. The analysis of excitation and luminescence spectra can contribute considerably to our knowledge of the geometry and the potential curves of the excited states. These measurements are also of immense importance for the investigation of the dynamics of excited states, and for the exploration of decay and energy-transfer processes. The time dependence of luminescence processes serves as a further means to provide accurate data on rate constants, both of radiative and nonradiative deactivation processes.

Luminescence experiments in the VUV, under selective optical excitation, have suffered greatly from intensity problems. Synchrotron radiation from storage rings has turned out to be a

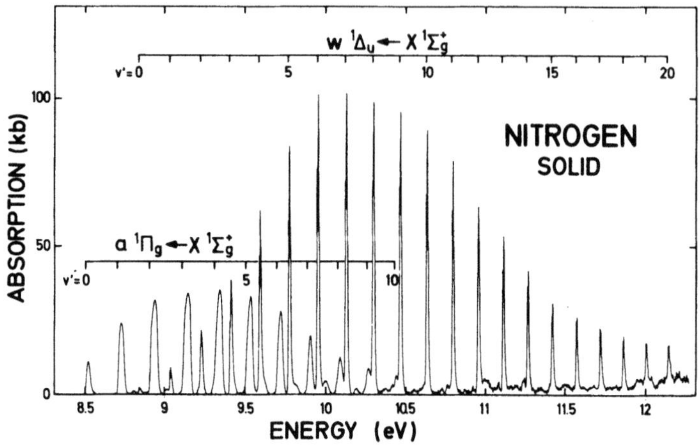

Fig. 16. Overview of the absorption spectrum from solid N_2 in the photon energy range from 8.5 eV to 12.3 eV. The two progressions corresponding to the gas-phase $a^1\Pi_g \leftarrow X^1\Sigma_g^+$ and $w^1\Delta_u \leftarrow X^1\Sigma_g^+$ transitions are labelled. (From [46].)

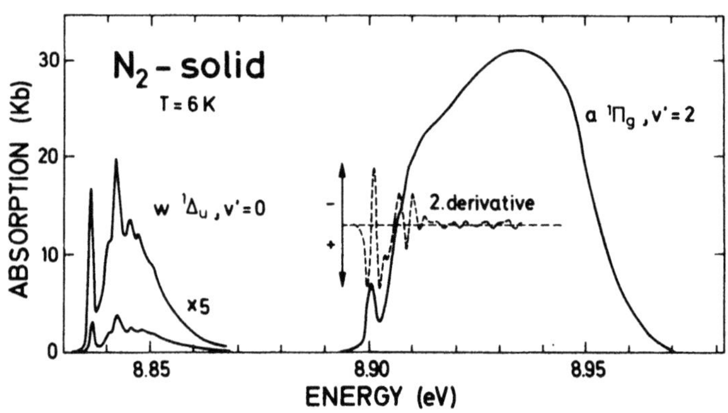

Fig. 17. Absorption spectrum of solid N_2 at 6 K in the range 8.83 eV to 8.89 eV. For the v' = 2 band of the $a^1\Pi_g$ system, the second derivative is also shown. (From [46].)

powerful excitation source for such experiments (see, e.g., [25, 49]) and a variety of problems have been tackled so far. For time-resolved and spectrally-resolved luminescence measurements, a multipurpose set-up is highly desirable. Fig. 18 shows such an experimental station which has a number of outstanding features [50,51]:

--- High flux and high resolution for excitation in the wavelength range from 30 nm to 300 nm. This is obtained by a beamline which includes several focusing mirrors that match a radiation fan of 50 mrad horizontal-width and 2.2 mrad vertical-width from the storage ring DORIS to a 2m normal-incidence monochromator.

--- Spectral analysis of luminescence covering an extremely large spectral range from 50 nm to 1000 nm. In the VUV, this is achieved with a specially designed toroidal grating monochromator with an extremely large f-number (1:2.8) [51].

--- Time resolution of spectrally-selected luminescence under state-selective excitation down to the 50 ps range. For these measurements, the excellent time characteristics of the

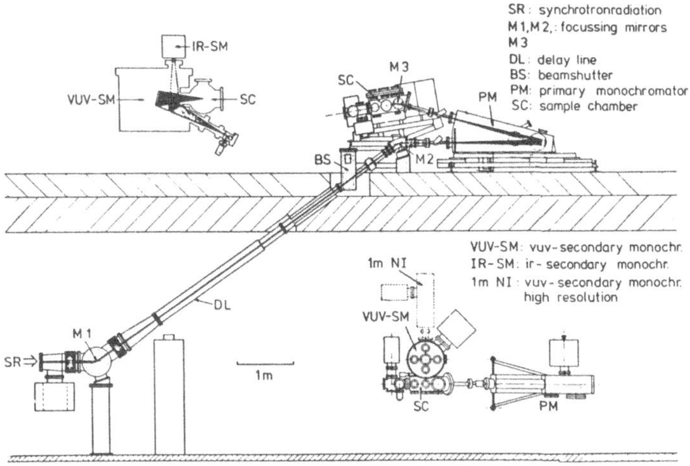

Fig. 18. Schematic drawing of the beam line, the monochromator and the secondary monochromator of the SUPERLUMI set-up, cut perpendicular to the storage ring plane. Details of the experiment and monochromator for VUV luminescence analysis are shown in the two inserts. (From [50,51].)

DORIS storage ring source (cf. Fig. 3) are of prime importance.

--- Sample preparation in a UHV environment, including growth of cryocrystals and *in situ* cleavage of crystals at low temperatures down to 4 K.

--- Interchangeable sample holders for solid and gas-phase samples.

--- Simultaneous recording of high-resolution absorption and excitation spectra.

In Fig. 18, a side-view of the set-up is shown, including schematics of the focusing mirrors, the beam line and the monochromator (PM). The insert shows the geometrical and optical arrangement for the VUV monochromator used in luminescence analysis. This monochromator has no entrance slit; instead, the luminescent part of the sample, typically 4 x 0.1 mm^2, serves as the source. In this way, intensity losses due to optical matching of the instrument to the sample, via additional elements, are avoided. This is of particular importance for $\lambda \lesssim 120$ nm. Luminescence in the 200 - 1000 nm range is analyzed with a commercial monochromator (VIS-SM).

In order to illustrate the capabilities of this instrument, two typical fluorescence spectra of Cl_2 are shown in Fig. 19 [52].

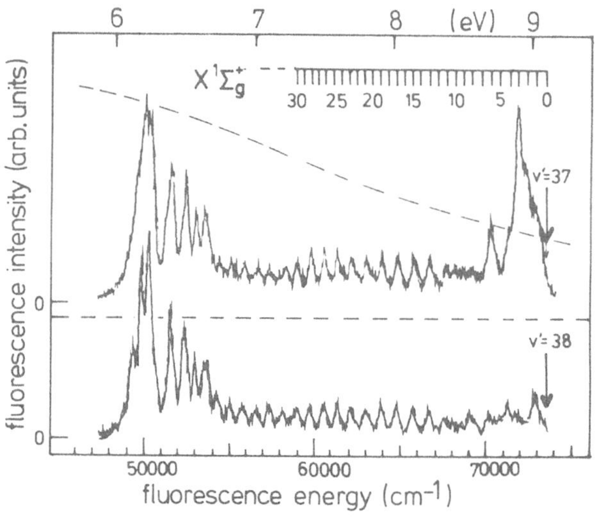

Fig. 19. Two typical fluorescence spectra for Cl_2. The excitation energies are indicated by arrows. The vibrational quanta of the final state ($X\,^1\Sigma_g^+$) are indicated in the upper part. (From [52].)

The excitation energies are indicated in the figure by arrows; these energies have been chosen in such a way that the v'=37 and v'=38 vibrations of the $1^1\Sigma_u^+$ state are reached respectively. Analysis of these and similar data clearly yields a double-well structure for the $^1\Sigma_u^+$ state, which results from an avoided crossing between an ionic valence state and a Rydberg state (cf. Fig. 20). The double-well structure is responsible for the irregular vibrational sequence in the luminescence spectrum between 73,000 cm^{-1} and 74,500 cm^{-1}, and for the remarkable difference in intensity of the two spectra in Fig. 19, a difference which either prevented an analysis in earlier investigations or led to wrong assignments. In general, this study is in excellent agreement with *ab initio* calculations [53] of the electronic structure of Cl_2.

Further examples of luminescence spectra, including the study of relaxation processes in condensed and matrix-isolated systems and time-resolved measurements, are discussed in the paper by Jortner, *et al.* [45] elsewhere in this volume.

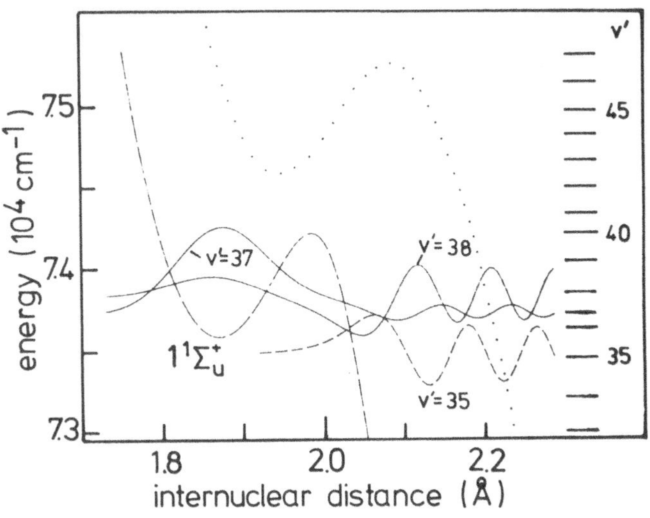

Fig. 20. Comparison between the $1^1\Sigma_u^+$ potential curve extracted from luminescence experiments (dashed curve) and the *ab initio* calculation (dotted curve) of Peyerimhoff and Buenker [53]. Typical wavefunctions for the v' = 35,37,38 levels of the $1^1\Sigma_u^+$ state are included. (From [52].)

Soft x-ray emission spectroscopy

In the present context, a few words concerning soft x-ray emission spectroscopy are appropriate. It is obvious that the high flux and the tunability of SR offer the opportunity to extend fluorescence spectroscopy into the x-ray region. However, only a rather few experiments, so far, have used both relatively high-resolution monochromators to produce a well-defined photon energy for spectral excitation, and a high-resolution monochromator for analysis of fluorescent radiation. Part of the reason for this paucity of data is the difficulty in operating high-intensity monochromators in the soft x-ray range (see Section 3). Up to now, in most experiments one had to be content with broad-band photon excitation. Yet even these experiments, when compared to electron-beam excitation, have the advantage that contamination, heating or decomposition of samples can be avoided to a large extent. We anticipate that, with the advent of wigglers and undulators, soft x-ray emission with high resolution for both excitation and analysis will become a very rewarding field of research. Apart from basic processes, such as multiple excitation and rearrangement processes during x-ray emission, valence-band to inner-orbital electron transitions provide unique information for molecular orbital and solid-state band analysis. Since emission spectra are primarily governed by dipole selection rules, information can be obtained on the distribution of a particular type of atomic-orbital character throughout the valence band (valence s + d $\rightarrow$ core p, or valence p $\rightarrow$ core s). Furthermore, because of the localized nature of the core hole, complementary x-ray emission can be obtained for the different constituent atoms of a compound. These possibilities are of particular importance in the soft x-ray region for the production of carbon, nitrogen and oxygen 1s core holes, whereby the valence p $\rightarrow$ 1s fluorescence spectra can be studied. In the next section we briefly describe the soft x-ray emission spectra of an organometallic compound and discuss the complementary information contained in such spectra, when compared to valence-band photoelectron spectra.

6. PHOTOEMISSION

Photoelectron emission ($\equiv$photoemission) spectroscopy has become a very popular application of synchrotron radiation and, without doubt, photoelectron spectroscopy is one of the areas where the high intensity, the polarization and the tunability over large photon energy ranges of SR have been of prime importance. Progress in instrumentation, such as efficient monochromators and angle-resolving spectrometers, has contributed to the widespread use of this technique, as well as to a thorough understanding of the photoemission process. Thus, SR has tremendously increased the usefulness of photoelectron spectroscopy as a probe of the electronic structure of atoms, molecules, surfaces, adsorbate states and interface states, and as a means of mapping energy bands of bulk solids [54-56]. The papers in this volume by S. Leach [57] and J. L. Dehmer, *et al.* [58] focus on aspects of this spectroscopy relevant for the investigation of the energetics and dynamics of molecules; hence, we concentrate rather narrowly on applications to molecular crystals.

The principle of a photoemission experiment, employing a solid sample, is illustrated in Fig. 21: Monochromatic light, with photon energy hν, impinges on the sample and the energy distribution curves (EDC's) of the photoemitted electrons are measured. The basic electronic-structure information lies in the probabili-

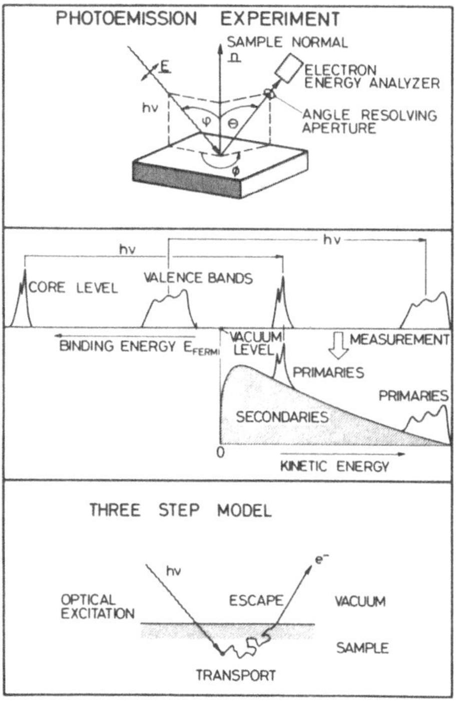

Fig. 21. Principle of a photoemission experiment from a solid. Upper panel, geometry of the experiment; middle panel, energetics of photoemission from valence bands and core levels and a typical photoelectron energy distribution curve (EDC) at photon energy hν; lower panel, three-step model of photoemission.

ty for a photon of energy $h\nu$ to excite an electron of initial energy E_i to a final state with energy E_f. The interpretation of photoemission data, so far, has been based mostly on the three-step model, as described by Berglund and Spicer [59]. Within this model, the photoemission process is approximated by three steps (cf. Fig. 21): (i), optical excitation; (ii), transport of the electrons to the surface, with a probability for undergoing elastic and inelastic scattering processes; and (iii), escape of the electrons into the vacuum, followed by detection. In a photoemission experiment, the influence of several parameters, such as the energy and polarization of the incident radiation or the direction and spin of the outgoing electrons, can be investigated.

The optical absorption of primary photons, in the VUV range, extends roughly 100 Å into the solid, depending upon the material and the photon energy. Further, the escape depth of the excited electrons is strongly dependent of their kinetic energy. Electron-electron scattering with valence electrons, and electron-phonon scattering with the crystal lattice, are the two most important mechanisms determining the mean free path and, hence, the escape depth [54]. It turns out that a plot of electron escape depth versus kinetic energy (or, equivalently, versus the incident photon energy), for a wide variety of different materials, yields a so-called "universal" curve [60]. The universal behavior is a fairly sharp decrease with increasing electron energy, a decrease which starts from an escape depth of 10,000 Å a few tenths of an electron volt above the Fermi energy or the vacuum level, and proceeds towards a flat minimum in the range 50 eV to 500 eV at an escape depth as low as $\sim$5 Å; thereafter, the curve shows an increase with increasing electron energy. Although the details of this curve differ for different materials, the general behavior is fairly similar for all kinds of materials. Thus, an important advantage of SR in photoemission experiments becomes immediately obvious: By choosing a photon energy around $h\nu = 100$ eV, the experiment can be made extremely surface sensitive and, therefore, it becomes possible to distinguish between bulk and surface properties and to identify surface states. Furthermore, the variation of polarization with respect to the plane of incidence (s- and p-polarization), and with respect to the crystallographic axis and bond directions of adsorbates, is another important advantage of SR photoemission experiments.

In addition to the usual technique of measuring an EDC with a fixed photon energy, there are other modes of photoemission spectroscopy which can only be applied with a continuum source. Among these are: (i), yield spectroscopy, where all of the electrons excited at a certain photon energy are collected; (ii), partial yield spectroscopy, or constant final state (CFS) spectroscopy, where all of the electrons excited into an energy interval around a given kinetic energy are collected; and (iii), constant initial state (CIS) spectroscopy, where $E_{kin} - h\nu$ is kept constant by scanning the source monochromator and the electron analyzer synchronously at the same rate. These different modes have been described previously [54].

In the following, we describe recent photoemission experiments on molecular crystals. These materials are characterized by small lattice energies indicative of the weak bonding forces between the units constituting the crystal. The individual molecules, held together by these weak van der Waals forces, are only slightly affected by the solid environment. Thus, the free-molecule electronic structure largely determines the electronic structure of the crystal. It follows that, in a first approximation (in the limit of vanishing intermolecular forces), the crystal may be thought of as an oriented gas. Interaction of the molecules in the crystalline state then perturbs the free-molecule electronic structure producing new, solid-state effects in the electronic structure (e.g., crystal-field splitting), which differentiate it from an oriented gas. Our discussion focuses on examples from two groups of molecular crystals: (i), cryodeposited films of condensed gases such as solid N_2, CO, CO_2; and (ii), organometallic compounds such as metal-phthalocyanines (M-PC's). In both cases the advantages of using SR for the photoemission experiments will become apparent implicitly.

6.1 Photoemission from Condensed Gases

In the time since the first photoelectron energy distribution measurements were made on solid rare gases and the results compared with "first-principles" band structure calculations for these prototype materials [61], a number of SR-based experiments have been performed on cryodeposited solidified gases, including solid N_2, O_2, CO_2 as well as matrices [62-65]. The electronic band structure, energy-transfer processes, as well as partial cross-sections and final state effects, have been investigated. The results for solid CO_2 serve as an example [64].

In the EDC's, one usually observes a one-to-one correspondence between gas-phase and solid-phase measurements. In Fig. 22, a family of EDC's , at different photon energies, is shown for solid CO_2. The three main features for direct emission from solid CO_2 are clearly visible. While the uppermost band, with the highest kinetic energy, is well-separated from the rest, several other bands coalesce to form a double-peaked structure at lower kinetic energies. The direct emission features are decomposed into four partly overlapping bands, as indicated in Fig. 23 for one particular photon energy. Thus, a clear one-to-one correspondence to the gas-phase photoelectron spectrum of CO_2 emerges (cf. Fig. 23). It is easy, therefore, to assign the peaks for solid CO_2, in the order of increasing energy, as photoemis-

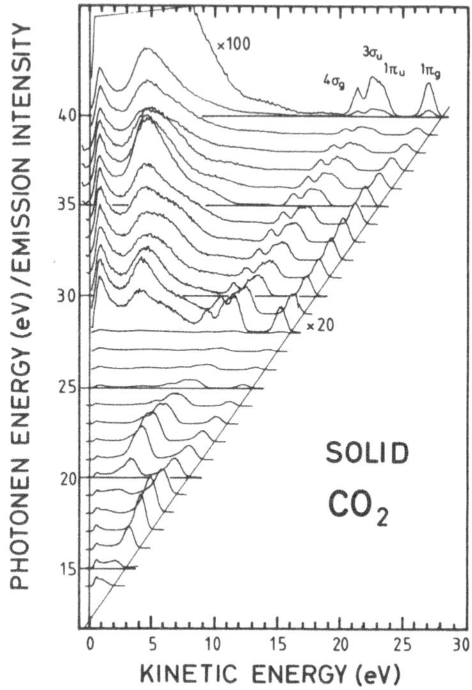

Fig. 22. Photoelectron energy distribution curves for polycrystalline solid CO_2 for excitation energies between 15 and 40 eV. In this plot, the same initial states follow inclined lines. The four valence molecular orbitals are denoted by the one-electron MO notation. (From [64].)

sion from valence bands formed by the $1\pi_g$, $1\pi_u$, $3\sigma_u$ and $4\sigma_g$ MO's of the CO_2 molecule. The resulting vertical binding energies E_B^V are 13.0, 16.7, 17.6 and 18.8 eV, respectively, with reference to the vacuum level ($E_{vac} = 0$) of the sample. From these data, a rigid gas-to-solid shift towards lower binding energies $\Delta E_R^V = 0.7 \pm 0.2$ eV is obtained for all four bands. The reduced ionization potentials in solid molecular crystals have been discussed previously within the framework of the Mott-Littelton model [62]. The analysis for a number of van der Waals insulators suggested that the dielectric polarization around the positive hole created in the photoemission process is the dominating relaxation mechanism. A hole polarization energy of $P_+ = 1.2$ eV was estimated for CO_2, within a simple macroscopic continuum model, as the energy difference between the screened hole and the free ion. This value

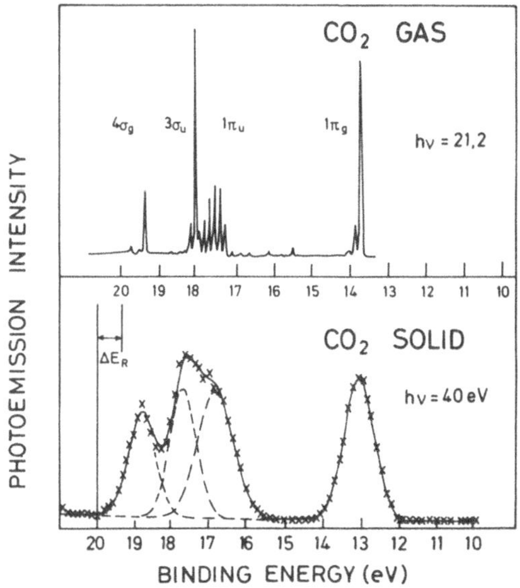

Fig. 23. Comparison of the gas-phase photoelectron spectrum (upper panel) with a photoelectron energy distribution curve for solid CO_2 (lower panel). $\Delta E_R^V = 0.8$ gives the rigid gas-to-solid relaxation shift by which the gas-phase spectrum has been shifted in order to obtain coincidence for the prominent features. The deconvolution of the EDC for solid CO_2 into four bands is also shown. (From [64].)

is in good qualitative agreement with the experimental gas-to-solid shift.

We next discuss the hν-dependence of the partial photoionization cross-sections for the valence bands. Inspection of Fig. 22 shows immediately that the intensity of each of the primary peaks goes through a maximum at ∿4.5 eV kinetic energy. The results of the detailed analysis are displayed in Fig. 24. From the energy distribution curves, such as those shown in Fig. 22, the relative partial ionization cross-sections $\sigma_i(h\nu)$ were obtained for the ith band by measuring, for each photon energy, the area A_i under each peak. This procedure is only valid for an angle-integrating geometry, where we can neglect the asymmetry parameter β_i which characterizes, in the molecular case, the angular distribution of the ith ionic state at a given photon energy. The above approximation is valid, however, since we are dealing with polycrystal-

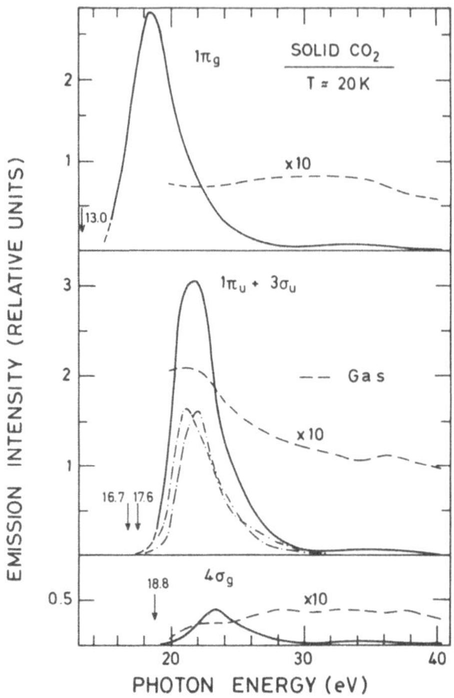

Fig. 24. Relative partial ionization cross-sections for the $1\pi_g$ derived, $1\pi_u$ + $3\sigma_u$ derived and $4\sigma_g$ derived valence bands of CO_2. The arrows mark the onset for photoemission from these bands. For comparison, the partial cross-sections for gaseous CO_2 are shown (crosses); these have been scaled for $h\nu$ = 30 eV for the $1\pi_g$ band, and enlarged by a factor of 10. (From [64].)

line, randomly-oriented samples and since the emitted photoelectrons can be regarded as an isotropic source.

In order to further discuss the intensities and partial photoionization cross-sections, we briefly recall those factors which influence the intensity $I(E,\omega)$ of photoemitted electrons having energy E and excited by photons of frequency ω from a solid sample [55]. This will enable us also to make a comparison to the optical spectrum $\varepsilon_2(\omega)$, the imaginary part of the dielectric constant [cf. Eq. (13)].

Within the framework of the three-step model for photoemission [59], the intensity distribution of primary electrons $I(E,\omega)$ can be factored as

$$I(E,\omega) = P(E,\omega)\cdot T(E)\cdot D(E) \tag{14}$$

according to the sequence of three events, namely: (i), the optical excitation of an electron, leading to a distribution of photoexcited electrons $P(E,\omega)$; (ii), the transport of the electron through the solid, characterized by a transmission function $T(E)$; and (iii), the escape of the electron through the sample surface into the vacuum, described by the escape function $D(E)$. The transport function $T(E)$ includes the possibility of inelastic scattering by the other electrons, or by elementary excitations, in the solid. It is a slowly-varying smooth function depending only upon the energy E of the photoelectrons, thus reflecting the fact that no single scattering process, characterized by one particular sharp threshold energy, dominates electron transport. In molecular crystals such as solid CO_2, excitation of several electronic and vibrational excited states can contribute to the scattering of primary excited photoelectrons. The energies for such excitations range from a few meV (phonons and librational motions) through several hundred meV (intramolecular vibrations), up to several eV (exciton formation). Hence, we do not expect a sharp onset of the electron-scattering mechanism since there is no single dominating process. As in the case of $T(E)$, $D(E)$ is a smooth function of E beyond the threshold, when the energy E is sufficient to surmount the potential barrier at the surface. While both $T(E)$ and $D(E)$ may distort the energy distribution, they are not expected to introduce structure or peaks in the intensity distribution $I(E,\omega)$.

The spectrum of photoexcited electrons, $P(E,\omega)$, is the remaining factor to be discussed. Within the three-step model, this distribution is given through the optical excitation of electrons from occupied valence-band states E_V into empty conduction-band states E_C:

$$P(E,\omega) \propto \sum_{v,c} \int_0^\infty d^3k \, |e\underset{\sim}{M}_{vc}|^2 \, \delta(E_c - E_v - \hbar\omega) \, \delta(E_c - E) \tag{15}$$

where $e\underset{\sim}{M}_{vc}$ is the dipole matrix element for the optical transition. It should be noted that the photoemission in this model becomes a very selective process since three δ functions (including the k-selection in the matrix element $|e\underset{\sim}{M}_{vc}|$) restrict the possible contributions. The expression for $P(E,\omega)$ is closely related to $\varepsilon_2(\omega)$, the imaginary part of the dielectric constant, as given by Eq. (13).

If we take the dipole matrix element to be constant, $P(E,\omega)$

is a measure of the total number of transitions possible at a photon energy ω, subject to energy and k-vector conservation (energy distribution of the joint density of states). We are thus led to attribute the strong resonances appearing for each photoemission band to the band structure of solid CO_2. The CO_2 valence bands are flat, as is characteristic for an insulator ($E_C \lesssim 1.5$ eV), and are expected to show limited dispersion. This is an immediate consequence of the prevailing tight-binding situation [66]. Thus, the maximum appearing for each partial ionization cross-section is due to the high density of final conduction-band states. This is schematically depicted in Fig. 25.

Our determination of the band structure from direct photoemission is supported by the observation of pronounced maxima of scattered electrons in each EDC (cf. Fig. 22), peaking at ∿4.5 eV

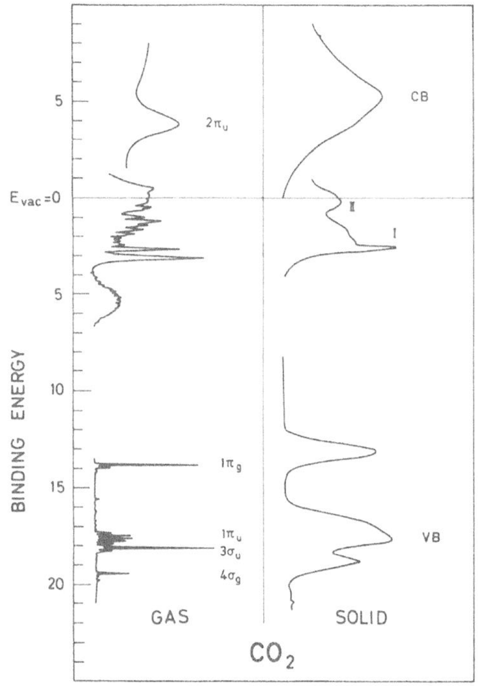

Fig. 25. Schematic energy scheme for the electronic band structure of solid CO_2 (right part). In order to deduce the molecular parentage, comparison is made to the valence MO's and to some excited states in gaseous CO_2 (see text). (From [64].)

kinetic energy. These maxima are clearly visible for $h\nu \gtrsim 30$ eV, where they are almost not obscured by primary emission; they reflect the high density of final states into which primary electrons are scattered inelastically. This peak of scattered electrons grows with increasing photon energy, since the number of successive scattering events which one primary electron can undergo increases with kinetic energy.

The electronic band structure of solid CO_2, as determined above, provides a key for the discussion of the hitherto unassigned optical absorption spectrum of solid CO_2 in the VUV region [67]. The imaginary part of dielectric function (cf. Fig. 25) shows a fairly simple irregular structure in the VUV, with sharp intense maxima at around 10.4 eV, 12.8 eV and 14.6 eV, and a broader structured region with its maximum at roughly 20.8 eV (see Fig. 7 in [45]). The total photoemission cross-section obtained by adding up the partial photoemission cross-sections yields roughly the density of valence-to-conduction band transitions. We are thus led to a simple interpretation of the optical spectrum. All structure appearing at photon energies $h\nu \lesssim 13$ eV has to be attributed to exciton states; even the maximum at $\sim$14.6 eV may be due to a metastable exciton state. The broad optical band centered at 20.8 eV is then due to interband transitions.

The excitons in solid CO_2 are of the Frenkel type, just as in solid N_2 and CO [30,45]; they are expected to be fairly localized with a parentage in intra-valence molecular excitations. An inspection of the excitation spectrum of CO_2 [68], and the schematic energy scheme of Fig. 25, suggests that the strong dominating gas-phase bands at 11.05 and 11.38 eV give rise to the strong exciton band at 10.4 eV in the solid. These intense broad gas-phase bands have previously been described as the lowest n = 3 members of the $1\pi_g \rightarrow np\sigma_u$ and $1\pi_g \rightarrow np\pi_u$ Rydberg series. The close correlation of these bands to the first exciton peak in solid CO_2, as suggested by the scheme in Fig. 25, leads us to postulate either valence character or a strong admixture of valence character to these bands for the gas phase. Similar results have recently been obtained for solid N_2, where the partial cross-sections for emission from $3\sigma_g$, $1\pi_u$ and $2\sigma_u$ derived valence bands show pronounced maxima at 3.4, 2.9 and 3.0 eV above threshold. These features have also been interrelated as being due to a high density of conduction-band final states [63].

Finally, we note that the high density of conduction-band final states for these molecular crystals can be traced back to molecular shape resonances which are well known for the gas phase (see the paper by J. L. Dehmer, *et al.* [58] in this volume).

6.2 Photoemission from Organic Molecular Crystals

Detailed photoemission experiments on organic solids, thin polycrystalline films or single crystals are of fairly recent vintage [69] in comparison to studies of metals and semiconductors. Certainly, the fact that these materials are frequently very poor conductors, with the accompanying problem of charging effects and additional problems associated with adequate sample and surface characterization, have prevented more detailed investigations. On the other hand, widespread interest in these materials, the practically unlimited choice in modelling materials with new and interesting properties, as well as recent discoveries of unexpected metal-like behavior of some organic materials make photoemission studies on organic solids a very rewarding theme.

The methods used to study photoemission from these systems are not different from those used in studies of conventional materials, and we refer the reader to pertinent references [54-56]. The sample preparation, particularly with more exotic materials including substances of biological relevance, requires UHV chambers separated from the electron analyzer chamber. Also, a number of standard techniques such as LEED, Auger spectroscopy and ion sputtering have recently been used in order to characterize the substrate and the organic film. It turns out that even a number of biologically important molecules, such as phthalocyanines (PC's), a number of dye stuffs and nucleic acid bases, can be prepared as thin films on single crystalline substrates by vacuum sublimation. While the emphasis in such studies is on the electronic structure of clean (bulk) materials, we note that studies of organic surfaces and of adsorbates on organic solids are still in their infancy.

Photoemission experiments on phthalocyanines serve as an example illustrating the present state of the art. The PC's are planar molecules with a geometry shown in the inset of Fig. 27. In metal-PC's, the central metal atom replaces the two central hydrogen atoms of the metal-free PC. The bonding is in most cases covalent. The electronic structure of the valence bands of the PC's and their metal adducts has attracted considerable attention, both from a theoretical and an experimental point of view, because of the similarity of the structure of the phthalocyanines to that of chlorophyll and blood pigments, and because of the important technological application of the phthalocyanines as dye stuffs, catalysts and as material for organic semiconductor applications [70].

Recent photoelectron energy distribution curve measurements on phthalocyanines have shown a quite complex structure for the valence bands, extending over a binding energy range from 5 to 35 eV (see, e.g., [71]). This complex structure, expected in view of the many molecular orbitals contributing to the density of states (in H_2PC, with D_{2h} symmetry, there are 186 valence electrons occupying 93 valence MO's), has been only partly disentangled by exploiting the differences in energy dependence of the partial cross-sections for different MO's, and by determining difference photoelectron spectra between the metal-PC and H_2PC. Experimentally, this latter method, applied to organometallic materials where a few metal atoms are dispersed in a quasi-matrix, is analogous to the determination of difference EDC's in photoemission experiments from adsorbates. An assessment of the differences in the EDC's from phthalocyanines and their metal adducts, by visual inspection, was attempted earlier for a number of phthalocyanines in the solid phase and also in the gas phase. However, in all these investigations neither the contribution of N 2p and C 2p partial densities of states to the valence bands, nor their energies with respect to the other MO's, could be determined. The results of a combined x-ray emission and photoemission experiment [71] are displayed in Fig. 26 for metal-free PC. The XPS result ($h\nu$ = 1487 eV) is compared with the photoelectron spectrum ($h\nu$ = 100 eV) and with the carbon and nitrogen K-shell soft x-ray emission spectra (see Section 5). Since the binding energies of the C 1s and N 1s levels in the PC's are known from XPS measurements, all data can be displayed on a common binding energy scale. In the binding energy range E_B = 15-30 eV, the soft x-ray 2p → 1s emission intensity is very low for both the carbon and the nitrogen emission. Thus, we conclude that the maxima labelled F, G and H in the EDC's are due to C and N 2s valence-band density of states. In addition, comparison with the XPS data and results for solid benzene shows that the maxima F, G and H are predominantly due to C 2s. (Note, in this context, that the ratio of carbon to nitrogen atoms is 4:1.) The 2s character of the features F to H is also evident from the relative enhancement of these peaks in the XPS spectra, as compared to the other valence-band maxima A to E, since at $h\nu$ = 1487 eV the partial cross-section for C and N 2s photoemission is roughly thirty times larger than that for 2p emission [72].

In the range of lower binding energies, E_B = 5-15 eV, we expect overlapping C and N 2p orbitals. MO's derived from N 2p are concentrated at binding energies around E_B = 8.6 eV (cf. Fig. 26), and photoemission from these MO's contributes strongly to peak B in the EDC's. We assume that peak B is (apart from its C 2p contribution) due mainly to emission from the nonbonding nitrogen lone-pair electrons which are located at the outer nitrogen atoms in the porphin ring. Since there is no other pronounced peak in the nitrogen soft x-ray emission spectrum, the remaining features (A, part of B, C, D and E) are of predominantly C 2p character. Most MO calculations agree that the maximum A in the EDC's, at around 5.9 eV binding energy, is due to a π-type MO of $4a_u$ symmetry in D_{2h}, with strong contributions from the C 2p orbitals situ-

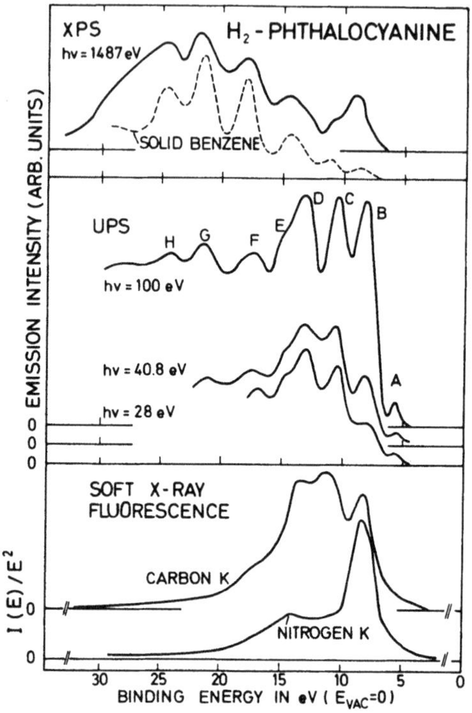

Fig. 26. Comparison of the x-ray photoelectron spectrum (uppermost panel), the photoelectron spectra for hν = 100, 40.8 and 28 eV (middle) and the carbon K 1s soft x-ray emission spectra (lower panel) for H_2-phthalocyanine. All spectra have been aligned on a common binding energy scale. (From [71].)

ated on the C-N ring. It is difficult at present to disentangle further the C 2p partial density of states, in the binding energy range 5-15 eV, in terms of its π and σ contributions.

As shown above, it is in general not an easy task to analyze the complicated partial density of states. On the other hand, one is particularly interested in the binding energies, possible splittings and relative strengths of the metal-derived orbitals. The possibilities offered by SR photoemission experiments for solving this problems are illustrated for Zn-PC in Figs. 27 and 28 [73-74], and for Cu-PC in Fig. 29 [75]. For Zn, the atomic ground-state configuration is $3d^{10}4s^2$ and the atomic photoelectron spectra show the d states well-separated from the s states.

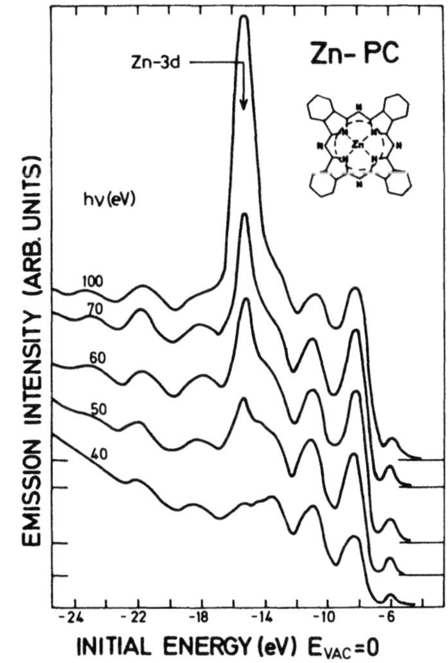

Fig. 27. A family of photoelectron energy distribution curves for Zn-phthalocyanine in the valence-band range for photon energies between 40 and 100 eV. The maximum due to the Zn derived 3d states is located at a binding energy 15.2 eV below the vacuum level E_{vac}. (From [74].)

This is carried through to the metallic state, where the Zn d bands are well-separated from the s-like and p-like bands [76]. Thus, Zn-PC appears to be a suitable material for studying both the 3d valence-orbital structure and the influence of the molecular environment on the metal d states. By using the characteristically different hν-dependence of s, p and d state photoelectron cross-sections in the 30-120 eV photon energy range (cf. Fig. 27), it was possible to determine the d partial density of states, which was then compared with results for atomic and metallic Zn (cf. Fig. 28). As shown within the one-electron approximation by Cooper and Fano [26], the partial cross-section of a subshell depends strongly on the overlap of the initial-state and final-state wavefunctions. In photoemission experiments, the higher-energy wavefunctions far above threshold are the continuumn-state wavefunctions. Thus, for d initial states, for example, delayed

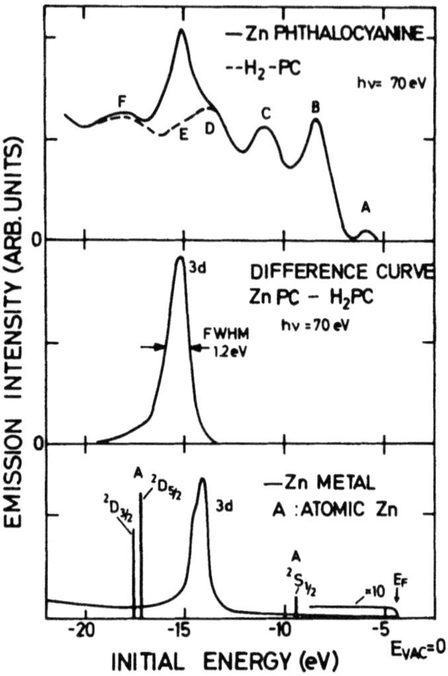

Fig. 28. Photoelectron energy distribution curves for Zn-PC (solid line) and H_2-PC (dashed line) are shown at $h\nu$ = 100 eV in the upper part. Counting rates have been set equal at $E_B \simeq 30$ eV, where no pronounced valence-band structure is observed. The difference spectrum (Zn-PC minus H_2-PC) $\Delta N(E)$ (middle part) is compared with the EDC's of atomic and metallic Zn (lower pannel). The binding energy scales of the upper and middle panels and of the atomic spectrum refer to $E_{vac} = 0$. For the solid spectrum, $e\phi = 4.4$ eV has been added to E_{FERMI} for this comparison. (From [74].)

maxima above threshold, due to an increasing overlap of d initial- and f final-state wavefunctions, are predicted in the atomic cross-sections, followed by Cooper minima which arise from a change of sign in the radial dipole matrix element associated with the d → f final-state transitions. These atomic effects often dominate the absorption cross-section for solids in the VUV

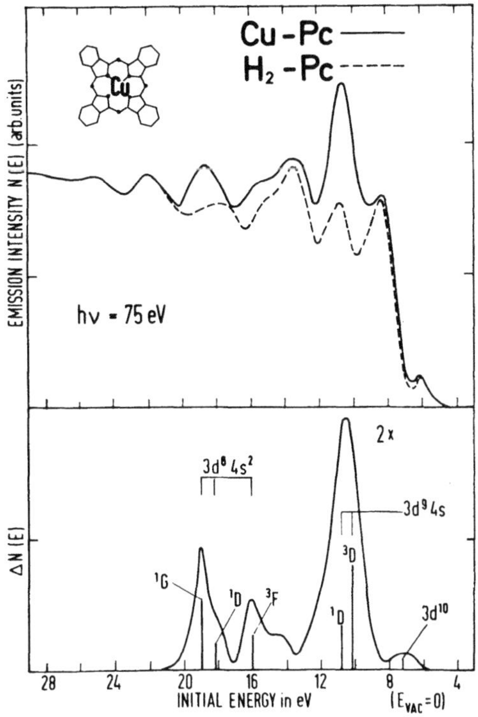

Fig. 29. Photoelectron energy distribution curves for Cu-phthalocyanine (solid curve) and H_2-PC (dashed curve) for a photon energy of $h\nu = 75$ eV at the 3p threshold show marked differences (upper panel) due to the Cu derived partial density of states. The difference spectrum at resonance (lower panel) can be assigned to a $3d^94s$ and a $3d^84s^2$ configuration. (From [75].)

range. They have now been studied in photoemission for a number of metals and compounds. The important point here is to note that, by tuning the photon energy, one can selectively enhance the visibility of partial density of states as demonstrated for the 3d levels in Zn-PC (cf. Fig. 27). Thus, these levels can be easily separated from the s-like and p-like partial density of states associated with the rest of the molecular framework, and an overall view of the pattern of valence bands in terms of the component partial density of states emerges. In this way, the location of the d bands in the series Fe-PC, Co-PC, Ni-PC, Cu-PC and Zn-PC has recently been determined [71,73-75].

Besides enabling the use of photoelectron spectroscopy in determining partial cross-sections over extended energy ranges, SR makes it possible to observe novel effects, as in the case of resonant photoemission at the onset of inner-shell absorption thresholds: In 1977, Guillot, *et al.* [77] observed a strong resonance enhancement in the photoelectron energy distribution curves from Ni. The intensity of a peak located at about 6 eV below the Fermi level increased dramatically (more than a factor of 10) when the photon energy was swept through the 3p threshold, at about 66 eV. This result was explained as a Fano-type resonance resulting from the interaction between a discrete excitation $3p^6 3d^9 4s \rightarrow 3p^5 3d^{10} 4s$ and the continuum channel $3p^6 3d^9 4s \rightarrow 3p^6 3d^8 4s\ \varepsilon\ell$ coupled via the Super-Coster-Kronig decay $3p^5 3d^{10} 4s \rightarrow 3p^6 3d^8 4s\ \varepsilon\ell$. In the meantime, a large number of experiments have been made using SR to further explore these resonances. Most of the 3d metals ($3p \rightarrow 3d$ resonance), rare-earth metals and compounds ($4d \rightarrow 4f$ resonance) and uranium ($5d \rightarrow 5f$ resonance) have been investigated.

Resonant photoemission, and the observation of two-electron-satellites in addition to one-electron emission, clearly demonstrate the importance of many-electron effects in photoemission. Aside from mandating fundamental investigations of these effects in order to clarify the photoemission process, resonance phenomena have offered novel possibilities for studying the electronic structure of more complicated materials. We consider Cu-PC as an example. A photoelectron energy distribution curve for Cu in Cu-PC, excited with photons around the Cu 3p threshold at $h\nu$ = 75 eV, is shown in Fig. 29 [75]. A sharp 3d electron satellite, with two peaks with initial energies of 5.5 and 8.5 eV below the center of the Cu 3d main peak, has been found which exhibits a strong resonant enhancement at threshold. The one-electron photoemission process, which excites the 3d electrons $3p^6 3d^{10} 4s + h\nu \rightarrow 3p^6 3d^9 4s + e^-$, results in the main lines ^3D, ^1D at $E_i = -10.5$ eV in the photoelectron distribution (cf. Fig. 29). At resonance, the $3p \rightarrow 4s$ excitation, followed by a Super-Coster-Kronig decay, creates a two-hole state with a $3d^8 4s^2$ final state: $3p^6 3d^{10} 4s + h\nu \rightarrow 3p^5 3d^{10} 4s^2 \xleftrightarrow{SCK} 3p^6 3d^8 4s^2 + e^-$. This process results in the resonant enhancement of the $3d^8 4s^2$ final state (cf. Fig. 29). The energetic separation of the satellite structure from the 3d emission peak (^3D, ^1D) is very close to the energy difference between $3d^8 4s^2$ and $3d^9 4s$ for atomic Cu^+, and was indeed corroborated by later resonant photoemission experiments for atomic Cu [78]. Thus, an atomic assignment of the main configurations contributing to the photoelectron spectrum for "quasi-atomic" Cu in Cu-PC, as shown in Fig. 29, is justified. Clearly, from this type of experiment, and from a comparison between the metal-PC and free atom results, important information can be obtained concerning relaxation (screening) and hybridization in the organic compound. For example, the importance of empty s-like states becomes evident by comparing the relative satellite intensity of the $3d^8$ configuration to the $3d^9$ intensity: For Zn-PC at resonance, the relative satellite intensity is about 30% in Zn-PC and 6% in Zn [73]; the corresponding relative satellite intensity is 36% in Cu-PC and 12% in Cu. One can associate the weakness of the satellite in Zn metal with the smaller number of empty 4s states in Zn as compared to Cu. In both Cu-PC and Zn-PC, the satellite intensity is enhanced in comparison to the metals, since 4s electrons are transferred from the metal site to the ligand.

As a final example of photoemission experiments, we discuss angle-resolved photoemission from oriented films of molecules. The determination of the angular distribution from adsorbates and ordered layers is of great interest for two inter-related reasons. First, one can obtain molecular orbital information, particularly about the symmetry of the MO's involved, from angle-resolved photoemission studies; second, the angular distribution of photoemission from an orbital of a molecule adsorbed on a surface contains, in principle, information concerning the orientation of the molecule on the surface [54,79]. An apparatus specifically designed for photoemission in the spectral range 5 eV $\lesssim h\nu \lesssim$ 50 eV is sketched in Fig. 30. The system consists of a beamline with three focusing mirrors, a 1 m Seya-Namioka monochromator with two interchangeable gratings, a modified VG ADES 400 photoelectron spectrometer with two angular rotations, and a sample preparation and transfer stage [80]. This photoemission set up has been used for a series of angle-resolved photoemission experiments with conventional He I or x-ray excitation, as well as for experiments exploiting the full advantages of SR. In Fig. 31, angle-resolved photoelectron energy distribution curves from thin films of Pb-PC on a Cu (100) single crystal are shown as an example [81]. The measured angular variations of the uppermost MO (peak A) follow the expected intensity variation for a p_z-like orbital (polar diagram inset). From this experimental observation, the molecular orientation can be derived: There is general agreement, among MO calculations for PC's, that the uppermost band A is largely formed by π orbitals of a_{1u}-type (in D_{4h} symmetry) extending over the inner ring of the PC's. In Pb-PC we also expect a contribution to the density of states from Pb 6p derived orbitals in this binding energy range. Thus, the observed p_z-type angular emission pattern is in excellent accord with the assumption of a parallel orientation of the planar Pb-PC molecules with respect to the substrate.

7. PERSPECTIVES

In the present overview we have briefly touched upon a few

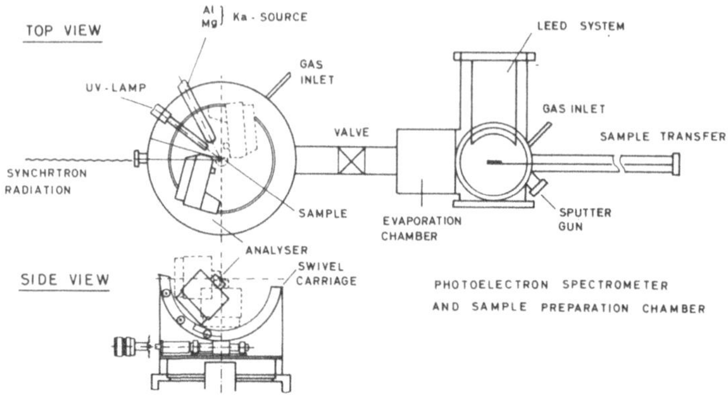

Fig. 30. Schematic photoelectron spectrometer with rotatable electron analyzer, sample preparation chamber and sample transfer rod. Auxilary equipment is indicated. (From [80].)

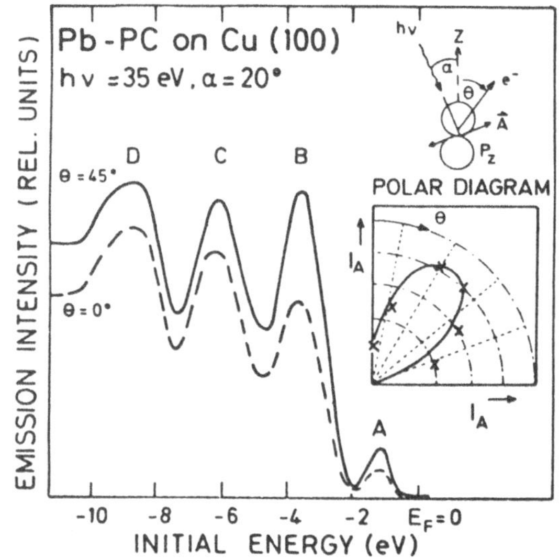

Fig. 31. Angle-resolved photoelectron energy distribution curves for an ordered Pb-phthalocyanine multilayer on a Cu (100) surface for two different angles of electron emission. In the polar diagram (inset), the expected angular variation for a p_z-like orbital (solid curve) is compared to the measured intensity variation for peak A (crosses, with an isotropic background subtracted). (After [81].)

of the many interesting and new possibilities offered by synchrotron radiation for the investigation of higher excited states of molecules and molecular crystals. We hope that the few examples discussed have shown that SR offers the unique ability to tune over extended regions of excitation energy in optical, luminescence or photoelectron emission experiments, so as to optimally probe binding energies ranging from those of loosely-bound valence electrons up to energies corresponding to the most tightly-bound levels. Fundamental quantities, such as binding energies, their shifts in molecular environments, potential curves and $h\nu$-dependent cross-sections may be determined over extended energy ranges. In addition to the static properties which determine the energetic structure of a system, dynamical processes, such as relaxation and energy-transfer processes, are receiving increasing attention (see, e.g., [44,45]). All of these studies provide a basis for further developments and refinements of theoretical models.

The above general understanding is basic to a large number of more obviously important phenomena, such as: (i), photon-induced chemical reactions in plasmas and in various regions of the atmosphere; (ii), the performance of molecules and clusters on surfaces in catalyzing chemical reactions; (iii), energy-transfer processes between molecules or molecular assemblies as an important step in biological processes; and (iv), the development and improvement of spectroscopic techniques used in applied research (e.g., for the development of VUV lasers). The electronic structure of molecular crystals is a subject of widespread interest. In particular, photoelectron emission and luminescence spectroscopy, when combined in their many forms with a tunable SR source, have proven to be versatile and powerful probes of the electronic structure of molecular crystals. Photoelectron spectroscopy encompasses a wide range of topics, such as atomic and molecular cross-section effects, SR/laser coincidence techniques [82], accurate determination of binding energies and material-induced binding energy shifts, valence-band structure (i.e., one-electron energy band versus momentum $E(\underset{\sim}{k})$ dispersions) for solids and for two dimensional surfaces, intrinsic surface states, adsorbate systems, surface reactions, etc. Information on these topics is of central interest because they are fundamental for our understanding of the chemical and physical properties of solids and surfaces. Moreover, the experimental results provide not only crucial tests for any theoretical model, but also have many interesting and important applications. For example, the topic of adsorbates and surface reactions covers basic mechanisms of catalysis, corrosion and oxidation.

SR is used today in more than 20 laboratories throughout the world. One can safely predict that there will be a steadily increasing output of new and interesting results in many scientific disciplines from these laboratories. Unfortunately, space considerations preclude mention of several important but developing uses of SR in this overview.

The development and use of superconducting wigglers, as well as improvement of simple and inexpensive permanent magnet wigglers, will greatly increase useful intensities by a factor of ten to a hundred. In certain spectral ranges, even these intensities will be greatly surpassed by SR from undulators. These future improvements of SR sources will be, by today's standard, spectacular. These greatly improved sources will undoubtedly be a driving force for new instrumentation and new techniques, and will permit entirely new classes of applications and lead to new breakthroughs.

In summary, during the last two decades SR has matured from being an unwanted by-product of high-energy electron accelerators to being a widely used source of radiaiton in the VUV and x-ray range which is in increasing demand. It is apparent that SR has an assured future for some time to come as a powerful tool in spectroscopy.

8. APPENDIX

In this Appendix, spectroscopic data (Tables A1-A8) for the various excited states of N_2 are compiled and comparison is made among data from the gas phase, solid phase and the matrix-isolated phase. This compilation is based on the Ph.D. thesis of P. Gürtler (Univ. of Hamburg, 1979). Data from the literature are given where appropriate.

A few general comments follow.

Energy E: From the measured spectra, the wavelength of a given transition was determined with an accuracy of 0.2 Å. These values have been converted with the factor $E[eV] = 12398.52/\lambda[\text{Å}]$. For the gas phase, the position of the band head was taken and, for the matrix-isolated phase or solid phase, the position of the zero-phonon line was taken if not stated otherwise.

Oscillator strength f: For the determination of f in absolute units, the area under a peak was evaluated by numerical integration. f was calculated according to $f = (mc^2/\pi e^2)(1/\lambda^2)\int\sigma(\lambda)\, d\lambda$.

Peak area $\int\sigma d\nu$: For a few systems, particularly for the matrix-isolated phase, only relative absorption spectra have been determined. In this case, the relative area of a peak is given in order to follow the intensity variation in a progression. The values are normalized in such a way that the area of the most intense peak of a progression is set to 1.

Halfwidth Γ: The halfwidth of a peak is given, in meV, as full width half maximum (FWHM).

Splitting S: For strong transitions, the peaks are split into several components. The splitting of these components is given in meV.

Vibrational constants: The position of a band in a vibrational progression is described by

$$E(v') = T_e + \omega_e (v' + \frac{1}{2}) - \omega_e x_e (v' + \frac{1}{2})^2$$

[See G. Herzberg, *Molecular Spectra and Molecular Structure*, Vol. III (Van Nostrand Reinhold, New York, 1966)]. The three constants T_e, ω_e and $\omega_e x_e$ are determined by a least-squares fit from the spectroscopic data. With these constants in hand, it is possible to determine the Morse potential of a given electronic state.

Table A1. $N_2\ a^1\Pi_g \leftarrow X^1\Sigma_g^+$ Transition.

	gas phase		solid phase			Ne-matrix		$^{15}N_2$-solid		
v'	E (eV)	f(x10^{-7})[a]	E (eV)	f(x10^{-7})[b]	Γ (meV)	E (eV)[c]	∫σdν	E (eV)[d]	∫σdν	Γ (meV)
0	8.549	1.43	8.498	27.2	27	8.537	0.23	e	e	e
1	8.756	2.21	8.701	81.4	36	8.744	0.69	8.719	0.55	33
2	8.959	2.78	8.903	118.4	43	8.947	0.93	8.922	0.89	41
3	9.159	2.92	9.106	137.5	44	9.147	1.00	9.117	0.96	43
4	9.356	2.54	9.305	133.8	42	9.344	0.97	9.310	1.00	42
5	9.547	2.41	9.502	104.2	38	9.536	0.71	9.497	0.84	38
6	9.738	2.10	9.694	80.8	32	9.725	0.61	9.678	0.64	34
7	9.922	1.53	9.882	44.5	27	9.913	0.43	9.855	0.43	29
8	10.103	1.26	10.069	36.1	--	10.097	0.29	10.034	0.37	--
9	10.288	1.01	10.248	22.6	--	10.274	0.19	10.205	0.21	--
10	10.462	0.86				10.451	0.11	10.373	0.14	--
11	10.635	0.68				10.625	0.08	10.542	--	--
12	10.803	0.63				10.792	--			

[a] Total f = 22.4 x 10^{-7}.

[b] Total f = 787 x 10^{-7}.

[c] For bands without pronounced zero-phonon lines, the energy was estimated from the position of the maximum and from the FWHM.

[d] The position of the maximum is given.

[e] The bands are disturbed by structures from CO impurities.

Table A2. N_2 $w^1\Delta_u \leftarrow X^1\Sigma_g^+$ transition.

	gas phase		Ne-matrix		solid phase		$^{15}N_2$-solid	
v'	E (eV)[a]	f(x10^{-8})[b]	E (eV)	$\int\sigma d\nu$	E (eV)	f(x10^{-7})[c]	E (eV)	$\int\sigma d\nu$
0	--				8.835	3	8.837	--
1	9.085	--			9.026	14	9.022	--
2	9.272	--	9.259	0.20	9.214	37	9.203	--
3	9.457	0.83	9.441	0.39	9.398	68	9.384	0.45
4	9.639	1.40	9.623	0.64	9.580	94	9.559	0.64
5	9.818	0.95	9.805	0.84	9.760	103	9.733	0.73
6	9.994	--	9.982	0.89	9.936	114	9.904	0.82
7	10.168	--	10.152	0.89	10.111	132	10.073	1.00
8	10.338	--	10.324	1.00	10.283	127	10.240	1.00
9	10.504	--	10.492	--	10.450	117	10.402	0.91
10	10.669	--	10.660	--	10.615	103	10.563	0.82
11	10.831	--	10.819	0.73	10.778	91	10.722	0.64
12			10.979	0.64	10.939	81	10.878	0.55
13			11.135	0.53	11.097	62	11.032	0.45
14			11.292	--	11.252	48	11.182	0.36
15			11.444	0.42	11.404	36	11.331	0.36
16			11.592	0.27	11.555	33	11.478	0.27
17					11.703	31	11.620	0.27
18					11.847	25		
19					11.986	15		
20					12.128	--		
21					12.266	--		

[a]Data are from Y. Tanaka, M. Orgawa and A. S. Jursa, J. Chem. Phys. 40, 3690 (1964).

[b]Data are from B. K. Ching, G. R. Cook and R. A. Becker, J. Quant. Spec. Rad. Transf. 7, 323 (1967).

[c]Total f = 13.3 x 10^{-5}.

Table A3. N_2 $b^1\Pi_u \leftarrow X^1\Sigma_g^+$ transition.

	gas phase		Ne-matrix		solid phase		
v'	E (eV)[a]	$\int\sigma d\nu$	E (eV)	$\int\sigma d\nu$	E (eV)	$\int\sigma d\nu$[b]	Γ (meV)[b]
0	12.500	0.06	12.494	0.09	12.418	0.67	40
1	12.579	0.33	12.573	0.21	12.499	0.87	43
2	12.668	0.55	12.664	0.41	12.581	0.84	45
3	12.761	0.75	12.757	0.67	12.677	1.00	55
4	12.856	1.00	12.850	0.90	12.774	0.97	55
5	12.953	0.92	12.946	1.00	12.879	0.97	54
6	13.052	0.72	13.042	0.96	12.977	0.78	52
7	13.150	0.64	13.137	0.93	13.079	0.63	50
8	13.248	0.66	13.233	0.83	13.179	0.49	48
9	13.343	0.55	13.328	0.70	13.272	0.37	47
10	13.436	0.36	13.422	0.52	13.371	0.23	45
11	13.528	0.25	13.512	0.35	13.465	0.15	43
12			13.602	--	13.558	0.13	46
13					13.649	0.10	--
14					13.732	--	--
15					13.819	--	--
16					13.893	--	--
17					13.962	--	--

[a]Deperturbed values are from M. W. Leoni, Thesis (Eidgen. Techn. Hochschule, Zurich, 1972).

[b]Γ and $\int\sigma d\nu$ are difficult to evaluate because of an overlap of vibrational bands and underlying continua.

Table A4. N_2 $b'\ ^1\Sigma_u \leftarrow X^1\Sigma_g^+$ transition

v'	gas phase		Ne-matrix	
	E (eV)[a]	$\int\sigma d\nu$	E (eV)	$\int\sigma d\nu$
5	13.314	0.10	13.287	--
6	13.390	0.25	13.377	0.11
7	13.490	0.53	13.465	0.45
8	13.577	1.00	13.550	1.00
9	13.663	1.71	13.632	--
10	13.748	2.67	13.710	--

[a]Deperturbed values are from M. W. Leoni, Thesis (Eidgen. Techn. Hochschule, Zürich, 1972).

Table A5. N_2 $C^3\Pi_u \leftarrow X^1\Sigma_g^+$

v'	gas phase	Ne-matrix	solid phase	
	E (eV)[a]	E (eV)	E (eV)	Γ (meV)
0	11.032	11.027	10.978	30
1	11.279	11.273	11.225	--

[a]Values are from S. G. Tilford, J. T. Vanderslice and P. G. Wilkinson, Astrophys. J. 142, 1203 (1965).

Table A6. Solid N_2; Vibrational Satellites of the $a^1\Pi_g$ Transition.

v'	$^{14}N_2$			$^{15}N_2$
	E_o (eV)[a]	E_{max} (eV)[b]	f(x10^{-7})	E_{max} (eV)[b]
0	8.786	8.801	3.2	8.795
1	8.990	---	---	8.994
2	9.192	---	---	---
3	---	---	---	---
4	---	---	---	---
5	---	9.809	---	---
6	9.977	9.996	8.9	9.956
7	10.160	10.179	6.0	10.128
8	---	10.367	5.0	10.308
9	10.521	10.544	---	10.482
10	---	10.713	---	---

[a] E_o position of the zero-phonon line.
[b] E_{max} position of the maximum of the band.

Table A7. Solid N_2; Vibrational Satellites of the $w^1\Delta_u$ Transition.

v'	$^{14}N_2$		$^{15}N_2$
	E (eV)	f(x10^{-7})	E (eV)
4	9.870	---	---
5	10.051	---	---
6	10.228	---	10.183
7	10.401	---	10.351
8	10.574	3.9	10.518
9	10.741	2.8	---
10	10.906	3.0	---
11	11.070	---	---

Table A8. N_2; Vibrational Constants for the $a^1\Pi_g$, $w^1\Delta_u$, $b^1\Pi_u$, $b'^1\Sigma_u^+$ and $C^3\Pi_u$ transitions.

Transition		gas phase	Ne-matrix	solid phase
$a^1\Pi_g$	T_e (eV)	8.445	8.432	8.389
	ω_e (meV)	210	210	211
	$\omega_e x_e$ (meV)	1.68	1.70	1.55
$w^1\Delta_u$	T_e (eV)	8.798	8.785	8.740
	ω_e (meV)	194[a]	193	192
	$\omega_e x_e$ (meV)	1.46	1.37	1.33
$b^1\Pi_u$	T_o (eV)[b]	12.500	12.494	12.418
	ω_e (meV)	93[c]	92	91
	$\omega_e x_e$ (meV)	--	--	--
$b'^1\Sigma_u^+$	T_o (eV)[d]	13.314	13.287	--
	ω_e (meV)	87[c]	85	--
	$\omega_e x_e$ (meV)	--	--	--
$C^3\Pi_u$	T_e (eV)	10.909	10.904	10.855
	ω_e (meV)	247[e]	246	247
	$\omega_e x_e$ (meV)	--	--	--

[a] Data are from Y. Tanaka, M. Orgawa and A. S. Jursa, J. Chem. Phys. 40, 3690 (1964).

[b] The vibrational bands can not be described by a Morse potential. Therefore, T_o is, in this case, the energy of the $v''=0 \rightarrow v'=0$ transition, and an averaged ω_e is given.

[c] Depurturbed values are from M. W. Leoni, Thesis (Eidgen. Techn. Hochschule, Zurich, 1972).

[d] In a Ne-matrix, the progression is visible from $v'=5$ onwards. Therefore, the energy of the $v''=0 \rightarrow v'=5$ transition is given along with an averaged ω_e.

[e] Data are from S. G. Tilford, J. T. Vanderslice and P. G. Wilkinson, Astrophys. J. 142, 1203 (1965).

ACKNOWLEDGMENT

In preparing this paper, we have frequently relied upon the work of our colleagues at HASYLAB. We have tried to acknowledge their efforts and achievements throughout this paper and again wish to thank them here for their advice, help and criticism. In particular, we are indebted to J. H. Fock and V. Saile, with whom over the past years much of the work reported here was done, for stimulating discussions. Our special thanks go to Helga Lehmann and Susanne Weigert for preparing the manuscript. This work was supported in part by the Bundesminiterium fur Forschung und Technologie (BMFT) from funds for Synchrotron Radiation Research.

REFERENCES

1. A. A. Sokolov and I. M. Ternov, *Synchrotron Radiation* (Pergamon, Oxford, 1968).
2. C. Kunz, ed., *Synchrotron Radiation, Techniques and Applications* (Topics in Current Physics Vol. 10, Springer-Verlag, Heidelberg, 1979).
3. H. Winick and S. Doniach, eds., *Synchrotron Radiation Research* (Plenum Press, New York, 1980).
4. E. E. Koch, ed., *Handbook on Synchrotron Radiation*, Vol. 1 (North Holland, Amsterdam, 1983).
5. P. L. Hartman and D. M. Tomboulian, Phys. Rev. 91, 1577 (1953).
6. D. Iwanenko and J. Pomeranchuk, Phys. Rev. 65, 343 (1944).
7. J. P. Blewett, Phys. Rev. 69, 87 (1946).
8. F. R. Elder, A. M. Gurewitsch, R. V. Langmuir and H. C. Pollock, Phys. Rev. 71, 829 (1947).
9. D. Iwanenko and A. A. Sokolov, DAN (USSR) 59, 1551 (1948).
10. J. Schwinger, Phys. Rev. 70, 1912 (1949).
11. An account of the historical developments, including more references, may be found in [4, Chap. 1].
12. See, for example, J. E. Spencer and H. Winick [3, p. 663].
13. See, for example, C. Pellegrini [3, p. 797].
14. E. E. Koch, ed., *X-Ray and VUV Synchrotron Radiation Instrumentation*, Proc. Int. Conf. Hamburg, Aug. 1982, Nucl. Instr. Methods Phys. Res. 208 (1983).
15. J. D. Jackson, *Classical Electrodynamics*, 2nd Ed. (Wiley, New York, 1975), p. 672 ff.
16. See, for example, G. N. Watson, *Bessel Functions* (MacMillan, New York, 1945), p. 188.
17. E. E. Koch, D. E. Eastman and Y. Farge [4, p. 41 ff].
18. G. Brown, K. Halbach, J. Harris and H. Winick, in: [14].
19. D. J. Thompson and M. W. Pool, eds., *European Synchrotron Radiation Facility, The Machine* (European Science Foundation, Strasbourg, 1979).

20. V. Rehn and W. J. Choyke, Nucl. Instr. Methods 177, 173 (1980).
21. J. H. Weaver, C. Krafka, D. W. Lynch and E. E. Koch, *Optical Properties of Metals*, Parts I and II, Physics Data, Vols. 18-1, 18-2 (Fachinformationszentrum Karlsruhe, 1981).
22. R. L. Johnson [4, p. 178]; V. Saile and J. B. West, in: [14].
23. T. Matsushita and H. Hashizume [4, p. 261].
24. Z. Hussain, E. Umbach, D. A. Shirley, J. Stohr and J. Feldhaus, Nucl. Instr. Methods 195, 115 (1982).
25. E. E. Koch and B. F. Sonntag [2, p. 269].
26. U. Fano and J. W. Cooper, Rev. Mod. Phys. 40, 441 (1968).
27. E. E. Koch and A. Otto, Int J. Radiat. Phys. Chem. 8, 113 (1976).
28. V. Saile, P. Gürtler, E. E. Koch, A. Kozevnikov, M. Skibowski and W. Steinmann, Appl. Opt. 15, 2559 (1976).
29. V. Saile, Ph.D. Dissertation (Univ. of Munchen, 1976).
30. P. Gürtler, Ph.D. Dissertation (Univ. of Hamburg, 1979).
31. P. Gürtler, V. Saile and E. E. Koch, Chem. Phys. Lett. 48, 245 (1977).
32. J. H. Fock, P. Gürtler and E. E. Koch, Chem. Phys. 47, 87 (1980).
33. E. Miescher, Y. T. Lee and P. Gürtler, J. Chem. Phys. 68, 2753 (1978).
34. P. Gürtler, V. Saile and E. E. Koch, Chem. Phys. Lett. 51, 386 (1977).
35. K. E. Ockenga, P. Gürtler, S. S. Hasnain, V. Saile and E. E. Koch, Extended Abstract, Vol. II, 6th VUV Conference (Charlottesville, Virginia, June, 1980), p. II-80.
36. See, for example, M. B. Robin, *Higher Excited States of Polyatomic Molecules*, Vols. 1 and 2 (Academic Press, New York, 1974, 1975).
37. P. Hochmann, H.-t. Wang, W. S. Felps, S. P. Foster and S. P. McGlynn, in: *Chemical Spectroscopy and Photochemistry in the Vacuum Ultraviolet*, C. Sandorfy, P. J. Ausloss and M. B. Robin, eds. (Reidel, Dordrecht, 1974), p. 385.
38. E. Lindholm, Ark. Fys. 40, 97 (1969); 40, 111 (1969); 40, 125 (1969).
39. H. J. Maria, J. L. Meeks, P. Hochmann, J. F. Arnett and S. P. McGlynn, Chem. Phys. Lett. 19, 309 (1973). See also P. Hochmann, P. H. Templet, H.-t. Wang and S. P. McGlynn, J. Chem. Phys. 62, 2588 (1975); S. Chattopadhyay, P. Hochmann and S. P. McGlynn, J. Chem. Phys. 65, 3341 (1976); S. P. McGlynn, S. Chattopadhyay, P. Hochmann and H.-t. Wang, J. Chem. Phys. 68, 4738 (1978); P. Hochmann, G. L. Findley and S. P. McGlynn, Z. Naturforsch. 35a, 595 (1980).
40. D. Spence, Phys. Rev. A15, 883 (1977); J. Chem. Phys. 66, 669 (1977).
41. R. B. Caton and A. E. Douglas, Can. J. Phys. 48, 432 (1970).
42. M. Ogawa and S. Ogawa, J. Mol. Spectrosc. 41, 393 (1972).

43. See, for example, F. Bassani and M. Altarelli [4, p. 463].
44. N. Schwentner, E. E. Koch and J. Jortner, *Electronic Excitations in Condensed Rare Gases* (Springer Tracts in Modern Physics, in press).
45. J. Jortner, E. E. Koch and N. Schwentner, paper in this volume.
46. E. Boursey, V. Chandrasekharan, P. Gürtler, E. E. Koch, P. Kunsch and V. Saile, Phys. Rev. Lett. 41, 1516 (1978).
47. P. Gürtler and E. E. Koch, Chem. Phys. 49, 305 (1980).
48. P. Gürtler and E. E. Koch, J. Mol. Struc. 60, 287 (1980).
49. U. Hahn, N. Schwentner and G. Zimmerer, Nucl. Instr. Methods 152, 261 (1978).
50. H. Wilcke, W. Böhmer and N. Schwentner, Nucl. Instr. Methods in Phys. Res. 204, 533 (1983).
51. P. Gürtler, E. Roick, G. Zimmerer and M. Pouey, in: [14].
52. T. Möller, B. Jordan, P. Gürtler, G. Zimmerer, D. Haaks, J. Le Calve and M. C. Castex, Chem. Phys., in press.
53. S. D. Peyerimhoff and R. J. Buenker, Chem. Phys. 57, 279 (1981).
54. For recent reviews, see I. Lindau and W. E. Spicer [3, p. 159] and N. V. Smith and F. J. Himpsel [4, p. 905].
55. M. Cardona and L. Ley, *Photoemission in Solids*, Vol. I (Springer-Verlag, Berlin, 1978).
56. L. Ley and M. Cardona, *Photoemission in Solids*, Vol. II (Springer-Verlag, Berlin, 1979).
57. S. Leach, paper in this volume.
58. J. L. Dehmer, D. Dill and A. C. Parr, paper in this volume.
59. C. N. Berglund and W. E. Spicer, Phys. Rev. A136, 1030 (1964).
60. I. Lindau and W. E. Spicer, J. Electron Spectrosc. Relat. Phenom. 3, 409 (1974).
61. N. Schwentner, F. J. Himpsel, V. Saile, M. Skibowski, W. Steinmann and E. E. Koch, Phys. Rev. Lett. 34, 528 (1975).
62. F. J. Himpsel, N. Schwentner and E. E. Koch, Phys. Stat. Sol. (b) 71, 615 (1975).
63. H. J. Lau, J. H. Fock and E. E. Koch, Chem. Phys. Lett. 89, 281 (1982).
64. J. H. Fock, H. J. Lau and E. E. Koch, Chem. Phys., submitted.
65. See, for example, N. Schwentner and E. E. Koch, Phys. Rev. B14, 4687 (1976); R. Nürnberger, F. J. Himpsel, E. E. Koch and N. Schwentner, Phys. Stat. Sol. (b) 81, 503 (1977).
66. C. Kittel, *Introduction to Solid State Physics* (Wiley, New York, 1967), p. 621.
67. E. E. Koch and M. Skibowski, Chem. Phys. Lett. 14, 37 (1982).
68. C. Friedh, L. Åsbrink and E. Lindholm, Chem. Phys. 27, 169 (1978).
69. For a summary of experiments up to 1977, see W. D. Grobman and E. E. Koch [56, p. 261].

70. See, for example, references cited in F. Gutmann and L. E. Lyons, *Organic Semiconductors* (Wiley, New York, 1967); M. Pope and C. E. Swenberg, *Electronic Processes in Organic Crystals* (Clarendon Press, Oxford, 1982).
71. E. Tegeler, M. Iwan and E. E. Koch, J. Electron Spectrosc. Rel. Phenom. 22, 297 (1981).
72. N. I. Nefedov, N. P. Sergushin, I. M. Band and M. B. Trzhuskovskaya, J. Electron Spectrosc. Rel. Phenom. 2, 383 (1973).
73. M. Iwan, E. E. Koch, T. C. Chiang and F. J. Himpsel, Phys. Lett. 76A, 177 (1980).
74. E. E. Koch, M. Iwan, K. Hermann and P. S. Bagus, Chem. Phys. 59, 249 (1981).
75. M. Iwan, E. E. Koch, T. C. Chiang, D. E. Eastman and F. J. Himpsel, Solid State Comm. 34, 57 (1980).
76. F. J. Himpsel, D. E. Eastman, E. E. Koch and A. R. Williams, Phys. Rev. B22, 4604 (1980).
77. J. C. Guillot, Y. Ballu, J. Paigné, J. Lecombt, K. P. Jain, P. Thiry, R. Pinchaux, Y. Petroff and L. M. Falicov, Phys. Rev. Lett. 39, 1632 (1977).
78. R. Bruhn, E. Schmidt, H. Schröder and B. Sonntag, J. Phys. B15, L 441 (1982).
79. N. V. Smith [55, p. 235].
80. C. A. Feldman, R. Engelhardt, T. Permien, E. E. Koch and V. Saile, in: [14].
81. T. Permien, R. Engelhardt, C. A. Feldmann and E. E. Koch, Chem. Phys. Lett. 98, 527 (1983).
82. V. Saile, Appl. Opt. 19, 4115 (1980).

NONLINEAR OPTICS AND LASER SPECTROSCOPY IN THE VACUUM ULTRAVIOLET

Stephen C. Wallace

Departments of Chemistry and Physics
University of Toronto
Toronto, Ontario M5S 1A1, Canada

1. INTRODUCTION

Over the past ten years enormous strides have been made towards producing novel sources of VUV and XUV radiation using lasers. This work has been motivated by many factors, including studies of highly-excited atomic states; interest in nonlinear optical phenomena; the development of x-ray lasers; and, the long-term prospect of applications of high energy radiation to photolithography in the production of integrated circuits. The spectroscopy community and, particularly, those in molecular spectroscopy, have been less active in fully exploiting the level of sophistication now available using laser techniques for studies of excited states in the VUV and XUV spectral regions. First, we now have the capability of producing high intensity [1-7], continuously-tunable, coherent and narrow line width VUV radiation, spanning the 100 nm - 200 nm region, with laboratory-scale dye laser systems. The intensity attainable [2,3] in the VUV ranges up to 10 kW per pulse (determined by the available dye laser system); in units of photons per 0.1 nm per second this intensity is $\sim 8 \times 10^{17}$, or approximately 10^7 times greater than that obtained following dispersion of currently available sources [8] of synchrotron radiation. Second, by using a sequence of harmonic generation steps or higher-order ($\geq$ third-order) nonlinear processes, coherent radiation has been produced at wavelengths [9] as short as 38 nm. More recently, this work has been extended, using dye lasers, to produce tunable radiation in the 60-80 nm spectral region. Third, there is the pioneering work of Harris [11], in which anti-Stokes scattering from metastable species is used to convert visible-wavelength lasers to wavelengths as short as 20 nm. And, finally, there is a wide repertoire of nonlinear laser methods for

S. P. McGlynn et al. (eds.), Photophysics and Photochemistry in the Vacuum Ultraviolet, 105–131.

studies of the spectroscopy and dynamics of excited states in the VUV: These include multiphoton absorption, multiphoton ionization [12,13], harmonic generation [14-16] and laser-assisted energy transfer [17]. All of these techniques will be discussed more fully below.

The above list does not highlight the use of actual VUV lasers because it was prepared from the perspective of molecular spectroscopy and dynamics, where narrow linewidths, time resolution and continuous tunability are the most important factors. Thus, although the first report [18] of the molecular hydrogen laser was the beginning of the rapid and successful development of VUV lasers, such devices have not found a wide range of applicability in spectroscopy. This is principally a result of either (i), an intrinsic lack of continuous tunability, as in the H_2 laser; or (ii), the difficulty in achieving suitably narrow linewidths, as in the excimer lasers.

This review discusses laser sources in the VUV in a broad context. First, we briefly review VUV lasers. The next section then deals with third-order nonlinear optical methods for producing tunable, coherent VUV radiation. A short discussion of anti-Stokes Raman effects is also included. Finally, we outline some examples of the application of tunable VUV laser sources to spectroscopy and molecular dynamics.

2. VACUUM ULTRAVIOLET LASERS

VUV lasers fall into two categories: First, there are the low-pressure, gas-discharge type lasers which use small molecules such as CO, H_2, HD or D_2 and in which the laser emission is between rovibronic levels of two bound potential energy surfaces. These levels are excited in a travelling-wave configuration (Blumlein), where the optical emission closely follows the excitation pulse, as the latter propagates through the laser medium, so that no optical cavity is required. In such lasers there is, therefore, no phase coherence in the output, which should be correctly classified as amplified spontaneous emission. Performance and other relevant parameters are summarized in Table 1.

The second category, the excimer lasers of the rare gases, uses an optical cavity. Excitation is either by an electron-beam pump to produce excimers [19] (Xe_2*, Kr_2*, Ar_2*) or by a high-pressure preionized discharge to generate exciplexes (ArF*) as the upper laser level. The laser transition is then a bound-continuum transition as indicated in Fig. 1. The observed con-

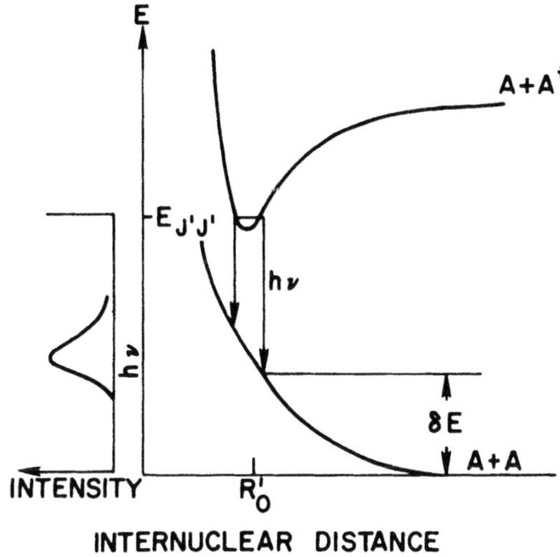

Fig. 1. Schematic potential energy curves and emission spectrum for the bound-continuum transition in excimer lasers.

tinuum emission will be a superposition of the Franck-Condon factors from the population distribution of vibrational levels in the upper state. Thus, these molecular-dissociation lasers have a broad gain profile and can be tuned. However, the gain per unit bandwidth is not large enough for really narrow line-width operation as in a dye laser, although such tunable excimer lasers have been successfully built [20,21]. More recently, a hybrid device [22], which uses the ArF gain medium as an amplifier, has produced narrow-bandwidth, tunable output in the 1930 Å region. This, then, is a fairly complex system using a tunable dye laser pulse to produce 1930 Å radiation via nonlinear optical methods (*vide infra*) and then amplifying this radiation in a series of ArF discharge modules.

The kinetics of excimer laser systems (the term excimer is now used for both excimer and exciplex) has been the subject of interesting and lively discussion over the past few years, a discussion much too involved to reproduce in detail here. The upper laser level is generally accepted to be populated by three-body collisions for the excimers, or by a "harpooning" mechanism for the exciplex. However, the detailed kinetics are very complex and there are often several pathways for both the formation and destruction of the upper state of the laser trans-

Table 1. Vacuum Ultraviolet Lasers

System	Wavelength (nm)	Excitation Source	Peak Power (MW)	Pulse Duration (ns)	Linewidth (cm^{-1})	Reference	Remarks
CO	181.1-197	gas discharge	10^{-5}	1.5	≤0.1	23	discrete spectrum
		electron beam	1.5×10^{-5}	3	≤0.1	24	
ArF	193	gas discharge	10	10-20	1000	25,26	commercially available, 600 cm^{-1} tuning range
		gas discharge	∿2	20	150	27	
		electron beam	1.6×10^{3}	55	1000	28	
Xe_2	173	electron beam	2	5	600	19	1000 cm^{-1} tuning range
		electron beam	2×10^{-2}	5	150	20	
		electron beam	1	50	10	21	
Kr_2	146	electron beam	$\leq 10^{-2}$	50	600	29	
Ar_2	126	electron beam	$\leq 10^{-2}$	50	600	30	

Table 1. Continued

System	Wavelength (nm)	Excitation Source	Peak Power (MW)	Pulse Duration (ns)	Linewidth (cm^{-1})	Reference	Remarks
H_2	125-161	gas discharge	$\leq 10^{-1}$	1-2	≤ 0.1	18,31,35	discrete spectrum
H_2,HD, D_2,p-H_2	109.8-161	electron beam	$\leq 10^{-1}$	1-2	≤ 0.1	37,38	
H_2	161	electron beam	0.3	1	≤ 0.1	39	
F_2	157	gas discharge	~ 10	~ 10		40	discrete spectrum, commercially available
Xe IV	154.82, 155.08	gas discharge	10^{-3}	≤ 1	≤ 50	41	
Kr IV	175.61, 195.03	gas discharge	10^{-3}	100	≤ 0.1	42	

ition. Many detailed studies of the individual kinetic processes, as well as measurement of the absorption and gain cross-sections, will be necessary to fully model the behaviour of these lasers.

The 157 nm molecular fluorine laser falls into both of the above categories in that the laser emission corresponds to a bound-bound transition but excitation is in a discharge in high pressure He. This laser, therefore, has a narrow-line emission spectrum consisting of two lines around 157 nm. It is important to note that both the F_2 and ArF lasers are excited in a simple, volume-preionized electric discharge device so that these lasers operate reliably at high repetition rates for long periods of time. In addition, these systems are commercially available.

3. NONLINEAR OPTICAL METHODS FOR PRODUCING TUNABLE VUV RADIATION

Third-order, nonlinear optical phenomena are used to transform visible-wavelength dye laser radiation, at frequency ω_i, into coherent VUV radiation at frequency ω_h, in atomic and molecular media. Such effects include both third-harmonic ($\omega_h = 3\omega_1$) and sum-frequency ($\omega_h = \omega_1 + \omega_2 + \omega_3$) or difference-frequency ($\omega_h = \omega_1 + \omega_2 - \omega_3$) schemes. As noted in the introduction, by using various combinations of sum and difference frequency mixing it is now possible to span the 100 nm to 200 nm region using several different nonlinear materials with visible-wavelength dye lasers providing input frequencies. This VUV radiation has the spatial and spectral properties intrinsic to the input laser radiation. Thus, from the perspective of high resolution spectroscopy or laser-induced fluorescence, harmonic generation schemes offer the highest brightness source (both spectral and spatial) in the VUV. Linewidths of $\Delta\nu \leq 0.1$ cm^{-1} are now routinely achieved with pulsed dye lasers leading to a resolution in the VUV of $\sim 5 \times 10^5$, equalling that of the best VUV spectrometers. Moreover, with recent advances in single-mode operation of pulsed dye lasers, we anticipate linewidths of 0.01 cm^{-1} to be readily achievable in the near future; and, with the demonstration of cw harmonic generation [43], we can anticipate several orders of magnitude further improvement in spectral resolution. Table 2 contains relevant examples of fixed-frequency, third-harmonic generation results which have led ultimately to the production of 38 nm radiation. The performance of tunable sources in the 100-200 nm region is summarized in Table 3, where the entries are selected on the basis of both performance and experimental ease and feasibility. (For a summary of earlier work see [44].) We now outline the theory on which these processes are based.

Table 2. VUV Harmonic Generation - Fixed Frequency

Nonlinear Material	Wavelength (nm)	Peak Power (kW)	Pulse Duration (ps)	Linewidth (cm^{-1})	Reference	Remarks
Cd	177.3	7	50	≤ 5	45	
	152.0	7.0×10^{-2}	50	≤ 5	45	
Xe	118.2	300	25	≤ 5	46	
Kr,He,Ne	88.7	---	25	≤ 5	47,48	
He	53.2	300	30	≤ 5	9,49	28th harmonic of Nd:Yag
	38.0	3.0×10^{-3}	30	≤ 5	9,49	

Table 3. Nonlinear Mixing Schemes* for Generation of Tunable Coherent VUV Radiation (106-200 nm)

Nonlinear Material	Wavelength Range (nm)	Resonant Enhancement	Dye Laser(s)	Output (kW)	Reference	Third Order Process
Xe	200-150	Yes	2	1	50	difference frequency
Xe	206-160	No	1	0.01-0.05	51	difference frequency
Hg	180-160	Yes	2	5-10	52	difference frequency
Mg	160-135	Yes	2	1	4,53,54	sum frequency
Kr	130-110	No	1	0.001-0.020	51	sum frequency
Hg	125.0	Yes	2	5	58	sum frequency
Kr:Ar	123.5-120	No	1	0.05	55,56	third harmonic
Zn	140-106	Yes	2	0.001-0.010	57	sum frequency

*In order to achieve these conversion efficiencies, megawatt dye lasers (pumped by either Nd:Yag or excimer lasers) are necessary.

3.1 Theory of Third-Order Nonlinear Optical Phenomena

In nonlinear optics, the theoretical framework falls naturally into two parts: First, one must calculate a nonlinear susceptibility χ which describes the microscopic (i.e., atomic) contribution to the polarization induced by the propagation of the laser beam at frequence ω_i. For the general case, this polarization is written as a power series in the electric field $E(\omega_i)$ so that:

$$P = \chi^{(1)}_{\omega_i} E(\omega_i) + \chi^{(2)}_{2\omega_i} E(\omega_i) E(\omega_i)$$

$$+ \chi^{(3)}_{3\omega_i} E(\omega_i) E(\omega_i) E^*(\omega_i) + \ldots \quad (1)$$

In isotropic media, only the odd-order susceptibilities are nonzero (because of inversion symmetry) and, hence, the lowest-order nonlinear term is the third-order susceptibility $\chi^{(3)}$. If, as in the gas phase, the number density N is low enough that local field effects are small, then the macroscopic nonlinear polarization induced by the laser field is simply $N\chi^{(3)}$ and this polarization may be introduced into the wave equation as a nonlinear source term. For the second part of the calculation, one reverts to classical electrodynamics in order to describe the evolution of the new third-harmonic wave $\omega_h = 3\omega_i$. The important physical feature of this latter stage is the phase matching between the fundamental (ω_i) and harmonic wave ($\omega_h = 3\omega_i$). The relevant parameters in the expression for the intensity of the harmonic radiation are as follows:

$$I_{3\omega_i} \propto N^2 |\chi^{(3)}|^2 I_{\omega_i} f(\Delta k) \quad (2)$$

where I_{ω_i} is the laser intensity at ω_i, Δk is the wave-vector mismatch ($k = 2\pi n/\lambda$) and f is a function determined by the actual geometrical conditions of the experiment (a comprehensive analysis of phase matching for various third-order effects may be found in [59]). For plane-waves, $f(\Delta k)$ is maximized at $\Delta k=0$ (i.e., $n_{\omega_h} = n_{\omega_i}$); but, for a focused beam, where there is a phase slip between the fundamental and harmonic waves through the focus, f is maximized for $n_{\omega_h} < n_{\omega_i}$ and $f = 0$ for $n_{\omega_h} = n_{\omega_i}$! Since all materials show significant dispersion in their transparent regions, it might appear to be difficult to achieve perfect phase matching. However, Harris [45] recognized that, in atomic systems, by choosing the harmonic radiation to lie in the region of anomalous dispersion on the high energy side of a resonance line then $n_{\omega_h} < n_{\omega_i}$ and, hence, by simply adding a second normally dispersive gas, one can compensate for the phase mismatch. This is schematically illustrated in Fig. 2. Thus, even though the nonresonant susceptibility per unit volume is low in gases (because the nonlinear material may be made of arbitrary dimension and has perfect optical homogeneity, as well as a high threshold for laser-induced breakdown), significant fractions of the incident fundamental radiation can be converted to the third harmonic. In rubidium vapour, for example, 3% conversion efficiencies have been reported [60]. Finally, even though difference-frequency mixing ($\omega_h = 2\omega_1 - \omega_2$) is inherently photon-energy inefficient, it has an important advantage for phase matching. For focused beams, $\Delta k=0$ as for plane-waves and the phase matching function $f(\Delta k)$ is symmetric about $\Delta k=0$. This considerably relaxes phase matching constraints and increases [50,51] the range of continuous wavelength significantly.

So far we have not considered the importance of the magnitude or the frequency dependence of the susceptibility $\chi^{(3)}_{3\omega}$. Although it was recognized early in the development of nonlinear optics [53,54] that the third-order nonlinear susceptibilities

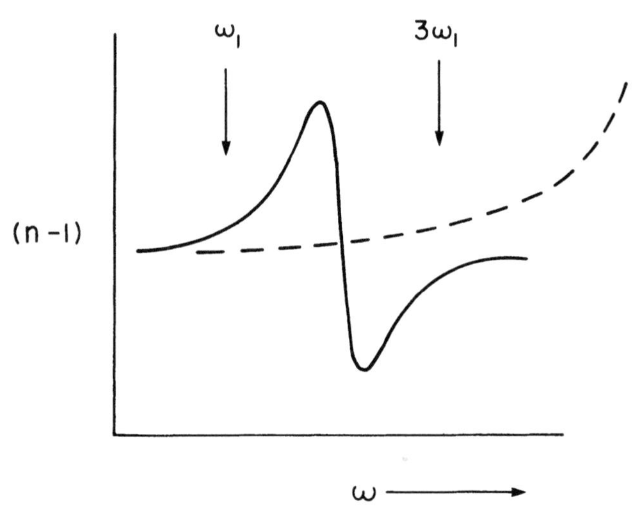

Fig. 2. Schematic representation of the compensation of refractive index for phase matching of third-harmonic generation in atomic vapors.

were frequency dependent (in a manner analogous to the resonant Raman effect), the potential for resonant enhancement in harmonic generation lay untapped until third-order effects were explored in gases. The important feature, which may be extracted from the perturbation expansion [61,62] of the susceptibility $\chi^{(3)}_{3\omega}$, is the existence of terms corresponding to a two-photon resonant enhancement process, a process which permits a large increase in $\chi^{(3)}$ without an associated linear absorption. Such effects were analyzed theoretically by Harris and Bloom [63] and first observed experimentally in strontium vapor by Hodgson, Sorokin and Wynne [64]. Resonant enhancement effects can increase the susceptibility by as much as 10^6-10^7 in favourable cases; hence, it is feasible to produce high intensity VUV radiation starting with tunable dye lasers and atomic vapors.

A detailed "first-principles" analysis of third-order susceptibilities for atoms and molecules will not be given in this article. However, it is possible to illustrate most of the important factors governing harmonic generation by selecting examples from several different experimental studies. Consider the third-harmonic process: We first approximate the susceptibility by a single term (there are a total of 24), namely the one which contains all three resonant enhancement factors in the denominator:

$$\chi^{(3)}_{3\omega_i} \simeq \sum_{a,b,c} \frac{\langle g|u|c\rangle\langle c|u|b\rangle\langle b|u|a\rangle\langle a|u|g\rangle}{(\Omega_{cg}-3\omega_i)(\Omega_{bg}-2\omega_i)(\Omega_{ag}-\omega_i)} \tag{3}$$

where Ω_{cg} is the energy of state c; ω_i is the laser frequency; and the $\langle m|u|n\rangle$ are the dipole matrix elements between states m and n. Clearly, in any real atomic system only one factor in the denominator can be made to approach zero (at which stage a linewidth is introduced phenomenologically by adding a term $-i\Gamma$). But, if we choose this factor to be $(\Omega_{bg} - 2\omega_i)$, the summation may be contracted to a single index:

$$\chi^{(3)}_{3\omega_i} \simeq \sum_{c} \frac{\langle g|u|c\rangle\langle c|u|b\rangle\langle b|u|a\rangle\langle a|u|g\rangle}{(\Omega_{cg}-3\omega_i)(\Omega_{bg}-2\omega_i-i\Gamma)(\Omega_{ag}-\omega_i)} \tag{4}$$

The two-photon transition strength to state b is approximated by a single sequence of matrix elements $\langle b|u|a\rangle\langle a|u|g\rangle$. This factor accounts for the principal resonant enhancement of $\chi^{(3)}$ in all four-wave sum-mixing schemes since, in contrast to its single-photon analogue (i.e., the resonant Raman effect), there is no linear absorption associated with it. By itself, the two-photon resonant enhancement term provides only a fixed frequency source (corresponding to the Doppler width of the transition to state b) of harmonic radiation. However, by using two laser frequencies, ω_1 fixed so that $2\omega_1$ gives rise to two-photon resonant enhancement and ω_2 variable, one may produce tunable harmonic radiation at the sum or difference frequency $2\omega_1 \pm \omega_2$.

This is illustrated [65] for the case of Mg in Fig. 3, where $2\omega_1$ is set to be equal to the 3s4s state and ω_2 is chosen to span the np Rydberg series. In this case the susceptibility reduces to a single term:

$$\chi^{(3)}_{2\omega_1+\omega_2} \simeq \frac{\langle g|u|3snp\rangle\langle 3snp|u|3s4s\rangle\langle 3s4s|u|3s3p\rangle\langle 3s3p|u|3s^2\rangle}{(\Omega_{3snp}-(2\omega_1+\omega_2)-i\Gamma_{3snp})(\Omega_{3s4s}-2\omega_1-i\Gamma_{3s4s})(\Omega_{3s3p}-\omega_i)} \quad (5)$$

and we have both two- and three-photon resonant enhancement terms in $\chi^{(3)}_{2\omega_1+\omega_2}$. This is the situation which is normally encountered. The experimental results from Mg vapour for this particular choice of energy levels are shown in Fig. 4. The experimental variable is the intensity of the harmonic radiation at the sum frequency $\omega_h = 2\omega_1 + \omega_2$ which, as noted above [Eq. (2)] is proportional to $|\chi^{(3)}|^2$. Fig. 4 clearly illustrates the resonant enhancement which is due to the np Rydberg series; however, the actual line shapes also show the importance of phase matching. These linewidths are not determined by the bandwidth of the dye lasers at ω_1 and ω_2 but by the dispersion around each p level, which acts to change the phase matching as one tunes through each line. Moreover, as the principal quantum number increases, the oscillator strength decreases and the variations in the refractive index become smaller, a phenomenon which is indeed observed.

In practice, large resonant enhancement terms at ω_h are to be avoided. First, they give rise to large intensity variations in the sum frequency. This is illustrated for the case of strontium [16] in Fig. 5. Second, these three-photon resonant enhancement terms also correspond to one-photon linear absorption at ω_h and, therefore, the intrinsic absorption of the harmonic radiation limits the conversion efficiency. For strontium at 191.97 nm [67], an absorption cross-section $\sigma = 3.5 \times 10^{-18}$ cm^2 results in an attenuation factor of 20 at a pressure of 6 torr and the conversion efficiency thereafter becomes independent of the number density of the nonlinear medium.

For the case of difference-frequency mixing ($\omega_h = 2\omega_1 - \omega_2$)

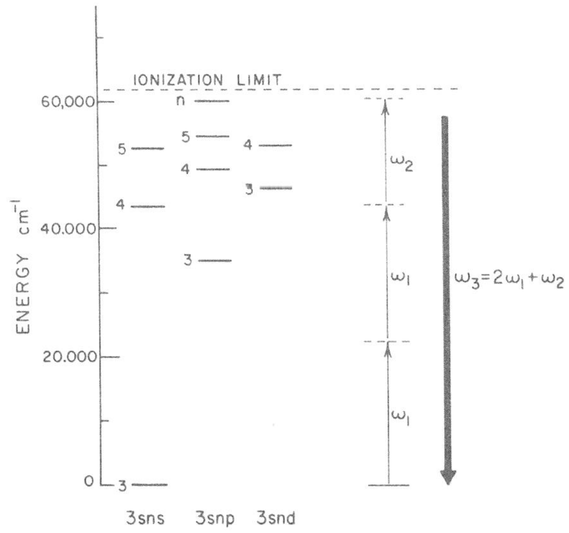

Fig. 3. Energy level diagram for Mg showing the four-wave sum-mixing process $\omega_h = 2\omega_1 + \omega_2$ with resonant enhancement at $2\omega_1$, by the 3s4s state, and ω_h, by the np Rydberg series.

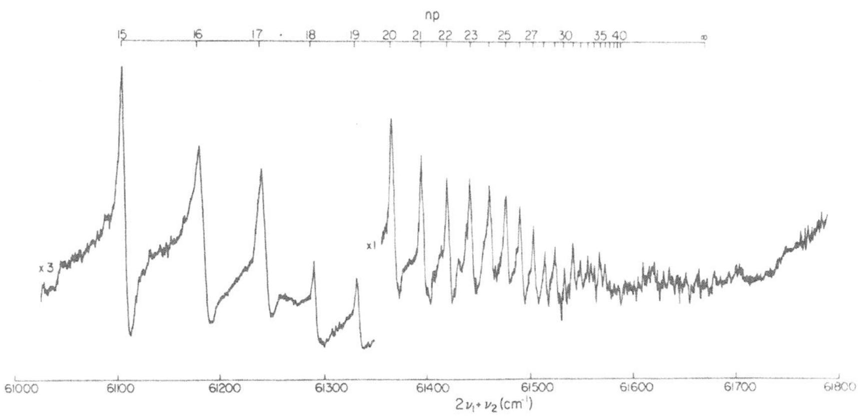

Fig. 4. Intensity of harmonic radiation in Mg vapor produced under the double resonant-enhancement conditions of Fig. 3.

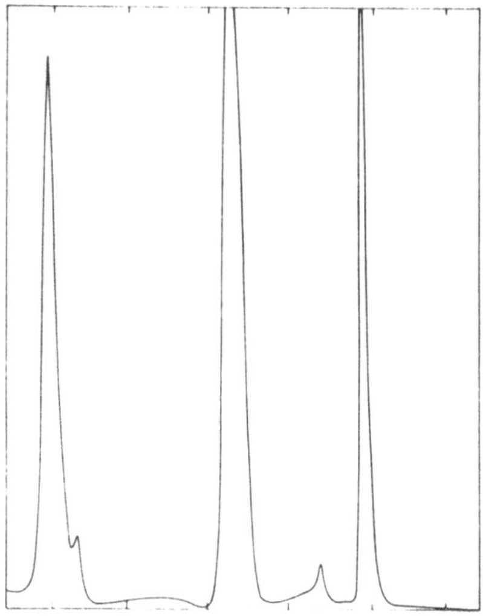

Fig. 5. Frequency dependence of four-wave sum-mixing for Sr vapor in the region of autoionizing states.

it becomes feasible and, indeed, practical to simultaneously engineer two-photon resonant enhancement, phase matching and optical transparency. These are the optimum conditions to achieve high conversion efficiency and wide tunability for harmonic generation in the VUV. Recent results on Xe [50] and Hg [52] clearly illustrate this point: 5-10 kW peak powers are readily obtained over a large spectral region using moderate-intensity input powers. It is likely that this type of system will ultimately prove to be the system-of-choice for any commercially available VUV laser source.

3.2 Saturation Phenomena

At this point it is important to discuss the effects of saturation on the harmonic conversion efficiency since it is only in the low intensity limit that $I_{\omega_h} \propto I_{\omega_i}{}^3$. There are fundamental limitations in the scaling of both resonantly-enhanced and nonresonant harmonic generation to high powers and such limitations have been observed in all attempts to obtain high harmonic-conversion efficiencies in gases. These saturation phenomena are an intrinsic feature since they arise through a variety of different nonlinear effects, effects closely related to harmonic generation itself.

It is possible to distinguish two regimes, depending upon the number density, where a group of physical processes predominates. At low number densities, multiphoton ionization [68] causes a significant depletion of species in the nonlinear medium. However, it is also essential, in this case, to include the effects of dynamical Stark shifts and power broadening [69] in order to obtain good agreement [68] with experiment [70]. At higher number densities, saturation effects become significant at power densities where optical Stark shifts are insignificant. In this latter case, conversion efficiency is limited by the loss of phase matching [71], a loss arising from multiphoton absorption. Multiphoton absorption may also lead to significant multiphoton ionization, resulting in further Stark broadening (by the ions) of the two-photon intermediate state which provides the resonant enhancement in $\chi^{(3)}$. This phenomenon also contributes to a reduction in conversion efficiency. The effect of the electric field generated by the ions produced in multiphoton ionization is quite dramatic since it gives rise to strong second-harmonic generation [50]. Finally, for the case of nonresonant harmonic generation, the limiting factor is again the loss of phase matching; in this case, however, the loss is attributable to field-dependent changes in the refractive index, changes which are described by the real part of $\chi^{(3)}$ (This is often referred to as the quadratic Kerr effect.) Thus, in order to compensate for the lower susceptibility in the nonresonant case, the use of a higher laser power simply induces a different limiting saturation condition.

It is not always easy to identify separately all of the physical processes giving rise to saturation in a particular case since the data necessary to compute the medium response are not available. The media chosen for inclusion in Table 3 represent the (current) best choices for total photon output and not, necessarily, for conversion efficiency. Many other systems have been studied; but, for the purposes of laser spectroscopy in the VUV these do not provide the optimum photon flux because of the limiting saturation effects described above.

3.3 XUV Harmonic Generation

The present record for producing the shortest wavelength laser radiation was achieved by generating [9] the seventh harmonic of 2660 Å (this is the fourth harmonic of Nd:Yag) in helium. The relevant energy level diagram is shown in Fig. 6: One observes that there is no appreciable resonant enhancement of the susceptibility. Such higher-order effects are only observed with extremely intense laser radiation (in this case, 10^{15} W/cm^2) at the fundamental wavelength. In order to avoid dielectric breakdown, high-intensity pulses are always of picosecond duration, an imposition which limits the ultimate spectral resolution and leads to a rather small number of photons per pulse, even though the peak power (as shown in Table 2) looks reasonable.

More recently, a tunable source below 100 nm has been developed using injection-locked [22] ArF lasers as the source of fundamental radiation (*vide supra*). With Kr, H_2 or Ar as the nonlinear medium, tunable third-harmonic radiation was produced in the 64 nm region. In later work [10], a second tunable dye laser was introduced in order to obtain radiation in the 79 nm region using H_2 as the nonlinear medium. Although the photon fluxes reported to date are still quite modest (corresponding to 1-10 W), further improvements can be expected as better schemes for differential pumping are introduced.

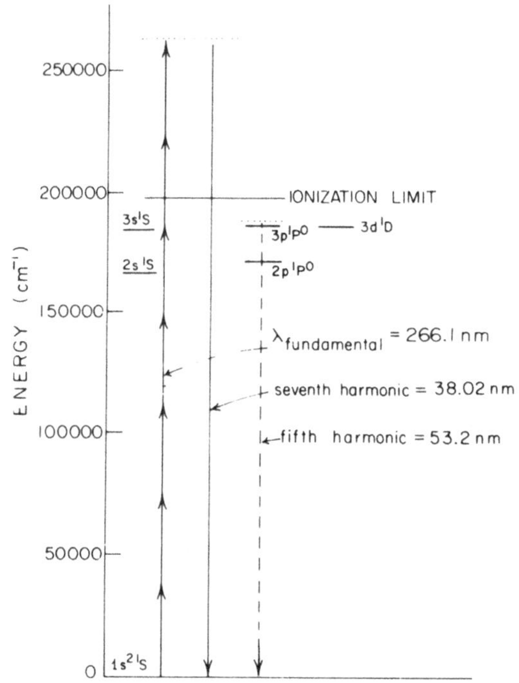

Fig. 6. Energy level diagram for He, showing fifth- and seventh-order harmonic generation processes.

3.4 Tunable XUV Sources Using the Anti-Stokes Raman Effect

Another approach to producing tunable XUV radiation has been developed by Harris and co-workers [72]. This technique uses the energy stored in highly-excited metastable species in an electric discharge to produce anti-Stokes Raman scattering of incident laser photons. The output frequency is the sum of the two energies; hence, XUV radiation is produced with visible lasers. The anti-Stokes radiation is polarized and has a linewidth which is the convolution of the Doppler width of the metastable species with the laser linewidth. Even though the radiation is isotropic in spatial distribution (because this is an incoherent process), the spectral brightness (at 56.9 nm) for the case of helium metastables exceeds the intensity of the 58.4 nm resonance line by 100 times. Experiments [11b] are currently in progress to use this type of source as a pump to produce a 20 nm laser.

If a population inversion exists for the metastable upper level, stimulated anti-Stokes Raman emission can be achieved, thus resulting in the production of coherent radiation. This type of device has been recently reported [72] for atomic iodine following selective photodissociation of NaI by a KrF excimer laser to produce an $I^*(5p\ ^2P^o_{1/2})$ inversion. With an input laser pulse at 206 nm, tunable radiation is produced around 178 nm in the VUV. This approach presents interesting possibilities if a high-lying metastable state can be inverted efficiently.

3.5 Laser-Produced Plasmas as an Extreme-UV Continuum Source

Recent studies [74] of the ablation region of laser-produced plasmas, using plane targets, have yielded intense emission continua in the 1-2 nm region resulting from free-free and free-bound transitions. Estimates of the brightness of this source indicate that it is comparable in intensity to existing synchrotron radiation sources. Given current advances in laser technology (e.g., slab-glass Nd lasers), we expect this to be a useable laboratory-scale source within a few years.

4. APPLICATIONS TO CHEMICAL PHYSICS

Perhaps the ultimate test of the successful development of tunable VUV laser sources is their application to spectroscopy and molecular dynamics. Thus far these applications fall into two categories: (i) photophysical studies of the excited states of small molecules (CO, NO) which absorb in the VUV; and (ii) laser-induced fluorescence studies of product state distributions in chemical dynamics for both atomic and molecular species.

4.1 Photophysics of Small Molecules

The first laser-induced fluorescence study [75] in the VUV was on the photophysical properties of the first excited state of CO($A^1\Pi$, v=0) in the region of strong interstate perturbation by neighbouring triplet states ($d^3\Delta$ and $e^3\Sigma^-$). The experimental arrangement used is shown schematically in Fig. 7. The intensity (and the net absorption) of the exciting light was monitored through a VUV monochromator by one photomultiplier while the fluorescence was viewed perpendicular to the exciting radiation, through a quartz filter. The measured fluorescence intensity was consistent with that estimated from the incident VUV flux and, even though the collection optics and spectral response of the particular VUV photomultiplier available were far from optimum, the signal-to-noise ratio was good enough that fluorescence decays were recorded in a single (internal) scan of the sampling oscilloscope time-window. Excitation spectra were obtained by simple charge-integration using an electrometer and a typical example is shown in Fig. 8 for pressures of 0.4 and 4 torr. As the carbon monoxide begins to become optically thick at higher pressures (≥4 torr), very weakly allowed singlet-triplet transitions begin to be observed (cf. Fig. 8). Two contrasting examples of fluorescence decays are given in Fig. 9. The first trace is for an unperturbed level of the $A^1\Pi$ state Q(24), while the second is for a weak, triplet-like level (interacting with the $d^3\Delta$ state) R(29). By measuring the radiative lifetimes of the "singlet-like" levels of the $A^1\Pi$ state as a function of rotational quantum number, we are able to quantitatively monitor the strength of the interaction between singlet and triplet states in isolated CO molecules.

The unusual time dependence shown in the second trace of Fig. 9 is a result of collision-induced triplet-singlet energy-transfer from the triplet-like levels of the $d^3\Delta$ to the $A^1\Pi$ state. The sequence of events is indicated in Fig. 10. Although the initial absorption is very weak because of the triplet-like character of the upper state, these levels can be probed with a tunable laser source. Following excitation, collisions with other CO molecules result in population transfer to nearby levels in $A^1\Pi$. Since these states have close to unperturbed lifetimes, similar to trace (a) of Fig. 9, the actual time evolution of the fluorescence in trace (b) is a direct measure of the triplet-singlet energy transfer.

Recent studies of a much more complex system, namely NO in the $B'^2\Delta$ state, have also been reported [76] using the same VUV system. These authors [76] find significant deviations from the lifetimes measured in previous studies using electron-impact excitation. A detailed analysis of interstate perturbations in this system is currently in progress.

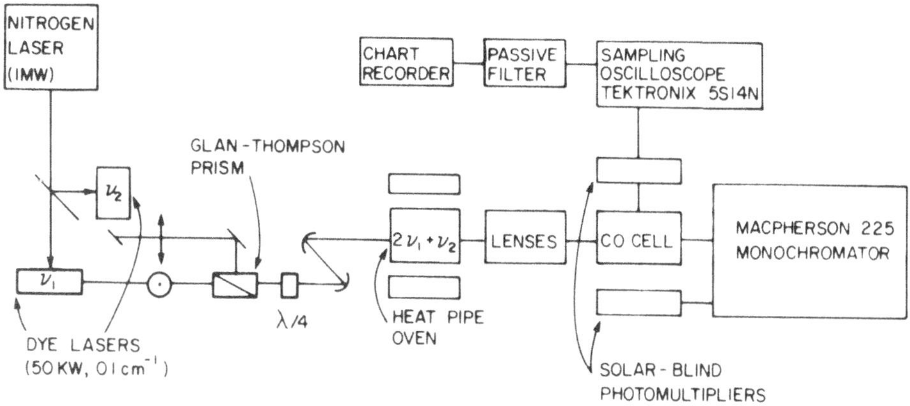

Fig. 7. Block diagram of the experimental arrangement for generation of tunable VUV radiation in Mg vapor and for photophysical studies of CO.

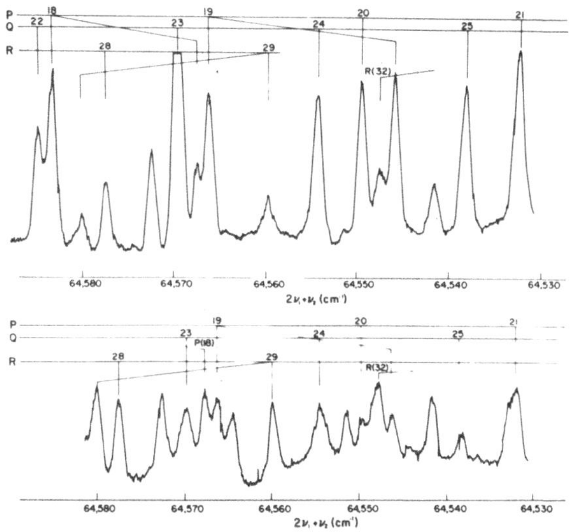

Fig. 8. Excitation spectra of CO at 0.4 torr (upper trace) and 4 torr (lower trace).

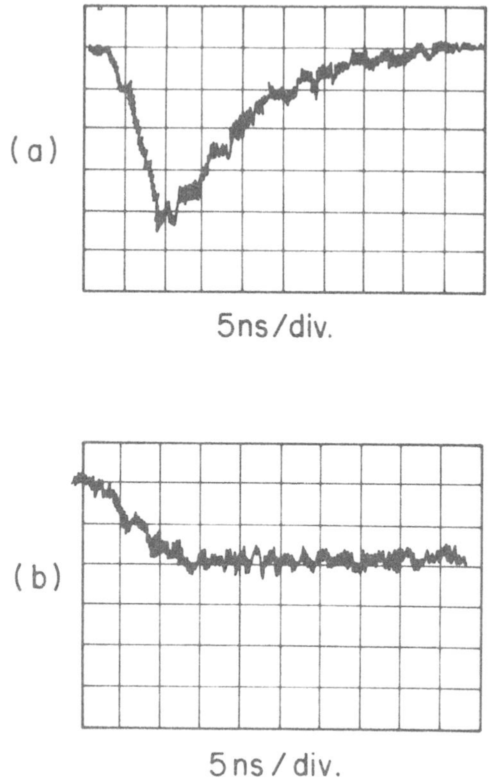

Fig. 9. Fluorescence decay curves for CO. See text for discussion.

4.2 Laser-Induced Fluorescence Detection

In chemical dynamics we would like to be able to detect atomic and molecular species produced in processes such as photodissociation, crossed-beam reactive scattering and surface scattering. Moreover, it is also important to monitor the nascent energy distribution of the product species, including translation, rotation and vibration. Tunable VUV laser sources have now been applied to studies in all of these areas and have yielded important and novel insights in molecular dynamics.

The first VUV laser-induced fluorescence measurements on reactive scattering were in a study [77] of the total cross-sections for $H + Br_2 \rightarrow HBr + Br$ as a function of hydrogen atom kinetic energy. The crossed molecular-beam apparatus used for this work is shown in Fig. 11. The source of atomic hydrogen

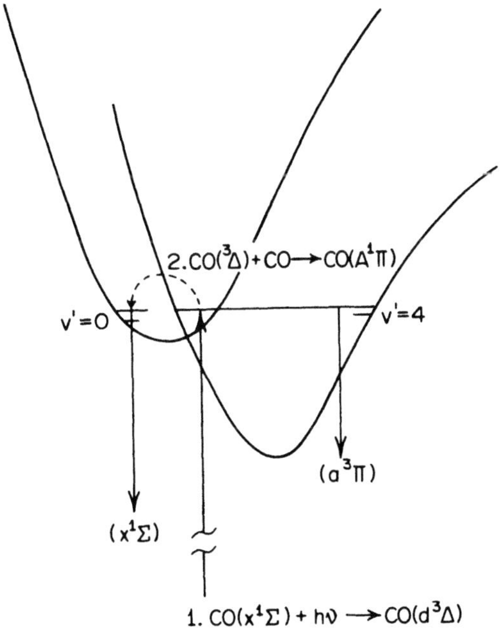

Fig. 10. Potential energy curves for CO showing the process of collision-induced, singlet-triplet energy transfer.

is a supersonic expansion from a tungsten oven, heated to 2800 °C, with variable kinetic energy achieved by inverse seeding techniques. The other reactant is introduced as an effusive beam from a collimated hole structure. The crossing region is illuminated by the VUV laser beam and fluorescence is detected perpendicular to the three intersecting beams. The VUV laser source used was the same as in [1] and [75] and, even with these modest dye lasers, the estimated lower limit of detection for ground state Br atoms was $10^6/cm^3$. Recent results with higher dye laser intensities has significantly increased the VUV flux and, hence, reduced this detection limit by several orders of magnitude. Cross-section data for the reaction of D, H + Br_2 are given in Fig. 12. These data are consistent with a very small barrier to reaction, namely $\leq$1 kcal/mole. Moreover, cross-sections for D and H become superimposable when plotted against relative velocity, as would be expected from a simple classical picture of the reaction of light and heavy hydrogen in the absence of any barrier to reaction.

More recently, reactions producing excited states of Br

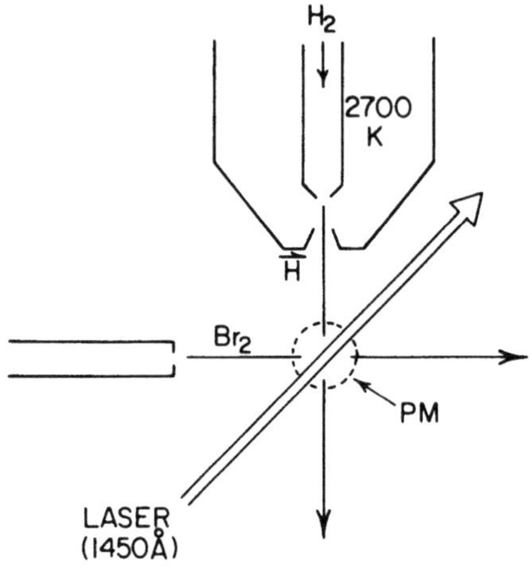

Fig. 11. Schematic diagram of the crossed-beam, reactive-scattering apparatus used to study the total cross-section for the process $H + Br_2 \rightarrow HBr + Br$.

have been investigated [78], as well as total cross-sections for $H + HBr \rightarrow H_2 + Br$ [79], again as a function of H atom kinetic energy.

Rotationally-inelastic scattering of CO from a LiF (001) surface has been studied [80] under molecular-beam conditions using VUV laser-induced fluorescence in the 1480 Å region (CO $A^1\Pi$, v=2). As noted above, the substitution of a Nd:Yag pumped dye laser significantly improved the VUV flux so that it was possible to achieve sensitivities of $10^5/cm^3$ per quantum state. Current studies are in progress to extend these techniques to molecular hydrogen [81].

Laser-induced fluorescence studies of the nascent rotational-energy distribution of CO formed following predissociation of formaldehyde have recently been performed [82]. Again, the VUV source was based on sum-frequency mixing in Mg vapor. This work clearly showed that the CO is formed promptly and highly rotationally hot (up to J=63). Further, these studies provide an important experimental comparison for recent *ab initio* results on formaldehyde photodissociation pathways [83].

Finally, laser-induced fluorescence studies of the translational distribution of H atoms following photodissociation of

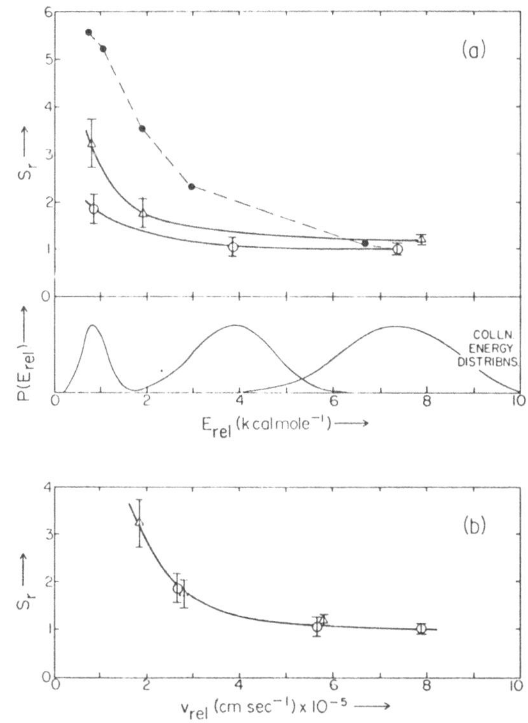

Fig. 12. Total cross-sections for the reaction H(D) + Br_2 → HBr(DBr) + Br plotted as a function of relative kinetic energy and relative velocity.

small molecules have recently been reported[84]. These measurements are an important test of the ultimate resolution of harmonic generation based VUV laser sources. In addition, they represent the final link in demonstrating the feasibility of the use of VUV laser-induced fluorescence in the complete characterization of the energy states (translational, rotational and vibrational) of atoms or small molecules produced in molecular studies of elementary systems.

5. FUTURE GOALS

The remarkable progress in development of laser sources in the VUV opens up an important spectral region to the techniques of laser spectroscopy and, hence, to the laser revolution in chemistry. The possibilities for fundamental studies in molecular dynamics are particularly attractive because of the potential for sensitive detection of nascent product distributions of species such as atomic or molecular hydrogen, carbon monoxide and atomic oxygen and halogens. The extension to coverage of wavelengths below 30 and 100 nm is still in progress but will be achieved as certain technological barriers are overcome. Current advances in laser technology, as yet unavailable commercially, will ultimately lead to laboratory-scale, intense (but incoherent) XUV sources based upon lasers. Finally, as we more completely understand the physics of highly-excited, inner-shell states of atomic systems, there remains the possibility of developing actual lasers operating in the soft x-ray region.

ACKNOWLEDGMENT

This work was supported in part by the Natural Science and Engineering Research Council of Canada, The Petroleum Research Fund administered by the American Chemical Society and the J. P. Bickell Foundation.

REFERENCES

1. S. C. Wallace and G. Zdasiuk, Appl. Phys. Lett. 28, 449 (1976).
2. J. Hager and S. C. Wallace, Chem. Phys. Lett., in press.
3. R. Wallenstein, to be published.
4. H. Juninger, H. B. Puell, H. Scheingraber and C. R. Vidal, IEEE J. Quantum Electronics JQE-16, 1132 (1980).
5. J. Bokor, R. R. Freeman, R. L. Panock and J. C. White, Opt. Lett. 6, 182 (1981).
6. F. S. Tomkins and R. Mahon, Opt. Lett. 6, 179 (1981).
7. W. Jamroz, P. E. LaRocque and B. P. Stoicheff, to be published.
8. K. Radler and J. Berkowitz, J. Opt. Soc. Am. 68, 1181 (1978).
9. J. Reintjes, C. Y. She and R. C. Eckardt, IEEE J. Quantum Electronics JQE-14, 581 (1978).
10. H. Egger, T. Srinivasan, K. Boyer, H. Pummer and C. K. Rhodes, in: *Proceedings of Laser Techniques for Extreme Ultraviolet Spectroscopy (Boulder, CO, 1982)*, R. R. Freeman and T. J. McIlrath, eds. (American Institute of Physics, New York, 1983).
11. (a) S. E. Harris, Appl. Phys. Lett. 31, 498 (1977);

(b) S. E. Harris, R. W. Falcone, J. F. Young, K. D. Pedrotti, J. E. Rothenber and J. R. Willison in: *Proceedings of Laser Techniques for Extreme Ultraviolet Spectroscopy (Boulder, CO, 1982)*, R. R. Freeman and T. J. McIlrath, eds., (American Institute of Physics, New York, 1983).
12. P. M. Johnson, J. Chem. Phys. 64, 4143 (1976).
13. J. Danon, H. Zacharias, H. Rothke and K. H. Welge, J. Chem. Phys. 76, 2399 (1982).
14. K. K. Innes, B. P. Stoicheff and S. C. Wallace, Appl. Phys. Lett. 29, 715 (1976).
15. S. C. Wallace and K. K. Innes, J. Chem. Phys. 72, 4805 (1980).
16. F. Valleé, S. C. Wallace and J. Lukasik, Opt. Comm. (in press).
17. J. Lukasik and S. C. Wallace, Phys. Rev. Lett. 42, 270 (1981).
18. R. T. Hodgson, Phys. Rev. Lett. 25, 494 (1970).
19. S. C. Wallace, R. T. Hodgson and R. W. Dreyfus, Appl. Phys. Lett. 23, 672 (1973).
20. S. C. Wallace and R. W. Dreyfus, Appl. Phys. Lett. 25, 498 (1974).
21. D. J. Bradley, M. H. R. Hutchison and C. C. Ling in: *Tunable Lasers and Applications*, A. Morrdian, T. Jaeger and P. Stockseth, eds., (Springer Verlag, New York, 1976). p. 40.
22. (a) H. Egger, M. Rothschild, D. Muller, H. Pummer, T. Srinivasan, J. Zavelovish and C. K. Rhodes in: *Laser Spectroscopy V: Proceedings of 5th International Conference of Laser Spectroscopy*, A. R. W. McKellar, T. Oka and B. P. Stoicheff, eds., (Springer Verlag, New York, 1981), p. 446.
(b) H. Pummer, T. Srinivasan, H. Egger, K. Boyer, T. S. Luk and C. K. Rhodes, Opt. Lett. 7, 93 (1982).
23. R. T. Hodgson, J. Chem. Phys. 55, 5378 (1971).
24. G. A. Kenney-Wallace, unpublished data.
25. R. Burnham and N. Djeu, Appl. Phys. Lett. 29, 707 (1976).
26. V. N. Ischenko, V. N. Lisitsyn and A. M. Razhev, Appl. Phys. 12, 55 (1977).
27. T. R. Loree, K. B. Butterfield and D. L. Barker, Appl. Phys. Lett. 32, 171 (1978).
28. J. M. Hoffman, A. K. Hays and G. C. Tisone, Appl. Phys. Lett. 28, 538 (1976).
29. P. W. Hoff, J. C. Swingle and C. K. Rhodes, Appl. Phys. Lett. 23, 245 (1973).
30. W. M. Hughes, J. Shannon and R. Hunter, Appl. Phys. Lett. 24, 488 (1974).
31. R. W. Waynant, J. D. Shipman, R. C. Elton and A. W. Ali, Appl. Phys. Lett. 17, 383 (1970).
32. R. W. Waynant, Phys. Rev. Lett. 28, 533 (1972).

33. V. S. Antonov, I. N. Kryazev, V. S. Letokhov and V. G. Movshev, Zh. Etf. Pis. Red. 17, 545 (1973).
34. R. W. Waynant, Appl. Phys. Lett. 22, 419 (1973).
35. R. W. Waynant, J. Shipman, R. C. Elton and A. W. Ali, Proc. IEEE 59, 679 (1971).
36. S. A. Borgstron, Opt. Comm. 11, 105 (1974).
37. R. T. Hodgson and R. W. Dreyfus, Phys. Lett. 38A, 213 (1972).
38. R. T. Hodgson and R. W. Dreyfus, Phys. Rev. A9, 2635 (1974).
39. R. W. Dreyfus, Appl. Phys. Lett. 29, 348 (1976).
40. H. Pummer, K. Hohla, M. Diegelmann and J. P. Reilly, Opt. Comm. 28, 104 (1979).
41. R. W. Waynant, Appl. Phys. Lett. 22, 419 (1973).
42. J. D. Marling and D. B. Lang, Appl. Phys. Lett. 31, 181 (1977).
43. R. R. Freeman, G. C. Bjorklund, N. P. Economou, P. F. Liao and J. E. Bjorkholm, Appl. Phys. Lett. 33, 739 (1978).
44. S. C. Wallace, Advances in Chemical Physics 47, 153 (1981).
45. A. H. Kung, J. F. Young, G. C. Bjorklund and S. E. Harris, Phys. Rev. Lett. 29, 985 (1972).
46. A. H. Kung, J. F. Young and S. E. Harris, Appl. Phys. Lett. 22, 301 (1973).
47. S. E. Harris, J. F. Young, A. H. Kung, D. M. Bloom and G. C. Bjorklund in: *Laser Spectroscopy I: Proceedings of 1st International Conference on Laser Spectroscopy*, R. G. Brewer and A. Mooradian, eds., (Plenum Press, New York, 1974), p. 59.
48. J. Reintjes, C. Y. She and R. C. Eckardt, IEEE J. Quantum Electronics JQE-14, 581 (1978).
49. J. Reintjes and C. Y. She, Opt. Comm. 27, 469 (1978).
50. J. Hager and S. C. Wallace, Chem. Phys. Lett., in press; and unpublished results.
51. R. Hilbig and R. Wallenstein, Appl. Opt. 21, 913 (1982).
52. J. Hager, F. Northrup, J. Williamson and S. C. Wallace, to be published.
53. S. C. Wallace and G. Zdasiuk, Appl. Phys. Lett. 28, 449 (1976).
54. F. Northrup and S. C. Wallace, unpublished results.
55. R. Hilbig and R. Wallenstein, IEEE J. Quantum Electronics JQE-17, 1566 (1981).
56. R. Mahon, T. J. McIlrath and D. W. Koopman, Appl. Phys. Lett. 33, 305 (1978).
57. W. Jamroz, P. E. LaRocque and B. P. Stoicheff, to be published.
58. F. S. Tomkins and R. Mahon, Opt. Lett. 6, 179 (1981).
59. G. J. Bjorklund, IEEE J. Quantum Electronics JQE-11, 289 (1975).
60. H. Puell, K. Spanner, W. Falkenstein, W. Kaiser and C. R. Vidal, Phys. Rev. A14, 2240 (1976).

61. J. A. Armstrong, N. Bloembergen, J. Ducuing and P. S. Pershan, Phys. Rev. 127, 1918 (1962).
62. B. J. Orr and J. Ward, Mol. Phys. 20, 513 (1971).
63. S. E. Harris and D. M. Bloom, Appl. Phys. Lett. 27, 229 (1974).
64. R. T. Hodgson, P. P. Sorokin and J. J. Wynne, Phys. Rev. Lett. 32, 343 (1974).
65. S. C. Wallace, unpublished data.
66. P. P. Sorokin, J. J. Wynne, J. A. Armstrong and R. T. Hodgson, Annals N.Y. Acad. Sci. 267, 30 (1976).
67. H. Scheingraber, H. Puell and C. R. Vidal, Phys. Rev. A18, 2585 (1978).
68. A. T. Georges, P. Lambropoulos and J. H. Marberger, Phys. Rev. A15, 300 (1977).
69. J. N. Elgin and G. H. C. New, Opt. Comm. 16, 242 (1976).
70. J. F. Ward and A. V. Smith, Phys. Rev. Lett. 35, 653 (1975).
71. H. Puell, H. Scheingraber and C. R. Vidal, Phys. Rev. A22, 1165 (1980).
72. L. Zych, J. Lukasik, J. F. Young and S. E. Harris, Phys. Rev. Lett. 40, 1493 (1978).
73. J. C. White and D. Henderson, Appl. Phys. B28, 125 (1982).
74. E. Jannitti, P. Nicolosi and G. Tondello in: *Proceedings of Laser Techniques for Extreme Ultraviolet Spectroscopy (Boulder, CO, 1982)*, R. R. Freeman and T. McIlrath, eds. (American Institute of Physics, 1983).
75. A. C. Provorov, B. P. Stoicheff and S. C. Wallace, J. Chem. Phys. 67, 5393 (1977).
76. J. R. Banic, R. H. Lipson, T. Efthimiopoulos and B. P. Stoicheff, Opt. Lett. 6, 461 (1981).
77. J. W. Hepburn, D. Klimik, K. Liu, J. C. Polyani and S. C. Wallace, J. Chem. Phys. 69, 4311 (1978).
78. J. W. Hepburn, K. Liu, R. G. Macdonald, F. J. Northrup and J. C. Polyani, J. Chem. Phys. 75, 3353 (1981).
79. J. W. Hepburn, D. Klimik, K. Liu, R. G. Macdonald, F. J. Northrup and J. C. Polyani, J. Chem. Phys. 74, 6226 (1981).
80. J. W. Hepburn, F. J. Northrup, G. L. Ogram, J. C. Polyani and J. M. Williamson, Chem. Phys. Lett. 85, 127 (1982).
81. J. Hager, F.J. Northrup, S. C. Wallace and J. Williamson, unpublished.
82. P. Ho and A. Smith, Chem Phys. Lett., in press.
83. J. D. Goddard and H. F. Schaeffer, III, J. Chem. Phys. 70, 5117 (1979).
84. K. Welge, to be published.

MULTIPHOTON IONIZATION AND THIRD-HARMONIC GENERATION IN ATOMS AND MOLECULES

John C. Miller and R. N. Compton

Chemical Physics Section
Health and Safety Research Division
Oak Ridge National Laboratory
Oak Ridge, Tennessee 37830, USA

1. INTRODUCTION

The advent of tunable dye lasers has totally revolutionized the fields of atomic and molecular physics and chemistry. The very narrow bandwidth and extensive tunability of dye lasers have made a quantum change in the precision and physical content of spectroscopic and photophysical studies. The other major property of lasers, high photon flux, has not only made traditional studies easier but has spawned the entirely new field of multiphoton spectroscopy. Although two-photon processes were investigated theoretically by Maria Goepert-Mayer in her thesis work in 1930 [1], multiphoton absorption in the visible spectral region was not observed until 1961, after lasers became available. Currently, so-called nonlinear spectroscopy is a vast, rapidly growing area - - in fact, it is one of the most exciting branches of chemical physics today.

The two fields considered here, multiphoton spectroscopy and nonlinear optics (frequency conversion) began shortly after the development of the laser. Kaiser and Garrett [2] observed two-photon absorption for the first time in crystals in 1961 and Abella [3] made the first such observation for gaseous species (Cs) in 1962. The first report of multiphoton ionization (MPI) was the eleven-photon ionization of Xe by a Q-switched ruby laser, seen by Voronov and Delone [4] in 1965. Shortly thereafter, in the same year, the multiphoton photodetachment of I^- was reported by Hall, *et al.* [5]. Although MPI of molecular background gases was reported in 1970 [6], the first serious molecular MPI studies were those of Lineberger and Patterson [7] on the photodetachment of C_2^- in 1972, and Collins, *et al.* [8]

S. P. McGlynn et al. (eds.), Photophysics and Photochemistry in the Vacuum Ultraviolet, 133–161.

in 1973 on Cs_2. In the intervening years, multiphoton studies of atoms and molecules have become so frequent as to require the publication of a yearly bibliography [9]. Several useful reviews of atomic [10-14] and molecular [15-17] MPI may be found in the literature.

In the case of nonlinear optics, frequency conversion via second-harmonic generation was observed in quartz by Franken, *et al.* [18] in 1961. Third-harmonic generation (THG) was first studied in the rare gases by Ward and New [19] in 1969. The current record [20] for frequency up-conversion is held by seventh-order harmonic generation in He to produce light of 38.0 nm. Various nonlinear optical techniques, collectively referred to as n-wave mixing, have allowed frequency conversion of visible lasers to regions from the far infrared to the vacuum ultraviolet. The review by Shen [21] is a good introduction to nonlinear optics.

In this paper, we present results of a number of experiments performed at the Oak Ridge National Laboratory (ORNL), involving both MPI and THG. In the next section (2), we discuss MPI experiments emphasizing mass spectroscopy and photoelectron spectroscopy, and MPI of alkali atoms and dimers. Section 3 concerns experiments involving the interference of MPI and THG in the rare gases. And, in this last section we also discuss vacuum ultraviolet (VUV) spectroscopy using photons produced by THG in rare gases as the source.

2. RESONANTLY-ENHANCED MULTIPHOTON IONIZATION

2.1 MPI Mass Spectroscopy and Photoelectron Spectroscopy

The use of tunable dye lasers to produce resonantly-enhanced multiphoton ionization (REMPI) provides a powerful new method for investigating atomic and molecular energy levels. This method is particularly useful in discovering and characterizing certain optically forbidden transitions. For example, new electronic states have been observed in iodine [22], benzene [23] and ammonia [24] using the REMPI technique. And, REMPI is ideal for studying Rydberg transitions in molecules and is experimentally easier than the traditional use of far ultraviolet radiation in conventional spectroscopy.

In addition, the use of two or more tunable dye lasers to produce sequential REMPI spectra has significant advantages in signal intensity, spectral simplification and the ability to probe different Franck-Condon regions of a molecule [25]. Studies of mass-selective two-color (i.e., two different dye laser beams) photoionization of benzene [26,27] and its clusters [28] have provided new information on high-Rydberg states and ionization limits for benzene, and structural information on small benzene clusters. Dual-laser experiments have also been used to obtain new information on excited-state lifetimes [28], ionization potentials [28], spectroscopic constants for free radicals [29,30] and dissociation kinetics [31,32]. Such studies have also provided new information on triplet states of carbon disulfide [33] and toluene [34]. This is by no means an exhaustive listing of past and present applications of MPI techniques; however, this roster serves to illustrate the vitality and potential of this new field of molecular spectroscopy.

The addition of mass spectroscopy to studies of MPI of polyatomic molecules has shown that considerable fragmentation can accompany ionization. In particular, REMPI of benzene can lead to extensive fragmentation, with C^+ being the dominant ion, while little or no parent ion is observed [28,35-37]. Although at least six photons are energetically required to produce C^+ from C_6H_6, the reported power indices range from 2 to 3.5 [35]. The ORNL spectrum [36] of benzene fragmentation is shown in Fig. 1, where the power dependence is easily seen. The degree of fragmentation is independent of wavelength for benzene; however, for other polyatomics (pyrrole, furan and ammonia [36,47]) the degree of fragmentation is highly dependent upon the wavelength (resonant intermediate state). In some regions of the spectrum, the parent ion dominates.

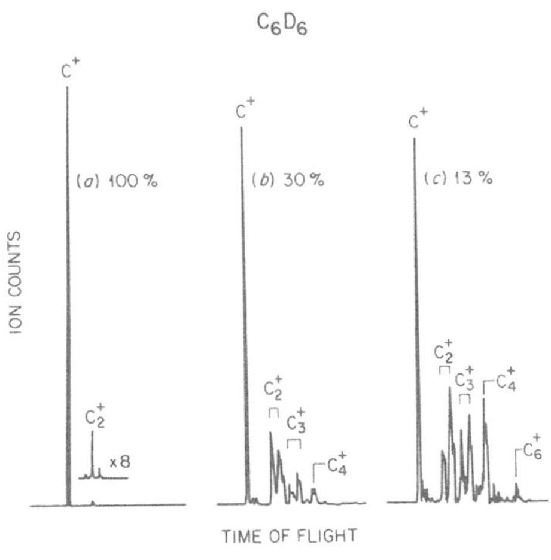

Fig. 1. Multiphoton ionization time-of-flight spectrum of perdeuterobenzene as a function of laser power.

Two mechanisms have been invoked to rationalize the production of small fragment ions from benzene and other molecules. In the first, multiple absorption of photons via *neutral* valence, Rydberg and autoionizing states leads to superexcited states which subsequently autoionize and dissociate to produce the observed fragmentation. The second mechanism assumes that the *parent ion* is initially produced by MPI, and then absorbs additional photons to reach dissociative or predissociative ionic states. Evidence that fragmentation occurs for benzene via absorption in the parent ion comes from the two-laser experiments of Beosl, *et al.* [27,37] and the nonresonant MPI photoelectron spectra of Meek, *et al.* [38], as well as from the ORNL experiments to be discussed below. Conversely, two-laser experiments on azulene [39] indicate that fragmentation results from photoabsorption through autoionizing states. Several theoretical treatments of MPI fragmentation, based upon statistical phase-space models, have recently appeared [40-43]. Recent measurements of photoelectron and ion kinetic-energy distributions at ORNL provide direct answers to the above questions and will be discussed below.

Fig. 2 shows the apparatus used at ORNL for recording mass and photoelectron spectra of molecules following MPI. Full descriptions of the apparatus may be found elsewhere [36,44-46]. Briefly, positive ions and electrons are detected separately and simultaneously (i.e., on the same laser shot) in a time-of-flight (TOF) mass spectrometer and spherical-sector electrostatic energy analyzer, respectively. In a separate experiment, the kinetic energy of positive fragment ions can be determined in the energy analyzer by reversing all of the fields, with ion masses being identified by TOF through the analyzer [45]. Charged particles are detected via channeltrons and the spectra displayed using either a boxcar integrator and x-y recorder, or a multichannel analyzer. The gas beam was effusive in the studies described here but is currently being upgraded to a pulsed supersonic nozzle. Two laser systems - - an N_2 laser pumped dye laser and an excimer laser pumped dye laser - - have been used, with the dye laser beam being focused by a 38 mm lens (power densities: 10^9-10^{11} W/cm^2) and crossed with the molecular beam. Laser bandwidths were on the order of 0.02 nm. Fig. 1 shows an example of a TOF spectrum for benzene. The MPI spectra of higher-pressure gases are recorded using the proportional counter shown in Fig. 12 and discussed in a later section.

Resonantly-enhanced multiphoton ionization photoelectron spectra (REMPI-PES) have been recorded for Xe [44], I_2 [45], NO [46], NH_3 [47], H_2S [48], CH_3I (ORNL, unpublished) and C_6H_6 [45]. Such a spectrum is shown for Xe in Fig. 3, in which the excimer-pumped dye laser is tuned to the $6s[3/2]^o$, J=1 state [49]. These data are of somewhat higher resolution but are

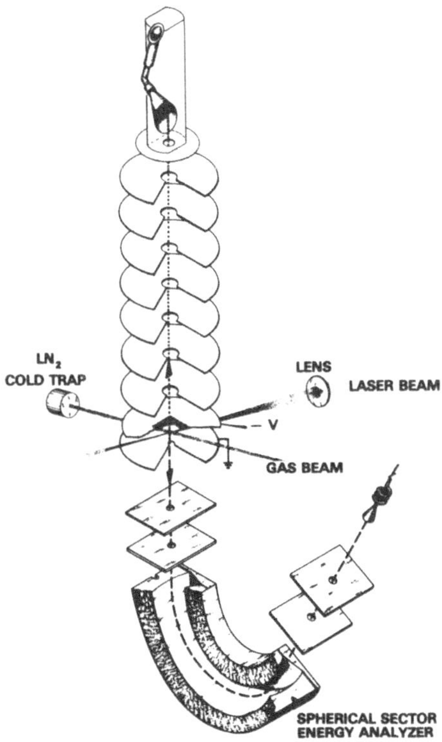

Fig. 2. Schematic diagram of the multiphoton ionization mass spectrometer and photoelectron energy analysis apparatus.

qualitatively the same as those published earlier using an N_2 dye laser [44]. The higher resolution in these data is attributed to three factors: (i), lower operating pressure afforded by the higher power of the excimer laser; (ii), reduced rf noise on the analyzer itself; and (iii), a digital data-acquisition system which allows detection of a very low ion density in the focal region, thus reducing the energy spread due to space charge. The interesting point to note for Xe is that MPI results in leaving Xe^+ in both possible fine-structure states, despite the fact that enhancement occurs through the $6s[3/2]^o$, J=1 core excitation.

MPI-PES contains a great deal of information about resonant intermediate states as well as pathways of ion fragmentation. In a recent study [46,49] of NO, where fragmentation does not occur, we demonstrated that the distribution of NO^+ vibration-

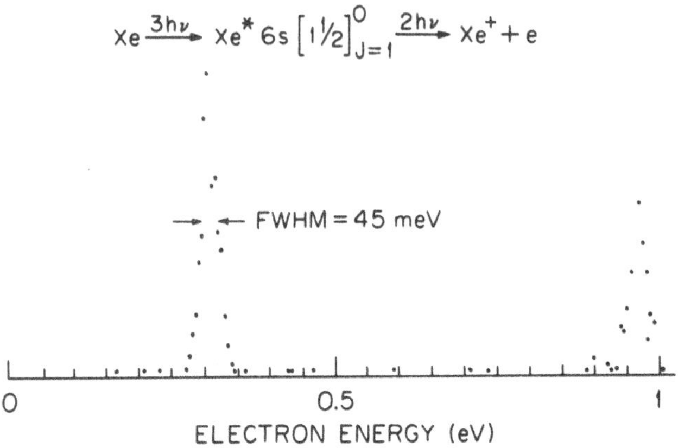

Fig. 3. Three-photon resonant, five-photon ionization photoelectron spectrum of Xe. The two peaks correspond to leaving the Xe^+ in the $^2P_{3/2}$, $^2P_{1/2}$ final states.

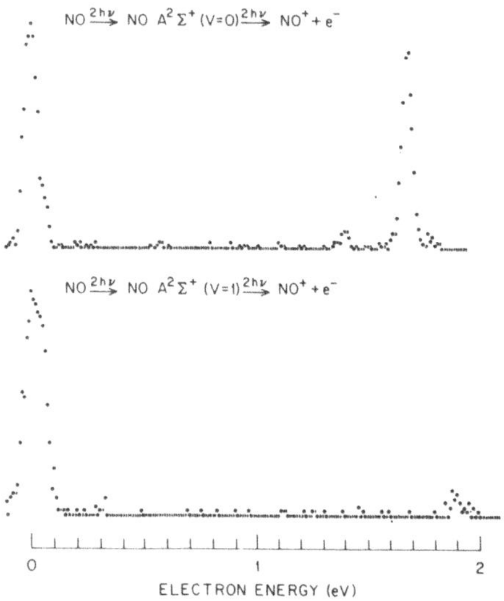

Fig. 4. Resonantly-enhanced multiphoton ionization photoelectron spectrum for four-photon ionization resonant with the v=0 and v=1 vibrational levels of the $A^2\Sigma^+$ state of NO.

al levels was determined primarily by the Franck-Condon factors connecting the resonant intermediate state ($A^2\Sigma^+$ or $C^2\Pi$) and the ground ionic state $X^1\Sigma^+$. Fig. 4 gives MPI-PES results for the case in which two photons are resonant with either the v=0 or v=1 levels of $A^2\Sigma^+$, and shows the higher quality data obtainable from the excimer-pumped dye laser [49]. The peaks at high energy correspond to direct ionization, leaving the ion in the v=0 and v=1 vibrational levels, respectively. This single peak is expected from Franck-Condon considerations since the $A^2\Sigma^+$ state is a Rydberg state converging upon the $X^1\Sigma^+$ ionic ground state; and, hence, these two potential curves are almost identical. We also note weaker photoelectron peaks corresponding to leaving the ion in v=2,3,4, etc. These peaks are probably due to electronic autoionization.

The intense peak at near-zero energy is of particular interest. We attribute this peak to vibrational autoionization resulting from the admixture of another resonant intermediate state at the third-photon energy level. Since *two* additional photons are required to ionize the resonant intermediate state, resonances (or near resonances) in the third-photon region can allow access to different autoionizing states than those accessible by direct two-photon ionization from the A state. Thus, the large zero-energy peak is attributed to vibrational autoionization from electronic states induced by transitions involving the three-photon excited state. Zero-energy electrons are not observed when the laser is tuned to the $C^2\Pi$ state *or* the v=3 level of the $A^2\Sigma^+$ state since the third photon is capable of ionization. Further evidence for this interpretation comes from the data of Fig. 5, where we have superimposed the dye laser beam with a beam from the direct output of the XeCl excimer pump (308 nm). The ionization signal resulting from the dye laser alone being resonant with the intermediate v=0 level of the $A^2\Sigma^+$ state is indicated. The large peak near zero energy is due to one-photon ionization of the $A^2\Sigma^+$ (v=0) state by the 308 nm excimer beam. No measureable increase of the zero-energy peak is observed when the excimer beam is co-propagated with the dye laser beam. This is taken as further evidence for the vibrational autoionization mechanism discussed above (i.e., when the "third-photon step" is eliminated by ionizing with one higher-energy photon, the slow electrons are not observed). The peak at $\sim$2.5 eV is a result of three-photon ionization of NO by the 308 nm excimer beam. The signal is resonantly enhanced by an accidental two-photon resonance with the $H^2\Sigma^+$ (v=1) state. The electrons occur at $3h\nu - IP - \omega_o$ ($h\nu$ = 4.02 eV, IP = 9.264 eV and $\omega_o = \omega_e - \omega_e x_e$ = 0.293 eV; ω_o is the first vibrational spacing of the ion). This is in accord with Franck-Condon factors, as discussed above. Finally, further evidence for the proposed mechanism for the zero-energy electrons comes from studies of MPI-PES of ^{15}NO, where the ratio of slow electrons to fast elec-

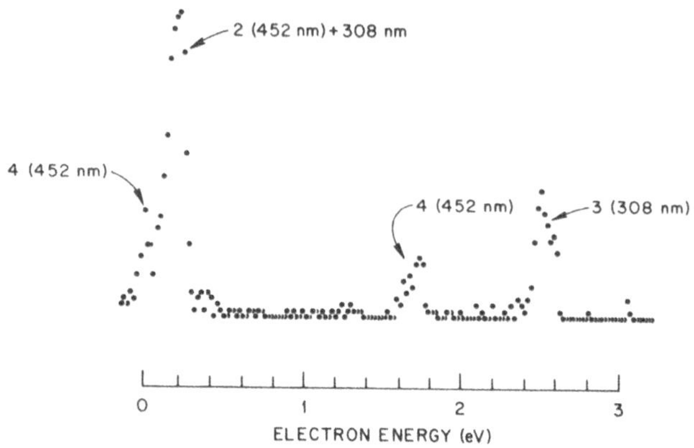

Fig. 5. Multiphoton-ionization photoelectron spectrum of NO using dual laser beams. One laser is tuned to the v=0 vibrational levels of the $A^2\Sigma^+$ state of NO (λ=452 nm); the second laser consists of the XeCl excimer beam at 308 nm. The origin of each photoelectron peak is labeled.

trons increases greatly over that shown in Fig. 5 for ^{14}NO. This result implicates a mechanism other than direct ionization.

I_2, NH_3 and H_2S are three simple systems which show intense fragment-ion peaks resulting from MPI. In each case, the PES shows directly that the initial ionization step results in the formation of the parent ion, I_2^+, NH_3^+ and H_2S^+. The fragment ions must be produced from the further absorption of photons by the parent ion. Furthermore, in the case of I_2, we were able to measure the kinetic energy of the I^+ ion [45], and this corresponds to the reaction $h\nu + I_2^+ \rightarrow I^+ + I$.

For the larger polyatomics, the ion kinetic energies did not provide direct evidence for this model. However, in the case of NH_3 it was possible to show that the extensive fragmentation is due to photon absorption into the first excited state of NH_3^+, a state which is dissociative [47]. This is shown in Figs. 6 and 7, where we have presented a mass spectrum for one particular wavelength and plotted the percent of fragment ions as a function of excess energy above the ground state of the ion. The data points represent fragment percentages for up to eight vibrational levels of five different electronic intermediate states. The energy onset for fragmentation corresponds exactly with the onset of the first excited state of NH_3^+, as shown in the one-photon PES of Fig. 8. In addition, the MPI-PES for NH_3 confirms the above argument [47]. The PES of the

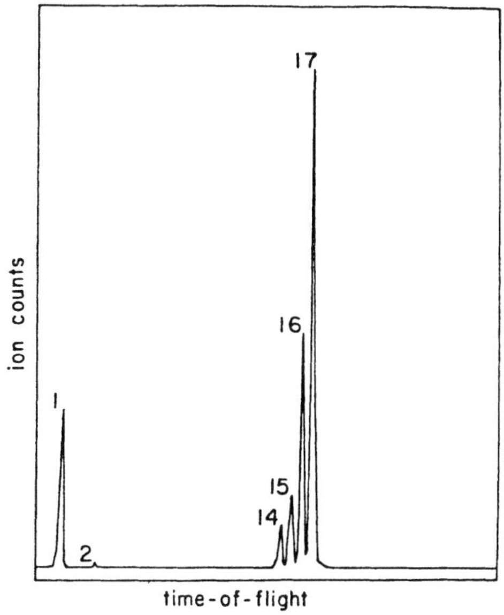

Fig. 6. Mass spectrum produced by multiphoton ionization and fragmentation of NH_3 molecules with 386.610 nm laser light as measured by the time-of-flight method. The resonantly-excited level is $\tilde{D}$-8. 1:H^+, 2:$H_2{}^+$, 14:N^+, 15:NH^+, 16:$NH_2{}^+$, 17:$NH_3{}^+$.

two larger molecules studied at ORNL, CH_3I and benzene, also show that fragmentation follows creation of the parent ion. *A major conclusion of our MPI-PES studies is that, in every case studied thus far, the fragmentation is due to absorption in the ionic manifold.* No evidence for absorption in the autoionization manifold has been observed in any molecule save NO, where a *minor* ionization pathway involves absorption of one photon from an autoionizing state [46]. The detailed physics of the ionic photodissociation processes is still under investigation.

2.2 Multiphoton Ionization of Cesium

The physical mechanisms leading to MPI of alkali atoms have received a great deal of theoretical and experimental attention. A wide variety of ionization phenomena have been observed, with the important processes which lead to ionization being dependent upon gas density, laser wavelength and laser power. For example, so-called "energy pooling" and "hybrid resonance" [50,51] effects have been observed in alkali excitation studies. It is even possible to perform experiments in which the laser field

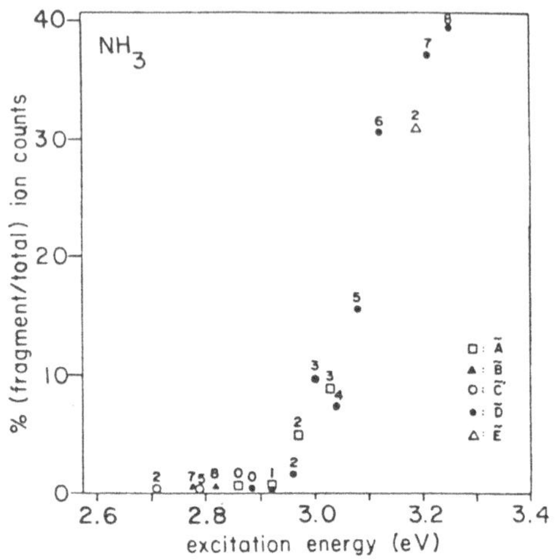

Fig. 7. Extent of fragmentation as a function of excitation energy for a fixed laser power. Square, $\tilde{A}$; closed triangles, $\tilde{B}$; open circles, $\tilde{C}$; closed circles $\tilde{D}$; open triangle, $\tilde{E}$. The numbers indicate the vibrational quanta (ν_2) excited in the particular electronic state.

alters a chemical reaction or makes the reaction energetically possible. Many recent studies have shown that intense photon fields, nonresonant with either of the separated collision partners, can induce energy transfer [52,53], collisional ionization [54,55] and charge transfer [56]. White [57] has recently observed the absorption of *two* photons during the collision between Ba and Tℓ ground-state atoms, resulting in the simultaneous excitation of both atoms to excited electronic states. These studies fall under the general heading of "Collisions in the Field of a Laser" and are in the early stages of development. When the laser is tuned to a resonance line of an alkali atom, direct MPI can be observed at low pressure. For somewhat higher alkali densities (10^{13}-10^{15} cm^{-3}), however, electrons which have been heated by superelastic collision with excited atoms can represent the dominant ionization mechanism. Despite the attention devoted to MPI studies of alkali atoms, the relative importance of the various possible ionization mechanisms, and the details of the physics of these processes, are only now being unraveled. Na has received the most attention because of its relative ease of handling, and because of the fact that the Na D lines occur at a convenient laser frequency. In what follows, we discuss some of our recent results on MPI of Cs [58].

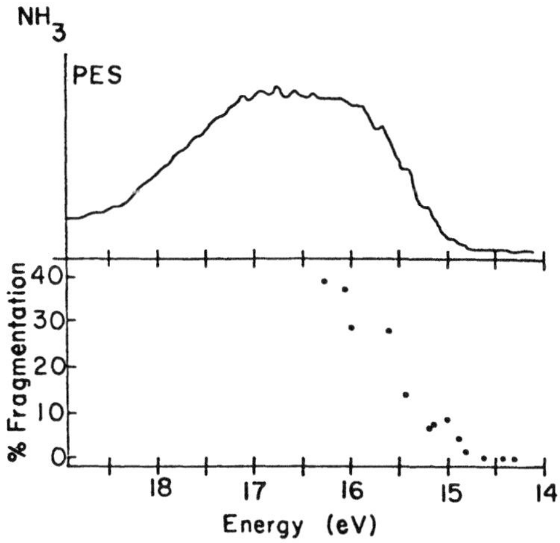

Fig. 8. Upper panel: photoelectron spectrum of the $(\ldots 1e^3 3a_1^2)\,^2E \leftarrow (\ldots 1e^4 3a_1^2)\,^1A_1$ transition (NH_3^+ $\tilde{A} \leftarrow \tilde{X}$) in NH_3. Lower panel: extent of fragmentation plotted versus five times the incident photon energy (to ensure coincidence of the two energy scales). [Adapted from M. J. Weis and G. M. Lawrence, J. Chem. Phys. 53, 214 (1970).]

Multiphoton ionization of low-density atomic beams of Cs has been studied under the experimental conditions shown in Fig. 9. The Cs MPI spectrum exhibits a rich array of atomic and molecular features in the near infrared. Fig. 10 gives a composite of two dye laser scans, including both the MPI signal and the dye laser power. Three photons are required to ionize Cs over the entire wavelength region shown. The second photon is in near resonance (<100 Å) with the two-photon allowed $6d^2D_{3/2,1/2}$ states of Cs. This near resonance probably accounts for the very strong ionization signals observed in the infrared. Starting at $12{,}000\ cm^{-1}$, and moving toward higher wavenumbers, the atomic $6p^2P_{3/2}$ transition is observed, followed by the $7d^2D_{3/2}$-$6p^2P_{1/2}$ hybrid resonance. The broad, continuum-like MPI signal observed between these two resonances appears to be molecular in origin but may also be due to nonresonant MPI, a process which is greatly enhanced by both the 6p and 6d states (near resonant with the first and second photons, respectively). Two-photon resonant, three-photon ionization via the intermediate $7d^2P_{3/2,5/2}$ states is observed next, followed by the second $6p^2P_{1/2}$ fine-structure peak. The $^2P_{1/2}$ intensity is considerably stronger than that due to $^2P_{3/2}$, despite the fact that the statistical ratio should

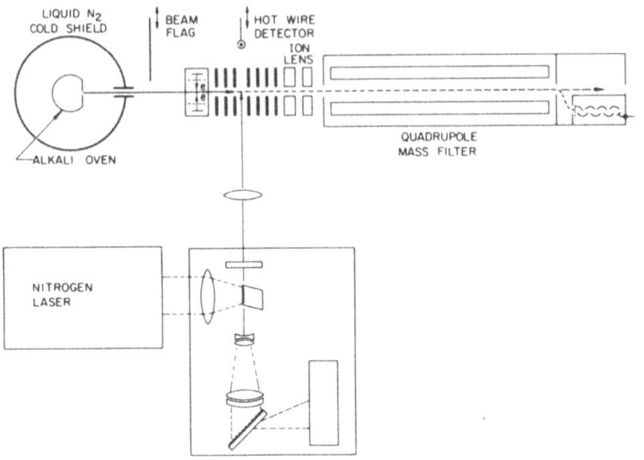

Fig. 9. Apparatus used to study multiphoton ionization of alkali beams. Vacuum chamber is all stainless steel (bakable to 400°C) and is pumped by a 500 liter/s vacuum pump.

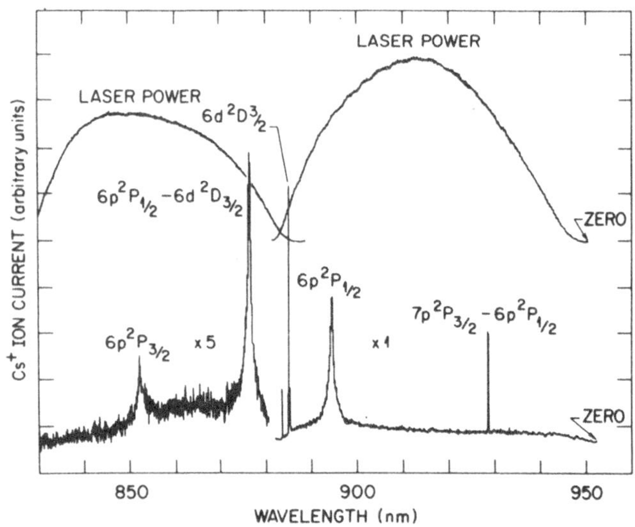

Fig. 10. Near-infrared multiphoton ionization spectrum (Cs^+ only) of Cs beam at ∿100°C. The laser power is also shown.

be 2:1, favoring the $^2P_{3/2}$ state. It is clear that MPI through both of these states is greatly affected by the presence of the 6d states. The last, very sharp feature shown in Fig. 10 is due to the dipole-forbidden but quadrupole-allowed $7p^2P_{3/2}$-$6p^2P_{1/2}$ transition. The $6p^2P_{1/2}$ atoms are produced in a hybrid resonance.

MPI signals were recorded for many of the one-photon resonant np states (n=7-12) and two-photon resonant s and d states. Fig. 11 shows the observed ionization signal as one approaches the two-photon ionization limit at 15,703.35 cm^{-1}. The ionization step occurring to the red of the correct ionization limit is due to field (electric and RF from quadrupole) and collisional ionization of high-Rydberg states. The intensity of the ionization signal versus principal quantum number n closely follows the n^{-8} dependence predicted by simple theoretical considerations (n^{-3} dependence for Rydberg oscillator strengths followed by an n^{-5} dependence for ionization). Red or blue Stark-shifts are observed, depending upon whether the first nonresonant photon energy lies above or below the intermediate 3p state.

The Cs_2^+ dimer ion was observed at the 7p, 8p and 9p atomic resonances. Since the dissociation limit for Cs_2^+ lies between the 8p and 9p levels, all 9p atoms, and those 8p atoms with kinetic energies greater than 700 cm^{-1}, can participate in the chemi-ionization reaction $Cs^* + Cs \rightarrow Cs_2^+ + e$.

The exact origin of the dimer-ion signal at the 7p resonance is, at present, unkown. Chemi-ionization by atoms with high kinetic energy is unlikely because of the small number of atoms in the tail of the Boltzman distribution. In addition, the dimer-ion signal at the 7p level is $\sim$10 times stronger than the signal at the 9p level. The high laser powers ($\sim 10^9$ W/cm^2) used in these experiments may open additional ionization channels, such as ionization by superelastically heated electrons, energy pooling (e.g., $Cs^* + Cs^* \rightarrow Cs_2^+ + e$) or laser-induced chemi-ionization (i.e., $Cs^* + Cs \xrightarrow{h\nu} Cs_2^+ + e$). Two-laser and photoelectron measurements are underway in order to distinguish between these possible mechanisms.

3. MULTIPHOTON IONIZATION AND THIRD-HARMONIC GENERATION IN RARE GASES

3.1 Generation of VUV Radiation

Tunable VUV radiation is an essential element in studies of absorption and photoionization of most atoms and molecules. The energy regime of interest begins at 200 nm (50,000 cm^{-1}) and extends to about 30 nm, where soft x-rays begin. The re-

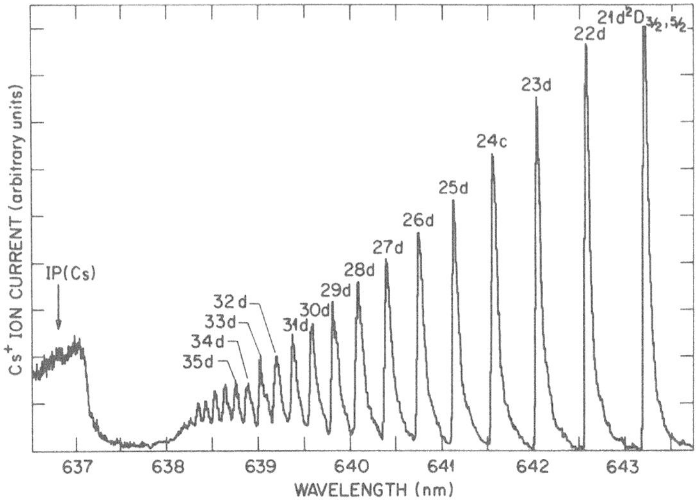

Fig. 11. Multiphoton ionization signals (Cs^+ only) in the region near the two-photon ionization limit (15,703.35 cm^{-1}).

gion from 200-100 nm (6.20-12.40 eV) is termed the VUV, while the region below 100 nm is usually termed the extreme ultraviolet (XUV).

The 200-100 nm spectral region is especially important since the ionization potential (IP) of most atoms lies above 6.20 eV, and since the IP of almost all organic and inorganic molecules lies between 6.2 and 12.4 eV. Although the spectroscopy of the lowest-lying states of most atoms and molecules is studied in the visible and near UV regions, most Rydberg states lie in the VUV. Due to the difficulty of obtaining and using VUV radiation, photoionization and Rydberg absorption studies have not been as frequent or as detailed as corresponding studies in the visible and near UV.

The traditional source of tunable photons in the VUV has been the rare-gas or hydrogen discharge lamp, coupled with a vacuum monochromator. More recently, synchrotron sources have provided relatively more intense, continuum VUV radiation which, however, also requires a monochromator for high resolution. Currently, there is intense interest in developing laser-based sources of VUV light. The aim is to extend, to this energy regime, the unique laser properties of high intensity, very narrow bandwidth, spatial and temporal coherence, and short pulsetimes. These factors, along with the extensive tunability of dye lasers, have already revolutionized the fields of atomic and molecular structure, spectroscopy and photophysics.

Extension of tunable, coherent light sources into the VUV, however, poses considerable problems since frequency conversion crystals become opaque in this region. For THG into the VUV, the rare gases have most often been used as the nonlinear medium. Studies of THG in Kr and Xe have characterized the properties of the nonlinear process, primarily with the aim of producing a bright source at the Lyman α and β lines for hydrogen detection schemes [59]. Tunable THG has been demonstrated in Xe in the ranges of 140.3-146.9 nm and 125.0-129.6 nm, and in Kr in the ranges of 120.3-123.6 nm and 110.0-116.5 nm, with conversion efficiencies of 10^{-5}-10^{-6} and fluxes of 10^{7}-10^{10} photons/pulse. Output power, linewidth and tuning range are, of course, dependent upon the properties of the laser supplying the fundamental frequencies and upon whether or not phase matching, by adding a second, positively-dispersing gas, is attempted.

In addition to THG, however, tight focusing of high-powered lasers into gases leads to other nonlinear processes. In particular, MPI occurs when the pump-laser wavelength is such that three or four photons are resonant with atomic levels. This MPI, which is interesting in its own right, is also an important loss mechanism for THG in rare gases. Conversely, the presence of VUV light from THG has a marked effect on the MPI spectra. Our experiments, monitoring both THG and MPI as functions of wavelength and pressure, are discussed below, following a description of the experimental apparatus. Finally, a number of applications of the VUV for molecular spectroscopy are outlined.

The apparatus shown in Fig. 12 is a further extension of a cell initially constructed for MPI studies [36] and previously modified for THG experiments [60]. It consists of three separately-evacuable chambers, each of which functions as a proportional counter. Electrons produced in such a counter are accelerated toward an electrode maintained at positive bias voltages (10-600 V) and undergo ionizing collisions with other gas molecules, thus causing an electron avalanche. The amplification produced is proportional to the ratio of electric field to pressure (E/P). The gain is limited finally by dielectric breakdown. The first chamber (I) consists of a stainless-steel Varian six-way cross connected to an ion pump and is capable of 10^{-8} torr baseline vacuums. The beam of a N_2 laser pumped dye laser is focused in chamber I by a 3.8 cm focal-length lens to a spot ≤ 20 μm, giving a power density on the order of 5×10^{9} W/cm^2. Kr and Xe were used as nonlinear media and any electrons produced by MPI in the tripling gas were detected with the biased flat-plate electrode. The VUV light, produced by THG in the focal spot, is collimated with a MgF_2 lens and passes through a MgF_2 window into the second chamber (II). Again, any elec-

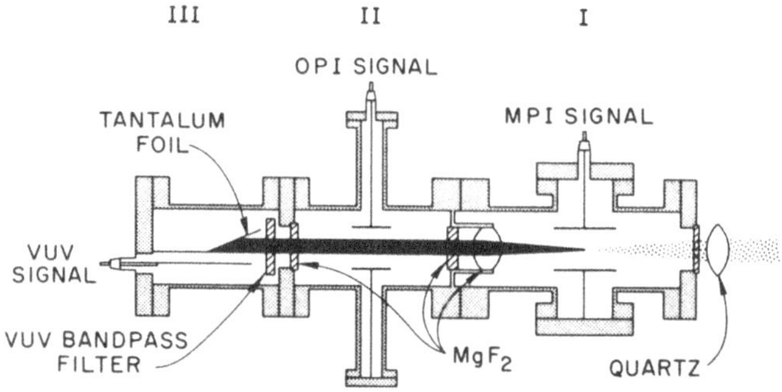

Fig. 12. Schematic diagram of the apparatus used for studies of one-photon absorption and ionization, and of two-photon ionization, using third-harmonic generation in rare gases. Chamber I is the frequency tripling cell; chamber II is the ionization or absorption cell; and chamber III is a vacuum ultraviolet light detector. Each chamber functions as a proportional counter for detecting electrons. In addition, chamber I also functions as a cell for conventional multiphoton ionization studies. See text for details.

trons resulting from one-photon ionization (OPI) or MPI are amplified in the sample gas and detected at the electrode. The VUV beam exits chamber II through an MgF_2 window and passes through a dielectric VUV bandpass filter to remove the blue pump light. This filter can also be placed between chambers I and II when necessary. Chamber III is a VUV photon detector consisting of a Varian nipple containing a tantalum foil and a single flat-plate electrode. Those photons with energies exceeding the work function of tantalum ($\sim$4.1 eV) eject electrons which are detected in the proportional counter. Signals from charge-sensitive preamplifiers connected to any two of the three chambers were averaged in a dual-channel box-car integrator and were displayed, versus wavelength, on a dual-channel x-y recorder and an oscilloscope.

The absolute intensity of the third-harmonic light is on the order of 10^7-10^8 photons/pulse, depending upon the wavelength. The bandwidth of the THG will be at most one-third the bandwidth of the input laser (nominally $\sim$0.1 Å). An additional factor of $\sqrt{3}$ reduction in width results from the I^3 dependence of the THG if a Gaussian pump beam is assumed. The expected bandwidth is thus 0.02 Å at 1470 Å ($\sim$0.9 cm^{-1}). (Use of an etalon would reduce the laser bandwidth to 0.01 Å and that of the THG to 0.002 Å (0.09 cm^{-1}) at 1470 Å. This is on the order of the

Doppler broadening for small molecules.)

For the experiments described below, chamber II was removed from the apparatus (cf. Fig. 12) and chamber III was attached directly to chamber I.

Aron and Johnson [61] reported the first REMPI spectrum of Xe at pressures ranging from 1 to 100 torr. They noted the complete absence of any ionization signal when tuning through the $6s[3/2]^o$, J=1 three-photon-allowed intermediate state. At low pressure (10^{-6} torr), we found that MPI through the 6s state is very intense and, in fact, dominates the spectrum [44]. In later publications [60,62], we further showed that, for a laser beam focused by a 3.8 cm lens to a power density of $\sim 5 \times 10^9$ W/cm^2, the ionization signal produced via the 6s state rapidly disappears at pressures above $\sim$0.3 torr. In addition, third-harmonic light generated in the direction of the pump beam can be detected as the ionization disappears. The excitation lineshape of both the MPI and THG signals are similar and were found to shift and broaden asymmetrically to shorter wavelength as the pressure was increased. Although the intensities of the ionization and third-harmonic light are entirely relative, the dramatic decrease of ionization with increase in THG suggests that the two processes are interfering (i.e., the two processes are in "competition"). THG light should be produced in any negatively-dispersing medium and the excitation lineshape is expected to shift to shorter wavelengths as the pressure increases, because of phase matching considerations. For wavelengths far off resonance, the normal theory for THG, involving the nonlinear susceptibility ($\chi^{(3)}$), applies. However, quenching of the MPI signal is not easily rationalized in terms of our present understanding of such phenomena. In recent theoretical studies of these results, Payne, *et al.* [63,64] found it necessary to include terms involving the coherent, collective excitation of Xe atoms. This model [63,64] shows that at resonance a cancellation exists between the amplitude for exciting the 6s level with three blue photons from the laser, versus that for exciting the level with one VUV photon produced by THG. In other words, at resonance, the one-photon Rabi frequency and the three-photon Rabi frequency are equal but 180° out of phase.

Multiphoton ionization and third-harmonic generation near the $6s'[1/2]^o$, J=1 resonance in Xe [62] show behavior similar to that of the $6s[3/2]^o$, J=1 state discussed above, as do the $5s[3/2]^o$, J=1 and $4s[3/2]^o$, J=1 levels of Kr and Ar, respectively. Plots of ionization and THG for Xe at the 6s' state are shown for pressures from 0.5 to 25 torr in Fig. 13 and at higher pressures (50 to 310 torr) in Fig. 14. The ionization signal is seen to broaden and shift at high pressures in a fashion similar to that observed near the $6s[3/2]^o$, J=1 region. The ionization

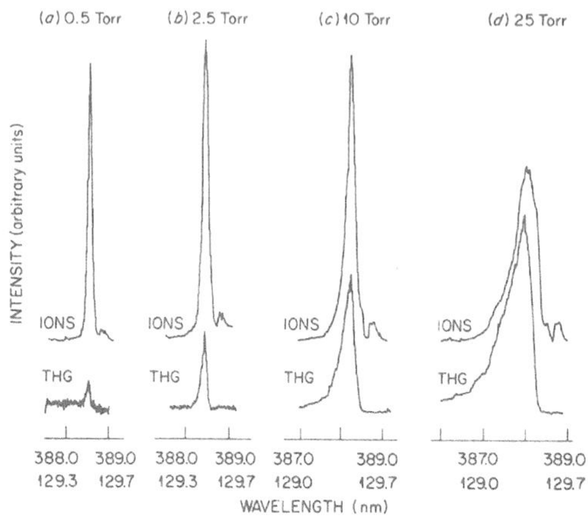

Fig. 13. Multiphoton ionization spectra and third-harmonic generation lineshapes near the 6s' resonance in Xe, recorded at various pressures from 0.5 to 25 torr. The lower wavelength scale corresponds to third-harmonic generation (one-third of the laser wavelength).

signal does not disappear at all wavelengths, but the resonance ionization does become unobservable. Also, molecular effects (Xe_2) are seen in the high pressure data (Fig. 14). It is important to note that only four photons are required to ionize in this wavelength region and this may account for the fact that the nonresonant ionization signal does not disappear.

The ionization signal which is observed to broaden and shift in the wavelength regions near the 6s and 6s' resonant intermediate states can be ascribed to MPI using one third-harmonic photon plus two or one laser photons, respectively. The fact that both the MPI and THG signals have similar lineshapes at low pressures strongly supports this hypothesis. However, it does not explain the absence of a signal at the resonance line position. This would require some other quenching mechanism such as that given by the model of Payne, *et al.* [63,64].

It is also evident from Fig. 14 (cf. also Fig. 3 of [60] and Fig. 10 of [62]) that MPI is an important loss mechanism for THG and ultimately limits the scaling of THG with higher-powered lasers. Sharp peaks in the MPI spectrum (which also appear as sharp dips in the VUV) occur at four-photon atomic resonances. For pressures such that the VUV occurs at these resonance wavelengths, absorption of one VUV plus one blue photon can also populate these states. One additional blue photon then leads to ionization. Similar absorptions into

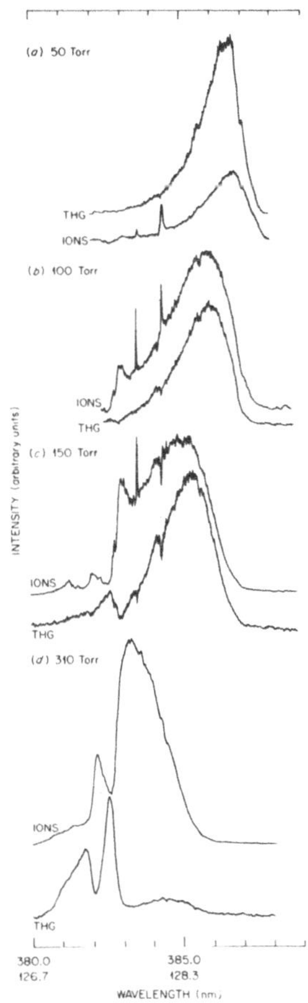

Fig. 14. Multiphoton ionization spectra and third-harmonic generation lineshapes near the 6s' resonance in Xe, recorded at various pressures from 50 to 310 torr. The upper wavelength scale corresponds to the laser; the lower scale corresponds to third-harmonic generation (one-third of the laser wavelength).

molecular states are important at higher pressures. Finally, near the 6s' level, nonresonant two-photon ionization (one VUV plus one blue) also limits the THG.

At the present time, new experiments using the higher power of an excimer laser pumped dye laser are underway to test the above models of MPI at high power. Such experiments will include unfocused laser MPI, polarization experiments, focal-length dependencies and two-color laser experiments. Some very recent two-laser experiments are described below.

Fig. 15 gives the results of a two-laser experiment, where laser 1 is tuned into three-photon resonance with the 6s level and laser 2 is tuned to a particular 6s → 7p level which can be easily saturated. It can be seen that the addition of laser 2, at pressures where the laser 1 signal at the 6s is just barely observable, results in the restoration of a strong, sharp signal exactly at the resonance position. Furthermore, as the pressure is increased the two-laser MPI signal decreases and again disappears. The spectra at high pressures are dominated by the so-called hybrid resonances [51]. These involve the excitation of dimers (or collision complexes), with either three blue photons or one VUV photon, which then dissociate to 6s atoms that are ionized via laser 2. Fig. 16 shows the laser 2 spectrum while laser 1 is tuned to the 6s transition. Clearly, the addition of laser 2 has allowed for more efficient detection of 6s atoms so that resonant MPI is easily observable. The interference effect can thus be followed to higher pressure.

Finally, very recent experiments by Glownia and Sander [65] provide additional support for the competition of THG and MPI. These authors [65] directed counterpropagating beams of circularly-polarized light into a cell similar to that of Fig. 12. Under these conditions MPI, resonant with the 6s and 6s', is possible using photons from both beams; however, THG cannot occur because of phase matching considerations. In the absence of THG, the 6s and 6s' MPI spectra are strong, sharp and unshifted at all pressures.

3.2 VUV Spectroscopy

In contrast to the many studies involving *generation* of THG radiation, there have been only a few involving the *use* of tunable VUV as a spectroscopic tool. Wallenstein, *et al.* [66] have traced the Lyman-α absorption of H and D atoms and a fluorescence excitation band of CO using THG in Kr and Ar generated with a Nd:Yag pumped dye laser system. Egger, *et al.* [67] and Rothschild, *et al.* [68,69], in an elegant series of experiments, have performed very high resolution studies of H_2 and D_2 Rydberg states in the XUV near 83 nm. These experiments used Xe to tri-

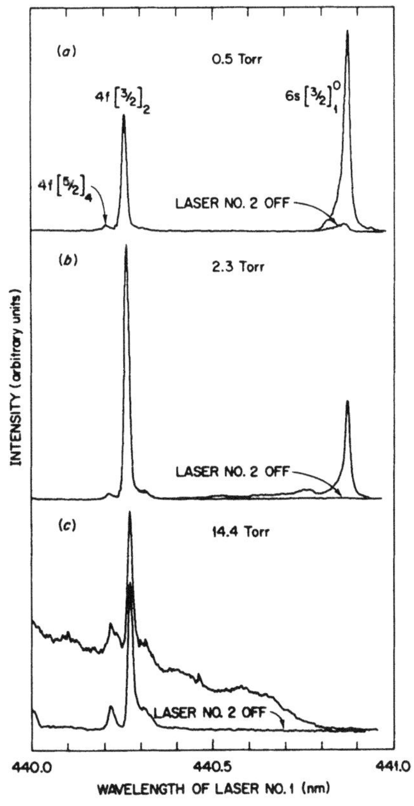

Fig. 15. Results of a two-color multiphoton ionization experiment where Laser No. 2 is set at the $6s[3/2]_1^o \rightarrow 7p[3/2]_2$ transition and Laser No. 1 is scanned in the region of the $6s[3/2]_1^o$, as a function of pressure.

ple the frequency of a high-resolution KrF laser system. Zacharias, *et al.* [70,71] have studied resonant two-photon ionization of H and D atoms and of CO in a mass spectrometer, where the VUV photon was produced by THG in Kr using a Nd:Yag pumped dye laser. The ionizing photon was one of the harmonics of the pump laser. Finally, the THG process itself depends upon resonances in molecules and may be used as a spectroscopic probe of energy levels. Such studies have been performed for NO [72,73], CO [74] and benzene [72].

The third-harmonic light generated in the above experiments is sufficiently intense to be useful in spectroscopic studies of gaseous species in the near ultraviolet region (100-200 nm). Table 1 gives the useful wavelength regions for THG in rare gas-

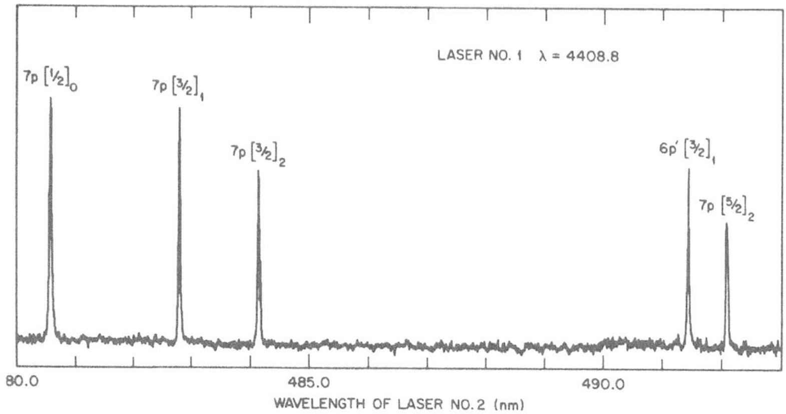

Fig. 16. Results of a two-color multiphoton ionization experiment in which Laser No. 1 is set at the three-photon $6s[3/2]_1^o$ transition and Laser No. 2 is scanned through various 7p and 6p' levels.

es. The gaps can probably be filled in by using molecular gases as the tripling medium, or by using various four-wave mixing schemes. The apparatus shown in Fig. 12 was used for these studies. Further details may be found in a recent paper [75].

By monitoring the THG intensity in chamber III (cf. Fig. 12) with a sample gas in chamber II, a VUV absorption spectrum can be obtained. Fig. 17 presents an absorption spectrum of the $A^1\Pi$ (v=3) $\leftarrow X^1\Sigma^+$(v=0) band of CO, which is a red-degraded band the head of which appears at 144.73 nm. This band is partially overlapped by the $e^3\Sigma^-$ (v=5) $\leftarrow X^1\Sigma^+$ (v=0) band at lower energies. Unfortunately, the spectrum does not show the full resolution of the system due to the mostly-unresolved, closely-spaced P,Q,R triplet in each band. The Doppler width for CO at room temperature is 0.0034 Å. Thus, an etalon in the present laser would produce sub-Doppler resolution for CO, as was observed by Hilbig and Wallenstein [59]. Absorption spectra were also taken for I_2 and NO in this spectral region which, while complicated, were similar to published results. The NO absorption results are shown in Fig. 18 and will be discussed below.

Direct photoionization spectra of benzene, CH_3I, NO and I_2 were studied in the Xe 6s' and Kr 5s wavelength regions. These spectra were easily obtained and were identical to previously reported spectra. Ionization thresholds for a number of molecules have been studied, from which accurate IPs have been determined [e.g., iodobenzene (IP = $8.75^{+0.01}_{-0.04}$ eV) o-xylene (IP =

Table 1. VUV Tuning Range For Nitrogen Laser Pumped Dye Laser Using Xenon and Krypton As The Tripling Gases

Rare Gas	Laser Tuning Range, nm	VUV Tuning Range nm	cm^{-1}	eV
Xe 6s	420.8-440.6	140.3-146.9	68063-71275	8.44-8.84
Xe 6s'	375.0-388.8	125.0-129.6	77160-88000	9.57-9.92
Kr 5s	360.9-370.5	120.3-123.5	80972-83126	10.04-10.31
Kr 5s'[a]	330.0-349.5	110.0-116.5	85837-90909	10.64-11.27

[a]Accessible only with excimer-pumped dye laser or doubled, red dye laser.

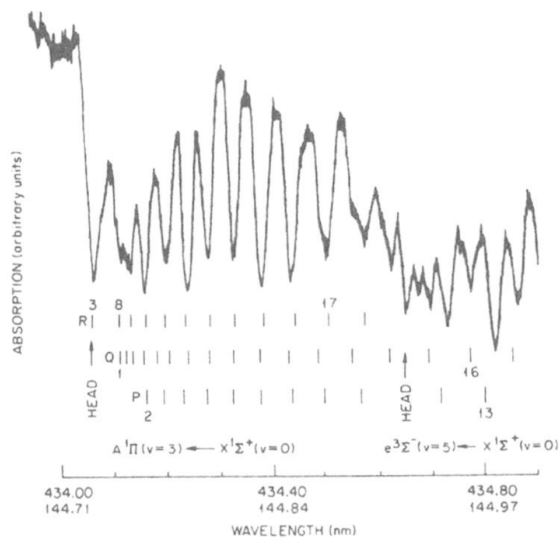

Fig. 17. A portion of the VUV absorption spectrum of CO recorded using third-harmonic generation in Xe. The upper wavelength scale is the laser wavelength in air; the lower scale is the VUV wavelength corrected for the refractive index of air.

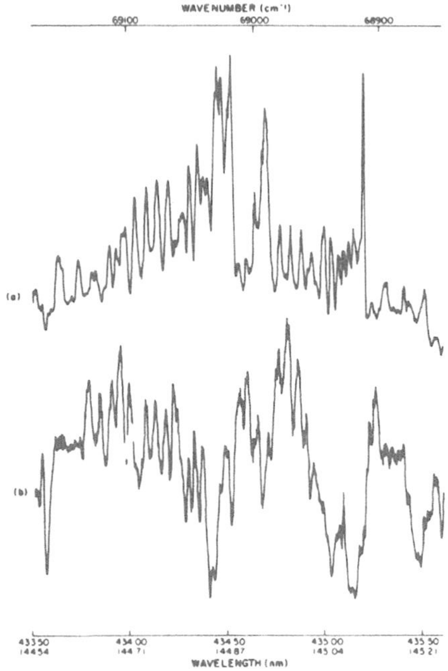

Fig. 18. The two-photon (one VUV plus one blue) ionization spectrum (upper) and one-photon absorption spectrum (lower) of ^{14}NO in the region of the $F^2\Delta$ (v=3) ← $X^2\Pi_{3/2,1/2}$ transition. The upper wavelength scale and the wavenumber scale have been corrected for the refractive index of air.

$8.54^{+0.01}_{-0.04}$ eV)].

Third-harmonic generation results quite naturally in a colinear beam of VUV and blue light. Therefore, by removing the VUV bandpass filter separating chambers I and II (e.g., the same configuration as in Fig. 12), it is possible to perform two-photon, two-color ionization experiments. Fig. 18 displays the two-photon (one VUV plus one blue photon) ionization spectrum (upper trace) and one-photon VUV absorption spectrum (lower trace) of NO in the region of the $F^2\Delta$ (v=3) ← $X^2\Pi$ transition. The transitions involved are shown in Fig. 19. The two leading peaks of each bandhead represent transitions arising from the $^2\Pi_{3/2,1/2}$ fine-structure splitting of the ground state. An indication of the spectral simplification provided by this technique can be seen by comparing the sharp, rotationally-resolved two-photon ionization spectrum with the absorption spectrum: Although there is a one-to-one correspondence in some areas of the spec-

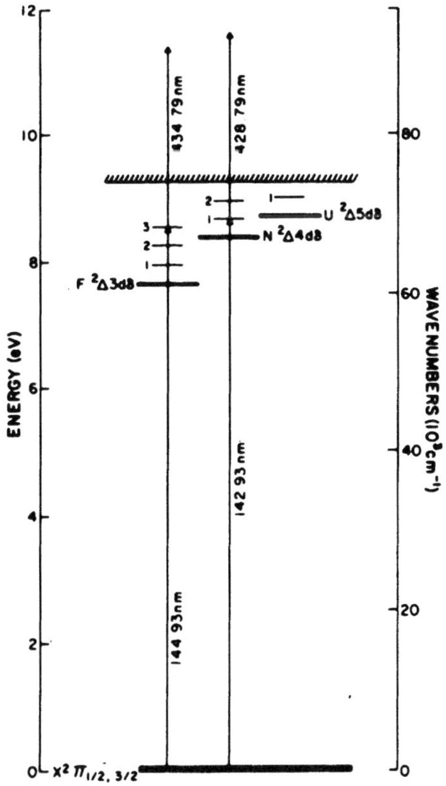

Fig. 19. Partial energy level diagram of NO, showing the $F^2\Delta$ state and the two-photon transition responsible for the ionization signal shown in Fig. 5.

trum, most regions of the absorption spectrum are vastly more complicated.

None of the $^2\Sigma$ and $^2\Pi$ Rydberg states known to occur in this wavelength region are observed, nor are any valence states. Valence states are observed in MPI spectra of NO only when they are mixed with nearby Rydberg states [76]. The absence of the $^2\Sigma$ and $^2\Pi$ states is easily explained: All states in this energy region are crossed by a repulsive $^2\Sigma^+$ state which dissociates to ground-state atoms. Due to the $\Delta\Lambda$ =0, $\pm$1 selection rule for perturbation and predissociation, the $^2\Sigma$ and $^2\Pi$ states can predissociate and, therefore, have a shortened lifetime. In these cases, the predissociation step is faster than the ionization step and these states are not seen in our experiments. The $^2\Delta$ states, however, cannot predissociate via the $^2\Sigma^+$ state and, therefore, have lifetimes long enough for ionization to occur. Higher vibrational levels of the $^2\Delta$ states do, however, show some signs of predissociation. It should be noted that these same kinetic considerations occur for four-photon MPI experiments resonant with the $^2\Delta$ and $^2\Pi$ states, yet these states are observed in the studies of Zakheim and Johnson [76]. However, the detection of four-photon MPI requires a focused laser since this is a higher-order process and, therefore, the one-photon ionization step occurs with much higher power density (this step is probably saturated). Thus, ionization *can* compete with predissociation. In contrast, in the present experiments only an unfocused beam is used and predissociation dominates.

Clearly, increased laser power density obtained either by a higher-powered dye laser or by focusing the emerging THG and blue beam would result in more states being observed. Thus, besides simplifying overlapping bands, an idea of relative lifetimes of states can be obtained. In fact, if the ionizing beam could be delayed in time relative to the VUV beam, exact lifetimes could be obtained.

ACKNOWLEDGMENTS

This research was sponsored by the Office of Health and Environmental Research, U.S. Department of Energy, under contract W-7405-eng-26 with the Union Carbide Corporation.

REFERENCES

1. M. Goepert-Mayer, Naturwissenschaften 17, 932 (1929); Ann. Phys. (Leipzig) 9, 273 (1931).
2. W. Kaiser and C. G. B. Garrett, Phys. Rev. Lett. 7, 229 (1961).
3. J. D. Abella, Phys. Rev. Lett. 9, 453 (1962).
4. G. S. Voronov and N. B. Delone, JETP Lett. 1, 66 (1965).
5. J. L. Hall, E. J. Robinson and L. M. Branscomb, Phys. Rev. Lett. 14, 1013 (1965).
6. S. C. Chin, Can. J. Phys. 48, 1314 (1970).
7. W. C. Lineberger and T. A. Patterson, Chem. Phys. Lett. 13, 40 (1972).
8. C. B. Collins, B. W. Johnson, D. Popescu, G. Muna, M. E. Pascu and I. Popescu, Phys. Rev. A8, 2197 (1973).
9. J. H. Eberly and B. Karizewski, eds., *Multiphoton Bibliography 1970-1976* (UCRL-13728); also, supplements for 1977, 1978, 1979 and 1980, J. H. Eberly, J. W. Gallagher and E. C. Beatty, eds. (NBS LP-92).

10. J. S. Bakos, in: *Advances in Electronics and Electron Physics*, vol. 36, L. Marton, ed., (Academic Press, New York, 1974), p. 57.
11. N. B. Delone, Sov. Phys.-Usp. 18, 169 (1975).
12. P. Lambropolous, Adv. At. Mol. Phys. 12, 87 (1976); A. T. Georges and P. Lambropolous, in: *Advances in Electronics and Electron Physics*, vol. 54, L. Marton and C. Marton, eds. (Academic Press, New York, 1980), p. 191.
13. P. Braunlich, in: *Progress in Atomic Spectroscopy, Part B*, W. Hanle and H. Kleinpoppen, eds. (Plenum Press, New York, 1979).
14. G. Mainfray and C. Manus, Appl. Opt. 19, 3934 (1980).
15. D. H. Parker, J. O. Berg and M. A. El-Sayed, in: *Advances in Laser Chemistry*, A. H. Zewail, ed. (Springer, West Berlin, 1978), p. 320.
16. P. M. Johnson, Acc. Chem. Res. 13, 20 (1980); P. M. Johnson and C. E. Otis, Ann. Phys. Chem. 32, 139 (1981).
17. V. A. Antonov and V. S. Letokhov, Appl. Phys. 24, 89 (1981).
18. P. A. Franken, A. E. Hill, C. W. Peters and G. Weinreich, Phys. Rev. Lett. 7, 118 (1961).
19. J. F. Ward and G. H. C. New, Phys. Rev. 185, 57 (1969).
20. J. Reintjes, C. Y. She, R. C. Eckardt, N. E. Kavangelen, R. H. Andrews and R. C. Elton, Appl. Phys. Lett. 30, 480 (1977).
21. Y. R. Shen, Rev. Mod. Phys. 48, 1 (1976).
22. G. Petty, C. Tai and F. W. Dalby, Phys. Rev. Lett. 34, 1207 (1975); F. W. Dalby, G. Petty-Sil, M. H. Pryce and C. Tai, Can. J. Phys. 55, 1033 (1977).
23. P. M. Johnson, J. Chem. Phys. 62, 4562 (1975).
24. G. C. Nieman and S. D. Colson, J. Chem. Phys. 68, 5656 (1978).
25. A. D. Williamson and R. N. Compton, Chem. Phys. Lett. 62, 295 (1979).
26. M. A. Duncan, T. G. Dietz and R. E. Smalley, J. Chem. Phys. 75, 2118 (1981).
27. U. Boesl, H. J. Neusser and E. W. Schlag, Chem. Phys. Lett. 87, 1 (1982).
28. J. B. Hopkins, D. E. Powers and R. E. Smalley, J. Phys. Chem. 85, 3739 (1981).
29. T. G. Digiuseppe, J. W. Hudgens and M. C. Lin, Chem. Phys. Lett. 82, 267 (1981); J. Phys. Chem. 86, 36 (1981).
30. J. Danon, H. Zacharias, H. Rottke and F. H. Welge, J. Chem. Phys. 76, 2399 (1982).
31. B. H. Rokney and E. R. Grant, Chem. Phys. Lett. 79, 15 (1981); R. J. S. Morrison, B. H. Rokney and E. R. Grant, J. Chem. Phys. 75, 2643 (1981).
32. G. Radhakrishnan, D. Ng and R. C. Estler, Chem. Phys. Lett. 84, 260 (1981).

33. R. Rianda, D. J. Moll and A. Kupperman, Chem. Phys. Lett. 73, 469 (1980).
34. T. G. Dietz, M. A. Duncan and R. E. Smalley, J. Chem. Phys. 76, 1227 (1982).
35. L. Zandee and R. B. Bernstein, J. Chem. Phys. 71, 1359 (1979); R. B. Bernstein, J. Phys. Chem. 86, 1178 (1982).
36. C. D. Cooper, A. D. Williamson, J. C. Miller and R. N. Compton, J. Chem. Phys. 73, 1527 (1980).
37. V. Boesl, H. J. Neusser and E. W. Schlag, Z. Naturforsch. A33, 1546 (1978); J. Chem. Phys. 72, 4327 (1980).
38. J. T. Meek, R. K. Jones and J. P. Reilly, J. Chem. Phys. 73, 3503 (1980).
39. D. M. Lubman, R. Naaman and R. N. Zare, J. Chem. Phys. 72, 3034 (1980).
40. J. Silberstein and R. D. Levine, Chem. Phys. Lett. 74, 6 (1980).
41. F. Rebentrost, K. L. Kompa and A. Ben-Shaul, Chem. Phys. Lett. 77, 394 (1981).
42. F. Rebentrost and A. Ben-Shaul, J. Chem. Phys. 74, 3255 (1981).
43. W. Dietz, H. J. Neusser, V. Boesl, E. W. Schlag and S. H. Lin, Chem. Phys. 66, 105 (1982).
44. R. N. Compton, J. C. Miller, A. E. Carter and P. Kruit, Chem. Phys. Lett. 71, 87 (1980).
45. J. C. Miller and R. N. Compton, J. Chem. Phys. 75, 2020 (1981).
46. J. C. Miller and R. N. Compton, J. Chem. Phys. 75, 22 (1981).
47. J. H. Glownia, S. J. Riley, S. D. Colson, J. C. Miller and R. N. Compton, J. Chem. Phys. 77, 68 (1982).
48. John C. Miller, R. N. Compton, T. E. Carney and T. Baer, J. Chem. Phys. 76, 5648 (1982).
49. John C. Miller and R. N. Compton, Chem. Phys. Lett. 93, 453 (1982).
50. V. S. Kushawaha and J. J. Leventhal, Phys. Rev. 22, 2468 (1980), and references therein.
51. C. B. Collins, J. A. Anderson, D. Popescu and I. Popescu, J. Chem. Phys. 74, 1053 (1981), and references therein.
52. R. W. Falcone, W. R. Green, J. C. White, J. F. Young and S. E. Harris, Phys. Rev. A15, 1333 (1977).
53. P. L. Cahuzac and P. E. Toschek, Phys. Rev. Lett. 40, 1087 (1978).
54. A. V. Hellfeld, J. Caddick and J. Weiner, Phys. Rev. Lett. 40, 1369 (1978).
55. P. Polak-Dingels, J.-F. Delpech and J. Weiner, Phys. Rev. Lett. 44, 1663 (1980).
56. W. R. Green, J. Lukasik, J. R. Willison, M. D. Wright, J. F. Young and S. E. Harris, Phys. Rev. Lett. 42, 970 (1979).
57. J. C. White, Opt. Lett. 6, 242 (1981).

58. W. Christian, R. N. Compton, J. A. Stockdale, J. C. Miller, C. D. Cooper, X. Tang and P. Lambropoulos, to be published.
59. See, for example, R. Hilbig and R. Wallenstein, IEEE J. Quantum Electron QE17, 1566 (1981), and references therein.
60. J. C. Miller, R. N. Compton, M. G. Payne and W. R. Garrett, Phys. Rev. Lett. 45, 114 (1980).
61. K. Aron and P. M. Johnson, J. Chem. Phys. 67, 5099 (1977).
62. J. C. Miller and R. N. Compton, Phys. Rev. A25, 2056 (1982).
63. M. G. Payne, W. R. Garrett and H. C. Baker, Chem. Phys. Lett. 75, 468 (1980).
64. M. G. Payne and W. R. Garrett, Phys. Rev. A26, 356 (1982).
65. J. H. Glownia and R. K. Sander, Phys. Rev. Lett. 49, 21 (1982).
66. R. Wallenstein, Opt. Comm. 33, 119 (1980).
67. H. Egger, R. T. Hawkins, J. Bokor, H. Pummer, M. Rothschild and C. K. Rhodes, Opt. Lett. 5, 282 (1980).
68. M. Rothschild, H. Egger, R. T. Hawkins, H. Pummer and C. K. Rhodes, Chem. Phys. Lett. 72, 404 (1980).
69. M. Rothschild, H. Egger, R. T. Hawkins, J. Bokor, H. Pummer and C. K. Rhodes, Phys. Rev. A23, 206 (1981).
70. H. Zacharias, H. Rottke and K. W. Welge, Opt. Comm. 35, 185 (1980).
71. H. Zacharias, H. Rottke, J. Danon and K. W. Welge, Opt. Comm. 37, 15 (1981).
72. K. K. Innes, B. P. Stoicheff and S. C. Wallace, Appl. Phys. Lett. 29, 715 (1976).
73. S. C. Wallace and K. K. Innes, J. Chem. Phys. 72, 4805 (1980).
74. J. H. Glownia and R. K. Sander, Appl. Phys. Lett. 40, 645 (1982).
75. J. C. Miller, R. N. Compton and C. D. Cooper, J. Chem. Phys. 76, 3967 (1982).
76. D. Zakheim and P. M. Johnson, J. Chem. Phys. 68, 3644 (1978).
77. D. H. Parker, in: *Ultra Sensitive Laser Spectroscopy*, D. S. Klinger, ed. (Academic Press, New York, 1983).
78. T. J. McIlrath and R. R. Freeman, eds., *Laser Techniques for Extreme Ultraviolet Spectroscopy* (American Institute of Physics, New York, 1982).

NOTE ADDED IN PROOF

Ref. [77,78] represent two very recent reviews that provide more details relevant to the studies described in this paper.

ELECTRON-IMPACT SPECTROSCOPY OF MOLECULES

H. H. Brongersma*, H. A. van Sprang† and G. J. Verhaart‡

Gorlaeus Laboratoria
University of Leiden
2300 RA Leiden, The Netherlands

1. INTRODUCTION

Electrons and ions have been used for the excitation of matter ever since the Franck-Hertz experiment in 1914 (see [1]). Although it is not useful to speak of beam experiments in those early years, it soon became obvious that the use of a monochromatic electron or ion beam would open up new avenues for experimental and theoretical advancement. In 1925, Goudsmit and Uhlenbeck [2] discovered electron spin and were thereby able to explain the multiplet structures in their experimental spectra. (This is especially important for low-energy electron-impact experiments, where the spin also determines transition probabilities.) A third important development, the Bethe theory (1930), must be mentioned: This theory provided a basis for the analysis of the interaction between high-energy electrons and matter [3]. The initial formalism for explaining results was now complete; it became operational when, after World War II, technical improvements allowed for the study of single-particle collision processes. Better vacuum systems, electron optics and solid-state electronics permitted an enormous increase in experimental results. These results concerned not only the use of beams for gas-phase spectroscopic research

*Permanent address: Dept. of Physics, Eindhoven University of Technology, P.O. Box 513, 5600 MB Eindhoven, The Netherlands
†Present address: Philips Research Labs, Eindhoven, The Netherlands
‡Present address: Optical Media Laboratory, Eindhoven, The Netherlands

S. P. McGlynn et al. (eds.), Photophysics and Photochemistry in the Vacuum Ultraviolet, 163–189.

but many other applications as well. Indeed, applications in solid-state research became especially important: Ion implantation, electron diffraction and excitation of matter (e.g., phosphors) may be mentioned. In addition, interest in plasma (or electron) chemistry and reactive etching increased.

In the present paper we focus on the spectroscopic use of electron beams. The two main topics are the Trapped-Electron (TE) method, which provides information on negative-ion states and low-lying excited neutral states, and the Electron-Impact-Induced Light Emission (EIILE) method, which is concerned with those excited neutral and positive-ion states which decay radiatively. (It is easily seen that the information from both approaches is largely complementary.)

2. GENERAL THEORY

2.1 Introduction

The encounter between an electron and a molecule may result in several processes, some of which are indicated below:

$e^-(E_i)$	$+ AB$		$\rightarrow AB$	$+ e^-(E_i)$	elastic scattering	(1)
$e^-(E_i > E_{ex})$	$+ AB$		$\rightarrow AB^*$	$+ e^-(E_i - E_{ex})$	direct excitation	(2)
$e^-(E_i = E_r)$	$+ AB$	$\rightarrow AB^-(^*)$	$\rightarrow AB$	$+ e^-(E_i)$	resonant elastic scattering	(3a)
			$\rightarrow AB^*$	$+ e^-(E_i - E_{ex})$	resonance excitation	(3b)
			$\rightarrow A(^*)$	$+ B^-(^*)$	dissociative attachment	(3c)
			$\rightarrow AB^-$	$+ h\nu$	radiative attachment	(3d)
$e^-(E_i > IP)$	$+ AB$		$\rightarrow AB^+(^*)$	$+ 2e^-$	ionization	(4)
			$\rightarrow AB^{**}$	$+ e^-$	superexcited state	(5)

where AB: a polyatomic molecule (A and B can be atoms as well as polyatomic groups).

E_i: energy of the incident electron.
E_{ex}: an excitation energy (electronic, vibrational or rotational) of the molecule AB.
E_r: energy, relative to the ground state of AB, of the negative-ion state $AB^-(*)$.
IP: ionization potential of the molecule AB.

Excitation by electrons differs from excitation by photons. A transition to a particular state by absorption of a photon can occur only when the photon energy matches the energy gap between the ground and excited states. Direct excitation by electrons (2) is, in principle, possible whenever the electron energy is equal to or larger than this energy gap since the electron transfers only part of its kinetic energy to the molecule. Resonance excitation (3), on the contrary, requires specific impact energies, but these are in general not equal to the excitation energy.

Excited states formed in processes (2), (3b), (4) and (5) may decay through emission of radiation. Study of this radiation provides information about both the excitation process and the states involved in the decay.

In the scheme given below, we summarize some of the processes observed experimentally. For simplicity we have assumed that the decay processes lead to molecules or ions in their ground states.

Non-dissociative processes:

$AB^* \rightarrow AB + h\nu$
$AB^{+*} \rightarrow AB^+ + h\nu$
$AB^{**} \rightarrow AB^+ + h\nu + e^-$

Dissociative processes:

$AB^* \rightarrow A^* + B \rightarrow A^+ + h\nu + B + e^-$
$AB^{+*} \rightarrow A^{+*} + B \rightarrow A^+ + h\nu + B$
$AB^{**} \rightarrow A^{+*} + B + e^- \rightarrow A^+ + h\nu + B + e^-$

2.2 Direct Excitation

The type of transition which is observed depends upon the energy of the incident electrons. At higher impact energies, optically allowed transitions become increasingly important in the excitation spectrum.

The character of a transition may be inferred from the energy dependence of the excitation cross-section, which is usually termed the excitation function. For a transition involving an exchange of spin angular momentum between projectile and target, the excitation function increases rapidly in intensity just above the threshold and reaches its maximum within a few eV. For transitions without spin exchange, a maximum is reached at energies of two to four times the threshold energy (the value of four being reached when the transition is also dipole-allowed).

2.3 Resonance Excitation and Dissociative Attachment

A molecule may capture an electron to form a temporary negative ion. This process can occur only when the energy of the incident electron (at infinite separation) matches the energy gap between the molecular ground state and an energy level of the negative-ion state. These latter states are often referred to as resonances. The populated vibrational levels of a specific negative-ion state are determined by Franck-Condon factors between the neutral ground state and the negative-ion state. Some typical potential energy curves for negative-ion states are shown in Fig. 1. In this and the following section, we present (i), a survey of processes resulting from the formation of a specific negative-ion state; and (ii), a classification of resonances.

A molecular negative-ion state may have a wide variety of decay channels [cf. the above processes (3a-3d)]. Formation of a stable parent negative ion (the minimum in curve 1, Fig. 1) is seldom observed since radiative attachment [process (3d)] is an extremely inefficient process and since stabilization by a collision with a third particle is not observed under the conditions prevailing in beam experiments.

When a negative-ion state is formed below its dissociation limit (cf. curves 2, 3, 5 and 6 of Fig. 1), autoionization is the only available decay channel. The molecule may be left either in its ground state [process (3a)] or in an excited state [process (3b)]. Resonances give rise to enhancements in the excitation cross-sections of molecular excited states. The width of a resonance (i.e., the range of electron energies for which a specific negative-ion state can be formed) is related to the resonance lifetime through the uncertainty principle. When this lifetime is sufficiently long, a resonance may show vibrational structure.

For a negative-ion state formed above its dissociation limit (cf. curves 1 and 3 of Fig. 1), the relative abundance of neutral excited states and dissociation products will depend

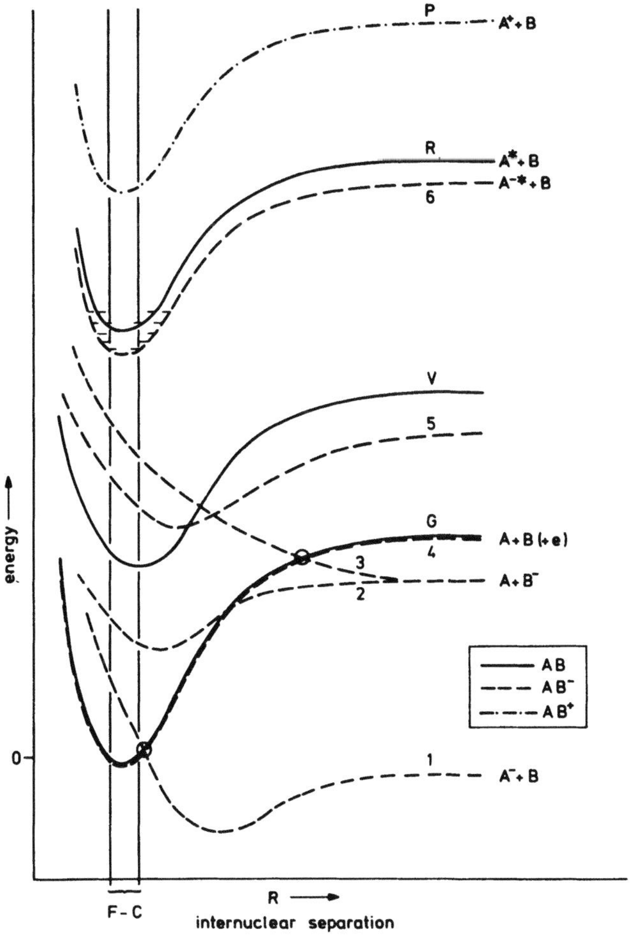

Fig. 1. Representative potential curves for some types of resonances. F-C: Franck-Condon region.

upon the ratio between the autodetachment lifetime and the time required for the nuclei to reach a separation at which the negative-ion state is stable against autodetachment (such separations are indicated by the circles in Fig. 1).

2.4 Classification of Resonances

A negative-ion state may be viewed as a state of the neutral molecule (the parent state) to which one electron has been added. Energetically, a negative-ion state may lie either above or below its parent state. When the negative-ion state is above its parent state, it is termed a shape resonance. In this case the potential forms a penetrable barrier; hence, the shape of the potential is responsible for the trapping of the electron. Shape resonances show a preference for decay into their parent states. Most molecules exhibit one or more shape resonances associated with the ground state, each of which gives rise to enhanced excitation of ground-state vibrational levels. Such a negative-ion state can be represented by curve 2 of Fig. 1; curve 5 represents a so-called core-excited shape resonance associated with the excited state V.

When a resonance lies energetically below its parent state, it is termed a Feshbach resonance. Core-excited Feshbach resonances, associated with Rydberg states, are mostly about half an eV below their parents. These resonances may be represented by two electrons in a Rydberg orbital, trapped in the field of a positive-ion core (the grandparent state). Since Rydberg orbitals extend far outside the molecular core, they have little influence on molecular bonding. This is reflected in the fact that Feshbach resonances, parent states and grandparent states often possess approximately equal vibrational frequencies. Curves 6, R and P of Fig. 1 represent a typical family.

Because decay into the parent state is energetically impossible, and decay into another state of the neutral molecule is not a fast process, core-excited Feshbach resonances are mostly rather "sharp" (i.e., long-lived).

The above classification of resonances as either shape or Feshbach is most useful when considering resonance excitation. Feshbach resonances usually give rise to sharply-structured enhancements in the excitation cross-section, while shape resonances appear as broad humps. The difference is nicely illustrated in [4].

When considering dissociative attachment, a classification of resonances into shape and Feshbach types loses much of its meaning. Consider, for example, the states represented by curves 1 and 3 in Fig. 1. It is clearly not useful to classify 1 as (nuclear-excited) Feshbach and 3 as shape since both possess broad maxima for negative-ion formation.

For molecules with large dissociative-attachment cross-sections, a classification of negative-ion states with respect to the orbital which accomodates the extra electron has been proposed [5] (see, also, Section 4.2 below). This orbital may be of valence type and, possibly, may have an anti-bonding character; it may also be of Rydberg type, in which case it will not influence internuclear bonding (e.g., core-excited

Feshbach resonances). Resonances have been observed in some methane derivatives (CH_4 [6], CF_4 and CF_3Cl [5 and below] and CF_3H [7]): These resonances are parentally associated with the ground state and a Rydberg-type description seems appropriate. And, it is possible that their potential energy curves are almost coincident with those for the neutral ground states (e.g., curve 4 of Fig. 1) [7].

2.5 Electron-Impact-Induced Light Emission (EIILE)

In these experiments the intensity of the emitted radiation is expressed in terms of the emission cross-section:

$$\sigma_{em} = 4\pi \frac{s(\omega)\ P}{\omega\ \phi\ k(\lambda) n\ L} \tag{1}$$

where $s(\omega)$ is the signal intensity for a solid angle of observation ω; ϕ is the electron flux; L is the observation length along the beam; $k(\lambda)$ is the quantum yield of the detection system; P is the polarization factor; and n is the target gas density. A radiating level has a characteristic lifetime τ which is inversely proportional to the total transition probability, including all of the radiative transition probability. Theoretical aspects of these transition probabilities, as well as special problems posed by ionic states with regard to the build up of space charge, have been discussed in [8].

When vibronic transitions are considered, it is customary to define the bandstrength as:

$$P_{v'v''} = |\langle v' | e\hat{R}(R) | v'' \rangle|^2 \tag{2}$$

with

$$e\hat{R}(R) = \langle \psi_u | M | \psi_\ell \rangle \tag{3}$$

where $e\hat{R}(R)$ is the electronic transition moment as a function of the internuclear distance R; and where $\psi_{u,v'}$ and $\psi_{\ell,v''}$ are the vibronic wavefunctions of the upper and lower states, respectively. Introduction of the Born-Oppenheimer approximation separates electronic and vibrational motions and permits the use of independent vibrational (Franck-Condon factors) and electronic transition moments. An example is the r-centroid approximation of Fraser [9].

When the energy dependence of the emission cross-section is considered, the Bethe theory is used in the formalism of Inokuti [10], leading to:

$$\sigma_{em} = \frac{4\pi a_o^2 R}{E_i} M_{em}^2 \ln \frac{C_{em} E_i}{R} \tag{4}$$

where M_{em}^2 is the "emission equivalent" of the electronic-excitation matrix element M_n^2 (i.e., $M_{em}^2 = \phi_\lambda M_n^2$); ϕ_λ is the efficiency for the emission process; a_o is the Bohr radius; R is the Rydberg constant; and C_{em} is a constant specific to the target gas. From Eq. (4), it is obvious that a plot of $\sigma_{em}E_i/4\pi a_o^2$ versus $\ln E_i$ is a straight line if the Bethe theory is valid. This represents a so-called Fano plot [11], where M_{em}^2 is the slope and C_{em} the intercept with the $\ln E_i$ axis.

If $C_{em} \approx 1$ and $M_{em}^2 > 0$, optically allowed processes dominate the formation of the decaying state; for $C_{em} >> 1$ and $M_{em}^2 \simeq 0$, however, forbidden processes are dominant. Examples are given in Section 4.5.

3. EXPERIMENTAL

3.1 General

The experimental possibilities for producing intense electron beams have improved significantly as a result of research on applications such as TV electron guns and electron microscopes. The rapidly growing field of surface science has led to the development of many new types of energy analyzers. However, many of these advances have not found widespread application to electron-impact studies of molecules.

In commercial electron guns, for example, a very high degree of sophistication has been reached. These guns have often been optimized by numerical solution of the Poisson equation for the electrostatic field using the electrode potentials as boundary conditions. In this way, space charge, fringing fields, aberrations and possible errors due to misalignment, etc. are automatically taken into account. Research institutes generally cannot afford the development of such elaborate computer codes, however.

A factor which is very often neglected is the significant influence space charge may have on the energy resolution. In "well-designed", aberration-free lenses, for instance, the electrons are accurately focused. This leads to high-density electron clouds in the cross-over region. It can be shown that, if such a cross-over occurs after acceleration of the electrons along the z-axis, Coulomb interactions among the electrons travelling with the beam (in the center-of-mass frame) may lead to a large energy spread of the electrons with respect to a target molecule at rest. For transport of intense monoenergetic beams, it is thus important to avoid cross-overs. If such avoidance is not feasible, the energy at which these cross-overs occur has to be carefully chosen.

3.2 Equipment

3.2a General

Three different experimental set-ups have been used for obtaining the results presented below. With the trapped-electron (TE) equipment, elastic and inelastic scattering of low-energy (0-30 eV) electrons is studied. (Some of the features of this set-up are briefly discussed in the next section.) The ion-cyclotron-resonance (ICR) equipment, which has been described previously [12], is used as a mass spectrometer for analyzing ions formed by low-energy electron impact. Only long-lived ($> 10^{-3}$ s) ions are detected. In the set-up for studies of electron-impact-induced light emission (EIILE), molecules are bombarded by an electron beam of 0-600 μA, having an energy selectable up to 2000 eV (energy spread 0.5 eV). Apparatus details have been presented elsewhere [13].

3.2b The Trapped-Electron (TE) Spectrometer

The TE spectrometer, as described previously [14], consists of an electron monochromator, a collision chamber and an energy analyzer for the scattered and unscattered electrons.

In the electron source, electrons are emitted by a cathode and energy selected in a trochoidal monochromator. An axial magnetic field, which is necessary for the operation of the monochromator, also collimates the electron beam. A disadvantage of this monochromator is that it only selects electrons with respect to the energy associated with the velocity component parallel to the magnetic field. Since a high radial energy corresponds, in a magnetic field, to a large Larmor radius, it is possible to intercept such electrons. Either a large, narrow slit or a channelplate having long, narrow channels can be used [4]. With such a device it is possible not only to improve the energy resolution but also to replace the easily poisoned oxide cathode by a tungsten filament, as well as to increase the beam intensity at the same time.

In Fig. 2, a schematic cross-section is given of the energy-analysis and detection systems for scattered electrons. Depending upon the electrode potentials, either the inelastically-scattered electrons (TE mode) or the unscattered electrons (transmission mode) are detected.

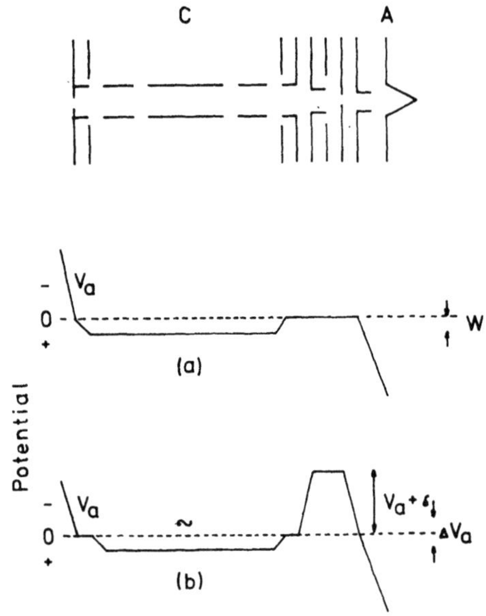

Fig. 2. Selection/detection system: electrode configuration and potential distribution along the axis of the system.
(a) trapped-electron mode
(b) transmission mode
V_a: accelerating voltage; W: well depth;
ΔV_a: modulation amplitude

In the TE mode (Fig. 2a), a potential well (of depth W) is formed along the axis of the collision chamber. Inelastically-scattered electrons, possessing a residual energy of less than eW, are collected and measured at C. Selection involves a diffusion mechanism in which the magnitude of the axial magnetic field and the target gas pressure are important quantities. Elastic collisions with target gas molecules often cause a change in the direction of the velocity vector of the electrons. (This is the basis of the selection mechanism.) The movement of the scattered electron consists of a helical path up and down the collision chamber. Again and again the electron will be energy selected at the end of the collision chamber, whether it has a sufficient axial velocity component to escape from the well or not. Therefore, electrons having an axial velocity component larger than eW will always escape from the well. Positive and negative ions, which may be formed in the collision chamber, can also be measured at the collector C. As a result of the larger masses of these ions, however, their motion will hardly be affected by the magnetic field.

The transition mode is a highly sensitive method for determining the impact energies at which temporary negative ions are formed since the formation of such ions often causes abrupt changes in the electron transmission as a function of energy (i.e., "resonances" appear). In the transmission mode, a retarding potential at the end of the collision chamber prevents the inelastically-scattered electrons from reaching the collector A. Most of the elastically-scattered electrons are also prevented from reaching the detector since scattered electrons have their axial velocity vector reoriented and thus possess insufficient axial momentum to overcome the potential barrier. In order to increase sensitivity to changes in the transmitted current as a function of the incident energy, the energy derivative of the current is generally measured.

4. EXPERIMENTAL RESULTS

4.1 Introduction

In this section, we discuss various experiments illustrating some of the possibilities of electron-impact studies. In Sections 4.2 and 4.3, attention is focused mainly upon the trapped-electron (TE) experiments, sometimes combined with ion-cyclotron-resonance (ICR) experiments. In Section 4.4, transmission electron spectra for layers of organic molecules are considered. Sections 4.2-.4 refer to low-energy electron-impact phenomena; Section 4.5, however, concerns high-energy impact followed by light emission. In Sections 4.2, 4.3 and 4.5, spectra resulting from electron impact on chlorofluoromethanes are used as examples. These results give an idea of the kinds of information which can be obtained from these experimental techniques.

4.2 Electron Transmission Spectroscopy

In Fig. 3, electron transmission spectra (cf. Section 3) are shown for CF_4 and CF_3Cl [5]. The strong vibrational progressions are attributed to a relatively long-lived negative-ion state since it is not likely that a direct process would exhibit such a sharp threshold as to cause structure in the derivative of the transmitted current. The observed vibrational energies are very close to known C-F vibrational energies in the neutral molecules (cf. Table 1). It has been proposed that the extra electron is in a Rydberg-type orbital since any valence orbital would be strongly anti-bonding (see Section 2.4). Similar phenomena have been observed in strongly polar molecules [17,18] but it is surprising that a highly symmetrical molecule such as CF_4 has a nonzero electron affinity too.

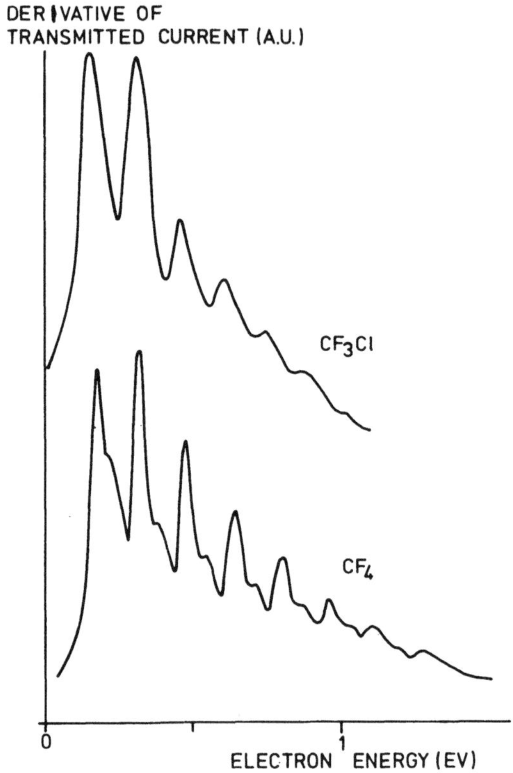

Fig. 3. Derivative of transmitted electron current.

Table 1. Vibrational Energies (eV) in the Negative Ions (Observed in Transmission) and the Neutral Molecules

	E_{vib} in anion	E_{vib} in neutral*	$E^{\dagger}$
CF_4	0.159 ± 0.002	0.159 [15]	0.32
CF_3Cl	0.142 ± 0.004	0.150 [16] (a)	0.29
		0.137 [16] (s)	

* "a" ≡ antisymmetric; "s" ≡ symmetric
† Energy of first fully resolved level in transmission

4.3 Dissociative-Attachment Processes

4.3a Chlorofluoromethanes

Dissociative-attachment processes in chlorofluoromethanes have been reported in earlier mass-spectrometric studies [19, 20] and in two recent studies [5,21]. Some results are summarized in Tables 2 and 3.

Illenberger, *et al.* [21] have used a set-up in which dissociative-attachment data down to zero eV impact energy are obtained directly, with a trochoidal electron monochromator in the electron gun and a quadrupole mass filter as an ion analyzer. Verhaart, *et al.* [5] have used an ICR apparatus for analyzing the ions formed. A distinct advantage of ICR is that the double-resonance technique [22] can be used to establish whether the ions are formed by a direct process or as a result of ion-molecule reactions. However, a disadvantage of the ICR apparatus is that the electron gun is not equipped with a monochromator. Therefore, it was not possible to obtain accurate appearance potentials. Moreover, exact values for impact energies below 2 eV could not be obtained because of experimental difficulties. This disadvantage was partially overcome by combining the ICR results with data from the TE spectrometer. The latter spectrometer is equipped with a high-resolution trochoidal monochromator. Negative-ion peaks can be discerned from electron scattering for impact energies down to 0.5 eV.

With respect to the main products (Cl^- in the 0-2 eV region; F^- and a fragment ion formed by the loss of one fluorine atom, $(M-F)^-$, in the 3-5 eV region), there is excellent agreement between the qualitative results of both studies [5,21]. There are, however, some differences in the relative intensities as functions of energy. In [17], an $(M-F)^-$ producing resonance in CF_3Cl, expected in the 3-5 eV region, was not found. Instead, a second Cl^- formation process was observed. It was speculated [5] that CF_2Cl^- would be primarily formed but that it would dissociate into CF_2 and Cl^- within the minimum lifetime required for ions to be detected in the ICR instrument (10^{-3} s). This supposition seems to be confirmed by the Illenberger, *et al.* results [21]. In that experiment, in which the observation time is much shorter ($\sim 10^{-6}$ s), CF_2Cl^- is detected in low abundance; its appearance potential coincides with the second onset for Cl^- formation, indicating that CF_2Cl^- is the primary fragment, with a lifetime on the μs scale.

Cl_2^- formation in $CFCl_3$ is observed at low impact energies

Table 2. Trichlorofluoromethane: Negative Ions (energies in eV, $M = CFCl_3$)

	Peak Maxima			Appearance Potentials		
	electron impact			electron impact		K-impact[a]
Reference	[5]	[21]	[19]	[21]	[19]	[23]
M^-						-1.1
Cl^-	<0.5	0.0	0.0	0.0 ± 0.05	0.0	-0.4
F^-	3.0	3.0	3.0	1.6 ± 0.2	1.3	<2.2
$(M-F)^-$	3.5	3.3	3.3	2.7 ± 0.2	2.75	
Cl_2^-	1-2	1.6		0.6 ± 0.2		2.0
$(M-Cl)^-$						2.1
CCl_2^-						3.9
FCl^-						3.9

[a]The reported appearance potentials have been corrected for IP (K) = 4.3eV to allow comparison with electron impact.

Table 3. Dichlorodifluoromethane and Chlorotrifluoromethane: Negative Ions (energies in eV)

	$M = CF_2Cl_2$					$M = CF_3Cl$		
	Peak Maxima electron impact			Appearance Potentials electron impact	K^a	Peak Maxima electron impact		Appearance Potentials electron impact
Reference	[5]	[21]	[20]	[21]	[23]	[5]	[21]	[21]
M^-					-0.4			
Cl^-	0.7	0.55	0.8	0.15 ± 0.1	-0.3	1.4 5.0	1.3 4.8	0.7 ± 0.3 3.4 ± 0.3
F^-	3.2	3.1		1.8 ± 0.2	<3.1	4.3	4.1	3.0 ± 0.3
$(M-F)^-$	3.7	3.55		2.8 ± 0.2			4.2	3.5 ± 0.3
Cl_2^-		0.65		0.2 ± 0.1	1.3			
$(M-Cl)^-$					1.7			
FCl^-		2.85		2.1 ± 0.2	<3.1		3.9	3.0 ± 0.3

[a] K-impact; the reported appearance potentials have been corrected for IP (K) = 4.3eV to allow comparison with electron impact.

in both of the above experiments [5,21]. Illenberger, *et al.* [21] have measured an appearance potential of 0.6 ± 0.2 eV for this process. Since the process is predicted to be forbidden at this energy from literature data for the heats of formation of the fragments, these authors [21] suggest that an ion-molecule reaction might be responsible for Cl_2^- formation. However, such a reaction is energetically not feasible either. Moreover, Cl_2^- has been shown to be a primary product by an ICR double-resonance experiment [24]. The consequences of this result for the previously accepted value of ΔH_f (CFCl) have been discussed in [25].

Besides those ions discussed above, Illenberger, *et al.* also report the detection of Cl_2^- in CF_2Cl_2, and FCl^- in both CF_2Cl_2 and CF_3Cl [21]. These ions were not found by other authors [5,7].

Another study of negative-ion formation in $CFCl_3$ and CF_2Cl_2, employing a beam of neutral potassium atoms, has been reported by Dispert and Lacmann [23] and their results are summarized in Tables 2 and 3. It is not expected that results from a neutral-atom impact experiment will be directly comparable with those from an electron-impact experiment. In electron impact, a molecule captures an electron with a specific energy to form a temporary negative-ion state, which subsequently dissociates into the observed products. In neutral-atom impact, such a state is formed by ionizing the atom. This process does not require a specific impact energy because the positive ion is available to carry away excess energy. Hence, it is not surprising that appearance potentials reported by Dispert and Lacmann [23] are higher than those reported for electron-impact processes [21]. Furthermore, the presence of the positive ion is expected to considerably disturb the potential curves of the initially formed negative-ion states [26]: This is reflected in the fact that fragments not observed in electron impact are found in neutral-atom impact, most notably $(M\text{-}Cl)^-$; however, $(M\text{-}F)^-$ fragments, always found in electron impact, are not observed in neutral-atom impact.

4.3b Tetrachloromethane

Fragment negative-ion formation in CCl_4 has been studied in the experimental set-up described in [5,7]. A survey of the results is given in Table 4, together with values from previous studies. There is general agreement that Cl^- is formed at zero impact energy. Reported values for the appearance potential of Cl_2^-, however, differ by two eV [28,29]. ICR data also suggest that the appearance potential is near zero impact energy. A double-resonance experiment clearly shows that Cl_2^- is formed in a one-step process. The observation that Cl_2^- is formed near

Table 4. Tetrachloromethane: Negative Ions (energies in eV)

	Peak Maxima			Appearance Potentials			
	electron impact			electron impact			K-impact[a]
Reference	[7]	int.[b]	[27]	this work	[28]	[29]	[23]
CCl_4^-							-2.0
Cl^-	<0.5		0.0	<0.1	0.0	0.0	-0.6
	5.8	1000	6.0		5.6		
Cl_2^-	~1.0		~1.5	<0.3	0.8	2.3	1.8
	5.8	18	6.3				
CCl_3^-	1-2		~1.5	<1.0	1.4		1.7
	5.7	5					
CCl_2^-	5.8	10	6.3	4.8			2.5

[a] The reported appearance potentials have been corrected for IP (K) = 4.3eV to allow comparison with electron impact.

[b] Relative intensity at the peak maximum.

zero electron energy could not be reconciled with literature data for heats of formation, which led to the proposal of a new value [25].

4.3c Reactions of Fragments

Dissociative-attachment processes have been monitored almost exclusively via the ions formed. However, the neutral fragments may give rise to the selective formation of certain products.

In a study by Verhaart, *et al.* [7,25], it has been shown that, on impact by electrons with near-zero energy, the following reactions occur in the chlorofluoromethanes:

$$e^- + CCl_4 \rightarrow CCl_3\bullet + Cl^-$$

$$\rightarrow CCl_2 + Cl_2{}^-$$

$$CCl_3\bullet + CCl_3\bullet \rightarrow C_2Cl_3$$

$$CCl_3\bullet + CCl_2 \rightarrow C_2Cl_4 + Cl\bullet$$

$$e^- + CFCl_3 \rightarrow CFCl_2\bullet + Cl^-$$

$$CFCl_2\bullet + CFCl_2\bullet \rightarrow C_2F_2Cl_4$$

$$e^- + CF_2Cl_2 \rightarrow CF_2Cl\bullet + Cl^-$$

$$CF_2Cl\bullet + CF_2Cl\bullet \rightarrow C_2F_4Cl_2$$

The selective product formation obtained in this fashion is usually not found in reactions induced by high-energy electrons or by photons.

4.4 Molecular Thin Films

The comparison of light-absorption data for low-pressure gas-phase (isolated) molecules with data for the corresponding solid films provides a powerful tool for distinguishing Rydberg states, which become significantly distorted in a solid, from valence excited states (Robin's "Rydberg criterion" [30]). Thin solid films of organic molecules have previously been investigated by electron-impact techniques [31]. The difficulties associated with film preparation, surface charging and radiation damage make such studies rather delicate. Moreover, the large cross-sections for electron scattering give rise to multiple-scattering processes which significantly complicate spectral interpretation.

Recently, Sanche [32] considerably improved the possibilities for electron-impact studies by using a monochromatic electron source in combination with a magnetic field to align the electrons. Much sharper structures are now observed in his transmission spectra. These spectra are divided into two regions: The collective regime is below 1.5 eV. Here, the de Broglie wavelength of the incident electron is large in comparison with molecular dimensions and no correlation is found with processes occuring in isolated molecules. Above about 1.5 eV, spectral structures generally correlate quite strikingly with electronic excitation energies of isolated molecules. These structures are a result of the relative importance of elastic and inelastic scattering. Since inelastically-scattered electrons do not have enough energy to leave the film again, they contribute negligibly to the transmitted current.

Sanche's results [32] seem to indicate that the Rydberg criterion is applicable here since no transitions are observed in the 8-9 eV Rydberg region of benzene. The fact that the 8.3 and 10.2 eV energy levels in pyridine are observed by Sanche makes it tempting to question the pure Rydberg assignment of these levels [33].

To illustrate that the results for isolated molecules and films do not always agree completely, the thin-film results of Sanche, *et al.* for 1,4-cyclohexadiene [32] and norbornadiene [34] are compared, in Figs. 4 and 5, with threshold excitation spectra by Brongersma [35].

A fundamental difference between the spectra compared in Figs. 4 and 5 is that in thin-film transmission spectra the current is plotted as a function of the incident energy, while near-threshold excitation spectra are plotted as a function of energy loss. As van Veen [36] has shown for thiophene, furan and pyrrole, the selection rules and, thus, the appearance of an excitation spectrum is strongly dependent upon the energy eW of the electron after scattering. As an illustration of this effect, van Veen's results [36] for thiophene are presented in Fig. 6. It is also clear from this figure that a transition which is not observed, for a given value of W, may very well be strongly resonance-enhanced for another value. This makes the applicability of the Rydberg criterion doubtful for electron impact.

Without further knowledge of the excitation functions and of the presence or absence of electron-exciton complex formation, it is difficult to decide which value of W is the best choice for comparison of excitation spectra with transmission data. Sanche's choice [32] in comparing his data with van Veen's threshold excitation spectra [36], namely W = 0.0 eV for

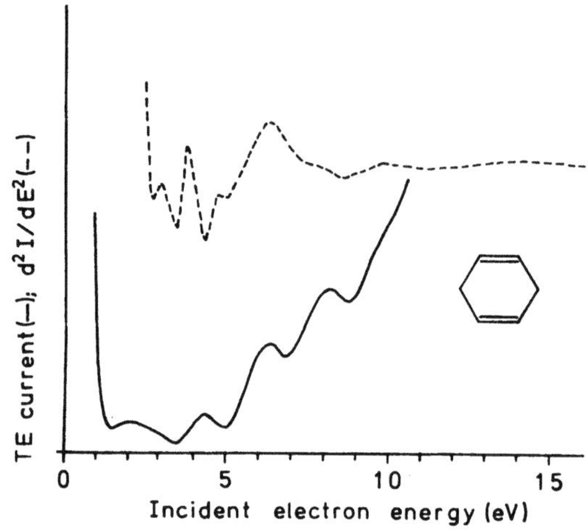

Fig. 4. 1,4 Cyclohexadiene. A comparison of TE (—) [35], with thin-film results (---) of Sanche [35].

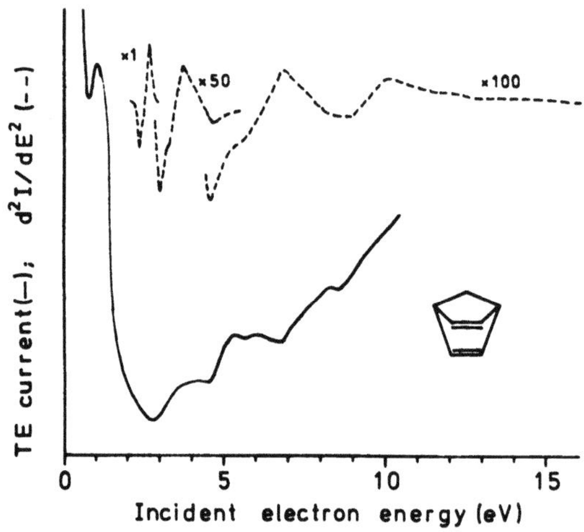

Fig. 5. Norbornadiene. A comparison of TE (——) [35] with thin-film results (---) of Sanche [35].

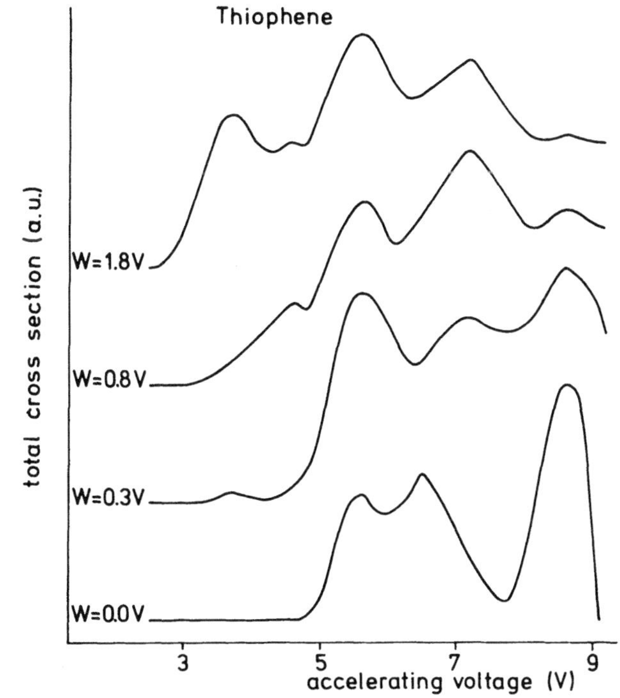

Fig. 6. Influence of the excess energy above threshold (eW) on the excitation spectrum of thiophene.

thiophene and W = 0.8 eV for pyrrole and furan, seems rather arbitrary.

Although the understanding of energy transfer in thin solid films has made much progress, it remains an intruiging field for more research.

4.5 Electron-Impact-Induced Light Emission (EIILE) Results

4.5a The HF Spectrum

The electron-impact emission spectrum of HF was first recorded by Gewurtz, *et al.* [37] in 1975. A weak band system was discovered between 3500 and 4600 Å (350 - 460 nm) which could be ascribed to the HF^+ ($A^2\Sigma^+ \rightarrow X^2\Pi$) emission. In 1978, van Sprang and de Heer [38] studied the most intense emission bands corresponding to HF^+ ($A^2\Sigma^+$,v'=0-3 $\rightarrow$ $X^2\Pi_i$,v"=0) vibrational transitions: Radiative lifetimes for the various vibrational levels were obtained and the collisional cross-section for

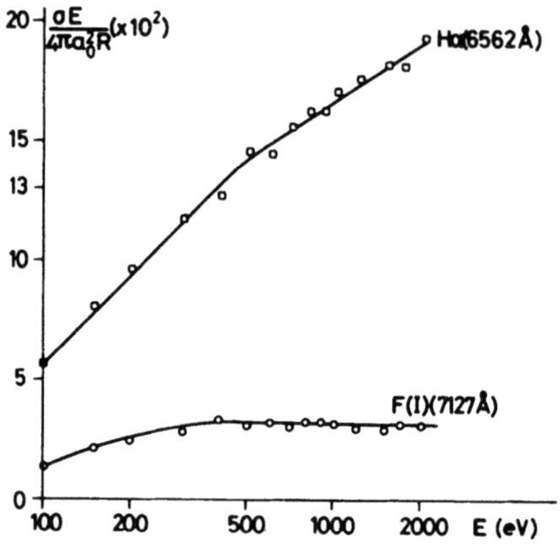

Fig. 7. Fano plot (100-2000 eV) for H_α and $F(^2P_o-^2P)$ radiation from HF.

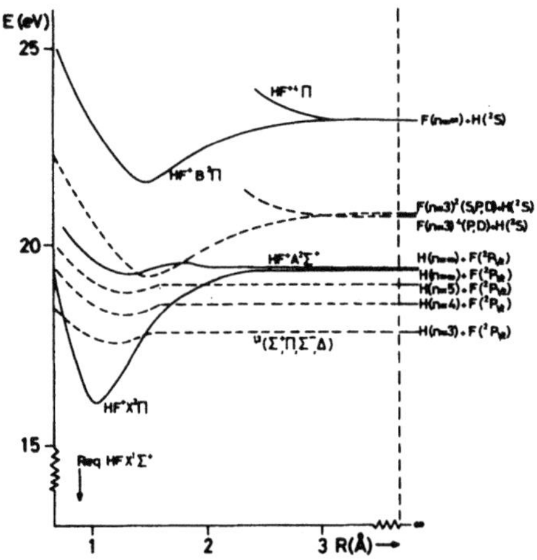

Fig. 8. Potential energy diagram for some ionic HF^+ levels.

quenching of HF^+ $A^2\ \Sigma^+$ by HF was determined to be 350 $Å^2$. Comparison of measured lifetimes with calculated values showed that, for HF^+, the electronic transition moment depends upon the internuclear distance. Clearly, for the observed transition the Born-Oppenheimer approximation failed.

The power of emission spectroscopy is amply demonstrated by a discussion of fragment emissions in HF. Excited hydrogen and fluorine atoms must be the sources of these emissions. In Fig. 7, the Balmer H_α and F(I) emission cross-sections at higher energies are given as Fano plots. It has been found that, for H_α, $M_{em}{}^2 = 3.0 \times 10^{-2}$ while $C_{em} = 1.8$, which implies that the formation of excited hydrogen atoms is dominated by optically allowed processes. The $F(^2P_o - {}^2P)$ radiation has $M_{em}{}^2 = 0$ and $C_{em} = \infty$, thus indicating optically forbidden processes to be dominant. Which mechanisms, however, are important in fragment

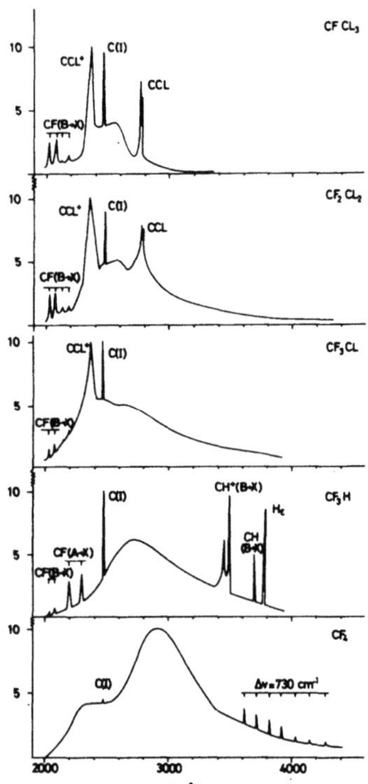

Fig. 9. Emission spectra of $CFC\ell_3$, $CF_2C\ell_2$, $CF_3C\ell$, CF_3H and CF_4 from 2000-4400 Å.

formation? In order to answer this, a potential energy diagram for HF^+ is given in Fig. 8. Here, full lines denote ionic curves while dashed lines represent Rydberg states converging to the various ionic limits (see, e.g., [39]). From this diagram, it appears that excited hydrogen atoms originate from those Rydberg states converging to $A^2\Sigma^+$, states which will be populated (above their dissociation limit from the HF $X^1\Sigma^+$ ground state) by a one-electron excitation. Such electron processes are optically allowed. The experimental threshold for H(n=3) formation at 17.8 ± 0.3 eV is in complete agreement with the dissociation limit of the corresponding Rydberg orbital at 18 eV.

For the fluorine ^2P_Q - ^2P emission, the Rydberg orbitals converging on $B^2\Pi$ are important. These orbitals are populated from $X^1\Sigma^+$ by a two-electron process which is optically forbidden (see, e.g., [40]). The threshold energy for F(n=3) formation was found experimentally to be 20.7 ± 0.3 eV, which compares well with the corresponding dissociation limit of 20.6 eV. These results have led to a better understanding of the origin of the various emission bands and lines of HF.

4.5b Chlorofluoromethanes (Freons)

The main interest in electron-impact data for freons has

Table 5. Experimental Thresholds for the Continuous Emissions

Compound	Experimental Threshold (eV)	Corresponding[a] IP	Upper State[b]
CF_4	19.5 ± 0.5	18.5	B^2E
CF_3	17.3 ± 0.5	17.2	C^2E
CF_3Cl	16.2 ± 0.5	15.8	B^2A_2
CF_2Cl_2	15.7 ± 0.5	15.9	E^2B_1
$CFCl_3$	15.2 ± 0.5	15.1	D^2F

[a] Ionization potential corresponding best to the experimental threshold.

[b] State corresponding to the ionization potential of third column.

been concentrated on the possible implications of low-energy electron-impact phenomena for atmospheric pollution. This section contains complementary information for higher impact energies.

The first emission spectrum of a freon, CF_4, was published by Hesser and Dressler [41], while van Sprang, *et al.* [42] published the first emission spectra of CF_3H, CF_3Cl, CF_2Cl_2 and $CFCl_3$. These spectra are shown in Fig. 9. Apart from the known emission bands and fragmentation lines, all compounds show continuous emission in the 2000-4000 Å (200-400 nm) range.

Table 5 contains the experimental thresholds for the emission continua, while the corresponding Fano plots are given in Fig. 10. These plots clearly show the optically allowed character of the emissions. The experimental thresholds are all in the range above the first ionization potentials and, as can be

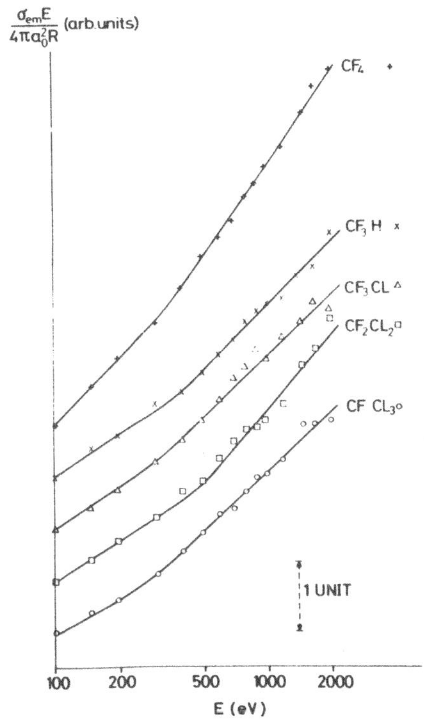

Fig. 10. Relative Fano plots for the continuous emissions of the various compounds.

seen in Table 5, a rather good correspondence can be found between these thresholds and higher ionization potentials of the various compounds. This correspondence permits an assignment of the upper electronic states to the ionic states shown in Table 5. For all molecules except CF_2Cl_2, the lower level is the ionic ground state. (For CF_2Cl_2, the E^2B_1-X^2B_2 transition is forbidden. Hence, the lower level in CF_2Cl_2 is A^2B_1.) The fact that the lower levels are dissociative is, of course, the reason for the continuous emission. The wavelength region of the continuum emission is in agreement with the corresponding energy difference of the ionic levels. Finally, the formation pathways for the various fragments could not be found because of a lack of theoretical data on the molecular states.

ACKNOWLEDGMENT

This work was sponsored by SON with financial support by ZWO, the Netherlands Organization for the Advancement of Pure Research.

REFERENCES

1. J. Franck and G. Hertz, Phys. Zeitschrift 20, 132 (1919).
2. G. E. Uhlenbeck and S. Goudsmit, Naturwissenschaften 13, 132 (1919).
3. H. A. Bethe, Ann. Rev. der Phys. 5, 235 (1930).
4. G. J. Verhaart and H. H. Brongersma, Chem. Phys. 52, 431 (1980).
5. G. J. Verhaart, W. J. van der Hart and H. H. Brongersma, Chem. Phys. 34, 161 (1978).
6. P. Burrow and J. A. Michejda, Argonne Natl. Lab. Rep., 113 (1976-1977).
7. G. J. Verhaart, Ph.D. thesis, Leiden University (1980).
8. G. R. Mohlmann and F. J. de Heer, Phys. Scripta 16, 51 (1977).
9. P. A. Fraser, Can. J. Phys. 32, 515 (1954).
10. M. Inokuti, Rev. Mod. Phys. 43, 297 (1971).
11. U. Fano, Phys. Rev. 95, 198 (1954).
12. W. J. van der Hart and H. A. van Sprang, J. Am. Chem. Soc. 99, 32 (1977).
13. C. I. M. Beenakker and F. J. de Heer, Chem. Phys. 6, 291 (1974).
14. E. H. van Veen and F. L. Plantenga, Chem. Phys. Lett. 38, 493 (1976).
15. P. J. H. Woltz and A. H. Nielson, J. Chem. Phys. 20, 307 (1952).
16. E. K. Plyler and N. Acquista, J. Res. Natl. Bur. Stand. 48, 92 (1952).

17. J. P. Ziesel, I. Nenner and G. J. Schulz, J. Chem. Phys. 63, 1943 (1975).
18. K. Rohr and F. Linder, J. Phys. B9, 2521 (1976).
19. R. K. Curran, J. Chem. Phys. 34, 2007 (1961).
20. W. M. Hickman and D. Berg, J. Chem. Phys. 29, 517 (1958).
21. E. Illenberger, H. U. Scheunemann and H. Baumgartel, Chem. Phys. 37, 21 (1979).
22. T. A. Lehman and M. M. Bursey, *Ion Cyclotron Resonance Spectroscopy* (Wiley, New York, 1976).
23. H. Dispert and K. Lacmann, Int. J. Mass Spectrom. and Ion Phys. 28, 49 (1978).
24. W. J. van der Hart and G. J. Verhaart, unpublished results.
25. G. J. Verhaart, H. A. van Sprang and H. H. Brongersma, Chem. Phys. 51, 389 (1980).
26. A. W. Kleyn, private communication.
27. F. H. Dorman, J. Chem. Phys. 44, 3856 (1966).
28. R. M. Reese, V. H. Dibeler and F. L. Mohler, J. Res. 57, 367 (1956).
29. J. J. de Corpo, D. A. Bafus and J. L. Franklin, J. Chem. Phys. 54, 1592 (1971).
30. M. B. Robin, *Higher Excited States in Polyatomic Molecules*, Vol. I (Academic Press, New York, 1974), p. 76.
31. K. Hiraoka and W. H. Hamill, J. Chem. Phys. 59, 5749 (1973).
32. L. Sanche, J. Chem. Phys. 71, 4860 (1979).
33. B. Ö. Jonsson, E. Lindholm and A. Skerbele, Int. J. Mass Spectrom. Ion Phys. 3, 385 (1969).
34. L. Sanche, G. Bader and L. Caron, J. Chem. Phys. 76, 4016 (1982).
35. H. H. Brongersma, Ph.D. thesis, Leiden University (1968).
36. E. H. van Veen, Chem. Phys. Lett. 41, 535 (1976).
37. S. Gewurtz, H. Lew and P. Flaieck, Can. J. Phys. 53, 1097 (1975).
38. H. A. van Sprang and F. J. de Heer, Chem. Phys. 33, 73 (1978).
39. J. Berkowitz, W. A. Chupka, P. M. Guyon, J. H. Holloway and R. Spohr, J. Chem. Phys. 54, 5165 (1971).
40. H. A. van Sprang, H. H. Brongersma and F. J. de Heer, Chem. Phys. 38, 277 (1979).
41. J. E. Hesser and K. Dressler, J. Chem. Phys. 47, 3443 (1967).
42. H. A. van Sprang, H. H. Brongersma and F. J. de Heer, Chem. Phys. 35, 51 (1978).

ELEMENTS OF QUANTUM DEFECT THEORY. I. INTRODUCTION AND FORMALISM

A. R. P. Rau

Department of Physics and Astronomy
Louisiana State University
Baton Rouge, Louisiana 70803, USA

1. INTRODUCTION TO QUANTUM DEFECT THEORY

1.1 Introduction

The term quantum defect theory (QDT) has been conventionally used to describe an analysis which interrelates phenomena in the discrete and continuum state regions of an atomic system consisting of a positive ion and an electron [1,2]. The analysis rests heavily on analytical properties of Coulomb wavefunctions. In view of the complexities associated with these functions, it is useful for pedagogic purposes to illustrate the basic ideas of QDT in simpler contexts. This is how we will begin, with Sections 1.1-1.3 of this paper treating examples in which there is no long-range potential.

When one considers the eigenstates of a Hamiltonian, particularly a Hamiltonian which contains both bound discrete levels (the realm of spectroscopy) and continuum levels (the realm of collisions), it is an elementary truism that all of these levels are members of a single family, namely, the spectrum of the Hamiltonian. It should be expected, therefore, that physical phenomena over different ranges of the spectrum can be interrelated and that, often, knowledge of some characteristics in the discrete region can be extrapolated to gain information about the continuum region (and *vice versa*). In particular, states around zero energy -- whether they are just below, in the discrete, or just above, in the continuum -- should be intimately connected. The study of scattering lengths and effective-range theory [3], in which low-energy scattering data are linked to parameters of shallow bound states, presents a canonical illustration. Let us

S. P. McGlynn et al. (eds.), Photophysics and Photochemistry in the Vacuum Ultraviolet, 191–215.

start even more simply, however, with a problem in only one dimension. Thus, consider the one-dimensional potential well shown in Fig. 1. Continuum states of even parity are described by

$$\begin{aligned} \psi(x) &= N_E \cos(k_o x - \delta), & x &\le -a \\ &= A \cos kx, & |x| &< a \\ &= N_E \cos(k_o x + \delta), & x &\ge a \end{aligned} \tag{1}$$

where k_o and k are, respectively, the wave-vectors outside and inside the well, with $k_o^2 = 2\mu E/\hbar^2$ and $k^2 = 2\mu(E+V_o)/\hbar^2$; N_E is an overall normalization constant (defined for normalization per unit energy, say); and A and δ are parameters determined by enforcing boundary conditions of continuity at $x = a$. Thus, δ (the *phase shift*) is determined by

$$k \tan ka = k_o \tan(k_o a + \delta). \tag{2}$$

For the discrete part of the spectrum (i.e., for $E<0$ and $k_o = i\kappa_o$), the solutions in Eq. (1) remain valid as even parity functions except that, on writing the asymptotic form

$$\psi(x) \sim \frac{1}{2} N_E \left[e^{i\delta} e^{ik_o x} + e^{-i\delta} e^{-ik_o x} \right], \quad x \sim \infty \tag{3}$$

the boundary conditions require that the coefficient of the rising exponential vanish:

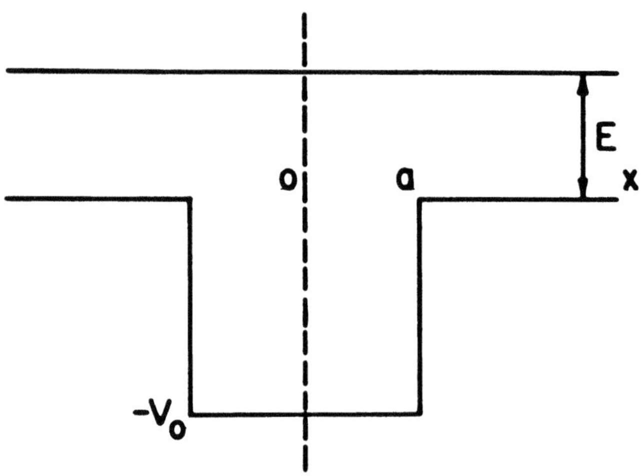

Fig. 1. A one-dimensional square well.

$$e^{-i\delta} = 0. \tag{4}$$

This requirement (written alternatively as $\tan\delta = -i$) gives, when combined with Eq. (2), the eigenvalue condition

$$k_B \tan k_B a = \kappa_{oB} \tag{5}$$

where $\kappa_{oB}^2 = 2\mu E_B/\hbar^2$ and where $E_B(>0)$ is the binding energy of the state.

This very elementary account exemplifies several features relevant to our later discussion. *First*, note that the form of the wavefunction inside the well is common to both discrete and continuum parts of the spectrum; in particular, the wavefunctions in these two spectral regions are identical when $E \simeq 0$, regardless of the sign of E. This is simply a realization of the fact that the wave-vector inside the well is determined primarily by the well depth when E is small. Also, this means that the structure of the wavefunction is insensitive, in general, to small changes in energy in a domain where the wavefunction is controlled primarily by the strong forces of an attractive potential. In atomic and molecular systems, the "small-distance" regime is a region of strong forces (due to the nucleus and other electrons) and can cause changes in atomic properties on a coarse energy scale only. In contrast, the region of weak forces, at larger distances (e.g., the region of zero potential in the above model problem), can lead to sensitive dependences on the energy, particularly near thresholds separating discrete and continuum states. It is in this latter region that the wavelength is determined primarily by the asymptotic kinetic energy of the particle. Thus, as illustrated by this initial example, it is useful to separate inner regions (where strong forces prevail) from outer regions (where weaker forces prevail) in following the variation with energy (and, therefore, the variation with respect to spectral position) of atomic properties. *Second*, our simple example demonstrates clearly that bound states are described, in general, by a condition such as Eq. (4), a condition motivated by the requirement that the asymptotic form of the wavefunction [Eq. (3)] remain finite. This is again a feature that will play a repeated and major role in our later discussion. The form $e^{-i\delta}$, termed a *Jost function* [4], is the most suitable parameter for viewing the entire spectrum of a Hamiltonian. Considered as a function of the complex wave-vector, the zeroes of the Jost function on the positive imaginary axis ($k_o = i\kappa_o$) locate the positions of the bound states.

1.2 s Wave in a Spherically-Symmetric Three-Dimensional Problem

The step from the one-dimensional square well to a three-dimensional problem -- such as a spherical potential $V(r) = -V_o$ for $r<r_o$, $V(r) = 0$ for $r \geq r_o$ -- is straightforward: Namely, one works in a partial-wave formalism so that the nontrivial problem is purely in one variable, the radial variable r. Consider first the s wave. Working with $F(r) \equiv rR(r)$ instead of with the radial wavefunction $R(r)$ [so that $F(r)$ is normalized with respect to integration over dr], the radial Schrödinger equation takes precisely a one-dimensional form, except that the range of r is $(0,\infty)$ rather than $(-\infty,\infty)$. We can, therefore, immediately write down the analogue of Eq. (1), namely,

$$
\begin{aligned}
F(r) &= A \sin kr \qquad\qquad , \quad r < r_o \\
&= N_E \sin(k_o r + \delta), \quad r \geq r_o \quad .
\end{aligned}
\tag{6}
$$

The only (but significant) difference between Eqs. (1) and (6) is that, because F(0) must vanish, the admissible solution is now a sine function at small r (and, therefore, corresponds to what would be the odd parity analogue of Eq. (1) in the one-dimensional problem). Other features, however, remain unchanged. Thus, we now have

$$k \cot kr_o = k_o \cot(k_o r_o + \delta) \tag{7}$$

and the bound states are again determined by Eq. (5), so that

$$k_B \cot k_B r_o = -\kappa_{oB} \quad . \tag{8}$$

For $E \approx 0$, the radial wavefunction inside the well changes little in going through the threshold. Outside the well, in the discrete spectrum with $\kappa_{oB} \simeq 0$, it is clear that the wavefunction $F(r) = N_E e^{-\kappa_{oB} r}$ is a slowly-varying linear function of r when $r \gtrsim r_o$. This is also true for the continuum function of Eq. (6), with $k_o \gtrsim 0$, since it follows from Eq. (7) that δ must vanish proportionally to r_o. Thus,

$$\delta = -k_o a \quad . \tag{9}$$

The parameter a is called the *scattering length* [3] and takes its name from the fact that the low-energy cross-section can be written $\sigma = (4\pi/k_o^2) \sin^2 \delta \simeq \pi (2a)^2$. In terms of the radial wavefunction, Fig. 2 illustrates the physical significance of the scattering length: Namely, the near zero energy radial function is proportional to $(r - a)$ for $r \gtrsim r_o$. The left-hand and middle plots of Fig. 2 correspond to a potential that is attractive within r_o; the right-hand plot, however, corresponds to a potential that is repulsive within r_o. Only the middle plot

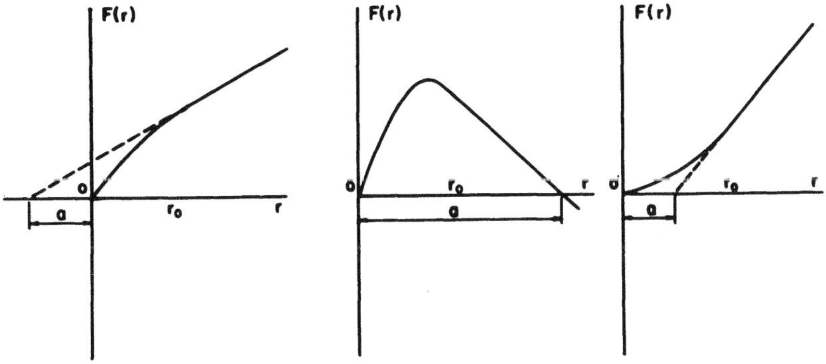

Fig. 2. Radial wavefunction with $\ell = 0$ and $E \simeq 0$ for a short-range potential. Note the straight line segments beyond the range of the potential $r > r_o$. The scattering length is denoted by a.

represents a situation in which the potential is capable of supporting a bound state near zero energy, since it is only in this case that the wavefunction at $r \gtrsim r_o$ can be matched to an exponential fall off $e^{-\kappa_{oB} r}$. In this latter case, when there is a shallow bound state, it follows that

$$a = 1/\kappa_{oB} \tag{10}$$

so that the scattering length is identical to what one refers to as the radius of the bound state. Note also the interconnections among Eqs. (5), (9) and (10): With δ small, Eq. (4) reduces to $(1 - i\delta) = 0$, or $\delta = -i$, as one indeed concludes from Eqs. (9) and (10). Finally, note that expanding Eq. (7), in the vicinity of $k_o \simeq 0$, to higher orders than that shown in Eq. (9) gives the effective-range expansion in powers of $k_o{}^2$. The scattering length is simply the first term in such an expansion [3].

1.3 A Two-Channel Generalization of the s-Wave Problem

As a next step in our development of an analysis applicable to atomic systems possessing one optical electron in various stages of excitation, consider photodetachment of a negative ion [5]. In particular, if the electron in question is initially in an $\ell = 1$ state, after detachment it leaves as an s or d wave because of dipole selection rules. At low energies, just above the threshold for photodetachment, it can be shown that the s wave is dominant [6]. And, the radial function of this s wave can be described in terms of a scattering length, as in the previous section. Let us consider the simplest system, namely, one which possesses a single bound negative-ion state. Such a system is exemplified by F^- or I^-, with the configuration and state labels $p^6\ {}^1S_0$. After detachment, the neutral atom is described by $p^5\ {}^2P_{1/2,3/2}$. The fine-structure splitting results in two detachment thresholds, with the neutral atom being left behind in one or the other of the two fine-structure levels. And, this fine-structure splitting gives rise to a two "channel" problem. (By "channel" we mean that the energy is unspecified, but that the remaining discrete quantum numbers of angular momentum and spin of the two system components, as well as any coupling scheme between these components, are specified.) In this first example of a multichannel system, all of the orbital quantum numbers are fixed (L = 1 for the atom, ℓ = 0 for the electron, J = 1 for the combined system) and the only difference between the two channels is in the angular momentum of the core (i.e., j = 1/2 or 3/2), a difference which arises from the coupling of orbital and spin angular momentum in the neutral atom.

The probability amplitude for photodetachment is given in terms of the matrix element

$$\langle p^5(^2P)jm_j,\ k_j s j_e m_e | P_q^{[1]} | p^6(^1S),\ J'=M'=0\rangle \tag{11}$$

where the ket is labeled with the quantum numbers of the negative ion; the electric-dipole operator for the absorbed photon is indexed with a polarization parameter q (q = ±1 for circularly-polarized light, q = 0 for linearly-polarized light); and the bra carries the quantum numbers of the dissociated state. The s-electron's wave-vector k_j indicates the energy ($\frac{1}{2}k_j^2$) of the outgoing electron, which is also, of course, the difference between the photon energy and that detachment energy which leaves the neutral atom in the state j. (*Note*: We use atomic units from here on out.) If the continuum wavefunction is normalized per unit energy, the absolute square of Eq. (11) gives the cross-section for photodetachment. Note, finally, that $m_j + m_e = q$ in our example.

The above problem also illustrates a general feature of many atomic problems, namely, that it may be convenient to work with alternative channels depending upon the r range of interest. Thus, the channels described above may be referred to as "fragmentation channels" because the quantum number j is relevant to distinguishing the alternative situations in the large-r dissociation limit, where the electron or residual atom is detected. This is a region of jj-coupling between the j of the neutral atom and j_e of the electron. On the other hand, the initial negative-ion wavefunction is concentrated at small r and is usually well described by LS-coupling. It is appropriate, therefore, to introduce another set of channels for the e + atom complex at small r, where $r \gtrsim r_0$ and where r_0 is the size of the negative ion. In this range, all of the electrons (including the one that finally escapes) are close together and LS-coupling is more appropriate for the complex, as well as for the initial negative-ion state. Two LS-coupled channels descriptive of an s-electron + ^2P-atom complex are given by $^{1,3}P$ (i.e., the channels are characterized by singlet and triplet labels); and, we call these "eigenchannels" because they are eigenstates for the absorption of the photon. The picture we have, then, is that one introduces a complete set (two, in this case) of LS-coupled states for the e + atom complex into Eq. (11) by writing

$$\sum_S \langle p^5(^2P)jm_j,\ k_j s j_e m_e | p^5(^2P)s(^{2S+1}P),\ J=1,\ M=q\rangle$$

$$\times \langle p^5(^2P)s(^{2S+1}P),\ J=1,\ M=q|P_q^{[1]}|p^6(^1S_0),\ J'=M'=0\rangle\ . \quad (12)$$

In our example, since J'= M'= 0 for the initial negative-ion state, the total quantum numbers J and M of the e + atom complex are uniquely fixed in terms of the quantum numbers of the photon. In a more general case, however, alternative values for these quantities may be allowed and Eq. (12) would then include a summation over J and M. To make the subsequent analysis less restrictive, we will refrain from specifying J and M in what follows and will equate them to 1 and q, respectively, only at the end.

The two factors in Eq. (12) have the following ready interpretation for the detachment process. The right-hand factor expresses the absorption of the photon in LS-coupling, giving rise to an e + atom complex at small r. The left-hand factor then expresses the dissociation of this complex from $r \gtrsim r_0$ to the fragmentation limit at $r \to \infty$. This dissociation involves a so-called "frame transformation" [2,7] which, in our example, is just a change from one coupling scheme (LS) to another (jj). The photon-absorption factor is given by conventional spectroscopic procedures; for our purpose, we represent this factor by d and note that because of electric-dipole selection rules the initial "feed" into the small-r e + atom complex is into S = 0 only. Nevertheless, both S = 0,1 are involved in a complete description of the complex in the small-r region, and in the later development of this complex into the j = 1/2,3/2 states. The left-hand factor in Eq. (12) describes this development; this factor can be viewed as half of a scattering since this part of the photoabsorption (i.e., separation of an electron from the atom) is just like the second half of an elastic e + atom scattering event. In order to study the development of the complex in detail, we begin with the ket which represents the e + atom complex at small r and in LS-coupling. In the small-r range, according to Section 1.2, the radial wavefunction is proportional to (r - a), where a is the scattering length for low-energy scattering of electrons from the atom. In the present case, however, we need two scattering lengths, depending upon whether S = 0 or 1, and we introduce a_S (S = 0,1) as the two parameters which describe the wavefunction. As a result, we find

$$|p^5(^2P)s(^{2S+1}P),\ JM\rangle = X^S_{JM}\ (r - a_S) \tag{13}$$

where we have represented the angular and spin parts of the wavefunction by X^S_{JM}. Note that, as in Section 1.2, the entire wavefunction is independent of energy in this region.

The fragmentation states at large r, on the other hand, differ from Eq. (13) in both the angular and the radial wavefunctions. In the large-r region, we find

$$|p^5(^2P)jm_j,\ k_j s j_e m_e,\ JM\rangle$$

$$= \phi^j_{JM}\ (2/\pi k_j)^{\frac{1}{2}}\ [c_j \sin k_j r - d_j \cos k_j r] \tag{14}$$

where the angular and spin parts of the wavefunction are now represented by ϕ^j_{JM}, and where the radial wavefunction is an energy-normalized $\ell = 0$ function. The full wavefunction at small r is a superposition of the basis functions from Eq. (13):

$$\psi = \sum_S X^S_{JM}\ (r - a_S)\ A^{(J)}_{S\mathcal{j}} \tag{15}$$

where $\mathcal{j}$ denotes that particular channel of interest in which the measurement is made. The superposition coefficients $A^{(J)}_{S\mathcal{j}}$ are determined by extrapolating Eq. (15) into the large-r region; expressing it in terms of the fragmentation channel functions [Eq. (14)]; and then applying boundary conditions. To express Eq. (15) appropriately, we first expand X^S_{JM} in terms of the ϕ^j_{JM}; this is a standard LS → jj recoupling transformation [8]:

$$X^S_{JM} = \sum_j \phi^j_{JM}\ U^{(J)}_{jS}\ . \tag{16}$$

The U is an orthogonal 2 x 2 matrix

$$U^{(1)}_{jS} = \begin{array}{c|cc|} & S = 1/2 & 3/2 \\ j = 1/2 & \sqrt{2/3} & \sqrt{1/3} \\ & & \\ 3/2 & -\sqrt{1/3} & \sqrt{2/3} \end{array} \qquad (17)$$

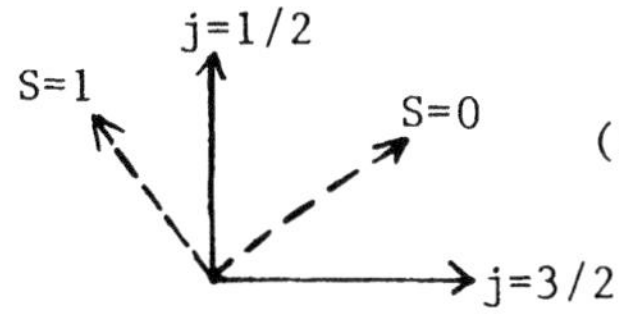

and has the geometrical interpretation of a rotation through 35°, as shown above. From Eqs. (15) and (16), therefore, we find

$$\psi = \sum_j \phi^j_{JM} \left[r \sum_S U^{(J)}_{jS} A^{(J)}_{S\mathcal{j}} - \sum_S U^{(J)}_{jS} a_S A^{(J)}_{S\mathcal{j}} \right] \qquad (18)$$

which extrapolates to the large-r form

$$\psi = \sum_j \phi^j_{JM} k_j^{-\frac{1}{2}} \left[\sin k_j r \sum_S k_j^{-\frac{1}{2}} U^{(J)}_{jS} A^{(J)}_{S\mathcal{j}} \right.$$

$$\left. - \cos k_j r \sum_S k_j^{\frac{1}{2}} U^{(J)}_{jS} a_S A^{(J)}_{S\mathcal{j}} \right] \qquad (19)$$

thus effectively fixing the coefficients c_j and d_j of Eq. (14) in terms of $A^{(J)}_{S\mathcal{j}}$ and other parameters.

To apply the boundary conditions, we re-write Eq. (19) as

$$\psi = \sum_j \phi^j_{JM} (2/\pi k_j)^{\frac{1}{2}} (1/2i) \left[e^{ik_j r} k_j^{-\frac{1}{2}} \sum_S B^{(J)}_{jS} A^{(J)}_{S\mathcal{j}} \right.$$

$$\left. - e^{-ik_j r} k_j^{-\frac{1}{2}} \sum_S B^{(J)*}_{jS} A^{(J)}_{S\mathcal{j}} \right] \qquad (20)$$

where

$$B^{(J)}_{jS} = U^{(J)}_{jS} (1 - ik_j a_S) (\pi/2)^{\frac{1}{2}} . \qquad (21)$$

The photodetachment boundary condition is the so-called "ingoing-wave" boundary condition [9] in which, of the two j values, only the $\mathcal{j}$ of interest should contribute to the $e^{ik_j r}$ term in Eq. (20); and, this term should have unit coefficient. Thus,

$$\sum_S B^{(J)}_{jS} A^{(J)}_{S\mathcal{j}} = k_j^{\frac{1}{2}} \delta_{j\mathcal{j}} \qquad (22a)$$

or, alternatively,

$$A^{(J)}_{S\mathcal{j}} = \left(B^{(J)^{-1}}\right)_{S\mathcal{j}} k^{\frac{1}{2}}_{\mathcal{j}} \quad . \tag{22b}$$

Note that the ingoing-wave boundary condition for a channel $\mathcal{j}$ which is open (i.e., a channel for which the kinetic energy $\frac{1}{2}k_{\mathcal{j}}^2$ is positive so that the electron can reach infinity) means that there are only ingoing waves in the other channel (when this channel is closed, the choice $k = -i\kappa$ means that it has only an exponentially-decreasing wavefunction). Thus, it is only in the channel of interest that a plane-wave with wave-vector $k_{\mathcal{j}}$ can be formed to represent the electron ejected into that state. Also note that with Eq. (22b) inserted into Eq. (20), we have the form

$$\psi \propto \left[- e^{-ik_{\mathcal{j}}r} \sum_{S} B^{(J)*}_{\mathcal{j}S} \left(B^{(J)^{-1}}\right)_{S\mathcal{j}} + e^{ik_{\mathcal{j}}r}\right] . \tag{23}$$

The coefficients B are the generalization of $e^{i\delta}$ from Section 1.1. Thus, B is the Jost matrix of this problem.

The cross-section for the $\mathcal{j}$ channel can now be written explicitly. With the initial "feed" into S = 0 only and with J taking only one value, the expression is particularly simple [cf. Eq. (12)]:

$$|\langle jm_{\mathcal{j}} \; 1/2 \; m_e | 1q\rangle \; A^{(1)}_{0\mathcal{j}} \; d|^2 \quad . \tag{24}$$

In a more general context, with multiple J and M values contributing to any particular $\mathcal{j}$ channel, Eq. (24) would include a coherent superposition of these various amplitudes before squaring. Note that the cross-section of Eq. (24) carries information on the spin polarization of the electron and the atom. For example, if q = 1 (circular polarization) and $\mathcal{j} = 1/2$, the only allowed values are $m_{\mathcal{j}} = m_e = +1/2$ (i.e., there is 100% polarization). If no spin measurements are made, however, we must sum over the m quantum numbers; this leaves behind the factor $d^2|A^{(1)}_{0\mathcal{j}}|^2$ in Eq. (24).

For convenience, we collect the results for the spin-averaged cross-sections below.

Just above the first threshold, j = 3/2:

Here, the j = 3/2 channel is open while the j = 1/2 channel is closed ($k_{1/2} = -i\kappa_{1/2}$). The B matrix is

$$(\pi/2)^{\frac{1}{2}} \begin{vmatrix} \sqrt{2/3}\ (1 - \kappa_{1/2}a_1) & \sqrt{1/3}\ (1 - \kappa_{1/2}a_0) \\ -\sqrt{1/3}\ (1 - ik_{3/2}a_1) & \sqrt{2/3}\ (1 - ik_{3/2}a_0) \end{vmatrix}$$

On evaluating $A_{03/2}$ from Eq. (22b), we find

$$\sigma_{3/2} = (2/\pi)d^2\ (2/3)\ (1 - \kappa_{1/2}a_1)^2\ k_{3/2}/\text{denom} \qquad (25)$$

where

$$\text{denom} \equiv [1-(1/3)\kappa_{1/2}(a_0+2a_1)]^2 + k_{3/2}^2[(1/3)(2a_0+a_1) - \kappa_{1/2}a_0a_1]^2 .$$

Above the second threshold, j = 1/2:

In this energy region both channels are open and B is given by

$$(\pi/2)^{\frac{1}{2}} \begin{vmatrix} \sqrt{2/3}\ (1 - ik_{1/2}a_1) & \sqrt{1/3}\ (1 - ik_{1/2}a_0) \\ -\sqrt{1/3}\ (1 - ik_{3/2}a_1) & \sqrt{2/3}\ (1 - ik_{3/2}a_0) \end{vmatrix}$$

The cross-sections are:

$$\sigma_{3/2} = (2/\pi)d^2\ (2/3)\ (1 + k_{1/2}^2a_1^2)\ k_{3/2}/\text{denom}' \qquad (26a)$$

$$\sigma_{1/2} = (2/\pi)d^2\ (1/3)\ (1 + k_{3/2}^2a_1^2)\ k_{1/2}/\text{denom}' \qquad (26b)$$

where

$$\text{denom}' \equiv [1-k_{1/2}k_{3/2}a_0a_1]^2 + (1/9)[k_{1/2}(2a_1+a_0) + k_{3/2}(2a_0+a_1)]^2.$$

Note that all of the above cross-sections involve both a_0 and a_1 even though the photoabsorption initially creates only the singlet state. The following physical picture accounts for this: In the evolution of the e + atom complex into a j of interest, there is a contribution to the wavefunction which results from reflection back into small r from the other j and then rescattering out into j. Thus, the reflected piece, when expressed at small r, involves both singlet and triplet components.

The above discussion constitutes a complete treatment of the photodetachment problem and, moreover, exemplifies those features of QDT which I wish to emphasize in this paper. In addition to the energy (which enters parametrically through the wave-vectors), the cross-sections are completely described by a few remaining parameters (four, in this case). First, there is d (the dipole absorption strength), which is common to all of the cross-section expressions. Next, there are the coefficients of the U matrix [Eq. (17)], which constitute only one parameter, namely, the angle of rotation from LS- to jj-coupling. (The frame transformation involved in describing alternative channels appropriate to different ranges of r typically introduces such purely geometrical factors.) Finally, the full dynamics of the short-range interaction -- a dynamics which enters through the radial functions -- is described in terms of a few parameters (two, in this case; one scattering length for each channel). Thus, the entire spectral information, over all energies, is described in terms of a very small set of parameters.

In our negative-ion example, there is only a continuum of states. However, even in problems with long-range forces (e.g., the Coulomb potential instead of the zero long-range potential we have considered), when the spectrum is much richer and possesses a large number of bound states below the lowest threshold as well as a large number of autoionizing levels in between the thresholds for the two j, all features of the spectrum can still be encapsulated in terms of a few parameters which are only weakly dependent on energy. For a long-range potential problem, these spectral features will include the positions of the bound states and the oscillator strengths for reaching these states, as well as the positions, intensities and profiles of the autoionization structure (see Section 2, below, and the papers by K. T. Lu and C. H. Greene in this volume).

1.4 Single-Channel QDT for a Coulomb Potential

We turn now to the problem of primary interest in atomic systems which possess one optical electron moving in the field of a positive ion: Namely, the Coulomb potential.

In the case of the hydrogen atom, the potential is $V(r) = -1/r$, for all r, and one finds the discrete eigenvalues $E_n = -1/2n^2$ and a Coulomb continuum at $E \geq 0$. In any other atom, however, an excited electron sees a more complicated potential, but one with the common feature that the potential at large r is once again hydrogenic:

$$\begin{aligned} V(r) &= v_s(r) - \frac{1}{r}\ , \quad r < r_o \\ &= -\frac{1}{r}\ , \quad\quad\quad r > r_o \quad . \end{aligned} \tag{27}$$

At small r, when the electron feels the influence of all of the other electrons in the system, the potential is complicated, stronger than $-1/r$ (in fact, as $r \to 0$, it approaches $-Z/r$, where Z is the charge on the nucleus) and, in general, neither spherically-symmetric nor local (because of the exchange interaction with the other electrons). In Eq. (27), therefore, $v_s(r)$ is merely a schemmatic representation of the short-range interactions. Moreover, the boundary r_o is not as sharply defined as it was in the previous sections; however, r_o is generally $\sim$ 5-10 au and is characterized by the fact that, beyond r_o, the wavefunctions of all of the other electrons fall off exponentially and exchange interactions are generally negligible.

Highly-excited bound states, which reside mainly in the region $r > r_o$ and which are held by the hydrogenic coulomb tail, are substantially the same as those of hydrogen, except that the penetration into the core region, with its stronger attractive forces, leads to an extra binding. This fact is the basis of the Rydberg formula:

$$\varepsilon_n = - 1/2(n - \mu)^2 \tag{28}$$

where μ is the so-called quantum defect. The quantum defect varies slowly with n and approaches a constant limiting value as n increases. Moreover, since μ depends on the amount of core penetration, it is strongly ℓ-dependent and usually negligible for $\ell > 2$.

Fig. 3 illustrates how the radial wavefunction for an electron with quantum defect μ can be viewed as being pulled in to smaller r, relative to hydrogen, as a result of the stronger attraction in the core. Thus, $\pi\mu$ represents a phase shift. As shown in the lower part of Fig. 3, at $r \gtrsim r_o$ the wavefunction can be written as

$$F(\nu,r) = f(\varepsilon,r) \cos \pi\mu - g(\varepsilon,r) \sin \pi\mu \tag{29}$$

where $\nu = (-2\varepsilon)^{-\frac{1}{2}} = 1/\kappa$ and where f and g are the regular and irregular Coulomb functions, respectively (these are the analogues of the sine and cosine functions used in Eq. (14) when there was zero potential in this region of r). The Coulomb functions are assumed normalized per unit energy. And, each of these functions is a superposition of an exponentially-rising and an exponentially-falling function at infinity (exactly as are the sine and cosine functions of Eq. (14) in the $\varepsilon < 0$ range, where sine and cosine become hyperbolic functions). In terms of the basis functions $\{f^+,f^-\}$ defined by

$$f^{\pm} \sim e^{\mp r/\nu} \, e^{\pm \nu \ell n \, r} \, , \quad r \to \infty \tag{30}$$

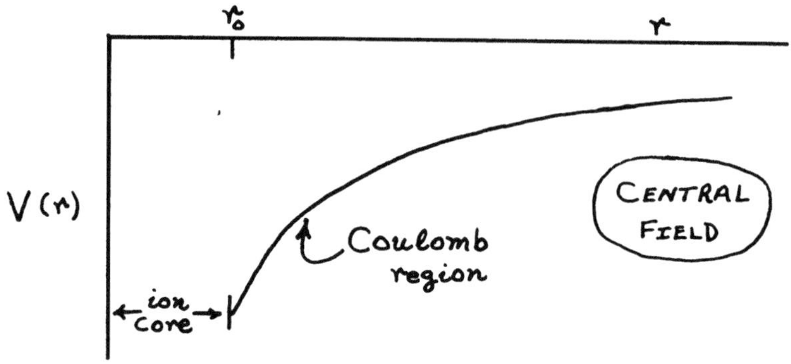

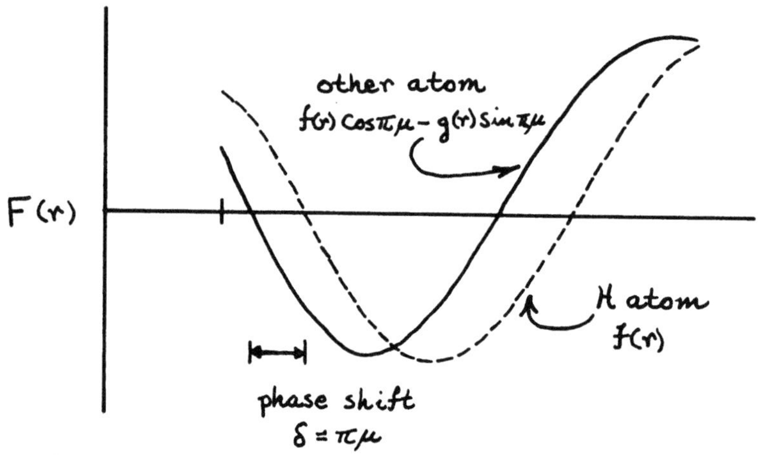

Fig. 3. Radial wavefunction of an excited electron in the outer Coulomb region of an atom. Adapted from U. Fano, J. Opt. Soc. Am. 65, 979 (1975).

the pair {f,g} can be expressed as

$$f(\varepsilon,r) = (\nu/\pi)^{\frac{1}{2}} \left[\sin\beta\, D^{-1} f^{-} - \cos\beta\, D f^{+}\right] \tag{31a}$$

$$\varepsilon < 0$$

$$g(\varepsilon,r) = -(\nu/\pi)^{\frac{1}{2}} \left[\cos\beta\, D^{-1} f^{-} + \sin\beta\, D f^{+}\right] \tag{31b}$$

where β and D are parameters that depend upon the energy and angular momentum [1,2,10]. In particular, for the Coulomb field

$$\beta = \pi(\nu-\ell). \tag{32}$$

These functions are somewhat complicated and we pause to note a few of their features. The logarithmic phase in Eq. (30) is characteristic of the long-range Coulomb field and is inconsequential. D is a so-called rescaling parameter which makes the exponentially-rising and exponentially-falling terms more compatible in magnitude; it is a monotonic function of the energy and depends upon the tunneling integral over the classically forbidden region, at large r, for $\varepsilon < 0$. All of these QDT parameters have similar ready interpretations in terms of phase integrals in a WKB framework [10].

In the absence of a core, when μ in Eq. (29) is zero, the solution is simply the regular hydrogenic function f since, in this case, the solution is valid at all r, including the origin, rather than just at $r \gtrsim r_0$. In addition, at large r we find from Eqs. (30) and (31a) that the bound state eigenvalues are given by the vanishing of the coefficient of f^-:

$$\sin \beta = 0. \tag{33}$$

With β as in Eq. (32), this gives

$$\nu-\ell = n_r + 1 \ , \quad n_r = 0,1,... \tag{34a}$$

or, since $\nu = (-2\varepsilon)^{-\frac{1}{2}}$,

$$\varepsilon = - 1/2(n_r + \ell + 1)^2 \tag{34b}$$

which are the usual hydrogenic energy levels. In a nonhydrogenic system, however, Eq. (29) represents the solution for $r \gtrsim r_0$, with μ determined by matching the interior ($r < r_0$) solution, which is regular at the origin, to Eq. (29) through logarithmic derivatives. Eq. (29), when extrapolated to large r, gives the following form for the vanishing of the coefficient of f^-:

$$\sin \pi(\nu+\mu) = 0. \tag{35}$$

Thus, the bound state positions are now given by Eq. (28).

The above discussion points out the significance of $\pi\mu$ as the phase difference between $F(\nu,r)$ and $f(\varepsilon,r)$ induced by the presence of a core region. μ can be determined, therefore, either by semi-empirically fitting a few bound-state eigenvalues to Eq. (28) or, *ab initio*, by starting with a regular solution at the origin, integrating over the core potential in Eq. (27) by some suitable procedure, and then matching logarithmic derivatives to the right-hand side of Eq. (29).

The matching of logarithmic derivatives of the solutions in the inner and outer regions implies a certain deficiency in the use of the functions {f,g} in Eq. (29). This is because, as emphasized in previous sections, the inner region is dominated by the strong attractive potential and, therefore, one would expect solutions developed in the inner region to be substantially independent of energy. Thus, for instance, a regular solution could be defined by a boundary condition at the origin:

$$f^o \to r^{\ell+1} \quad , \; r \to 0 \; . \tag{36}$$

And, this boundary condition can be taken to be energy independent since the angular momentum potential totally overwhelms the asymptotic kinetic energy $|\varepsilon|$ in this small-r domain. Such a regular function, indicated by a superscript "o", differs from f by an energy-dependent normalization constant. With Eq. (36) as the starting form, and in the presence of $v_s(r)$ [we assume $v_s(r)$ is less dominant at the origin than is the angular potential], integration from r = 0 to r_o would still give an $F(\nu,r)$ at r_o which is only weakly dependent upon energy. On the other hand, however, we must match to the energy-normalized functions {f,g} at r_o; and, since these functions are defined with respect to their large-r behavior (i.e., defined in a region of weak potential so that ε is now important), strong energy dependences are introduced, particularly near $\varepsilon = 0$. Such a procedure makes μ an unsuitable parameter for extrapolations in energy and, in particular, for going from the discrete to the continuum spectrum. Section 1.3 gives a specific illustration of this: The small-r solution in Eq. (15) possesses no energy dependences; however, when this solution was expressed in terms of the large-r sines and cosines in Eq. (19), it acquired coefficients proportional to $k^{\pm\frac{1}{2}}$, which are strong, nonanalytic threshold singularities. Therefore, for the extrapolation in energy that is central to QDT, an alternative procedure is called for. (Note that in Section 1.3 we skirted this question by dealing with negative ions which, of course, do not have an entire set of discrete states. Thus, extrapolation between discrete and continuum states was not very relevant.)

For the purposes of energy extrapolation, QDT employs an alternative pair of Coulomb functions $\{f^o,g^o\}$ [10]. The regular function f^o is defined through the boundary condition of Eq. (36). As stated above, this function can differ from f only by an energy-dependent normalization factor. Thus,

$$f^o(\varepsilon,r) = B^{-\frac{1}{2}}(\varepsilon) \; f(\varepsilon,r) . \tag{37}$$

Since f^o is smooth in energy, whereas f is not, all of the nonsmooth, nonanalytic dependences are contained in the parameter B;

in particular, B does not extrapolate smoothly through $\varepsilon = 0$. The smooth irregular function g^o, however, cannot be defined as easily as f^o through the corresponding energy-independent boundary condition at the origin:

$$g^o \to -\frac{1}{\pi(\ell + \frac{1}{2})} r^{-\ell} . \tag{38}$$

[The factor in front is added to ensure that the Wronskian $W(f^o,g^o)$ is equal to $W(f,g)$, which is $2/\pi$]. A solution to a differential equation cannot be specified uniquely by such a boundary condition at a point of singularity; therefore, g^o (unlike f^o) cannot be independently defined but must be constructed from the nonanalytic function g. The construction of a well-defined g^o for QDT applications is algebraically involved [10] but straightforward. First of all, it is clear from Eq. (37), and from the equality of the Wronskians mentioned above, that we must have the form

$$g^o(\varepsilon,r) = B^{\frac{1}{2}}(\varepsilon)\, g(\varepsilon,r) - G(\varepsilon)\, f^o(\varepsilon,r) . \tag{39}$$

The appearance of the second term in Eq. (39), in contrast to Eq. (37), is because the first term by itself is not smooth and analytic; thus, a further subtraction is necessary before one obtains a smooth irregular function. $G(\varepsilon)$, therefore, is another of the nonanalytic parameters of QDT. (Note, of course, that addition of a piece proportional to f^o disturbs neither the Wronskian nor the behavior of g^o near the origin in Eq. (38).) Using Eqs. (37) and (39) in Eq. (29), we find

$$F(\nu,r) = N(\varepsilon)\left[f^o(\varepsilon,r)\cos\delta^{(o)} - g^o(\varepsilon,r)\sin\delta^{(o)}\right] \tag{40}$$

where $N(\varepsilon)$ and $\delta^{(o)}$ are defined in terms of μ and the QDT parameters by

$$\begin{aligned} N(\varepsilon)\cos\delta^{(o)} &= B^{\frac{1}{2}}(\varepsilon)\cos\pi\mu - B^{-\frac{1}{2}}(\varepsilon)\, G(\varepsilon)\sin\pi\mu \\ N(\varepsilon)\sin\delta^{(o)} &= B^{-\frac{1}{2}}(\varepsilon)\sin\pi\mu \quad . \end{aligned} \tag{41}$$

Thus, in QDT applications Eq. (40) is a more useful form than Eq. (29). The parameter $\delta^{(o)}$, defined by equating logarithmic derivatives of an interior solution, smooth in ε, with logarithmic derivatives of the smooth pair $\{f^o,g^o\}$, will likewise be smooth in energy; it is the candidate for energy extrapolations and, in particular, for going from $\varepsilon < 0$ to $\varepsilon > 0$. Moreover, $\delta^{(o)}$ is defined entirely in terms of the short-range interactions and is insensitive to ε. In the continuum ($\varepsilon > 0$), once $\delta^{(o)}$ is obtained by extrapolation Eq. (41) can be inverted to yield the new counterpart of $\pi\mu$, namely, the scattering phase shift due to the short-range interaction $v_s(r)$. (This must be added to the Coulomb phase shift, already contained in {f,g}, to give the total phase shift.) Finally, for completeness we record the long-range form of the {f,g} functions when $\varepsilon = \frac{1}{2}k^2 > 0$; these are the analogues of Eq. (31) and are expressed as

$$
\begin{aligned}
f(\varepsilon,r) &\sim \quad (2/\pi k)^{\frac{1}{2}} \sin[kr + \frac{1}{k}\,\ln r + \eta(k)] \\
& \qquad\qquad\qquad\qquad\qquad\qquad \varepsilon > 0 \qquad (42) \\
g(\varepsilon,r) &\sim - (2/\pi k)^{\frac{1}{2}} \cos[kr + \frac{1}{k}\,\ln r + \eta(k)] \;.
\end{aligned}
$$

The above analysis of QDT separates clearly the contributions from short-range and long-range interactions. In particular, we see that strong energy dependences of physical quantities such as the phase shift in the continuum or the positions of bound states, come entirely from the long-range field. And, since this is a standard field common to all atomic systems, these dependences can be characterized once and for all by a few parameters, namely $B(\varepsilon)$, η, β, D and G. Once this characterization is performed, any individual system of interest can be handled (most often numerically or semi-empirically) by focusing only on the region $r \lesssim r_o$. To illustrate the energy insensitivity of radial functions in the region $r \lesssim r_o$, we point to Table 1, drawn from Starace [11], which records characteristics of $F(\nu,r)/N(\varepsilon)$ for various s states of the uranium atom. Notice how remarkably constant the entries are in spite of a change in ε by four orders of magnitude. Had this table included continuum s orbitals as well, the same constancy would have been seen to stretch across the threshold.

1.5 Other Long-Range Potentials

In principle, QDT analysis, presented in detail above for the Coulomb field, need not be restricted to the specific nature of the long-range field. In fact, Sections 1.2 and 1.3 can be seen as a similar treatment of the particularly simple case that arises when $\ell = 0$ and when there is no long-range field at all. (The values of the various parameters in this case are D = 1, η = 0, $\beta = \pi/4$, G = 0 and B = k or κ.) Extensions have also been made to other long-range fields: Canonical examples are an attractive ("dipole") or repulsive $1/r^2$ field, or a polarization potential $-1/r^4$. In all of these cases, the same philosophy -- namely, that it is useful to separate inner and outer regions, and to distinguish the different influence these regions exert on atomic properties -- guides the development of a QDT analysis. Such treatments now exist for many potentials and the pairs of functions {f,g} and $\{f^o,g^o\}$, together with the expressions for

Table 1. Behavior of Uranium ns Radial Wavefunctions $P_{ns}(r)/r$ for Small Radial Distances r.*

n	Energy (Ry)	r_{max1}	r_{node1}	r_{min1}	r_{node2}	$\frac{P(r_{min1})}{P(r_{max1})}$
1s	-7409.2	0.01097	---	---	---	---
2s	-1289.2	0.00844	0.02199	0.05922	---	-1.9589
3s	-326.64	0.00823	0.02110	0.04771	0.08230	-1.6643
4s	-83.16	0.00817	0.02089	0.04620	0.07812	-1.6223
5s	-18.49	0.00816	0.02083	0.04583	0.07714	-1.6128
6s	-2.96	0.00815	0.02082	0.04572	0.07691	-1.6095
7s	-0.35	0.00815	0.02082	0.04581	0.07686	-1.6093

*Positions (in Bohr radii) are of the first maximum (r_{max1}), the first minimum (r_{min1}) and the first two nodes ($r_{node1,2}$) of $P_{ns}(r)$. Adapted from [11].

the QDT parameters, have been catalogued [10]. Some of these problems (e.g., the dipole potential) also possess a rich spectrum of bound states; and, the interrelations provided by QDT in these cases are particularly useful and interesting. Further generalizations, particularly those appropriate to the situation when the sub-systems of the complex are not restricted to an electron and a residual atom or molecule but, rather, involve two more general fragments, are also subject to the same treatment. Thus, molecular dissociation involving a spectrum which consists of a set of discrete vibrational states and a dissociation continuum, all referred to the long-range R potential between the two partners, has been considered in a QDT analysis [12]. (QDT applied to molecular dissociation is discussed at length in the paper by C. H. Greene in this volume.)

2. MULTICHANNEL QUANTUM DEFECT THEORY IN ATOMS

2.1 Introduction

The treatment given in Section 1.4, involving a single ionization threshold and a single channel, is applicable to an alkali atom. In other atoms, however, there are multiple ionization thresholds, typically split by spin-orbit interactions (just as in the example of Section 1.3), and this leads to further complications. For example, when there are alternative channels it is not even clear how to assign a particular state to a given channel or ionization limit. We passed over such complications in Section 1.3 because, even though there were two ionization thresholds ($j = 1/2$ and $3/2$), we were concerned only with continuum states. In such a case, since when the electron reaches infinity either it or the residual atom can be monitored, applications of energy conservation to the known photon energy is sufficient to fix the wave-vector as either $k_{1/2}$ or $k_{3/2}$. When there are discrete states, however, this constraint no longer holds and it is unclear how to assign a quantum defect, with respect to either I_1 ($j = 3/2$) or I_2($j = 1/2$), to any particular level in the bound state region. Moreover, there is the interesting possibility that levels will lie between I_1 and I_2 and, therefore, will involve a superposition of discrete and continuum properties. The multichannel generalization of Section 1.4 turns out, however, to be a powerful technique for handling this entire set of phenomena -- for example, positions of the discrete levels and the interaction between the channels; the positions, widths and shapes of the autoionization levels; and the photoelectron distribution between the two channels in the continuum -- all in terms of a few basic QDT parameters. In Section 2.2 we develop such a QDT description; this description will be illustrated by atomic and molecular examples in the papers by K. T. Lu and C. H. Greene later in this volume. In particular, K. T. Lu will focus on a semi-empirical implementation of QDT, one which involves a graphical representation of experimental data to extract the significant parameters [13]. These parameters can then serve as input for extrapolation to other spectral ranges which might otherwise be unavailable. In addition to this practical aspect, the importance of multichannel QDT (MQDT) is vested in its condensation of a large number and variety of data into those few significant parameters which should be the target for other physical studies.

2.2 Photoabsorption by the Rare Gases

For concreteness, we consider the same system as in Section 1.3, but now in the context of an atom rather than a negative ion: That is, we consider a rare gas with a $p^6\ {}^1S_0$ configuration which absorbs a photon to reach one of the singly-excited or ionized states built on the p^5 core of the positive ion. Again, there are two alternative states of the ion, ${}^2P_{1/2,3/2}$. Further, an immediate difference from Section 1.3 is that when an electron moves in a Coulomb field there is now no longer any suppression of higher partial waves, even if the electron is slow (i.e., even if $|\varepsilon| \simeq 0$). In fact, a characteristic of the Coulomb problem is that right around $\varepsilon \simeq 0$, very high ℓ values can be attained either in the discrete (n and ℓ large) or in the continuum. (That photoionization near threshold is independent of ℓ, or that the Rutherford cross-section is infinite (which attests to contributions from arbitrarily large ℓ), are illustrations of this fact.) Thus, the excited electron can just as well be in the d wave as in the s wave, and the former is no longer suppressed. In the fragmentation limit, there are five possible J = 1 states exhibiting various alternative angular momentum quantum numbers for the electron and the ion; jj-coupling holds in the fragmentation limit and, therefore, these states can be enumerated as

$$p^5(^2P_{3/2})d_{5/2},\ p^5(^2P_{3/2})d_{3/2},\ p^5(^2P_{3/2})s_{1/2},$$
$$p^5(^2P_{1/2})d_{3/2},\ p^5(^2P_{1/2})s_{1/2}\ . \qquad (43)$$

Each of the above states represents a distinct channel; thus, we are faced with a five-channel problem, with three channels converging to $I_1(^2P_{3/2})$ and two converging to $I_2(^2P_{1/2})$. (In the context of negative-ion photodetachment, the d channels are absent because of the Wigner threshold law [6] and the problem reduces to two channels, each built separately on the two thresholds.)

The fragmentation channels given in Eq. (43), and labeled by quantities directly observable at infinity, are appropriate for the description at large r and are often called i-channels (i = 1,...,5). In the large-r region, the separation between I_1 and I_2 is dominant because the interaction of the electron with the ion is negligible. When we turn to the small-r region, however, the situation is drastically different: Since this region is one of strong attractive forces, the electron picks up a large kinetic energy which completely dwarfs the separation between thresholds (which is at most $\sim$ 1 eV). The "eigenchannels" relevant to this region are, therefore, different from the fragmentation channels and, as in Section 1.3, are more appropriately described by LS-coupling, particularly in the lighter rare gases. These channels are termed the α-channels (α = 1,...,5) and are described as

$$(p^5d)\,^1P,\ (p^5d)\,^3D,\ (p^5s)\,^3P,\ (p^5d)\,^3P,\ (p^5s)\,^1P\ . \qquad (44)$$

Any i-channel is a superposition of the five α-channels; and, the interaction between i-channels implied by this superposition means that various properties, such as phase shifts and cross-sections in the continuum or quantum defects in the discrete, are no longer simple monotonic functions of energy, as they were in the case of a single channel. (Section 1.3 contained a small illustration of this in Eqs. (25) and (26) for the cross-sections; the plots presented in the paper by K. T. Lu in this volume give more dramatic examples.)

The role of the short-range α-channels in effecting interactions between the various long-range i-channels lends itself to the following picture: Any asymptotic i-channel state can be described as being scattered by the short-range interaction into one of the asymptotic j-channels. Thus, states of the outer electron, whether in the discrete or in the continuum, can be viewed as having been formed by continuous scatterings off the core, thereby giving rise to a unified collision-theoretic picture of both collisions and spectra. A reaction matrix R_{ij} expresses this scattering from the core according to

$$R_{ij} = \sum_{\alpha} U_{i\alpha} \tan \pi\mu_\alpha \, U^{\dagger}_{\alpha j} \ . \tag{45}$$

$U_{j\alpha}$ represents the frame transformation from j to α, with scattering from the core inducing a phase shift $\pi\mu_\alpha$, while $U_{i\alpha}$ represents the frame transformation from α to i. In the present rare-gas application, the frame transformation $U_{j\alpha}$ is largely the LS to jj transformation, as in Section 1.3. However, $U_{j\alpha}$ may be regarded simply as an antisymmetric matrix that is expected to be slowly dependent upon energy, and its 10 parameters treated as open coefficients to be obtained by data fitting. In this way, $U_{j\alpha}$ can be interpreted as transforming from jj-coupling at large r, to whatever intermediate-coupling scheme that prevails at small r. The $\pi\mu_\alpha$ are the *eigenphase shifts*[eigenvalues of the R matrix in Eq. (45)]; they are the counterparts of the a_S of Section 1.3.

The description of the discrete levels in MQDT proceeds by assigning, to each level, alternative quantum numbers defined with respect to the various ionization limits. Thus, in our example, each discrete level is assigned two effective quantum numbers, ν_1 and ν_2:

$$\begin{aligned} E_n &= I_1 - \frac{1}{2\nu_1^2} \\ &= I_2 - \frac{1}{2\nu_2^2} \ . \end{aligned} \tag{46}$$

Eq. (46) gives an implicit relation between ν_2 and ν_1, namely

$$\nu_1 = \nu_1(\nu_2) = \nu_2[1 - 2(I_2 - I_1)\nu_2^2]^{-\frac{1}{2}} \ . \tag{47}$$

As in Section 1.3, at any energy E the eigenfunction is written as a superposition of the various α-channels, with amplitudes A_α, and is then expressed at large r in terms of the i-channels:

$$\psi = \sum_i \phi_i \, [f(\nu_i,r) \sum_\alpha U_{i\alpha} \cos \pi\mu_\alpha \, A_\alpha - g(\nu_i,r) \sum_\alpha U_{i\alpha} \sin \pi\mu_\alpha \, A_\alpha] \, , \quad r > r_o \, . \tag{48}$$

Once again, ϕ_i represents the complete wavefunction in the i-channel, except for the radial part of the outer electron; it includes the core wavefunction and all of the angular and spin parts of the outer electron, coupled in a manner appropriate to the dissociation region. Eq. (48), with the asymptotic forms of {f,g} thrown in, is ripe for application of the boundary conditions. However, these boundary conditions take slightly different forms depending upon the spectral region of interest.

Discrete region below $I_1(j = 3/2)$:

Here, f and g are written as in Eq. (31) and the exponentially-rising parts set equal to zero. This gives

$$\sum_\alpha U_{i\alpha} \sin \pi(\nu_i + \mu_\alpha) \, A_\alpha = 0 \, , \quad i = 1,\ldots,5 \, . \tag{49}$$

This is the multichannel generalization of Eq. (35); it represents a homogeneous system of linear equations for A_α and nontrivial solutions exist only if

$$F(\nu_1,\nu_2) = \det |U_{i\alpha} \sin \pi(\nu_i + \mu_\alpha)| = 0 \, . \tag{50}$$

Simultaneous solution of Eqs. (47) and (50) locates the positions of the bound states. The significant point about Eq. (50) is that it represents a set of curves periodic in (ν_1,ν_2) because it depends only upon ν_i (modulo 1); thus, it lends itself to a simple graphical analysis, where $F(\nu_1,\nu_2) = 0$ is drawn as a single curve in a unit (ν_1,ν_2) square (with opposite sides identified). Any horizontal line has two, and any vertical line three, intersections with the $F(\nu_1,\nu_2) = 0$ curve. The curves defined by Eq. (47) run at a slant across the (ν_1,ν_2) square and each of these curves has, in general, five intersections with $F(\nu_1,\nu_2) = 0$, which locate the positions of the five levels. That the various channels interact strongly, and that the effective quantum numbers or quantum defects along a series can display very nonmonotonic behavior, is shown by the fact that $F(\nu_1,\nu_2) = 0$ exhibits sharp twists and bends over certain ranges of ν_1 and ν_2 (modulo 1). The paper by K. T. Lu in this volume gives several examples of these plots and provides a detailed discussion of the graphical aspects of MQDT.

Autoionization region between $I_1(j = 3/2)$ *and* $I_2(j = 1/2)$:
In this region the 3/2 channels are open and the 1/2 channels closed. For the latter, Eq. (49) still holds, but now for only two values of i. The other three conditions are obtained, as in Section 1.3, by imposing ingoing-wave boundary conditions on the open channels:

$$\sum_\alpha U_{i\alpha} \sin \pi(-\tau_\rho + \mu_\alpha) A_{\alpha\rho} = 0$$
$$\sum_\alpha U_{i\alpha} \cos \pi(-\tau_\rho + \mu_\alpha) A_{\alpha\rho} = T_{i\rho} \tag{51}$$

where we have introduced the eigenstates ρ which are superpositions of the open i-channels with the same eigenphase shift $\pi\tau_\rho$, and where T is an orthogonal matrix that is discussed further in the following papers by K. T. Lu and C. H. Greene. The form of these equations remains substantially unchanged from that of Eq. (49): This is an expression of the remark which followed Eq. (22) in Section 1.3, namely, that the ingoing-wave boundary condition for an open channel other than the one of interest becomes, on extrapolation to $k = -i\kappa$, the condition that exponentially-rising pieces in closed channels vanish. Further discussion of the solutions of Eq. (51), and of the expressions for the continuum region above I_2, will be taken up by Lu and Greene.

In summary, MQDT provides a facile analysis of a wide range of experimental data in the discrete, autoionization and continuum regions of excitation of an electron interacting with a positive-ion core, an analysis which can be encapsulated in terms of the μ_α, the frame transformation matrix $U_{i\alpha}$ and dipole absorption strengths d_α into the α-channels (in Section 1.3, we had only one dipole absorption strength). And, as discussed in Section 1.4, if one carries out *ab initio* calculations in the core region to provide the μ_α, it is convenient to work with $\{f^o, g^o\}$ and the smooth $\delta_\alpha^{(o)}$, since the latter is more suitable for extrapolation in energy.

ACKNOWLEDGMENTS

I thank Dr. C. W. Clark for his comments on this write-up. My presentation of quantum defect theory owes much to Professor U. Fano, from whom I learned the subject.

REFERENCES

1. M. J. Seaton, Proc. Phys. Soc. 88, 801, 815 (1966); Rep. Prog. Phys. 46, 167 (1983).
2. U. Fano, Phys. Rev. A2, 353 (1970); J. Opt. Soc. Am. 65, 979 (1975); K. T. Lu, in: *Proceedings of the Daresbury Study Weekend*, W. Eissner, ed. (March, 1982); J. J. Wynne and J. A. Armstrong, Comm. At. Mol. Phys. 8, 155 (1979).
3. H. A. Bethe, Phys. Rev. 76, 38 (1949); see, for example, A. Messiah, *Quantum Mechanics*, Vol. I (North-Holland, New York, 1961), p. 406.
4. R. Jost, Helv. Phys. Acta 20, 256 (1947); R. G. Newton, *Scattering Theory of Waves and Particles* (McGraw Hill, New York, 1966), Section 12.1.2.
5. A. R. P. Rau and U. Fano, Phys. Rev. A4, 1751 (1971).
6. E. P. Wigner, Phys. Rev. 73, 1002 (1948).
7. E. S. Chang and U. Fano, Phys. Rev. A6, 173 (1972).
8. U. Fano and G. Racah, *Irreducible Tensorial Sets* (Academic Press, New York, 1959), Chapter 13.
9. L. D. Landau and E. M. Lifshitz, *Quantum Mechanics: Non-Relativistic Theory*, 3rd Ed. (Pergamon, New York, 1977), Section 136.
10. C. H. Greene, U. Fano and G. Strinati, Phys. Rev. A19, 1485 (1979); C. H. Greene, A. R. P. Rau and U. Fano, Phys. Rev. A26, 2441 (1982).
11. A. F. Starace, in: *Photoionization and Other Probes of Many-Particle Interactions*, F. J. Wuilleumier, ed. (Plenum, New York, 1976).
12. R. Colle, J. Chem. Phys. 74, 2910 (1981); A. Giusti, J. Phys. B 13, 3867 (1980).
13. K. T. Lu and U. Fano, Phys. Rev. A2, 81 (1970).

ELEMENTS OF QUANTUM DEFECT THEORY. II. A UNIFIED THEORY OF RYDBERG AND AUTOIONIZING STATES*

K. T. Lu

Chemistry Division
Argonne National Laboratory
Argonne, Illinois 60435, USA

1. INTRODUCTION

The purpose of this paper is to present a quantum defect theory (QDT) basis for a unified approach to Rydberg and autoionizing states in atoms and molecules. A scattering-theoretic outline of QDT was given by A. R. P. Rau in the previous paper in this volume; the present paper focuses on atomic aspects of QDT and provides comment on the transition from atoms to molecules; finally, the MQDT of diatomic molecules will be discussed in detail by C. H. Greene in the following paper in this volume.

The early treatment of QDT, involving a single ionization threshold and a single channel, is applicable to valence Rydberg spectra of alkali atoms. A single parameter, namely the *quantum defect* $\mu = n - \nu$ (where ν is the effective quantum number), is sufficient to characterize the short-range non-Coulombic interaction between an excited Rydberg electron and a closed-shell ion core. The quantum defect is a smooth function of energy throughout the channel. For a Rydberg atom, the eigenenergy scales as ν_n^{-2}, while the oscillator strength and radius scale as ν_n^{-3} and ν_n^{2}, respectively: That is, the higher the excitation in ν_n, the smaller the binding energy, the weaker the oscillator strength and the larger the radius. In other atoms, or in inner-shell excitation of alkali atoms, there are multiple ionization thresholds (due to spin-orbit interactions, for example) that serve to enrich the spectral structure. In this case, exchange and corre-

*Dedicated to Professor Ugo Fano on the occasion of his seventieth birthday.

S. P. McGlynn et al. (eds.), Photophysics and Photochemistry in the Vacuum Ultraviolet, 217–243.

lation effects in the interaction between the excited electron and ion-core electrons generate channel interactions, leading to the perturbations and mixing of the Rydberg channels. The quantum defect in the mixed-channel case is no longer a smooth function of energy; instead, μ varies from level to level. Moreover, oscillator strength and radius, as well as other physical quantities, no longer follow the simple ν_n scaling laws discussed above. One-photon absorption from the ground state of a rare gas (Ne, Ar, Kr and Xe) gives rise to five Rydberg channels, each with odd parity and J = 1 final states, of the form $np^6\ (^1S_0) \rightarrow np^5\ (^2P)\varepsilon\ell$. The spin-orbit interaction splits the np^5 ion core into two states, $^2P_{1/2}$ and $^2P_{3/2}$. As shown in Fig. 1, three of the five series converge to the first ionization threshold I_1, $np^5\ (^2P_{3/2})$, while the remaining two series converge to the second ionization threshold I_2, $np^5\ (^2P_{1/2})$. Below the first ionization threshold, all five channels are closed (closed channels consist of discrete levels); between the two thresholds, however, two channels are closed while three channels are open (open channels consist of continuum states). Channel interaction mixes all channels: In the discrete region, energy-level perturbation and anomalous oscillator strength redistribution are typical examples of channel interaction; however, a particularly dramatic consequence of channel interaction occurs between the two ionization thresholds, namely, the appearance of Beutler-Fano autoionization resonances. Multichannel quantum defect theory [1,2] has evolved into a powerful technique descriptive of a host of Rydberg and autoionization phenomena. The positions and oscillator strengths of the discrete levels; the positions, widths and shapes of the autoionization resonances; and the photoelectron branching ratio in the continuum can all be described in terms of a few basic quantum defect parameters, namely, the eigen-quantum defect μ_α, the frame transformation matrix $U_{i\alpha}$ or the channel mixing angle θ_α, and the dipole matrix element d_α. These parameters can be obtained either from data fitting or from *ab initio* calculations.

In Section 2, we first develop the basic principles and concepts of QDT and then illustrate them by a graphical method [3]. Since the quantum defect plot (i.e., the (μ,ν_2) plot of Section 2) does not rely on any specific properties of the quantum defect parameters, such as energy independence of μ_α, θ_α and d_α, it can be used as a guide for refining the theory and for generating a better understanding of the physical processes involved [4]. For example, when the graphical method is applied to the analysis of alkaline-earth atomic spectra, it is necessary to relax the energy dependence of the eigen-quantum defect μ_α, as well as that of the channel mixing angle θ_α [5,6]. The energy-dependent mixing angle serves to characterize radial correlation between hybridized Rydberg and doubly-excited channels [7]. In fact, this characterization was the main reason for studying doubly-excited alkaline earths in hyperspherical coordinates, where quantum num-

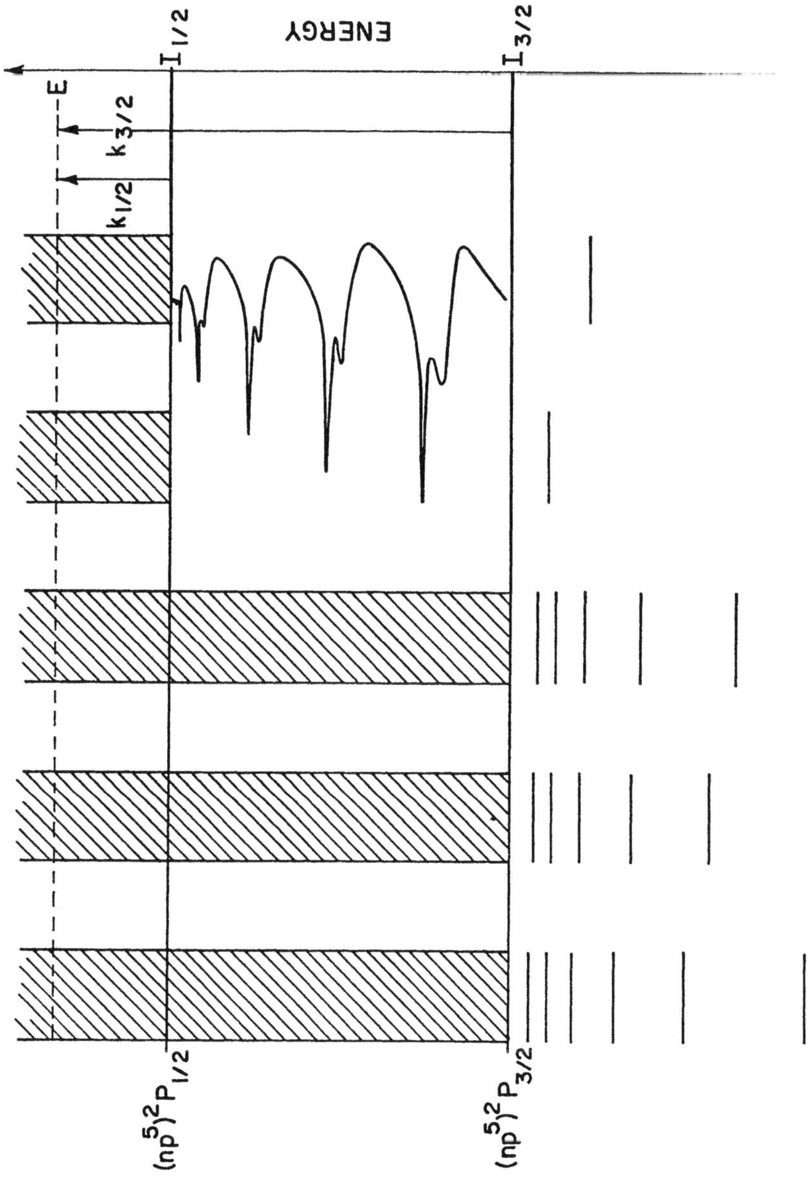

Fig. 1. Schematic energy diagram for the rare gases.

bers for radial correlation occur naturally [8].

The QDT of Seaton was developed in order to adapt collision concepts to the description of an (N+1)-electron atomic system, in terms of two-body interactions between the optical electron and the remaining N-electron ion core. For a certain class of e + atom (ion) problems, ones for which the interaction potential outside the ion core can be treated as pure Coulombic, analytic solutions of the Schrödinger equation have been determined for arbitrary energy and radii by the eigenvalue problem. From a scattering point of view, the asymptotic wavefunction of the outer electron is far separated from the remaining ion core in configuration space; they are coupled by the Coulomb field and form a dissociation channel (i-channel). This solution is a linear combination of two independent Coulomb wavefunctions which determine the regular energy spacings of successive levels of a series. The mixing coefficients of the wavefunction are determined by the normalization factor and by matching the boundary conditions at the ion core, where the electron and ion core form a compound-state channel (α-channel). These mixing coefficients consist of two types of parameters, namely, the eigen-quantum defect μ_α and an orthogonal matrix $U_{i\alpha}$, which are the eigenvalues and eigenvectors, respectively, of the reaction matrix R. The eigen-quantum defect μ_α measures the extra phase shift in the wavefunction due to the short-range non-Coulombic many-body interaction between the outer electron and ion core of the compound-state α-channel; μ_α is a physical observable and can be inferred from graphical analysis of the data. The orthogonal matrix $U_{i\alpha}$ transforms between two configuration spaces, the i-channel dissociation zone and the α-channel compound-state zone. The picture of scattering is now complete, whether the processes are in the discrete or in the continuum. An electron coupled to the ion core in the dissociation i-channel is scattered by the ion core, where channel interaction takes place, into one of the dissociation i'-channels. The excited electron can exchange angular momentum and energy with the core only through collisions in the course of radial motions. To this extent, channel interaction takes into account both angular momentum mixing as well as radial correlation. The essential aspect of eigenchannel quantum defect theory (EQDT) [1,2] is the recognition of two distinctive regions of configuration space where the interaction dynamics of the excited electron with the core is quite different. Whether the core is atomic or molecular is secondary. (This may explain the striking similarity of the Rydberg and autoionization spectra of the molecule methyl iodide (CH_3I) with those of the atom Xe [9].) This important concept was introduced and elaborated by Ugo Fano in his pioneering work on the frame transformation theory of the hydrogen molecule [1,10]; and, this concept serves as the base for transformation from an atom to a molecule.

Configuration mixing calculations of wavefunctions by the variational technique, via trial electronic wavefunctions, have been very successful for low-lying valence states. However, in calculating highly-excited states or continuum problems, one encounters several difficulties: The large number of configurations required, orthogonality between states of the same symmetry and numerical convergence are major problems. It is, therefore, of interest and importance when a method is available which not only makes short-cuts in the calculations possible but also can handle the data analysis readily. QDT, in fact, provides such a procedure. Since QDT aims at calculating quantum defect parameters, it bypasses the difficulties inherent in evaluating wavefunctions [11,12]. Moreover, QDT takes reasonably accurate account of exchange and correlation effects in the interaction between the valence and core electrons in the heavy elements; thus, the method is particularly relevant when these effects are large enough to cause hybridization between different channels. We discuss briefly the *ab initio* calculation of quantum defect parameters in Section 3.

QDT has been further developed and applied to express physical quantities, such as branching ratios of photoelectrons, angular distribution of photoelectrons [13], spin polarizations [14], Landé g-factors [15], diamagnetic and Stark shifts [16], Raman processes [17] and hyperfine structure [18], all in terms of the basic quantum defect parameters μ_α, d_α and $U_{i\alpha}$. Moreover, all of these quantities can be evaluated easily once the basic quantum defect parameters are known. QDT has also been developed for multichannel relativistic problems [19] and for dielectronic recombinations [20]. In Section 4 we discuss briefly the application of QDT to (i) the Stark effect, (ii) diamagnetic shifts, (iii) iso-valence homologues, and (iv) the transformation of atoms into molecules, including applications to iso-chromophore studies of polyatomic molecules.

Several review articles on MQDT have appeared recently [21-23]; the reader is referred to these articles for missing subjects and references.

2. EIGENCHANNEL QUANTUM DEFECT THEORY

2.1 Basic Theory

An (N+1)-electron atomic system may be described in terms of the two-body interaction between the excited electron and the rest of the atomic core A. The many-body effect is treated through the coupled differential equation (we use atomic units throughout)

$$\left[-\frac{1}{2}\frac{d^2}{dr^2} + \frac{\ell_i(\ell_i + 1)}{2r^2} - z^2\varepsilon_i \right] \psi_i + \sum_j V_{ij}\, \psi_j = 0$$

where $z = Z - N$ ($z = 1$ for neutral e + A systems).

Suppose the interaction potential between the electron and particle A has the form

$$V_{ij} = v(r) + B(r,\theta,\phi) - \frac{1}{r} \tag{1}$$

such that it satisfies the following conditions at different configuration zones:

compound-state zone

$$V_{ij} \to v(r) \qquad r < r_o$$

intermediate zone

$$V_{ij} \to -\frac{1}{r} \qquad r > r_o$$

dissociation zone

$$V_{ij} \to B(r,\theta,\phi) - \frac{1}{r} \qquad r \to \infty$$

where $B(r,\theta,\phi)$ can be external potentials. If analytic wavefunctions exist in the dissociation zone for the potential $B(r,\theta,\phi) - \frac{1}{r}$, then a parameter (i.e., the quantum defect) can be introduced as a mixing coefficient to represent the effect due to the short-range non-Coulombic interaction; and, this parameter can be determined by matching the boundary condition at r_0 with the compound-state zone. This is the underlying principle of QDT. (In the following discussion, we adopt the analytic treatment of [2].)

For e + positive atomic-ion systems, the potential $B(r,\theta,\phi)$ vanishes and the interaction potential can be divided into two configuration zones. In the compound-state zone, the electron and positive ion form a compound system and the potential is dominated by a short-range non-Coulombic spherical interaction. In the far zone, however, the electron is dissociating from the ion core and the electron and ion core interact mainly by the attractive Coulomb field. A "channel" is defined as a set of discrete and continuum states (of the e + ion complex) which differ only in the energy of the excited electron, and is specified by the orbital, spin and fine-structure quantum numbers of the ion; the orbital and spin angular momentum of the electron; and the coupling of the electron to the ion core. For example, as discussed in Section 1, one-photon absorption from the ground state of a rare gas leads to five channels consisting of odd parity, J = 1 final states, $np^6\,(^1S_0) \to np^5\,(^2P)\varepsilon\ell$. The energy ε of the outer electron equals $1/2\nu_i^2$ for discrete states and $k^2/2$ for continuum states. ν_i is an effective quantum number defined at each total energy E_n, with respect to the two different ionization limits:

$$\begin{aligned} E_n &= I_1 - \frac{1}{2\nu_1^2} \\ &= I_2 - \frac{1}{2\nu_2^2} \; . \end{aligned} \tag{2}$$

In the i-channel dissociation zone ($r > r_o, r \to \infty$) the interaction is dominated by the pure Coulomb potential between the electron and the ion core and the coupling can be described by a jj scheme: $[^2P_{3/2}]\varepsilon d_{5/2}$, $[^2P_{3/2}]\varepsilon d_{3/2}$, $[^2P_{3/2}]\varepsilon s_{1/2}$, $[^2P_{1/2}]\varepsilon d_{3/2}$ and $[^2P_{1/2}]\varepsilon s_{1/2}$. In the α-channel compound zone ($r \lesssim r_o$), however, the non-Coulombic electrostatic potential dominates and, therefore, the coupling is characterized by an LS scheme: $(p^5\varepsilon d)^1P$, $(p^5\varepsilon d)^3D$, $(p^5\varepsilon s)^3P$, $(p^5\varepsilon d)^3P$ and $(p^5\varepsilon s)^1P$. In the dissociation zone, analytic Coulomb wavefunctions are known, namely, $f(\nu_i,r)$ and $g(\nu_i,r)$; however, the e + ion short-range non-Coulomb interaction in the compound zone, an interaction resulting from collisions between the electron and the ion core, causes channels to be mixed. We must therefore describe the eigen-wavefunction, at a given energy E, as a mixture of all of the i-channels:

$$\psi(r,E) = A \sum_{i,j} \Phi_i [f(\nu_i,r)\delta_{ij} - g(\nu_i,r)R_{ij}]h_j , \quad r > r_0 \tag{3a}$$

where Φ_i represents the wavefunctions of the ion core, of spin, and of angular momentum coupling in the i-channel; the h_j are the mixing coefficients; and A is an antisymmetrization operator. R_{ij} is the reaction matrix ($R_{ij} = \sum_\alpha U_{i\alpha} \tan\pi\mu_\alpha U^\dagger_{\alpha j}$) and $U_{i\alpha}$ and μ_α are the eigenvectors and eigenvalues, respectively, of the R-matrix. $U_{i\alpha}$ is an orthogonal matrix which transforms between the two frames, i-channel and α-channel. The eigenvalue μ_α measures the additional quantum defect due to the short-range non-Coulombic interaction of the e + ion compound-state α-channel. An alternative form of this eigen-wavefunction is

$$\psi(r,E) = A \sum_i \Phi_i [f(\nu_i,r) \sum_\alpha U_{i\alpha} \cos \pi\mu_\alpha -$$

$$- g(\nu_i,r) \sum_\alpha U_{i\alpha} \sin \pi\mu_\alpha] A_\alpha , \quad r > r_o \; . \tag{3b}$$

When the total energy E falls below the first ionization limit I_1, all of the channels are closed. The requirement that the eigen-wavefunction of Eq. (3b) be finite asymptotically then leads to the relation

$$\sum_\alpha U_{i\alpha} \sin \pi(\nu_i + \mu_\alpha) A_\alpha = 0 \tag{4a}$$

and the compatibility condition

$$F(\nu_1,\nu_2) = \det|U_{i\alpha} \sin \pi(\nu_i + \mu_\alpha)| = 0. \tag{4b}$$

The pair of Coulomb wavefunctions {f,g} are chosen such that $U_{i\alpha}$ and μ_α are smooth functions of energy. Thus, $F(\nu_1,\nu_2) = 0$ represents a set of smooth curves *periodic* in a (ν_1,ν_2) plot, with $U_{i\alpha}$ and μ_α as parameters. Energy eigenvalues (i.e., discrete levels) lie at the intersections between the $F(\nu_1,\nu_2) = 0$ curves and curves represented by

$$\nu_1 = \nu_1(\nu_2) = \nu_2[1 - 2(I_2 - I_1)\nu_2^2]^{-\frac{1}{2}}. \tag{5}$$

The quantum defect μ is identified as $\mu = \nu_1 - n$ or, equivalently, $\mu = \nu_1$ (modulo 1).

The wavefunction in Eq. (3) is normalized by

$$\bar{\psi}_n(r,E) = \psi_n(r,E)/N_n$$

such that

$$\int_{r_o}^{\infty} |\bar{\psi}_n|^2 d\tau = 1$$

and where the normalization factor is

$$N_n = \left[\nu_{1,n}^3 + \nu_{2,n}^3 \frac{d(-\nu_1)}{d\nu_2}\bigg|_n\right]^{\frac{1}{2}}.$$

The normalized wavefunction for a given state n can be expressed as a superposition of all of the i-channels:

$$\bar{\psi}_n(r,E) = \sum_i \Phi_i \, P_i^{(n)}(r,\nu_i)\, Z_i^{(n)} \tag{6a}$$

where $P_i^{(n)}(r,\nu_i)$ is the wavefunction of the outer electron. The mixing coefficients $Z_i^{(n)}$ for the nth state in the i-channel is related to the mixing coefficients $A_\alpha^{(n)}$ in the α-channel by

$$Z_i^{(n)} = (-1)^{\ell_i+1} \nu_i^{3/2} \sum_\alpha U_{i\alpha} \cos \pi(\nu_{i,n} + \mu_\alpha) A_\alpha^{(n)}/N_n. \tag{6b}$$

These equations connect the channel approach to the traditional spectroscopic interpretation in terms of configuration interaction, as will be discussed below.

The discrete oscillator strength f_n can be written as

$$f_n = 2(E_n - E_o) \left| \sum_\alpha d_\alpha A_\alpha^{(n)} \right|^2 / N_n^2 \tag{7}$$

where d_α is the dipole transition matrix element which connects the initial state to the α-eigenchannel:

$$d_\alpha = \int_{r_o}^{\infty} \psi_\alpha^\dagger (\sum_i z_i)\, \psi_o \, d\tau . \tag{8}$$

Thus, the complete bound-state physical quantities, such as energy levels and oscillator strengths, are expressed in terms of three sets of parameters -- the dipole matrix elements d_α, the eigen-quantum defect μ_α and the transformation matrix $U_{i\alpha}$. The total number of parameters for an n-channel problem is $n + n + \frac{1}{2}n(n-1)$.

In the autoionization region ($I_1 < E < I_2$), some of the channels become open (denoted $i\varepsilon P$) and $\nu_i = i/k_i$ is imaginary, whereas others remain closed (denoted $i\varepsilon Q$) and ν_i remains real. The condition that the closed-channel wavefunction, expressed as a linear combination of $f(\nu_i,r)$ and $g(\nu_i,r)$, remain finite at $r \to \infty$ yields only part of the relations contained in Eq. (4a), namely, those relations with $i\varepsilon Q$. However, one recovers the rest of these relations in a modified form by introducing a boundary condition for the open channels ($i\varepsilon P$). The asymptotic behavior of wavefunctions in this region requires oscillatory standing waves in the open channels and exponentially-damped waves in the closed channels. This can be achieved by introducing eigenstates ρ, such that each eigenstate ρ is a superposition of the standing waves of open channels with the same eigen-phase shift $\pi\tau_\rho$:

$$\psi_\rho \xrightarrow[r \to \infty]{} \sum_{i\varepsilon P} \Phi_i (2/\pi k_i)^{\frac{1}{2}} \sin(\theta_i + \pi\tau_\rho)\, T_{i\rho}$$

$$- \sum_{i\varepsilon Q} \Phi_i \, v(\nu_i, r) \sum_\alpha U_{i\alpha} \cos \pi(\nu_i + \mu_\alpha)\, A_\alpha \tag{9}$$

where

$$\begin{aligned} &\sum_\alpha U_{i\alpha} \sin \pi(\nu_i + \mu_\alpha)\, A_\alpha = 0, && i \in Q \text{ (closed channels)} \\ &\sum_\alpha U_{i\alpha} \sin \pi(\mu_\alpha - \tau_\rho)\, A_\alpha = 0, && i \in P \text{ (open channels)} \\ &\sum_\alpha U_{i\alpha} \cos \pi(\mu_\alpha - \tau_\rho)\, A_\alpha = T_{i\rho} && i \in P \text{ (open channels)} \end{aligned} \tag{10}$$

and where $T_{i\rho}$ is an orthogonal matrix, while $v(\nu_i,r)$ is an exponentially-decreasing function of r.

An essential point of this treatment is that the relations of Eq. (10) differ from those of Eq. (4a) merely by the substitution, for $i \varepsilon P$,

$$\nu_1 \longleftrightarrow -\tau_\rho \text{ (modulo 1).} \tag{11}$$

This means that the variable ν_1 of each level of the discrete spectrum can be represented in the form $\nu_1 = n - \tau_\rho$, where $\pi\tau_\rho$ represents the phase shift of the collision eigenstate ψ_ρ for the autoionization region. Thus, a correspondence between bound-state levels and continuum eigenstates for multichannel systems is established, which is analogous to that for single-channel problems.

With the above extension in mind, the compatibility condition becomes

$$F(-\tau_\rho, \nu_2) = \det \left| U_{i\alpha} \sin \pi(-\tau_\rho + \mu_\alpha) \right| = 0, \ \rho \ \varepsilon \ P \ . \tag{12}$$

Eq. (12) represents functional curves in a $(-\tau_\rho, \nu_2)$ plot. Moreover, the oscillator strength density of autoionization resonances can be written as

$$\frac{df}{dE} = \sum_\rho \frac{df^{(\rho)}}{dE} = 2(E - E_o) \sum_\rho \left| \sum_\alpha d_\alpha A_\alpha^{(\rho)} \right|^2 / N_\rho^2 \tag{13}$$

where $N_\rho^2 = \sum_{i \varepsilon P} \left| \sum_\alpha U_{i\alpha} \cos \pi(-\tau_\rho + \mu_\alpha) A_\alpha^{(\rho)} \right|^2$. The oscillator strength densities in the autoionization region are expressed in terms of the same set of parameters d_α, μ_α and $U_{i\alpha}$ as in the discrete region, with the quantities τ_ρ and ν_2 treated as continuously changing variables. Thus, it is possible to relate the oscillator strength density $df^{(\rho)}/dE$ to the discrete oscillator strength f_n by a normalization factor with $\nu_{2,n}$ substituted for ν_2:

$$\frac{df^{(\rho)}}{dE}(\nu_2)\Big|_n = f_n(\nu_2) \left[\nu_{1,n}^3 + \nu_{2,n}^3 \frac{d(-\nu_1)}{d\nu_2} \Big|_n \right] . \tag{14}$$

This relation successfully connects the oscillator strength in the discrete with that in the continuum. Figs. 3(a) and 3(b), discussed below, demonstrate this fact.

2.2 Graphical Analysis of Perturbed Rydberg and Autoionization Spectra

In this section, we describe a graphical analysis of perturbed Rydberg and autoionization spectra based upon the analytic theory discussed in the previous section, and upon information extracted from that theory. We use Ar as a case study [4].

Initially, we do not distinguish between the "perturbed series" and the "perturbing level"; instead, we consider equally all experimental levels, with $J = 1$ and odd parity, below the first ionization limit I_1. To each of these levels we assign two alternative effective quantum numbers ν_1 and ν_2, where $\nu_1 = \nu_1(\nu_2)$ according to Eq. (5). The quantum defect μ is identified as $\mu = \nu_1 - n$, or $\mu = \nu_1$ (modulo 1), where n is an integer which may have the same value for two or more levels, as shown in Fig. (2a). Values of μ for all of the levels available for Ar are thus plotted, versus ν_2, in Fig. 2(a). To emphasize the connection with customary practice, the points (μ,ν_2) usually assigned to series converging to I_1 are marked by open circles, whereas the three points usually assigned to series converging to I_2 are marked by "χ". The solid and dashed lines joining the points are interpolated curves which represent all quantum defects as points lying on the plot of a multivalued continuous function $\mu(\nu_2)$, which has the analytic form given in Eq. (4b). All of the experimental points are located at the intersections between curves represented by Eq. (4b) and curves represented by Eq. (5). The compatibility conditions of Eq. (4b) consist of trigonometric functions of ν_1 and ν_2. To check out the periodicity, the data of Fig. 2(a) are plotted modulo 1 -- that is, with successive unit intervals of ν_2 folded onto one another as shown in Fig. 2(b). The points corresponding to each different interval are now seen to interpolate rather smoothly and to define a single multi-branched periodic curve $\mu(\nu_2)$.

Fig. 3 shows the composite graphical representation of all of the data on Rydberg series as well as on autoionization resonances. The eigen-phase shift τ_ρ, oscillator strength (in this case relative intensity) and oscillator strength density $df^{(\rho)}/dE$ are all plotted as functions of ν_2. Fig. 3(a) shows the correspondence between bound-state levels in terms of quantum defect μ and eigenphases τ_ρ in the continuum, a correspondence which is expressed by the analytic relationship of Eq. (11). The connection between oscillator strength in the discrete and oscillator strength density $df^{(\rho)}/dE$ in the continuum, established in Eq. (14), is demonstrated in Fig. 3(b). One should note that the superposition of the three unit sections of Fig. 3(a) reproduces Fig. 2(b), and that the addition of the three unit sections of Fig. 3(b) generates the autoionization resonance cross-section df/dE, according to Eq. (13), as shown in Fig. 4.

The basic quantum defect parameters μ_α, $U_{i\alpha}$ and d_α are obtained by fitting the analytic relations of Eqs. (4b) and (14) to the discrete energy levels and autoionization profiles [4]. The following features of Figs. 2 and 3 help to determine the parame-

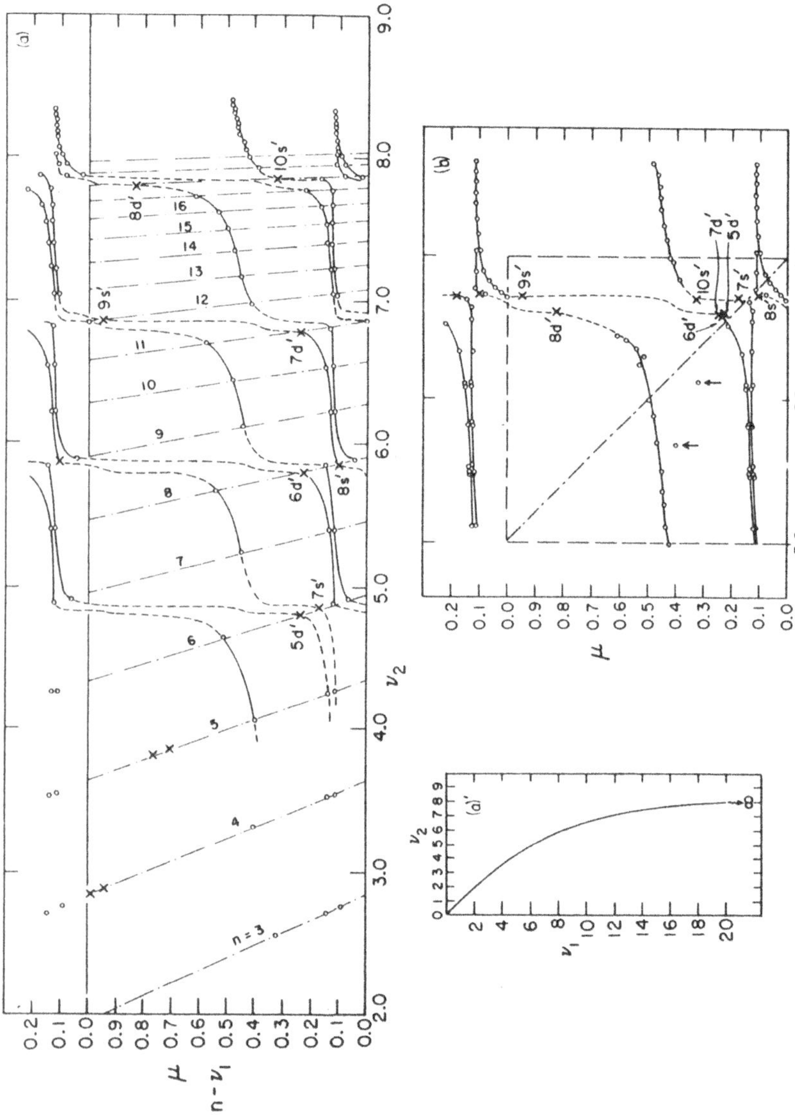

Fig. 2. Caption follows Fig. 3.

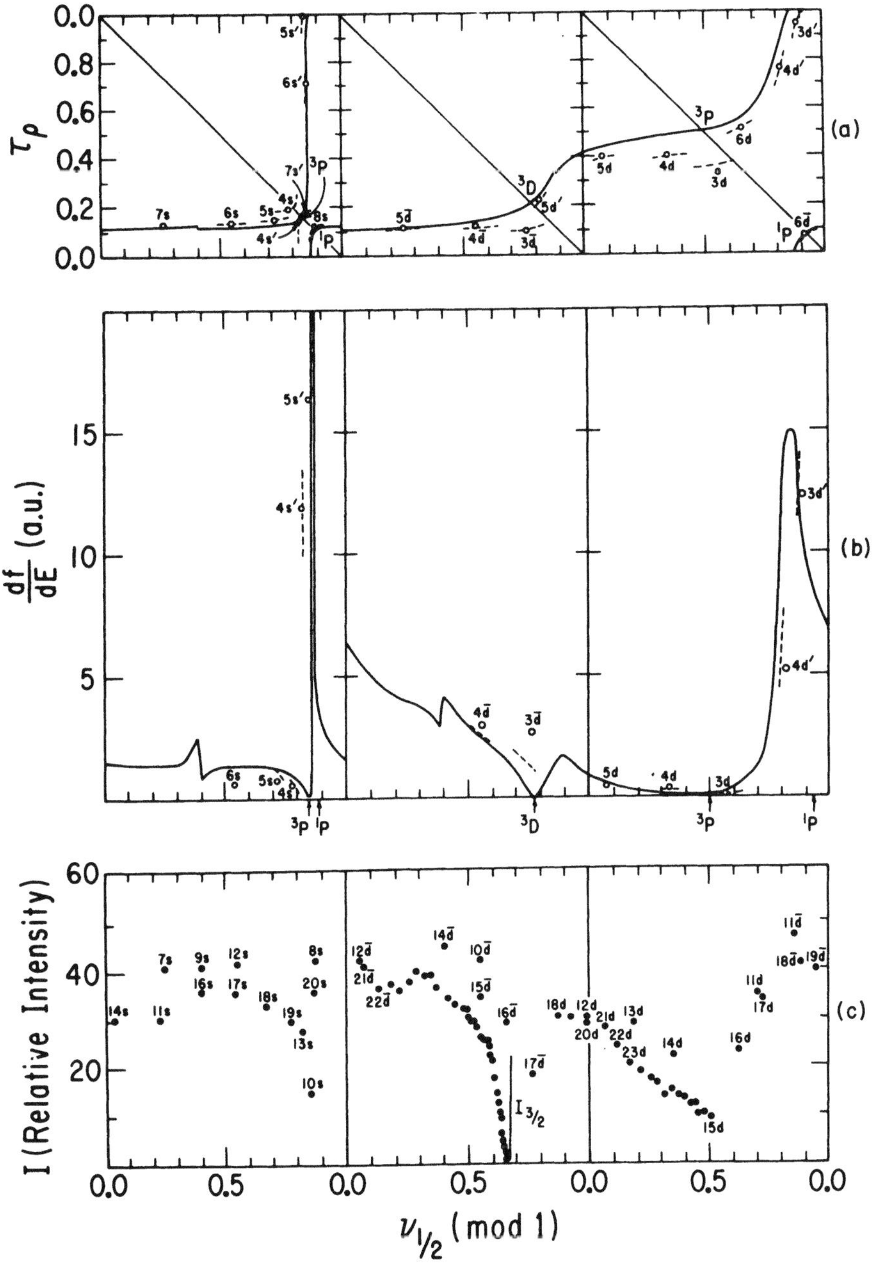

Fig. 3. Caption follows on next page.

Fig. 2. (a), The quantum defect (μ,ν_2) plot for Ar. The function $-\nu_1(\nu_2)$, defined by Eq. (5) of the text, is shown in the box (a)' and again, modulo 1, by a dot-dash line in (a). Numerical labels on these curves indicate values of the integer n from $\mu = n - \nu_1$. (b), The folded quantum defect (μ,ν_2) plot for Ar. The data of (a) with $\nu_2 > 4.0$ are replotted on a ν_2 (modulo 1) scale (i.e., successive intervals of ν_2 are folded onto one another). Some points with $\nu_2 < 4.0$ are included (↑) to show departure from the interpolated curves. (*Note*: $s' = (^2P_{1/2})s_{1/2}$, $d' = (^2P_{1/2})d_{3/2}$.) Taken from [3].

Fig. 3. Eigenphases τ_ρ and oscillator strength densities $df^{(\rho)}/dE$ for separate eigenchannels $\rho = 1,2,3$ of Ar. (*Note*: $\nu_{3/2}$, $\nu_{1/2}$ and $I_{3/2}$ correspond to ν_1, ν_2 and I_1, respectively, of the text; $s = (^2P_{3/2})s_{1/2}$, $s' = (^2P_{1/2})s_{1/2}$, $d = (^2P_{3/2})d_{5/2}$; $d' = (^2P_{1/2})d_{3/2}$; $\bar{d} = (^2P_{3/2})d_{3/2}$.) (a), solid lines represent $\tau_\rho(\nu_{1/2})$ for parameters fitted near $I_{3/2}$; dashed lines represent $\tau_\rho(\nu_{1/2})$ for parameters fitted at the $\nu_{1/2}$ value for low-lying lines; and open circles represent experimental data [$\tau_\rho = -\nu_{3/2,n}$ (modulo 1)]. (b), solid lines represent $df^{(\rho)}/dE$ for parameters fitted near $I_{3/2}$; dashed lines represent $df^{(\rho)}/dE$ for parameters fitted at $\nu_{1/2}$ values of low-lying levels; and open circles represent experimental data renormalized by Eq. (14) of the text. (c), Closed circles represent experimental data (relative intensities) [44]. Taken from [4].

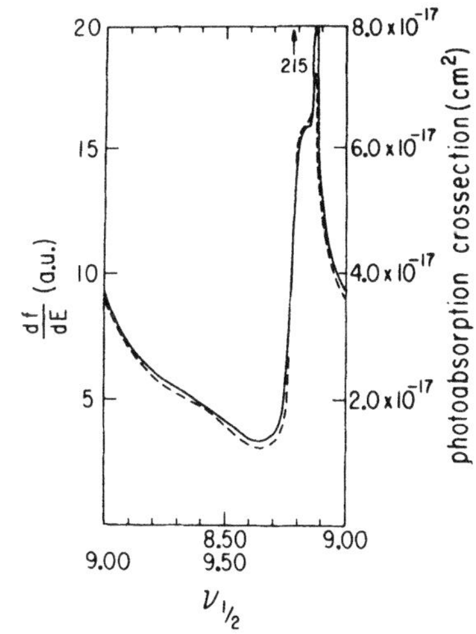

Fig. 4. Ar autoionization profile for $8.64 \leq \nu_{1/2} \leq 9.64$. (*Note*: Abscissa scale is modulo 1; $\nu_{1/2}$ corresponds to ν_2 of the text.) Dashed line represents the experimental data from [44] while solid line represents the theoretical fit (using the d_α listed in [4]).

ters of the theory [2,23]:

--- A special solution of the eigenvalue problem occurs at the intersections between the $F(\nu_1,\nu_2) = 0$ curves of Eq. (4b) and the diagonal line in the (ν_1,ν_2) plot (i.e., ν_1(modulo 1) = ν_2 (modulo 1) + n = μ_α). The eigen-quantum defect μ_α can be read easily from the plot; or, in other words, the discrete levels located at these intersections have the special α-channel character.

--- The slope of the curve $F(\nu_1,\nu_2) = 0$ at any point is represented by the ratio of squared overlap integrals of a state ψ with the closed channels "Q" and the open channels "P"; thus, the slope is always positive:

$$\frac{d(-\nu_1)}{d\nu_2} = \frac{|\langle Q|\psi\rangle|^2}{|\langle P|\psi\rangle|^2} \geq 0. \qquad (15)$$

This slope takes a special value at the intersections between $F(\nu_1,\nu_2) = 0$ and the diagonal line:

$$\frac{d(-\nu_1)}{d\nu_2} = \sum_{i=1,4} U_{i\alpha}^2 \Big/ \sum_{i=2,3,5} U_{i\alpha}^2$$

where we have adopted the following i-channel and α-channel labels:

i	*label*	α	*label*
1	$(^2P_{1/2})d_{3/2}$	1	$(p^5d)\,^3D$
2	$(^2P_{3/2})d_{5/2}$	2	$(p^5d)\,^1P$
3	$(^2P_{3/2})d_{3/2}$	3	$(p^5d)\,^3P$
4	$(^2P_{1/2})s_{1/2}$	4	$(p^5s)\,^3P$
5	$(^2P_{3/2})s_{1/2}$	5	$(p^5s)\,^1P$

It has been stressed [2] that data fitting based upon energy values alone cannot provide unique information about $U_{i\alpha}$, since the eigenphase τ_ρ and the corresponding slope $d\tau_\rho/d\nu_2$ depend on the invariant combinations of $U_{i\alpha}$. Additional physical observables [14], such as oscillator strength or autoionization resonances, that distinguish between i = 2,3,5 or between i = 1 and 4 are necessary to get the full $U_{i\alpha}$.

The periodic character of the eigenphase τ_ρ in the presence of the closed channels Q, as revealed in Figs. 2 and 3, can be described by the following relations:

--- τ_ρ is a monotonic increasing function of ν_2 such that $d\tau_\rho/d\nu_2 \geq 0$ and $\tau_{\rho+p} = \tau_\rho + \pi$, where p is the number of open channels. The slope of the eigenphase τ_ρ [Eq. (15)] is proportional to the time delay defined by

$$2\hbar \frac{d\tau_\rho}{dE} = 2\hbar \frac{d\tau_\rho}{d\nu_2} \cdot \frac{d\nu_2}{dE} = 2\hbar\nu_2^3\pi \frac{d(-\nu_1)}{d\nu_2} . \tag{16}$$

This relation has been shown to represent the lifetime of a resonant multichannel scattering process due to nonhydrogenic interactions. For simplicity, a two-channel example is given to demonstrate that the time delay indeed has the resonant structure:

$$\frac{d(-\nu_1)}{d\nu_2} = \frac{1}{\cos^2 \pi\nu_2} \cdot \frac{\Gamma}{(\tan \pi\nu_2 - \alpha)^2 + \Gamma^2} . \tag{17}$$

--- $\tau_\rho(\nu_2 + 1) = \tau_{\rho+q}(\nu_2)$, where q is the number of closed channels.

$$--- \sum_\rho \tau_\rho(\nu_2 + 1) = q\pi + \sum_\rho \tau_\rho(\nu_2).$$

The above three properties can be checked easily from Figs. 2 or 3 for a system with three open channels and two closed channels.

The time delay, or the slope of the eigenphase $(d\tau_\rho/d\nu_2)$, provides information on the strength of channel interactions represented by the avoided crossings in Figs. 2 and 3, and can be easily read off from the figures. For example, the channel interaction between the two s channels is weak, since the avoided crossing is small and the slope of τ_ρ in the first unit square of Fig. 3(a) is almost zero everywhere except at the location of closed channel s', $\nu_2 = 0.85$. This weak channel interaction is reflected in the oscillator strength density in Fig. 3(b) as an isolated resonance. The channel interactions among the three d channels are stronger, as represented by the large avoided crossings and the non-zero slope $d\tau_\rho/d\nu_2$ over a large section of values in ν_2. The corresponding oscillator strength densities also show a marked resonance extended over a large range of values of ν_2 in Fig. 3(b). These figures also indicate that little channel interaction exists among s channels and d channels.

The overall agreement between the intensity (or oscillator strength) of discrete levels with the oscillator strength densities of autoionization resonances is rather striking. A further demonstration of the power of the channel interaction picture is the interpretation of intensity anomalies along perturbed Rydberg spectra, illustrated by Fig. 3(c) [2]. The three intensity minima at 10s, 17$\bar{d}$ and 15d (cf. Fig. 3(c)) are located at the intersections between $F(\nu_1, \nu_2) = 0$ and the diagonal line. These intersections represent values of μ_α for α-eigenchannels with triplet character according to LS-coupling (cf. Fig. 3(a)). The other two intersections have singlet character eigenchannels, as demonstrated by the substantial intensity of the 8s and 6$\bar{d}$ levels. The above argument follows the LS dipole-allowed selection rule from the ground singlet state.

The channel interaction picture provides a connection with the traditional configuration mixing picture through the channel mixing coefficients A_α or Z_i in Eqs. (6a) and (6b). Levels located on flat or vertical portions of the curves in Fig. 3(a) represent states belonging to a pure channel. For example, 6s and 7s belong to the same channel i = 5, whereas the 5s' and 6s' levels belong to the channel i = 4: That is, the corresponding mixing coefficients Z_i for these levels are dominated by i = 5 and 4, respectively, with $Z_5 \simeq 1$ for 6s and 7s and $Z_4 \simeq 1$ for 5s' and 6s'. On the other hand, levels located on the curving portion of Fig. 3(a) represent states belonging to a mixed channel (i.e., the mixing coefficients A_α or Z_i are dominated by more than one α- or i-channel). For example, the mixing coefficients A_α of level

4d' have $A_1 = 0.640$, $A_2 = 0.63$ and $A_3 = 0.44$ [4]. Thus, a perturbed Rydberg series consists of levels located along a (τ_ρ, ν_2) curve in Figs. 2 and 3(a), with the channel mixing coefficients varying from one level to the other, while the periodicity due to channel interactions persists in the reduced (μ_1, ν_2) plot.

This semi-empirical graphical method has been adopted to analyze a vast amount of data from diverse experimental techniques. Atomic spectra analyzed by this method include the rare gases [24], alkaline earths [25-29], carbon-group elements [30], rare earths [26,31] and uranium [32].

3. *AB INITIO* CALCULATIONS OF QUANTUM DEFECT PARAMETERS

Various dynamical theories have been developed and adopted to calculate, *ab initio*, quantum defect parameters and physical observables.

--- Fano and Lee [11] have developed a variational approach for calculating eigenchannel quantum defect parameters, for a given energy, by solving the many-electron Schrödinger equation within only a finite volume surrounded by a boundary Σ. The wavefunction is expanded in terms of a set of basis functions according to the nature of the physical problem and according to the shape of the boundary Σ. The expansion coefficients can then be obtained variationally by diagonalizing a real Hermitian matrix which follows from the analytic Hartree-Fock-Roothaan scheme. Oscillator strengths and autoionization resonances of Ar have been calculated. A variational-type calculation of the parameters of the Ne-ion spectrum has also been carried out, with the parameters being expressed in terms of Slater integrals and spin-orbit interactions along an iso-electronic sequence [12].

--- A formulation of relativistic MQDT has been given by Lee and Johnson, with the relativistic random-phase approximation being applied to calculate the eigenchannel quantum defect parameters [19]. Two examples have been demonstrated: The first calculation is on oscillator strengths and phase shifts near the first $1s \rightarrow 2p$ autoionization resonances of the Be iso-electronic sequence [19]. Angular distribution and spin polarization of photoelectrons in the Beutler-Fano autoionization resonances for the rare gases Ar, Kr and Xe have also been calculated [33].

--- Greene [8] developed a "first-principle" calculation of Be autoionization resonances, based upon hyperspherical coordinates. Since the hyperspherical coordinates provide an additional labeling for a doubly-excited state $\psi(R,\alpha,E)$, where $R = [r_1^2 + r_2^2]^{\frac{1}{2}}$ and $\alpha = \tan^{-1}(r_1/r_2)$, radial correlation due to doubly-excited states can be formally incorporated. This calculation confirms the early findings that the energy-dependent mixing angle θ [5,6] in the frame transformation matrix serves the purpose of characterizing the radial correlation of doubly-excited channels [7].

4. APPLICATIONS

In this section we discuss applications of MQDT for (i) Stark shifts, (ii) diamagnetic shifts, (iii) iso-valence homologues and (iv) iso-chromophoric molecules.

For atoms in external electric or magnetic fields, the long-range potential $B(r,\theta,\phi)$ of Eq. (1) takes the form r^n, where $n = 1$ for the Stark effect and $n = 2$ for the diamagnetic effect.

4.1 Stark Effect

For nonhydrogenic atoms in a Stark field, the effective potential has the form

$$-\frac{1}{r} + v(r) + \vec{F}\cdot\vec{r} \ .$$

The Stark field dominates in the dissociation zone whereas $v(r)$ dominates in the compound-state zone. However, the symmetries in these two configuration spaces are different: Cylindrical symmetry obtains in the dissociation zone for the Stark field, while spherical symmetry obtains in the compound-state zone where the electrostatic field dominates. This difference of symmetry in different configuration spaces requires a transformation between two different coordinate systems. The pure Coulomb plus Stark potential is separable in parabolic coordinates and an analytic wavefunction is known. Thus, the eigenfunctions of a nonhydrogenic atom can be represented by a linear superposition of hydrogenic Stark wavefunctions in parabolic coordinates. The mixing coefficients are expressed in terms of a reaction matrix K and fixed by the boundary conditions in the compound-state zone [34]. The eigenvalue of the reaction matrix K is represented in the space of orbital angular momentum ℓ by a phase shift δ_ℓ:

$$\pi N^*_{\ell\varepsilon} K_{\ell\ell}(\varepsilon)\ N_{\ell\varepsilon} = -\tan\delta_\ell(\varepsilon) \ .$$

The transformation of the K-matrix from the spherically-symmetric frame of ℓ into the cylindrically-symmetric frame of parabolic coordinates is carried out by a local frame transformation $A^*_{\ell,m}(\beta_n,\nu)$ [34]:

$$K_{nn'}(\varepsilon) = \sum_\ell A^*_{\ell m}(\beta_{n'},\nu)\ K_{\ell\ell}(\varepsilon)\ A_{\ell m}(\beta_n,\nu)$$

where β_n is a separation parameter and n is the parabolic quantum number. The local frame transformation is performed in the region where the field strength F is not important, namely as $(\xi,\eta) \to 0$ and $r \to 0$.

The photoionization cross-section is proportional to

$$\sum_{n'n} D^*_{n'm}(\varepsilon)\ [1 + (\pi N^*_\varepsilon\ K(\varepsilon)\ N_\varepsilon)^2]^{-1}_{n'n}\ D_{nm}(\varepsilon)$$

where $D_{nm}(\varepsilon) = \sum_\ell N^*_{n\varepsilon}\ A^*_{\ell m}\ (\beta_n,\nu)\ d_{\ell m}(\varepsilon)$ and where $d_{\ell m}(\varepsilon)$ is the zero-field transition dipole moment. The photoionization cross-section for a nonhydrogenic atom in a Stark field is thus expressed in terms of the zero-field quantum defect and the frame transformation parameters. "First-principle" calculations on Na in a Stark field have been carried out [35].

4.2 Diamagnetic Shift

A unified channel approach for atomic diamagnetism is not yet possible since the Coulomb plus diamagnetic potential is not separable in any coordinate system [36]. Nonetheless, if the diamagnetic potential can be treated as a perturbation, the quantum defect method can be readily applied. The diamagnetic shift for perturbed Rydberg series is expressed in terms of zero-field QDT parameters [16] as

$$E_D = A\ \nu_1^4\ B^2\ [1 - b(\nu_1,\ \nu_2,\ \mu_\alpha,\ U_{i\alpha})]$$

$$+ C\ \nu_1^{11}\ B^4\ [1 - d(\nu_1,\ \nu_2,\ \mu_\alpha,\ U_{i\alpha})]$$

where A and C are constants; b measures the channel interaction; and d represents the channel interaction due to the diamagnetic ℓ-mixing. The functions b and d can be obtained from zero-field spectra and, therefore, the diamagnetic shift can be evaluated.

4.3 Iso-Valence Homologues

We use the systematic study of atomic spectra along a sequence of atoms possessing the same number of valence electrons (i.e., iso-valence homologues [37]) as an example to highlight the usefulness of the graphical method in sorting out notable spectral irregularities and in pointing to areas for further study.

The first successful applications of MQDT in the characterization of complex Rydberg and autoionization spectra were to the rare gases (Ne [12], Ar [4], Kr [5] and Xe [2]). The conspicuous irregularities of the rare-gas spectra were brought out by focusing on the spin-orbit interaction of the p^5 ion core, relative to the electrostatic interaction, from Ne to Xe [38]. Extension of these studies to the Be-group atoms reveals an interesting and general effect, namely, that hybridization due to electron correlation between a Rydberg state and a doubly-excited state takes place not only in the ground state or low-energy excited states, but also in the highly-excited and continuum spectra. One consequence of hybridization between these two contrasting orbital types is the anomalous intensity redistribution in the spectra of the alkaline earths. For Be and Mg, the intensity redistribution between a Rydberg channel and a doubly-excited channel extends from the discrete into the continuum [39]. For Ca, Sr and Ba, the dominating channels are of sp and dp type and are characterized by the same set of quantum defect parameters [25]; however, the intensity redistribution is localized in these spectra. It is interesting to note that Ca, Sr and Ba are chemically very similar, and dissimilar to Be; Mg, however, is more similar to Be than to the other atoms of this group.

The carbon-group atomic spectra provide another interesting case. Fig. 5 shows the quantum defect (μ, ν_2) plots for Si, Ge, Sn and Pb (J = 1 channels, upper row; J = 2 channels, lower row; data taken from [30]). The interacting channels are Rydberg s^2pd and s^2ps, and doubly-excited sp^3. The similarity of the quantum defect plots for Ge and Sn is striking: The eigen-quantum defects for the J = 1 states of Ge are μ_α = 0.90, 0.87, 0.80, 0.32, 0.06, whereas those for Sn are μ_α = 0.91, 0.85, 0.79, 0.33, 0.06. (These numbers are read off from the figure at the intersections between the quantum defect curves and the diagonal line.) Again, it is interesting to note that the chemical and material properties of Ge are most similar to those of Sn: Both are semiconducting and superconducting materials. The significance of the similarity between quantum defect correlations and correlations arising from the macroscopic properties of different elements is not yet understood.

4.4 Transformation of Atoms into Molecules

In this section we outline the steps in which atomic states are coupled and transformed into molecular states within the general framework of MQDT. Accordingly, we point out the difference and similarity of excited Rydberg and autoionization states in atoms as compared to molecules. The results are used to interpret the striking qualitative similarity of the Rydberg spectra and autoionization structure of CH_3I and Xe (members of an isochromophoric series [40]).

Step 1. Atomic states are coupled in the α-channel compound-state zone, in which the excited electron is close to the ion core. The electron's motion is rigidly coupled to the interatomic axis $\hat{R}$ and the squared orbital-momentum projection $\Lambda^2 = (\vec{\ell}\cdot\vec{R})^2$

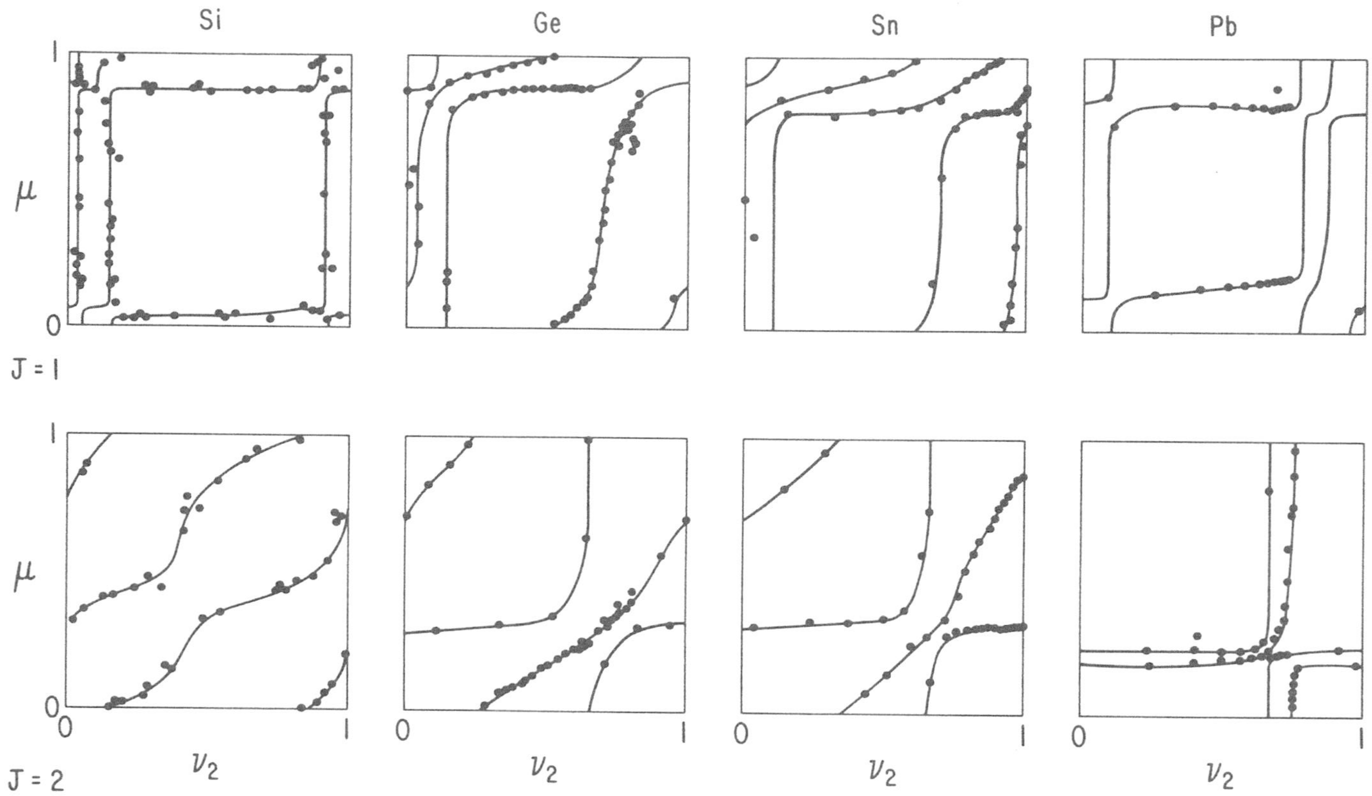

Fig. 5. The quantum defect (μ, ν_2) plots of Si, Ge, Sn and Pb for J = 1 channels (upper row) and J = 2 channels (lower row). Data taken from [30]. (Courtesy of Tang Jiang.)

is a constant of motion, whereas the squared momentum of molecular-ion rotation $\vec{N}^2$ does not have a definite value [1].

Step 2. In the i-channel dissociation zone, the electron is effectively coupled to the compound-state ion core (molecular-ion core) by the isotropic Coulomb attraction. The squared angular momentum of the electron $\vec{\ell}^2$ and of the core $\vec{N}^2$ are constants of the motion, as is $\vec{J} = \vec{\ell} + \vec{N}$.

Step 3. The transformation from the atomic (or laboratory) frame (r,θ,ϕ) to the molecular frame (r,θ',ϕ') is carried out by the wavefunction of the symmetric rotor $D^{(\ell)}$ [1]:

$$Y_{\ell\Lambda}(\theta',\phi') = \sum_m Y_{\ell m}(\theta,\phi)\, D^{(\ell)}_{m\Lambda}(-\pi-\Phi,\Theta,\pi)$$

where Θ and Φ are Euler angles.

The compound states are formed in the α-channel where the usual Born-Oppenheimer factorization of the molecular wavefunction applies. For excited electronic states and autoionization resonances, the electron is coupled to the isotropic Coulomb attraction of the ion core; thus, it makes little difference whether the ion core is atomic or molecular. This explains the qualitative similarity of the Rydberg and autoionization structures of Xe and CH_3I [9], shown in Figs. 6 and 7.

The ground state of a molecular ion in an iso-chromophoric sequence such as HI^+ and CH_3I^+ is $^2\Pi$, whereas the ground state of the corresponding atomic iso-chromophore Xe^+ is ^2P. Thus, the formation of an e + ion compound state (i.e., the α-channel) of CH_3I^* or HI^* is quite similar to that of Xe^*.

There are fundamental differences between CH_3I^* and Xe^*, however:

--- Since CH_3I has C_{3v} symmetry, the np and np' channels which were forbidden in Xe are allowed in CH_3I. In fact, an np' channel has been identified by both photoionization [41] and photoabsorption [42] methods.

--- The (Λ,S) coupling scheme for α-channels may not be a good assignment for CH_3I. This condition, as well as the preceding condition, generate more allowed channels in CH_3I as compared to Xe.

--- Most of the Rydberg channels of CH_3I are predissociated since the C-I dissociation energy is only 2.33 eV [43]. Thus, reliable intensity measurements, rather than higher resolution, are needed to make detailed assignments.

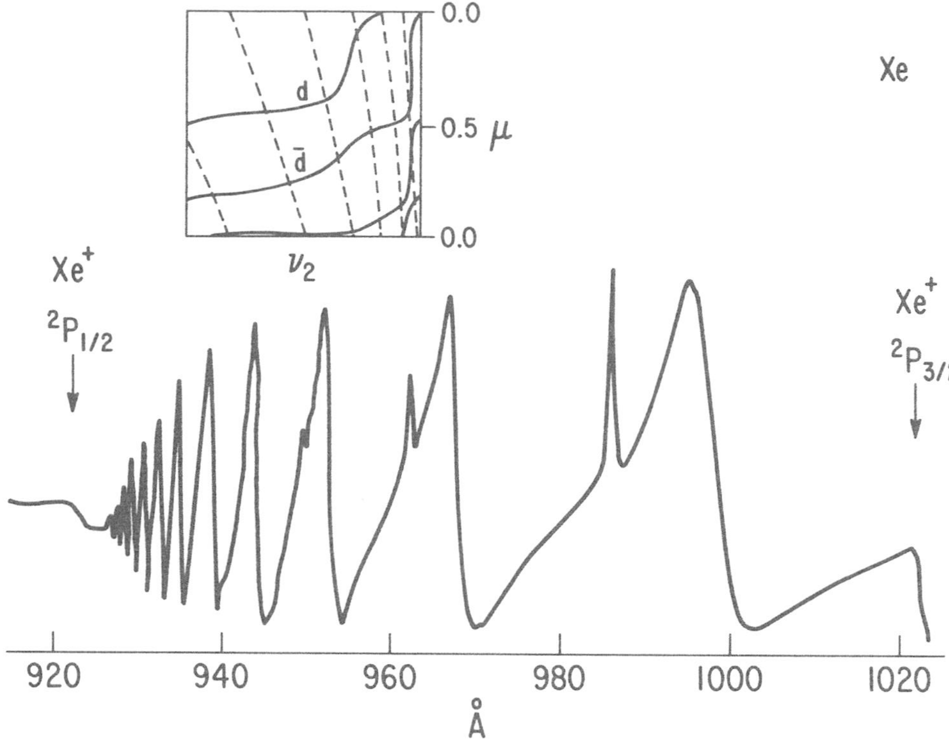

Fig. 6. The quantum defect (μ, ν_2) plot and autoionization resonances for Xe. (*Note*: $d = (^2P_{3/2})d_{5/2}$, $\bar{d} = (^2P_{3/2})d_{3/2}$.) Adapted from [2].

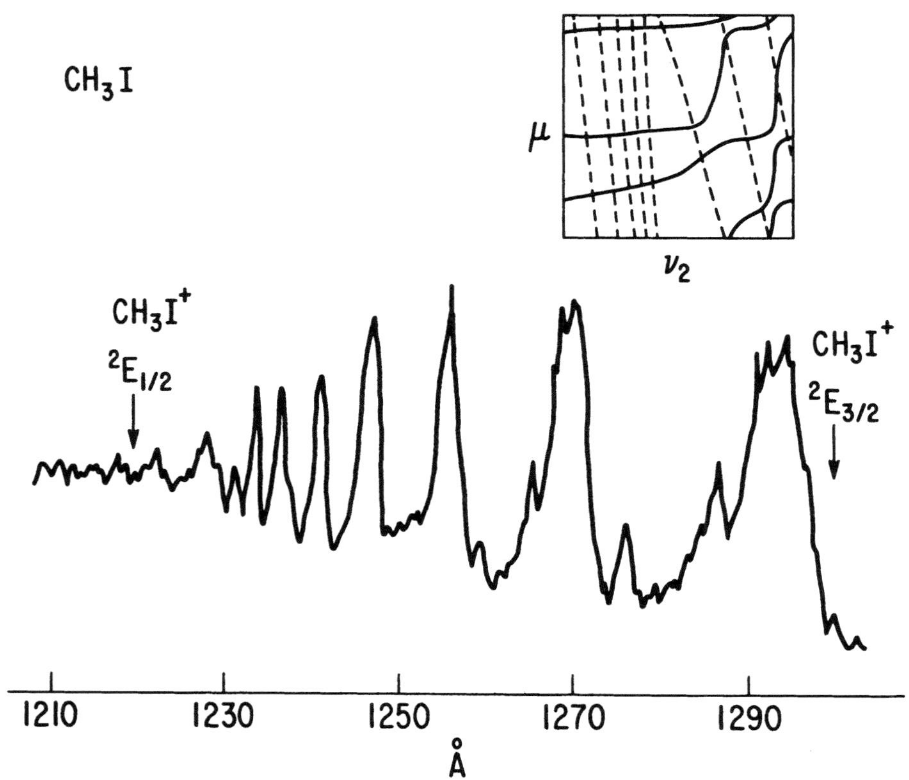

Fig. 7. The quantum defect (μ,ν_2) plot and autoionization resonances for CH_3I. Adapted from [9,42].

This qualitative understanding leads naturally into a future extension of MQDT to include rotational and vibrational coupling in order to treat excited states of polyatomic molecules. This study may develop into a powerful means for understanding chemical reactions involving excited species, on the one hand, and for disentangling the spectra of polyatomic molecules, on the other.

REFERENCES

1. U. Fano, Phys. Rev. A2, 353 (1970), and references therein.
2. K. T. Lu, Phys. Rev. A4, 579 (1971), and references therein.
3. K. T. Lu and U. Fano, Phys. Rev. A2, 81 (1970).
4. C. M. Lee and K. T. Lu, Phys. Rev. A8, 1241 (1973).
5. J. Geiger, J. Phys. B12, 2277 (1979).
6. J. A. Armstrong, P. Esherick and J. J. Wynne, Phys. Rev. A15, 180 (1977).
7. K. T. Lu, Proc. Roy. Soc. (London) A353, 431 (1977); Bull. Am. Phys. Soc. 25, 1133 (1980).
8. C. H. Greene, Phys. Rev. A23, 661 (1981).
9. J. A. Dagata, G. L. Findley, S. P. McGlynn, J. P. Connerade and M. A. Baig, Phys. Rev. A24, 2485 (1981).
10. C. S. Chang and U. Fano, Phys. Rev. A6, 173 (1972).
11. C. M. Lee, Phys. Rev. A10, 584 (1974); U. Fano and C. M. Lee, Phys. Rev. Lett. 31, 1573 (1973).
12. K. T. Lu and M. Mansfield, in: *Electron and Photon Interaction with Atoms*, H. Kleinpoppen and M. R. C. McDowell, eds. (Plenum, New York, 1975).
13. D. Dill, Phys. Rev. A7, 1976 (1973).
14. C. M. Lee, Phys. Rev. A10, 1598 (1974).
15. J. J. Wynne, J. A. Armstrong and P. Esherick, Phys. Rev. Lett. 39, 1520 (1977).
16. K. T. Lu and A. R. P. Rau, Phys. Rev. A, to be published.
17. K. T. Lu, in: Abstracts of the 5th International Conference on Atomic Physics (1976), p. 291; Invited Talk at the Xth International Meeting on Quantum Electronics, J. Opt. Soc. Am. Abstracts (1978), p. 674.
18. K. T. Lu, Bull. Am. Phys. Soc. 22, 557 (1977); R. Beigang, E. Matthias and A. Timmermann, Phys. Rev. Lett. 47, 326 (1981); 48, 290 (1982).
19. C. M. Lee and W. R. Johnson, Phys. Rev. A22, 979 (1980); W. R. Johnson and K. T. Cheng, Phys. Rev. A20, 978 (1979).
20. R. Bell and M. J. Seaton, J. Phys. B: Atom. Molec. Phys., to be published.
21. U. Fano, J. Opt. Soc. Am. 65, 979 (1975).
22. A very extensive review has been written by M. J. Seaton, Rep. Prog. Phys. 46, 167 (1983).

23. K. T. Lu, in: *Proceedings of the Daresburg Study Weekend on Recent Developments in Electron-Atom and Electron-Molecule Collision Processes*, W. Eissner, ed. (March, 1982).
24. J. Geiger, Z. Phys. A282, 129 (1977).
25. J. J. Wynne and J. A. Armstrong, IBM J. Res. Dev. 23, 490 (1979); J. A. Armstrong, P. Esherick and J. J. Wynne, Phys. Rev. A15, 180 (1977).
26. M. Aymar, P. Camus, M. Dieulin and C. Marillon, Phys. Rev A 18, 2173 (1978); M. Aymar, A. Debarre and O. Robaux, J. Phys. B: Atom. Molec. Phys. 13, 1089 (1980).
27. J. P. Connerade, M. W. D. Mansfield, G. H. Newsom, D. H. Tracy, M. A. Baig and K. Thimm, Phil. Trans. Roy. Soc. Lond. 290, 327 (1979).
28. T. F. Gallagher, W. Sandner and K. A. Safinya, Phys. Rev. A 23, 2969 (1981).
29. S. A. Bhatti, C. L. Crommer and W. E. Cooke, Phys. Rev. A24, 161 (1981).
30. C. B. Brown, S. G. Tilford and M. L. Ginter, (a) Si, J. Opt. Soc. Am. 65, 385 (1975); (b) Ge, J. Opt. Soc. Am. 67, 584 (1977); (c) Sn, J. Opt. Soc. Am. 67, 607 (1977); (d) Pb, J. Opt. Soc. Am. 67, 1240 (1977).
31. M. G. Kozlov and S. A. Kotochigova, Opt. and Spektrosk. (USSR) 45, 869 (1978); Opt. and Spectrosc. (USA) 45, 742 (1978).
32. K. T. Lu, Phys. Rev. A16, 2184 (1977).
33. W. R. Johnson, K. T. Cheng, K. N. Huang and M. LeDourneuf, Phys. Rev. A22, 989 (1980).
34. U. Fano, Phys. Rev. A24, 619 (1981).
35. D. Harmin, Phys. Rev. A26, 2656 (1982).
36. C. Clark, K. T. Lu and A. F. Starace, in: *Progress in Atomic Spectroscopy*, Part C, H. F. Beyer and H. Kleinpoppen, eds. (Plenum, New York, in press).
37. I acknowledge W. Martin for this terminology.
38. K. T. Cheng and W. R. Johnson, Phys. Rev. A, to be published.
39. K. T. Lu, J. Opt. Soc. Am. 64, 706 (1974).
40. H. -t. Wang, Ph.D. Thesis, Louisiana State University, Baton Rouge, 1976; H. -t. Wang, W. S. Felps, G. L. Findley, A. R. P. Rau and S. P. McGlynn, J. Chem. Phys. 67, 3940 (1977).
41. P. M. Dehmer and W. H. Chupka, in: Radiological and Environmental Research Division Annual Report, ANL-78-65, Part I (Argonne National Laboratory, 1978), p.20.
42. M. A. Baig, J. P. Connerade, J. Dagata and S. P. McGlynn, J. Phys. B: Atom. Molec. Phys. 14, 125 (1981).
43. G. Herzberg, *Molecular Spectra and Molecular Structure. III. Electronic Spectra and Electronic Structure of Polyatomic Molecules* (Van Nostrand Reinhold Co., New York, 1966).
44. K. Yoshino, J. Opt. Soc. Am. 60, 1220 (1970).

ELEMENTS OF QUANTUM DEFECT THEORY. III. DIATOMIC MOLECULES

Chris H. Greene

Department of Physics and Astronomy
Louisiana State University
Baton Rouge, Louisiana 70803, USA

1. INTRODUCTION TO MOLECULAR QDT

The preceding two papers by A. R. P. Rau and K. T. Lu describe the main features of QDT with illustrative applications to atomic Rydberg states and to the spectra of atomic negative ions. The interpretation of *molecular* spectra, however, requires considerably more sophistication since the relevant degrees of freedom include vibrational and rotational excitations in addition to electronic excitation. This complexity is illustrated schematically in Fig. 1 for H_2. The large number of fragmentation channels can, in principle, be handled simultaneously using the general form of QDT; however, the bookkeeping alone is a formidable task.

For the moment we restrict ourselves to those processes leading to ionization or Rydberg excitation only, namely,

$$h\nu + H_2(v,J) \rightarrow e^- + H_2{}^+(v^+,N^+) \tag{1}$$

ignoring, for simplicity, the "true" dissociation channels which lead to H + H and $H^+ + H^-$. The processes of Eq. (1) already include spectra of extremely high complexity since the vibrational, rotational and electronic degrees of freedom interact strongly. In particular, Rydberg series are found converging to each rovibrational energy level of $H_2{}^+(v^+,N^+)$. Levels belonging to different series interact strongly and perturb one another, and result in preionization (autoionization) when one or more channels are open. The potential energy of the outer electron in the large-r range is shown schematically in Fig. 2. As long as the electron

S. P. McGlynn et al. (eds.), Photophysics and Photochemistry in the Vacuum Ultraviolet, 245–259.

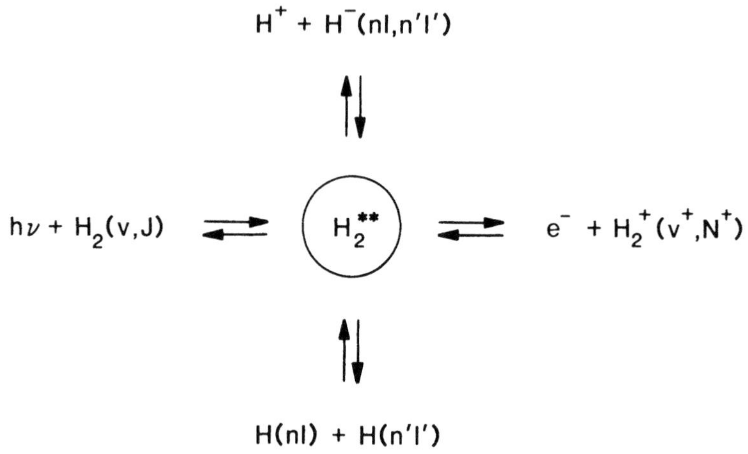

Fig. 1. Fragmentation channels for the H_2 molecule. These include neutral and ionic dissociation, ionization, and photon absorption or emission. Dissociative ionization is also possible at higher energies but is not explicitly shown here.

is outside the core radius r_o, it remains on one of these Coulomb potentials, each of which converges to a different H_2^+,v^+ energy level. After entering the core on one potential, however, the outer electron can scatter and re-emerge on any other potential since the motion is nonseparable and complicated within the core. It is just this nonseparability which causes the series perturbations and which is also responsible for the preionization of levels in the continuum.

Quantum mechanically, the amplitudes for these alternative rearrangement processes are represented by a reaction matrix $K_{v^+N^+,v^{+\prime}N^{+\prime}}$ in a basis of energy-normalized Coulomb functions $\{f_{v^+N^+},g_{v^+N^+}\}$ in each channel. Once the reaction matrix is known, the spectral properties of bound and autoionizing states are obtained in the same fashion as described in the preceding two papers in this volume. Thus, the H_2 photoabsorption spectrum can be accounted for, in principle, by fitting it to the K-matrix elements and to the corresponding dipole absorption amplitudes. In practice, however, this approach is far from satisfactory. The high density of H_2^+ rovibrational levels implies that a prohibitively large number of channels is required in order for this approach to be realistic.

Perhaps the most striking success of the frame transformation viewpoint has been its use in simplifying and clarifying

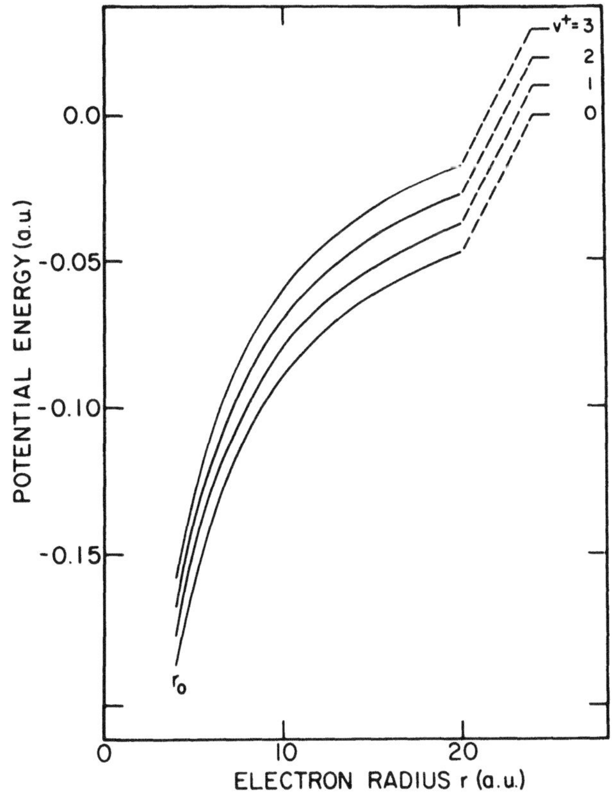

Fig. 2. Combined Coulomb and centrifugal potentials for an $\ell=1$ outer electron, as a function of the electron - ion separation r. Near the core boundary ($r = r_o$), the electronic wavelength is almost independent of the H_2^+ vibrational quantum number v^+.

problems associated with the large size of the reaction matrix, by breaking $K_{v^+N^+,v^{+\prime}N^{+\prime}}$ down into simpler and fewer significant parts. The reaction matrix is generally diagonal, or block-diagonal, when transformed into a representation for which the short-range Hamiltonian is diagonal. When the outer electron is inside the core ($r < r_o$), the good "quantum numbers" are the internuclear distance R and the orbital angular momentum projection Λ, rather than the conjugate quantum numbers v^+ and N^+ appropriate at large radii ($r > r_o$) [1-5]. That Λ is conserved at small r is quite familiar in molecular spectroscopy, and the "ℓ-uncoupling" which occurs at $r \to \infty$ reflects the ability of the $H_2^+(v^+,N^+)$ ion to rotate freely and independently of the distant outer electron

[6]. Similarly, R is very nearly a constant of the motion at small r since the electron experiences a strong Coulomb acceleration and has a high velocity. To an excellent approximation, then, once the outer electron enters the core region ($r < r_o$), the internuclear distance R does not change until the electron reemerges from the core. Formally, these frame transformation factors in $K_{v^+N^+,v^{+\prime}N^{+\prime}}$ are extracted by writing the exact expression

$$K_{v^+N^+,v^{+\prime}N^{+\prime}} = \sum_{\Lambda,\Lambda'} \int dR dR' (v^+|R)(N^+|\Lambda) \times K_{\Lambda,\Lambda'}(R,R')(R'|v^{+\prime})(\Lambda'|N^{+\prime}) \quad (2)$$

where $(v^+|R)$ are the $H_2{}^+(v^+,N^+)$ vibrational wavefunctions and $(N^+|\Lambda)$ are essentially Clebsch-Gordan coefficients, given in [4]. The physically reasonable approximations outlined above imply that $K_{\Lambda,\Lambda'}(R,R')$ is approximately local in R and diagonal in Λ:

$$K_{\Lambda,\Lambda'}(R,R') \simeq K_\Lambda(R)\ \delta_{\Lambda\Lambda'}\ \delta(R - R'). \quad (3)$$

Eq. (3) reduces Eq. (2) to the far simpler form

$$K_{v^+N^+,v^{+\prime}N^{+\prime}} \simeq \sum_{\Lambda} \int dR\ (v^+|R)(N^+|\Lambda) \tan \pi\mu_\Lambda(R)\ (R|v^{+\prime})(\Lambda|N^{+\prime}) \quad (4)$$

where we have introduced the quantum defect function, $\tan \pi\mu_\Lambda(R) \equiv K_\Lambda(R)$. As we will see, Eq. (4) holds to a remarkably high degree of accuracy, with the possible exception of states having very high vibrational quantum numbers.

The main novelty of this formulation is the continuous nature of the eigenchannel index $\alpha = \{\Lambda,R\}$, while the ionization channel index $i = \{v^+,N^+\}$ remains discrete to the extent that only *bound* vibrational states $(v^+|R)$ of the ion are relevant. A general solution to the Schrödinger equation is a superposition of all of the eigenchannel solutions $\psi_\Lambda(R')$ with as yet undetermined coefficients $A_\Lambda(R')$:

$$\Psi = \sum_\Lambda \int dR'\ \psi_\Lambda(R')\ A_\Lambda(R') \quad (5)$$

where, for $r > r_o$,

$$\psi_\Lambda(R') = \sum_{v^+N^+} [\phi_{N^+}(\hat{r},\hat{R})\ (v^+|R)]\ (v^+|R')\ (N^+|\Lambda) \times \{f_{v^+N^+}(r) \cos \pi\mu_\Lambda(R') - g_{v^+N^+}(r) \sin \pi\mu_\Lambda(R')\}. \quad (6)$$

Here $\phi_{N^+}(\hat{r},\hat{R})$ is a core wavefunction, which includes the H_2^+ electronic wavefunction and the spin and angular wavefunctions of the electrons and nuclei. The internuclear separation R plays two quite different roles in Eq. (6). Within the square brackets, R is a *coordinate* required to fully specify the $\{v^+,N^+\}$ wavefunction of H_2^+. The remaining factors in Eq. (6) contain R', which is an *eigenchannel quantum number* to be summed over as in Eq. (5). Accordingly, the factor $(v^+|R)$ in square braces occurs because it is a vibrational wavefunction, whereas the factor $(v^+|R')$ occurs because it is a factor of the frame transformation matrix $U_{i\alpha}$. The orbital angular momentum of the outer electron is almost purely $\ell = 1$ for the ungerade singlet states in the energy range near the H_2^+ 1sσ state. While ℓ is not denoted explicitly in Eq. (6), the factors containing $\{f_{v^+N^+}, g_{v^+N^+}\}$, $(N^+|\Lambda)$ and $\phi_{v^+N^+}$ depend upon ℓ implicitly. The transformation coefficients $(N^+|\Lambda)$ depend also on J (the total angular momentum of H_2), and the vibrational wavefunctions $(v^+|R)$ depend also on N^+.

The quantum defect functions $\mu_\Lambda(R)$ can be given a simple physical interpretation. Notice that in Fig. 2 the kinetic energy of the outer electron is, when near the core, nearly identical for each rovibrational channel $\{v^+,N^+\}$. This permits us to drop the v^+N^+ indices on $\{f,g\}$ and it now shows that $\pi\mu_\Lambda(R)$ is simply the body-frame phase shift of an electron escaping from H_2^+ at the fixed internuclear separation R. The requirement that

$$f(r)\cos\pi\mu_\Lambda(R) - g(r)\sin\pi\mu_\Lambda(R) \xrightarrow[r\to\infty]{} 0 \tag{7}$$

quantizes the allowed values of the electronic Rydberg states converging to the H_2^+ 1sσ state at a given R:

$$U_{n\Lambda}(R) = U_{1s\sigma}(R) - \frac{1}{2(n-\mu_\Lambda(R))^2} \; . \tag{8}$$

Thus, the quantum defect function $\mu_\Lambda(R)$ contains the same information as the adiabatic potential curves $U_{n\Lambda}(R)$, and these latter can be either taken from *ab initio* calculations or else fitted to obtain agreement with experimental results. Fig. 3 shows the $^1\Sigma_u$ and $^1\Pi_u$ quantum defect curves associated with H_2 Rydberg states converging to the H_2^+ ground-state potential $U_{1s\sigma}(R)$. In some sense, the difference $\mu_\Pi(R) - \mu_\Sigma(R)$ is a measure of the anisotropy of the electron + ion interaction since this difference would vanish for a spherically-symmetric potential.

Since Eq. (8) recovers the Born-Oppenheimer potential curves of H_2, it might appear that QDT reduces to more familiar adiabatic methods. In fact, however, the QDT approach proves to be far more accurate than the Born-Oppenheimer approach, accounting for

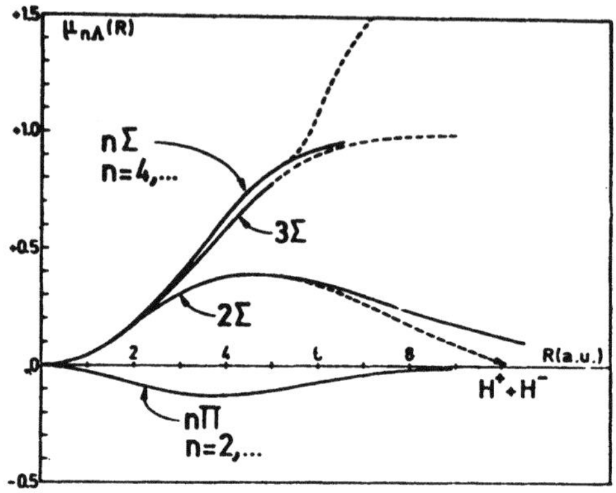

Fig. 3. Quantum defect curves $\mu_{n\Lambda}(R)$ for the ungerade Rydberg states of H_2. Taken from [5].

many so-called "nonadiabatic" effects such as predissociation. These nonadiabatic processes are accounted for in the present method precisely because the fixed-R quantization condition of Eq. (7) is *not* imposed on the wavefunction at $r \to \infty$. Instead, the asymptotic boundary conditions that are applied to the full wavefunction of Eq. (5) are the usual ones of MQDT: Namely, Ψ vanishes at $r \to \infty$ for those $\{v^+,N^+\}$ channels which are closed ($i \in Q$), whereas the wavefunction components in open $\{v^+,N^+\}$ channels ($i \in P$) satisfy continuum boundary conditions such as the collision eigenchannel condition or the incoming-wave boundary condition. (We follow closely the discussion and notation of Jungen and Dill [7].) In general, the infinite number of asymptotic channels $\{v^+,N^+\}$ must be truncated to a finite number N, whereby the large-r boundary conditions specify N conditions on the $A_\Lambda(R')$ of Eq. (5). This is an underdetermined system since the $A_\Lambda(R')$ represent an infinite number of unknowns. To make this problem well-defined, the usual approach is to expand $A_\Lambda(R')$ with respect to some complete set of vibrational functions:

$$A_\Lambda(R') = \sum_{v^{+\prime}N^{+\prime}} (\Lambda|N^{+\prime})(R'|v^{+\prime})A_{v^{+\prime}N^{+\prime}} \quad . \tag{9}$$

When this expansion is truncated to its first N terms, the homogeneous system for the N coefficients $A_{v^{+\prime}N^{+\prime}}$ is well-defined:

$$\sum_{i'} [\sin \pi\nu_i \; c_{ii'} + \cos \pi\nu_i \; s_{ii'}]A_{i'\rho} = 0, \quad i \in Q \tag{10}$$

$$\sum_{i'} [\sin \pi\tau_\rho \; c_{ii'} - \cos \pi\tau_\rho \; s_{ii'}]A_{i'\rho} = 0, \quad i \in P \tag{11}$$

where we have used the notation $i = \{v^+,N^+\}$ for the asymptotic channel indices and where i' represents the expansion indices of Eq. (9). A given i channel belongs to Q (or P) if it is closed (or open) asymptotically. The matrix $s_{ii'}$ is defined by

$$s_{ii'} = \sum_{\Lambda}(N^+|\Lambda)\left[\int dR'(v^+|R')\sin\pi\mu_\Lambda(R')(R'|v^{+\prime})\right](\Lambda|N^{+\prime}) \tag{12}$$

and a similar expression holds for $c_{ii'}$ with $\sin\pi\mu_\Lambda(R)$ replaced by $\cos\pi\mu_\Lambda(R)$. Note that $s_{ii'}$ and $c_{ii'}$ would vanish for $i \neq i'$ if $\mu_\Lambda(R)$ were to be independent of R and Λ. In this limiting case there would be no channel interaction and, hence, no series perturbations or preionization.

The collision eigen-phase shifts τ_ρ are determined by Eqs. (10) and (11) [as in the preceding paper by K. T. Lu], with the number of values of the index ρ equal to the number of open channels, N_P. Also, the normalization of the solution vectors $A_{i'\rho}$ is fixed by requiring the $N_P \times N_P$ matrix $T_{i\rho}$ to be orthogonal, where $T_{i\rho}$ is defined by

$$T_{i\rho} = \sum_{i'}[\cos\pi\tau_\rho\, c_{ii'} + \sin\pi\tau_\rho\, s_{ii'}]A_{i'\rho} \tag{13}$$

Once the $A_{i'\rho}$, τ_ρ and $T_{i\rho}$ are constructed, the observables corresponding to specific measurements can be calculated. For photoionization, the collision eigenchannel solutions $\{\rho\}$ must be superposed to satisfy incoming-wave boundary conditions. Next, the reduced dipole matrix elements D(i,J) can be calculated and from these one obtains the desired partial or total cross-sections, photoelectron angular distribution, spin polarization and the alignment of the H_2^+ residue [8,9]. The calculation of D(i,J) requires the eigenchannel reduced dipole matrix elements d_α (see the papers by Rau and Lu in this volume), which can be written for the present problem as $d_\Lambda(R)$. These must be found in advance, or else fitted to obtain agreement with experiment. Previous workers [7,10,11], as a starting approximation, took $d_\Sigma = d_\Pi = 1.86$ au, independent of R and Λ (which would be correct in the united-atom limit).

2. COMPARISON WITH EXPERIMENT

The $2p\sigma$ and $2p\pi$ energy levels of H_2 (usually denoted B $^1\Sigma_u^+$, $^1\Pi_u^+$) provide a good starting point to compare the QDT results of Jungen and Atabek [5] with experiment, and with conventional *ab initio* calculations [12] based on the adiabatic approximation. This is the energy range best suited to the adiabatic approach since it lies well below the ionization threshold of H_2 (or D_2)

and, therefore, the interactions with higher Rydberg states are comparatively weak. Fig. 4 shows the deviations of the calculated J = 0 levels from experiment; these deviations are quite small for both methods. Note that the QDT approach, which uses the *ab initio* quantum defect function calculated in [12], agrees somewhat better with experiment than does adiabatic theory. Moreover, QDT does not require separate evaluation of the so-called adiabatic and nonadiabatic corrections which, in this example, are quite sizeable. For higher-lying Rydberg states, such as the B'' $^1\Sigma_u^+$ state, Jungen and Atabek find the QDT results to be significantly better than adiabatic theory [5].

At energies above the first ionization threshold of H_2, all levels are preionized to some extent, a feature which is difficult to incorporate within the conventional treatment of the spectrum. In the QDT formulation, however, this continuum is automat-

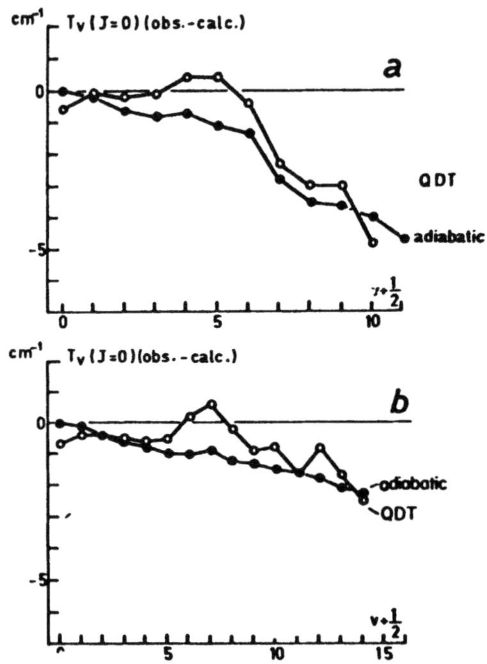

Fig. 4. Deviation between theoretical and experimental energy levels of the J = 0, B $^1\Sigma_u^+$ states of H_2(a) and D_2(b) as a function of the vibrational quantum number. The filled circles are the conventional adiabatic results, while the open circles are the QDT values. Taken from [5].

ically included. Fig. 5 compares the quantum defect calculations of Jungen and Raoult [11] with the relative photoionization cross-sections measured by Dehmer and Chupka [13]. Even subtle features in the cross-section, such as the barely noticeable $8p\pi, v = 5$ resonance, are reproduced by theory. The $6p\sigma, v = 6$ resonance, shown on the right-hand side of Fig. 5, is embedded among the np Rydberg levels converging to the $v^+ = 4$, $N^+ = 0,2$ levels of H_2^+ and it distributes its intensity among a large number of these Rydberg states. The same theoretical study [11] also predicted more detailed aspects of the photoionization spectrum, including the photoelectron angular distribution asymmetry parameter β and the vibrational branching ratios. These complicated and interesting features have not yet been observed experimentally, except for the branching ratios at the wavelength 752.866 Å, corresponding to the $6p\pi$, $v = 6$ peak:

Branching Ratios	$v^+ = 4$	$v^+ = 3$	$v^+ = 2$
MQDT [9], %	82	16	2
Experiment [13], %	82	15	3

The above agreement lends support to the approximations used, namely, that only p waves of the photoelectron are relevant in the final state and that $d_\Lambda(R)$ is independent of Λ and R within the radial range of the H_2 ground state.

One particularly interesting aspect of the QDT results is their ability to account for many so-called "nonadiabatic" effects usually associated with variations of the electronic wavefunction with internuclear distance. This is illustrated in Fig. 6. Levels calculated in the strict Born-Oppenheimer approximation, which ignores terms such as $(\Phi_{el}|\partial^2\Phi_{el}/\partial R^2)$, differ from experiment by roughly an order of magnitude more than the quantum defect results do. Physically, the ability of QDT to account for such effects can be understood in terms of the fact that the QDT wavefunction is essentially exact for $r > r_0$. In particular, this is precisely the region of configuration space where nonadiabatic processes dominate since, in this region, the outer electron moves more slowly the greater is r. To the extent that Eq. (3) is valid, then, the quantum defect method accounts for nonadiabatic effects *exactly*.

3. INCLUSION OF MOLECULAR DISSOCIATION

Because nonadiabatic processes are included in the theory, it is natural to adopt QDT to describe the dissociation continuum as an additional "i"-channel, and thereby to account for processes such as predissociation and dissociative recombination. These extensions have been developed very recently by Giusti-Suzor [14] and by Jungen [15]. These nonadiabatic processes generally be-

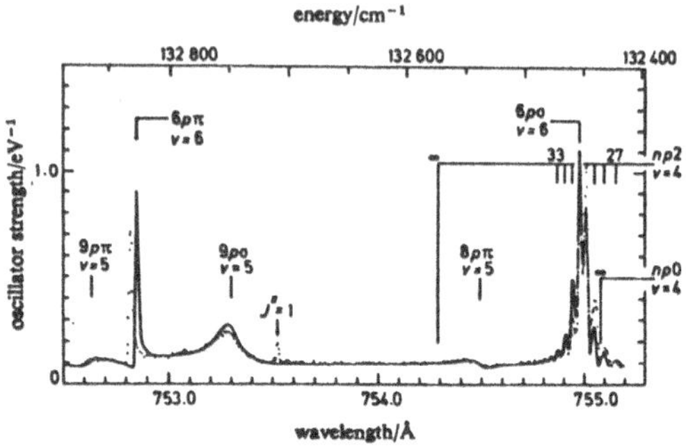

Fig. 5. Comparison of experimental and theoretical (QDT) oscillator strengths for the photoionization of H_2 (J = 0), in the spectral range of autoionizing levels. Taken from [5].

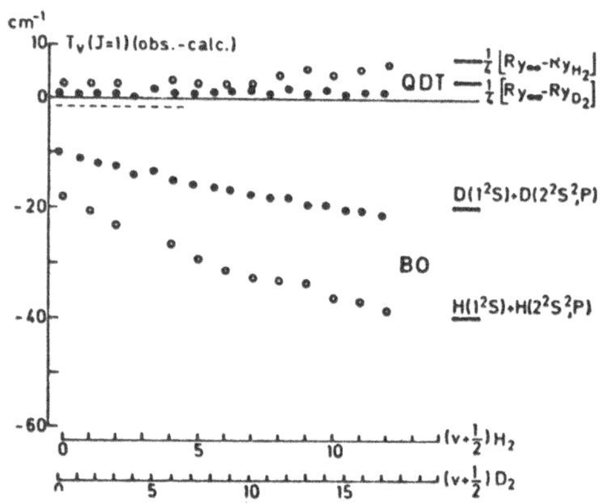

Fig. 6. Deviations between experimental and theoretical energies of the J = 1, C $^1\Pi_u^-$ states as a function of vibrational quantum number, for H_2 and D_2. The QDT results, as well as the values obtained in the strict Born-Oppenheimer (BO) approximation, are shown. Taken from [5].

long to one of two possible types:

Type (i). Nonadiabatic processes at large electronic radii, $r \gg R$. This type consists of interactions within a single electronic channel, such as the np Rydberg states treated by Jungen and coworkers [5,7,11].

Type (ii). Nonadiabatic processes at smaller radii, $r \simeq R$. This type usually arises from configuration interaction between different electronic channels. Accordingly, it involves additional nonadiabatic effects not described by an R-dependent quantum defect function $\mu_\Lambda(R)$.

Because of their qualitative difference, these two types of nonadiabatic effects require rather different theoretical treatments. We consider *Type* (i) first. The inclusion of vibrational continuum states in the treatment outlined above is rather natural since the summations over v^+ in Eqs. (9-11) are intended formally to include those unbound states. This fact was noted long ago by Chang, Dill and Fano [3], although it has been implemented in detail by Jungen only much more recently [15]. The eigenchannel formulation of the R-matrix theory given by Fano and Lee [16] has provided the decisive key to this extension of molecular QDT. For definiteness, consider those levels which lie below the first ionization threshold of H_2, but which are *above* the dissociation threshold $H(n = 1) + H(n = 2)$. In the usual MQDT approach, using only bound vibrational functions $(v^+|R)$ as in Eqs. (9-12), the energy levels E_n are considered as truly discrete. Jungen's elegant method [15] of including the possibility of dissociation differs in two key ways from the usual approach:

--- The solutions $(v^+|R)$ are calculated only within a finite range $R \leq R_o$, with R_o to be specified below.

--- The solutions $(v^+|R)$ are not required to vanish at large R, but instead satisfy the "collision eigenchannel" boundary condition at $R \geq R_o$:

$$(v^+|R) \propto f(R) \cos \pi\tau - g(R) \sin \pi\tau \tag{14}$$

where $\{f,g\}$ is the energy-normalized basis pair of wavefunctions calculated in the relevant internuclear potential curve.

The modified boundary condition of Eq. (14) now allows the homogeneous system of Eqs. (10,11) to be solved at *all* energies E (instead of only at certain energies E_n), for a suitable choice of the phase parameter τ. More explicitly, once τ is fixed a set of discrete levels $E_n(\tau)$ is generated in the usual MQDT approach. The lowest vibrational levels are hardly changed as τ varies since they are exponentially small at $R = R_o$. The higher $E_n(\tau)$

levels, however, possess some appreciable amplitude near $R = R_0$ and sweep over a considerable energy range as τ is varied from 0 to 1. For these energies, an inversion of the solutions $E_n(\tau)$ determines the single-valued phase shift as a function of energy, namely $\tau(E)$. Near a predissociated state, τ will rise rapidly by roughly unity; and, from the maximum rate of change $d\tau/dE$, the predissociation width can be obtained.

When ionization is also energetically allowed, the above approach permits the competition between photoionization and photodissociation to be studied. In the main, photoionization is significantly stronger than photodissociation when both are allowed, although exceptions to this rule occur, notably near resonances. Fig. 7 shows such an exception, namely, the $4p\sigma$, $v = 6$, $J = 0$ state of H_2. The experiment of Dehmer and Chupka [13] demonstrates this state clearly in the photoabsorption cross-section, but not in the photoionization signal. The difference is attributed to photodissociation and Jungen's approach [15] gives theoretical results in good agreement with experiment.

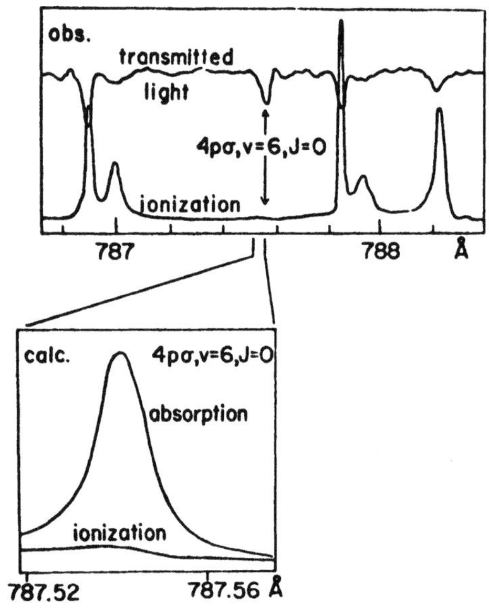

Fig. 7. Competition of preionization and predissociation in the photoabsorption spectrum of H_2, above the $v^+ = 1$ ionization threshold. The top figure shows experimental results [13], while the bottom figure is the QDT spectrum of [15]. Taken from [15].

It should be mentioned that the choice of R_0 is quite crucial in this formulation, and is dictated by two conflicting desires. The value of R_0 should be as small as possible from the point of view of obtaining rapid convergence, since the density of energy levels increases rapidly with R_0. On the other hand, R_0 must be large enough so that all of the relevant nonadiabatic processes are contained within $R < R_0$. In addition, for energies in the vicinity of a given H_2^+ energy level $E_{v^+N^+}$, the boundary should easily contain the entire vibrational wavefunction $(v^+|R)$ inside $R < R_0$. In practice, these constraints may require experimentation with alternative values of R_0 in order to find an optimal value.

For the second class [*Type* (ii)] of nonadiabatic processes considered by Giusti-Suzor [14], a repulsive electronic potential curve leading to dissociation, A + B, adiabatically crosses a family of Rydberg states, $AB^+ + e^-(n\ell)$. In the absence of electrostatic "interactions" (as in the Hartree-Fock model), these Rydberg potential curves are completely independent of the repulsive "dissociative" potential curve and, therefore, dissociative recombination, $e^- + AB^+(v^+) \rightarrow A + B$, cannot occur. In reality, these different types of potential curves have some residual coupling V(R) which is nearly independent of n in an energy-normalized representation of the Rydberg states. The reaction amplitudes for $e^- + AB^+(v^+)$ scattering are still represented by $K_{v^+N^+,v^{+\prime}N^{+\prime}}$ in Eq. (4); however, the additional dissociative channel "d" implies that an extra row and column must be added to the K-matrix. Dissociative recombination is then mediated by the amplitudes $K_{v^+N^+,d}$, while elastic A + B scattering involves the element $K_{d,d}$. To lowest order in perturbation theory, Giusti-Suzor [14] has shown that the new reaction amplitudes are given by $K_{d,d} \simeq 0$ and

$$K_{v^+N^+,d} \simeq \pi^{-1} \int dR\ (v^+N^+|R)\ V(R)\ F_d(R) \tag{15}$$

where $F_d(R)$ is the regular energy-normalized continuum wavefunction calculated in the dissociative-state NO potential curve at the desired total energy. Calculations by Giusti-Suzor [14] on NO and H_2 have shown that Rydberg levels converging to excited vibrational states of AB^+ induce a striking resonant pattern in the dissociative recombination cross-section σ_{dr}. Moreover, σ_{dr} often varies by more than an order of magnitude when the level of target vibration is changed by one quantum. The dramatic dependence on target vibration has also been seen experimentally by Allan and Wong [17]. The above formulation applies equally well to photoabsorption by the neutral AB molecule and it has successfully explained significant differences between the photoionization and photoabsorption cross-sections of NO, with the explanation being couched in terms of the competing photodissociation channel [14].

The above discussion illustrates the power of quantum defect methods. The extension of these methods to the related problem of dissociative attachment, $e^- + AB \rightarrow A^- + B$, appears ripe for development. The different long-range $e^- + AB$ potential (i.e., r^{-2} or r^{-4}) should pose no particular difficulty. However, the body-frame scattering parameters [e.g., $\mu_\Lambda(R)$] acquire an energy dependence of short-range origin (particularly near shape resonances) and the effects of this energy dependence remain to be explored.

ACKNOWLEDGMENT

The hospitality of Professor R. N. Zare, during a summer stay at Stanford when this work was completed, is greatly appreciated. Comments on the manuscript by Mr. M. Raoult of Orsay have been extremely informative. I am deeply grateful to Dr. A. Giusti-Suzor, Dr. Ch. Jungen and Mr. M. Raoult for lengthy conversations on these matters and for access to their unpublished results. This work was supported in part by the National Science Foundation.

REFERENCES

1. U. Fano, Phys. Rev. A2, 353 (1970).
2. G. Herzberg and Ch. Jungen, J. Mol. Spectrosc. 41, 425 (1972).
3. E. S. Chang, D. Dill and U. Fano, in: *Abstracts of Papers, Proceedings of the VIIIth ICPEAC*, B. C. Bocic and M. V. Kurepa, eds. (Institute of Physics, Belgrade, 1973), p. 536.
4. E. S. Chang and U. Fano, Phys. Rev. A6, 173 (1972).
5. Ch. Jungen and O. Atabek, J. Chem. Phys. 66, 5584 (1977).
6. G. Herzberg, *Molecular Spectra and Molecular Structure. I. Spectra of Diatomic Molecules* (Van Nostrand Reinhold, New York, 1950).
7. Ch. Jungen and D. Dill, J. Chem. Phys. 73, 3328 (1980).
8. C. M. Lee, Phys. Rev. A10, 1598 (1974); D. Dill, Phys. Rev. A6, 160 (1972).
9. C. H. Greene and R. N. Zare, Phys. Rev. A25, 2031 (1982); C. H. Greene and R. N. Zare, Ann. Rev. Phys. Chem. 33, 119 (1982).
10. M. Raoult, Ch. Jungen and D. Dill, J. Chim. Phys. 77, 599 (1980).
11. Ch. Jungen and M. Raoult, Faraday Discuss. Chem. Soc. 71, 253 (1981).
12. W. Kołos and L. Wolniewicz, J. Chem. Phys. 43, 2429 (1965); 48, 3672 (1968).
13. P. M. Dehmer and W. A. Chupka, J. Chem. Phys. 65, 2243 (1976); 66, 1972 (1976).

14. A. Giusti-Suzor, J. Phys. B 13, 3867 (1980); A. Giusti-Suzor, in: *Invited Papers at the XIIth ICPEAC*, S. Datz, ed. (North-Holland, Amsterdam, 1982), p. 381.
15. Ch. Jungen, in: *Invited Papers at the XIIth ICPEAC*, S. Datz, ed. (North-Holland, Amsterdam, 1982), p. 455.
16. U. Fano and C. M. Lee, Phys. Rev. Lett. 31, 1573 (1973); C. M. Lee, Phys. Rev. A10, 584 (1974).
17. M. Allan and S. F. Wong, J. Chem. Phys. 74, 1687 (1981).

NEGATIVE-ION STATES

R. N. Compton

Chemical Physics Section
Health and Safety Research Division
Oak Ridge National Laboratory
Oak Ridge, Tennessee 37830, USA

1. INTRODUCTION

Negative ions play important roles in many areas of photophysics and photochemistry. Negative ions can be produced directly by photodissociation of molecules through ion-pair production, or indirectly via charge exchange between photoexcited species and electronegative atoms or molecules. The most vivid illustration of the importance of negative ions in photophysics is to be found in the radiation emitted by the sun. The solar spectrum exhibits an intense self-absorption for wavelengths longer than 5000 Å. In 1939, Wildt [1] showed that H^- might exist in the solar atmosphere in sufficient concentrations to be the main source of opacity in the infrared region of the solar spectrum. In order to explain the absorption over the entire infrared region, it is necessary to take into account both photodetachment of H^- and the so-called free-free absorption in which an electron makes a transition between two continuum states in the field of an H atom.

In this review, we will discuss only gas-phase negative-ion properties and photo-reactions. Negative ions in the condensed phase have been extensively studied by spectroscopic, electron spin resonance and other techniques. Sanche [2] has recently suggested that shape resonances [3], formed by the capture of an incident electron by its centrifugal angular momentum barrier, could form tight-binding π bands in organic insulators. Electronic motion within these bands could account for electrical conductivity in organic solids. Many organic dyes used in dye lasers probably fluoresce and lase between negative-ion transi-

S. P. McGlynn et al. (eds.), Photophysics and Photochemistry in the Vacuum Ultraviolet, 261–295.

tions. Although there have been no reports of a negative-ion gas-phase laser, there seems to be no reason why such a device would not be possible. Since most small (3 or 4 atom) negative ions do not possess bound excited states, such a laser would probably involve large organic anions and therefore might afford some moderate tuning capabilities.

We will also omit, in this discussion, studies of negative ions on surfaces, although this is an interesting area of research. It has recently been shown [4] that compound [3] states of O_2^- and N_2^- can play significant roles in the scattering of electrons from surfaces covered with O_2 and N_2.

2. NEGATIVE-ION PROPERTIES

The potential energy surface of a negative ion essentially defines the physical properties of that ion. Obtaining experimental and theoretical information about these surfaces remains an arduous task. There is accurate information on electron affinities, vibrational frequencies and geometries for only a number of small negative ions. In a few cases, reasonably precise potential energy surfaces can be constructed.

The adiabatic electron affinity (EA) of a molecule is defined as the difference in energy between the neutral molecule in its electronic and vibrational ground state E^0, and the ion in its ground state E^- (i.e., $EA = E^0 - E^-$). Studies of electron photodetachment or radiative attachment have provided values for atomic EA's with accuracies often better than 0.01 eV. The first accurate measurements of halogen EA's are credited to Berry and Reimann [5]. On the other hand, very few accurate determinations of molecular EA's have been made because of the complexity of molecular negative ions. In a few cases (notably NO and O_2) where photodetachment and photoelectron spectroscopy [6] have been applied, EA's are known with accuracies better than 0.01 eV. In cases where the EA is negative and a well-defined vibrational progression is observed, the resonance scattering technique [7] has provided adiabatic EA's with accuracies of ±0.05 eV. In some cases, notably the studies of Richardson and co-workers [8] and Herbst, *et al.* [9], accurate lower limits to EA's for complex molecules have been determined from photodetachment thresholds.

The determination of energy thresholds for the transfer of an electron between heavy particles and target molecules is often used to infer molecular EA's. For example, studies of reactions of the type $X^- + M \rightarrow X + M^-$ have provided accurate EA's (±0.2 eV) for many molecules [10-15].

Helbing and Rothe [16] first employed the alkali collisional-ionization technique to determine molecular EA's. This method involves the measurement of the energy threshold for the charge-exchange reaction between an alkali atom and a molecule, namely, $A + M \rightarrow A^+ + M^-$. In one instance, the energy of the alkali ion was measured to infer the EA of CO_2 [17]. A number of groups are actively employing the threshold collisional-ionization technique for the purpose of determining molecular EA's (see the review by Baede [18]). This technique has proven especially useful for polyatomic molecules where other methods are not presently applicable. However, the method suffers from three major uncertainties: (i), strictly speaking, only lower limits to EA's are deduced; (ii), the thresholds are sensitive to the target-gas temperature; and (iii), in some cases the actual threshold is difficult to determine because of weak signals or because of the nature of the cross-section (or both). In cases where the target molecules form a gas at room temperature, the Doppler motion can have serious effects on the observed threshold.

The flowing afterglow technique has provided a wealth of information on negative-ion/molecule reaction rates. The EA's of some molecules have also been bracketed by observing whether charge transfer does or does not proceed with the negative ions of known EA (see, e.g., [19, 20]).

Electron affinity values which have been reported for atoms and small molecules (fewer than five atoms) range from $\sim$-2 to +4 eV. Fig. 1 illustrates this with a plot of measured EA's for molecules made from the atoms of the first (H) and second rows of the periodic table. The EA is plotted versus the number of valence electrons (total number of electrons minus the 2 K-shell electrons). Note that ions with 16 and 24 valence electrons are quite stable, reflecting the fact that greatest stability exists for "closed shells".

Even though the EA's for the molecules shown in Fig. 1 cover a limited range, there is evidence that some of the heavier hexafluoride molecules possess EA's of up to 10 eV! The first evidence of these surprising molecular negative-ion properties came from thermochemistry considerations. Bartlett [21] analyzed the oxidizing properties of the hexafluorides of the third transition metal series and found that the EA's increase in the order $WF_6 < ReF_6 < O_5F_6 < IrF_6 < PtF_6$ with EA (ReF_6) > 3.9 eV, EA (IrF_6) > 5.45 eV and EA (PtF_6) > 6.75 eV.

The EA's increase ($\sim$0.85 eV) with unit increase in atomic number of the transition metal. More recent data of Barberi and Bartlett [22] give EA (PtF_6) = 9.3 eV. Bartlett's recent estimate (private communication) of the EA (AuF_6) is 10 ± 0.5 eV! Compton and Reinhardt [23] have studied the reactions $Cs + (AuF_5)_2 \rightarrow AuF_6^- + AuF_4$ and also find evidence for an EA > 9 eV for AuF_6.

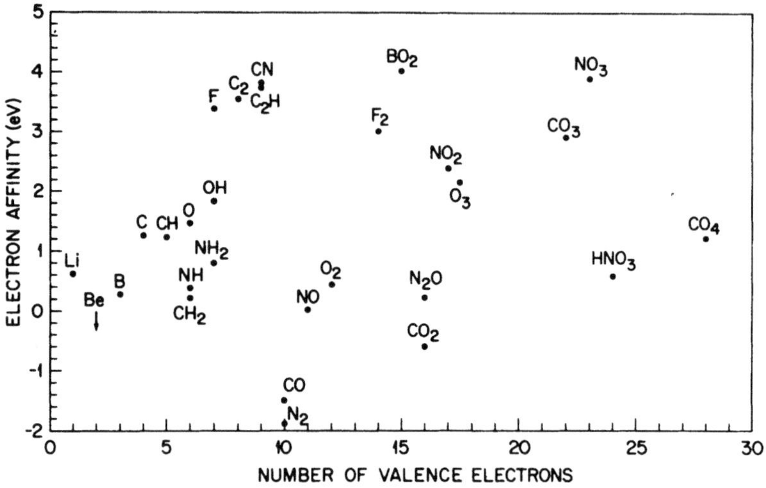

Fig. 1. Experimentally measured electron affinities for atoms and molecules made from atoms of the first (H) and second rows of the periodic table versus the number of valence electrons for the neutral. Notice that the values range from ∿-2 to +4 eV and that abrupt changes occur at a completed shell.

Jensen and Miller have studied electron attachment and compound formation in flames and were able to deduce the following rather large molecular EA's: BO_2 (4.07 eV) [24], WO_3 (3.64 eV) [24], HWO_4 (4.35 eV) [25], CrO_3 (4.04 eV) [26], ReO_4 (4.45 eV) [27] and $HMoO_4$ (4.25 eV) [28]. Although these EA's are not as large as those reported for some hexafluorides, they serve to illustrate that the EA's of many molecules can be larger than the accepted values for the halogen atoms.

Burgess, *et al.* [29] have studied the thermochemistry of alkali metal hexafluoromolybdates and hexafluorotungstates to derive the standard enthalpies of formation of KWF_6, $RbWF_6$, $CsWF_6$, $KMoF_6$, $RbMoF_6$ and $CsMoF_6$. From these measurements, they estimate the gas-phase EA's of WF_6 and MoF_6 to be 5.09 ± 0.05 eV and 5.36 ± 0.05 eV, respectively.

Charge-exchange reactions involving negative ions and hexafluorides, and alkali atoms and hexafluorides, have provided further evidence for very large EA's of many hexafluorides (see [30] and the literature cited therein). The large EA's and the complex electronic structure of these hexafluorides give rise to low-lying bound excited states. The extra electron in UF_6^-, for example, exhibits two excited states ($^2T_{2u}$ and $^2T_{1u}$) in the region of 1-4 eV [31]. Experimental studies of these excited states in the gas phase would be interesting.

3. UNIMOLECULAR ELECTRON ATTACHMENT

The attachment of a free electron to a molecule can be studied under single-collision conditions in the gas phase. The electron is said to be attached if it is "temporarily" held up for a time at least greater than its normal "transit time" across the molecule. In this case, one describes the electron-molecule system as a "resonance". Bardsley and Mandl [32] classified such resonances as (i) shape resonances, (ii) nuclear-excited Feshbach resonances, or (iii) electronically-excited Feshbach resonances. In the case of a shape resonance, the electron-molecule interaction potential possesses a barrier, most commonly a centrifugal barrier built in by the orbital angular momentum of the colliding particles. The electron is then temporarily held up at the electron-molecule separation distance where the "potential well" exists due to the barrier. Electrons can be temporarily bound through Feshbach resonances if electronic or nuclear motion of the target molecule is excited. Thus, an electron of energy E_e is attached to a molecule to form a molecular ion with total energy E*: That is,

$$e(E_e) + M(E) \underset{\tau(E^*)}{\overset{\sigma_a(E_e)}{\rightleftarrows}} M^-(E^*) \quad . \tag{1}$$

The width of the attachment resonance, Γ (i.e., the range of energies E_e over which attachment can occur), is related to the lifetime $\tau(E^*)$ through the uncertainty principle $\tau(E^*)\Gamma \simeq h$. In many instances, the width of a resonance is broader than the spacings between vibrational (and rotational) levels of the ion and a broad resonance is observed, the width of which is determined by the Franck-Condon factors between the neutral molecule and the ion. The lifetime of a resonance can be inferred from the measured width or, in some cases, measured directly.

The cross-section is related to the lifetime through the principle of detailed balance, namely,

$$\sigma_a(E_e)\tau(E^*) = \frac{\rho^-(E^*)}{v\rho^o(E)} \tag{2}$$

where v is the velocity of the electron and where ρ designates the density of states of the ion (-) or neutral (o). The lifetime of the ion against autodetachment, τ, is a function of the total internal energy E*. The total internal energy is the sum of the incident electron energy and the electron affinity EA of the molecule (i.e., $E^* = E_e + EA$). The lifetimes of these resonances can vary from $\sim 10^{-15}$ s (e.g., H_2^-), where it becomes difficult to speak of a resonance, to $> 10^{-3}$ s (e.g., $TCNQ^-$), where one is dealing with almost stationary states of the electron-molecule system.

When a complex molecule attaches a slow electron, the total energy E* is shared among the various degrees of freedom of the ion. Inspection of Eq. (2) shows that if $\rho^-(E^*)$ is large, the lifetime *and* cross-section can be large. For negative ions with lifetimes greater than 10^{-6} s, it is possible to study these resonances directly and even to measure their lifetimes as a function of total energy E*. Over the past ten years, we have determined the lifetimes of a large number of complex anions in a time-of-flight mass spectrometer. We measure the ratio of neutral molecules, arriving at a detector at a distance from the ion source, to the negative ions arriving at the same detector. The neutrals result from autodetachment of the metastable anions. The time during which autodetachment can occur is varied by changing the ion velocity or the distance of travel. As an example, Fig. 2 shows the data for electron attachment to a conjugated organic molecule, parabenzoquinone [33]. The PBQ^- ion has three excited states $^2B_{1u}$, ^2A_u and $^2B_{1u}$ which lie 3 to 4 eV above the $^2B_{3g}$ ground state. The second state leads to long-lived PBQ^-

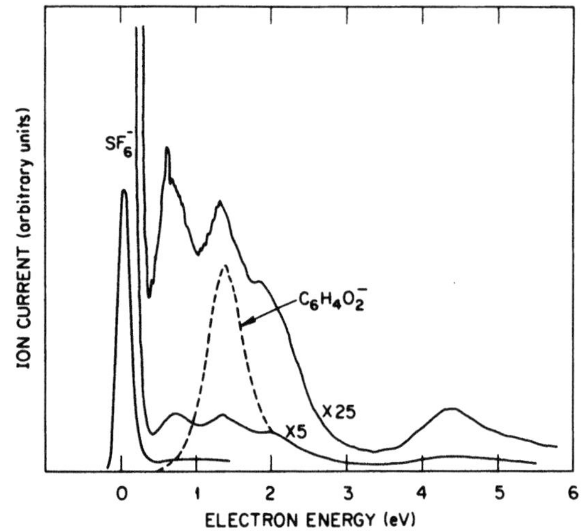

Fig. 2. Solid lines are the SF_6 scavenger spectra showing low-lying compound negative-ion states in p-benzoquinone (PBQ). The PBQ^- negative ion signal is also shown as a dashed curve. The energy dependence of the PBQ^- ion lifetime is shown in Fig. 3 [33].

ions, the lifetime of which varies with electron energy (see Fig. 3).

Electron attachment into π orbitals (shape resonances) of organic molecules has been extensively studied by the techniques of electron transmission spectroscopy and negative-ion mass spectrometry. The first study that showed vibrational structure in such resonances was due to Hasted and his colleagues [34]. Jordan and Burrow (see [35] for a review) have made significant recent progress in describing and categorizing shape resonances of organic molecules. They have shown that it is possible, by using electron transmission spectroscopy, to map out the spectrum of unoccupied molecular orbitals in much the same manner that photoelectron spectroscopy maps out the occupied orbitals. Some remarkably simple, but useful, conclusions have already been reached. For example, Younkin, *et al.* [36] have used the Koopmans theorem to relate the calculated energy of the lowest

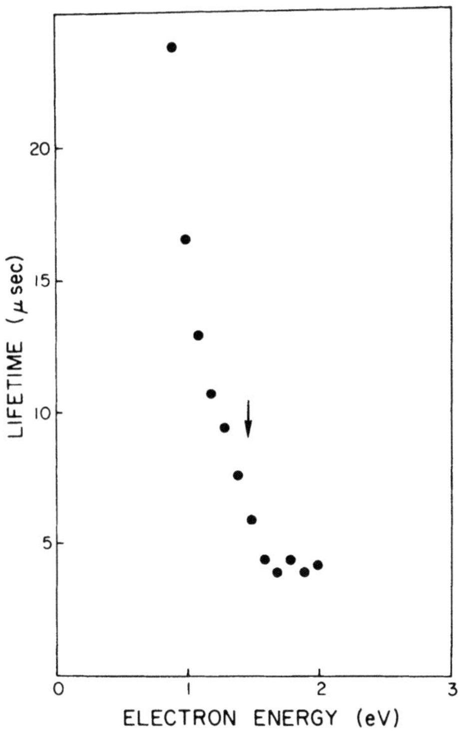

Fig. 3. The dependence of autodetachment lifetimes of PBQ^- on the bombarding electron energy. The energy dependence of the cross-section for the reaction $e + PBQ \rightarrow PBQ^-$ is shown in Fig. 2 [33].

Table 1. Summary of Experimental Evidence for Multiply-Charged Negative Ions

Ion Observed	Mass Spectrometer	Ref.	Comments
Cs^{-2}	Quadrupole	a	Cs^{-2} was only suggested later (private communication, L. M. Chanin); believed to be artifact.
F^{-2}	Quadrupole	b	Later attempts to reproduce F^{-2} were unsuccessful (W. S. Koski, private communication).
NO_2^{-2} $C_6H_4-(CH_2)_n^{-2}$ CO_2R ($R=H$ or CH_3)	Magnetic	c	Lifetime must be greater than 10^{-5} s. Three distinct types of doubly charged negative ions are described.
SO_4^{-2}, FeF_6^{-3} $HFeF_5^{-}$, $RbBr_2^{-2}$	45° Sector Magnetic	d	Ions were produced by field desorption.
Cl^{-2}, I^{-2}, Br^{-2}, SF_6^{-2}, SF_5^{-2}	Ion Cyclotron Resonance	e	Cl^{-2} is suggested but *harmonics* could not be ruled out. Mechanism believed to be $X^- + E \rightarrow X^{-2}$.
O^{-2}, Te^{-2}, Bi^{-2}, F^{-2}, Cl^{-2}, Br^{-2}, and I	Penning Source, 60° Sector Magnetic	f	Electric and magnetic deflection was employed. I^{-2} observation is placed in some doubt by L. Frees, E. Heinicke and W. S. Koski, Nucl. Inst. Meth. 159, 105 (1979).

Table 1. Continued

Ion Observed	Mass Spectrometer	Ref.	Comments
H^{-2} D^{-2}	Tandem Mass (1) 30-cm Wien filter (2) 90° sector magnetic	g	See also: M. L. Vestal, J. Chem. Phys. 65, 4331 (1976); J. Durup, J. Chem. Phys. 65, 4331 (1976); and R. Schnitzer and M. Anbar, J. Chem. Phys. 65, 4332 (1976).
O^{-2}, F^{-2}, Cl^{-2}, Br^{-2}, CN^{-2}	0 Megatron	h	See alternate interpretation (harmonics) by J. H. Fremlin, Nature 211, 1453 (1960).
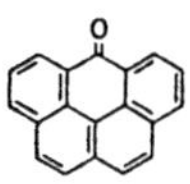 -2 I	Magnetic	i	

[a] D. G. Kuehn, D. E. Sutliff and L. M. Chanin, Appl. Phys. Lett. 33, 906 (1978).
[b] J. E. Ahnell and W. S. Koski, Nature 245, 30 (1973).
[c] J. H. Bowie and B. Jane Stapleton, J. Am. Chem. Soc. 98, 6480 (1976).
[d] M. Anbar and G. A. St. John, J. Am. Chem. Soc. 97, 7195 (1975).
[e] B. Jane Stapleton and J. H. Bowie, Aust. J. Chem. 30, 417 (1977).
[f] H. Bauman, E. Heinicke, H. J. Kaiser and K. Bethge, Nucl. Inst. Meth. 95, 389 (1971).
[g] R. Schnitzer and M. Anbar, J. Chem. Phys. 64, 2466 (1976).
[h] W. K. Stuckey and R. W. Kiser, Nature 211, 963 (1966).
[i] R. C. Dougherty, J. Chem. Phys. 50, 1896 (1969).

vacant molecular orbital to the adiabatic EA of a conjugated molecule. An excellent one-to-one correlation was found. Furthermore, calculations and experiment provide evidence for the so-called "pairing theorem" (see [36]) which states that the sum of the EA and the ionization potential is a constant for alternate hydrocarbons. (Younkin, *et al.* [36] give a value of 8.18 eV for this constant.) This relationship is true for excited states as well as the ground state of the ion. Jordan (private communication) has suggested this could be used as a method of identifying "π" ionization potentials from photoelectron spectra. Experimentalists in this area of collision physics are making significant new contributions in organic chemistry.

4. MULTIPLY-CHARGED NEGATIVE IONS

Doubly-charged negative ions were first reported in 1960 by Stuckey and Kiser [37]. Since that time there have been many claims of multiply-charged negative ions. Table 1 summarizes these data. Note that even triply-charged negative ions are reported in one field-desorption mass spectrometer experiment. Recently, a lengthy review article by Kiser [38] on doubly-charged negative ions concluded with the assertion: "Several conclusions may be drawn from the investigations above. Foremost among these is that doubly-charged negative ions exist." This conclusion is not unanimously adhered to by all researchers in the field, at least as far as the statement applies to long-lived ($\tau > 1\mu s$) doubly-charged atomic ions.

Spence, Chupka and Stevens [39] have very recently reexamined the possible existence of such species under experimental conditions which mimick, as closely as possible, those experiments which have reported positive identification of doubly charged I^{-2}, F^{-2}, Cl^{-2} and O^{-2}. These authors [39] allowed for an absolute identification of any ions which may result in false signals caused by impurities, and in which artifact peaks do not occur. They set upper limits on the ratio of X^{-2} to X^- of the order 10^{-7} whereas earlier experiments claim values of 10^{-1}-10^{-2}. In contrast to previous experiments, they find no evidence for doubly-charged atomic negative ions.

Baumann, *et al.* [40] have reported the most detailed experiments on doubly-charged ions. They reported O^{-2}, Te^{-2}, Bi^{-2}, F^{-2}, Cl^{-2}, Br^{-2} and I^{-2}. Ratios of X^{-2} to X^- of 10^{-3} to 10^{-4} were found. Frees, *et al.* [41] have attempted to reproduce these results for the case of I^{-2} using radioactive ^{131}I sources and they conclude: "The presence of such a peak has been previously attributed to I^{-2}; however, applications of radioactive tracer techniques show that this peak in our apparatus was not due to iodine, but to some impurity."

Hird and Ali [42] searched for O^{-2} from an rf discharge and set an upper limit of (0.07 ± 1.6) x 10^{-12} for the ratio of O^{-2}/O^{-}.

The present author and colleagues have searched for doubly-charged negative ions during the past 20 years using electron bombardment, surface ionization, ion-molecule reactions, Rydberg reactions and cesium charge-exchange reactions, and have found no evidence for doubly-charged negative ions. Mass analysis of the negative ions was also performed with a time-of-flight mass spectrometer, which should be free of many of the potential artifacts which are known to plague other mass spectrometers.

In the following paragraph we report an experimental observation which may bear on the subject of multiply-charged negative ions.

P. W. Reinhardt and R. N. Compton (unpublished) have recorded, for the past few years, a negative ion of mass 103 ± 2 amu under conditions in which gold and fluorine are present. In one experiment, a gold wire was heated in the presence of fluorine gas and an intense ion at mass 103 ± 2 amu was present. This ion was also recorded when $(AuF_5)_n$ was exposed to a hot wire (tungsten). Although impurity ions have not been ruled out, it is interesting to speculate that the ion is AuF_6^{-3}. The EA of AuF_6 is believed to be 10 ± 0.5 eV (N. Bartlett). An interesting, but unproven, possibility for metastability of AuF_6^{-3} is that the additional electrons are temporarily bound into a giant shape resonance. The potential shape is a combination of the long-range repulsive Coulomb potential and the short-range attraction due to the electron binding to the molecule (i.e., polarizability). Provided that the well is deep enough and the width of the well is wide enough, a long-lived (microseconds) metastable, multiply-charged negative ion could exist. AuF_6 does, in fact, require three additional electrons to complete a "shell". In addition, the large EA and the symmetry of AuF_6 make this system a good candidate for a multiply-charged long-lived shape resonance. The fact that the ion is only observed from a surface means that all of the molecular-anion constituents (3e's, 6F's and Au) come together "at once" to yield AuF_6^{-3} and thereby circumvent the difficulty of adding the extra electrons one at a time. All of these statements make for interesting speculation but much new experimental work is required. The possibility of a molecular resonance analogous to that of a nucleus which is metastable toward alpha decay would be exciting.

5. PHOTOABSORPTION BY NEGATIVE IONS

The interaction of electromagnetic radiation with negative

ions constitutes a large field of physics and chemistry. Recent review articles summarize most of the data [43-45] for the gas-phase negative ions. There is an even larger data set for negative ions in the condensed phase.

The cross-section for photodetachment from an atomic negative ion is given by

$$\sigma_d = \frac{g_o 32\pi^4 m^2 e^2 v\nu}{g_i 3hc^3} |\langle\chi_i|\vec{r}|\chi_f\rangle|^2 \quad (3)$$

where ν is the light frequency; $\langle\chi_i|\vec{r}|\chi_f\rangle$ is the dipole matrix element; and g_i and g_o are the statistical weights of the ionic and neutral states, respectively. Photodetachment cross-sections have been calculated from Eq. (3) for many atomic negative ions. The accuracy of the calculated cross-section is tested by applying known oscillator sum rules. The sum rule takes the form of an integral over the frequency and, in the case where there are no bound excited states, becomes [43]

$$\left(\frac{e^2}{ha_o}\right)^s \frac{mc}{\pi e^2} \int_{\nu_o}^{\infty} \frac{\sigma_d(\nu)}{\nu^s} d\nu = A_s \quad (4)$$

where A_s is a constant depending upon the wavefunction of the ground-state negative ion.

There are few known examples of bound excited states of negative ions. Herzberg and Lagerquist [46] observed a well-developed band system in the range from 4800-6000 Å, emitted from a high-intensity flash discharge in methane, which is not attributable to C_2 or C_2^+. Lineberger and Patterson [47] studied two-photon detachment of C_2^- and showed that the Herzberg-Lagerquist bands were due to $^2\Sigma_u^+ \rightarrow {}^2\Sigma_g^+$ transitions in C_2^-. This study of Lineberger and Patterson was a monumental contribution to negative-ion physics and to nonlinear laser spectroscopy as well.

Autodetaching states of negative ions can have a large influence on photodetachment cross-sections. The somewhat analogous autoionizing states of neutral atoms have been extensively studied in optical absorption spectra. The famous Beutler lines in the rare gases (except He) are classic examples. Similarly, autodetaching states can affect the continuum photodetachment cross-sections. Many "window" type resonances are observed in the alkali negative ions, both experimentally and theoretically [48]. One of the most interesting recent studies of the effects of shape resonances on photodetachment is the photodetachment of $He^-(1s2s2p)^4P^o$ via the $He^-(1s2p^2)^4P^e$ shape resonance [49]. The $1s2p^2\ ^4P^e$ state of He^- gives rise to a very large ($\sim$24 Å^2) and narrow peak 10 meV above the 2^3P^o threshold. The transition $1s2s2p\ ^4P^o \rightarrow 1s2p^2\ ^4P^e$ is dipole allowed and, since the state lies in the continuum of $1s2p\ ^3P^o$, the resonance contributes to the photodetachment cross-section. Fig. 4 gives the theoretical calculation [49] along with the recent experimental cross-sections [50,51]. Note that the peak in the theoretical cross-section is ~16 times the maximum value shown in Fig. 4.

Perhaps one of the most exciting new discoveries in negative-ion physics involves Li^-. There are two known excited states of Li^- which are bound relative to the parent excited neutral; $Li^-\ 1s2p^3\ ^5S^o$ and $Li^-\ 1s2s2p^2\ ^5P$ [52]. The $Li^-\ ^5S^o$ state radiates to $Li^-\ ^5P$ with a lifetime of 2.9 ns. Radiation at 3490 Å due to this transition has been identified in beam foil spectroscopy [53].

Recent studies of high-resolution photodetachment of polar anions [54] have revealed a series of very narrow, well-resolved resonances which are found just above the photodetachment thresholds for several negative-ion species. The authors [54] suggested the possibility that the resonant structures might be associated with bound excited dipole-states of the molecular anions. These excited states, although bound relative to the ground-state neutral, are so weakly bound that internal nuclear energy (rotation-vibration) allows for autodetachment. Garrett [55] found that the dipole field of the species studied in [54] was exactly in the correct range (3 to 3.5 D) to yield one or two excited dipole-states with binding energies so small that the

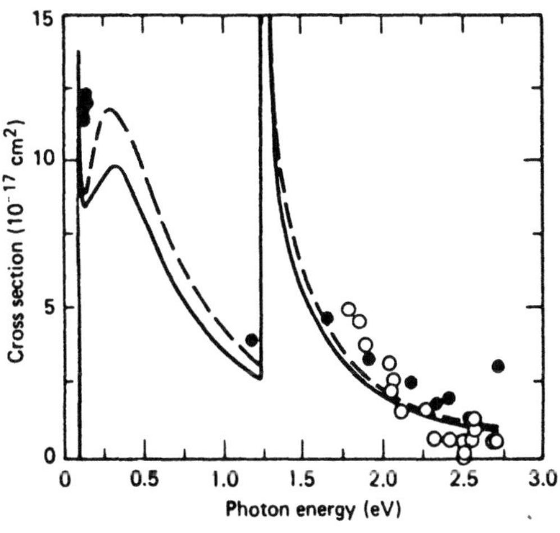

Fig. 4. Total absolute photodetachment cross-section for He^- ($^4P^o$); theory [49] and experiment [40-41].

bound excited states, with some finite ro-vibrational energy, would overlap with the continuum of one or more lower-lying ro-vibrational levels.

6. RECENT STUDIES OF LONG-LIVED He NEGATIVE IONS

Long-lived negative ions of helium have been of great theoretical curiosity and practical importance. He^- ions are routinely used as injection ions in a tandem accelerator.

Negative helium ions were first detected in a mass spectrometer in 1939 by Hiby [56]. This surprising observation required the He^- ions to be extremely long-lived ($\gtrsim$ 1 μs). In 1955, Holøien and Midtdal [57] argued that the $^4P^o_{5/2}$ state of (1s2s2p) $^4P^o$ He^- is stable against autodetachment and might explain Hiby's observation of long-lived He^- ions. They further calculated that $^4P^o$ He^- is bound relative to $(1s2s)^3S$ by 0.075 eV [57]. Table 2 provides a summary of more recent calaculations and one experimental measurement of the energy level of $^4P^o$ He^-.

Riviere and Sweetman [62], in 1960, were the first to show that He^- has a finite lifetime. The $^4P_{3/2,1/2}$ level can decay through spin-orbit and spin-spin interactions via coupling with the doublet P states, while $^4P_{5/2}$ He^- can decay only through the considerably weaker spin-spin interaction. Blau, Novick and Weinflash [63] established that the J=5/2 lifetime is 345 ± 90 μs and that a weighted average of the J=3/2 and 1/2 lifetime is 11

Table 2. Energy Levels for $^4P^o$ He^-

$E(1s2s)^3S - E(1s2s2p)^4P$ (eV)	$E(1s2s2p)^4P$ (eV)	Reference
$\gtrsim$0.075	19.744	[57] (Theor.)
$\gtrsim$0.033	19.786	[58] (Theor.)
0.067	19.752	[59] (Theor.)
0.080 ± 0.02	19.739	[60] (Exptl.)
0.0774 ± 0.0003	19.7417	[61] (Theor.)

μs. Table 3 presents a summary of measured lifetimes for He^- ions produced in a variety of collisions.

The theoretical values for the lifetime of a $^4P_{5/2}$ state of He^- are 303 [68], 505 [68], 266 [69] and 455 [70] μs, with the latter value of Estberg and LaBahn [70] expected to be the most accurate. This value is in good agreement with the experimental values shown in Table 3. Manson [71], using wavefunctions which are less accurate than Estberg and LaBahn, obtained 10^{-3} s for the $^4P_{3/2}$ state, but was unable to estimate the lifetime of $^4P_{1/2}He^-$. Manson further predicted that the energy for the fine-structure levels would increase with J, and that the energy excess Δ_{53} and Δ_{51} of the 5/2 level above the 3/2 and 1/2 levels would be 8.43 x 10^{-6} eV and 3.89 x 10^{-5} eV, respectively. Mader and Novick [72] used magnetic resonance techniques to determine Δ_{53} to be 3.41 x 10^{-6} eV and Δ_{51} to be 3.58 x 10^{-5} eV.

Donnally and Thoeming [73] were the first to demonstrate that He^- could be produced in a two-step process, namely,

$$He^+ + X \rightarrow He(1s2s)^3S + X^+ \tag{5}$$

$$He(1s2s)^3S + X \rightarrow He^-(1s2s2p)^4P + X^+ \tag{6}$$

where X is a target atom. The alkali atoms are the most efficient electron donors. Very large beam currents can be generated in this manner. Hooper, *et al.* [74], for example have been able to produce 70 mA pulses (10 ms pulse length) by charge exchange in sodium at 10.5 kV. Most tandem accelerator charge-exchange sources can inject microampere beams of He^-.

Table 3. Measured Lifetimes of He^-

Source of He^-	Lifetime (μs)	Reference
$He^+ + He$	18.2	[64]
$He^+ + N_2$	9^{+5}_{-3}	[65]
$He^+ + K$	11.5 ± 5, 345 ± 90	[63]
$He^+ + K$	10 ± 2, 16 ± 4, 500 ± 200	[66]
$He^+ + Ca$	10.5 ± 2	[67]

Until about 1975, it was generally assumed that long-lived He^- ($\tau > 10^{-6}$ s) ions exist only in the $(1s2s2p)^4P^o$ state. Three groups [75-78] have recently raised the question of whether or not He^- may also exist in a long-lived doublet state. This conjecture was based on the fact that He^+ can pick up two electrons from an atom or molecule with closed s shells (i.e., $He^+ + Ca \rightarrow He^- + Ca^{+2}$). More recent calculations [61] dispute these conjectures since no long-lived ^2P states of He^- were obtained.

Previously, Compton, *et al.* [67] measured the lifetime of He^- formed through collisions of He^+ and Ca vapor and found it to agree with earlier reported values for $He^-(1s2s2p)^4P^o_{3/2,1/2}$. This was taken as partial evidence that the He^- is formed in a quartet state. Also, the photodetachment cross-sections shown in Fig. 4 are in agreement with theoretical and other experimental photodetachment data for $He^-(^4P^o)$.

The energy level of He^- formed through collisions of He^+ and Ca was also determined by directly measuring the energy of the autodetached electron. Fig. 5 shows a sketch of the experimental geometry. Electrons are analyzed in the direction of the ion beam.

Fig. 6 shows a typical energy spectrum. Two peaks are observed, the one at low energy being due to collisionally-detached (or "cusp") electrons produced by background gas or from gas introduced into a scattering cell in order to enhance the "cusp" electron signal. Helium gas was introduced into the collision chamber in the particular experiment shown in Fig. 5. The autodetachment energy E_a is related to the He^- ion kinetic energy E_i through the expression

$$E = E_c + 2\sqrt{M_e/M_{He}}\,\sqrt{E_i E_a} + E_a \tag{7}$$

where $E_c = (M_e/M_i)E_i$ = energy of the "cusp"; and E is the electron energy in the laboratory system. Using autodetachment data over the range of incident energies from 10 to 120 keV, we obtain an autodetachment energy and, thereby, an energy level for He^- of 19.7 ± 0.1 eV, which agrees very well with the values shown in Table 2 for $He^-(^4P^o)$. Again, we find only evidence for long-lived quartet states of He^-.

The method of autodetachment spectroscopy described above would be applicable to any metastable negative ion (e.g., Be^-, Ca^-, N^-, etc.). We see considerable potential for this technique in studying metastable states of negative ions.

Finally, we mention that at low background pressures, a residual "cusp" electron peak is observed which is pressure inde-

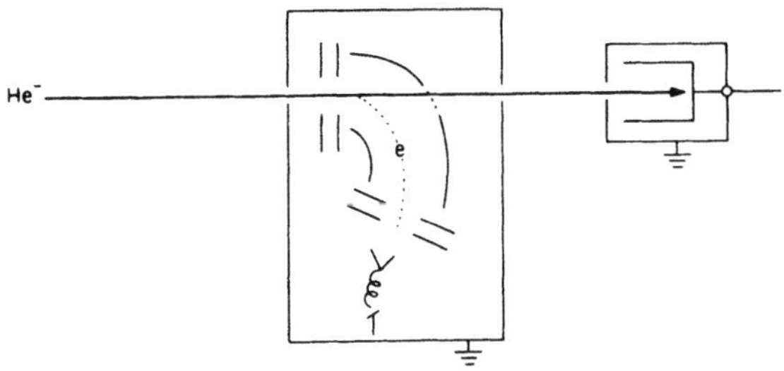

Fig. 5. Experimental sketch of double-focusing, hemisperical, electron-energy analyzer arrangement used to measure autodetached and collisionally-detached electrons from He^- ion beams.

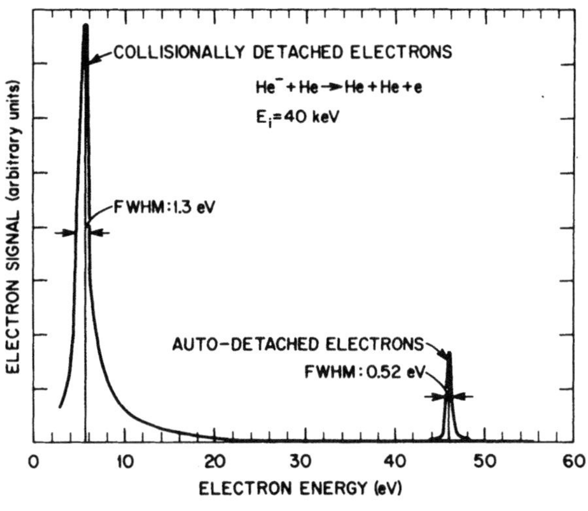

Fig. 6. Energy spectrum of autodetached electrons and collisionally-detached electrons for He^- ($^4P^o$).

pendent for pressures below $\sim 10^{-6}$ torr. This peak is somewhat more narrow in energy spread than the peak produced by collisional detachment; and, it results, in part, from photodetachment by room temperature blackbody radiation. The spread in energy and observed count rate is close to that calculated assuming a blackbody distribution convoluted with the known experimental [51] and theoretical [49] cross-sections.

Nicolaides, *et al.* [79] have recently studied bound states and decay mechanisms in He^-. They conclude that He^- has only two bound states, $2p^3\ ^4S^o$ and $1s2s2p\ ^4P^o$. Both $^4S^o$ and $^4P^o$ autodetach to the adjacent continuum via spin-dependent forces. They describe two other decay mechanisms: Nonrelativistic radiative autodetachment for $^4S^o$ and relativistic radiative autodetachment for $^4P^o$. Fig. 7 shows the autodetachment schemes involved. The outgoing electron energy is expected to be low and the photon takes away most of the excitation energy. Nicolaides, *et al.* [79] calculate a probability of radiative autodetachment for the $^4S^o$

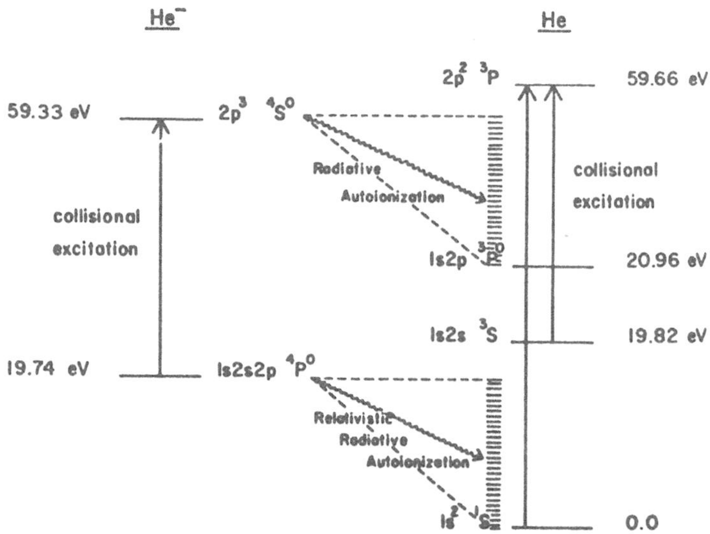

Fig. 7. Energy levels of He^- ($^4S^o$) and He^- ($^4P^o$) relative to the ground $1s2s\,^3S$ and $2p^2\ ^3P$ states of neutral helium. $He^-\ ^4S^o$ can decay only to the He $1s2p\ ^3P^o$ continuum with a maximum photon energy of 38.64 eV. The $^4P^o$ state can undergo relativistic radiative autoionization to the He $1s^2 + e^- + h\nu$ continuum by mixing with the $1s^2\ \varepsilon p$ and $1s2s2p\ ^2P^o$ continuum via spin-dependent interactions. Helium $2p^2\ ^3P$ can also decay to the $He^+\ 1s + e^- + h\nu$ continuum. However, its main decay channel is to the lower discrete $1s2p\ ^3P^o$ level. (Taken from [80].)

state of $1.14 \times 10^{10}\ s^{-1}$. Nicolaides and Komminos [80] have suggested that it may be possible to construct a tunable (over a narrow energy range) laser based upon radiative autodetachment of He^- $(^4S^o)$ at $\lambda \approx 323$ Å. The primary obstacle is a mechanism for population of $^4S^o$. One possibility is that an intense He^- ion beam (70 mA pulses) could be passed through an intense electron beam, thus relying upon electron collisions to populate the $^4S^o$ state:

$$e + He^-(^4P^o) \rightarrow He^-(^4S^o) \rightarrow He + e_{slow} + h\nu \ .$$

7. PRODUCTION OF NEGATIVE IONS BY PHOTON REACTIONS

The formation of ion pairs by photon absorption, namely,

$$h\nu + AB \rightarrow A^+ + B^-$$

is sometimes referred to as polar dissociation. This process can be particularly important in cases where the ion-pair limit lies below the ionization potential. The halogens I_2 and Br_2 are classic examples (e.g., the $I^+ + I^-$ limit is 0.49 eV below $I_2^+ + e$). In these cases, large cross-sections are observed below the ionization limit; however, once the photon energy approaches and exceeds the ionization limit, the ion-pair cross-section decreases dramatically. Fig. 8 shows the experimental data of Morrison, *et al.* [81] for I_2.

It is interesting to consider the polar dissociation of molecular clusters by photons. Although such experiments have not yet been reported, it is instructive to examine the energetics of such reactions. Let us use, as an example, the important case of photodissociation of water clusters into ion pairs. The energy required to ionize a single water molecule is 12.51 eV and the ionization potential for the dimer is 11.21 ± 0.09 eV [82]. A recent value for the threshold for photoelectron emission from liquid water is 10.06 eV [83]. On the other hand, the threshold for the polar dissociation of the dimer is 9.81 eV:

$$h\nu + (H_2O)_2 \rightarrow H_3O^+ + OH^-$$

$$E = D(H_2O \cdot H_2O) + D(H\text{-}OH) - D(H^+ - H_2O) - EA(OH) + IP(H)$$

$$= 0.17 + 5.113_6 - 7.18 - 1.89 + 13.6 = 9.81 \text{ eV} \ .$$

The significant lowering of energy comes primarily from the rather large proton affinity of water (7.18 eV) [82]. Also, since the ion-dipole interaction energies are generally greater than the van der Waals interactions, we would expect the energy for the polar-dissociation reaction to decrease as the cluster

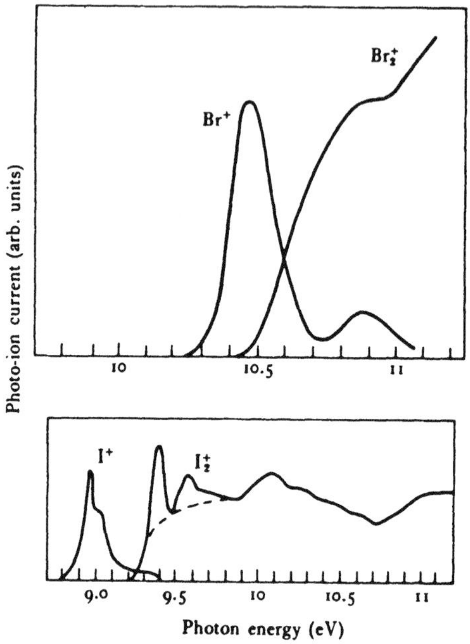

Fig. 8. Relative cross-sections for $Br^+ + Br^-$ and Br_2^+ ($I^+ + I^-$ and I_2^-) produced from photon collision with Br_2 (I_2) as observed by Morrison, *et al.* [81].

size increases. Fig 9 shows an approximate plot of photodissociation energy versus cluster size for the case of symmetric fission:

$$h\nu + (H_2O)_N \rightarrow H_3O^+ \cdot (H_2O)_n + OH^- \cdot (H_2O)_n$$

where N = 2n + 2.

In Fig. 9, dissociation energies of $OH^- \cdot (H_2O)_n$ were taken from Kebarle and co-workers [84,85], while those for $H_3O^+ \cdot (H_2O)_n$ were taken from Lau, *et al.* [86]. The major uncertainty in the data plotted is in the neutral bond dissociation energies. The important point to note from Fig. 9 is that photodissociation of $(H_2O)_N$, for N=12, could occur for photons with energies of $\lesssim$4 eV. The proton transfer reaction in liquid water,

$$(H_2O)_{liquid} \leftrightarrow H^+_{(aq)} + OH^-_{(aq)}$$

exhibits a low ΔH (0.628 eV) because of the effective solution and dielectric shielding of product ions by water. Natzle, *et al.* [87] have recently studied dissociative ionization of water induced by single-photon vibrational excitation. The quantum yield for production of $H^+_{(aq)}$ and $OH^-_{(aq)}$ increases from 2 x 10^{-9} to 4 x

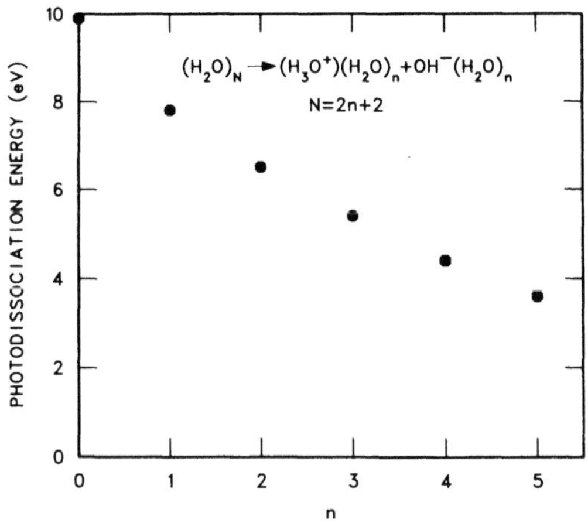

Fig. 9. Plot of the calculated minimum energy required to photodissociate a water cluster of size N into approximately equal size ion-pair fragments (i.e., symmetric fission). The energies are approximate since the neutral bond energies are not known accurately.

10^{-5} for energies between 0.94 and 2.25 eV. Fig. 10 shows the data taken from [87].

The quantum yield is seen to rise rapidly at the onset of a vibrational band and is relatively constant at $\sim 10^{-5}-10^{-6}$ through the band center. Given the photon intensities of present dye lasers, and the abilty of producing large molecular water clusters from nozzle beams, it is possible to study the production of ion pairs in water clusters for various cluster sizes. It may also be possible to obtain information on the minimum energy required to produce an ion pair for various cluster sizes, thereby accurately reproducing and extending the approximate data shown in Fig. 9.

8. COLLISION OF HIGH-RYDBERG ATOMS WITH ELECTRON-ATTACHING MOLECULES AND THE POSSIBLE INFLUENCE OF DIPOLE BOUND STATES

The Rydberg electron in a highly-excited Rydberg state can be viewed as almost free and slow: That is, the excited electron and the ion core are sufficiently separated that when a molecule collides with a Rydberg atom, the molecule can be viewed as reacting with either the electron *or* ion core separately. Fermi [88] used data on ultra-low-energy elastic electron scattering to analyze successfully the observed pressure shift of the high-Rydberg series absorption spectra of an alkali metal atom perturbed by a

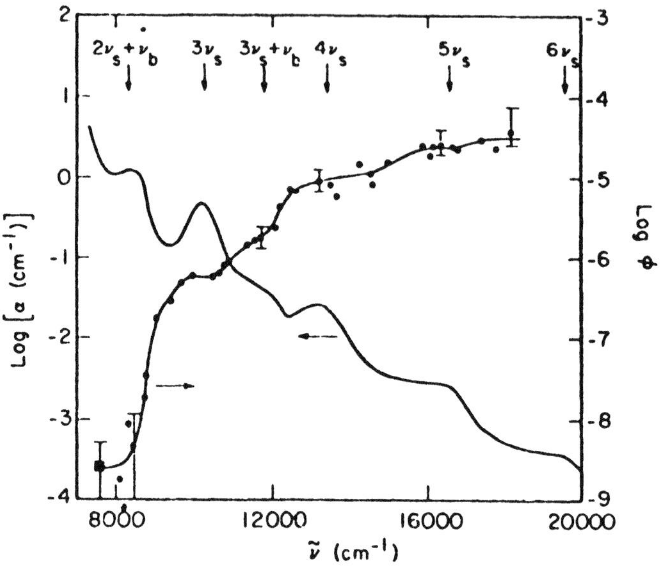

Fig. 10. Quantum yield (Φ) for photoionization of water at 283 ± 1 K as a function of photon energy ($\tilde{\nu}$). The absorption spectrum of water is shown as the dependence of the natural absorption coefficient (α) on wave number ($\tilde{\nu}$). [$I/I_0 = \exp(\alpha)$.] Band assignments are shown with subscripts b and s representing bending and stretching modes, respectively. (Taken from [87].)

foreign gas atom. Since Fermi's original contribution, there has been a considerable body of theoretical [89,90] and experimental [91-93] evidence to suggest that the rate constant for transfer of a loosely-bound Rydberg electron to a target molecule should be equal to the attachment rate for free electrons of the same energy. Perhaps the best experimental evidence for this equivalence comes from the unpublished data of Chupka [94]. Fig. 11 displays the electron charge-transfer rate from Rydberg-excited argon atoms to some hexafluoride molecules, as measured by Chupka, versus the thermal electron attachment rate for the same hexafluorides, as determined by Davis, *et al.*[95]. The solid line through the data points is the Fermi-Matsuzawa prediction [88-90] of equivalence between Rydberg reaction rates and slow-electron attachment rates. The data not only confirm the Fermi-Matsuzawa model but the ions observed by Chupka are in fact identical to those found in slow-electron attachment experiments (i.e., SF_6^-/SF_6, SeF_5^-/SeF_6, TeF_5^-/TeF_6). It is worth noting that Rydberg reaction rates and electron attachment rates for a large number of other molecules also fall on this line.

The Rice University group of Stebbings, Dunning and co-

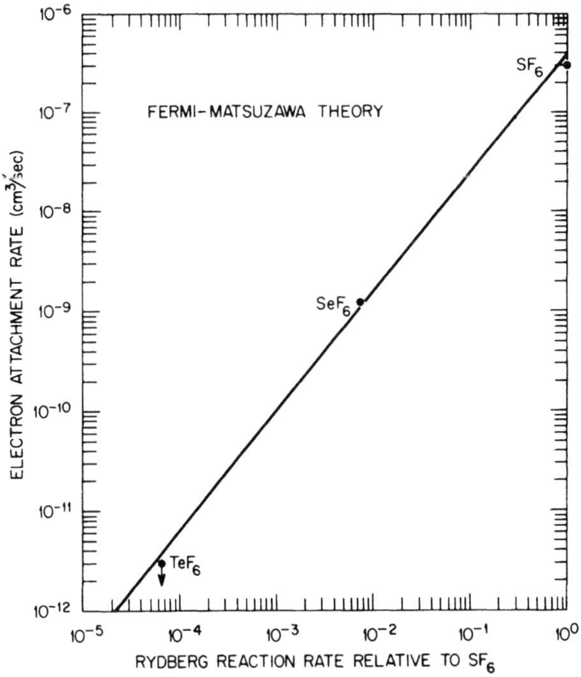

Fig. 11. Rydberg reaction rate leading to negative ions for various hexafluoride molecules, relative to SF_6, as measured by W. A. Chupka (unpublished results), versus the thermal-electron attachment rate as determined by Davis, *et al.* [95]. The Fermi- Matsuzawa [88-90] model predicts a straight line as shown.

workers has presented some equally compelling evidence for the free-electron model. The thermal-electron attachment rate for perfluoromethylcyclohexane (C_7F_{14}) is known to increase with electron energy [95]. Hildebrant, *et al.* [96] have actually observed an analogous increase of the ionization rate with *decreasing* n when xenon atoms in high-Rydberg states collide with C_7F_{14}. Thus the respective equivalence of the magnitude, energy dependence and ion products for both Rydberg charge-exchange rates and thermal attachment rates seem to be well founded. We hasten to point out, however, that the positive-ion core of the Rydberg state can play a role in the collision other than that of simply bringing up the slow electron. For example, we know that thermal electrons attach to CH_3NO_2 to produce a short-lived $CH_3NO_2^-$ ion with a lifetime of less than 10^{-6} s. However, the $CH_3NO_2^-$ formed through Rydberg charge exchange is stable [88], indicating that the recoiling positive ion relaxes vibrations initially excited in the negative ion. The very presence of the ion core can also make certain reactions possible, reactions which are apparently not available for slow electrons. For example, Chupka (unpublished results) has found that the reaction Ar** + HCl → ArH^+ + Cl^- proceeds. Such a reaction would not be predicted from the model of free-electron attachment since the thermal-electron attachment rate to HCl is less than 3 x 10^{-12} cm^3/s, and may well be zero [95]. Thus, HCl is an example where the Rydberg core must play a critical role; and, in fact, the reaction is only possible because of the ArH^+ + Cl^- channel. But does the Rydberg electron transfer to HCl^-, and the Ar^+ core then react to abstract a hydrogen atom? If so, we must postulate a potential energy surface for HCl^- that lies very close to the ground state. Such states have been postulated by Rohr and Linder [97] to explain their low-energy electron scattering data for the case of the hydrogen halides. The negative-ion surface is believed to result from the molecular dipole field. Such negative-ion states do not show up in limited basis set valence-type calculations. However, the properties of the dipole field are such that we are assured of bound states if the dipole moment is greater than ∿2 Debye [98, 99]. The dipole moment of HCl is only 1.08 Debye, which is less than the critical moment, but virtual states of the dipole field could provide a mechanism for electron transfer via the reaction

$$Ar^{**} + HCl \rightarrow (Ar^+ \ldots HCl^-) \rightarrow ArH^+ + Cl^- \ .$$

Under this proposal, the Fermi-Matsuzawa model would again apply. However, one could make an equally convincing argument that the Ar^+ core of Ar** reacts with HCl to form ArH^+, and that the Cl^- takes away the electron. Clearly, more experimental and, perhaps, theoretical work is necessary to solve this riddle.

There is a further example in which the thermal-electron attachment rate does not approximate the Rydberg reaction rate. Namely, highly-excited rare-gas Rydberg atoms (excited by electron impact) react rapidly [91,100] with acetonitrile (CH_3CN) to produce CH_3CN^-, whereas the thermal-energy electron attachment rate constant for CH_3CN is $\lesssim 1.25 \times 10^{-14}$ cm^3/s. This, incidentally is the lowest upper limit which has been placed on an electron attachment rate. Acetonitrile, like HCN, is a "closed-shell" molecule, with the valence-type empty orbitals being high in energy, and no valence-type bound states are expected. The dipole moment of CH_3CN is 3.92 Debye and well exceeds the minimum dipole moment required to form a bound state [98,99]. The dipole states are very fragile because of the low binding energy and would not survive the collisions inherent in an electron swarm experiment. Thus, it is not surprising that the attachment rate is immeasurably small. We should mention that the Rydberg reaction rate estimated by Stockdale, *et al.* [91] was not as large as that reported by Sugiura and Arakawa [100]. A possible explanation is that Stockdale, *et al.* [91] did not detect all of the CH_3CN^- produced. In this connection, Odom and Hill(unpublished results) have reacted sodium high-Rydberg states with CH_3CN and see *no* CH_3CN^- negative ions. They do observe a large cross-section for collisional ionization producing a free electron. More experiments are definitely required to resolve this discrepancy; and in its resolution, perhaps we will learn more about dipole states of the electron-molecule system.

The production and detection of dipole-bound negative-ion states offer a real challenge to experimentalists. The problem is to gently place the electron on the dipole and then to detect the ion, without ejecting the electron in the process of detection. Table 4 represents a list of molecules that are assured of weakly-bound dipole states. These compounds have vapor pressures that are suitable for use in mass spectrometer experiments. We have studied fast alkali collisions with each of these molecules and observe parent negative ions only in the case of nitromethane, which has a bound state resulting from an empty valence orbital (EA = 0.44 ± 0.2 eV) [101]. In this case, the dipole state should be considered as an excited state of $CH_3NO_2^-$.

Finally, we should mention that Matsuzawa [102] has suggested the use of high-Rydberg atoms to obtain information on electron-molecule resonances at ultra-low energies. Resonances in elastic electron-molecule scattering would cause an oscillation in the n dependence of the pressure shift. As an example, if one could detect oscillations of 0.005 cm^{-1} in the pressure shift at low pressures, where one can avoid multiple scattering effects,

Table 4. Some Molecules with Dipole Moments Sufficiently Large to Permanently Bind an Electron

Molecule	Dipole Moment (Debye)
Cyanomethane H_3CCN (acetonitrile)	3.92
Formamide H_2NCOH	3.25
Nitromethane H_3CNO_2	3.46
Cyanamide H_2NCN	4.27
Dimethyl sulfoxide $(CH_3)_2SO$	3.96
Vinylene carbonate $C_3O_3H_2$	4.5

then the high-Rydberg atom could be used to detect resonances of 10^{-3} eV.

9. NEGATIVE IONS OF MOLECULAR CLUSTERS

The clustering of neutral molecules to ions is a broad subject. Various techniques have been employed to study the interactions of an ion and a molecule which lead to clustering: For example, flowing and stationary afterglow, ion-cyclotron resonance and high-pressure mass spectrometry. These studies are directed toward understanding the nature of the forces affecting cluster stability. In many cases, the ion and clustering ligand have been symmetrically varied in order to access, separately, individual ionic or neutral properties such as ion size, polarizability, etc. In some cases, the cluster ions themselves have been studied via collisional-dissociation, photodetachment or photodissociation reactions. During the past few years, C. E. Klots and R. N. Compton have taken a different approach to cluster-ion studies by examining electron attachment to preexisting clusters of molecules. A supersonic nozzle expansion is employed to prepare the molecular cluster so that it is possible to study electron attachment to ultracold ($T_{trans} \simeq 1$ K, $T_{rot} \simeq 2$ K, $T_{vib} \simeq 20$-30 K) molecules and molecular clusters. We will briefly discuss some of these studies below.

9.1 Electron Attachment to Ultracold Molecules and Molecular Clusters

The cross-section for electron attachment to molecules can be especially sensitive to thermal effects. Henderson, Fite and Brackman [103] found, for example, that the maximum in the cross-section for the process $e + O_2 \rightarrow O^- + O$ shifts to lower energy and increases in magnitude upon heating the O_2. Also, the energy onset for O^- from CO_2 is temperature dependent: At room temperature the threshold is 0.25 eV below the expected thermodynamic threshold [104]. The most dramatic effect of internal energy on electron attachment cross-sections is found in the case of N_2O. For N_2O, at room temperature, a separate O^- peak appears at $\sim$0.5 eV [105]. Chantry [105] attributes this dramatic temperature effect to thermal excitations of N_2O bending modes. Such excitation would yield good Franck-Condon overlap with the bent configuration of the N_2O^- ground electronic state.

Klots and Compton [106,107] have largely eliminated these artifactual thresholds by employing sonic nozzle expansions to cool the gas adiabatically. For example, Fig. 12 shows the O^- signal from the reaction $e + N_2O \rightarrow O^- + N_2$ as a function of electron energy. The prominent peak at $\sim$0.5 eV in the room temperature data is absent in the nozzle experiment. Similarly, the on-

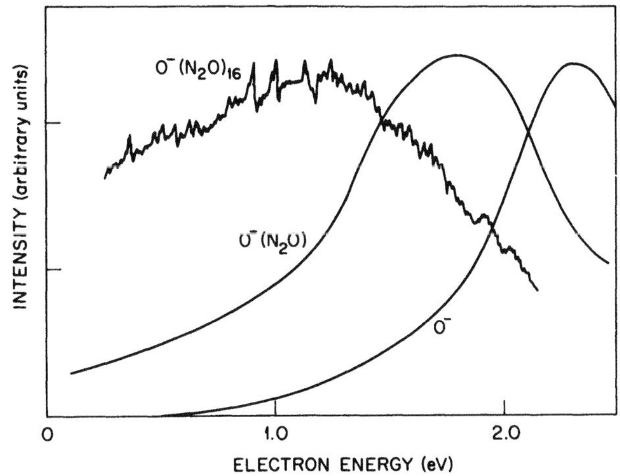

Fig. 12. A few of the negative ions observed from electron attachment to supersonic molecular beams of nitrous oxide, as a function of the electron energy. Ions with the formula $O^-(N_2O)_n$ with $0 \leq n \leq 3$ are observed [107].

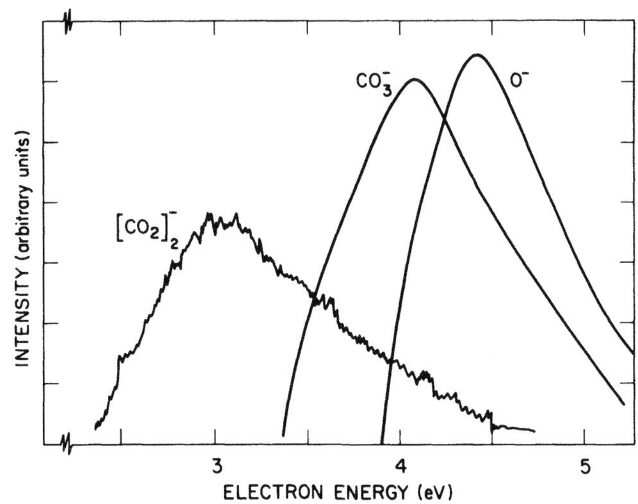

Fig. 13. A few of the negative ions observed from electron attachment to supersonic molecular beams of carbon dioxide as a function of the electron energy. Ions with the formula $(CO_2)_n$, with $0 \leq n \leq 6$, are observed. Larger ion clusters were most likely present [107].

set for O^- from CO_2, shown in Fig. 13, occurs at the proper thermodynamic threshold. Electron-beam experiments employing sonic nozzle expansions offer a number of advantages: (i), rotational (∿1 K), vibrational (∿20-50 K) and translational (∿1 K) cooling; (ii), translational cooling greatly reduces the Doppler profile from the energy dependence of a resonance or threshold; and (iii), increased molecular density for a given collision chamber pressure.

Van der Waals polymers can also be produced in sonic expansions. We have studied electron attachment to polymers of CO_2 [106,107], N_2O [107] and H_2O [108]. In the case of CO_2, the most intense cluster ions occur with general formula $O^-(CO_2)_n$, with $0 \leq n \leq 6$ (see Fig. 13). Long-lived ions with charge-to-mass ratios corresponding to $(CO_2)_n^-$, with $2 \leq n \leq 6$, were detected. Carbon dioxide is linear whereas CO_2^- is bent at about 135°. For bond angles less than $\sim 140^o$, the CO_2^- curve lies below that for CO_2. The experimental vertical [109] and adiabatic [110] EA's of CO_2 are -3.8 and -0.6 ± 0.2 eV, respectively. Linear CO_2^- autodetaches in $\sim 10^{-15}$ s. However, due to poor Franck-Condon factors between the bent CO_2^- and linear CO_2, the lifetime of ground state CO_2^- is ∿90 μs. The $(CO_2)_n^-$ ions appear to be stable ($\tau > 10^{-3}$ s). Rossi and Jordan [111] have employed the Hartree-Fock method to predict that the adiabatic EA of $(CO_2)_2$ is small but positive. Their most stable configuration consisted of a bent CO_2^- "solvated" by a nearly linear CO_2 (C_s symmetry). If this turns out to be true, these experimental results will provide the first observation of a solvated electron in the gas phase. Solvation energies can be large [e.g., > 0.6 eV in the case of $(CO_2)_2^-$] and could provide stability for a large number of anions. In this connection, Klots [112] has mixed H_2O and CO_2 in a nozzle expansion to produce the $(CO_2 \cdot H_2O)^-$ ion.

The formation of CO_2^- ions from CO_2 clusters is an example of "evaporative electron attachment". This is a particularly good illustration of collisional stabilization by a built-in third body. The attachment of low-energy electrons to an O_2 molecule has usually been envisaged in terms of a classical three body mechanism: An electron is captured by an O_2 monomer. Collision with a third body can stabilize the ion if the collision occurs before autodetachment takes place. We have found that an O_2^- can be formed with high efficiency from a single-step attachment to a preexisting oxygen dimer [107]. Spurred by these observations, Shimamori and Fessenden [113] have reexamined the mechanisms of electron attachment to oxygen in the gas phase. They find that electron attachment to oxygen at low temperature occurs much faster than predicted by the classical three-body mechanism. The estimated rate constants for attachment to van der Waals molecules are nearly two orders of magnitude larger than for O_2 itself. This does not imply that the cross-section for attachment is necessarily smaller, however. The cross-section for attachment to a single O_2 molecule exhibits strong maxima (actually, double maxima due to fine-structure splittings) at the vibrational levels of O_2^-. On the other hand, O_2^- from $(O_2)_2$ is a dissociative process and could show a rather broad energy dependence depending upon the potential energy surfaces involved.

As a further example of stabilization by collisions, let us consider the attachment of low-energy electrons to carbon tetrachloride. Thermal electron attachment to CCl_4 yields almost entirely Cl^-. Yet, when CCl_4 is bound initially to CO_2 in a van der Waals molecule, one obtains CCl_4^- in a good yield [106].

As a final example of electron attachment to van der Waals molecules, let us consider the case of methyl iodide, CH_3I [114]. In the low-pressure mass spectrometer, a methyl iodide molecule captures only electrons of near-zero energy to form I^- ions. When the molecule is part of a van der Waals molecule, this is no longer the case. Fig. 14 shows the signal of $I^- \cdot CH_3I$ for electron energies above 3 eV, together with the SF_6^- scavenger spectrum. The scavenger spectrum is a measure of the inelastic scattering cross-section for excitation of electronically excited

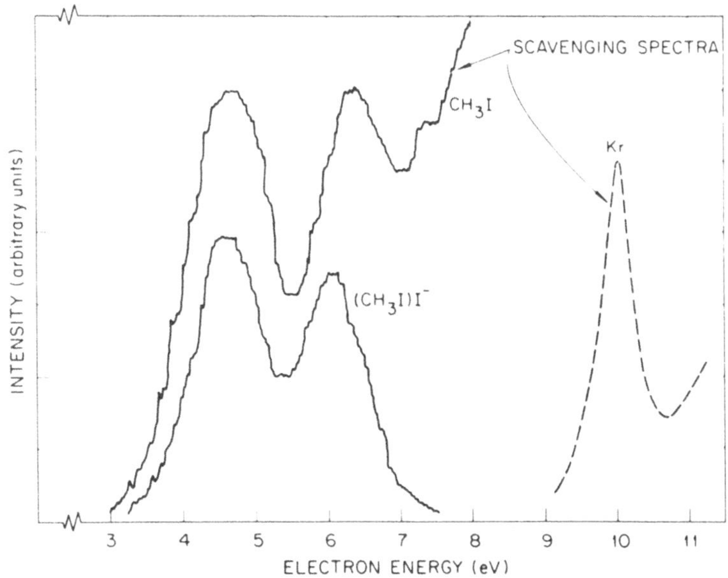

Fig. 14. Scavenging spectrum of methyl iodide monomer, and product of self-scavenging in the trimer, as functions of electron energy. The scavenger spectrum is essentially the spectrum of excited states produced by electron-impact excitation [114].

states of methyl iodide. The close similarity of the two spectra has prompted us to argue [114] that this should be viewed as involving an *intermolecular Feshbach resonance*. This is a mechanism whereby an incident electron excites one moiety of the cluster to an electronically excited state and the resulting slow electron is captured elsewhere in the cluster.

We originally began these cluster experiments in order to search for the "hydrated" electron in the gas phase. Thus far, we have been unsuccessful [108]. The large dipole moment of $(H_2O)_2$ (>2.4 D) [115] assures us that $(H_2O)_2^-$ is bound. However, the dipole moment of $(H_2O)_3$ is probably very small. Thus, unless $(H_2O)_2^-$ can be further solvated by water molecules, any connection with the hydrated electron is tenuous. Very recently, Armbruster, Haberland and Schindler [116] have detected negatively-charged water-cluster ions $(H_2O)_n^-$, for $n \geq 8$, in the gas phase. These authors [116] expanded steam and free electrons through a supersonic nozzle and observed $(H_2O)_n^-$, with $n \geq 8$. For $(H_2O)_{11}^-$, the signal is larger than the $OH^- \cdot (H_2O)_{10}$ signal. These studies represent a breakthrough in the quest for the gas-phase hydrated electron.

ACKNOWLEDGMENT

This research was sponsered by the Office of Health and Environmental Research, U. S. Department of Energy, under contract W-7405-eng-26 with the Union Carbide Corporation.

REFERENCES

1. R. Wildt, Astrophys. J. 89, 295 (1939).
2. L. Sanche, G. Bader and L. Caron, J. Chem. Phys. 76, 4016 (1982).
3. G. J. Schulz, Rev. Mod. Phys. 45, 378 (1973).
4. L. Sanche and M. Michaud, Phys. Rev. Lett. 47, 1008 (1981); Chem. Phys. Lett. 84, 497 (1981).
5. R. S. Berry and C. W. Reimann, J. Chem. Phys. 38, 1540 (1963).
6. M. W. Siegal, R. J. Celotta, J. L. Hall, J. Levine and R. A. Bennett, Phys. Rev. A6, 607 (1972); R. J. Celotta, R. A. Bennett, J. L. Hall, M. W. Siegel and J. Levine, Phys. Rev. A6, 631 (1972).
7. L. Sanche and G. J. Schulz, J. Chem. Phys. 58, 479 (1973); I. Nenner and G. J. Schulz, J. Chem. Phys. 62, 1747 (1975).

8. J. H. Richardson, L. M. Stephenson and J. I. Brauman, J. Chem. Phys. 59, 5068 (1973); Chem. Phys. Lett. 25, 321 (1974); J. Am. Chem. Soc. 97, 1160 (1975).
9. E. Herbst, T. A. Patterson and W. C. Lineberger, J. Chem. Phys. 61, 1300 (1974).
10. W. A. Chupka, J. Berkowitz and D. Gutman, J. Chem. Phys. 55, 2724 (1971).
11. J. Berkowitz, W. A. Chupka and D. Gutman, J. Chem. Phys. 55, 2733 (1971).
12. C. Lifshitz, B. M. Hughes and T. O. Tiernan, Chem. Phys. Lett. 7, 469 (1970).
13. T. O. Tiernan, B. M. Hughes and C. Lifshitz, J. Chem. Phys. 55, 5692 (1971).
14. B. M. Hughes, C. Lifshitz and T. O. Tiernan, J. Chem. Phys. 59, 3162 (1973).
15. C. Lifshitz, T. O. Tiernan and B. M. Hughes, J. Chem. Phys. 59, 3182 (1973).
16. R. K. B. Helbing and E. W. Rothe, J. Chem. Phys. 51, 1607 (1969).
17. R. N. Compton, P. W. Reinhardt and C. D. Cooper, J. Chem. Phys. 63 3821 (1975).
18. A. P. M. Baede, Adv. Chem. Phys. 30, 463 (1975).
19. L. M. Babcock and G. E. Streit, J. Chem. Phys. 75, 3864 (1981).
20. E. E. Ferguson, D. B. Dunkin and F. C. Feshenfeld, J. Chem. Phys. 57, 1459 (1972).
21. Neil Bartlett, Angew. Chem. Int. Edit. 7, 433 (1968).
22. P. Barberi and N. Bartlett, private communication.
23. R. N. Compton and P. W. Reinhardt, J. Chem. Phys. 72, 4655 (1980).
24. D. E. Jensen, Trans. Faraday Soc. 65, 2123 (1969).
25. D. E. Jensen and W. J. Miller, J. Chem. Phys. 53, 3287 (1970).
26. W. J. Miller, J. Chem. Phys. 57, 2354 (1972).
27. R. K. Gould and W. J. Miller, J. Chem. Phys. 62, 644 (1975).
28. D. E. Jensen and W. J. Miller, Thirteenth Symposium on Combustion, Abst. No. 363, University of Utah, Salt Lake City, August 23-29, 1970.
29. J. Burgess, I. H. Haigh, R. D. Peacock and P. Taylor, J. Chem. Soc. (Dalton), 1062 (1974).
30. R. N. Compton, P. W. Reinhardt and C. D. Cooper, J. Chem. Phys. 68, 2023 (1978).
31. M. Boring, J. H. Wood and J. W. Moskowitz, J. Chem. Phys. 61, 3800 (1974).
32. J. N. Bardsley and F. Mandl, Rep. Prog. Phys. 31, 471 (1968).
33. C. D. Cooper, W. T. Naff and R. N. Compton, J. Chem. Phys. 63, 2752 (1975).

34. M. J. W. Boness, I. W. Larkin, J. B. Hasted and L. Moore, Chem. Phys. Lett. 1, 292 (1967).
35. J. K. Jordan and P. D. Burrow, Acc. Chem. Res. 9, 341 (1978).
36. J. M. Younkin, L. J. Smith and R. N. Compton, Theor. Chim. Acta (Berl.) 41, 157 (1976).
37. W. K. Stucky and R. W. Kiser, Nature 211, 963 (1960).
38. R. W. Kiser, Top. Curr. Chem. 83, 89 (1979).
39. D. Spence, W. A. Chupka and C. M. Stevens, Phys. Rev. A 26, 654 (1982).
40. H. Baumann, E. Heinicke, H. J. Kaiser and K. Bethge, Nucl. Instrum. Methods 95, 389 (1971).
41. L. Frees, E. Heincke and W. S. Koski, Nucl. Instrum. Methods 159, 105 (1979).
42. B. Hird and S. P. Ali, J. Chem. Phys. 74, 3620 (1981).
43. H. S. W. Massey, *Negative Ions*, 3rd Ed. (Cambridge University Press, Cambridge, 1976).
44. B. M. Smirnow, *Negative Ions* (McGraw-Hill, New York, 1982).
45. H. Hotop and W. C. Lineberger, J. Chem. Phys. Ref. Data 4, 539 (1975).
46. G. Herzberg and A. Lagerquist, Can. J. Phys. 46, 2363 (1968).
47. W. C. Lineberger and T. A. Patterson, Chem. Phys. Lett. 7, 469 (1970).
48. T. A. Patterson, H. Hotop, A. Kasdan, D. W. Norcross and W. C. Lineberger, Phys. Rev. Lett. 32, 189 (1974); D. L. Moores and D. W. Norcross, Phys. Rev. A10, 1646 (1974).
49. A. V. Hazi and K. Reed, Phys. Rev. A24, 2269 (1981).
50. R. N. Compton, G. D. Alton and D. J. Pegg, J. Phys. B13, L651 (1980).
51. R. V. Hodges, M. J. Coggiola and J. R. Peterson, Phys. Rev. A 23, 59 (1981).
52. C. F. Bunge, Phys. Rev. Lett. 44, 1450 (1980); Phys. Rev. A22, 1 (1980).
53. S. Mannervik, G. Astner and M. Kisielinski, J. Phys. B13, L441 (1980); R. L. Brooks, J. E. Hardis, H. G. Berry, L. J. Curtis, K. T. Cheng and W. Ray, Phys. Rev. Lett. 45, 1318 (1980); A. Dennis and J. Desesquelles, J. Physique 42, L59 (1981).
54. A. H. Zimmerman and J. I. Brauman, J. Chem. Phys. 66, 5823 (1977); R. L. Jackson, A. H. Zimmerman and J. I. Brauman, J. Chem. Phys. 71, 2088 (1979).
55. W. R. Garrett, J. Chem. Phys. 73, 5721 (1980).
56. Von Julius W. Hiby, Ann. Phys. 34, 473 (1939).
57. E. Holøien and J. Midtal, Proc. Phys. Soc. London 68, 815 (1955).
58. E. Holøien and S. Geltman, Phys. Rev. 153, 81 (1967).
59. A. W. Weiss, Phys. Rev. 166, 70 (1968).
60. B. Brehm, M. A. Gusinow and J. L. Hall, Phys. Rev. Lett. 19, 737 (1967).

61. A. V. Bunge and C. F. Bunge, Phys. Rev. 19, 452 (1979).
62. H. G. Riviere and D. R. Sweetman, Phys. Rev. Lett. 5, 560 (1960).
63. L. M. Blau, R. Novick and D. Weinflash, Phys. Rev. Lett. 24, 1268 (1970).
64. D. J. Nichols, C. W. Trowbridge and W. D. Allen, Phys. Rev. 167, 38 (1968).
65. F. R. Simpson, R. Browning and H. B. Gilbody, J. Phys. B 4, 106 (1971).
66. R. Novick and D. Weinflash, in: *Proc. Int. Conf. Precision Measurements and Fundamental Constants* (Gaitherburg, Maryland), D. N. Langenberg and B. N. Baylor, eds. (Natl. Bur. Stand. (US) Spec. Pub. 343, 1970), p. 403.
67. R. N. Compton, G. D. Alton, A. D. Williamson and A. E. Carter, Abstract of Papers, XI Int. Conf. Physics of Electronic and Atomic Collisions (Kyoto, Japan, 1979), p. 68.
68. C. Laughlin and A. L. Stewart, J. Phys. B. 1, 151 (1968).
69. G. N. Estberg and R. W. LaBahn, Phys. Lett. A 28, 420 (1968).
70. G. N. Estberg and R. W. LaBahn, Phys. Rev. Lett. 24, 1265 (1970).
71. S. T. Manson, Phys. Rev. A 3, 147 (1971).
72. D. L. Mader and R. Novick, Phys. Rev. Lett. 29, 199 (1972).
73. B. L. Donnally and G. Thoeming, Phys. Rev. 159, 87 (1967).
74. E. B. Hooper, Jr., P. A. Pincosy, P. Poulsen and C. F. Burrell, Rev. Sci. Instrum. 51, 1066 (1980).
75. R. A. Baragiola, Abstracts of Papers, VIII Int. Conf. Physics of Electronic and Atomic Collisions (Belgrade, 1973), p. 827.
76. R. A. Baragiola and E. R. Salvatelli, J. Phys. B 8, 382 (1975).
77. E. H. Pedersen, F. R. Simpson and P. Hvelplund, J. Phys. B 7, L294 (1974); Phys. Rev. A11, 516 (1975).
78. K. F. Dunn, B. J. Gilmore, F. R. Simpson and H. B. Gilbody, J. Phys. B 11, 1797 (1978).
79. C. A. Nicolaides, Y. Kominos and D. R. Beck, Phys. Rev. 24, 1103 (1981).
80. C. A. Nicolaides and Y. Kominos, Chem. Phys. Lett. 80, 463 (1981).
81. J. D. Morrison, H. Hurzeler, M. G. Inghram and H. E. Stanton, J. Chem. Phys. 33, 821 (1960).
82. C. Y. Ng, D. J. Trevor, P. W. Tiedemann, S. T. Ceyer, P. L. Kronesbush, B. H. Mahan and Y. T. Lee, J. Chem. Phys. 67, 4235 (1977).
83. P. Delahay and K. Von Berg, Chem. Phys. Lett. 83, 250 (1981).
84. I. Dzidic and P. Kebarle, J. Phys. Chem. 74, 1466 (1970).
85. M. Arshadi, R. Yamdagni and P. Kebarle, J. Phys. Chem. 74, 1475 (1970); J. Phys. Chem. 74, 1483 (1970).

86. Y. K. Lau, S. Ikuta and P. Kebarle, J. Am. Chem. Soc. 104, 1462 (1982).
87. W. C. Natzle, C. B. Moore, D. M. Goodall, W. Frisch and J. F. Holzworth, J. Phys. Chem. 85, 2882 (1981).
88. E. Fermi, Nuova Cim. 11, 157 (1934).
89. M. Matsuzawa, J. Phys. Soc. (Japan) 32, 1088 (1972).
90. M. Matsuzawa, J. Phys. B 8, 2114 (1975).
91. J. A. D. Stockdale, F. J. Davis, R. N. Compton and C. E. Klots, J. Chem. Phys. 60, 4279 (1974).
92. W. P. West, G. W. Foltz, F. B. Dunning, C. T. Latimer and R. F. Stebbings, Phys. Rev. Lett. 36, 854 (1976).
93. G. W. Foltz, G. J. Latimer, G. F. Hildebrant, F. G. Kellert, K. A. Smith, W. P. West, F. B. Dunning and R. F. Stebbings, J. Chem. Phys. 67, 1352 (1977).
94. W. A. Chupka, unpublished results.
95. F. J. Davis, R. N. Compton and D. R. Nelson, J. Chem. Phys. 59, 2324 (1973).
96. G. F. Hildebrandt, F. G. Kellert, F. B. Dunning, K. S. Smith and R. F. Stebbings, J. Chem. Phys. 68, 1349 (1978).
97. K. Rohr and F. Linder, J. Phys. B 9, 2521 (1976).
98. O. H. Crawford, Mol. Phys. 20, 585 (1971).
99. W. R. Garrett, J. Chem. Phys. 69, 2621 (1978).
100. T. Sugiura and A. Awakawa, in: *Proc. Int. Conf. Mass Spectrom.* (University of Tokyo, Tokyo, 1970), p. 858.
101. R. N. Compton, P. W. Reinhardt and C. D. Cooper, J. Chem. Phys. 68, 4360 (1978).
102. M. Matsuzawa, J. Phys. B 10, 1543 (1977).
103. W. R. Henderson, W. L. Fite and R. T. Brackman, Phys. Rev. 183, 157 (1969).
104. G. J. Schulz and D. Spence, Phys. Rev. Lett. 22, 47 (1969).
105. P. J. Chantry, J. Chem. Phys. 51, 3369 (1969).
106. C. E. Klots and R. N. Compton, J. Chem. Phys. 67, 1779 (1977).
107. C. E. Klots and R. N. Compton, J. Chem. Phys. 69, 1636 (1978).
108. C. E. Klots and R. N. Compton, J. Chem. Phys. 69, 1644 (1978).
109. M. J. Boness and G. J. Schulz, Phys. Rev. A9, 1969 (1974).
110. R. N. Compton, P. W. Reinhardt and C. D. Cooper, J. Chem. Phys. 63, 3821 (1975).
111. A. R. Rossi and K. D. Jordan, J. Chem. Phys. 70, 4422 (1979).
112. C. E. Klots, J. Chem. Phys. 71, 4122 (1979).
113. H. Shimamori and R. W. Fessenden, J. Chem. Phys. 74, 453 (1981).
114. C. E. Klots and R. N. Compton, Chem. Phys. Lett. 73, 589 (1980).
115. T. R. Dyke, K. M. Mack and J. S. Muenter, J. Chem. Phys. 66, 498 (1977).

116. M. Armbruster, H. Haberland and H. G. Schindler, Phys. Rev. Lett. 47, 323 (1981).

SUPEREXCITED AND IONIC STATE RELAXATION PROCESSES IN VACUUM ULTRAVIOLET EXCITED POLYATOMIC MOLECULES

Sydney Leach

Laboratoire de Photophysique Moléculaire du CNRS
Bâtiment 213
Université Paris-Sud
91405 Orsay Cédex, France

1. INTRODUCTION

In certain energy regions, a molecular system can be excited to the ionization continuum (direct ion formation) and/or to excited states of the neutral species lying above the lowest ionization potential. These neutral superexcited states can decay by various processes, among which is autoionization which involves coupling with the ionization continuum (Fig. 1). In this paper, we discuss the relaxation processes of superexcited states and also of the ionic states formed by direct ionization or by autoionization. Photoabsorption of vacuum ultraviolet radiation is the principal excitation mechanism with which we will be concerned. However, it is of interest to note that superexcited states and ions can be formed as a result of interaction between neutral molecules and many different types of particles of sufficient energy, namely, photons, electrons, fast atoms or molecules, atoms and molecules in metastable excited states, and charged atoms or molecules [1].

In a discussion of fundamental processes (Section 2), we first consider the general question of direct and indirect processes (Section 2.1), and introduce the most appropriate language for the description of these processes. Next, the photoionization process is presented and the nature of autoionization is discussed (Section 2.2). Some generalities on photoionization cross-sections and manifestations of autoionization (Section 2.3) are followed by a discussion of the competing decay channels of superexcited states (Section 3). Methods for studying superexcited and ionic states and their relaxation processes are then present-

S. P. McGlynn et al. (eds.), Photophysics and Photochemistry in the Vacuum Ultraviolet, 297–340.

ed, namely, photon excitation sources (Section 4.1); high-resolution absorption spectroscopy in the vacuum ultraviolet (Section 4.2); ion-yield spectroscopy (Section 4.3); photoelectron spectroscopy and coincidence studies (Section 4.4); and fluorescence excitation spectra of neutral or ionic products (Section 4.5). The role of autoionization in dissociative ionization is singled out for particular discussion (Section 5).

Section 6 is devoted to the radiative and nonradiative decay of ionic states. The advantages of molecular ions for the study of radiationless transitions is presented in Section 6.1, and the measurement of ion fluorescence decay and yields in Section 6.2. A survey of radiationless transition theory and results of studies on molecular ions (Section 6.3) includes a discussion of the four cases of radiationless transitions: (i), the resonance limit; (ii), the statistical limit; (iii), the intermediate case; and (iv), predissociation. Final remarks single out some particularly interesting cases, implications and proposals for further work.

2. FUNDAMENTAL PROCESSES

2.1 Direct and Indirect Processes

In discussing those processes which result from the interaction between a photon, or other particle of sufficient energy, and a molecular system, we can use a quantum, a semi-classical or a classical treatment for dealing with the matter-radiation, matter-field or matter-matter interactions. The choice of treatment will depend upon the particular type of interaction process and upon the energies involved [2].

Before or after the interaction, the system is characterized by a specific quantum state describing its electronic and nuclear configuration. A transition from the initial to the final state of the system involves an intermediate period in which the system corresponds to a superposition of its asymptotic states. Experimental observations usually pertain to the initial and final states -- the asymptotic states. Elementary atomic and molecular processes have been classified in terms of the initial and final electronic and nuclear states by Berry [3] and by Berry and Leach [4].

An important parameter for a particular process is the duration of the intermediate compound state (ICS). When the duration of the ICS is comparable to the photon (or other particle) interaction time with the molecule, then the process is *direct*. We talk of *indirect* processes when the ICS has a lifetime comparable to at least several Bohr periods. In this case, there is a possibility of identifying the ICS experimentally. A useful view of an indirect process is to consider it as having an energetically allowed exit channel, but one for which initially the energy is not all in the appropriate degree of freedom [3]. The system evolves to its asymptotic region in the particular exit channel when enough energy has collected in one (or a few) degrees of freedom. The relative phases of the different modes may have to be restricted in order to achieve the asymptotic region, if more than one degree of freedom is involved. Thus, for example, in preionization (autoionization) or predissociation processes, there is a redistribution of energy from an initial storage in electronic and/or nuclear degrees of freedom, to an ionizing or dissociating mode.

2.2 The Photoionization Process

The (single photon, single ionization) photoionization process corresponds to the removal of a bound electron, nominally to infinity, following absorption of a photon of adequate energy. The probability that a molecular species in an initial (i) bound state will undergo transition to a final (f) electron continuum state is usefully defined in terms of a cross-section at the particular photon energy absorbed. The differential cross-section $d\sigma(\nu)$ is proportional to the ratio of the bound-continuum transition rate $P_{if}(\nu)$ to the exciting light intensity $I(\nu)$:

$$d\sigma(\nu) = P_{if}(\nu)h\nu/I(\nu) \tag{1}$$

The transition rates can be calculated using first-order time-dependent perturbation theory within the context of the semiclassical theory of radiation-matter interaction -- that is, where the radiation field is treated classically as a plane wave with vector potential $\underset{\sim}{A}$, so that $I(\nu) = (2\pi\nu^2/c)|\underset{\sim}{A}|^2$. The transition rate is given by the Wentzel [5] (Fermi) Golden rule expression:

$$P_{if}(\nu) = (2\pi/\hbar)|\langle f|H_{int}|i\rangle|^2\rho_f \tag{2}$$

where H_{int} is the interaction Hamiltonian, and where ρ_f is the density of continuum final states. H_{int} can be written as $H_{int} = H_{int}\,(E1) + H_{int}\,(M1) + H_{int}\,(E2) + \ldots$, where the three terms on the right-hand side are the electric-dipole, magnetic-dipole and electric-quadrupole interactions, respectively. We will be concerned mainly with electric-dipole transitions.

A key problem in calculations of the transition probability is the description of the wavefunctions used, especially the adequate form of the continuum wavefunction. Many different approaches have been employed [2].

The independent-particle approximation is a convenient, but not unique, approach to the description of atomic and molecular electronic states that take part in the photoionization process. Within this approach, an energy state will be an eigenstate of an approximate Hamiltonian; it can be described by its electronic configuration -- that is, by a listing of the single-particle states that are occupied and the values of angular momentum and vibrational quantum numbers (or approximations to them). Upon photoionization, the removal of a bound electron is accompanied by the quasi-simultaneous reorganization of the orbitals and energies of the remaining electrons. The amount of electron reorganization is a function of the rapidity of electron ejection, the process being adiabatic when the kinetic energy of the outgoing electron is relatively small. This can be considered in terms of the orbital relaxation time [6,7]. For example, with photons in the 10 eV range, the interaction time is on the order of 2 x 10^{-13} s, assuming 500 cycles of the radiation field. Since the orbital reorganization time is on the order of 10^{-15} s (in saturated hydrocarbons, for example), photoionization will correspond to an electronic adiabatic process. This is in contrast to electron-impact ionization, where electrons of 10 eV kinetic energy will have interaction times on the order of 10^{-16} s, making ionization a partially diabatic process.

Some of the discrete bound states, as described in the independent-particle approximation for example, will have the same energy as one or more of the continuum states and, therefore, will decay to the continuum if an adequate value of the matrix element of the interaction operator exists at this level of approximation. These *superexcited* (i.e., excited above the lowest ionization energy) states are Rydberg levels converging to limits above the lowest ionization potential; they are known as autoionizing (or preionizing) states and the decay process is termed autoionization (or preionization) [3,4] (Fig. 1). Other decay modes of the superexcited states will be discussed later.

Autoionization can occur only if the atomic and molecular ion core is excited with an energy at least equal to the binding energy of the detachable electron in the superexcited state. Some useful theories of this autoionization process -- for example, the multichannel quantum defect theory (MQDT) [4] -- are "collisionist" in approach, in which one conceives of an inelastic collision occuring between an excited electron and the ion core, leading to an exchange of energy and angular momentum. Although for atoms the energy exchange can only be electronic in nature, for molecules there also exists the possibility of conversion of vibrational and/or rotational energy of the ion core into kinetic energy of the outgoing electron. Thus, for molecules,

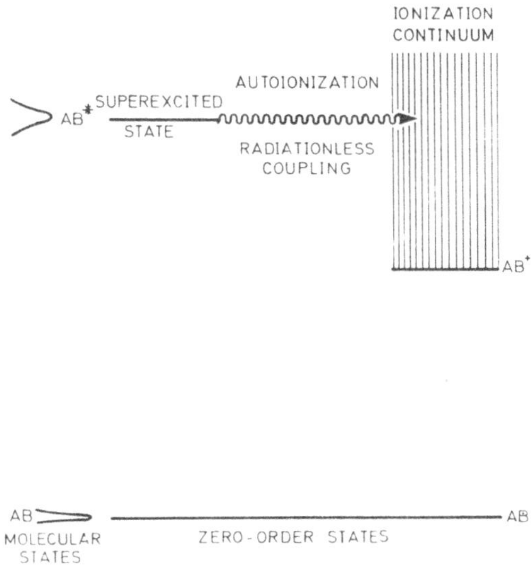

Fig. 1. Autoionization as a radiationless coupling between a zero-order superexcited state and an ionization continuum: $AB^* \rightsquigarrow AB^+ + e^-$.

there exist rotational, vibrational and electronic autoionization mechanisms. These processes can be competitive and can even interfere with each other.

Ion or electron currents are generally used to determine total (integrated) photoionization cross-sections σ_i. Absolute values of σ_i are not easy to obtain; most literature values are relative to standard calibrations. Electron measurements can be used to represent faithfully ionization processes, except when ion pairs are formed (no electrons ejected) or when multiple ionization occurs (multiple electron ejection). Much useful information has indeed been obtained from *photoelectron measurements*, including the angular distribution of the photoelectrons. This latter aspect has been reviewed elsewhere [4] and will not be discussed in the present paper.

Although we are concerned here with processes resulting from photon absorption, it is possible to study quantitatively a wide range of energy-dependent "simulated" photoabsorption and photoionization phenomena by the detection of fast scattered electrons in coincidence with ejected electrons (e, 2e process) and ions (e, e + ion process). The fast-electron differential scattering cross-sections are determined at negligibly small momentum transfers [8]. Photoabsorption cross-sections can be obtained by kinematic correction of the forward-scattering electron energy-loss spectrum according to the Bethe-Born theory. Partial photoionization cross-sections are then available via the ion branching ratios. Ion yields from ionization and dissociative ionization have been measured with this technique for a number of molecular species (e.g., O_2, CO_2, N_2O, NH_3, CH_4) [8].

2.3 Photoionization Cross-Sections

2.3a General Characteristics

The *partial photoionization cross-section* $\sigma_i(k)$ for producing a singly-charged ion in the kth level is ideally a step function of the photon energy, corresponding to a constant branching ratio [9,10]. Two kinds of general behavior are exhibited for atoms. Fig. 2 shows the photoionization cross-section for helium [11]. The sharp rise of $\sigma_i(k)$ at threshold is followed by a slow decrease to higher energies. The second type of behaviour is illustrated by the results of Ederer and Tomboulian [12] for neon, which exhibits a *maximum* in the cross-section function (Fig. 3). The threshold for the production of the $^2P_{3/2}$ and $^2P_{1/2}$ states of the $2s^22p^5$ Ne II ion occurs at about 575 Å with a *total* cross-section $\sigma_i \simeq 6.3 \times 10^{-18}$ cm^2. After passing through a maximum of 10^{-17} cm^2 at 375 Å, σ_i decreases, falling to 7.8×10^{-18} cm^2 at about 260 Å, the onset for formation of the $2s2p^6\ {}^2S_{1/2}$ state of Ne II.

The two types of function are related to the presence or absence of a node in the initial bound-state radial wavefunction of the photoejected electron. Minima can also occur when, at certain energy regions, the radial wavefunctions of the initial state and the continuum state tend to cancel in the matrix elements [13]. A new threshold step might be expected in cross-section curves as each new ionization level k' is reached with increasing photon energy. Above this threshold there will be a superposition of the partial photoionization cross-sections to form the total cross-section σ_i. This is seen in Fig. 3 at the $^2S_{1/2}$ level threshold, where the jump is on the order of 0.7×10^{-18} cm^2.

2.3b Autoionization Features for Atoms and Molecules

Discontinuities at photoionization thresholds above the lowest threshold are not always observed. Indeed, Samson [14] has found no evidence for the discontinuity at 260 Å in neon (Fig. 3). Fano and Cooper [15] have shown that autoionization processes can give rise to interference effects which can cause σ_i to increase, decrease or remain relatively unmodified at ionization thresholds For molecules, of course, these thresholds can correspond to rota-

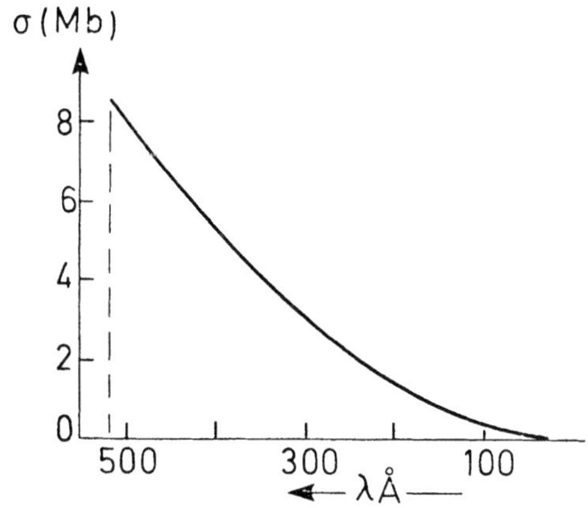

Fig. 2. Photoionization cross-section function for He in the 100 - 600 Å region [11].

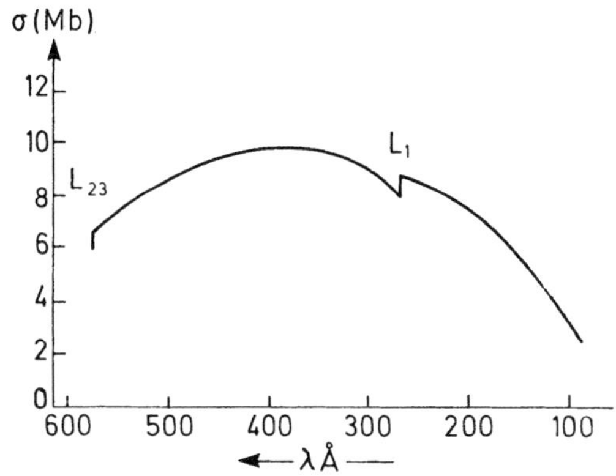

Fig. 3. Photoionization cross-section function for Ne in the 100 - 600 Å region [12]. The discontinuity at the L_1 threshold was not observed by Samson [14].

tional or vibrational, as well as electronic, state formation. Resonance peaks and dips (antiresonance minima) of various symmetrical or asymmetrical shapes can occur as a result of interferences between the probability amplitudes for direct ionization and indirect ionization (autoionization from superexcited levels). The shape factor has been parameterized by Fano [16] in a treatment based on isolated autoionization resonances. The "Fano profile" function $f(\varepsilon,q)$ is given by

$$f(\varepsilon,q) = \frac{(\varepsilon + q)^2}{(\varepsilon^2 + 1)}$$

where ε is the energy (in units of peak width) of the incident photon, and where q is related to the ratio of electric-dipole coupling strength from the ground (neutral) state ψ_0 to the compound state Ψ_E and to the continuum state ψ_E (where Ψ_E is a linear combination of the continuum state ψ_E and the discrete excited state ψ_s), as well as to the interaction between discrete and continuum states.

Fano profiles in autoionization features are illustrated in Fig. 4 by the photoionization cross-section function for atomic nitrogen in the 610-850 Å region, obtained by Dehmer, *et al.* [17]. The photoionization threshold for production of the $2s^22p^2\ ^3P$ state of N II is at 852.2 Å. The first excited state of the ion to which an optical dipole transition is allowed from the $2s^22p^3\ ^4S$ ground state of the neutral N I atom is the $2s2p^3\ ^5S$ state of

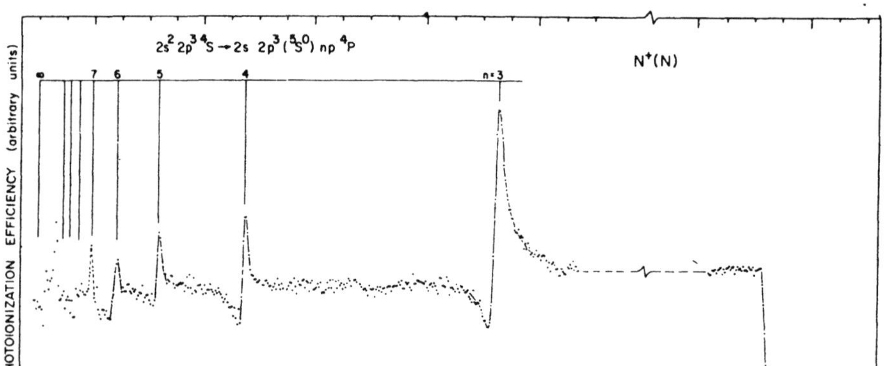

Fig. 4. Photoionization cross-section function for atomic nitrogen in the 610 - 850 Å region. Wavelength resolution FWHM = 0.83 Å [17].

N II at 609.3 Å. The asymmetrical peaks and antiresonance minima in Fig. 4 correspond essentially to autoionization into the continuum associated with the ^3P ion state from the n = 3,4,... Rydberg series levels leading to the $2s2p^3\ ^5S$ state of the ion.

The photoionization cross-section functions of molecules are very rich in autoionization features for several electron volts in energy above the first ionization threshold. This subject will be examined below in the context of competing decay channels. Mies [18] (see also Smith [19]) extended Fano's original treatment to the case of several continua, and also to the general case of several closely spaced autoionizing states and any number of continua. Jungen [20] has stressed that, due to overlap of resonances, photoionization cross-sections can have forms that are more complex than the relatively simple analytical forms that derive from Fano's original treatment. This is clear from the results of the theoretical calculation carried out for H_2 in the threshold region by Jungen and co-workers [20-24], based on Seaton's [25] and Fano's [26] quantum defect theory. The observed [27] and calculated cross-section curves [20-24] are in excellent agreement and well-demonstrate the existence of structure far more complex than a single Fano profile. In general, molecular high-resolution photoionization spectra will require analyses of the same order of refinement as those developed by Jungen and co-workers [20-24] for the study of the hydrogen molecule.

3. COMPETING DECAY CHANNELS OF SUPEREXCITED STATES

3.1 Generalities

A measure of the relative importance of nonionic relaxation channels of superexcited states is given by the photoionization efficiency, or quantum yield. The *photoionization efficiency* γ_I is defined as the number of ions produced per photon initially absorbed. For atoms, γ_I is generally unity above the first threshold; however, it can be less than unity in regions where superexcited states occur, if autoionization is a slower process than other decay processes such as fluorescence emission. Such behaviour has been noted for atomic oxygen [28,29]. Nevertheless, in most atomic cases, photoionization cross-sections can be considered as identical to those of photoabsorption.

The situation is markedly different for molecules, where there is a greater diversity of decay channels for superexcited states and a greater density of the latter since these now include rovibronic levels of electronic states. The primary act of photoabsorption can lead to direct ionization and also to the formation of superexcited states, the decay processes of which include not only autoionization and fluorescence, as is possible for atoms, but also other nonionic processes besides fluorescence, such as neutral dissociation, predissociation and intramolecular coupling to stable states. To a first approximation, each process can be considered as independent and competitive, and thus as having a distinct partial cross-section; however, strong interactions can also occur between these channels, leading to interference effects, just as we have seen in the case of direct and indirect ionization.

For convenience we list here, in reaction form, the various decay channels that can occur for the state reached on absorption of monochromatic photons of energy $h\nu$. The primary act of photon absorption may lead to the following processes, having specific cross-sections σ: It can form a stable ion state, with cross-section σ_k,

$$AB + h\nu \xrightarrow{\sigma_k} AB^+ + e^- \qquad (3)$$

or it may lead to ion fragmentation,

$$AB + h\nu \xrightarrow{\sigma_g} A^+ + B + e^- \qquad (4)$$

or to ion pair production,

$$AB + h\nu \xrightarrow{\sigma_p} A^+ + B^-. \qquad (5)$$

The primary act can also form a superexcited neutral state i,

$$AB + h\nu \xrightarrow{\sigma_i} AB^*. \qquad (6)$$

This superexcited state may decay by autoionization, with rates k_{if}, k_{if}' and k_{if}'' for the following three processes:

$$AB^* \xrightarrow{k_{if}} AB^+ + e^- \qquad (7)$$

$$AB^* \xrightarrow{k_{if}'} A^+ + B + e^- \qquad (8)$$

$$AB^* \xrightarrow{k_{if}''} A^+ + B^- \qquad (9)$$

Nonionic decay of the superexcited state is also possible (assuming only first-order processes):

Dissociation	$AB^* \rightsquigarrow A + B$	(10)
Predissociation	$AB^* \rightsquigarrow AB^{**} \rightsquigarrow A + B$	(11)
Fluorescence	$AB^* \rightsquigarrow AB + h\nu$	(12)
Internal Conversion	$AB^* \rightsquigarrow AB$	(13)

We define k_n as the total probability for nonionic radiative and nonradiative decay processes (10)-(13).

The total photon absorption cross-section σ_a is given by

$$\sigma_a = \sigma_k + \sigma_g + \sigma_p + \sigma_i \tag{14}$$

and the total ionization cross-section σ_I is

$$\sigma_I = \sigma_k + \sigma_g + \sigma_p + \sigma_i \left(\frac{k_{if} + k_{if}' + k_{if}''}{k_{if} + k_{if}' + k_{if}'' + k_n} \right) \tag{15}$$

For conditions where the ion-pair production processes (5) and (9) may be neglected (i.e., when we assume $\sigma_p = 0$ and $k_{if}'' = 0$), the photoionization efficiency γ_I is given simply by the ratio σ_I/σ_a:

$$\gamma_I = \frac{\sigma_I}{\sigma_a} = \frac{\sigma_k + \sigma_g + \sigma_i \left(\frac{k_{if} + k_{if}'}{k_{if} + k_{if}' + k_n} \right)}{\sigma_k + \sigma_g + \sigma_i} \tag{16}$$

Since for every ion formed a photoelectron is ejected, under these circumstances [i.e., neglect of processes (5) and (9)] γ_I will be equal to the photoelectron yield. However, if ion-pair production processes (5) and (9) are significant, the photoionization efficiency is no longer given as σ_I/σ_a. Since these processes do not produce photoelectrons, the ratio σ_I/σ_a will be greater than the true photoelectron yield γ_e, where

$$\gamma_e = \sigma_e/\sigma_a \tag{17}$$

In this equation, σ_a is given by Eq. (14) and σ_e is the sum

$$\sigma_e = \sigma_k + \sigma_g + \sigma_i \left(\frac{k_{if} + k_{if}'}{k_{if} + k_{if}' + k_{if}'' + k_n} \right) \tag{18}$$

3.2 Competing Decay Channels of Superexcited H_2

Some of the competing decay channels of superexcited states have been demonstrated for the hydrogen molecule: These include fluorescence [30], neutral predissociation [31] and ion-pair formation [32]. Intramolecular coupling between stable states has also been invoked in some of the decay processes. Results on these processes have been reviewed recently by Berry and Leach [4]. In spite of the extensive work on H_2, there is still insufficient information for a complete quantitative study of the rates and yields of the various processes over a large energy range.

3.3 Molecular Photoionization Efficiencies: Larger Molecules

In general, it is found that the ionization efficiency for molecules does not reach unity until several electron volts above the ionization threshold. This implies that nonionic channels must have decay rates that are comparable (to within a factor of 100) to autoionization rates over this energy region. There is little detailed work on the competitive processes for molecular species other than hydrogen.

When γ_I reaches unity, autoionization becomes the overwhelming decay process of the superexcited states. In this region, the photoionization cross-section curves show little or no structure, as is consistent with the linewidths associated with ultrafast autoionization decay rates ($\sim 10^{16}$ s^{-1}) and with the high densities of superexcited states at these energies. Fig. 5, which illustrates this behaviour, shows the absorption and photoionization cross-sections for H_2, from the work of Cook and Metzger [33]. Photoionization onset is at 804 Å, but the corresponding cross-section does not become comparable to that for photoabsorption until $\sim$680 Å.

Table 1 (taken from Berry and Leach [4]) presents the following data for a number of di-, tri- and poly-atomic molecules: Column 2 shows the adiabatic first ionization potential; column 3 gives the photon energy corresponding to ion formation with unit efficiency, estimated mainly from compilations given by Berkowitz [29]; and, column 4 gives ΔE, the energy difference between columns 3 and 2. ΔE, as defined, corresponds to the energy range above the first ionization limit in which superexcited states have non-negligible probabilities for nonionic decay. One sees

Table 1. Domain of Competitive Nonionic Decay Processes of Superexcited States in Some Molecular Species.

Molecule	Adiabatic First Ionization Potential (eV)	Lower Energy Limit[a] for $\phi_i = 1$ (eV)	Energy Difference ΔE[b] (eV)
H_2	15.43	18.23	2.80
N_2	15.58	20.66	4.08
O_2	12.07	18.09	4.02
CO	14.01	19.37	5.36
NO	9.26	17.71	8.45
H_2O	12.62	20.66	4.04
CO_2	13.77	19.37	5.60
NH_3	10.15	17.71	7.56
CH_4	12.62	15.49	2.87
C_2H_2	11.40	17.71	6.31
C_2H_4	10.51	16.75	6.24
C_2H_6	11.56	∿18	∿6.4
C_6H_6	9.26	16.53	7.27

[a] These limits have an accuracy of ∿0.2–0.5 eV.

[b] Difference between third and second columns equals the domain of competitive nonionic decay.

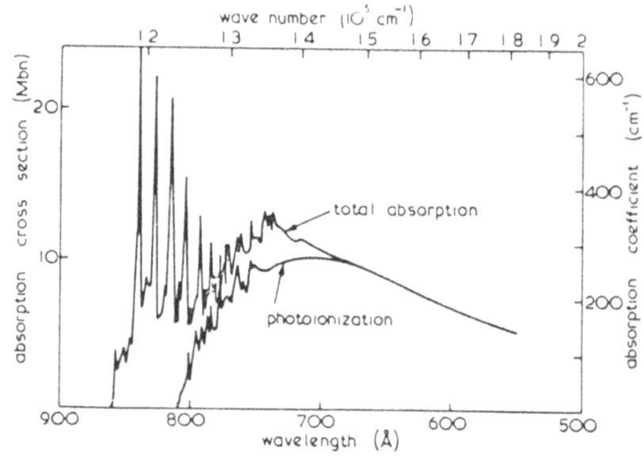

Fig. 5. Absorption cross-section and photoionization cross-section functions for H_2 in the 500 - 900 Å region [33].

that, for this series of molecules, ΔE ranges from $\sim$3 to $\sim$9 eV, but no systematic trends are obvious.

4. METHODS FOR STUDYING SUPEREXCITED AND IONIC STATES AND THEIR RELAXATION PROCESSES

We restrict ourselves to those processes which result from photon absorption. Before discussing the individual experimental techniques, it is useful to mention some of the photon sources that can be used for studying excitation and ionization processes.

4.1 Photon Excitation Sources

Much of the earlier work on photoionization processes was carried out using line emission sources of the gas-discharge or vacuum-spark varieties. Detailed structure in cross-section curves is not easily revealed with such sources. Continuum sources, particularly the high-pressure rare-gas continua, the Lyman continuum generated by a high current-density discharge through a low-pressure gas (such as in the Garton source) and the Ballofet-Romand-Vodar (BRV) vacuum-spark source (which has a low inductance, is triggered by an auxiliary sliding spark, and uses refractory high atomic-weight anode material) have been the principal VUV sources used for quantitative studies. Descriptions and references to these line and continuum sources are given by Samson [34] and by Zaidel' and Shreider [35]. References to recent developments of the very useful BRV plasma source are given by Berry and Leach [4]. In more recent years, the synchrotron radiation (SR) source has provided more reliable data on photoionization cross-sections and on other processes induced by photon absorption. SR sources can provide continuum radiation from the far-IR to the x-ray region and, with adapted monochromators, can give tunable radiation over this entire range. The use of wigglers in storage rings [36] can provide both wavelength shifting (to the blue) and intensity enhancement of synchrotron radiation.

A comparison of various VUV sources is made in Table 2. It should be noted that one difficulty in the use of continuum sources is to account for scattered light and, especially with synchrotron radiation, to eliminate higher spectral-order radiation.

Laser-generated plasmas can also be used to obtain continuum radiation over the 50-2000 Å region. Results have been obtained that show promise for these plasmas as sources for quantitative absorption spectroscopy in the far-UV. Work on suitable target materials is being actively pursued [37].

The rapid development of VUV lasers also shows promise for photoselective studies over a narrow tunability range in the 5-10 eV spectral region for one-photon excitation, and in the 15-30 eV region for multiphoton excitation [38]. The use of visible and near-UV laser sources, especially tunable ones, has given great impetus to the study of multiphoton ionization. This field is of interest not only as a tool for spectroscopy, where easily detected ions monitor intermediate real or virtual states which play a role in ion formation, but also for studies of fragmentation processes.

The use of undulators (multiple wigglers) [36,39] in storage rings shows promise for the development of a free-electron laser capable of high intensity and, perhaps, of going down to ∿500 Å in wavelength.

Knowledge of the relaxation processes of molecular superexcited states can be obtained from several complementary techniques, to which we now turn.

4.2 High-Resolution Absorption Spectroscopy in the Vacuum Ultraviolet

Continuum photon sources are required for such studies. Some of the experimental techniques appropriate for high-resolution VUV spectroscopy are given in books by Samson [34,44], Zaidel' and Shreider [35], Damany, *et al.* [40] and Berkowitz [41]. High-resolution VUV spectrographs and monochromators have been described in the literature [42,43]. Holographic gratings are now being used more extensively [44].

Table 2. Comparison Between Sources of Vacuum Ultraviolet Radiation

Source	Useful Energy Range (eV)	Line-Width (eV)	Pulse Length (Sec)	Intensity at Monochromator Exit Slit Photons/(s·Å)
A. Conventional Rare-Gas Continua				
He Continuum (50 torr)	12 - 21	---	---	$\sim 10^8$
Ne Continuum (100 torr)	12.4 - 16.8	---	---	$\sim 10^8$
Ar Continuum (200 torr)	8.0 - 11.8	---	---	$\sim 10^8$
Kr Continuum (200 torr)	6.9 - 9.9	---	---	$\sim 10^8$
Xe Continuum	6.2 - 8.4	---	---	$\sim 10^8$
B. Pulsed Discharge in Rare-Gas Continua				
Ar (2000 torr)	8.0 - 11.8	---	10^{-6}	$\sim 10^9 - 10^{10}$
Kr (2000 torr)	6.9 - 9.9	---	10^{-6}	$\sim 10^9 - 10^{10}$
C. Hydrogen Lamp				
Hinterreger Lamp	4 - 12	---	---	10^7

Table 2. Continued.

Source	Useful Energy Range (eV)	Line-Width (eV)	Pulse Length (Sec)	Intensity at Monochromator Exit Slit Photons/(s·Å)
D. Resonance Lamps				
He(I) Resonance	21.2	10^{-3}	---	$\sim 10^{10}$
He(II) Resonance	40.8	$\leq 10^{-2}$	---	$\sim 10^{9}$
Ne(I) Resonance	16.8	10^{-3}	---	$\sim 10^{10}$
Ne(II) Resonance	26.9	$\leq 10^{-2}$	---	$\sim 10^{9}$
E. Plasma Source				
BRV Source	4 - 250	---	2×10^{-7}	3×10^{12} (at 5 Hz) [6×10^{11} photons/pulse Å]
F. Synchrotron Radiation	10^{-1} - 1000	---	$\geq 0.4 \times 10^{-10}$	10^{11}

From VUV absorption spectra, one can measure the energies of the "superexcited" bands and determine the energies of their corresponding excited levels. Exact determination of zero-order band origins can be difficult because of interactions between the discrete superexcited level and autoionization (or other) continua. Nevertheless, one uses the information so obtained to classify Rydberg series and to establish quantum defects. Series limits enable one to determine ionic level energies. The study of line and band profiles enables one, in principle, to explore interaction phenomena, including interferences between direct ionization and superexcited state decay. [Recall [Eq. (14)] that the absorption cross-section is the sum of the partial cross-sections for all primary processes (3)-(6).]

A certain amount of work on the high-resolution VUV absorption spectroscopy of polyatomic molecules has been done recently. We mention, in particular, studies on CO_2 [45], H_2O [46,47], H_2S [48], C_2H_2 [49], CH_3Br [50], CH_3I [51] and C_6H_6 [50,52]. In some cases, these studies have led to the discovery of rotationally-structured Rydberg bands and to extensive analyses of Rydberg energy levels and states. Relatively little work has been done on Rydberg absorption line profiles for polyatomic molecules. A brief discussion of Rydberg band profiles has been given for some species; it was concluded that autoionization dominates other relaxation channels in high-n Rydberg states [50].

4.3 Ion-Yield Spectroscopy

Studies, as a function of incident photon energy, of the yield of parent and of fragment ions enable one to observe some autoionizing superexcited states and to follow the relative cross-sections for autoionization to stable or dissociative ionic states. Monochromatized SR and other continuum sources are used in such studies. Experimental techniques have been described in [34,35,40,41]. Examples of work done have been given in the book by Berkowitz [41]. More recent work has been carried out on a number of triatomic species (CO_2, COS, CS_2, N_2O, SO_2) and on other species, some of which will be mentioned and referenced in Section 5.1. We note also the studies by Baumgärtel and his co-workers on a number of small organic molecules [53], as well as recent molecular-beam photoionization studies of molecules and clusters [54].

4.4 Photoelectron Spectroscopy and Coincidence Studies

Since one electron is emitted for every singly-charged ion (excluding ion pairs) formed by photoionization with photons of energy $h\nu$, the ion appearance potentials I can be determined by measurement of the kinetic energy of the photoejected electrons, according to the relation $I = h\nu - mv^2/2$ (where m is the mass and v the velocity of the electron). For a molecular species, the ion will be formed not only in a particular electronic state, but also in well-defined vibrational and rotational levels. The energies of these ion levels are displayed in photoelectron spectral (PES) bands, as exemplified in Fig. 6. The energy resolution in PES depends upon the linewidth of the exciting radiation (see Table 1), the translational motion of the absorbing species, and the resolving power of the electron-energy analyzer. With present day techniques, resolutions down to $\sim$5 meV ($\sim$40 cm^{-1}) can be achieved with care, and to $\sim$2 meV ($\sim$16 cm^{-1}) with extreme care. Usual energy resolution is $\sim$20 meV. We note that PES employs a fixed photon excitation energy and measures the variable kinetic energy of photoelectrons. The angular distribution of the photoelectrons also yields valuable structural information.

Fig. 6 also illustrates another form of photoelectron spectroscopy of interest in photoionization studies, namely, threshold photoelectron spectroscopy (TPES). In TPES, a variable photon energy source is used but only zero kinetic energy photoelectrons (threshold photoelectrons ≡ TPE) are detected. The states detected therefore have the same energy as that of the absorbed photon. Threshold photoelectron spectroscopy can thus provide information on both direct-ionization and autoionization processes. Comparison with conventional photoelectron spectra enables one to separate the threshold photoelectron signal into its direct-ionization and autoionization components.

Several types of coincidence studies can be carried out in connection with photoionization phenomena; these are indicated in Fig. 6 and are classified in Table 3. We note in particular that studies can be made of coincidences between threshold photoelectrons and the products of superexcited state autoionization channels, such as parent ions, fragment ions, fluorescence of excited ions, etc. Further useful information can be had from the photoelectron energy distribution at specific excitation energies. Some explicit examples of coincidence studies will be mentioned later.

4.5 Fluorescence Excitation Spectra of Neutral or Ionic Products

Fluorescence excitation spectroscopy can provide information on autoionizing, dissociative and predissociative channels for superexcited state relaxation, when one or more of the autoionization or dissociation products is in an excited state of non-negligible quantum yield. Work has been done, using monochromatized SR as a variable-energy excitation source, on a number of small species. We mention in particular the studies of Judge and his co-workers on CO_2 [63-65,91]. However, as has been discussed elsewhere [4], it is essential to stress that extreme care must

Table 3. Types of Coincidence Studies.

Acronym	Coincidence Technique	Coincidence Between	Technique Described [Refs]	Principal Ionization Mechanism: Direct (D) Autoionization (A)	Channel Studied
PIFCO	PhotoIon-Fluorescence photon	Ion - $h\nu_F$	[55,56]	D	Ion-selected parent or fragment-ion fluorescence
PEFCO	PhotoElectron-Fluorescence photon	e^- - $h\nu_F$	[57,58]	D	Energy-selected parent or fragment-ion fluorescence
TPEFCO	Threshold PhotoElectron-Fluorescence photon	TPE - $h\nu_F$	[59,60]	D,A	Energy-selected parent or fragment-ion fluorescence
PEPICO	PhotoElectron-PhotoIon	e^- - ion	[61]	D	Energy-selected parent or fragment-ion formation
TPEPICO	Threshold PhotoElectron-PhotoIon	TPE - ion	[61,62]	D,A	Energy-selected parent or fragment-ion formation

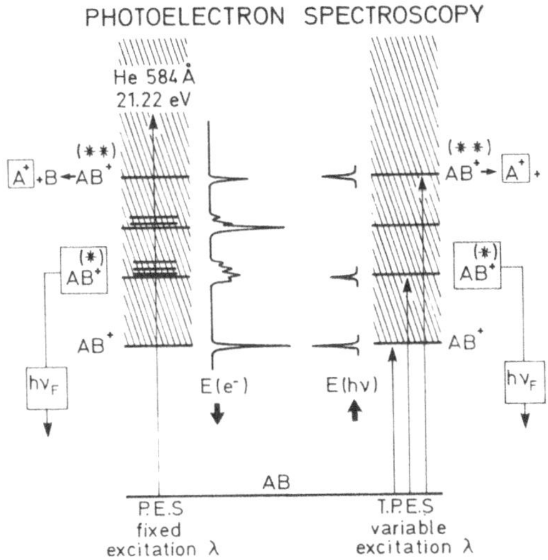

Fig. 6. Photoelectron spectroscopy (PES): fixed excitation wavelength, measurement of variable kinetic energy photoelectrons; and threshold photoelectron spectroscopy (TPES): variable excitation wavelength, measurement of fixed (zero) kinetic energy photoelectrons = threshold photoelectrons (TPE). Various types of coincidence experiments are possible (see text):

PIFCO = ion - $h\nu_F$ coincidences
PEFCO = e^- - $h\nu_F$ coincidences
TPEFCO = TPE - $h\nu_F$ coincidences
PEPICO = e^- - ion coincidences
TPEPICO = TPE - ion coincidences

be taken in interpreting fluorescence data to infer partial cross-sections whenever there exist radiationless transitions involving the fluorescent state.

5. FURTHER REMARKS ON MOLECULAR AUTOIONIZATION

5.1 Role of Autoionization in Dissociative Ionization

The optical resolution available in recent photoionization mass spectrometric studies has made it possible to examine the profiles of autoionization features not only for the parent ion, but also for the various fragment ion channels of small molecules. The results of systematic studies carried out by Berkowitz and his co-workers on diatomic and triatomic molecules, as well as on neopentane, have been brought together in a paper by Eland, *et al.* [66] which gives an empirical analysis of the autoionization resonances observed in the total and partial photoionization cross-sections. This analysis is based on the Fano single-resonance formulation, extended to the case of a single resonance interacting with several continua. It is clear that, in reality, there can be many overlapping resonances, as discussed for H_2 [20-24]; nevertheless, in the absence of a full multichannel quantum defect theory treatment, which would require much unavailable data on energy levels, the analysis [66] was carried out, *faute de mieux*, on a single-resonance basis. This approach can be considered as a simple parameterization, in which the observed profile data are expressed by an effective Fano q shape-parameter, a resonance width Γ, and the resonance and continuum coupling strengths. The parameterization for the series of species studied [66] gave consistent resonance energies and widths independent of the particular channel; peak shapes were found to vary systematically with the relative intensity of the resonance and the continuum in the channel examined; and, triatomic species of related electronic structure were found to have similar series of low $|q|$ resonances. Further work on this parameterization approach is discussed by Eland [67].

Many results show that *thresholds* for dissociative photoionization are often associated with autoionization processes, rather than with direct photoionization. This can be seen, for example, in comparing the ionization cross-section curve with photoelectron spectra taken at a nonautoionizing photon energy. Such comparisons often show that ionization events, including dissociative processes, can take place in a "Franck-Condon gap". The onset for many dissociative-ionization processes occurs at the thermodynamic threshold, which lies within the Franck-Condon gap, where the probability of ion formation is extremely low as determined by He I PES. This is the case for the formation of O^+ from N_2O [68-70]; O^+ from CO_2 [71-73]; S^+ from COS [74,75]; S^+ and CS^+ from CS_2 [75,76]; SO^+ from SO_2 [77,78]; $C_2{}^+$ from C_2N_2 [79]; HCO^+ from H_2CO [80,81]; $C_3H_3{}^+$ from $CH_3C{\equiv}CH$ [82]; and $C_3H_3{}^+$ from $H_2C{=}C{=}CH_2$ [83]. Some further examples are given by Murray and Baer [84].

5.2 Dissociative Ionization in CO_2

We consider, as an example, dissociative ionization in CO_2. The electronic energy levels of the $CO_2{}^+$ ion are given with the He I PES of CO_2 in Fig. 7. Fig. 8 shows the photoionization efficiency curves for $CO_2{}^+$, CO^+ and O^+ in the neighborhood of the dissociative-ionization thresholds in the 19 eV region, taken from

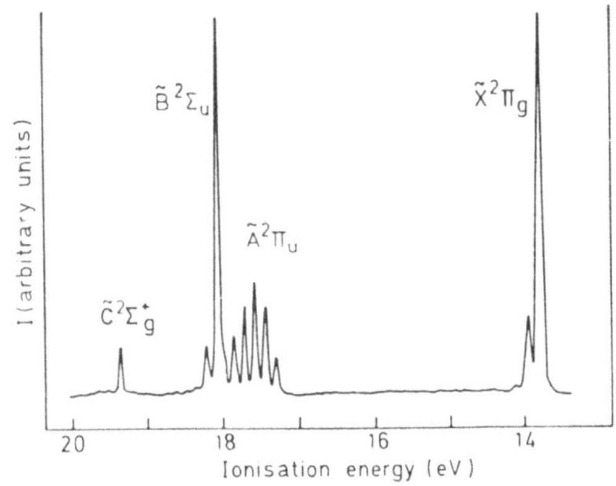

Fig. 7. He I photoelectron spectrum of CO_2.

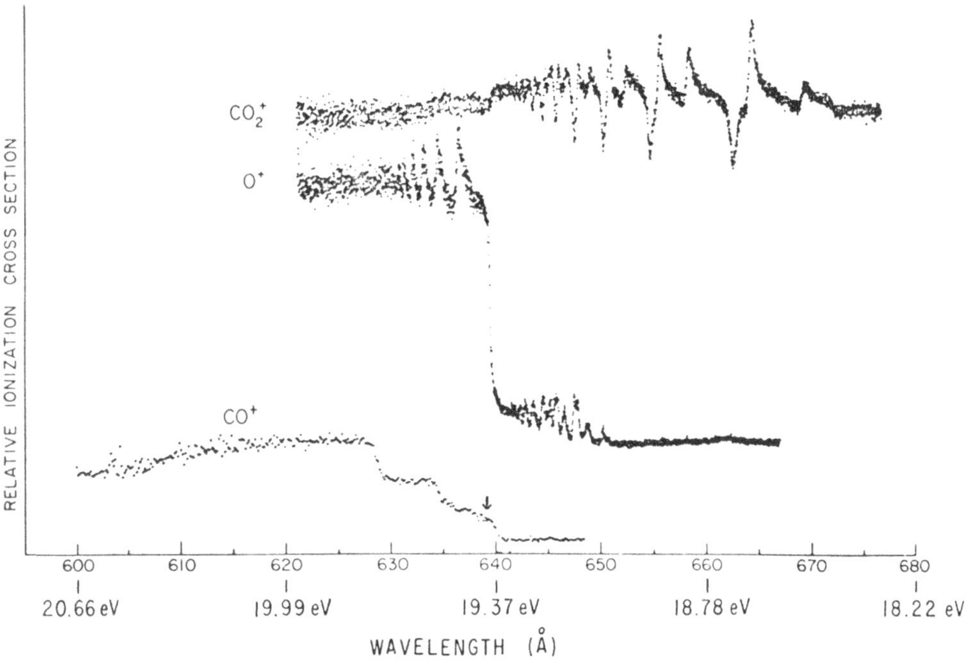

Fig. 8. Relative photoion yield curves of CO_2^+, O^+ and CO^+ in the vicinity of the dissociative-ionization thresholds [73]. Each curve is in different arbitrary units. The vertical arrow indicates the CO_2^+ $\tilde{C}^2\Sigma_g^+$ state threshold energy.

the work of Eland and Berkowitz [73]; each curve is in different arbitrary efficiency units, so that their relative intensities are not directly comparable. The He I photoelectron spectrum of CO_2 (Fig. 7) shows that the 19 eV region is not accessible by direct ionization. The threshold for O^+ is at 19.07 eV (650 Å) [$CO_2 \rightarrow O^+(^4S) + CO(X^1\Sigma^+, v=0)$], about 0.32 eV below the CO_2^+ $\tilde{C}^2\Sigma_g^+$ state threshold. The structure in the O^+ curve between 19.07 and 19.39 eV indicates clearly that the dissociation process occurs here via superexcited states -- namely, the Rydberg levels of the Tanaka-Jursa-Leblanc series [85] converging to the $\tilde{C}^2\Sigma_g^+$ state of CO_2^+ -- which are coupled to dissociative-ionization continua corresponding to lower-lying CO_2^+ states. The O^+ current increases sharply at the $\tilde{C}^2\Sigma_g^+$ state threshold; this state has been shown to be completely predissociative [86], thus accounting for the large increase in the partial cross-section. However, spin conversion rules are violated here, since the dissociation products $O^+(^4S) + CO(^1\Sigma^+)$ can only form quartet states.

The threshold for forming O^+ and the CO molecule in its v=1 state lies at 19.337 eV. Eland [86], in a study of the kinetic energy distribution of the O^+ ion, has concluded that at the CO_2^+ $\tilde{C}^2\Sigma_g^+$ state onset, predissociation to O^+ + CO creates about 85% of the CO $X^1\Sigma^+$ molecules in the v=1 state, with the remainder in v=0. The CO^+ ion is formed at its thermodynamic threshold at 19.466 eV, the process being $CO_2 \rightarrow CO^+ (X^2\Sigma^+, v=0) + O(^3P)$. At this energy, the observed CO^+/O^+ ratio, from the two dissociative channels, is 0.3 at 300 K. Rotationally-excited levels of the CO_2^+ $\tilde{C}^2\Sigma_g^+$, v=0 state, with J > 40, are above the CO^+ dissociation limit and predissociation is principally to CO^+ + O. Rather large impact parameters are involved in conserving angular momentum, which makes it somewhat surprising that the CO^+ dissociation channel dominates over the O^+ channel. However, this probably reflects the fact that $CO^+(^2\Sigma^+) + O(^3P)$ correlates with doublet states of CO_2^+ [87], and so could favor predissociation of CO_2^+ $\tilde{C}^2\Sigma_g^+$ with respect to the quartet state formed by CO + O^+. The structure in the CO^+ partial photoionization yield curve indicates that vibrationally-excited states open up new channels for CO^+ production.

An experimental difficulty in determining the exact branching ratios for forming parent and fragment ions is the discrimination of ions formed with, or given, different kinetic energies. Most mass spectrometric methods tend to discriminate against energetic ions; thus, carefully designed instrumentation is required [88].

Measurements of the $CO_2^+/CO^+/O^+/C^+$ ratios have been made at a number of wavelengths between 304 Å and 740 Å [41]. This has been extended down to 90 Å, with measurements also including double-ionization processes forming CO_2^{2+} and C^{2+} [88]. Cross-section data can be obtained by two methods: (i), the branching ratio method, in which the mass spectrum is accumulated at a given photon energy and the ratio of the number of ions collected for a particular species to the total number of ions produced is determined; and (ii), the ions per photon method, in which the wavelength is scanned for a fixed mass, with the photon intensity being monitored simultaneously with the ion intensity. The relative cross-sections determined by either method are put on an absolute basis by normalization at some calibration point. Maximum peak partial cross-sections are as follows: CO_2^+ (∿37 x 10^{-18} cm^2 at ∿600 Å); CO^+ (∿3.2 x 10^{-18} cm^2 at ∿450 Å); O^+ (∿3.8 x 10^{-18} cm^2 at ∿310 Å); C^+ (∿2.8 x 10^{-18} cm^2 at ∿340 Å); CO_2^{2+} (∿0.35 x 10^{-18} cm^2 at ∿230 Å); and C^{2+} (∿0.8 x 10^{-20} cm^2 at ∿130 Å).

Recently, the (e, e + ion) technique has been used to study ion fragmentation in CO_2 up to 80 eV [89], with results in good agreement with the photon data. Structure in all of the fragmentation curves obtained by the photon and (e, e + ion) techniques can be correlated with the multielectron transitions observed in the pseudo-photoelectron spectra obtained with the (e,2e) technique by Brion and Tan [90,91]. The total fragmentation cross-section has three principal peaks: (i), ∿8 x 10^{-18} cm^2 at ∿420 Å; (ii), ∿9.5 x 10^{-18} cm^2 at ∿330 Å; and (iii), ∿7.8 x 10^{-18} cm^2 at ∿230 Å [88]. The excitation spectrum of VUV fluorescence from CO_2 has been measured by Lee, *et al.* [91]; the spectrum is similar in shape to the total fragmentation curve determined by Masuoka and Samson [88]. The absolute fluorescence cross-sections appear to be about 5-10% of the total fragmentation values. The fluorescence is probably due to excited fragments produced by dissociative photoionization.

6. DECAY OF ION STATES

Ionic states formed can decay via three principal processes: (i), fluorescence; (ii), radiationless transition to bound states and (iii), radiationless transition to dissociation continua. In addition, collisions can modify these intramolecular processes and give rise to new, reactive, channels. We will be concerned only with intramolecular processes.

The theory of radiationless transitions, involving the coupling of two or more electronic states and concomitant vibrational redistribution, has been extensively developed over the past few years [92] but the models and applications relate mainly to neutral species. Relatively little has been done to study these processes in molecular ions, although such transitions have often been invoked, particularly to account for ionic fragmentation. Some of the advantages of molecular ions as objects for the study of radiationless transitions are described below.

6.1 Advantages of Molecular Ions for the Study of Radiationless Transitions

In experimental studies of radiationless transitions, the quantities determined are quantum yields and time-resolved decay patterns. Interpretation is in terms of coupling between the initially excited zero-order state and vibronic levels of lower-lying electronic states. With respect to neutral species, the study of molecular ions presents a number of advantages; these derive from two essential properties, one structural (the open-shell electronic configurational nature of many (odd-electron) molecular ions), and the other instrumental (the ease of ion detection and counting). We detail these advantages as follows:

Structural advantages:

Many molecular ions are open-shell species. The energy separation $\Delta E(D_1 - D_0)$ between the electronic ground state (doublet state D_0) and the first excited state of the same multiplicity (doublet state D_1) is often rather less than in the parent molecule $\Delta E(S_1 - S_0)$ [cf. Table 4]. The D_1 state will therefore tend to lie below the first dissociation limit of the ion; whereas in

Table 4. Energy of the First Electronic Transition of the Same Multiplicity as the Ground State for Some Ionic Species and Their Parent Neutral Molecules (Data from [93]).

Species	First Transition	Transition Energy (eV)
N_2	$a'\,^1\Sigma_u^- - X\,^1\Sigma_g^+$	8.40
N_2^+	$A\,^2\Pi_u - X\,^2\Sigma_g^+$	1.12
CO_2	$\tilde{A}\,^1B_2 - \tilde{X}\,^1\Sigma_g^+$	5.7
CO_2^+	$\tilde{A}\,^2\Pi_u - \tilde{X}\,^2\Pi_g$	3.53
H-C≡C-Cl	$^1? - {}^1\Sigma^+$	4.96
$H\text{-}C{\equiv}C\text{-}Cl^+$	$\tilde{A}\,^2\Pi - \tilde{X}\,^2\Pi$	3.32
1,3,5-$C_6F_3H_3$	$\tilde{A}\,^1A_2' - \tilde{X}\,^1A_1'$	4.78
1,3,5-$C_6F_3H_3^+$	$\tilde{B}\,^2A'' - \tilde{X}\,^2E''$	2.71

neutral (closed-shell) species, the first excited singlet state S_1 can very often have an energy close to or greater than that of the lowest dissociation threshold and, therefore, can possibly be subject to the occurrence of predissociation. Furthermore, the lower excitation energy of ions brings the first electronic transition to a more easily accessible spectral region from the viewpoint of excitation (selective or otherwise) and detection of photon emission.

Another structural advantage of molecular ions is particularly relevant to the applicability of radiationless transition theories, and provides a simplification with respect to closed-shell species. This simplification derives from the fact that, in closed-shell systems, the excited S_1 state can be coupled to vibronic levels of at least *two* lower-lying electronic states, namely, the first excited triplet state T_1 and the electronic ground state S_0; whereas in many open-shell molecular ions, the vertical excitation energy (with respect to the D_0 state) of the lowest-lying quartet state Q_1 will be greater than that of the first excited doublet state D_1, so that coupling leading to radiationless transitions will occur between D_1 and vibronic levels of only *one* lower-lying state, namely, the electronic ground state D_0 (cf. Fig. 9).

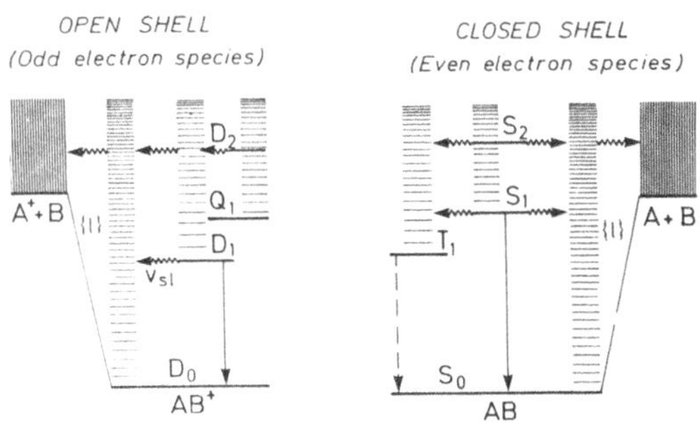

Fig. 9. Schematic comparison, between open-shell and closed-shell species, of electronic level manifolds, interelectronic state coupling and dissociation limits. $v_{s\ell}$ is the matrix element coupling the $|s\rangle$ vibronic level of the D_1 state to the isoenergetic $\{|\ell\rangle\}$ manifold of high vibrational levels of the D_0 ground state; similar matrix elements are posited for coupling between other open-shell electronic states, or between closed-shell states. See text for meaning of other symbols.

Instrumental advantages:

These advantages arise from the existence of an electric charge on the ion, which makes it relatively easy to detect, count and identify. Photoionization will also create a photoelectron (except in the case of ion-pair formation) having a well-defined kinetic energy. Charged-particle detection can proceed by means of ion multipliers, electron multipliers or channel plates, so that the number of ionic species can be counted; determination of the nature of these species requires some form of mass spectrometry, employing magnetic- and/or electric-field ion dispersion or mass-dependent time-of-flight techniques. It is thus possible to use coincidence methods to determine ion fluorescence lifetimes and quantum yields (ion-photon or electron-photon coincidences) and ion fragmentation yields (cf. Table 3). Coincidence techniques can permit the study of the behaviour of selected vibronic states; the techniques used for equivalent measurements in neutral species are much less direct [94].

6.2 Measurement of Ion Fluorescence Decay (τ) and Yields (ϕ_F)

Ion fluorescence decay behaviour and quantum yields have been measured using the three different coincidence techniques PIFCO, PEFCO and TPEFCO (cf. Table 3). The principles of these techniques, experimental details and results are to be found in the references listed in Table 5. We recall that the PIFCO technique measures the average $\overline{\tau}$ and $\overline{\phi}_F$, where the average is over the vibrational levels of a particular electronic state of the molecular ion. The PEFCO and TPEFCO techniques allow τ and ϕ_F measurements to be made at specific excitation energy levels within the energy resolution of the particular method, which is $\sim$100 meV for PEFCO and $\sim$40 meV for TPEFCO.

6.3 Survey of Radiationless Transition Theory and Results of Studies on Molecular Ions

The development and status of radiationless transition theory have been extensively reviewed in the last few years [92]. In what follows, we briefly survey these aspects of the theory useful in the study of molecular ions.

An electronic excited state $|S\rangle$ (D_1 in Fig. 9) is coupled radiatively to the ground state $|L\rangle$ (D_0 in Fig. 9). $|S\rangle$ and $|L\rangle$ are zero-order states (e.g., Born-Oppenheimer states). A (ro)vibronic level $|s\rangle$ of $|S\rangle$ is coupled nonradiatively, via matrix elements $v_{s\ell}$ of the nuclear kinetic energy operator (we consider only coupling between states of the same spin multiplicity), to one or more members of a set of (ro)vibronic level(s) $\{|\ell\rangle\}$ isoenergetic with $|s\rangle$. The (ion) molecular states are denoted by $|n\rangle$:

Table 5. Ion Fluorescence Decay and Yields: Methods of Measurement and Species Studied.

Species Studied	Coincidence Method [References]		
	PIFCO	PEFCO	TPEFCO
N_2^+	[55]		[99]
N_2O^+	[55]	[58]	[95]
CO_2^+	[55,93,96,97]	[58]	[59]
COS^+	[55]	[58]	
CS_2^+	[55]	[58]	
$ClCN^+$	[55]		
$BrCN^+$	[55]		[99]
ICN^+	[55]		
SO_2^+	[55,98]		[99]
$C_2H_2^+$	[100]		
$C_2N_2^+$	[100]		
NH_3^+	[99,100]		
CH_3CN^+	[100]		
$Cl-C\equiv C-H^+$	[93,102]		
$Cl-C\equiv C-D^+$	[93,102]		
$Cl-C\equiv C-Cl^+$	[93,102]		
$H-(C\equiv C)_2-H^+$	[103]	[104]	
$D-(C\equiv C)_2-D^+$	[103]	[104,105]	
$CH_3-(C\equiv C)_2-H^+$		[106]	
$CH_3-(C\equiv C)_2-D^+$		[106]	
$CD_3-(C\equiv C)_2-H^+$		[106]	
$CD_3-(C\equiv C)_2-D^+$		[106]	
$CH_3-(C\equiv C)_2-CH_3^+$	[103]	[106]	
$CD_3-(C\equiv C)_2-CD_3^+$		[106]	
$C_2H_5-(C\equiv C)_2-H^+$		[106]	

Table 5. Continued.

Species Studied	Coincidence Method [References]		
	PIFCO	PEFCO	TPEFCO
cis 1,2-$C_2H_2F_2^+$		[105,107]	
$Cl-(C{\equiv}C)_2-H^+$		[108]	
$Cl-(C{\equiv}C)_2-Cl^+$		[108]	
$N{\equiv}C-C{\equiv}C-C{\equiv}N^+$		[109]	
3-5 Octadiyne		[110]	
$C_6H_6^+$	[55,111]		
$C_6D_6^+$	[111]		
$C_6F_6^+$	[55,56,93]	[112a]	[60,113]
$C_6F_5H^+$	[55,56,93]	[112a]	
1,2,3,4-$C_6F_4H_2^+$	[56,93]	[112a]	
1,2,3,5-$C_6F_4H_2^+$	[56,93]	[112a]	
1,2,4,5-$C_6F_4H_2^+$	[56,93]	[112a]	[101]
1,2,4,5-$C_6F_4D_2^+$		[112a]	[101]
1,2,4-$C_6F_3H_3^+$	[55,56,93]	[112a]	
1,3,5-$C_6F_3H_3^+$	[56,93]	[112a]	[60,113]
1,3,5-$C_6F_3D_3^+$	[56,93]	[112a]	
1,3-$C_6F_2H_4^+$	[56,93]	[112a]	
1,3-$C_6Cl_2H_4^+$		[112b]	
1,4-$C_6Cl_2H_4^+$		[112b]	
1,3,5-$C_6Cl_3H_3^+$		[112b]	
1,3,5-$C_6Cl_3F_3^+$		[112b]	
1,3,5-$C_6Br_3F_3^+$		[112b]	

$$|n\rangle = \alpha_{ns}|s\rangle + \sum_{\ell} \beta_{n\ell}|\ell\rangle \tag{19}$$

where α_{ns} and $\beta_{n\ell}$ are the amplitudes of the zero-order $|s\rangle$ and $|\ell\rangle$ states.

Three cases are commonly distinguished in the theory: (i), the resonance limit case; (ii), the statistical limit case; and (iii), the case intermediate between the previous two limits.

6.3a The Resonance Limit

In this limit, the $\{|\ell\rangle\}$ levels are very sparse, the density of states being a fraction of, or a few, states per cm^{-1}. This means that each linear combination of interacting levels will correspond to different energy intervals and will have a distinct $v_{s\ell}$ value, so that each molecular eigenstate will emit with its own characteristic decay time. The molecular levels correspond to perturbed zero-order levels; the situation is the well-known one of spectral perturbations in small molecules [114,115], in which the unperturbed $|S\rangle \leftrightarrow |L\rangle$ dipole transition amplitudes will be distributed among the resulting molecular state transitions (cf. Fig. 10). If one of the zero-order transitions is forbidden, the molecular transitions (perturbed states) may exhibit "extra-lines" corresponding to transitions induced by s/ℓ mixing. In the resonance limit, the s/ℓ recurrence time is generally so short that the observed individual level decay rates will each have exponen-

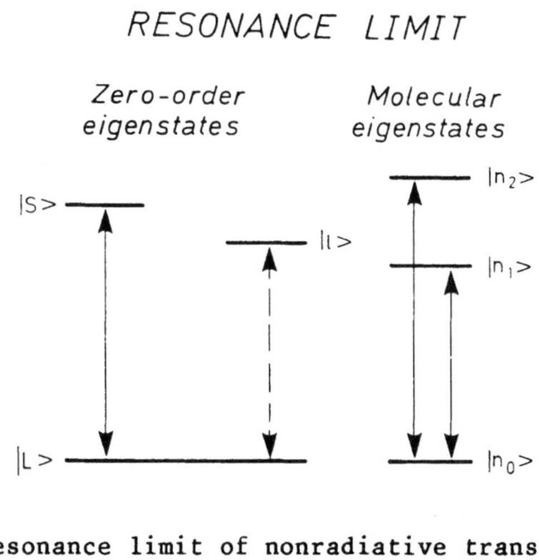

Fig. 10. Resonance limit of nonradiative transition theory: Zero-order and molecular eigenstates. See text for meaning of symbols.

tial character. If the $\{|\ell\rangle\}$ levels cannot be relaxed by collision or via dissociation, one expects the total fluorescence quantum yield to be unity.

Since the density of $\{|\ell\rangle\}$ states ρ_ℓ is so small in the resonance limit, ρ_ℓ is of little significance in the theory. The state mixing coefficients α_{ns} and $\beta_{n\ell}$, and the coupling matrix element $v_{s\ell}$, can be determined from high-resolution studies on the perturbed spectral lines. Useful information can also be obtained from lifetime measurements on individual molecular state transitions since these lifetimes will reflect interstate mixing [115,116].

In the case of two interacting levels, $|s\rangle$ and $|\ell\rangle$, the solution of the secular equation

$$\begin{vmatrix} E_s - E & v_{s\ell} \\ v_{s\ell} & E_\ell - E \end{vmatrix} = 0 \qquad (20)$$

gives the perturbed energies, namely, the molecular eigenstate energies E_{n_1} and E_{n_2}:

$$2E_{n_2} = (E_s + E_\ell) + \left[4|v_{s\ell}|^2 + (E_s - E_\ell)^2\right]^{1/2} \qquad (21)$$

$$2E_{n_1} = (E_s + E_\ell) - \left[4|v_{s\ell}|^2 + (E_s - E_\ell)^2\right]^{1/2} . \qquad (22)$$

The shifts of the zero-order levels from their mean values are therefore equal in magnitude and opposite in sign. More complex matrix equations and solutions arise if more than two zero-order levels interact [117].

An excellent example of the resonance limit is the CO_2^+ ion, for which radiationless coupling between the rovibronic levels of the $\tilde{B}^2\Sigma_u^+$ and $\tilde{A}^2\Pi_u$ states is manifested in spectral perturbations and effects on level lifetimes and apparent fluorescence quantum yields [96,118].

6.3b The Statistical Limit

In this other limiting case, the density of those $\{|\ell\rangle\}$ states isoenergetic with $|s\rangle$ is very high, so that $\{|\ell\rangle\}$ forms a uniform quasi-continuum. Typical values are $\rho_\ell = 10^{10} - 10^{20}$ states/cm^{-1}. To a first approximation, the coupling strengths are given by their average values $\langle v_{s\ell}\rangle$. The radiative width γ_s^r of the $|s\rangle$ state is supplemented by a nonradiative term γ_s^{nr}, which derives from s/ℓ coupling, so that the fluorescence quantum yield ϕ_F is given by

$$\phi_F = \gamma_s^r/(\gamma_s^r + \gamma_s^{nr}) \tag{23}$$

and will therefore tend to be less than unity in the statistical limit. Furthermore, in this limit, the nonradiative rate $k_s^{nr} = \gamma_s^{nr}/\hbar$ is related to $v_{s\ell}$ and ρ_ℓ by a Fermi golden-rule expression when $v_{s\ell}$ varies slowly enough to be represented by its average value [92]:

$$k_s^{nr} = (2\pi/\hbar)\rho_\ell v_{s\ell}^2 \tag{24}$$

and the decay of the $|s\rangle$ state will then be monoexponential:

$$\tau_s = \hbar/(\gamma_s^r + \gamma_s^{nr}) = (k_s^r + k_s^{nr})^{-1} \tag{25}$$

In the statistical limit, one might expect the nonradiative rates k^{nr} to be governed by the well-known *energy gap law* [92]. If the energy gap E_{SL} is large in comparison to the vibrational quantum $\hbar\omega_i$ of the acceptor mode i, and if we assume only one such acceptor mode (i.e., if we assume that geometry changes between the $|s\rangle$ and $|\ell\rangle$ potential surfaces are affected solely via mode i), then the nonradiative rate k_s^{nr} is given by [92]

$$\hbar k_s^{nr} \simeq (2\pi\hbar\omega_i/E_{SL})^{\frac{1}{2}}({v_{s\ell}^{el}}^2/\hbar\omega_i)\exp(-x_i)\exp(-\Omega E_{SL}/\hbar\omega_i) \tag{26}$$

where
$$\Omega \equiv \ln(E_{SL}/x_i\hbar\omega_i) - 1 \tag{27}$$

$$x_i \equiv (M_i\omega_i/2\hbar)\ (Q_{eq}^s - Q_{eq}^\ell)^2 \tag{28}$$

and where $v_{s\ell}^{el}$ is the electronic part of $v_{s\ell}$. x_i is a dimensionless measure of the shift in equilibrium positions for normal mode i, M_i being the corresponding reduced mass. Since, in general, $x_i \ll 1$ and Ω is a slowly-varying function of E_{SL}, the k_s^{nr} rate will then be simply an exponentially-decreasing function of the energy gap E_{SL}. Frequency changes (i.e., $\omega_i^S \neq \omega_i^L$) and the contributions of other accepting modes affect this result only slightly [92].

The electronic energy gap E_{SL} cannot, of course, be varied experimentally for an isolated molecule. However, we can compare the k^{nr} rates for the members of a class of related species with different E_{SL}, but for which the electronic part of $v_{s\ell}$ is expected to be approximately constant. This was done by Leach, *et al.* [93] using the mean values $\overline{\Phi}_F(\tilde{B})$ and $\overline{\tau}(\tilde{B})$ measured by the PIFCO technique for the occupied vibronic levels of eight fluorobenzene cations. Although many intrinsic factors (discussed by Dujardin, *et al.* [56]) can change from one fluorobenzene cation to another, an approximate exponential decrease of $\overline{k}^{nr}$ with E_{SL} was observed experimentally [93].

There is another "energy gap" law for large molecules, which is valid for moderate to large values of E_{SL} [92,119]. This predicts an increase of $k^{nr}(E_{SL} + E_v^S)$ when E_v^S increases, where E_v^S is the vibrational energy in the $|S\rangle$ state; several observations confirming this prediction have been made for neutral molecules. PEFCO measurements by Maier and Thommen [112] have demonstrated that such a law is valid for vibrational energy levels of the $\tilde{B}$ state of a number of polyfluorobenzene cations. This has been confirmed by Dujardin, *et al.* using the TPEFCO technique [60,113]; however, the superior energy resolution of the latter method enabled these authors to show that for $E_v^S < 3100$ cm^{-1} in $C_6F_6^+$, k^{nr} no longer depends monotonically on E_v^S but is sensitive to the particular vibrational mode excited in the $\tilde{B}$ state. This mode-selective behaviour has been reproduced by model calculations on $C_6F_6^+$ [113]. Comparison with experiment also shows that, for combination levels, the excited vibronic state gradually loses its memory of initial mode composition as E_v^S increases, and exhibits an increasing degree of collisionless vibrational redistribution [113].

We remark that, in the TPEFCO method, ions are formed both by direct and by autoionization processes. For sym-trifluorobenzene and hexafluorobenzene, considerable autoionization occurs from superexcited states to high vibrational levels of the ion ground state. This is seen both in Franck-Condon regions and in Franck-Condon gaps. Furthermore, in some cases there exists competition between threshold autoionization and ion-core nonradiative relaxation processes. These results, as well as methods of determining the branching ratios for forming the fluorescent state, are discussed by Dujardin, *et al.* [60].

6.3c The Intermediate Case

The density of $\{|\ell\rangle\}$ states, in this case, is intermediate between those in the resonance and statistical limits. These states can be considered to consist of two classes, $\{|\ell_c\rangle\}$ and $\{|\ell_d\rangle\}$, which are coupled to $|s\rangle$ with matrix elements v_{sc} and v_{sd}, respectively, where $\langle v_{sc}\rangle \gg \langle v_{sd}\rangle$ (cf. Fig. 11). The $\{|\ell_c\rangle\}$ states form a nondissipative set of states with respect to $|s\rangle$, as in the small-molecule (i.e., resonance) limit, whereas the $\{|\ell_d\rangle\}$ states can be dissipative with respect to $|s\rangle$ and may be numerous enough to form a quasi-continuum, as in the statistical limit. At this stage we disregard interactions, via anharmonic coupling, among and between $\{|\ell_c\rangle\}$ and $\{|\ell_d\rangle\}$ states.

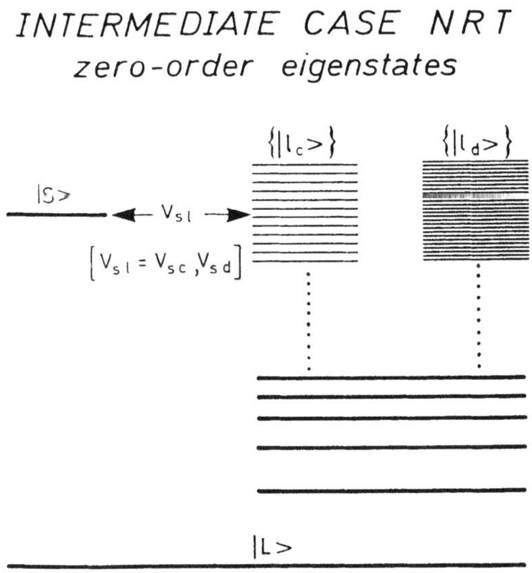

Fig. 11. Intermediate case of nonradiative transition theory: Zero-order states and couplings. See text for meaning of symbols.

Excited molecular ions formed, for example, by photoionization, have wavefunctions that are superpositions of the quasi-molecular (ion) functions $\{|n\rangle\}$ [cf. Eq. (19)]. We assume that the $\{|\ell\rangle\}$ states cannot undergo dipole transitions to $|L\rangle$ (except for very weak infrared dipole transitions between $|\ell\rangle$ states). Excitation of all $\{|n\rangle\}$ for which $\alpha_{ns} \neq 0$ leads to formation of the excited ion in its nonstationary state $|s\rangle$. In order to determine the emission characteristics of the excited ion, one may use the effective Hamiltonian formalism in which the effects of real- and quasi-continua provide purely imaginary contributions to the $|s\rangle$ and $\{|\ell\rangle\}$ energies.

We denote the energies of the zero-order states $|s\rangle$ and $\{|\ell_c\rangle\}$ as E_s and E_c, respectively, and the widths of these states as γ_s and $\{\gamma_c\}$, respectively. The effective Hamiltonian is defined by [92]

$$H_{eff} = \begin{pmatrix} E_s - i\gamma_s/2 & \cdots & v_{sc} & \cdots \\ \vdots & \ddots & & 0 \\ v_{sc} & & E_c - i\gamma_c/2 & \\ \vdots & 0 & & \ddots \end{pmatrix} \qquad (29)$$

where we limit the $\{|\ell\rangle\}$ states to $\{|\ell_c\rangle\}$, strongly-coupled to $|s\rangle$, and where the damping matrix γ is assumed to be diagonal.

We note that, in principle, there are several contributions to the widths γ_s and γ_c and, therefore, to the corresponding decay rates $k_s(=\gamma_s/\hbar)$ and $k_c(=\gamma_c/\hbar)$:

$$k_s = k_s^r + k_s^{coll}\,P + k_s^{nr} \tag{30}$$

$$k_c = k_c^r + k_c^{coll}\,P + k_c^{nr} \tag{31}$$

where k^r and k^{nr} refer to intramolecular radiative and nonradiative rates, respectively, and where $k^{coll}\,P$ is the collisional deactivation rate at pressure P.

Diagonalization of H_{eff} gives the complex eigenvalues $\bar{E}_n = E_n - i\gamma_n/2$; the corresponding eigenvectors can also be obtained and are of the form

$$|n\rangle = \alpha_{ns}|s\rangle + \sum_{\ell_c} \beta_{n\ell_c}|\ell_c\rangle \tag{32}$$

where $$\beta_{n\ell_c} = \left[v_{sc}/(\bar{E}_n - \bar{E}_c)\right]\alpha_{ns} \tag{33}$$

and $$\alpha_{ns}^2 = \left\{1 + \sum_{\ell_c}\left[|v_{sc}|^2/\bar{E}_c - \bar{E}_n)^2\right]\right\}^{-1} \quad . \tag{34}$$

The fluorescence decay properties can be obtained by exploring the probability $|\langle s|n(t)\rangle|^2$ of finding the system in state $|s\rangle$ at time t (assuming excitation at t = 0). If we recall our assumption that the dipole transition strengths of transitions from the $\{|\ell\rangle\}$ states are negligible, the time-resolved emission intensity and the total quantum yield are given by:

$$I_{em}(t) = k_s^r\,|\langle s|n(t)\rangle|^2 \tag{35}$$

$$\phi_F = \int_0^\infty I_{em}(t)\,dt \tag{36}$$

We consider the emission dynamics in the "strong-coupling" limit [92,120], in which the average interaction $\langle v_{sc}\rangle$ is supposed to significantly exceed $\varepsilon_c(=\rho_c^{-1})$, the average spacing between successive $|\ell_c\rangle$ levels. In the approximation of constant v_{sc}, equal spacing of successive energy levels of $|\ell_c\rangle$ and constant γ_c, it has been shown that $I_{em}(t)$ and ϕ_F can be described by simple analytical expressions in the strong-coupling limit [92,120].

When the emission can be resolved into two exponential components with decay constants τ_1 and τ_2:

$$I_{em}(t) = |A_1 \exp(-t/\tau_1) + A_2 \exp(-t/\tau_2)| \, k_s^r/(A_1 + A_2) \quad (37)$$

one obtains the following expressions:

$$A_2/A_1 = 1/2(N + 1) \quad (38)$$

$$\tau_1^{-1} = \Delta/h + k_s^r \quad (39)$$

$$\tau_2^{-1} = k_s^r/2(N + 1) + k_c^r(1 - 1/2(N + 1)) \quad (40)$$

$$\phi_F = k_s^r(A_1\tau_1 + A_2\tau_2)/(A_1 + A_2) \quad (41)$$

where $N/\pi = \Delta\rho_c/2 = \pi v_{sc}^2 \rho_c^2$, corresponds to the number of $\{|\ell_c\rangle\}$ states effectively coupled to $|s\rangle$. N/π can also be interpreted as the number of $|n\rangle$ states which contain a significant weight $|\alpha_{ns}|^2$ of $|s\rangle$.

Methods for deriving values of the coupling parameters, density of states and decay rates are described, and experimental results are reported, for the intermediate-case chloroacetylene ions by Dujardin, *et al.* [102] and Leach, *et al.* [93] in experiments carried out using the PIFCO technique. For example, for $Cl\text{-}C{\equiv}C\text{-}H^+$, the nonradiative coupling between the $\tilde{A}^2\Pi$ and $\tilde{X}^2\Pi$ states was found to have a coupling parameter $v_{sc} = 4 \times 10^{-5}$ cm^{-1}, whilst the $|\ell_c\rangle$ levels were found to have a nonradiative rate $k_c = 1.8 \times 10^6$ s^{-1}. This relatively high rate probably reflects intramolecular vibrational energy redistribution among those high vibrational levels of the $\tilde{X}^2\Pi$ state which are isoenergetic with $\tilde{A}^2\Pi$. Another contributing factor could be fast infrared decay of these high vibrational levels.

6.3d Further Remarks on Radiationless Transition Studies of Molecular Ions

In addition to the three cases discussed above, a further case corresponds to ionic dissociation -- that is, coupling between the $|s\rangle$ state and a dissociation continuum. In some cases the predissociation process can be in competition with ionic fluorescence (e.g., SO_2^+ [98,99]; cis-1,2-$C_2H_2F_2^+$ [105,107]).

Other interesting cases are those for which one might expect to see ionic fluorescence from a particular excited electronic state, when that state lies well below the lowest dissociation limit of the ion, yet experiment shows that the fluorescence quantum yields are vanishingly small ($\phi_F < 10^{-5}$). Some species of this type have been studied by the PIFCO technique ($C_6H_6^+$ [55,111]; NH_3^+ [99,100]; $C_2H_2^+$ [100]; $C_2N_2^+$ [100]). The reasons for the fast nonradiative relaxation are specific to each ion.

In reviewing radiationless transition theory, it is evident that the concomitant data available on specific rovibronic energy levels, mainly from high-resolution optical spectroscopy, is a great asset in making the resonance limit very clear cut. Although detailed information of this type may become available in the future for a few intermediate-case species, no such hope can be held out for the statistical-limit species. The approximations of average coupling strengths $\langle v_{s\ell} \rangle$ and uniform $\{|\ell\rangle\}$ level spacing, often used to obtain tractable analytical models in intermediate- and statistical-limit cases, introduce inevitable, but little desirable, coarse-graining effects. Such coarse-graining is over-simplificative in that analysis of the experimental emission decay behaviour is directed towards the extraction of one or a few exponential components from what naturally must be a vast conglomeration of individual exponential decay processes. However, it has been found that in the statistical limit, one, and in the intermediate case, two or three, clear-cut exponential decay components account for the observed emission characteristics. This is good evidence that the behaviour of the vast majority of individual levels is close to that of the average representative levels entering into the theoretical treatment.

Some other interesting problems arise out of the study of radiationless transitions in gas-phase molecular ions and are singled out for further work:

--- Deeper analysis is required of the molecular (ion) eigenstates prepared in the photoionization process and, in particular, of the extent of their coherent excitation. In the PIFCO and PEFCO work, the exciting photon energy (He I or Ne I) is much greater than that of the ion eigenstates studied. The photoelectron kinetic energy is then large and, therefore, photoelectron ejection will be a sufficiently rapid process for an important number of ion eigenstates to be formed in a coherent fashion. This was implicitly assumed in our previous discussion. In this respect, it is of interest, especially in the intermediate case, to compare the emission decay behaviour of ion excited states prepared directly by photoionization in PIFCO and PEFCO experiments, with that of states formed by photon absorption from (previously created) ground-state ions. (The energy resolution in the TPEFCO technique is insufficient for such a comparison.)

--- More work is required on the question of effective level densities. This is a trivial problem, and of limited utility, in the resonance-limit case where detailed spectroscopic analysis can provide this information. However, the coarse-grained model used in the intermediate- and statistical-limit cases introduces real difficulties in estimating the validity of derived values of ρ_ℓ. The Haarhoff approximation [121] underestimates the total $\{|\ell\rangle\}$ level density, in particular because of the neglect of anharmonicity (and because of some neglected rotational effects, particularly in asymmetric top species [122]); therefore, quantitative comparison with derived effective level densities remains somewhat ambiguous, although relative level densities may be compared with profit in a series of related compounds.

--- The factorization assumption $v_{s\ell} = v_{s\ell}^{el} \cdot v_{s\ell}^{vib}$ requires further examination in cases where the $\{|\ell\rangle\}$ states are high vibrational levels of an electronic (ground) state.

--- The multiexponential emission decays observed for the chloroacetylene ions require an intermediate-case treatment [93, 102]. Meaningful analysis could be made using the strong-coupling approximation; with the weak-coupling approximation no real solutions could be obtained. The strong-coupling approximation is nominally applicable in the intermediate case only if $v_{sc} > \rho_c^{-1}$ [120,122]. However, derived values for v_{sc} are only slightly greater than ρ_c^{-1} for the chloroacetylene ions studied [93,102]. This probably reflects the imperfections of a coarse-grained approach and merits further attention.

--- In the chloroacetylene case, nonradiative intramolecular vibrational redistribution is invoked to account for a considerable part of the width of the $\{|\ell_c\rangle\}$ states strongly coupled to $|s\rangle$ [93,102]. This working hypothesis also requires more study, particularly since it presupposes (i), an adequate density of $\{|\ell_d\rangle\}$ states to which the $\{|\ell_c\rangle\}$ states (or, alternatively, the molecular (ion) eigenstates $\{|n\rangle\}$) are coupled in this process; and (ii), a sufficient variance in the distribution of $v_{s\ell}^2$ [123].

--- Our results have led us to consider the possibility of $\{|\ell\rangle\}$ relaxation via fast infrared radiative decay. A theoretical discussion of this problem [102] enables one to rationalize quite high infrared radiative rates for high vibrational levels of the electronic ground state, especially in molecular ions. Direct experimental evidence is highly desirable in this respect. The existence of very rapid infrared emission rates for vibrationally-hot molecular ions could indeed have important effects on cold plasma systems and planetary ionospheres; it could, for example, significantly modify temperature/altitude modelling of the

Martian atmosphere [124].

Finally, we emphasize that the continuing study of radiationless transitions in molecular ions will provide important information necessary in understanding the intramolecular dynamics occurring in fragmentation processes.

REFERENCES

1. S. Leach, J. Chim. Phys. 77, 585 (1980).
2. T. N. Rescigno, V. McKoy and B. Schneider, eds., *Electron-Molecule and Photon-Molecule Collisions* (Plenum, New York, 1979).
3. R. S. Berry, Adv. Electron. Electron Phys. 51, 137 (1980).
4. R. S. Berry and S. Leach, Adv. Electron. Electron Phys. 57, 1 (1981).
5. G. Wentzel, Z. Phys. 40, 474 (1926); 41, 828 (1927).
6. R. S. Berry, Proc. Conf. Int. Phenom. Ionis. Gaz, 6th (1963), Vol. 1. p. 13.
7. H. W. Meldner and J. D. Perez, Phys. Rev. A4, 1388 (1971).
8. C. E. Brion and A. Hamnett, Adv. Chem. Phys. 45, 2 (1981).
9. E. P. Wigner, Phys. Rev. 73, 1002 (1948).
10. S. Geltman, Phys. Rev. 102, 171 (1956).
11. J. F. Lowry, D. H. Tomboulian and D. L. Ederer, Phys. Rev. 137, A 1054 (1965).
12. D. L. Ederer and D. H. Tomboulian, Phys. Rev. 133, A 1525 (1964).
13. J. W. Cooper, Phys. Rev. 128, 681 (1962).
14. J. A. R. Samson, Adv. Atom. Mol. Phys. 2, 192 (1966).
15. U. Fano and J. W. Cooper, Phys. Rev. 137, A 1364 (1965).
16. U. Fano, Phys. Rev. 124, 1866 (1961).
17. P. M. Dehmer, J. Berkowitz and W. A. Chupka, J. Chem. Phys. 60, 2676 (1974).
18. F. Mies, Phys. Rev. 175, 164 (1968).
19. A. L. Smith, Phil. Trans. R. Soc. London Ser. A 268, 169 (1970).
20. Ch. Jungen, J. Chim. Phys. 77, 27 (1980).
21. Ch. Jungen and D. Dill, J. Chem. Phys. 73, 3338 (1980).
22. M. Raoult, Ch. Jungen and D. Dill, J. Chem. Phys. 77, 599 (1980).
23. M. Raoult and Ch. Jungen, J. Chem. Phys. 79, 3388 (1981).
24. Ch. Jungen and M. Raoult, Faraday Disc. Chem. Soc. 71, 253 (1981).
25. M. J. Seaton, Proc. Phys. Soc. Lond. 88, 801 (1966).
26. U. Fano, J. Opt. Soc. Am. 65, 979 (1975).
27. P. M. Dehmer and W. A. Chupka, J. Chem. Phys. 65, 2243 (1976).

28. P. M. Dehmer, J. Berkowitz and W. A. Chupka, J. Chem. Phys. 59, 5777 (1973).
29. J. A. R. Samson and V. E. Petrosky, J. Electron Spectrosc. Relat. Phenom. 3, 461 (1974).
30. J. Y. Roncin, H. Damany and Ch. Jungen, Vac. Ultraviolet Radiat. Phys., Proc. Int. Conf. 4th (1974), p. 52.
31. P. M. Cuyon, J. Breton and M. Glass-Maujean, Chem. Phys. Lett. 68, 314 (1979).
32. K. E. McCulloh and J. A. Walker, Chem. Phys. Lett. 25, 439 (1974); W. A. Chupka, P. M. Dehmer and W. T. Jivery, J. Chem. Phys. 63, 3929 (1975).
33. G. R. Cook and P. H. Metzger, J. Chem. Phys. 41, 321 (1964).
34. J. A. R. Samson, *Techniques of Vacuum Ultraviolet Spectroscopy* (Wiley, New York, 1967).
35. A. N. Zaidel' and E. Ya. Shreider, *Vacuum Ultraviolet Spectroscopy* (Ann Arbor-Humphrey Sci. Publ., Ann Arbor, Michigan, 1970).
36. J. E. Spencer and H. Winick, in: *Synchrotron Radiation Research*, H. Winick and S. Doniach, eds. (Plenum, New York, 1980), Chap. 21, p. 663 *et seq.*
37. P. K. Carroll, E. T. Kennedy and G. O'Sullivan, Appl. Opt. 19, 1454 (1980).
38. J. Jortner and S. Leach, J. Chim. Phys. 77, 7 (1980).
39. C. Pellegrini, in: *Synchrotron Radiation Research*, H. Winick and S. Doniach, eds. (Plenum, New York, 1980), Chap. 22, p. 717 *et seq.*
40. N. Damany, B. Vodar and J. Romand, eds., *Some Aspects of Vacuum Ultraviolet Radiation Physics* (Pergamon, Oxford, 1974).
41. J. Berkowitz, *Photoabsorption, Photoionization and Photoelectron Spectroscopy* (Academic Press, New York, 1979).
42. A. E. Douglas and J. G. Potter, Appl. Opt. 1, 727 (1962).
43. F. Launay, Le Courrier du C. N. R. S. 12, 10 (1974).
44. J. A. R. Samson, in: *Methods of Experimental Physics*, 13A, D. Williams, ed. (1976), p. 204.
45. C. Cossart-Magos, S. Leach, M. Eidelsberg, F. Launay and F. Rostas, J. Chem. Soc. Faraday Trans. (2) 78, 1477 (1982).
46. P. Gürtler, V. Saile and E. E. Koch, Chem. Phys. Lett. 51, 386 (1977).
47. J. P. Connerade, M. A. Baig, S. P. McGlynn and W. R. S. Garton, J. Phys. B: At. Molec. Phys. 13, 705 (1980).
48. M. A. Baig, J. Hormes, J. P. Connerade and S. P. McGlynn, J. Phys. B: At. Molec. Phys. 14, L725 (1981).
49. M. Herman and R. Colin, Physica Scripta 25, 275 (1982).
50. J. P. Connerade, M. A. Baig and S. P. McGlynn, J. Phys. B: At. Mol. Phys. 14, L67 (1981).
51. M. A. Baig, J. P. Connerade, J. Dagata and S. P. McGlynn, J. Phys. B: At. Mol. Phys. 14, L25 (1981).

52. K. E. Ockenga, P. Gürtler, S. S. Hasnain, V. Saile and E. E. Koch, Proc. 6th Int. Conf. VUV Rad. Phys., Vol. 2 (Univ. Virginia, Charlottesville, 1981), paper 80.
53. See, for example, W. Lohr, H. W. Jochims and H. Baumgärtel, Ber. Bunsenges. Phys. Chem. 79, 901 (1975).
54. C. Y. Ng, Adv. Chem. Phys. 52, 263 (1983).
55. J. H. D. Eland, M. Devoret and S. Leach, Chem. Phys. Lett. 43, 97 (1976).
56. G. Dujardin, S. Leach and G. Taieb, Chem. Phys. 46, 407 (1980).
57. M. Bloch and D. W. Turner, Chem. Phys. Lett. 30, 344 (1975).
58. J. P. Maier and F. Thommen, Chem. Phys. 51, 319 (1980).
59. E. W. Schlag, R. Frey, B. Gotchev, W. B. Peatman and H. Pollack, Chem. Phys. Lett. 51, 406 (1977).
60. G. Dujardin, S. Leach, O. Dutuit, T. Govers and P. M. Guyon, J. Chem. Phys. 79, 644 (1983).
61. T. Baer, in: *Gas Phase Ion Chemistry*, Vol. 1, M. T. Bowers, ed. (Academic Press, New York, 1979), p. 153.
62. T. Baer, P. M. Guyon, I. Nenner, T. R. Govers, A. Tabché-Fouhaillé, R. Botter and L. F. A. Ferreira, J. Chem. Phys. 70, 1585 (1979).
63. R. W. Carlson, D. L. Judge and M. Ogawa, J. Geophys. Res. 78, 3194 (1973).
64. R. W. Carlson, D. L. Judge and M. Ogawa, J. Phys. B: Atom. Molec. Phys. 6, L343 (1973).
65. L. C. Lee, R. W. Carlson and D. L. Judge, J. Phys. B: Atom. Molec. Phys. 9, 855 (1976).
66. J. H. D. Eland, J. Berkowitz and J. E. Monahan, J. Chem. Phys. 72, 253 (1980).
67. J. H. D. Eland, J. Chim. Phys. 77, 613 (1980).
68. V. H. Dibeler, J. A. Walker and S. K. Liston, J. Res. N. B. S. A71, 371 (1967).
69. J. Berkowitz and J. H. D. Eland, J. Chem. Phys. 67, 2740 (1977).
70. I. Nenner, P. M. Guyon, T. Baer and T. R. Govers, J. Chem. Phys. 72, 6587 (1980).
71. V. H. Dibeler and J. A. Walker, J. Opt. Soc. Am. 57, 1007 (1967).
72. K. E. McCulloh, J. Chem. Phys. 59, 4250 (1973).
73. J. H. D. Eland and J. Berkowitz, J. Chem. Phys. 67, 2782 (1977).
74. V. H. Dibeler and J. A. Walker, Adv. Mass Spectrom. 4, 767 (1968).
75. J. H. D. Eland and J. Berkowitz, J. Chem. Phys. 70, 5151 (1979).
76. P. Coppens, J. C. Reynaert and J. Drowart, J. Chem. Soc. Faraday Trans. (2) 75, 292 (1979).
77. V. H. Dibeler and S. K. Liston, J. Chem. Phys. 49, 482 (1968).

78. M. J. Weiss, T. C. Hsieh and G. G. Meisels, J. Chem. Phys. 71, 567 (1979).
79. J. H. D. Eland, in: *Electron Spectroscopy: Theory, Techniques and Applications*, Vol. 3, C. R. Brundle and A. D. Baker, eds. (Academic Press, New York, 1979), p. 231.
80. P. M. Guyon, W. A. Chupka and J. Berkowitz, J. Chem. Phys. 64, 1419 (1976).
81. M. Vaz Pires, C. Galloy and J. C. Lorquet, J. Chem. Phys. 69, 3242 (1978).
82. A. C. Parr and F. A. Elder, J. Chem. Phys. 49, 2659 (1968).
83. A. C. Parr, A. J. Jason and R. Stockbauer, Int. J. Mass Spectrom. Ion Phys. 26, 23 (1978).
84. P. T. Murray and T. Baer, Int. J. Mass Spectrom. Ion Phys. 30, 165 (1979).
85. Y. Tanaka, A. S. Jursa and F. J. LeBlanc, J. Chem. Phys. 32, 1199 (1960).
86. J. H. D. Eland, Int. J. Mass Spectrom. Ion Phys. 9, 397 (1972).
87. S. Leach, J. Chim. Phys., Special Issue on Radiationless Transitions in Molecules, 74 (1970).
88. T. Masuoka and J. A. R. Samson, J. Chim. Phys. 77, 623 (1980).
89. A. P. Hitchcock, C. E. Brion and M. J. Van der Wiel, Chem. Phys. 45, 461 (1980).
90. C. E. Brion and K. H. Tan, Chem. Phys. 34, 141 (1978).
91. C. E. Brion and K. H. Tan, J. Electron Spectrosc. Relat. Phenom. 15, 241 (1979).
92. (a) K. F. Freed, Top. Appl. Phys. 15, 23 (1976); (b) P. Avouris, W. M. Gelbart and M. A. El Sayed, Chem. Revs. 77, 793 (1977); (c) S. H. Lin, ed., *Radiationless Transitions* (Academic Press, New York, 1980).
93. S. Leach, G. Dujardin and G. Taieb, J. Chim. Phys. 77, 705 (1980).
94. See, for example, H. Okabe, *Photochemistry of Small Molecules* (Wiley Interscience, New York, 1978).
95. R. Frey, B. Gotchev, W. B. Peatman, H. Pollak and E. W. Schlag, Chem. Phys. Lett. 54, 411 (1978).
96. S. Leach, M. Devoret and J. H. D. Eland, Chem. Phys. 33, 113 (1978).
97. S. Leach, P. Stannard and W. M. Gelbart, Molec. Phys. 36, 1119 (1978).
98. G. Dujardin and S. Leach, J. Chem. Phys. 75, 2521 (1981).
99. G. Dujardin and S. Leach, unpublished results.
100. M. Devoret, Thèse 3ème Cycle, Orsay (1976).
101. V. Winkoun, G. Dujardin and S. Leach, unpublished results.
102. G. Dujardin, S. Leach, G. Taieb, J. P. Maier and W. M. Gelbart, J. Chem. Phys. 73, 4987 (1980).
103. G. Dujardin, S. Leach and G. Taieb, unpublished results.

104. J. P. Maier and F. Thommen, J. Chem. Phys. 73, 5616 (1980).
105. J. P. Maier, Acc. Chem. Res. 15, 18 (1982).
106. P. Forster, J. P. Maier and F. Thommen, Chem. Phys. 59, 85 (1981).
107. J. P. Maier and F. Thommen, J. Chem. Soc., Faraday Trans. (2) 77, 845 (1981).
108. J. P. Maier, O. Marthaler, L. Misev and F. Thommen. Roy. Soc. Chem., Faraday Disc. 71, 181 (1981).
109. J. P. Maier, L. Misev and F. Thommen, J. Phys. Chem. 86, 514 (1982).
110. J. P. Maier, L. Misev and F. Thommen, Helv. Chim. Acta 64, 1985 (1981).
111. O. Braitbart, E. Castellucci, G. Dujardin and S. Leach, J. Phys. Chem. 87, 000 (1983).
112. (a) J. P. Maier and F. Thommen, Chem. Phys. 57, 319 (1981); (b) J. P. Maier and F. Thommen, J. Chem. Phys. 77, 4427 (1982).
113. G. Dujardin and S. Leach, J. Chem. Phys. 79, 658 (1983).
114. G. Herzberg, *Electronic Spectra of Polyatomic Molecules* (Van Nostrand, New York, 1966).
115. S. Leach, in: *Molecular Energy Transfer*, R. D. Levine and J. Jortner, eds. (Halsted Press, New York, 1975), p. 163.
116. J. Dufayard, J. M. Negre and O. Nedelec, J. Chem. Phys. 61, 3614 (1974).
117. D. Cossart and T. Bergeman, J. Chem. Phys. 65, 5462 (1977); D. Cossart, J. Physique 41, 489 (1980); T. Bergeman and D. Cossart, J. Mol. Spectrosc. 87, 119 (1981).
118. M. Johnson, J. Rostas, S. Leach and R. N. Zare, to be published.
119. A. Nitzan and J. Jortner, J. Chem. Phys. 55, 1355 (1971).
120. F. Lahmani, A. Tramer and C. Tric, J. Chem. Phys. 60, 4431 (1974).
121. P. C. Haarhoff, Mol. Phys. 7, 101 (1963).
122. A. Tramer and R. Voltz, in: *Excited States*, Vol. 4, E. C. Lim, ed. (Academic Press, New York, 1979), p. 281.
123. C. Tric, Chem. Phys. 14, 189 (1976).
124. J. C. McConnell, in: *Physics and Chemistry of Upper Atmospheres*, B. M. McCormac, ed. (Reidel, Dordrecht, 1973), p. 309.

PHOTOIONIZATION DYNAMICS OF SMALL MOLECULES

J. L. Dehmer

Argonne National Laboratory
Argonne, IL 60439, USA

Dan Dill

Department of Chemistry
Boston University
Boston, MA 02215, USA

A. C. Parr

Synchrotron Ultraviolet Radiation Facility
National Bureau of Standards
Washington, D.C. 20234, USA

1. INTRODUCTION

The last decade has witnessed remarkable progress in characterizing dynamical aspects of the molecular photoionization process. The general challenge is to gain physical insight into those processes occuring during photoexcitation and eventual escape of the photoelectron through the anisotropic molecular field, in terms of various observables such as photoionization cross-sections and branching ratios, photoelectron angular distributions and even newer probes mentioned below. Much of the progress in this field has mirrored earlier work in atomic photoionization dynamics [1] where many key ideas were developed (e.g., channel interaction, quantum defect analysis, potential barrier phenomena and experimental techniques). However, additional concepts and techniques were required to deal with the strictly molecular aspects of the problem, particularly the anisotropy of the multicenter molecular field and the interaction among rovibronic modes.

S. P. McGlynn et al. (eds.), Photophysics and Photochemistry in the Vacuum Ultraviolet, 341–408.

From among the great variety of successful approaches to the study of molecular photoionization dynamics, one can identify four broad classes which together form the core of recent activities in this field: *First*, the extensive measurements of total photoionization cross-sections from the VUV to the x-ray range by a variety of means(see, e.g., [2,3] and original literature cited below) have continuously provided fresh impetus to account for a variety of phenomena displayed in molecular oscillator strength distributions. *Second*, in response to observations [4-17] of broad, intense features in photoionization spectra of small molecules, the interpretation and study of shape resonances in molecular photoionization began [18-32] and grew into a sizable subfield [33-134]. Benefitting greatly from the timely development of realistic, independent-electron models [67,94,123,135,136] for molecular photoionization, studies in this area not only accounted for the features in the total photoionization spectra, but also established and predicted several manifestations in other physical observables as discussed below. *Third*, multichannel quantum defect theory (MQDT) was adapted [137-150] to molecular photoionization, thus providing a framework for the quantitative and microscopic analysis of autoionization phenomena. This powerful theoretical framework has been successfully applied to a number of prototypical diatomic molecules, yielding both insight into detailed dynamics of resonant photoionization and some specific predictions for experimental testing by means discussed in the next item. *Fourth*, technical advances, especially the development of intense synchrotron radiation sources [151-156], have made it feasible to perform triply-differential photoionization measurements (see, e.g., [75, 157-163]) on gas-phase atoms and molecules. By "triply-differential" we mean that photoelectron measurements are made as a function of three independent variables - - incident photon wavelength, photoelectron energy and photoelectron ejection angle. Variable wavelength permits the study of photoionization at and within spectral features of interest; photoelectron energy analysis permits separation and selection of individual (ro)vibronic ionization channels; and measurement of photoelectron angular distributions accesses dynamical information (i.e., relative phases of alternative, degenerate ionization channels) not present in integrated cross-sections. This level of experimental detail approaches that at which theoretical calculations are made and, hence, permits us to isolate and study dynamical details which are otherwise swamped in integrated or averaged quantities. We emphasize that, although the current trend is toward the use of synchrotron radiation for variable-wavelength studies, a variety of light sources have been successfully used to study photoionization dynamics. For example, in the shape-resonance literature cited above, many of the pioneering measurements were carried out with laboratory-scale sources. Likewise, although most current measurements of vibrational branching ratios [132-134,164-170] and angular distributions [166-170] within autoionizing resonances employ synchrotron radiation, many early and ongoing studies with traditional light sources have made significant observations of the effects of autoionization on vibrational branching ratios [171-183] and angular distributions [175,184-193].

In this paper we review the recent progress outlined above, with emphasis on photoionization dynamics exhibited by shape and autoionizing resonance phenomena. Prototypes, such as H_2 and N_2, will be used in order to best illustrate the fundamental aspects and the variety of manifestations of these processes. Finally, in Section 4, we present an overview of several other approaches to the study of molecular photoionization dynamics in order to show how they complement those topics emphasized here and to convey, to some extent, the richness of the field.

2. SHAPE RESONANCES IN MOLECULAR PHOTOIONIZATION

2.1 Overview

Shape resonances are quasibound states in which a particle is temporarily trapped by a potential barrier, through which it may eventually tunnel and escape. In molecular fields, such states can result from so-called "centrifugal barriers", which block the motion of otherwise free electrons in certain directions, trapping them in a region of space with molecular dimensions. Over the past few years, this basic resonance mechanism has been found to play a prominent role in a variety of processes in molecular physics, most notably in photoionization and electron scattering. As discussed more fully in later sections, the expanding interest in shape-resonant phenomena arises from a few key factors:

First, shape resonance effects are being identified in the spectra of a growing and diverse collection of molecules and now appear to be active somewhere in the observable properties of most small (nonhydride) molecules. Examples of those processes which exhibit shape resonance effects are x-ray and VUV absorption spectra, photoelectron branching ratios and photoelectron angular distributions (including those vibrationally resolved), Auger-electron angular distributions [82], elastic electron scattering [194-196], vibrational excitation by electron impact [81,195-196], and so on. Thus, concepts and techniques developed in this connection can be used extensively in molecular physics.

Second, being quasibound inside a potential barrier on the perimeter of the molecule, shape resonances are localized, have enhanced electron density in the molecular core, and are uncoupled from the external environment of the molecule. This localization often produces intense, easily studied spectral features, while suppressing the nearby continuum and/or Rydberg structure and, as discussed more fully below, has a marked influence on vibrational motion. In addition, localization causes much of the conceptual framework developed for shape resonances in free molecules to apply equally well [68] to photoionization and electron scattering, and to other states of matter such as adsorbed molecules [29,31,197-198], molecular crystals [11,20] and ionic solids [199].

Third, resonant trapping by a centrifugal barrier often imparts a well-defined orbital momentum character to the escaping electron. This can be directly observed (e.g., by angular distributions of scattered electrons [194-196] or photoelectron angular distributions from oriented molecules [29,31,197,198]) and shows that the centrifugal-trapping mechanism has physical meaning and is not merely a theoretical construct. Recent case studies have revealed trapping of $\ell = 1$ to $\ell = 5$ components of continuum molecular wave functions. The purely molecular origin of the great majority of these cases is illustrated by the N_2 prototype discussed in Section 2.2.

Fourth, the predominantly one-electron nature of the phenomenon lends itself to theoretical treatment by realistic, independent-electron methods [67,94,123,135-136], with the concomitant flexibility in terms of complexity of molecular systems, energy ranges and alternative physical processes. This has been a major factor in the rapid exploration of this area. Continuing development of computational schemes holds the promise of elevating the level of theoretical work on molecular ionization and scattering in order to test and quantify the independent-particle results and to treat other circumstances such as weak channels, multiply-excited states, etc., where the simpler schemes become invalid.

The earliest, and still possibly the most dramatic, examples of shape resonance effects in molecules are the photoabsorption spectra of the sulfur K- [4,14a] and L-shells [5,8,11,14a] in SF_6. The sulfur L-shell absorption spectra of SF_6 and H_2S, shown in Fig. 1, illustrate the types of phenomena that originally drew attention to this area. In Fig. 1, both spectra are plotted on a photon energy scale referenced to the sulfur L-shell ionization potential (IP), which is chemically shifted by a few eV in the two molecular environments but which lies near $h\nu \simeq 175$ eV. The ordinates represent relative photoabsorption cross-sections and have been adjusted so that the integrated oscillator strength for the two systems is roughly equal in this

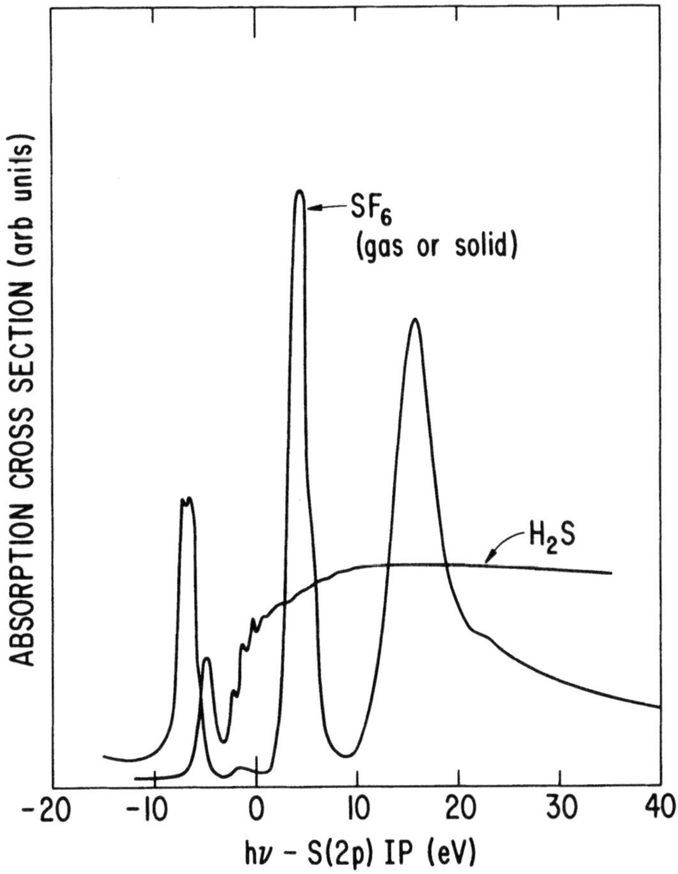

Fig. 1. Photoabsorption spectra of H_2S (adapted from [8]) and SF_6 (adapted from [11]) near the sulfur $L_{2,3}$ edge.

spectral range (absolute calibrations are not known). The H_2S spectrum is used here as a "normal" reference spectrum since hydrogen atoms normally do not contribute appreciably to shape resonance effects and, in this particular context, can be regarded as weak perturbations on the inner-shell spectra of the heavy atom. Indeed, the photoabsorption spectrum exhibits a valence transition, followed by partially resolved Rydberg structure which converges to a smooth continuum. The gradual rise at threshold is attributable to the delayed onset of the "$2p \rightarrow \varepsilon d$" continuum which, for second-row atoms, will exhibit a delayed onset prior to the occupation of the 3d subshell. This is the qualitative behavior one might well expect for the absorption spectrum of a core level.

In sharp contrast to this, the photoabsorption spectrum of the same sulfur 2p subshell in SF_6 shows no vestige of the "normal" behavior just described. Instead, three intense, broad peaks appear, one below the ionization threshold and two above, and the continuum absorption cross-section is greatly reduced elsewhere. Moreover, no Rydberg structure is apparent although an infinite number of Rydberg states must necessarily be associated with any molecular ion. (Actually, Rydberg states were detected [33], using photographic methods, superimposed on the weak bump below the IP; obviously, however, these states are extremely weak in this spectrum.) This radical reorganization of the oscillator strength distribution has been interpreted [20] in terms of potential barrier effects in SF_6, resulting in three shape-resonantly enhanced final-state features of a_{1g}, t_{2g} and e_g symmetry (in order of increasing energy). Another shape-resonant feature (t_{1u} symmetry) is prominent in the sulfur K-shell spectrum [4] and, in fact, is responsible, via channel interactions, for the weak feature just below the IP in Fig. 1. Hence, four prominent features occur in the photoexcitation spectrum of SF_6 as a consequence of potential barriers caused by the molecular environment of the sulfur atom. Another significant observation [11] is that the SF_6 curve in Fig. 1 represents both gaseous and solid SF_6, within experimental error bars. This is definitive evidence that the resonances are eigenfunctions of the potential well inside the barrier and are effectively uncoupled from the molecule's external environment.

In what follows, we describe the underlying physics of shape resonances in molecular fields and go on to discuss several other manifestations of this phenomenon, manifestations which have been brought to light since the initial interpretation [20] of the features in Fig. 1.

2.2 Basic Properties

The central concept in shape resonance phenomena is the single-channel, barrier-penetration model familiar from introductory quantum mechanics. In fact, the name "shape resonance" means simply that the resonance behavior arises from the "shape" (i.e., the barrier and associated inner and outer wells) of a local potential. The basic shape resonance mechanism is illustrated schematically [200] in Fig. 2. There, an effective potential for an excited and/or unbound electron is shown to have an inner well at small distances, a potential barrier at intermediate distances and an outer well (asymptotic form not shown) at large separations. In the context of molecular photoionization, this would be a one-dimensional abstraction of the effective potential for the photoelectron in the field of a molecular ion. Accordingly, the inner well would be formed by the partially screened nuclei in the molecular core and would therefore be

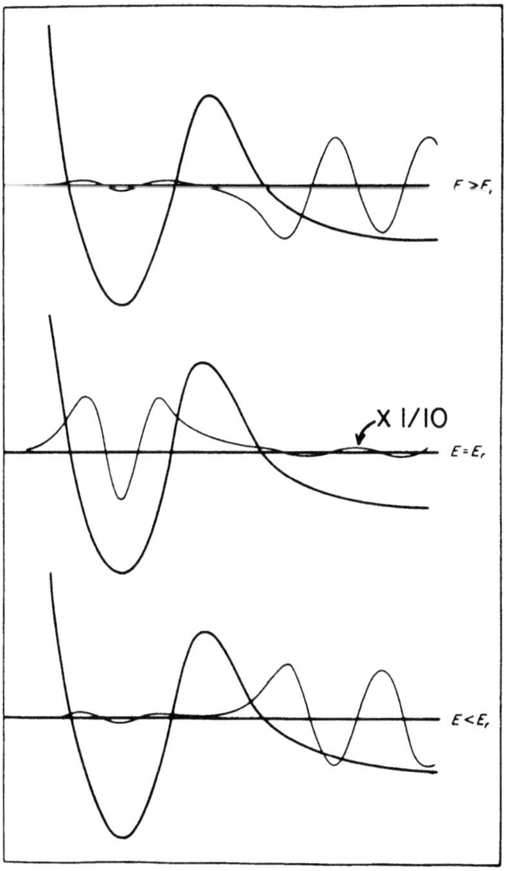

Fig. 2. Schematic of the effect of a potential barrier on an unbound wavefunction in the vicinity of a quasibound state at E = E_r (adapted from [200]). In the present context, the horizontal axis represents the distance of the excited electron from the center of the molecule.

highly anisotropic and would overlap much of the molecular charge distribution (i.e., the initial states of the photoionization process). The barrier, in all well-documented cases, is a so-called centrifugal barrier (other forces such as repulsive exchange forces, high concentrations of negative charge, etc. may also contribute, but have not yet been documented to be pivotal in the molecular systems studied to date). This centrifugal barrier derives from a competition between repulsive centrifugal forces and attractive electrostatic forces and usually resides on the perimeter of the molecular charge distribution where the centrifugal forces can compete effectively with electrostatic forces. Similar barriers are known for d and f waves in atomic fields [1]; however, the ℓ (orbital angular momentum) character of resonances in molecular fields tends to be higher than those of constituent atoms owing to the larger spatial extent of the molecular charge distribution (e.g., see the discussion given below in connection with N_2 photoionization). The outer well lies outside the molecule where the Coulomb potential ($\sim -r^{-1}$) of the molecular ion again dominates the centrifugal terms ($\sim r^{-2}$) in the potential. We stress that this description has been radically simplified to convey the essential aspects of the underlying physics. In reality, effective barriers to electron motion in molecular fields occur for particular ℓ components of particular ionization channels and restrict motion only in certain directions. Again, a specific example is described below.

Focusing now on the wavefunctions in Fig. 2, we see the effect of the potential barrier on the wave mechanics of the photoelectron. For energies below the resonance energy $E < E_r$ (lower part of Fig. 2), the inner well does not support a quasibound state (i.e., the wave function is not exponentially decaying as it enters the classically forbidden region of the barrier). Thus, the wavefunction begins to diverge in the barrier region and emerges in the outer well with a much larger amplitude than that in the inner well. When properly normalized at large r, the amplitude in the molecular core is very small; hence, we say this wavefunction is essentially an eigenfunction of the outer well although small precursor loops extend inside the barrier into the molecular core.

At $E = E_r$, the inner well supports a quasibound state. The wavefunction exhibits exponential decay in the barrier region so that if the barrier extended to $r \to \infty$, a true bound state would lie very near this total energy. Therefore, the antinode that was not supported in the inner well at $E < E_r$ has traversed the barrier to become part of a quasibound waveform which decays monotonically until it reemerges in the outer-well region, much diminished in amplitude. This "barrier penetration" by an antinode produces a rapid increase in the asymptotic phase shift by $\sim\pi$ radians and greatly enhances the amplitude in the inner well over a narrow band of energy near E_r. Therefore, at $E = E_r$ the wavefunction is essentially an eigenfunction of the inner well although it decays through the barrier and reemerges in the outer well. The energy halfwidth of the resonance is related to the lifetime of the quasibound state and to the energy derivative of the rise in the phase shift in well-known ways. Finally, for $E > E_r$ the wavefunction reverts to being an eigenfunction of the outer well since the behavior of the wavefunction at the outer edge of the inner well is no longer characteristic of a bound state.

Obviously, this resonant behavior will cause significant physical effects: The enhancement of the inner-well amplitude at $E \simeq E_r$ results in good overlap with the initial states, which reside mainly in the inner well. Conversely, for energies below the top of the barrier but not within the resonance halfwidth of E_r, the inner amplitude is diminished relative to a more typical barrier-free case. This accounts for the strong modulation of the oscillator strength distribution in Fig. 1. Also, the rapid rise in the phase shift induces shape resonance effects in the photoelectron angular distribution. Another important aspect is that eigenfunctions of the inner well are localized inside the barrier and are substantially uncoupled from the external environment of the molecule. As mentioned above, this means that shape-resonant phenomena often persist in going from the gas phase to the condensed phase (cf., Fig..1) and, with suitable modification, shape resonances in molecular photoionization can be mapped [68] into electron-scattering processes and vice versa. Finally, note that this discussion was focused on total energies, measured from the bottom of the outer well to the top of the barrier, and that no explicit mention was made of the asymptotic potential that determines the threshold for ionization. Thus, valence or Rydberg states in this range can also exhibit shape-resonant enhancement, even though they have true bound state behavior at large r, beyond the outer well.

We now turn, for the remainder of this section, to the specific example of the well-known σ_u shape resonance in N_2 photoionization, which was the first documented case [28] in a diatomic molecule and has since been used as a prototype in studies of various shape resonance effects. To identify the major final-state features in N_2 photoionization at the independent-electron level, we show (Fig. 3) the original calculation [28-30] of the K-shell photoionization spectrum, performed with the multiple-scattering model. This calculation agrees qualitatively, with all major features in the experimental spectrum [13,44,87], except for a narrow band of double excitation features, and with subsequent, somewhat more accurate, calculations using L^2 techniques [47]. The four partial cross-sections in Fig. 3 represent the four dipole-allowed channels for K-shell (IP = 409.9 eV) photoionization. Here, we have neglected the localization [54] of the K-shell hole, since it does not greatly affect the integrated cross-sections, and the separation into u and g symmetry both helps the present discussion and is rigorously applicable to the subsequent discussion of valence-shell excitation. (Note that the identification of shape-resonant behavior is generally easier in inner-shell spectra since the problems of overlapping spectra, channel interaction and zeros in the dipole matrix element are all reduced relative to valence-shell spectra.)

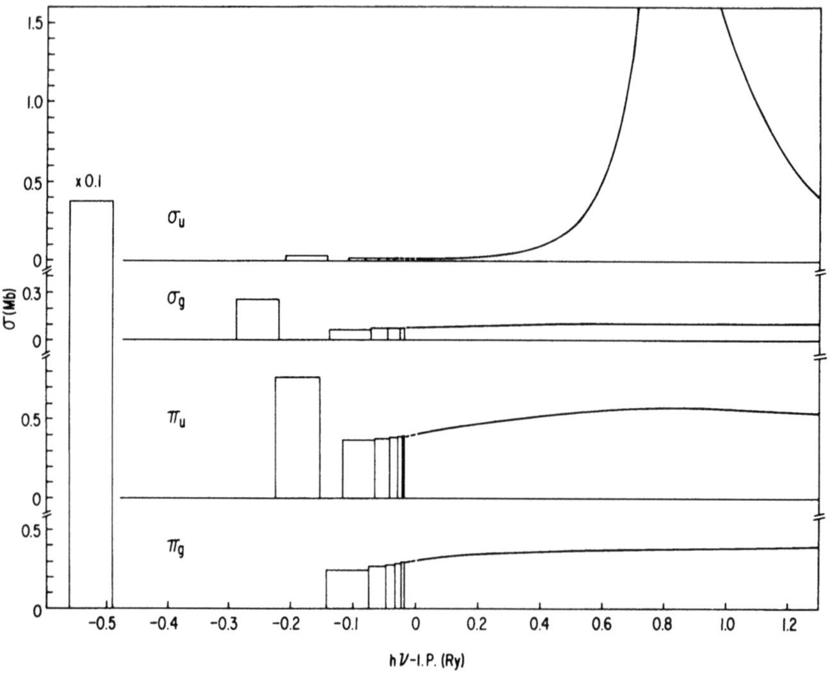

Fig. 3. Partial photoionization cross-sections for the four dipole-allowed channels in K-shell photoionization of N_2. Note that the energy scale is referenced to the K-shell IP (409.9 eV) and is expanded two-fold in the discrete part of the spectrum.

The most striking spectral feature in Fig. 3 is the first member of the π_g sequence, which dominates every other feature in the theoretical spectrum by a factor of ∿30. (Note that the first π_g peak has been reduced by a factor of 10 in order to fit in the frame.) The concentration of oscillator strength in this peak is a centrifugal-barrier effect in the d-wave component of the π_g wavefunction. The final state in this transition is a highly localized state, about the size of the molecular core, and is the counterpart of the well-known [195, 196] π_g shape resonance in e-N_2 scattering at 2.4 eV. For the latter case, Krauss and Mies [201] demonstrated that the effective potential for the π_g elastic channel in e-N_2 scattering exhibits a potential barrier resulting from the centrifugal repulsion acting on the dominant $\ell = 2$ lead term in the partial-wave expansion of the π_g wavefunction. In the case of N_2 photoionization, there is one less electron in the molecular field to screen the nuclear charge so that this resonance feature is shifted [68] to lower energy and appears in the discrete region of the spectrum. It is in this sense that we refer to such features as "discrete" shape resonances. The remainder of the π_g partial cross-section consists of a Rydberg series and a flat continuum. The π_u and σ_g channels both exhibit Rydberg series, the initial members of which correlate well with partially resolved transitions in the experimental spectrum below the K-shell IP.

The σ_u partial cross-section, on the other hand, was found to exhibit behavior rather unexpected for the K-shell of a first-row diatomic. Its Rydberg series was extremely weak and an intense, broad peak appeared at $\sim$1 Ry above the IP in the low-energy continuum. This effect is caused by a centrifugal barrier acting on the $\ell = 3$ component of the σ_u wavefunction. The essence of the phenomenon can be described in mechanistic terms as follows: The electric-dipole interaction, localized within the atomic K shell, produces a photoelectron with angular momentum $\ell = 1$. As this p wave escapes to infinity, the anisotropic molecular field can scatter it into the entire range of angular momentum states contributing to the allowed σ- and π-ionization channels ($\Delta\lambda = 0, \pm 1$). In addition, the spatial extent of the molecular field, consisting of two atoms separated by 1.1 Å, enables the $\ell = 3$ component of the σ_u continuum wavefunction to overcome its centrifugal barrier and to penetrate into the molecular core at a kinetic energy of $\sim$1 Ry. This penetration is rapid, with a phase shift of $\sim\pi$ occurring over a range of $\sim$0.3 Ry. These two circumstances combine to produce a dramatic enhancement of photoelectron current at $\sim$1 Ry kinetic energy, with predominantly f-wave character.

The specifically molecular character of this phenomenon is emphasized by comparison with K-shell photoionization in atomic nitrogen and in the united-atom case, silicon. In contrast to N_2, there is no mechanism for the essential p-f coupling and neither atomic field is strong enough to support resonant penetration of high-ℓ partial waves through their centrifugal barriers. (With substitution of "d" for "f", this argument applies equally well to the d-type resonance in the discrete part of the spectrum.) Note that the π_u channel also has an $\ell = 3$ component but does not resonate. This underscores the directionality and symmetry dependence of the trapping mechanism.

To place the σ_u resonance in a broader perspective and to show its connection with high-energy behavior, we present, in Fig. 4, an extension of the calculation of Fig. 3 to much higher energy. Again, the four dipole-allowed channels in $D_{\infty h}$ symmetry are shown. The dashed line is two times the atomic nitrogen K-shell cross-section. Note that the modulation about the atomic cross-section, caused by the potential barrier, extends to $\sim$100 eV above threshold before the molecular and atomic curves seem to coalesce.

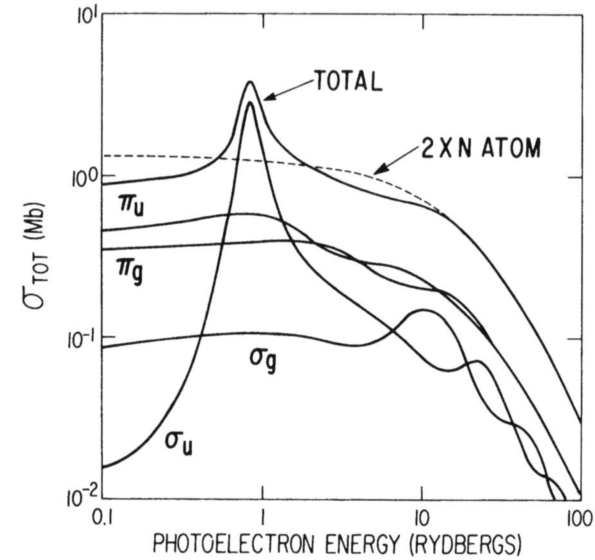

Fig. 4. Partial photoionization cross-sections for the K shell of N_2 over a broad energy range. The dashed line represents twice the K-shell photoionization cross-section for atomic nitrogen, as represented by a Hartree-Slater potential.

At higher energies, a weaker modulation appears in each partial cross-section. This weak modulation is a diffraction pattern, resulting from scattering of the photoelectron by the neighboring atom in the molecule, or, more precisely, by the molecular field. Structure of this type was first studied over 50 years ago by Kronig [202] in the context of metal lattices. It currently goes by the acronym EXAFS (extended x-ray absorption fine structure) and is used extensively [151,152] for local structure determination in molecules, solids and surfaces. The net oscillation is very weak in N_2 since the light atom is a weak scatterer. More pronounced effects are seen, for example, in K-shell spectra [203] of Br_2 and $GeCl_4$. Our reason for showing the weak EXAFS structure in N_2 is to emphasize that the low-energy resonant modulation (called "near-edge" structure in the context of EXAFS) and high-energy EXAFS evolve continuously into one another and emerge naturally from a single molecular framework, although high-energy EXAFS is usually treated from an atomic point of view.

Fig. 5 shows a hypothetical experiment which clearly demonstrates the ℓ character of the σ_u resonance. In this experiment, we first fix the nitrogen molecule in space and orient the

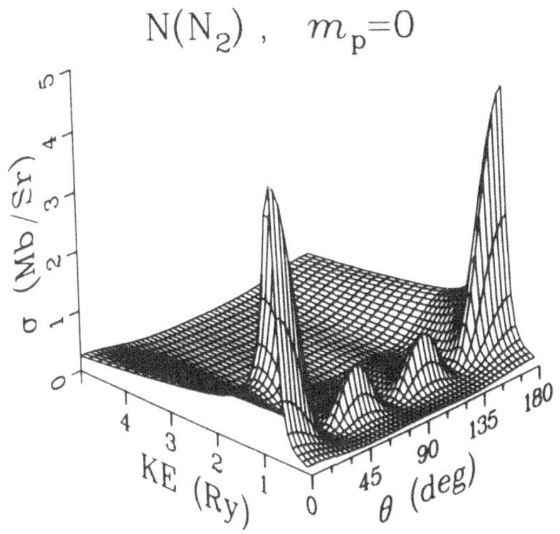

Fig. 5. Fixed-molecule photoelectron angular distribution for kinetic energies 0-5 Ry above the K-shell IP of N_2. The polarization of the ionizing radiation is oriented along the molecular axis in order to excite the σ continua; and the photoelectron ejection angle, θ, is measured relative to the molecular axis.

polarization direction of a photon beam, tuned near the nitrogen K edge, along the molecular axis. This orientation will cause photoexcitation into σ final states, including the resonant σ_u ionization channel. Fig. 5 shows the angular distribution of photocurrent as a function of both excess energy above the K-shell IP and angle of ejection θ, relative to the molecular axis. The enhanced photocurrent at the resonance position, KE $\simeq$ 1 Ry, is very apparent in Fig. 5. Moreover, the angular distribution exhibits three nodes, with most of the photocurrent exiting the molecule along the molecular axis and none at right angles to it. This is an f-wave (ℓ = 3) pattern and clearly indicates that the resonant enhancement is caused by an ℓ = 3 centrifugal barrier in the σ_u continuum of N_2. Thus, the centrifugal barrier has observable physical meaning and is not merely a theoretical construct. Note that the correspondence between the dominant, asymptotic partial wave and the trapping mechanism is not always valid, especially when the trapping is on an internal or off-center atomic site where the trapped partial wave can be scattered by the anisotropic molecular field into alternative asymptotic partial waves, as is the case in, for example, BF_3 [114] and SF_6. Finally, note that the hypothetical experiment discussed above has been approximately realized by photoionizing molecules adsorbed on surfaces. The shape-resonant features tend to survive adsorption and, owing to their observable ℓ character, can even provide evidence [197,198] as to the orientation of the molecule on the surface.

The final topic in the discussion of basic properties of shape resonances involves eigenchannel contour maps [107], or, "pictures" of unbound electrons. This is the continuum counterpart of contour maps of bound-state electronic wavefunctions, maps which have proven so valuable as tools in quantum-chemical visualization and analysis. Indeed, the present example helps achieve a physical picture of the σ_u shape resonance and the general technique promises to be a useful tool for future analyses of resonant trapping mechanisms and other observable properties. The key to this visualization (given in Eq. (5) below) is the construction of those particular combinations of continuum orbital momenta that diagonalize the interaction of the unbound electron with the anisotropic molecular field. These combinations, known as eigenchannels, are the continuum analogues of the eigenstates in the discrete spectrum (i.e., the bound states).

The electronic eigenvalue problem in the molecular continuum is inhomogeneous --- that is, there is a solution at every energy. Moreover, there are, in general, alternative solutions possible at a given energy, depending upon how the inhomogeneity is chosen. Typically, calculations are done in terms of real, oscillatory radial functions for the alternative possible orbital momenta. The result is the K-matrix normalized partial-wave expansion of the continuum electronic wavefunction, which takes the asymptotic form [67,135,204]

$$\Psi_L \sim (\pi k)^{-1/2}\ r^{-1} \textstyle\sum_{L'} [\sin\theta_{\ell'}\ \delta_{LL'} + K_{LL'}\ \cos\theta_{\ell'}]\ Y_{L'}(\hat{\mathbf{r}}) \qquad (1)$$

where $L=(\ell,m)$, the photoelectron orbital momentum and its projection along the molecular z axis; k^2 is the electron kinetic energy in Rydbergs; the electron distance r from the molecular center is in Bohrs; $\theta_\ell = kr - \ell(\pi/2) + \omega$, where the Coulomb phase $\omega = -(Z/k)\ln(2kr) + \arg\Gamma[\ell+1-i(Z/k)]$; molecular-ion charge is $Z=1$. The coefficients $K_{LL'}$ of the cosine terms form a real, symmetric matrix known as the K matrix. The K matrix tells us about the coupling, due to the nonspherical molecular potential, between different angular momenta and, as such, summarizes the electron-molecule interaction in a compact way. That single orbital momentum L (we refer here to ℓ and m collectively as orbital momentum) for which there occurs a sine term specifies the inhomogeneity, and each choice of L gives a row of the K matrix. By determining all rows in this way, we obtain the full K matrix and thereby a complete set of functions at the given energy.

The wavefunction (1) specifies what might be called a calculational boundary condition. Its form allows us to obtain the K-matrix while working in terms of real radial functions. There are two other types of boundary conditions [67,135,204], however -- physical boundary conditions, appropriate for representing physical observables, and eigenchannel boundary conditions, appropriate for analysis of the continuum wavefunction itself. The sets of wavefunctions for these alternative three types of boundary conditions are interrelated by unitary transformations.

Physical boundary conditions are introduced to obtain physical observables. One transforms to complex radial functions so that the directional character of the continuum electron at large distances, where it is detected, can be represented. This so-called S-matrix boundary condition, and its use in representing electron-molecule scattering and molecular photoionization, is discussed, for example, in [67,135].

Physical boundary conditions, however, are not well-suited to analysis of the continuum wavefunction itself. What is needed, rather, are the "normal modes" of the interaction of the continuum electron with the molecule --- that is, the eigenchannels of the electron-molecule complex. These are obtained by diagonalizing the K-matrix

$$\tan(\pi\mu_\alpha)\delta_{\alpha\alpha'} = \sum_{LL'} U_{\alpha L} K_{LL'} U_{L'\alpha'} \qquad (2)$$

to give the eigenvectors Ψ_α and eigenphases μ_α. The coefficients $U_{L\alpha}$ give the composition of the alternative eigenvectors Ψ_α in terms of the K-matrix-normalized functions Ψ_L:

$$\Psi_\alpha = \sum_L \Psi_L U_{L\alpha} \ . \qquad (3)$$

Using Eq. (2) and the fact that the matrix U is unitary,

$$\delta_{LL'} = \sum_\alpha U_{L\alpha} U_{\alpha L'} \ , \qquad (4)$$

we can rewrite the eigenvectors as [204]

$$\Psi_\alpha \sim (\pi k)^{-1/2} r^{-1} \sum_L [\sin\theta_\ell + \tan(\pi\mu_\alpha)\cos\theta_\ell] Y_L(\hat{r}) U_{L\alpha} . \qquad (5)$$

Eq. (5) is the key result of this discussion. Because molecules are not spherical, an electron of a particular angular momentum is, in general, "rescattered" into a range of angular momenta. This rescattering is indicated in Eq. (1) by the sum over L', and the coefficients $K_{LL'}$ give the relative amplitudes of the various rescatterings. The eigenchannel functions (5), on the other hand, correspond to those special combinations of incident angular momenta which are unchanged by the anisotropic potential of the molecule, i.e., the normal (eigen) modes of the electron-molecule interaction.

Comparison of Eq. (5) with Eq. (1), then, shows why the eigenchannel representation (5) is more suitable for analysis of the continuum molecular electronic wavefunction: First, as we have seen, the mixing of different orbital momenta is greatly simplified. Second, the radial wavefunctions for different angular momenta all have the same mixing coefficient $\tan(\pi\mu_\alpha)$. Third, if an eigenchannel is dominated by a particular orbital momentum, the eigenchannel wavefunction (5) has the characteristic angular pattern of the corresponding spherical harmonic. Last, and perhaps most important, because quasibound, shape-resonant states generally resonate in a single eigenchannel α, the eigenchannel representation gives us the most direct image of these resonant states.

As discussed earlier, the σ_u resonance is accompanied by a corresponding rise by $\sim\pi$ radians in one component of the eigenphase sum

$$\mu_{sum} = \sum_\alpha \mu_\alpha \; . \tag{6}$$

This resonant component, in turn, consists almost entirely of the single partial wave $\ell = 3$,

$$U_{L\alpha_{res}} \simeq \delta_{\ell 3} \; . \tag{7}$$

At these kinetic energies only one other orbital momentum, $\ell = 1$, contributes appreciably to the photoelectron wavefunction. (Orbital momenta $\ell = 0,2,4$, etc. do not contribute because they are of even parity, and $\ell = 5$ and higher odd orbital momenta are kept away from the molecule by their centrifugal barriers.) This means that there is only one other appreciable eigenchannel. The eigenphase component of this channel, primarily $\ell = 1$, is nearly constant throughout the resonant region. Thus, within about 20 eV above the ionization threshold, the N_2 odd-parity continuum can be analyzed in terms of just two eigenchannels, a nonresonant p-like channel and a resonant f-like one.

In Fig. 6, we have plotted the p-like eigenchannel wavefunction for two kinetic energies, one before (top) and one at (bottom) the resonance, which in this calculation falls at $\sim$1.2 Ry. The molecule is in the yz plane, along the z axis. The surface contours have been projected onto the plane to show more clearly the angular variation of the wavefunction. The single nodal plane characteristic of p waves is clearly seen in this projection. It is remarkable that, despite the complex ℓ-mixing induced by the anisotropic molecular potential, this eigenchannel has such a well-defined ($\ell = 1$) orbital momentum character. We stress, however, that such emergence of a single orbital momen-

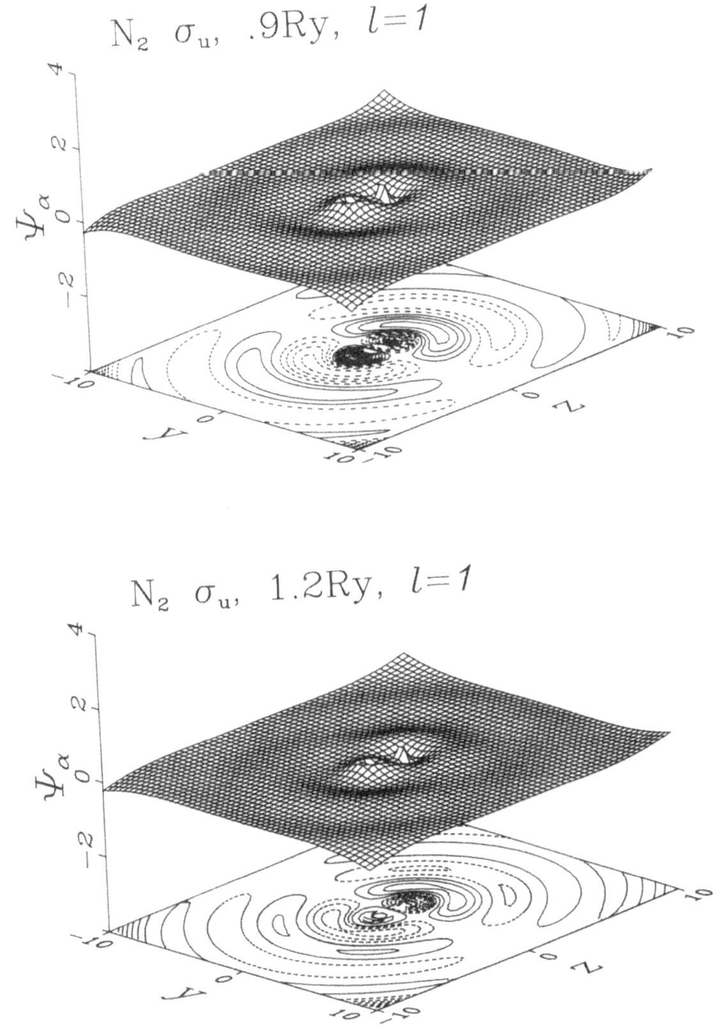

Fig. 6. p-wave-dominated eigenchannel wavefunctions for two electron kinetic energies in the σ_u continuum of N_2. The molecule is in the yz plane, along the z axis, centered at y = z = 0. Contours mark steps of 0.03, from 0.02 to 0.29; positive = solid, negative - dashed.

tum is not universal and that the angular pattern can be expected, in general, to be more complex. The cusps in the surface mark the nuclear positions. The only apparent change from one surface to the other is the slight shortening of wavelength as the kinetic energy increases. This smooth contraction of nodes continues monotonically through the resonance energy to higher energies. We conclude from Fig. 6 that the p eigenchannel is indeed nonresonant.

The f-like eigenchannel, plotted in Fig. 7, exhibits strikingly different behavior. In this case the surfaces (the contours of which have, at large distances, the three nodal planes characteristic of f orbitals) clearly show the resonant nature of the f eigenchannel. Note again the definite emergence of a single (ℓ = 3) orbital momentum character over the whole wavefunction. Below and above (not shown) the resonance energy, the probability amplitude is similar to the p-wave amplitude. But, at the resonance energy, there is an enormous enhancement of the wavefunction in the molecular interior; the wavefunction now resembles a molecular bound-state probability amplitude distribution. It is this enhancement, in the region occupied by the bound states, that leads to the very large increase in oscillator strength indicative of the resonance, and to the other manifestations discussed earlier and in the next section.

These eigenchannel plots are discussed more fully elsewhere [107]; however, several points should be noted before leaving the subject. First, the N_2 example that we have chosen is somewhat special in that there is a near one-to-one correspondence between the eigenchannels and single values of orbital angular momentum. Orbital angular momentum is, however, not a "good" quantum number in molecules and, generally, we should not always expect such clear nodal patterns. More typically, earlier work [67,96,113,114] indicates that several angular momenta often contribute to the continuum eigenchannels (although a barrier in only one ℓ component will be primarily responsible for the temporary trapping that causes the enhancement in that component, and in components coupled to it), thus implying that the resulting eigenchannel plots will be correspondingly richer. Second, Eqs. (1) and (5) are asymptotic expressions. The orbital momentum composition of these wavefunctions is more complicated in the molecular interior, as seen, for example, in Figs. 6 and 7. Nonetheless, continuity, and a dominant ℓ component may, as in the case of N_2, cause the emergence of a distinct ℓ pattern, even into the core region. Third, even though these ideas were developed [107] in the context of molecular photoionization, the continuum eigenchannel concept carries over without any fundamental change to electron-molecule scattering. This is a further example of the close connections [68] between shape resonances in molecular photoionization and electron-

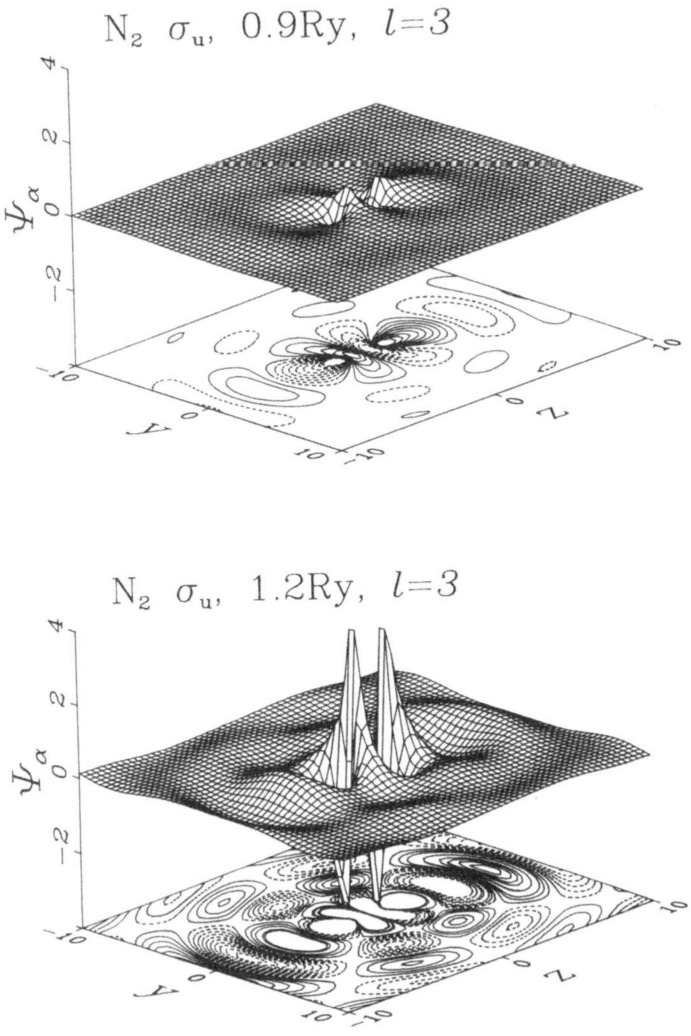

Fig. 7. f-wave-dominated eigenchannel wavefunctions for nonresonant (top) and resonant (bottom) electron kinetic energies in the σ_u continuum of N_2. The molecule is in the yz plane, along the z axis, centered at y = z = 0. Contours mark steps of 0.03, from 0.02 to 0.29; positive = solid, negative = dashed. The lack of contour lines near the nuclei for 1.2 Ry is because of the 0.29 cutoff.

molecule scattering. Finally, while we have used one-electron wavefunctions here, obtained within the multiple-scattering model, we emphasize that the eigenchannel concept is a general one, and we look forward to its use in the analysis of more sophisticated, many-electron molecular continuum wavefunctions.

2.3 Shape-Resonance-Induced Non-Franck-Condon Effects

2.3a Theoretical Predictions

Molecular photoionization at wavelengths unaffected by autoionization, predissociation or ionic thresholds has been generally believed to produce Franck-Condon (FC) vibrational intensity distributions within the final ionic state, and v-independent photoelectron angular distributions. We now discuss a recent prediction [69,81] that shape resonances represent an important class of exceptions to this picture. These ideas are illustrated with a calculation of the $3\sigma_g \rightarrow \varepsilon\sigma_u$, $\varepsilon\pi_u$ photoionization channel of N_2, which accesses the same σ_u shape resonance as that discussed above, at approximately $h\nu = 30$ eV, or $\sim$14 eV above the $3\sigma_g$ IP. The potential energy curves for N_2 are shown in Fig. 8 in order to orient this discussion and for later reference in Sections 3 and 4. The process we are considering involves photoexcitation of N_2 X $^1\Sigma_g^+$, in its vibrational ground state, with photon energies starting from the first IP and extending beyond the region of the shape resonance at $h\nu \simeq 30$ eV. This process ejects photoelectrons, leaving behind N_2^+ ions in energetically accessible states. Fig. 9 shows a typical photoelectron spectrum, measured at $h\nu = 21.2$ eV, which exhibits photoelectron peaks corresponding to production of N_2^+ in its X, A and B states (i.e., the three lowest ionic states in Fig. 8). Since we are interested in ionization of the $3\sigma_g$ electron, an event which produces the X $^2\Sigma_g^+$ ground state of N_2^+, we are concerned with the photoelectron band in the range 15.5 eV $\leq$ IP $\leq$ 16.5 eV in Fig. 9. The physical effects we seek involve the relative intensities and angular distributions of the v = 0-2 vibrational peaks in the X $^2\Sigma_g^+$ electronic band and, more specifically, the departures of these observables from behavior predicted by the FC separation.

The breakdown of the FC principle arises from the quasibound nature of the shape resonance which, as we discussed in Section 2.2, is localized in a spatial region of molecular dimensions by a centrifugal barrier. This barrier and, hence, the energy and lifetime (width) of the resonance are sensitive functions of internuclear separation R and vary significantly over a range of R corresponding to the ground-state vibrational motion. This is illustrated in the upper portion of Figs. 10 and 11, where the dashed curves represent separate, fixed-R calculations of the

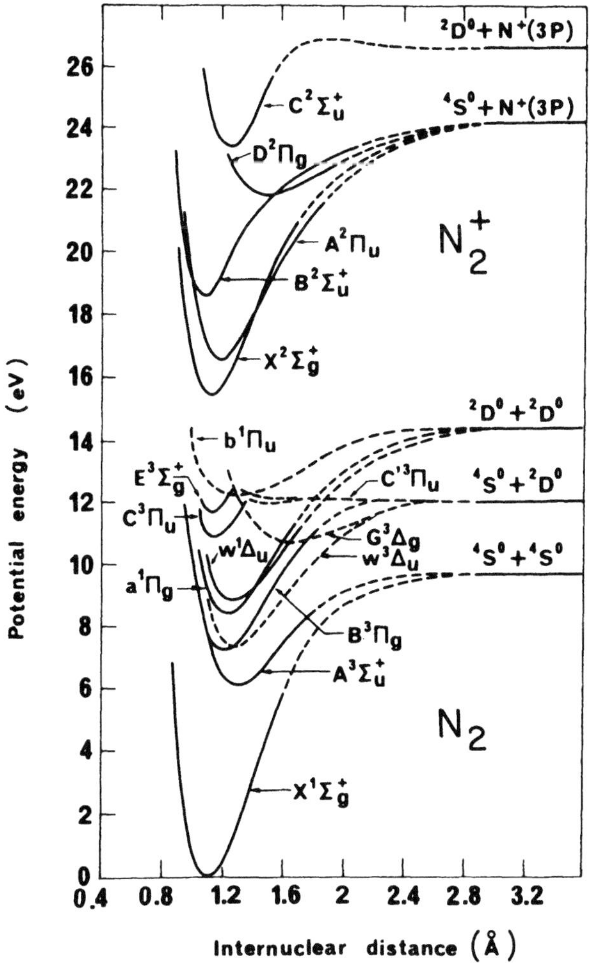

Fig. 8. Potential energy curves for N_2 and N_2^+.

partial cross-section and asymmetry parameter (respectively) for N_2 $3\sigma_g$ photoionization over the range $1.824a_0 \leq R \leq 2.324a_0$, which spans the N_2 ground-state vibrational wavefunction.

Of central importance in Fig. 10 is the clear demonstration that resonance positions, strengths and widths are sensitive functions of R. In particular, for larger separations, the inner well of the effective potential acting on the $\ell = 3$ component of the σ_u wavefunction is more attractive, and the shape resonance shifts to lower kinetic energy, becoming narrower and higher. Conversely, for lower values of R, the resonance is

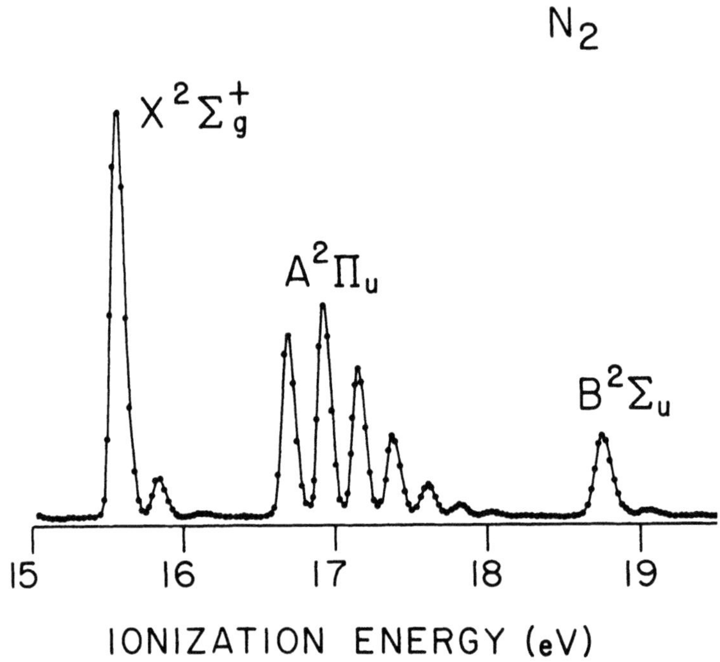

Fig. 9. Photoelectron spectrum of N_2 at hν = 21.2 eV.

pushed to higher kinetic energy and is weakened. This indicates that nuclear motion exercises great leverage on the spectral behavior of shape resonances and that small variations in R can significantly shift the delicate balance between attractive (mainly Coulomb) and repulsive (mainly centrifugal) forces which combine to form the barrier. In the present case, variations in R, corresponding to the ground-state variation in N_2, produce significant shifts of the resonant behavior over a spectral range several times the fullwidth at half maximum of the resonance calculated at $R = R_e$. By contrast, nonresonant channels are relatively insensitive to such variation in R, as was shown by results [98] on the $1\pi_u$ and $2\sigma_u$ photoionization channels in N_2.

Thus, in the vicinity of a shape resonance, the electronic transition moment varies rapidly with R. This parametric coupling was estimated [69,205], in the adiabatic-nuclei approximation, by computing the net transition moment for a particular vibrational channel as an average of the R-dependent dipole amplitude, weighted by the product of the initial- and final-state vibrational wavefunctions at each R:

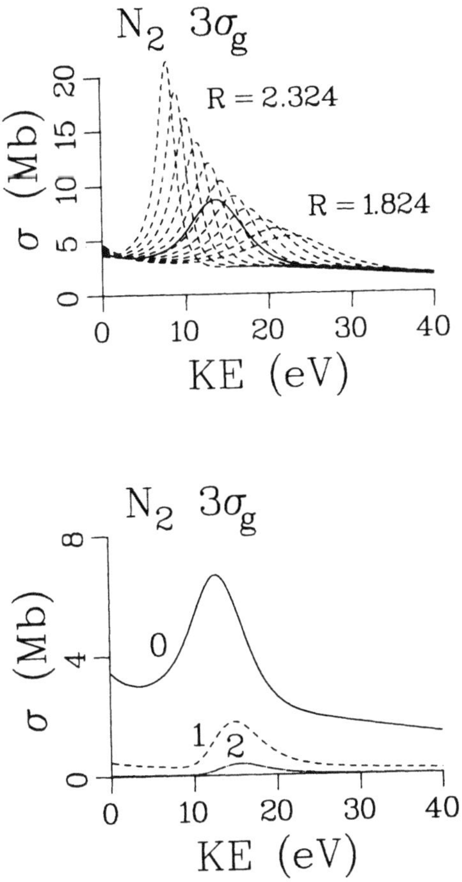

Fig. 10. Cross-sections, σ, for photoionization of the $3\sigma_g$ (v_i = 0) level of N_2. Top: fixed-R (dashed curves) and R-averaged, vibrationally-unresolved (solid curve) results. Bottom: results for resolved, final-state vibrational levels, v_f = 0-2.

$$D_{v_f v_i} = \int dR \chi_{v_f}^{\dagger}(R) D^{-}(R) \chi_{v_i}(R) \quad . \tag{8}$$

The vibrational wavefunctions were approximated by harmonic-oscillator functions and the superscript " – " denotes that incoming-wave boundary conditions have been applied and that the transition moment is complex. Note that even when the final vibrational levels v_f of the ion are unresolved (summed over), vibrational motion within the initial state v_i = 0 can cause the

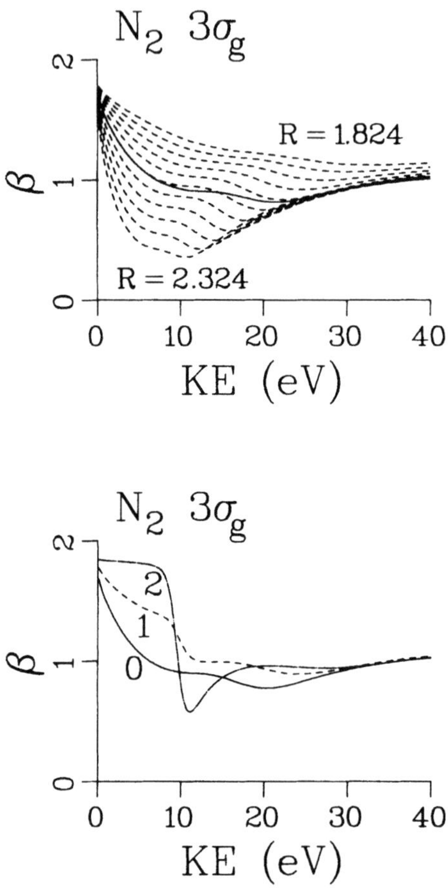

Fig. 11. Asymmetry parameters, β, for photoionization of the $3\sigma_g$ (v_i = 0) level of N_2. Top: fixed-R (dashed curves) and R-averaged, vibrationally-unresolved (solid curve) results. Bottom: results for resolved, final-state vibrational levels, v_f = 0-2.

above equation to yield results significantly different from the $R = R_e$ result, since the R dependence of the shape resonance is highly asymmetric. This gross effect of R averaging can be seen in the upper half of Fig. 10, by comparing the solid line (R-averaged result, summed over v_f) and the middle dashed line ($R = R_e$). Hence, even for the calculation of gross properties of the whole, unresolved electron band, it is necessary to take into account vibrational motion effects in channels exhibiting shape resonances. As we stated earlier, this is generally not a critical issue in nonresonant channels.

Effects of nuclear motion on individual vibrational levels are shown in the bottom half of Figs. 10 and 11. Looking first at the partial cross-sections in Fig. 10, we see that the resonance position varies over a few volts depending on the final vibrational state and that higher levels are relatively more enhanced at their resonance position than is $v_f = 0$. This sensitivity to v_f arises because transitions to alternative final vibrational states preferentially sample different regions of R. In particular, $v_f = 1,2$ sample successively smaller R, governed by the maximum overlap with the ground vibrational state, which causes the resonance in those vibrational channels to peak at higher energy than that for $v_f = 0$. The impact of these effects on branching ratios is clearly seen in Fig. 12, where the ratio of the higher v_f intensities to the intensity of $v_f = 0$ is plotted in the resonance region. There we see that the ratios are slightly above the FC factors (9.3%, $v_f = 1$; 0.6%, $v_f = 2$) at zero kinetic energy, go through a minimum just below the resonance energy in $v_f = 0$, then increase to a maximum as individual $v_f > 0$ vibrational intensities peak and, finally, approach the FC factors again at high kinetic energy. Note that the maximum enhancement over the FC factors is progressively more pronounced for higher v_f, i.e., 340% and 1300% for $v_f = 1,2$, respectively.

The effects on $\beta(v_f)$, shown in the lower portion of Fig. 11, are equally dramatic. Especially at and below the resonance position, the β's vary greatly for different final

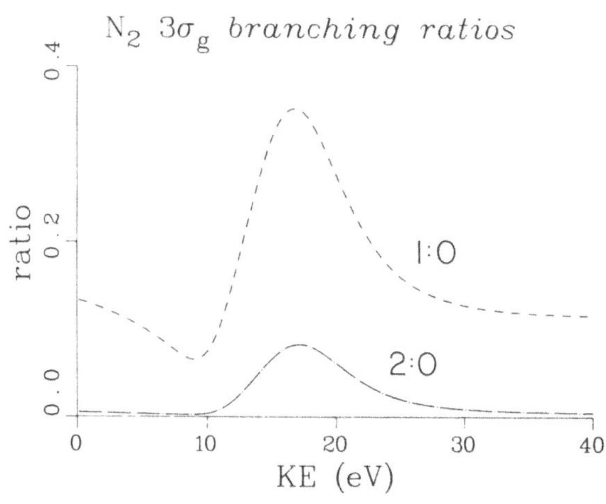

Fig. 12. Vibrational branching ratios $\sigma(v_f)/\sigma(v_f=0)$ for photoionization of the $3\sigma_g$ level of N_2.

vibrational levels. The $v_f = 0$ curve agrees well with the solid curve (in the upper half) since the gross behavior of the vibrationally-unresolved electronic band will be governed by the β of the most intense component. The $R = R_e$ curve has been found to agree well with recent wavelength-dependent measurements [75,78] and the agreement is improved by the slight damping caused by R averaging. More significant for the present discussion, however, is the v_f dependence of β. Carlson first observed [175,184] that, at 584 Å, the $v_f = 1$ level in the $3\sigma_g$ channel of N_2 had a much larger β than the $v_f = 0$ level, even though there was no apparent autoionizing state at this wavelength. This is in semiquantitative agreement with the present calculation which gives $\beta(v_f = 0) \simeq 1.0$ and $\beta(v_f = 1) \simeq 1.5$. Although the agreement is not exact, we feel this demonstrates that the "anomalous" v_f dependence of β in N_2 stems mainly from the σ_u shape resonance, a resonance which acts over a range of the spectrum many times its own $\sim$5 eV width. The underlying cause of this effect is the shape-resonance-enhanced R dependence of the dipole amplitude, just as for the vibrational partial cross-sections. In the case of $\beta(v_f)$, however, both the R dependence of the phase and of the magnitude of the complex dipole amplitude play a crucial role, whereas the partial cross-sections depend only upon the magnitude.

The above prototypical study demonstrates several new aspects of photoionization channels exhibiting shape resonances: First, the time delay in photoelectron escape, associated with a shape resonance, enhances the coupling between nuclear motion and electronic motion, invalidating the FC factorization of the two modes. In time-independent language, the great asymmetric and nonmonotonic sensitivity of the transition amplitude to internuclear separation requires a folding of the transition amplitude with the vibrational motion of the molecule, at the very least in an adiabatic scheme. When resonance lifetimes and the vibrational period are comparable, nonadiabatic effects (analogous to those observed [195,196] for the 2.4 eV π_g resonance in e-N_2 scattering) are likely to be important as well. Second, shape resonance effects are large for both vibrational intensities and angular distributions, but were largely overlooked in earlier work because these effects tend to lie in an inconvenient wavelength range for laboratory light sources. Synchrotron radiation has since solved this problem, as discussed in the next section. Third, it is significant to note that the effects of the shape resonance, as described above, act over tens of volts of the spectrum, several times the half-width of the resonance, and that σ and β probe the effects differently, i.e., σ and β exhibit maximal effects in different energy regions. This underscores the well-known difference in dynamical information contained in the two physical observables. Fourth, a long-standing "anomalous" v_f dependence in the photoelectron angular distributions of the $3\sigma_g$ channel of N_2 has been resolved. Finally, the phenomena described here for one channel of N_2 should be very widespread since shape resonances now appear to affect one or more inner- and outer-shell channels in most small (non-hydride) molecules.

2.3b Triply Differential Photoelectron Measurements

In order to fully examine dynamical aspects discussed in earlier sections, especially the last section, it is essential to perform measurements of photoelectron intensity as a function of three independent variables --- wavelength of the incident light, to select the spectral features; photoelectron kinetic energy, to select the ionization channel of interest; and ejection angle, to measure angular distributions. For convenience, we refer to this level of experiment as "triply-differential" photoionization experiments. Over the last few years, electron spectroscopy has been combined with synchrotron radiation sources to achieve successful triply-differential measurements in molecules, including vibrational state resolution. Presently several groups (e.g., [75,157-163]) in the U.S., France, Germany and England are involved in this type of experiment, each with its own specific experimental configuration and special emphasis, but each fulfilling the requirements for full triply-differential studies. Here we will review, as an example, the experimental aspects of a new instrument [163] at the Synchrotron Ultraviolet Radiation Facility (SURF-II), at the U.S. National Bureau of Standards, in order to focus on some of the experimental considerations in triply-differential photoionization studies. After a review of experimental techniques, we show how earlier measurements at SURF-II [99] and elsewhere [79] were used to test the predictions of non-FC effects in N_2 photoionization.

The new triply-differential electron spectrometer system at NBS consists of a high-throughput, normal-incidence monochromator [206] (Fig. 13) and a pair of 10 cm mean-radius hemispherical electron spectrometers in an experimental chamber (Fig. 14). To avoid later confusion, we note that an earlier configuration using the same monochromator with a single, rotatable 5 cm mean-radius spectrometer [159] has been used for the past five years in several triply-differential photoionization studies and is, in fact, the instrument used to obtain the data presented later in Figs. 15, 27 and 28. The special emphasis with both generations of instruments has been the same, namely, to optimize the photon and electron resolution in order to probe detailed dynamics within shape resonance and autoionization structure in the near-normal-incidence range ($h\nu \leq 35$ eV). The new instrument further optimizes several aspects of the electron spectrometer system to greatly extend the sensitivity and/or resolution com-

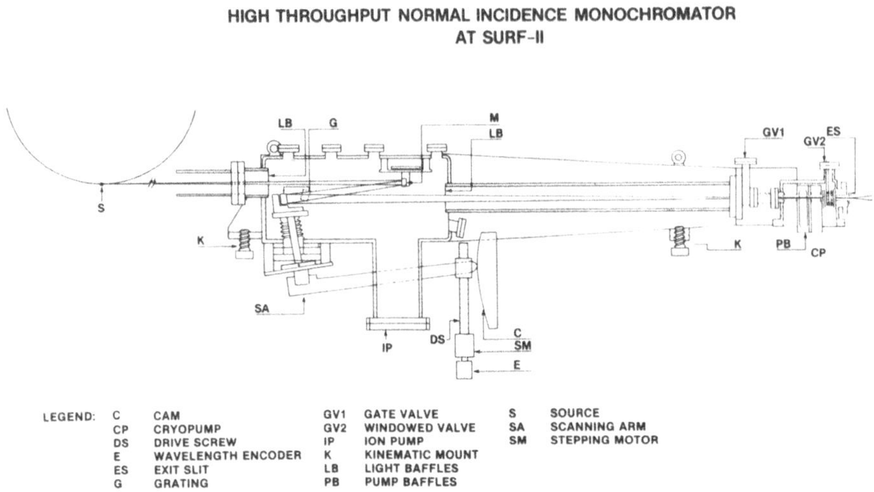

Fig. 13. Schematic diagram of high-throughput, normal-incidence monochromator (taken from [206]).

pared to earlier measurements, for reasons discussed in Section 3.

The high-flux, 2-meter monochromator shown in Fig. 13 has been specifically matched to the characteristics of the SURF-II storage ring: First and foremost, it uses the small vertical dimension of the stored electron beam (∿80 μm) as the entrance aperture of the monochromator, a unique feature which eliminates loss of incident flux on the entrance slit. Second, it is attached directly to the exit port of the storage ring, resulting in a very large capture angle of 65 mrad in the horizontal plane This arrangement produces approximately 10^{10} photons $sec^{-1}Å^{-1}$per mA of circulating current (typical initial current ≃15 mA) at 1000 Å and has been used for triply-differential measurements down to ∿375 Å. Together with a 1200 line/mm grating and a 200 μm (100 μm) exit slit, this configuration yields a photon resolution of ∿0.8 Å (0.4 Å). Plans have been made to improve the resolution with a higher-dispersion grating. The dispersed light is channeled by a 2 mm i.d. capillary tube for a distance of ∿25 cm into the interaction region of the experimental chamber. The low pumping conductance of this capillary tube is very effective in reducing the gas load on the monochromator and storage ring during experiments in which the gas pressure in the experimental chamber can be as high as 10^{-4} torr.

The new electron spectrometer system is shown schematically

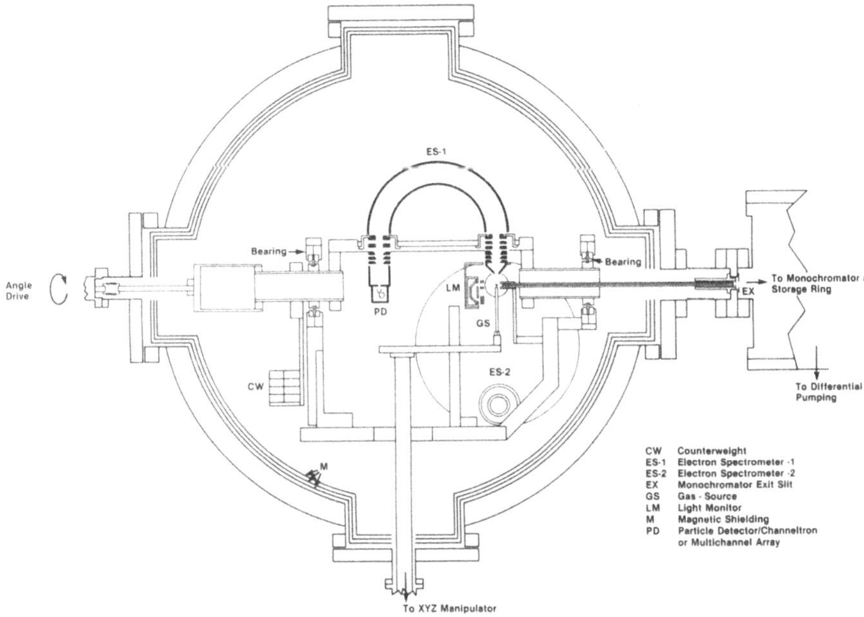

Fig. 14. Schematic diagram of dual electron spectrometer system (taken from [163]).

in Fig. 14. The chamber is a 76 cm diameter, 92 cm long stainless-steel vacuum chamber. It is pumped by a 500 liter/sec turbomolecular pump and a 2000 liter/sec closed-cycle helium cryopump to provide maximum flexibility in studying the whole range of gaseous targets. Low magnetic fields of <500 μgauss are maintained throughout the chamber by three layers of high-permeability magnetic shielding.

The system is designed to operate with either one rotating (ES-1) or two (ES-1 and ES-2) electron spectrometers. In either configuration the electron spectra can be recorded as a function of ejection angle relative to the principle axis of polarization. This leads to the determination of the photoelectron branching ratios and angular distributions according to the following expression, which applies to dipole excitation of free molecules with elliptically-polarized light [159,207]:

$$\frac{d\sigma_v}{d\Omega} = \frac{\sigma_v}{4\pi}\left[1 + \frac{\beta_v}{4}\,(3P\cos 2\theta + 1)\right] \tag{9}$$

where,

β_v = asymmetry parameter for vibrational state v.

σ_v = total cross-section for transition to vibrational state v.

P = polarization of the light with the horizontal component being the major axis.

θ = angle of ejection of photoelectrons with respect to the horizontal direction.

The number of electrons ejected per unit light flux per unit solid angle, $dn_v/d\Omega$, is proportional to the differential cross-section and, hence, we can write:

$$\frac{dn_v}{d\Omega} = N_v \left[1 + \frac{\beta_v}{4} (3P \cos 2\theta + 1)\right]. \qquad (10)$$

Measurement of P, θ and the number of electrons as a function of θ enables a determination of β_v and N_v. N_v, when normalized over a relevant set of possible alternative ionization channels, gives the branching ratio for the particular transition. The measurement of P is accomplished with a triple-reflection polarizer based upon the considerations of Horton, *et al.* [208]. The incoming light flux is monitored by a 90% transparent tungsten photocathode on the input aperture of the polarization monitor. After three reflections, the light is intercepted by a second tungsten photodiode. The ratio of these two photodiode signals, at 0° and 90° with respect to the major polarization axis, determines the polarization.

Each electron analyzer is a 10 cm mean-radius version of our previous instrument [159] and maintains the same electron-lens system --- a three-aperture zoom lens to focus the electrons into the hemispherical dispersive element and a similar one to refocus the energy-analyzed electrons on the exit slit. There are no entrance or exit apertures in the equatorial plane of the hemisphere and, therefore, the aperture in the entrance cone determines the basic resolution. This entrance aperture, and an aperture in front of the channeltron, are changeable to ensure control of geometric constraints on the resolution. The resolution obtainable, while yet maintaining a good signal, is expected to be on the order of 20 meV. This resolution is a significant improvement over that typically used now and will allow for the extension of studies of non-Franck-Condon effects to small polyatomic molecules.

The electrical aspects of the electron spectrometer are maintained much as in the previously described apparatus [159].

Briefly, the fixed voltages are controlled by highly regulated conventional power supplies and the variable voltages are under computer control. A 16 bit digital-to-analog converter (DAC) with a basic increment of 0.0005 V controls the ramping of offset voltages. The variable focus voltages are controlled by isolated power sources run by the computer. The computer (LSI-11) resides in a CAMAC crate and also controls the grating drive, angular position, light-detecting system, electron-counting system and other experimental chores. Both analyzers are ramped off the 16 bit DAC but have their own separately controlled power supplies for lens voltages. The two identical instruments will allow a determination of the branching ratios and asymmetry parameters without rotation (i.e., the electron intensity at two angles can be measured simultaneously). The device can be calibrated by reference to a gas with a known cross-section [209] and asymmetry parameter [192,210]. The calibration features for the two instruments are incorporated into a computer program that analyzes and reduces the data. Area detectors have been purchased and will soon be integrated into the experiment, thus significantly increasing the sensitivity of the system.

The gas-inlet system is mounted on an XYZ manipulator for external alignment of the gas jet to optimize signal intensity and resolution. Exit nozzles of the system are interchangeable and provide both effusive beams and supersonic jets by use of pinhole apertures of diameter 7-50 μm. The positioning of the supersonic source is of particular importance and necessitates the positioning capability of this inlet system. With the larger hemispherical dispersive element, the use of two analyzers, better gas-source technology, incorporation of area detectors and enhanced pumping, we expect a significant improvement in basic sensitivity of the instrument (a very conservative estimate would be >100x) as compared to our present 5 cm radius single-analyzer system.

The same LSI-11 computer used for automation is also used for data reduction. The basic data consist of electron counts as a function of wavelength, ramp voltage and angle. The ramp voltage is converted, using known quantities, to electron kinetic energy. The electron counts are then normalized for correction factors that depend upon kinetic energy, such as the transmission functions of the instruments and any angular correction factors. Upon obtaining a suitably normalized set of data, the photoelectron spectra are fitted to a gaussian basis set, using spectroscopic values for vibrational energy spacings, and treating peak height, peak width and overall position as free parameters. The areas of the respective peaks are used to infer the values of the branching ratios and asymmetry parameters from the equation given above. The calculated curve and normalized data are plotted to aid in the evaluation of the quality of fit. In addition, this fitting program outputs statistical parameters which can be used to estimate accuracy.

2.3c Experimental Confirmation of Non-FC Effects

The theoretical predictions of Section 2.3a were soon tested in two separate experiments, as indicated in Figs. 15 and 16. In Fig. 15, the branching ratio for production of the v=0 and 1 vibrational levels of N_2^+ X $^2\Sigma_g^+$ is shown. The dash-dot line is the original prediction [69] from Fig. 12. The solid dots are recent measurements [99] in the vicinity of the shape resonance at hν ≃ 30 eV. The conclusion drawn from this comparison is that the observed variation of the vibrational branching ratio relative to the FC factor over a broad spectral range qualitatively confirms the prediction; however, subsequent calculations [94,109] with fewer approximations have achieved much better

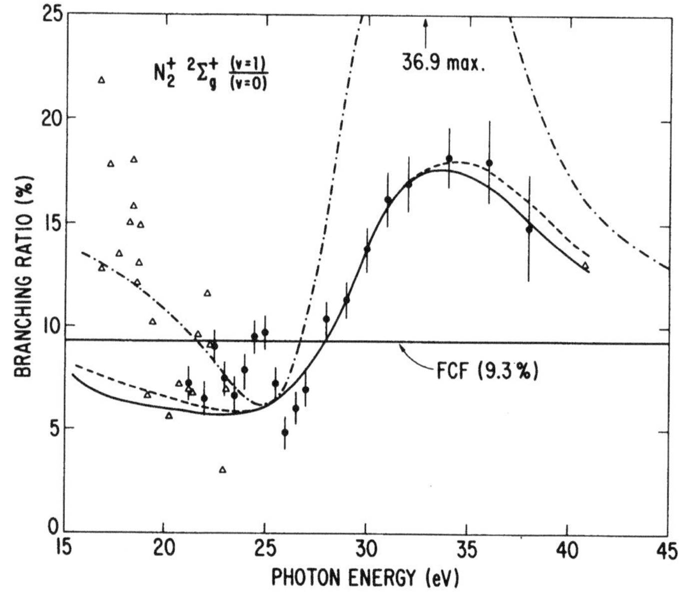

Fig. 15. Branching ratios for production of the v = 0,1 levels of N_2^+ X $^2\Sigma_g^+$ by photoionization of N_2. References: •, [99]; Δ, [181]; -·-·-, multiple-scattering model prediction from [69]; ——, frozen-core Hartree-Fock dipole-length approximation from [109]; ----, frozen-core Hartree-Fock dipole-velocity approximation from [109].

agreement based on the same mechanism for breakdown of the FC separation. The dashed and solid curves in Fig. 15 are results based on a Schwinger variational treatment [109] of the photoelectron wavefunction. The two curves represent a length and velocity representation of the transition matrix element, both of which are in excellent agreement with the data. This is an outstanding example of interaction between experiment and theory, proceeding as it did from a novel prediction, through experimental testing, and final quantitative theoretical agreement in a short time. Also shown in Fig. 15 are data in the 15.5 eV $\leq h\nu \leq$ 22 eV region which represent earlier results [181] obtained using laboratory line sources. The apparently chaotic behavior arises from unresolved autoionization structure, the detailed study of which is discussed in Section 3.3.

Fig. 16 shows angular distribution asymmetry parameters, β, for the v=0,1 levels of $N_2^+ X^2\Sigma_g^+$ over roughly the same energy region as that discussed above. These data were obtained at the Synchrotron Radiation Center at the University of Wisconsin by Carlson and co-workers [79]. In the region above $h\nu \simeq$ 25 eV, these data also show qualitative agreement with the predicted [69] v dependence of β caused by the σ_u shape resonance. In this case, the agreement is somewhat improved in later calcula-

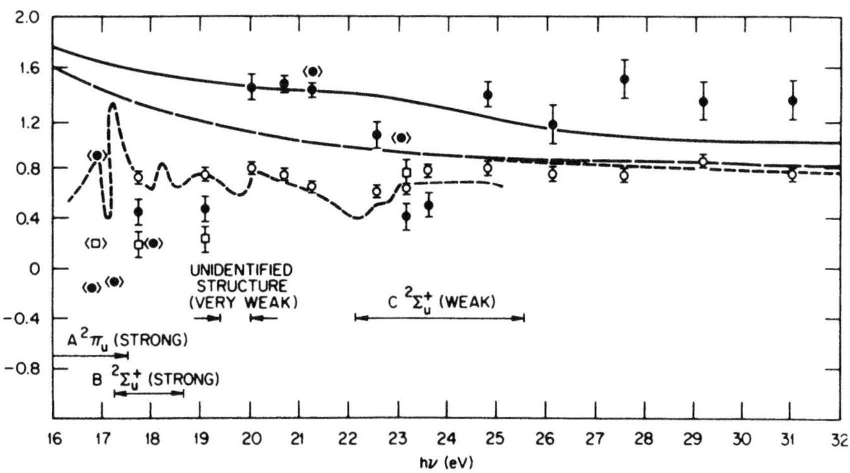

Fig. 16. Photoelectron asymmetry parameters for the v = 0,1 levels of $N_2^+ X\ ^2\Sigma_g^+$. References: o, v = 0 data from [79]; •, v = 1 data from [79]; long-dashed curve, v = 0 prediction from [69]; solid curve, v = 1 prediction from [69]; other data described in [79], from which this figure was taken.

tions [109], mainly for v=1: however, the change is less dramatic than for the branching ratios.

2.4 Progress and Prospects

In summary, we have used prototypical studies on N_2 to convey the progress made in the study of shape resonances in molecular photoionization, most of which has occurred over the last ten years or so. This included the identification of shape-resonant features in photoionization spectra of molecules and the accrual of substantial physical insight into their properties, many of which are peculiar to molecular fields. One very recent example has been the prediction and experimental confirmation of the role of shape resonances in producing non-Franck-Condon effects in vibrational branching ratios and photoelectron angular distributions.

What this discussion has failed to convey so far is the already extensive body of work that has developed around these basic themes. Even in an early interpretation [20] of shape resonance effects in x-ray spectra, it was obvious that the phenomena would be widespread since more than ten molecules, or local molecular moieties, were observed (see, e.g., [8,20]) to exhibit shape-resonant behavior. At the present, it is not difficult to identify over two dozen examples of molecules exhibiting the effects discussed above, in one or more final-state symmetries. (The following examples are found in [2-134] and the inner-shell cases are listed according to molecule in the bibliography given by Hitchcock [3].) These include simple diatomics (N_2, O_2, CO, NO), polyatomics with subgroups related to the first-row diatomics (HCN, C_2N_2, CH_3CN), triatomics (CO_2, CS_2, OCS, N_2O, SO_2) and more highly coordinated molecules and local molecular environments (SF_6, SO_4^{-2}, SF_5CF_3, SF_2O_2, SF_2O, BF_3, SiF_4, $SiCl_4$, SiF_6^{-2}, SiO_2, NF_3, CF_4). There is no doubt that many cases have been overlooked in this listing and that many more examples will be identified in the future as the exploration of molecular photoionization dynamics continues, particularly with the increasing utilization of synchrotron radiation sources.

Several examples from the above body of literature serve both to emphasize some of the interesting complications that can arise, and to caution against assuming that manifestations of shape resonances will always conform to the independent-electron concepts used above to explain the fundamentals of the subject in connection with N_2 photoionization: (i) For example, in the iso-electronic molecule CO, much of what was said for N_2 would carry over with suitable modifications to account for the loss of inversion symmetry. However, for photoionization of the outermost 5σ orbital (the counterpart of the $3\sigma_g$ in N_2) the vibrational branching ratios [76] and β's [80] *for the weaker* $v \geq 2$

channels deviate qualitatively from calculations [112]. This has been postulated [112] to arise from channel interaction with weak, doubly-excited autoionizing states in the region of the shape resonance. In any case, some departures from the simplified picture drastically alter the otherwise anticipated behavior of these weak vibrational channels. (ii) In the case of O_2 photoionization, an analogous σ_u shape resonance is expected [84,111,119] but its identification in the photoionization spectrum has been complicated by the existence of extensive autoionization structure in the region of interest. Recent work [132], using variable-wavelength photoelectron measurements and an MQDT analysis of the principal autoionizing Rydberg series, has sorted out this puzzle, with the result that the σ_u shape resonance was established to be approximately where expected but was not at all clearly identifiable without extensive analysis. (iii) In the case of CO_2, a σ_u shape resonance of completely different origin was expected [85,96] in photoionization of the $4\sigma_g$ orbital, leading to the C $^2\Sigma_g^+$ state of CO_2^+. This resonance, however, was not apparent in partial cross-section measurements on this channel [56], for reasons still not definitely known. Nevertheless, predictions [85,113] of a shape-resonant feature in the corresponding β were confirmed [101] and results from several laboratories [85,92,96,101,102,108,110,113,122] have now converged to reasonable agreement for this observable. Future experimental work on vibrational branching ratios [113] and v-dependent β's [113] would greatly aid in the further study of this puzzle. (iv) The most dramatic display of shape resonance phenomena is in SF_6, a molecule which exhibits four prominent resonantly-enhanced features (a_{1g}, t_{1u}, t_{2g} and e_g symmetries) in its inner-shell spectrum [4,5,20]. However, the role of the t_{2g} and e_g shape resonances in valence-shell spectra is poorly established. The experimental and theoretical evidence is too involved to summarize here; however, we do note that strong evidence [129], associated with the behavior of the t_{2g} resonance, exists for strong channel interaction among valence-shell photoionization channels. Moreover, the failure to clearly observe [55,74] the e_g resonance also suggests substantial departures from the elementary ideas described above. (v) Finally, returning to the case of N_2, we note that the photoionization of the $2\sigma_g$ should also access the σ_u shape resonance. However, for this inner-valence orbital, extensive vibronic coupling causes a breakdown [211] of the single-particle model, leading to the observation [91] of tens of "satellite" vibronic states in the photoelectron spectrum instead of a single peak due to ionization of the $2\sigma_g$ orbital. Nevertheless, if the intensity of this complicated structure is summed and plotted versus photon energy, the resonant enhancement reemerges.

These five cases are excellent examples of the additional challenges that can arise in the study of shape resonance phenomena. These challenges should not diminish the simplicity and power of the fundamental shape resonance dynamics; rather, they show how the fundamental framework showcases more complicated (and interesting) photoionization dynamics which, in turn, require a more sophisticated framework for full understanding.

Another form of progress is measured by the applicability of ideas to other observables or, more broadly, to other subfields: (i) We have already touched upon the close connection [68] between shape resonance phenomena in molecular photoionization and electron-molecule scattering. The mapping of the set of shape resonances from one case to the other by a shift of $\sim$10 eV on the kinetic energy scale has been discussed elsewhere [68]. (ii) Shape resonances in adsorbed molecules are now used rather extensively [197,198] as a probe of the geometry and electronic properties of adsorption sites. (iii) As discussed in connection with inner-shell spectra of SF_6, free-molecule concepts concerning localized states carry over to the condensed phase. In such cases a local "molecular" point of view can often provide more direct physical insight into photoexcitation dynamics of solids than does a band-structure approach. It is interesting to note that shape-resonant features in solid SiO_2, interpreted [20] long ago, are now used to monitor the oxidation of silicon in the fabrication of metal-oxide semiconductors. (iv) As noted above, shape resonances are often low-energy precursors to EXAFS structure occurring from $\sim$100 eV to thousands of eV above inner-shell edges. In fact, such resonant features are very sensitive to local structure (cf. Section 2.3) and may be very useful for local structure determination. (v) An intimate connection also exists with antibonding valence states in quantum chemical language. This was dramatically demonstrated over ten years ago, when Gianturco, *et al.* [21] interpreted the shape resonances in SF_6 using unoccupied virtual orbitals in an LCAO-MO calculation. This connection is a natural one since shape resonances are localized within the molecular charge distribution and therefore can be realistically described by a limited basis set suitable for describing the valence MO's. However, the scattering approach used in the shape resonance picture is necessary for analysis of various dynamical aspects of the phenomenon. (vi) Finally, shape resonances have been used as characteristic features in the analysis of such diverse subjects as electron optics [67] of molecular fields and hole localization [54] in inner-shell ionization, and as the cause for molecular alignment in photoionization, leading to anisotropy in the angular distributions [82] of Auger electrons from the decay of K-shell holes.

Looking to the future, there are several enticing prospects for significant progress. For instance, the present set of known and characterized resonance features is only the tip of the iceberg. The joint theoretical and experimental efforts needed to extend our knowledge of these useful states to other molecules and to other detection channels should be very fruitful and should present new issues to be resolved. Another theme, so natural in research, is the study of those cases in which our prevailing ideas break down. Several examples were cited above but this is expected to be a major growth area in molecular photoionization dynamics, owing to the growth both in detailed experimental studies and in the variety of computational methods capable of treating inter- and intra-channel coupling and non-Born-Oppenheimer effects. In this context, the study of weak channels, such as weak vibrational channels, should be most useful in highlighting the departures from the independent-particle, adiabatic framework used above. Another obvious example is the potential for more active investigations of shape-resonantly localized excitations in other contexts, such as adsorbed molecules, solids and biological molecules. In these and unforseen ways, the expansion, refinement and unification of recent developments in the study of shape resonances in molecular fields will provide a stimulating theme in molecular physics in the coming years.

3. MOLECULAR AUTOIONIZATION

3.1 Overview

Autoionization is an intrinsically multichannel process in which an excited discrete state from one channel couples to the underlying electronic continua of one or more other channels to effect ionization. It has been known since Fano's original work [1,212,213], almost fifty years ago, that this process produces characteristic asymmetric Fano-Beutler profiles in the photoionization cross-section. Since then, there have been extensive studies of autoionization structure in photoionization spectra of atoms [1] and molecules (see, e.g., [2,120,214-228]). In addition, the manifestations of autoionization in dynamical parameters such as photoionization branching ratios and photoelectron angular distributions have been recognized and have recently evolved into a major focal point for current studies of molecular photoionization dynamics (see, e.g., [64,93,116,125,132-134, 137-150,164-193,229-230]).

A more physical description of the autoionization process is helpful in discussing the alternative decay mechanisms possible in molecules: In most cases, autoionizing states consist of an excited Rydberg electron bound to an excited ion (also called core), primarily by Coulomb attraction. (The case of two highly correlated electrons bound to an ion is another important example, but one which requires special treatment [231] and will not be discussed here.) A necessary condition for decay of this state by ionization is that the excitation energy of the ion core be greater than the binding energy of the Rydberg electron. Then, barring alternative decay paths, autoionization will take place by means of a close collision, between the Rydberg electron and the core, in which excitation energy of the core is transferred to the excited electron to overcome its binding energy and to permit its escape from the ionic field. Note that although a Rydberg electron spends only a very small fraction of time within the molecular ion, such close encounters are essential for autoionization, since only when the Rydberg electron is nearby can it participate fully in the dynamics of the core and exchange energy efficiently with it.

A molecular ion core can store the energy needed to ionize a Rydberg electron in any of its three modes -- electronic, vibrational or rotational. The most direct means of storing electronic energy is the production of a hole in a molecular orbital other than the outermost occupied MO by, for example, promoting one of the inner electrons into a Rydberg orbital to initially prepare the autoionizing state. In addition, various degrees of vibrational and rotational excitation can accompany photoexcitation of Rydberg states. It is the existence of, and interplay among, the alternative energy modes which lead to the unique properties of molecular autoionization.

As a concrete example of rotational and vibrational autoionization, we shall discuss photoionization of H_2. The alternative rovibrational ionization channels for para-H_2 (J=0) are shown schematically in Fig. 17. The ground ionic state, H_2^+ X $^2\Sigma_g^+$, is the only bound electronic state in this spectral range; hence, the possibility of electronic autoionization is eliminated. In Fig. 17, the vertical, shaded bars represent various vibrational channels of H_2^+ X $^2\Sigma_g^+$, labeled by v^+=0-5. Pairs of continua are associated with each v^+, reflecting the two rotational continua N^+ = 0,2 produced by photoionization of para-H_2. Converging to each of these (and higher) rovibrational thresholds is a Rydberg series supported by the Coulomb field of the H_2^+ ion. A small subset of these Rydberg states is indicated, for later reference, by horizontal lines at the observed spectral locations, with each state being placed directly under the threshold to which it converges. Any of these optically allowed Rydberg states can autoionize by coupling with accessible open channels. In this case, autoionization would proceed by transferring energy stored in rotation or vibration of the ion core to the photoelectron. If more than one continuum is available, the decay will proceed into each with a branching ratio determined by the detailed dynamics of the decay process. Moreover, the angular distribution of the photoelectrons escaping in

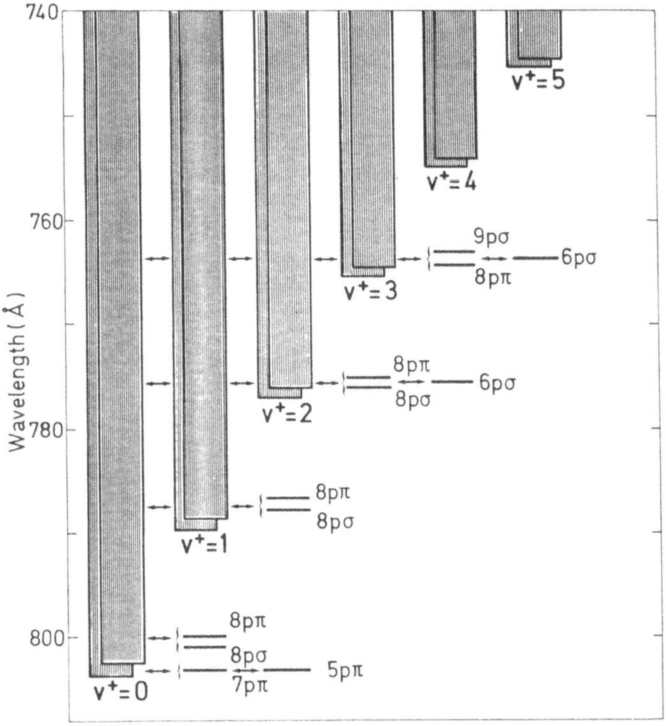

Fig. 17. Schematic illustration of vibrational-rotational autoionization in cold para-H_2 (J = 1, negative parity final states). Continua are indicated by vertical hatching. For each given v^+ of the ion H_2^+ there are two continua corresponding to rotational quantum number N^+ = 0 and 2 of the ion (J = 1). Selected discrete Rydberg levels are indicated below the vibrational ionization limit with which they are associated (taken from [148]).

each channel will reflect further details of the dynamics, including relative phases of degenerate photoelectron wavefunctions.

Based on the picture, so far, of a set of rovibrational thresholds and Rydberg series converging to each threshold, one might expect a dense pattern of autoionizing levels, but one which would straightforwardly yield to spectroscopic analysis in terms of characteristic rovibrational spacings and the known behavior of Rydberg series. However, this simple picture of a rich, but fundamentally uncomplicated, spectrum ignores all in-

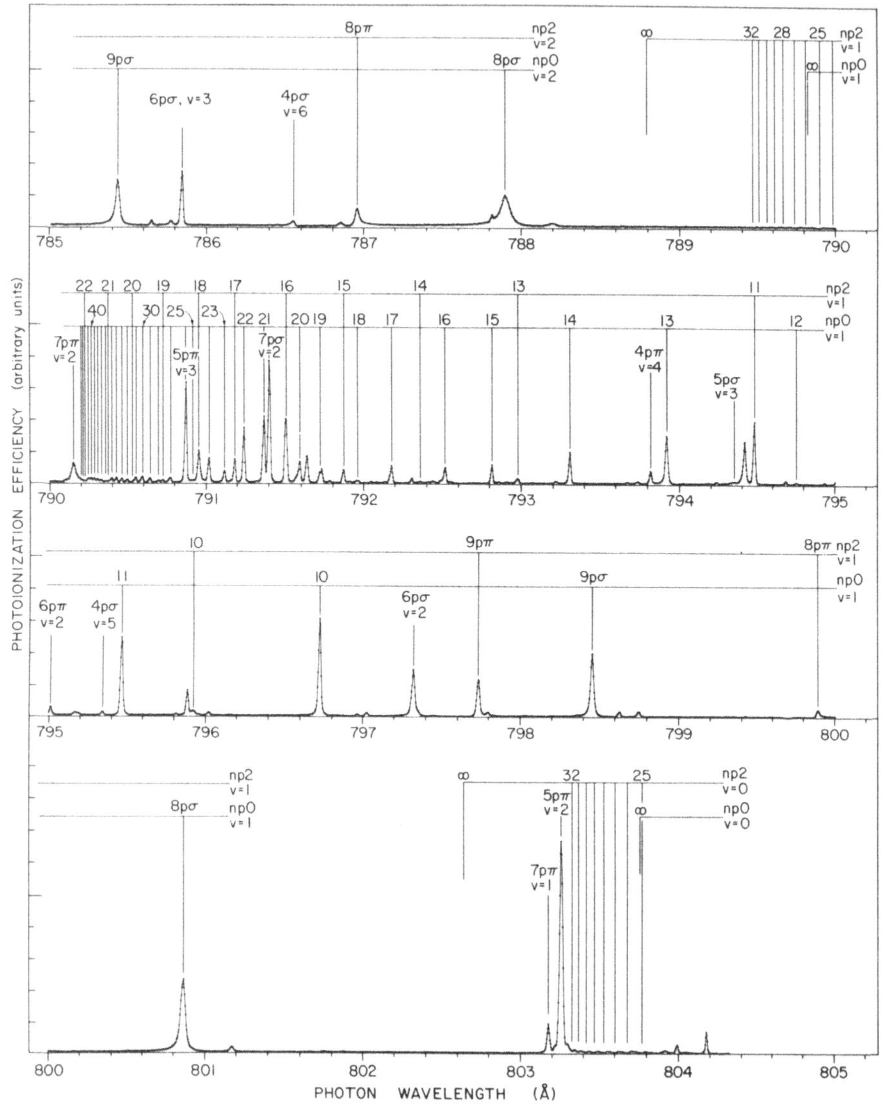

Fig. 18. A portion of the photoionization cross-section of para-H_2 at 78^0K (taken from [220]).

teraction between the Rydberg electron and the molecular core, and is wrong. This is demonstrated dramatically in Figs. 18 and 19, which show the total photoionization spectrum in two spectral regions (chosen for later discussion) covered in Fig.

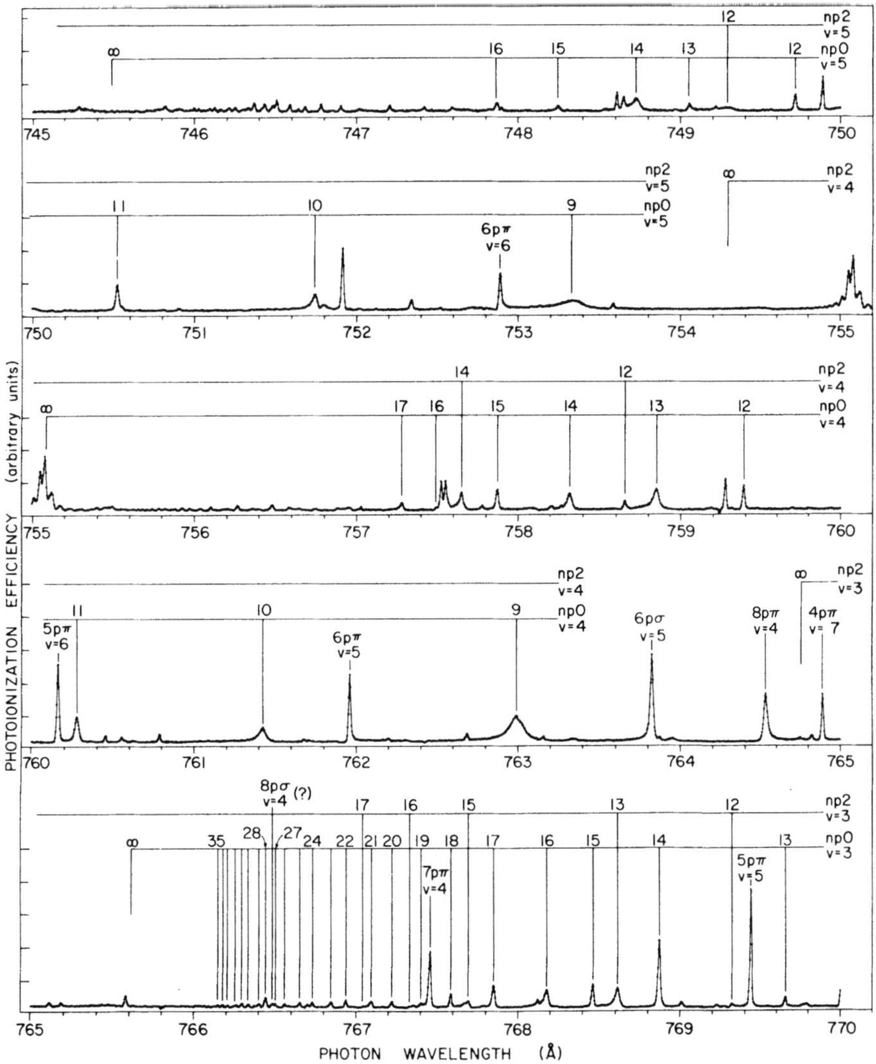

Fig. 19. A portion of the photoionization cross-section of para-H_2 at 78°K (taken from [220]).

17 --- one region from the first IP to 785 Å and the other from 770-745 Å. Careful inspection reveals that, for any Rydberg series, the spacings, intensities and profiles differ strongly from a simple Rydberg pattern. This is especially true near "interlopers", i.e., Rydberg states falling in the midst of a Rydberg series but converging to a higher limit. For such cases, level shifts and intensity redistribution frequently modify the entire host Rydberg series. These modifications arise from mutual interactions mediated by short-range forces and have been accounted for in detail in this prototypical system.

In the following two sections, we discuss two examples which represent the state-of-the-art in theoretical and experimental studies of molecular autoionization dynamics. The most accurate and penetrating theoretical analysis [137-139,143-145,147,148] has been carried out on parts of the H_2 spectrum using MQDT. Two representative cases will be discussed: Rotational autoionization, and a prediction of vibrational branching ratios and photoelectron angular distributions resulting from vibrational autoionization above the $v^+=3$ limit at ∿764.8 Å. These predictions have not yet been tested, although equivalent experiments on electronic autoionization in N_2 have recently been performed [168,169]. These latter results will constitute our example of experimental progress. In Section 3.4 we summarize the progress and prospects in this area, emphasizing that while experiment and theory are approaching one another, it is crucial that this progress converge to direct tests at the triply-differential level as soon as possible.

3.2 MQDT Treatment of Autoionization in H_2

Multichannel quantum defect theory and its application to molecular photoionization have been described in detail elsewhere [137-150], including elsewhere in this volume. Hence, we only briefly summarize the important attributes of MQDT and then turn to two examples of its application to photoionization in H_2.

MQDT is a theoretical framework which *simultaneously* treats the interactions between and within *whole* excitation channels. The input to an MQDT calculation consists chiefly of a small set of physically meaningful parameters (quantum defects and dipole amplitudes) which characterize the short-range interactions between the excited electron and the core, are slowly-varying functions of energy relative to rovibronic structure in the spectrum, and can generally be obtained from the positions and intensities of low-lying states in the spectrum. Also used are known transformation properties of molecular wavefunctions, e.g., transformations between Hund's coupling cases, and the asymptotic boundary conditions pertinent to a particular spectral range. Given this input, straightforward matrix mechanics yields, at each excitation energy, the spectral composition of the total final-state wavefunction in terms of the short-range, body-frame basis set, known dipole strengths and the asymptotic eigenphase shifts of the observable ionization channels. These quantities are then related, by now standard formulas, to such observables as the total photoionization cross-section, vibrational branching ratios and photoelectron angular distributions.

We wish to emphasize that this theoretical framework is not only elegant, but also reflects very accurately the internal mechanics of the excited complex. Hence, the quality of the computed observable depends solely upon the quality of the input. When accurate, empirical quantum defects are used, adiabatic and nonadiabatic corrections to the Born-Oppenheimer approximation are included automatically, to all orders. Hence, given physical input from low-lying excited states, one can use MQDT to generate accurate predictions for experiments throughout the extremely complex high-excitation regions. This is to be contrasted to the normal perturbation approach, which would require treating each state separately, with explicit adiabatic and non-adiabatic corrections and with little hope of treating higher-order interactions within and between whole exictation channels.

A striking example of the power and accuracy of MQDT in a complex situation is provided by the rich, rotational/vibrational autoionization structure in the 174.3 cm^{-1} spectral range between the N^+=0 and 2 rotational thresholds associated with the lowest v^+=0 ionization potential of H_2. The results of a calculation [144] of this structure are given in Fig. 20. The level positions indicated across the top of the figure are those one obtains with discrete boundary conditions (i.e., by eliminating open channels from the linear system). This level of analysis determines the initial assignments of spectral features. Alternatively, photoabsorption directly into the continuum is shown across the bottom of Fig. 20. This represents the single-channel level of approximation used throughout the discussion of shape resonances in Section 2. If rotational autoionizing channels only are introduced, the np2 levels of the Rydberg series converging to the upper (N^+=2) threshold of H_2^+ X $^2\Sigma_g^+$ (v^+=0) autoionize and distort the continuum into a Rydberg series of Fano-Beutler profiles. This is shown in the middle frame of Fig. 20. Finally, if vibrational autoionization channels are also introduced, the 5pπ(v=2) and 7pπ(v=1) levels autoionize and strongly distort the rotationally autoionizing levels as well, shifting intensity from above to below the vibrational interlopers.

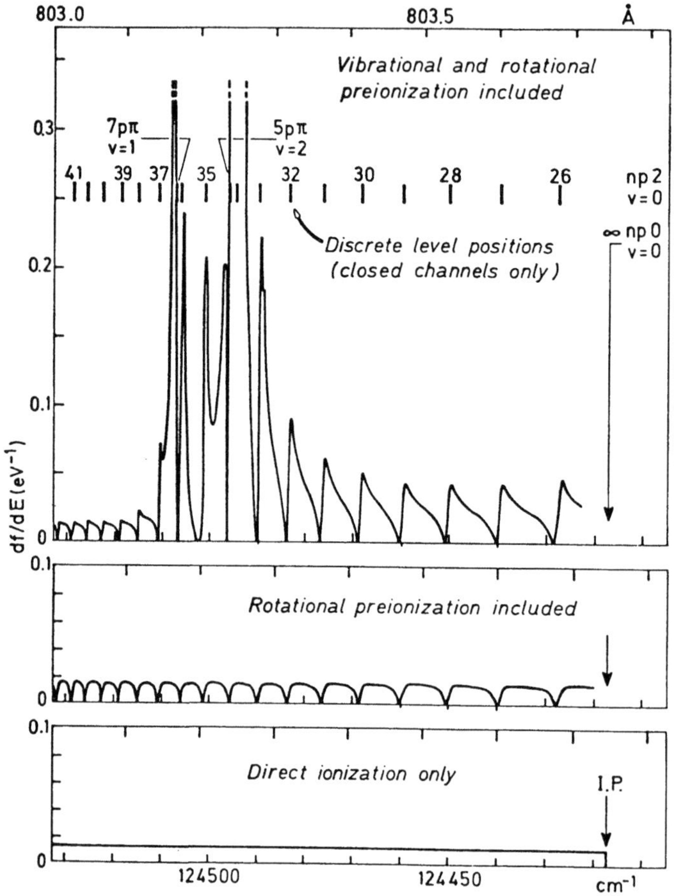

Fig. 20. MQDT calculation of photoionization of H_2 X $^1\Sigma_g^+$ (J" = 0, v" = 0) near the ionization threshold (taken from [144]).

We see, therefore, that the effect of vibrational autoionization on the ionization cross-section is profound, and that it affects the whole range shown (corresponding to ∿100 cm^{-1}). Indeed, if the fine variations in the cross-section are neglected, the whole spectrum can be viewed as one "giant" resonance of ∿50 cm^{-1} width, which causes a global transfer of intensity from the high-energy to the low-energy side of the vibrational peaks. This transfer leads to further modifications of the fine structure. For example, for n=26 and 27 the intensity minima still correspond nearly to the discrete np2 levels; for higher n, however, the profiles become progressively distorted until, near n=

32 to 35, it is the intensity maxima which coincide with the discrete level positions. In other words, there no longer exists a simple relationship between the extrema of the ionization curve and the positions of the autoionizing levels. In view of these complexities, it is clear that vibrational and rotational autoionization cannot be meaningfully treated as separate processes in this spectral region. A key feature of MQDT, however, is that it is based on no such assumed separability, i.e., it

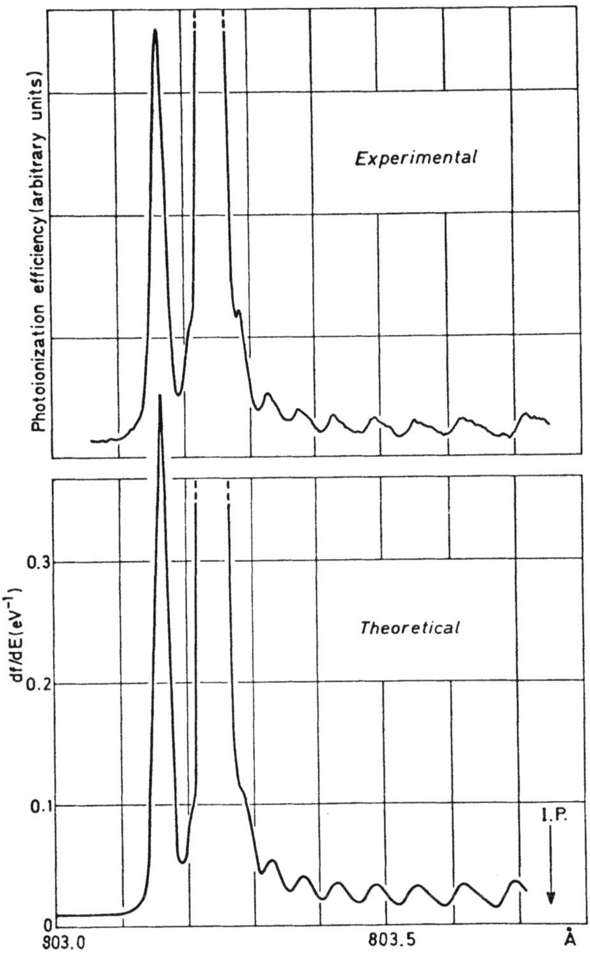

Fig. 21. MQDT results from Fig. 20, broadened to a resolution of 0.022 Å and compared with data from [220] (taken from [144]).

is applicable independently of coupling strengths between alternative decay mechanisms.

In Fig. 21, the calculated spectrum is compared with the high-resolution photoionization spectrum [220]. The calculated spectrum from Fig. 20 was convoluted with a triangular apparatus function of width 0.022 Å to mimic the finite experimental resolution. The comparison in Fig. 21 shows essentially exact agreement and reflects, more clearly than words, the state-of-the-art in computational simulation of detailed photoionization dynamics.

Triply-differential cross-sections have also been computed [148] near each of the levels shown explicitly in Fig. 17. However, we shall skip to the highest set of levels, above the $v^+=3$ limit, to discuss MQDT predictions of vibrational branching ratios and β's. This spectral range, between 762.5 and 765 Å in Fig. 19, is most attractive for future experimental examination since it produces four photoelectron peaks, corresponding to $v^+=$ 0-3, with sufficiently large photoelectron energies to be measured with existing electron-energy analyzers.

Fig. 22 shows the calculated total and vibrational partial cross-sections (the rotational sublevels have been summed over) in this spectral range, together with total photoionization data [220]. In this case the calculations are not folded with the instrument function and are, accordingly, sharper and higher. The horizontal arrow indicates the height that would be obtained from such a convolution. Also, the vertical arrows show more precise peak positions from the high-resolution absorption spectrum [139]. Given these qualifications, the agreement in the total cross-section is, again, quite satisfactory. The vibrational partial cross-sections, for which no experimental data are available, are shown at the bottom of Fig. 22. There, we see that the widths of the vibrational autoionization peaks are the same as in the total cross-section, as expected, but that the profiles vary drastically. Other details of the behavior of these partial cross-sections are displayed more clearly in the vibrational branching ratios, discussed below.

Fig. 23 again shows the total photoionization cross-section, together with the total β (summed over v^+) and includes the vibrational branching ratio and β_{v^+} for each vibrational ionization channel v^+. The total β curve is observed to dip strongly within the resonances. This is a consequence of the discrete wavefunction component mixing strongly into the ionization continuum. Classically speaking, one would say that the "quasibound" photoelectron spends more time near the core so that angular momentum exchange is enhanced. In this case the $N^+=$ 2 ionization channel becomes dominant near the center of the re-

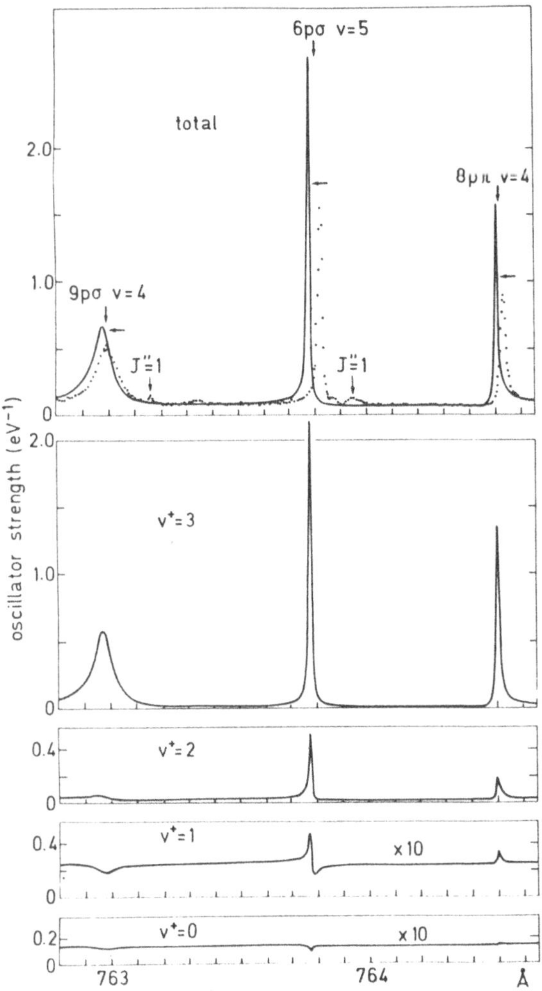

Fig. 22. Total and partial oscillator strengths for photoionization of H_2 near the $v^+ = 3$, $N^+ = 2$ ionization threshold (764.755 Å). Experimental points from [220]; figure taken from [148].

sonance and the value of β is depressed correspondingly, showing that directional information carried by the incoming photon is largely transferred to molecular rotation in the subsequent electron-core collisions. This general behavior is also reflected in the partial β_{v^+} curves although, in addition, a strong v^+ dependence in the magnitudes and shapes of these curves is

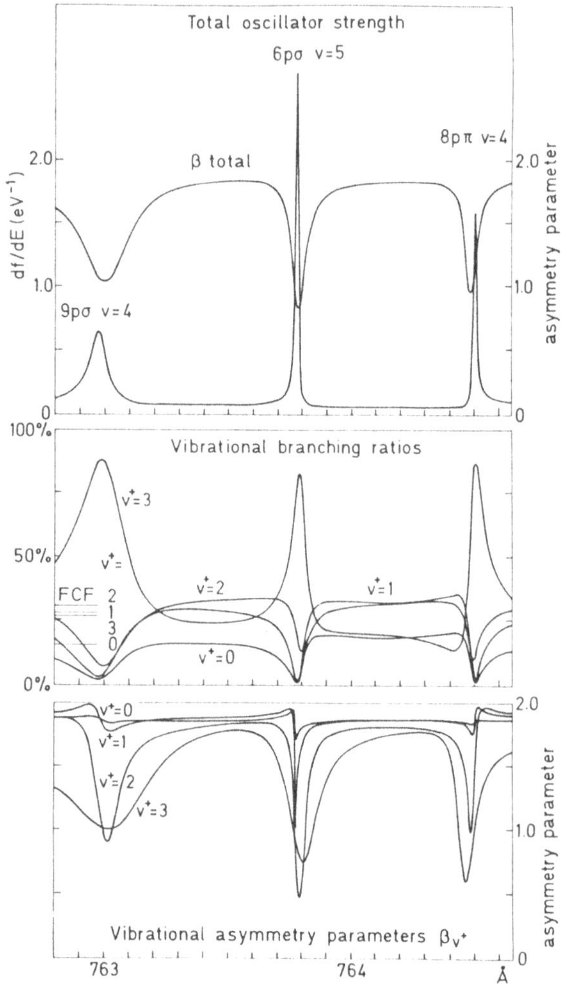

Fig. 23. Asymmetry parameter β (total and vibrationally-resolved) and vibrational branching ratios for photoionization of H_2 near the $v^+ = 3$, $N^+ = 2$ ionization threshold (764.755 Å). Figure taken from [148].

observed. Note that the spectral extent of the variations induced in β_{v^+} by vibrational autoionization is considerably larger than the halfwidths of the resonances themselves. This is a significant advantage in triply-differential experimental studies, which are difficult to perform with narrow photon bandwidth, for intensity reasons, and are only now being planned with bandwidths of 0.1-0.2 Å.

The middle frame in Fig. 23 shows the vibrational branching ratios in this energy region, along with the FC factors for direct ionization. As in the case of shape-resonant photoionization, striking non-FC behavior is observed in the vicinity of the vibrationally autoionizing states. And, as observed for β_{v^+}, the spectral extent of the autoionization effects is greater than the autoionizing resonance halfwidth, when displayed in terms of vibrational branching ratios, an observation which is significant in the context of experimental tests of these predictions. These calculations also predict that the branching ratio in the open channel possessing the highest v^+ (corresponding to autoionization with the lowest possible $|\Delta v|$) is strongly enhanced at the expense of all other channels, which are strongly depressed from their FC factors. Thus, the well-known qualitative rule [229], which states that a vibrationally-autoionized level decays preferentially with the smallest possible change of vibrational quantum number, is globally confirmed in these calculations, although for certain wavelengths (e.g., 764.4 Å in Fig. 23) the exact opposite may be true.

The above results tend to lull one into the feeling that we understand photoionization in H_2 completely, needing only to extend the range of the MQDT treatment to any energy region of

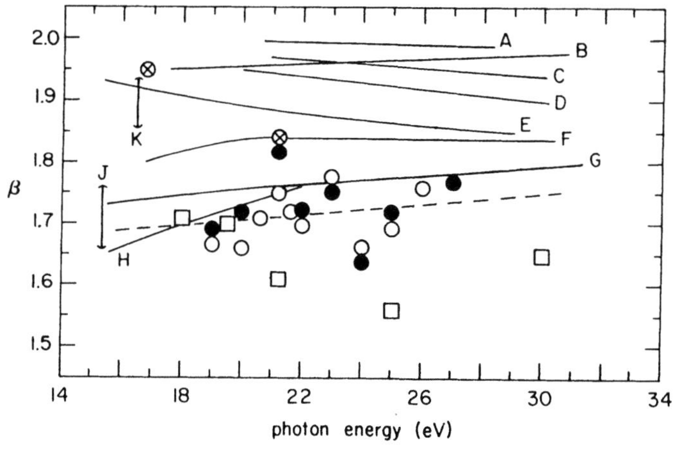

Fig. 24. Various experimental and theoretical results for the asymmetry parameter, β, for photoionization of H_2 and D_2. Complete citations for the various data are given in [232], from which this figure was taken.

interest. Nevertheless, it is imperative to perform experimental tests at the triply-differential level since it is in the more detailed quantities, such as vibrational branching ratios and angular distributions, that we are most likely to observe shortcomings in our detailed understanding of this important fundamental system. That such measurements have *not* been performed is often surprising to some. To emphasize the dearth of detailed data on this point, we show, in Fig. 24, a recent summary [232] of β measurements on H_2 (and D_2), together with other theoretical treatments. The gap between Figs. 23 and 24 is enormous. Indeed, the measurements are difficult; however, optimization of current technology should make this goal attainable. This is, one of the main motivations for the new-generation instrument described in Section 2.3b.

3.3 Triply-Differential Measurements Within Autoionization Resonances

Recall the scattered data in the 15.5 eV ≤ hν ≤ 20 eV region of the vibrational branching ratio results for N_2 (cf. Fig. 15). As indicated in the discussion of these data, the scatter was

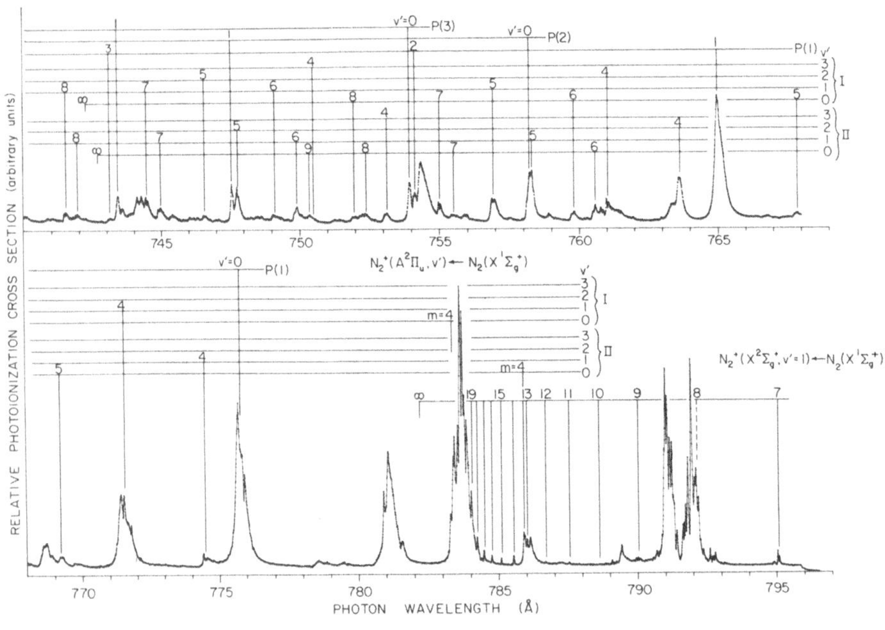

Fig. 25. A portion of the photoionization spectrum of N_2 at 78°K (taken from [222]).

produced by unresolved autoionization structure leading to the A and B states of N_2^+ (see Fig. 8). The high-resolution, total photoionization spectrum [222] of N_2 in that spectral range is given in Figs. 25 and 26, showing the rich autoionization structure in the N_2 spectrum. This structure serves to emphasize the difference in precision needed in studying the two types of resonant photoionization processes --- shape resonances and autoionizing resonances. The relatively broad Hopfield absorption and emission Rydberg series [233], converging to the v=0 level of N_2^+ B $^2\Sigma_u^+$ at 661.2 Å, was chosen for the initial triply-differential study [168] of molecular autoionization structure. This choice was guided by two considerations: First, this series represents a relatively isolated Rydberg series, the lower members of which are sufficiently broad to permit a systematic, detailed investigation of autoionization profiles with available instrumentation. Second, this was considered a good prototype for investigating the full rotational-vibrational-electronic interplay in the autoionization process. In this example, the autoionizing states consist of a Rydberg electron bound to an electronically excited N_2^+ B $^2\Sigma_u^+$ (v=0) core. Ionization occurs when the Rydberg electron collides with the core, thus enabling an exchange of the large electronic excitation energy of the core, together with smaller amounts of energy from the nuclear

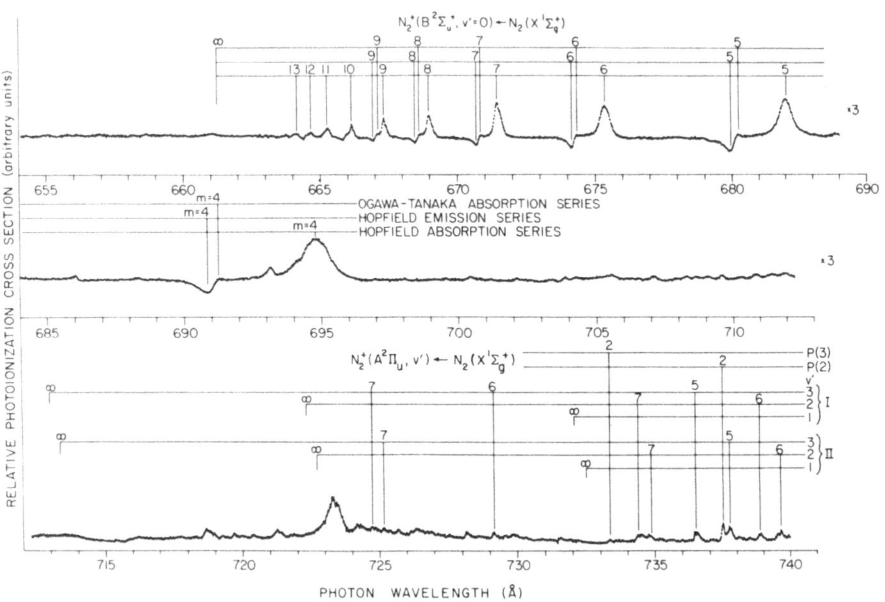

Fig. 26. A portion of the photoionization spectrum of N_2 at 78°K (taken from [222]).

modes.

Although several members of the Hopfield absorption and emission series have been studied [169] by triply-differential photoelectron spectroscopy, we focus here on the lowest (and broadest) m=3 members of these series at λ = 723.3 Å (absorption) and 715.5 Å (emission). The spectroscopic assignments for these series are summarized elsewhere [168]. We note, however, that the window and main absorption series, together with a weaker absorption series occurring between them, seem to be most consistent with the designations $nd\pi_g$, $nd\sigma_g$ and $ns\sigma_g$, although these assignments have not been unequivocally established.

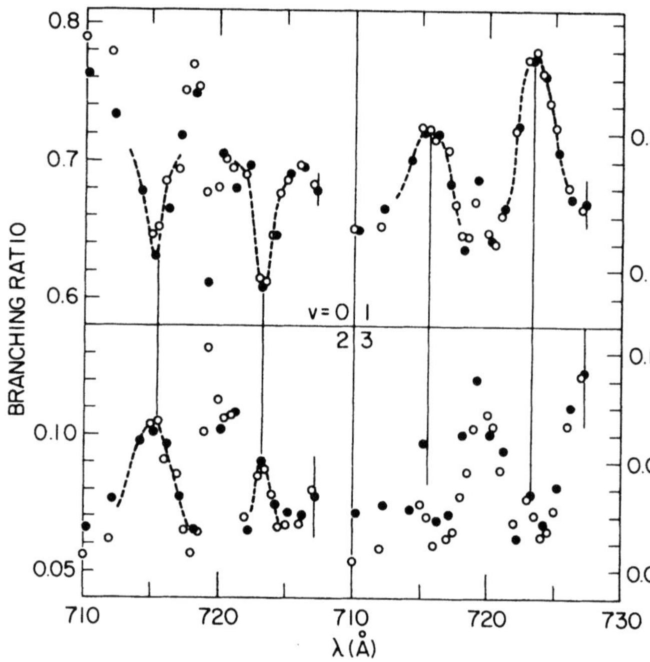

Fig. 27. Vibrational branching ratios for production of N_2^+ X $^2\Sigma_g^+$ (v = 0-3) in the range 710 Å $\leq \lambda \leq$ 730 Å. Vertical lines at 715.5 and 723.3 Å denote the positions of the first members of the Hopfield "emission" and absorption series approaching the N_2^+ B $^2\Sigma_u^+$ (v = 0) limit. Typical error bars are indicated on the last point in each frame. The dashed line is hand drawn to guide the reader's eye. Open and closed circles represent two independent runs. Figure taken from [168].

In Fig. 27, we present the vibrational branching ratios for formation of the ground-state ion N_2^+ X $^2\Sigma_g^+$ by photoionization in the range 710 Å ≤ λ ≤ 730 Å. (Here we define the vibrational branching ratio as the ratio of the intensity of a particular vibrational level to the sum over the whole vibrational band.) In Fig. 28, the asymmetry parameter β is given for the same processes. In both figures, the positions of the Hopfield emission and absorption features at 715.5 Å and 723.3 Å, respectively, are indicated by solid lines joining the upper and low-

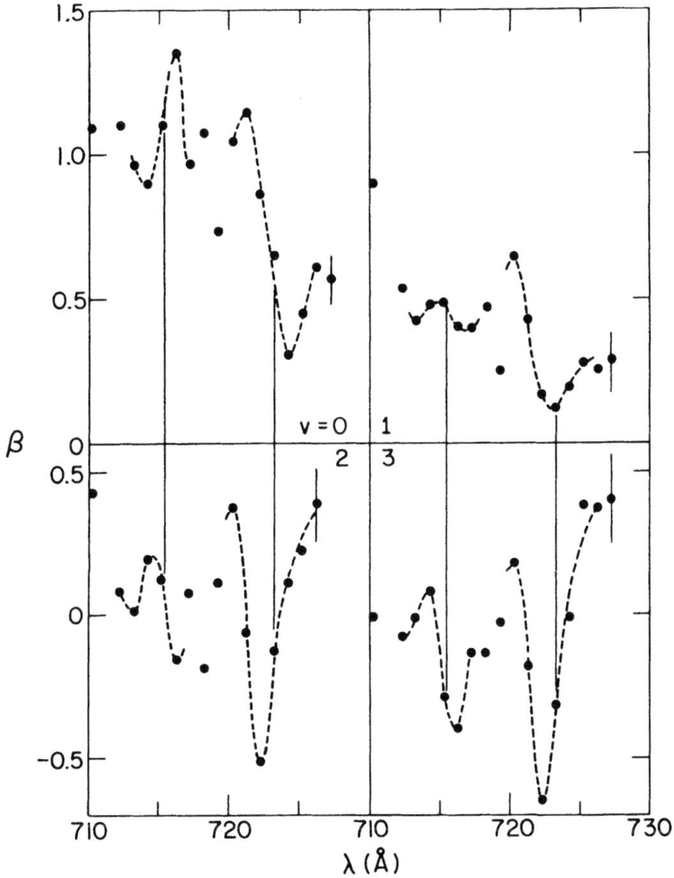

Fig. 28. Photoelectron asymmetry parameters corresponding to producing N_2^+ X $^2\Sigma_g^+$ (v = 0-3) in the range 710 Å ≤ λ ≤ 730 Å. Other conventions as in Fig. 27. Figure taken from [168].

er frames. In the vicinity of these features, a hand-drawn dashed curve is constructed only to guide the eye and should not be taken too seriously. Typical error bars for the data in both figures are shown on the last point in each frame. Duplicate branching-ratio measurements (the open circles were taken at the magic angle and the solid dots were deduced from the angular distribution measurements) show the reproducibility of the data. Note that an early branching-ratio study of this region of the N_2 photoionization spectrum was reported by Woodruff and Marr [164,165], but without angle dependence and with insufficient wavelength resolution to characterize the profiles of the Hopfield resonances.

Focusing first on the vibrational branching ratios in Fig. 27, we see three major qualitative features:

(i) The v=0 branching ratio exhibits pronounced dips at the locations of the two major autoionization features, whereas the higher vibrational channels, most notably v=1, show enhancements. Hence, the quasibound autoionizing states mediate a transfer of dipole amplitude from the v=0 channel to the much weaker v=1,2 and 3 channels, a transfer which involves simultaneous electronic deexcitation and vibrational excitation of the ion core. This transfer is primarily directed to the v=1 channel and is much diminished by v=3. Such an enhancement of vibrational channels possessing small Franck-Condon factors relative to the most intense channel is reminiscent of the effects of shape resonances in those few cases studied so far. Comparison with vibrational autoionization in H_2, where ionization channels with the minimum (negative) Δv usually dominate [148,229], is not straightforward since $\Delta v=0$ is permitted in this case; and, at any rate, transitions in which vibrational excitation occurs are favored. Establishing the systematics of this diverse set of observations is, obviously, a most timely problem.

(ii) Despite the great contrast between window and absorption profiles in the photoabsorption and photoionization spectra, the profile in Fig. 27 are of similar shape and both exhibit either an enhancement or depletion, depending upon the channel.

(iii) Definite "interloper" structure occurs between the two major resonances, with variable shape and strength. Both the weak absorption peak near the window resonances and other weak structures (one peak in the photoionization spectrum at 718.8 Å correlates well with the main interloper structures in Fig. 27) may play a role here.

The angular distribution results of Fig. 28 also exhibit structure at the positions of the two major resonances and in between. Implicit in the spectral variations in β is information on both the vibrational branching ratios and the relative phases of the alternative vibrational ionization channels. Specifically, the competition between asymptotic phases produces large asymmetric variations in β at the resonance positions, variations which differ from one final vibrational level to another. For instance, the β curve near the Hopfield "emission" line exhibits a peak for v=0 which evolves into a dip for v=3. Near the Hopfield absorption profile, the position of the minimum in β, although not extremely well-defined by these data, clearly shifts from the long-wavelength side of the resonance position to the short-wavelength side.

We conclude by observing that, although this example is one of only a few measurements of its kind at this time (see the next section), the experimental developments described in Section 2.3b, and those occurring in other laboratories, will make available more systematic, triply-differential studies in many molecules in the next few years. It is, therefore, most timely to extend calculations, such as those described in Section 3.2, to the full rovibronic description of molecular autoionization, in order to help guide experimental progress and to extract the wealth of detailed dynamical information contained in the anticipated growth of triply-differential data. Fortunately, theoretical progress in this direction is already substantial, as summarized in the next section.

3.4 Progress and Prospects

In addition to the prototype H_2 and N_2 cases discussed above, there have been a large number of molecules for which autoionization structure in the total photoionization cross-section has been mapped out (see, e.g., [2,120,214-228]). These spectra contain a great deal of dynamical information in themselves and create an indispensable roadmap for choosing particular cases for detailed study. More recently, additional triply-differential measurements have been performed on broad autoionization features in O_2 [166], CO [167], CO_2 [170] and C_2H_2 [125,131]. In addition, very detailed vibrationally-resolved partial cross-sections for production of $O_2{}^+$ B $^4\Sigma_g^-$ and B $^2\Sigma_g^-$ have been measured [132-134] systematically throughout the complex autoionization structure in the $h\nu$ = 18-21 eV range. As mentioned above (Section 2.4), this work addressed, and has rather clearly resolved, the spectroscopy and dynamics arising from overlapping shape and autoionizing resonances in this region, using a combination of experimental and theoretical methods.

Progress on the theoretical side has also been substantial. The MQDT analysis of electronic autoionization in O_2 just mentioned is a good example of utilizing MQDT in a more complicated situation, where the basic input is not as complete as in the case of H_2 and must, therefore, be approximated or empirically derived. This type of application exploits the powerful theoretical framework of MQDT to reflect complicated dynamics in terms of a few parameters, which can even be deduced by forcing the calculated spectrum to mimic experiment. This permits us to access the information contained in these data, at least until means [150] to compute input parameters for more complicated molecules are adequately developed. Other examples of successful MQDT analyses include NO [149] and N_2 [146, 149]; both of these molecules exhibit a very interesting hybrid autoionization mechanism in which the net result is vibrational autoionization, but for which all of the strength derives from indirect electronic coupling with an interloper converging to a higher electronic threshold.

Clearly, experimental and theoretical programs to probe more deeply into molecular autoionization dynamics are proceeding in a healthy manner. However, special emphasis should be placed on making progress toward the (still missing) direct comparison at the triply-differential level. Technical difficulties exist on both experimental and theoretical sides but the progress cited above would indicate that this prime goal is within present capabilities. To be specific, the shortest path toward a direct comparison would appear to be an experimental test of the theoretical predictions in Fig. 23, and theoretical analyses of the experimental data in Figs. 27 and 28. We expect these and many more examples to emerge in the next few years.

4. CONCLUDING REMARKS

In order to place the preceeding sections in a broader perspective, we briefly indicate how a variety of techniques are presently being used to probe aspects of molecular photoionization dynamics which are complementary to those emphasized here: (i) Multiphoton ionization is a recently developed technique which has a great potential for expanding our understanding of photoionization dynamics in totally new directions. When used in conjunction with photoelectron detection, such as that described in Section 2.3b, vibrational intensities and photoelectron angular distributions can be measured (see, e.g., [234-241]) to provide windows onto the dynamics of the multiphoton process, just as they do in the single-photon case. In addition, when the multiphoton process proceeds via resonances with excited neutral states, such as the excited valence states of N_2 in Fig. 8, very high-resolution ($\Delta\lambda < 0.05$ cm^{-1}) spectroscopy and dynamics of these *intermediate* resonances can be examined. (ii) Another recently developed technique [242,243] is polarization of fluorescence from molecular ions formed by photoionization. This approach accesses information on the orientation of the molecular ion and the relative strengths of degenerate photoelectron channels in the photoionization process. To give a concrete example, photoionization of N_2 to form the B $^2\Sigma_u^+$ state of N_2^+ (see Fig. 8) leads to the rightmost photoelectron band in Fig. 9. This band consists of electrons in degenerate $\varepsilon\sigma_g$ and $\varepsilon\pi_g$ ionization channels which cannot be separated by straightforward electron spectroscopy or by β measurements. Nevertheless, by observation of the polarization of the B-X fluorescence, information on the branching ratio for these degenerate photoelectron channels can be obtained [242]. This provides a unique test of dynamical information which exists, but usually remains implicit, in theoretical calculations. This technique has recently been extended [243] to the study of autoionization structure in the photoionization of CO_2. Another, apparently dissimilar, technique for examining relative strengths of degenerate photoionization channels is the measurement of angular distributions of photoions from dissociative photoionization [244]. The common link between these techniques is that they do not detect (and, hence, integrate over the angular distribution of) the photoelectrons, thus eliminating the interference effects between degenerate channels and isolating their relative strengths in the observed parameters. (iii) An important extension of the triply-differential studies discussed above is the measurement of spin polarization of the photoelectrons. Spin-polarization measurements access additional dynamical information described in detail elsewhere [158]. Full quadruply-differential measurements have already been performed on atoms [158] and spin-polarization measurements have been made on molecules [245]. (iv) A variety of dissociative phenomena occur either in competition with or subsequent to molecular photoionization, and are an important part of the broader picture. For example, MQDT has been applied [149] to the competition between predissociation and autoionization in photoexcitation of NO, an application which represents an extension of MQDT analysis to dissociation channels and thereby serves as a prototype for an extensive class of processes always present to some degree in molecular photoexcitation spectra. Another subject with a rich literature (see, e.g. [246-254]) is fragmentation of molecular ions formed by photoionization. Usually studied by photoelectron-photoion coincidence techniques, this subfield focuses on the decay of excited molecular ions into alternative dissociation channels. (v) A class of phenomena which appear to be very common in inner-valence-shell spectra results from the breakdown of the single-particle model brought on by extensive vibronic coupling among the high density of states in the inner-valence region [211]. This breakdown manifests itself in a high density of satellites in photoelectron spectra (consisting of admixtures of single-hole states, two-hole one-particle states and certain higher-order combinations) to the extent that the main photoelectron line associated with a single inner-valence hole can even be difficult to recognize. This has been treated successfully, e.g., in the case of the "$2\sigma_u$" spectrum [91] of N_2, by many-body Green's function techniques [211], and may be expected to be an important dynamical effect in the photoionization of molecular levels possessing IP's in the $h\nu \approx 30$-50 eV range.

In closing, we emphasize that, although extensive, this survey is not at all comprehensive. Hopefully, it does succeed in indicating both the present richness of the field of molecular photoionization dynamics and the opportunities for significant progress in the future.

ACKNOWLEDGMENT

This work was supported by the U.S. Department of Energy, the Office of Naval Research, NATO Grant No. 1939, and National Science Foundation Grant CHE-8203267.

REFERENCES

1. U. Fano and J. W. Cooper, Rev. Mod. Phys. 40, 441 (1968).
2. E. E. Koch and B. F. Sonntag, in: *Synchrotron Radiation: Techniques and Applications*, C. Kunz, ed. (Springer-Verlag, Heidelberg, 1979), p. 269.
3. A bibliography of inner-shell spectra is given by A. P. Hitchcock, J. Electron Spectrosc. 25, 245 (1982).
4. R. E. LaVilla and R. D. Deslattes, J. Chem. Phys. 44, 4399 (1966).
5. T. M. Zimkina and V. A. Fomichev, Dokl. Akad. Nauk SSSR 169, 1304 (1966) [Sov. Phys. Dokl. 11, 726 (1966)].
6. V. A. Fomichev, Fiz. Tverd. Tela. 9 3167 (1971) [Sov. Phys. Solid State 9, 2496 (1968)].
7. M. Nakamura, M. Sasanuma, S. Sato, M. Watanabe, H. Yamashita, Y. Iguchi, A. Ejiri, S. Nakai, S. Yamaguchi, T. Sagawa, Y. Nakai and T. Oshio, Phys. Rev. 178, 80 (1969).
8. T. M. Zimkina and A. C. Vinogradov, J. Phys. (Paris) Colloq. 32, 3 (1971).
9. W. Hayes and F. C. Brown, J. Phys. B 4, L85 (1971).
10. W. Hayes, F. C. Brown and A. B. Kunz, Phys. Rev. Lett. 27, 774 (1971).
11. D. Blechschmidt, R. Haensel, E. E. Koch, U. Nielsen and T. Sagawa, Chem. Phys. Lett. 14, 33 (1972).
12. M. J. van der Wiel and Th. M. El-Sherbini, Physica (Utrecht) 59, 453 (1972).

13. G. R. Wight, C. E. Brion and M. J. van der Wiel, J. Electron Spectrosc. 1, 457 (1972/73).
14. (a) R. E. LaVilla, J. Chem. Phys. 57, 899 (1972); (b) 58, 3841 (1973).
15. G. R. Wight and C. E. Brion, J. Electron Spectrosc. 3, 191 (1974).
16. G. R. Wight and C. E. Brion, J. Electron Spectrosc. 4, 313 (1974).
17. G. R. Wight and C. E. Brion, J. Electron Spectrosc. 4, 327, 335 (1974).
18. V. I. Nefedov, Zh. Strukt. Khim. 11, 292 and 299 (1970) [J. Struct. Chem. 11, 272 and 277 (1970)].
19. V. A. Fomichev and R. L. Barinskii, Zh. Strukt. Khim. 11, 875 (1970) [J. Struct. Chem. 11, 810 (1970)].
20. J. L. Dehmer, J. Chem. Phys. 56, 4496 (1972).
21. F. A. Gianturco, C. Guidotti and U. Lamanna, J. Chem. Phys. 57, 840 (1972).
22. B. Cadioli, U. Pincelli, E. Tosatti, U. Fano and J. L. Dehmer, Chem. Phys. Lett. 17, 15 (1972).
23. U. Fano, Comments At. Mol. Phys. 3, 75 (1972).
24. R. L. Barinskii and I. M. Kulikova, Zh. Strukt. Khim. 14, 372 (1973) [J. Struct. Chem. 14, 335 (1973)].
25. J. L. Dehmer, Phys. Fenn. 9S, 60 (1974).
26. V. P. Sachenko, E. V. Polozhenstev, A. P. Kovtun, Yu. F. Migal, R. V. Vedrinski and V. V. Kolesnikov, Phys. Lett. A 48, 169 (1974).
27. L. N. Mazalov, F. Kh. Gel'mukhanov and V. M. Chermoshentsev, Zh. Strukt. Khim. 15, 1099 (1974) [J. Struct. Chem. 15, 975 (1974)].
28. J. L. Dehmer and D. Dill, Phys. Rev. Lett. 35, 213 (1975).
29. D. Dill, J. Siegel and J. L. Dehmer, J. Chem. Phys. 65, 3158 (1976).
30. J. L. Dehmer and D. Dill, J. Chem. Phys. 65, 5327 (1976).
31. J. W. Davenport, Phys. Rev. Lett. 36, 945 (1976).
32. J. L. Dehmer and D. Dill, in: *Proceedings of the International Conference on Inner Shell Ionization Phenomena-Invited Papers*, W. Mehlhorn and R. Brenn, eds. (Universitat Freiburg, Freiburg, 1976), p. 221.
33. M. Nakamura, Y. Morioka, T. Hayaishi, E. Ishiguro and M. Sasanuma, in: *Third International Conference on Vacuum Ultraviolet Radiation Physics*, Y. Nakai, ed. (Physical Society of Japan, Tokyo, 1971), paper 1pA1-6.
34. Th. M. El-Sherbini and M. J. van der Wiel, Physica (Utrecht) 59, 433 (1972).
35. R. E. LaVilla, J. Chem. Phys. 63, 2733 (1975).
36. W. Eberhardt, R. P. Haelbich, M. Iwan, E. E. Koch and C. Kunz, Chem. Phys. Lett. 40, 180 (1976).
37. A. Hammett, W. Stoll and C. E. Brion, J. Electron Spectrosc. 8, 367 (1976).

38. L. C. Lee, R. W. Carlson and D. L. Judge, J. Phys. B 9, 855 (1976).
39. J. A. R. Samson and J. L. Gardner, J. Electron Spectrosc. 8, 35 (1976).
40. M. Tronc, G. C. King, R. C. Bradford and F. H. Read, J. Phys. B 9, L555 (1976).
41. G. R. Wight, M. J. van der Wiel and C. E. Brion, J. Phys. B 9, 675 (1976).
42. J. W. Davenport, Int. J. Quantum Chem. Symp. 11, 89 (1977).
43. A. P. Hitchcock and C. E. Brion, J. Electron Spectrosc. 10, 317 (1977).
44. R. B. Kay, Ph. E. van der Leeuw and M. J. van der Wiel, J. Phys. B 10, 2513 (1977).
45. G. C. King, F. H. Read and M. Tronc, Chem. Phys. Lett. 52, 50 (1977).
46. E. W. Plummer, T. Gustafsson, W. Gudat and D. E. Eastman, Phys. Rev. A 15, 2339 (1977).
47. T. N. Rescigno and P. W. Langhoff, Chem. Phys. Lett. 51, 65 (1977).
48. J. A. R. Samson, J. L. Gardner and G. N. Haddad, J. Electron Spectrosc. 12, 281 (1977).
49. J. A. R. Samson, G. N. Haddad and J. L. Gardner, J. Phys. B 10, 1749 (1977).
50. W. H. E. Schwarz, T. C. Chang and J. P. Connerade, Chem. Phys. Lett. 49, 207 (1977).
51. A. Bianconi, H. Peterson, F. C. Brown and R. Z. Bachrach, Phys. Rev. A 17, 1907 (1978).
52. C. E. Brion and K. H. Tan, Chem. Phys. 34, 141 (1978).
53. F. C. Brown, R. Z. Bachrach and A. Bianconi, Chem. Phys. Lett. 54, 425 (1978).
54. D. Dill, S. Wallace, J. Siegel and J. L. Dehmer, Phys. Rev. Lett. 41, 1230 (1978); 42, 411 (1979).
55. T. Gustafsson, Phys. Rev. A 18, 1481 (1978).
56. T. Gustafsson, E. W. Plummer, D. E. Eastman and W. Gudat, Phys. Rev. A 17, 175 (1978).
57. A. P. Hitchcock and C. E. Brion, Chem. Phys. 33, 55 (1978).
58. A. P. Hitchcock, C. E. Brion and M. J. van der Wiel, J. Phys. B 11, 3245 (1978).
59. P. W. Langhoff, A. E. Orel, T. N. Rescigno and B. V. McKoy, J. Chem. Phys. 69, 4689 (1978).
60. D. G. McCoy, J. M. Morton and G. V. Marr, J. Phys. B 11, L547 (1978).
61. N. Padial, G. Csanak, B. V. McKoy and P. W. Langhoff, J. Chem. Phys. 69, 2992 (1978).
62. T. N. Rescigno, C. F. Bender, B. V. McKoy and P. W. Langhoff, J. Chem. Phys. 68, 970 (1978).
63. M. Sasanuma, E. Ishiguro, H. Masuko, Y. Morioka and M. Nakamura, J. Phys. B 11, 3655 (1978).

64. T. Baer, P. M. Guyon, I. Nenner, A. Tabche-Fouhaille, R. Botter, L. F. A. Ferreira and T. R. Govers, J. Chem. Phys. 70, 1585 (1979).
65. D. M. Barrus, R. L. Blake, A. J. Burek, K. C. Chambers and A. L. Pregenzer, Phys. Rev. A 20, 1045 (1979).
66. C. E. Brion, K. H. Tan, M. J. van der Wiel and Ph. E. van der Leeuw, J. Electron Spectrosc. 17, 109 (1979).
67. J. L. Dehmer and D. Dill, in: *Electron-Molecule and Photon-Molecule Collisions*, T. Rescigno, V. McKoy and B. Schneider, eds. (Plenum, New York, 1979), p. 225.
68. J. L. Dehmer and D. Dill, in: *Symposium on Electron-Molecule Collisions*, I. Shimamura and M. Matsuzawa, eds. (University of Tokyo Press, Tokyo, 1979), p. 95.
69. J. L. Dehmer, D. Dill and S. Wallace, Phys. Rev. Lett. 43, 1005 (1979).
70. A. P. Hitchcock and C. E. Brion, J. Electron Spectrosc. 15, 401 (1979).
71. A. P. Hitchcock and C. E. Brion, Chem. Phys. 37, 319 (1979).
72. A. P. Hitchcock and M. J. van der Wiel, J. Phys. B 12, 2153 (1979).
73. P. W. Langhoff, A. Gerwer, C. Asaro and B. V. McKoy, Int. J. Quantum Chem. 13, 645 (1979).
74. H. J. Levinson, T. Gustafsson and P. Soven, Phys. Rev. A 19, 1089 (1979).
75. G. V. Marr, J. M. Morton, R. M. Holmes and D. G. McCoy, J. Phys. B 12, 43 (1979).
76. R. Stockbauer, B. E. Cole, D. L. Ederer, J. B. West, A. C. Parr and J. L. Dehmer, Phys. Rev. Lett. 43, 757 (1979).
77. M. Tronc, G. C. King and F. H. Read, J. Phys. B 12, 137 (1979).
78. S. Wallace, D. Dill and J. L. Dehmer, J. Phys. B 12, L417 (1979).
79. T. A. Carlson, M. O. Krause, D. Mehaffy, J. W. Taylor, F. A. Grimm and J. D. Allen, J. Chem. Phys. 73, 6056 (1980).
80. B. E. Cole, D. L. Ederer, R. Stockbauer, K. Codling, A. C. Parr, J. B. West, E. D. Poliakoff and J. L. Dehmer, J. Chem. Phys. 72, 6308 (1980).
81. J. L. Dehmer and D. Dill, in: *Electronic and Atomic Collisions*, N. Oda and K. Takayanagi, eds. (North-Holland, Amsterdam, 1980), p. 195.
82. D. Dill, J. R. Swanson, S. Wallace and J. L. Dehmer, Phys. Rev. Lett. 45, 1393 (1980).
83. H. Friedrich, B. Pittel, P. Rabe, W. H. E. Schwarz and B. Sonntag, J. Phys. B 13, 25 (1980).
84. A. Gerwer, C. Asaro, B. V. McKoy and P. W. Langhoff, J. Chem. Phys. 72, 713 (1980).
85. F. A. Grimm, T. A. Carlson, W. B. Dress, P. Agron, J. O. Thomson and J. W. Davenport, J. Chem. Phys. 72, 3041 (1980).

86. T. Gustafsson, Chem. Phys. Lett. 75, 505 (1980).
87. A. P. Hitchcock and C. E. Brion, J. Electron Spectrosc. 18, 1 (1980).
88. A. P. Hitchcock and C. E. Brion, J. Electron Spectrosc. 19, 231 (1980).
89. A. P. Hitchcock, C. E. Brion and M. J. van der Wiel, Chem. Phys. 45, 461 (1980).
90. R. M. Holmes and G. V. Marr, J. Phys. B 13, 945 (1980).
91. S. Krummacher, V. Schmidt and F. Wuilleumier, J. Phys. B 13, 3993 (1980).
92. P. W. Langhoff, T. N. Rescigno, N. Padial, G. Csanak and B. V. McKoy, J. Chim. Phys. 77, 589 (1980).
93. P. Morin, I. Nenner, P. M. Guyon, O. Dutuit and K. Ito, J. Chim. Phys. 77, 605 (1980).
94. G. Raseev, H. Le Rouzo and H. Lefebvre-Brion, J. Chem. Phys. 72, 5701 (1980).
95. B. Ritchie and B. R. Tambe, J. Phys. B 13, L225 (1980).
96. J. R. Swanson, D. Dill and J. L. Dehmer, J. Phys. B 13, L231 (1980).
97. M. Tronc, G. C. King and F. H. Read, J. Phys. B 13, 999 (1980).
98. S. Wallace, Ph.D. thesis, Boston University (1980).
99. J. B. West, A. C. Parr, B. E. Cole, D. L. Ederer, R. Stockbauer and J. L. Dehmer, J. Phys. B 13, L105 (1980).
100. T. A. Carlson, M. O. Krause, F. A. Grimm, J. D. Allen, D. Mehaffy, P. R. Keller and J. W. Taylor, J. Chem. Phys. 75, 3288 (1981).
101. T. A. Carlson, M. O. Krause, F. A. Grimm, J. D. Allen, D. Mehaffy, P. R. Keller and J. W. Taylor, Phys. Rev. A 23, 3316 (1981).
102. F. A. Grimm, J. D. Allen, T. A. Carlson, M. O. Krause, D. Mehaffy, P. R. Keller and J. W. Taylor, J. Chem. Phys. 75, 92 (1981).
103. T. Gustafsson and H. J. Levinson, Chem. Phys. Lett. 78, 28 (1981).
104. A. P. Hitchcock and C. E. Brion, J. Electron Spectrosc. 22, 283 (1981).
105. P. W. Langhoff, S. R. Langhoff, T. N. Rescigno, J. Schirmer, L. S. Cederbaum, W. Domcke and W. von Niesson, Chem. Phys. 58, 71 (1981).
106. P. W. Langhoff, B. V. McKoy, R. Unwin and A. M. Bradshaw, Chem. Phys. Lett. 83, 270 (1981).
107. D. Loomba, S. Wallace, D. Dill and J. L. Dehmer, J. Chem. Phys. 75, 4546 (1981).
108. R. R. Lucchese and B. V. McKoy, J. Phys. Chem. 85, 2166 (1981).
109. R. R. Lucchese and B. V. McKoy, J. Phys. B 14, L629 (1981).
110. N. Padial, G. Csanak, B. V. McKoy and P. W. Langhoff, Phys. Rev. A 23, 218 (1981).

111. G. Raseev, H. Lefebvre-Brion, H. Le Rouzo and A. L. Roche, J. Chem. Phys. 74, 6686 (1981).
112. J. A. Stephens, D. Dill and J. L. Dehmer, J. Phys. B 14, 3911 (1981).
113. J. R. Swanson, D. Dill and J. L. Dehmer, J. Phys. B 14, L207 (1981).
114. J. R. Swanson, D. Dill and J. L. Dehmer, J. Chem. Phys. 75, 619 (1981).
115. A. Tabche-Fouhaille, I. Nenner, P. M. Guyon and J. Delwiche, J. Chem. Phys. 75, 1129 (1981).
116. R. Unwin, I. Khan, N. V. Richardson, A. M. Bradshaw, L. S. Cederbaum and W. Domcke, Chem. Phys. Lett. 77, 242 (1981).
117. T. A. Carlson, M. O. Krause and F. A. Grimm, J. Chem. Phys. 77, 1701 (1982).
118. J. J. Delaney, I. H. Hillier and V. R. Saunders, J. Phys. B 15, 1477 (1982).
119. P. M. Dittman, D. Dill and J. L. Dehmer, J. Chem. Phys. 76, 5703 (1982).
120. T. Hayaishi, S. Iwata, M. Sasanuma, E. Ishiguro, Y. Morioka, Y. Iida and M. Nakamura, J. Phys. B 15, 79 (1982).
121. E. Ishiguro, S. Iwata, Y. Suzuki, A. Mikuni and T. Sasaki, J. Phys. B 15, 1841 (1982).
122. R. R. Lucchese and B. V. McKoy, Phys. Rev. A 26, 1406 (1982).
123. R. R. Lucchese, G. Raseev and B. V. McKoy, Phys. Rev. A 25, 2572 (1982).
124. L. E. Machado, E. P. Leal, G. Csanak, B. V. McKoy and P. W. Langhoff, J. Electron Spectrosc. 25, 1 (1982).
125. A. C. Parr, D. L. Ederer, J. B. West, D. M. P. Holland and J. L. Dehmer, J. Chem. Phys. 76, 4349 (1982).
126. S. H. Southworth, C. M. Truesdale, P. Kobrin, D. Lindle, W. D. Brewer and D. A. Shirley, J. Chem. Phys. 76, 143 (1982).
127. S. Wallace, D. Dill and J. L. Dehmer, J. Chem. Phys. 76, 1217 (1982).
128. C. E. Brion and K. H. Tan, J. Electron Spectrosc. 23, 1 (1981).
129. J. L. Dehmer, A. C. Parr, S. Wallace and D. Dill, Phys. Rev. A 26, 3283 (1982).
130. P. M. Dittman, D. Dill and J. L. Dehmer, Chem. Phys., in press.
131. P. R. Keller, D. Mehaffy, J. W. Taylor, F. A. Grimm and T. A. Carlson, J. Electron Spectrosc. 27, 223 (1982).
132. P. Morin, I. Nenner, M. Y. Adam, M. J. Hubin-Franskin, J. Delwiche, H. Lefebvre-Brion and A. Giusti-Suzor, Chem. Phys. Lett., in press.
133. P. Morin, I. Nenner, P. M. Guyon, L. F. A. Ferreira and K. Ito, Chem. Phys. Lett., in press.

134. I. Nenner, P. Morin, L. F. A. Ferreira, P. M. Guyon, K. Ito, K. Kollman and O. Dutuit, to be published.
135. D. Dill and J. L. Dehmer, J. Chem. Phys. 61, 692 (1974).
136. P. W. Langhoff, in: *Electron-Molecule and Photon-Molecule Collisions*, T. N. Rescigno, B. V. McKoy and B. Schneider, eds. (Plenum, New York, 1979), p. 183.
137. U. Fano, Phys. Rev. A 2, 353 (1970).
138. D. Dill, Phys. Rev. A 6, 160 (1972).
139. G. Herzberg and Ch. Jungen, J. Mol. Spec. 41, 425 (1972).
140. O. Atabek, D. Dill and Ch. Jungen, Phys. Rev. Lett. 33, 123 (1974).
141. U. Fano, J. Opt. Soc. Am. 65, 979 (1975).
142. Ch. Jungen and O. Atabek, J. Chem. Phys. 66, 5584 (1977).
143. D. Dill and Ch. Jungen, J. Phys. Chem. 84, 2116 (1980).
144. Ch. Jungen and D. Dill, J. Chem. Phys. 73, 3338 (1980).
145. Ch. Jungen, J. Chim. Phys. 77, 27 (1980).
146. A. Giusti-Suzor and H. Lefebvre-Brion, Chem. Phys. Lett. 76, 132 (1980).
147. M. Raoult, Ch. Jungen and D. Dill, J. Chim. Phys. 77, 599 (1980).
148. M. Raoult and Ch. Jungen, J. Chem. Phys. 74, 3388 (1981).
149. A. Giusti-Suzor, in: *Electronic and Atomic Collisions*, S. Datz, ed. (North-Holland, Amsterdam, 1982), p. 381.
150. G. Raseev and H. Le Rouzo, Phys. Rev. A 27, 268 (1983).
151. *Synchrotron Radiation: Techniques and Applications*, C. Kunz, ed. (Springer-Verlag, Berlin, 1979).
152. *Synchrotron Radiation Research*, H. Winick and S. Doniach, eds. (Plenum, New York, 1980).
153. V. Schmidt, App. Opt. 19, 4080 (1980).
154. E. M. Rowe, Physics Today 34(5), 28 (1981).
155. F. J. Wuilleumier, in: *Atomic Physics*, vol. 7, D. Kleppner and F. M. Pipkin, eds. (Plenum, New York, 1981), p. 491.
156. R. P. Madden and A. C. Parr, App. Opt. 21, 179 (1982).
157. M. G. White, R. A. Rosenberg, G. Gabor, E. D. Poliakoff, G. Thornton, S. H. Southworth and D. A. Shirley, Rev. Sci. Instrum. 50, 1268 (1979).
158. U. Heinzmann, App. Opt. 19, 4086 (1980).
159. A. C. Parr, R. Stockbauer, B. E. Cole, D. L. Ederer, J. L. Dehmer and J. B. West, Nucl. Instr. Meth. 172, 357 (1980).
160. M. O. Krause, T. A. Carlson and P. R. Woodruff, Phys. Rev. A 24, 1374 (1981).
161. H. Derenbach, R. Malutzki and V. Schmidt, Nucl. Intsr. Meth. 208, 845 (1983).
162. P. Morin, M. Y. Adam, I. Nenner, J. Delwiche, M. J. Hubin-Franskin and P. Lablanquie, Nucl. Instr. Meth. 208, 761 (1983).
163. A. C. Parr, S. H. Southworth, J. L. Dehmer and D. M. P. Holland, Nucl. Instr. Meth. 208, 767 (1983).

164. G. V. Marr and P. R. Woodruff, J. Phys. B 9, L377 (1976).
165. P. R. Woodruff and G. V. Marr, Proc. R. Soc. Lond. A 358, 87 (1977).
166. K. Codling, A. C. Parr, D. L. Ederer, R. Stockbauer, J. B. West, B. E. Cole and J. L. Dehmer, J. Phys. B 14, 657 (1981).
167. D. L. Ederer, A. C. Parr, B. E. Cole, R. Stockbauer, J. L. Dehmer, J. B. West and K. Codling, Proc. R. Soc. Lond. A 378, 423 (1981).
168. A. C. Parr, D. L. Ederer, B. E. Cole, J. B. West, R. Stockbauer, K. Codling and J. L. Dehmer, Phys. Rev. Lett. 46, 22 (1981).
169. J. B. West, K. Codling, A. C. Parr, D. L. Ederer, B. E. Cole, R. Stockbauer and J. L. Dehmer, J. Phys. B 14, 1791 (1981).
170. A. C. Parr, D. L. Ederer, J. L. Dehmer and D. M. P. Holland, J. Chem. Phys. 77, 111 (1982).
171. P. H. Doolittle and R. I. Schoen, Phys. Rev. Lett. 14, 348 (1965).
172. J. Berkowitz and W. A. Chupka, J. Chem. Phys. 51, 2341 (1969).
173. J. E. Collin and P. J. Natalis, Int. J. Mass Spectrom. Ion Phys. 2, 231 (1969).
174. J. L. Bahr, A. J. Blake, J. H. Carver, J. L. Gardner and V. Kumar, J. Quant. Spectrosc. Radiat. Transfer 11, 1839 (1971); 11, 1853 (1971); 12, 59 (1972).
175. T. A. Carlson, Chem. Phys. Lett. 9, 23 (1971).
176. J. L. Gardner and J. A. R. Samson, J. Electron Spectrosc. 2, 153 (1973).
177. K. Tanaka and I. Tanaka, J. Chem. Phys. 59, 5042 (1973).
178. J. L. Gardner and J. A. R. Samson, Chem. Phys. Lett. 26, 240 (1974).
179. G. Caprace, J. Delwiche, P. Natalis and J. E. Collin, Chem. Phys. 13, 43 (1976).
180. P. Natalis, J. Delwiche, J. E. Collin, G. Caprace and M. -T. Praet, Chem. Phys. Lett. 49, 177 (1977).
181. J. L. Gardner and J. A. R. Samson, J. Electron Spectrosc. 13, 7 (1978).
182. J. H. D. Eland, J. Chem. Phys. 72, 6015 (1980).
183. V. Kumar and E. Krishnakumar, J. Electron Spectrosc. 22, 109 (1981).
184. T. A. Carlson and A. E. Jonas, J. Chem. Phys. 55, 4913 (1971).
185. R. Morgenstern, A. Niehaus and M. W. Ruf, in: *Electronic and Atomic Collisions*, L. M. Branscomb, *et al.*, eds. (North-Holland, Amsterdam, 1971), p. 167.
186. T. A. Carlson and G. E. McGuire, J. Electron Spectrosc. 1, 209 (1972/73).
187. A. Niehaus and M. W. Ruf, Z. Phys. 252, 84 (1972).

188. W. H. Hancock and J. A. R. Samson, J. Electron Spectrosc. 9, 211 (1976).
189. D. M. Mintz and A. Kuppermann, J. Chem. Phys. 69, 3953 (1978).
190. S. Katsumata, Y. Achiba and K. Kimura, J. Electron Spectrosc. 17, 229 (1979).
191. M. H. Kibel, F. J. Leng and G. L. Nyberg, J. Electron Spectrosc. 15, 281 (1979).
192. J. Kreile and A. Schweig, J. Electron Spectrosc. 20, 191 (1980).
193. J. Kreile, A. Schweig and W. Thiel, Chem. Phys. Lett. 79, 547 (1981).
194. J. N. Bardsley and F. Mandl, Rep. Prog. Phys. 31, 472 (1968).
195. G. J. Schulz, Rev. Mod. Phys. 45, 422 (1973).
196. N. F. Lane, Rev. Mod. Phys. 52, 29 (1980).
197. T. Gustafsson and E. W. Plummer, in: *Photoemission and the Electronic Properties of Surfaces*, B. Feuerbacher, B. Fitton and R. F. Willis, eds. (J. Wiley, New York, 1978), p. 353.
198. T. Gustafsson, Surface Science 94, 593 (1980).
199. T. Åberg and J. L. Dehmer, J. Phys. B 6, 1450 (1973).
200. M. S. Child, *Molecular Collision Theory* (Academic Press, New York, 1974), p. 51.
201. M. Krauss and F. H. Mies, Phys. Rev. A 1, 1592 (1970).
202. R. de L. Kronig, Z. Physik 70, 317 (1931); 75, 191 (1932).
203. B. Kincaid and P. Eisenberger, Phys. Rev. Lett. 34, 1361 (1975).
204. See, for example, R. G. Newton, *Scattering Theory of Waves and Particles* (McGraw-Hill, New York, 1966), p. 457.
205. D. M. Chase, Phys. Rev. 104, 838 (1956).
206. D. L. Ederer, B. E. Cole and J. B. West, Nucl. Instr. Meth. 172, 185 (1980).
207. J. A. R. Samson and A. F. Starace, J. Phys. B 8, 1806 (1975).
208. V. G. Horton, E. T. Arakawa, R. N. Hamm and M. W. Williams, App. Opt. 8, 667 (1969).
209. G. V. Marr and J. B. West, At. Data Nucl. Data Tables 18, 497 (1976).
210. D. M. P. Holland, A. C. Parr, D. L. Ederer, J. L. Dehmer and J. B. West, Nucl. Instr. Meth. 195, 331 (1982).
211. L. S. Cederbaum, W. Domcke, J. Schirmer and W. von Niessen, Phys. Scr. 21, 481 (1980).
212. U. Fano, Nuovo Cimento 12, 156 (1935).
213. U. Fano, Phys. Rev. 124, 1866 (1961).
214. V. H. Dibeler and J. A. Walker, J. Opt. Soc. Am. 57, 1007 (1967).
215. V. H. Dibeler and S. K. Liston, Inorg. Chem. 7, 1742 (1968).

216. W. A. Chupka and J. Berkowitz, J. Chem. Phys. 51, 4244 (1969).
217. V. H. Dibeler and J. A. Walker, Int. J. Mass Spectrom. Ion Phys. 11, 49 (1973).
218. K. E. McCulloh, J. Chem. Phys. 59, 4250 (1973).
219. P. M. Dehmer and W. A. Chupka, J. Chem. Phys. 62, 4525 (1975).
220. P. M. Dehmer and W. A. Chupka, J. Chem. Phys. 65, 2243 (1976).
221. J. Berkowitz and J. H. D. Eland, J. Chem. Phys. 67, 2740 (1977).
222. P. M. Dehmer and W. A. Chupka, Argonne National Laboratory Report ANL-77-65 (1977), p. 28.
223. P. Gurtler, V. Saile and E. E. Koch, Chem. Phys. Lett. 48, 245 (1977).
224. P. M. Dehmer and W. A. Chupka, Argonne National Laboratory Report ANL-78-65 (1978), pp. 4, 10, 13, 19.
225. J. Berkowitz, *Photoabsorption, Photoionization and Photoelectron Spectroscopy* (Academic Press, New York, 1979), Chap. 6.
226. P. M. Dehmer, P. S. Dardi and W. A. Chupka, Argonne National Laboratory Report ANL-79-65, p. 7.
227. Y. Ono, E. A. Osuch and C. Y. Ng, J. Chem. Phys. 76, 3905 (1982).
228. C. Y. R. Wu and C. Y. Ng, J. Chem. Phys. 76, 4406 (1982).
229. R. S. Berry and S. E. Nielsen, Phys. Rev. A 1, 383 and 395 (1970); 4, 865 (1971).
230. C. Duzy and R. S. Berry, J. Chem. Phys. 64, 2431 (1976).
231. U. Fano, Physics Today 29(9), 32 (1976).
232. S. Southworth, W. D. Brewer, C. M. Truesdale, P. H. Kobrin, D. W. Lindle and D. A. Shirley, J. Electron Spectrosc. 26, 43 (1982).
233. J. J. Hopfield, Phys. Rev. 35, 1133 (1930); 36, 789 (1930).
234. J. C. Miller and R. N. Compton, J. Chem. Phys. 75, 22 (1981).
235. J. C. Miller and R. N. Compton, J. Chem. Phys. 75, 2020 (1981).
236. J. H. Glownia, S. J. Riley, S. D. Colson, J. C. Miller and R. N. Compton, J. Chem. Phys. 77, 68 (1982).
237. J. Kimman, P. Kruit and M. J. van der Wiel, Chem. Phys. Lett. 88, 576 (1982).
238. J. C. Miller, R. N. Compton, T. E. Carney and T. Baer, J. Chem. Phys. 76, 5648 (1982).
239. M. G. White, M. Seaver, W. A. Chupka and S. D. Colson, Phys. Rev. Lett. 49, 28 (1982).
240. J. C. Miller and R. N. Compton, Chem. Phys. Lett. 93, 453 (1982).
241. S. T. Pratt, E. D. Poliakoff, P. M. Dehmer and J. L. Dehmer, J. Chem. Phys. 78, 65 (1983).

242. E. D. Poliakoff, J. L. Dehmer, D. Dill, A. C. Parr, K. H. Jackson and R. N. Zare, Phys. Rev. Lett. 46, 907 (1981).
243. E. D. Poliakoff, J. L. Dehmer, A. C. Parr and G. E. Leroi, J. Chem. Phys. 77, 5243 (1982).
244. J. L. Dehmer and D. Dill, Phys. Rev. A 18, 164 (1978).
245. U. Heinzmann, B. Osterheld, F. Schafers and G. Schonhense, J. Phys. B 14, L79 (1981).
246. C. J. Danby and J. H. D. Eland, Int. J. Mass Spectrom. Ion Phys. 8, 153 (1972).
247. B. Brehm, J. H. D. Eland, R. Frey and A. Kustler, Int. J. Mass Spectrom. Ion Phys. 12, 197 (1973).
248. R. Stockbauer, J. Chem. Phys. 58, 3800 (1973).
249. R. Stockbauer and M. G. Inghram, J. Chem. Phys. 62, 4862 (1975).
250. D. M. Mintz and T. Baer, J. Chem. Phys. 65, 2407 (1976).
251. R. Stockbauer and M. G. Inghram, J. Chem. Phys. 65, 4081 (1976).
252. R. Stockbauer, Int. J. Mass Spectrom. Ion Phys. 25, 89 (1977).
253. T. Baer, in: *Gas Phase Ion Chemistry*, M. T. Bowers, ed. (Academic Press, New York, 1979), Chap. 5.
254. R. Stockbauer, Adv. Mass Spectrom. 8, 79 (1980).

PHOTODISSOCIATION DYNAMICS OF GAS-PHASE SMALL MOLECULES

M. N. R. Ashfold

School of Chemistry
Bristol University
Bristol, BS8 1TS, United Kingdom

1. INTRODUCTION

Any detailed discussion of the vacuum ultraviolet (VUV) photophysics of simple polyatomic molecules must involve consideration of the strong inter-relation between photochemistry and spectroscopy. The present paper concentrates largely on just one photophysical aspect -- photodissociation (i.e., molecular fragmentation as a result of photon absorption) -- since this represents one of the more probable, more interesting and, indeed, more studied consequences of the interaction of VUV photons with gas-phase small molecules. Thus, in the specific examples considered below, neither fluorescence nor photoionization significantly compete with photodissociation as the dominant decay process from the parent-molecule excited states prepared by photon absorption.

Traditionally, photodissociation has been represented most simply in terms of the process

$$AB + h\nu \rightarrow A + B \ .$$

While there was a time, not so long ago, when the limit of a photochemist's aspiration involved merely the identification of the product fragments A and B, the combined efforts of present-day experiment and theory now render possible a fairly complete picture of the detailed dynamical aspects of the dissociation. In general, the absorbed VUV photon will carry energy $h\nu$ which is greater than that required to break the A-B bond, $D_0^o(A-B)$; often, in fact, the excess energy will be sufficient to allow formation of one (or even both) of the fragments in electronically excited

S. P. McGlynn et al. (eds.), Photophysics and Photochemistry in the Vacuum Ultraviolet, 409–465.

states, which may then spontaneously fluoresce. In pre-laser days, the observation of this photofragment fluorescence represented the only possible route to a study of photodissociation dynamics under anything even approaching collision-free conditions. This explains, in part at least, the traditional emphasis placed on studies of small molecule photodissociation as a result of VUV (rather than less energetic near-UV or even visible) photoexcitation. Important photochemical and spectroscopic information may be derived from measurements of the branching ratios for the various energetically allowed decomposition pathways which follow photon absorption; studies of this kind provide a useful point of entry into a survey of the VUV photophysics of a few selected gas-phase small molecules.

Having characterized some, or preferably all, of the fragmentation channels, further insight into the dynamics of any particular dissociation process requires completion of the detailed energy balance between reactants (parent molecule plus the photon) and the resulting photofragments. Overall energy conservation requires that the excess energy be partitioned into relative translational motion of the recoiling fragments and/or into their respective internal quantum states: That is,

$$E_{int}^{AB} + h\nu = E_{int}^{A} + E_{int}^{B} + \frac{\underset{\sim}{P}^2}{2\mu_{AB}} + D_o^o(A-B)$$

where the relative translational energy of the products is expressed in terms of the relative linear momentum $\underset{\sim}{P}$. Angular momentum conservation can influence the pattern of energy disposal through the requirement that

$$\underset{\sim}{J}_{AB*} = \underset{\sim}{J}_A + \underset{\sim}{J}_B + \underset{\sim}{L}$$

where $\underset{\sim}{J}_{AB*}$, $\underset{\sim}{J}_A$ and $\underset{\sim}{J}_B$ are rotational angular momenta of the photoexcited molecule and the resulting photofragments, respectively; and where $\underset{\sim}{L}$ is the relative orbital angular momentum of the recoiling fragments. As we proceed, the importance of angular momentum conservation in constraining the partitioning of excess energy amongst the various product degrees of freedom will become increasingly apparent.

Product-state detection methods are now sufficiently advanced as to enable precise determination of the primary energy disposal pathways in a number of photodissociations. For example, translational energy disposal may be derived through measurement of the fragment velocity distributions using molecular beams and time-of-flight methods. Internal vibrational, and even rotational, state distributions have been obtained by monitoring and analyzing (i), the wavelength-resolved spontaneous fluorescence from electronically excited product species; or (ii), the laser-induced fluorescence (LIF) from the ground state photofragments, under collision-free conditions. Nozzle expansion techniques are beginning to find a role in narrowing the thermal spread of energies E_{int}^{AB} and rotational states J_{AB*} from which a given dissociation proceeds. Nevertheless, any attempt to provide a detailed interpretation of the observed energy partitioning is likely to remain somewhat speculative in the absence of reliable information concerning the properties of that excited state of the parent molecule from which dissociation occurs.

Thus, we must necessarily concern ourselves with some of the various types of study that attempt to probe the nature of the excited state (or states) populated through VUV photon absorption. One natural starting point involves analysis of the VUV absorption spectrum of the parent molecule. This, as we shall see, tends to be no easy matter since, even in simple molecules, many of the electronic transitions at these energies give rise to broad continua that can only be assigned through recourse to accurate theoretical calculations. Even in those instances where population of the excited state gives rise to a structured spectrum, it is unusual for the predissociation rate out of the initially prepared state to be slow enough that the rotational structure can be resolved sufficiently to permit assignment of the orbital symmetry and excited-state geometry. The fact that both intravalence and Rydberg excitations can be expected in this wavelength region further complicates spectral analysis. Preparation of excited states possessing substantial Rydberg character involves electronic promotions to large, diffuse molecular orbitals which, to all intents, may be envisaged as centred on an "atom-like" molecular core; thus, details of the molecular geometry may be expected to have relatively little effect on the orbital character.

Adopting the opposite viewpoint, it follows that the more spatially extensive the Rydberg orbital, the less influence it will have on molecular bonding; and, the photoexcited molecule will therefore tend towards the equilibrium geometry associated with the molecular ion. If the ion itself is stable with respect to dissociation, it is not unreasonable to suppose that the higher-lying Rydberg states of the neutral molecule will also be comparatively stable. Thus, Rydberg transitions frequently manifest themselves as sharp, well-defined features in VUV absorption spectra (cf. Fig. 1), especially in those situations in which the Rydberg state geometry differs appreciably from that associated with neighboring, nearly isoenergetic (dissociative) intravalence states since, in these circumstances, Franck-Condon considerations may restrict the predissociation rate from the initially prepared Rydberg state.

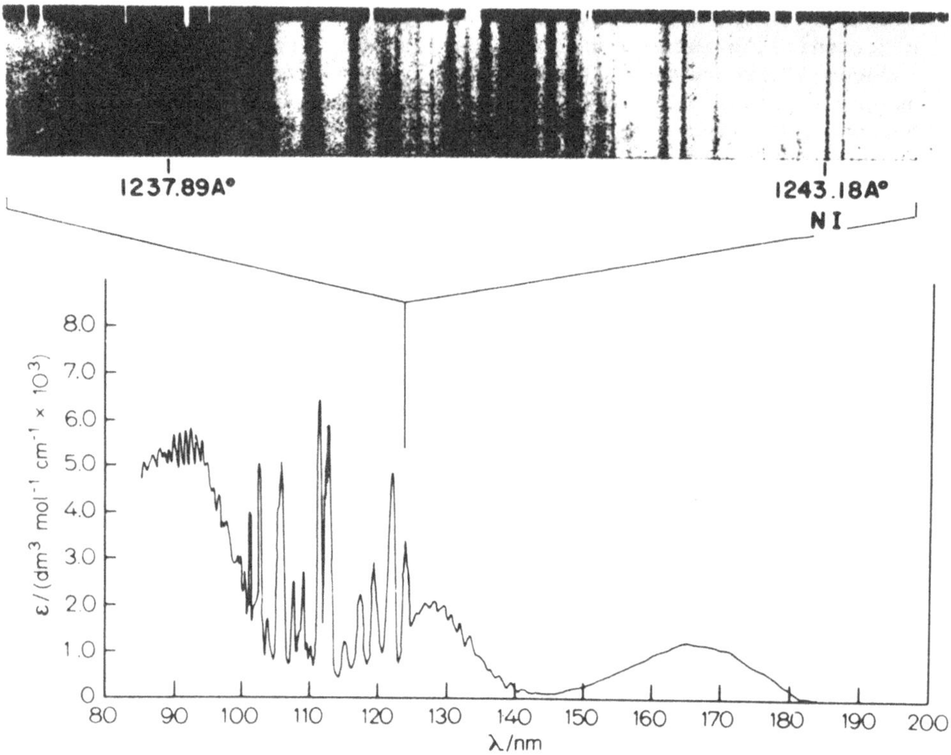

Fig. 1. Vacuum ultraviolet absorption spectrum of H_2O (after [1]), together with an illustration of the resolved rotational structure associated with the 0-0 band of the $\tilde{C}\ ^1B_1 \leftarrow \tilde{X}$ transition at ∿124 nm (after [2]).

Complementary data for the excited states of the parent molecule can frequently be obtained through monitoring the wavelength dependence of the yields of the various primary photofragments. Judicious application of state-to-state correlation diagrams, linking the parent molecule and the various product channels, can aid assignment of excited-state symmetries. In what follows, we shall see that analysis of the collision-free photofragment angular distributions, following dissociation brought about by an anisotropic photoexcitation (through use of a linearly-polarized laser pulse, for example), can provide information about the excited-state symmetry, geometric configuration and lifetime with respect to fragmentation. In cases where an electronically excited species is produced, much the same information can be derived through analysis of the polarization of the resulting photofragment fluorescence.

Further elucidation of the VUV absorption spectra of small molecules can be anticipated as a result of the ever increasing experimental and theoretical interest being shown in multiphoton ionization (MPI) processes. A much enhanced ion yield is obtained whenever the overall MPI occurs via a long-lived resonant intermediate excited state of the neutral parent molecule. Thus, MPI spectra tend to reveal the structure due to Rydberg states without interference from those (predominantly dissociative) intravalence states that lie in the same wavelength region and that normally complicate the more conventionally obtained VUV absorption spectrum [3-5].

In the remainder of this article, we shall concentrate on studies of the photofragmentation of a few specific molecules which, we hope, will serve to demonstrate the applicability of many of the above techniques to VUV photodissociation investigations. In no way do we intend that this work attempt to compete with any of the now numerous and comprehensive recent reviews of the VUV photochemistry and spectroscopy of small molecules [6-8], or of their photofragmentation dynamics [9-13]. Rather, we hope to provide (i), a feel for the kind of experimental information that is now obtainable, both in practice and in principle; (ii), the significance of these results in interpreting some of the finer details regarding the dynamics of photodissociation processes; and (iii), some indication of the likely, and the desirable, directions that this kind of research should take.

The first of three illustrative case studies to be discussed below involves photofragmentation brought about using near-UV excitation wavelengths; this selection largely reflects the continuing shortage of readily available, and sufficiently intense, tunable sources of monochromatic VUV radiation.

2. ICN PHOTODISSOCIATION IN THE NEAR ULTRAVIOLET

This temporary excursion into the near-UV can be justified by the fact that this system, probably more than any other, has served to illustrate many of the various experimental and theoretical types of investigation currently available. That some of the studies have lead to conflicting conclusions merely serves to demonstrate that in photodissociation processes -- as in so many other fields -- that which might at first appear straightfoward can, when probed in greater depth, reveal many hidden subtleties.

The first absorption feature in the electronic spectrum of ICN comprises an apparently structureless, weak continuum (the so-called $\tilde{A}$-continuum) which spans the wavelength region 220-300 nm (cf. Fig. 2). The current consensus associates this absorption with a purely intravalence $\pi \rightarrow \pi^*$ transition from the linear

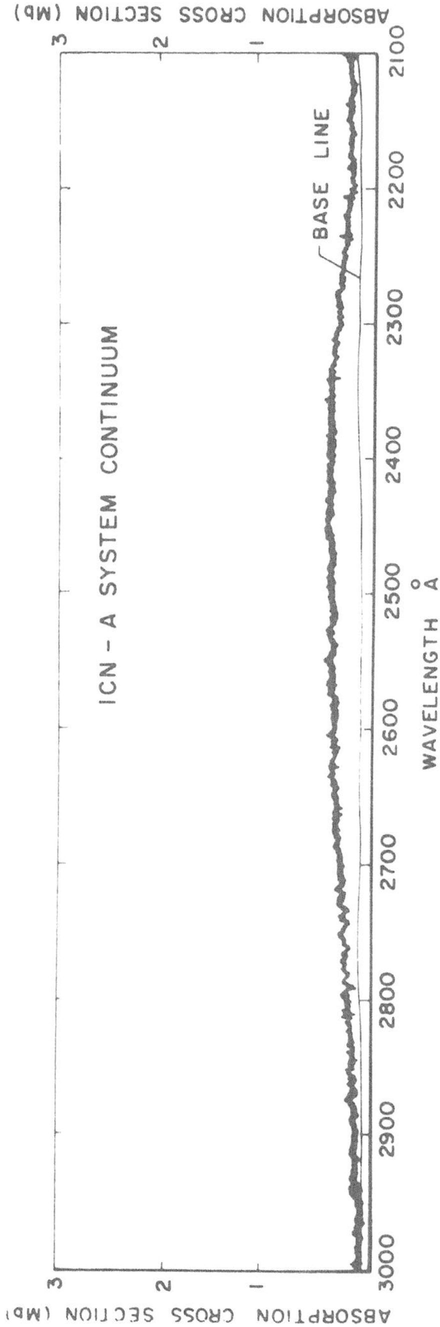

Fig. 2. Near ultraviolet absorption of ICN. (Taken from [14].)

Table 1. $\pi^3\pi$ Excited-State Symmetries for Linear and Bent Geometries of ICN

Point Group	Molecular Term			
$C_{\infty v}$ (linear)	Δ		Σ^+	Σ^-
	↙	↘	↓	↓
C_s (bent)	A'	A''	A'	A''

ground state of ICN, the electronic configuration of which may be represented as†

$$....(1\pi)^4(6\sigma)^2(2\pi)^4 \ .$$

If we assume that a Russell-Saunders (Λ,S) description is the appropriate scheme for the coupling of electron spin and orbital angular momenta in ICN, this electronic promotion will result in the possible population of singlet and triplet Δ, Σ^+ and Σ^- states. For a sixteen electron molecule such as ICN, Walsh's rules [15] predict that the excited states may be bent, in which case further complications will likely be introduced by Renner-Teller splitting of the degenerate Δ states. Table 1 provides a correlation between the excited $\pi^3\pi$ electronic state representations for both linear and bent molecular configurations. However, it may be more appropriate to consider the excited states of a relatively heavy molecule such as ICN in terms of a core ($2\pi^3$), within which there is strong spin-orbit coupling, and an outer electron (3π). The projection of the total electronic angular momentum of each system (core and outer electron) along the molecular axis is quantized and is represented by Ω_c and ω, respectively. The axial projection of the *total* angular momentum Ω is then derived from the coupling of Ω_c and ω. A schematic correlation between the excited states derived using (Λ,S) and (Ω_c,ω) coupling schemes for the $\pi^3\pi$ configuration of the linear ICN molecule is presented in Fig. 3. Note that the latter coupling rules yield a total of ten possible excited states. Little of this potential complexity is apparent from just a simple glance at the ICN absorption spectrum, however.

---

†*Note added in proof*: Unpublished CI calculations [64] presented at this NATO-ASI strongly suggest that the $\tilde{A}$-continuum in ICN arises, in fact, via a $\pi \rightarrow \sigma^*$ transition, thus vindicating one earlier spectroscopic study [43] in which this assignment was offered on the basis of a comparision with other small halogen-containing molecules.

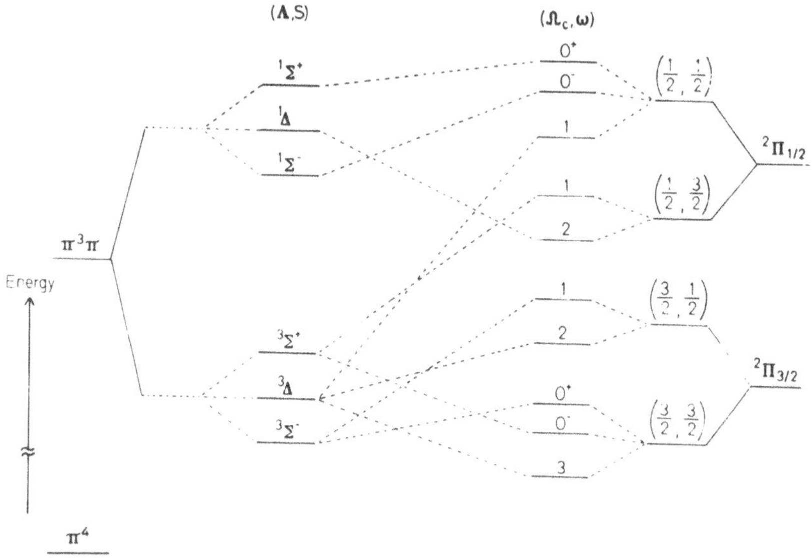

Fig. 3. Schematic correlation between (Λ,S) and (Ω_c,ω) coupling for a $\pi^3\pi$ configuration. Energy differences are not to scale.

The I-CN bond dissociation energy is 304 ± 5 kJ mol^{-1}. Thus, the following product channels are *energetically* accessible in the near-UV photofragmentation of ICN:

$$\mathrm{ICN}\begin{cases}\rightarrow \mathrm{I}(^2P_{3/2}) + \mathrm{CN}(X^2\Sigma^+) & \lambda < 398 \text{ nm}\\ \rightarrow \mathrm{I}^*(^2P_{1/2}) + \mathrm{CN}(X^2\Sigma^+) & \lambda < 306 \text{ nm}\\ \rightarrow \mathrm{I}(^2P_{3/2}) + \mathrm{CN}^*(A^2\Pi_i) & \lambda < 292 \text{ nm}\\ \rightarrow \mathrm{I}^*(^2P_{1/2}) + \mathrm{CN}^*(A^2\Pi_i) & \lambda < 239 \text{ nm}\ .\end{cases}$$

ICN was one of the first polyatomic molecules to be investigated by the technique of *photofragment spectroscopy*, as pioneered by Wilson and co-workers [16]. In these experiments, the pulsed output of a linearly-polarized, frequency-quadrupled Nd-YAG laser was used to photodissociate a molecular beam of ICN molecules at 266 nm. The resultant CN and I fragment densities, which reached the ionization region of a mass spectrometer located a fixed distance along an axis orthogonal to the crossed laser/molecular-beam interaction region, were recorded separately and selectively as a function of (i), the time after the 10 ns photodissociating

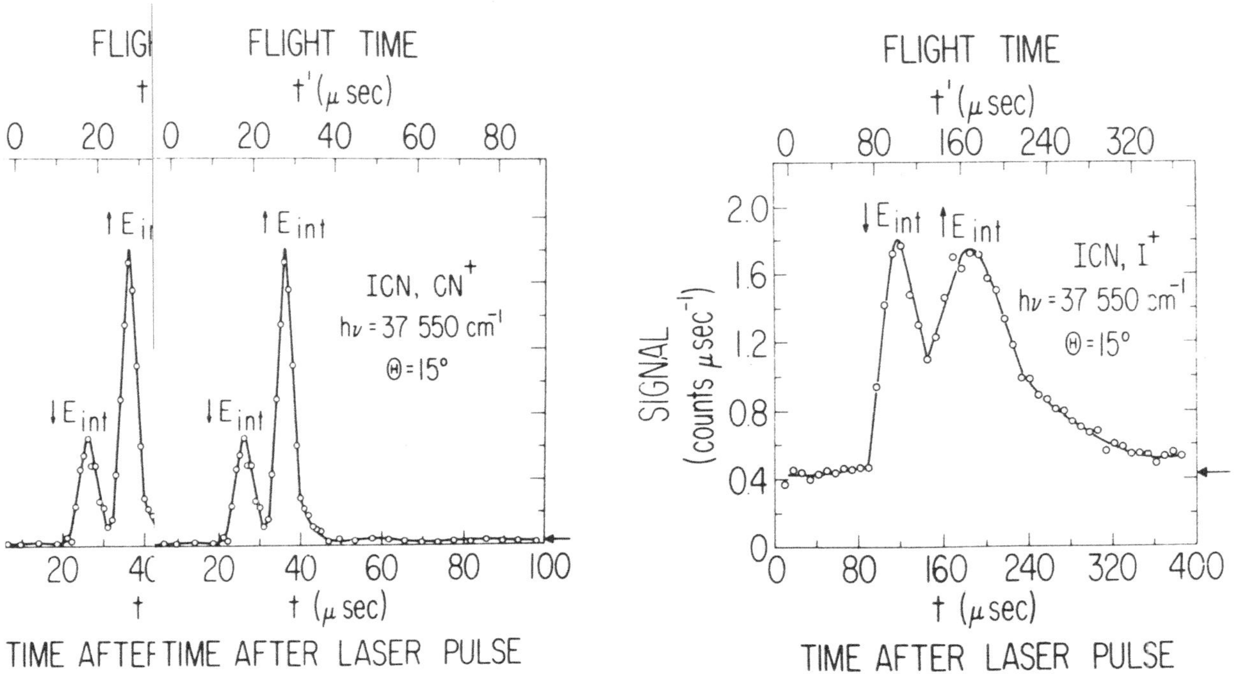

Fig. 4. Time-of-flight distributions of (a) CN and (b) I fragments (detected at a laboratory recoil angle of 15°) in the 266 nm photodissociation of ICN. The early time $\downarrow E_{int}$ peak corresponds to fragments with lower internal energy; the later $\uparrow E_{int}$ peak corresponds to fragments carrying a higher level of internal excitation. (Taken from [17].)

laser pulse (Fig. 4); and (ii), the direction of fragment recoil relative to that of the electric vector of the linearly-polarized laser beam (Fig. 5) [17]. This latter variation was achieved, with the fixed-geometry configuration, by using a quartz half-plate to rotate the $\underset{\sim}{\varepsilon}$-vector of the linearly-polarized laser beam.

Two peaks are observed in the flight-time distributions of both products. The more translationally excited set of fragments which give rise to the early time peak must, through total energy conservation, carry less internal excitation. Note that the heavier atomic iodine products display substantially longer flight times than the CN fragments, as is to be expected from linear momentum conservation in the center-of-mass frame. The peaks in

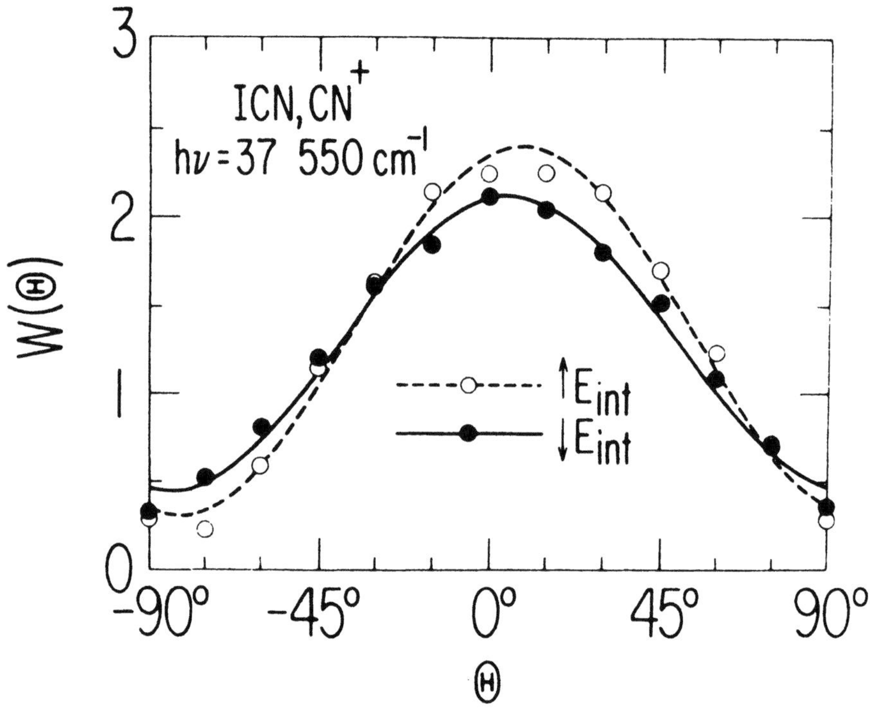

Fig. 5. Angular distribution of the recoiling CN fragments produced in the 266 nm photodissociation of ICN. The fragmentation channels yielding high ($\uparrow E_{int}$) and low ($\downarrow E_{int}$) internal energy products both exhibit laboratory angular distributions peaking at $\theta_{lab} \simeq 0^o$. (Taken from [17].)

the I atom time-of-flight spectrum are less well-resolved because the parent molecules from which the dissociation occurs possess, themselves, a distribution of velocities. Thus, the fragment recoil velocity distributions are necessarily superimposed on this range of parent molecular velocities and -- when the relevant fragment is heavy and, therefore, relatively slow moving -- this parent molecular velocity distribution will, to some extent, smear out the detail in the laboratory recoil velocity distribution of the fragments. Since D_0^o(I-CN) is known, and since the average internal energy E_{int}^{ICN} of the parent molecule is readily estimable, it should be possible to calculate how the excess energy provided by the monochromatic photoexcitation is partitioned between the product internal and translational degrees of freedom. Such a calculation requires that the fragment velocity distribution measured in the laboratory frame be transformed into a translational energy distribution in the center-of-mass frame of the molecule (cf. Fig. 6). Then, the average internal energy associated with each peak in the center-of-mass translational energy distribution can be estimated and possible product states deduced. In this particular instance (i.e., photodissociation of ICN at 266 nm), the low internal energy peak ($\downarrow E_{int}$) can only be associated with formation of I atoms and CN fragments both in their ground electronic states, with vibrational and rotational excitation within the CN(X) fragments accounting for the spread of translational energies observed. This channel accounts for ∿40% of the total fragmentation yield. Interpretation of the "slow" peak ($\uparrow E_{int}$) in the fragment time-of-flight spectrum is less clear cut; Fig. 5 reveals that the associated internal energy could allow any of three possible fragmentation channels, namely: (i), I and CN in their ground electronic states, with considerable vibrational and rotational excitation in the molecular fragment; (ii), excited I*($^2P_{1/2}$) atoms and ground-state, vibrationally unexcited CN; or (iii), electronically excited CN*$(A^2\Pi_i)_{v=0}$, together with ground-state I atoms. Alternative experimental techniques have subsequently been applied in order to determine the relative importance of these various decomposition channels (see below).

As mentioned previously, photofragment spectroscopy can also be used to provide information regarding the nature of the photoexcited state of the parent molecule. When, as in this case, linearly-polarized light is used, those molecules with electric-dipole transition moments μ aligned parallel to the electric vector ε of the polarized light will be excited preferentially. Thus, provided that (i), the subsequent photofragmentation occurs before significant molecular rotation has had the opportunity to wash out the anisotropically oriented ensemble of photoexcited molecules and (ii), the relative translational energy of the separating fragments is much greater than the rotational energy of the parent molecule, this anisotropy should be reflected in the

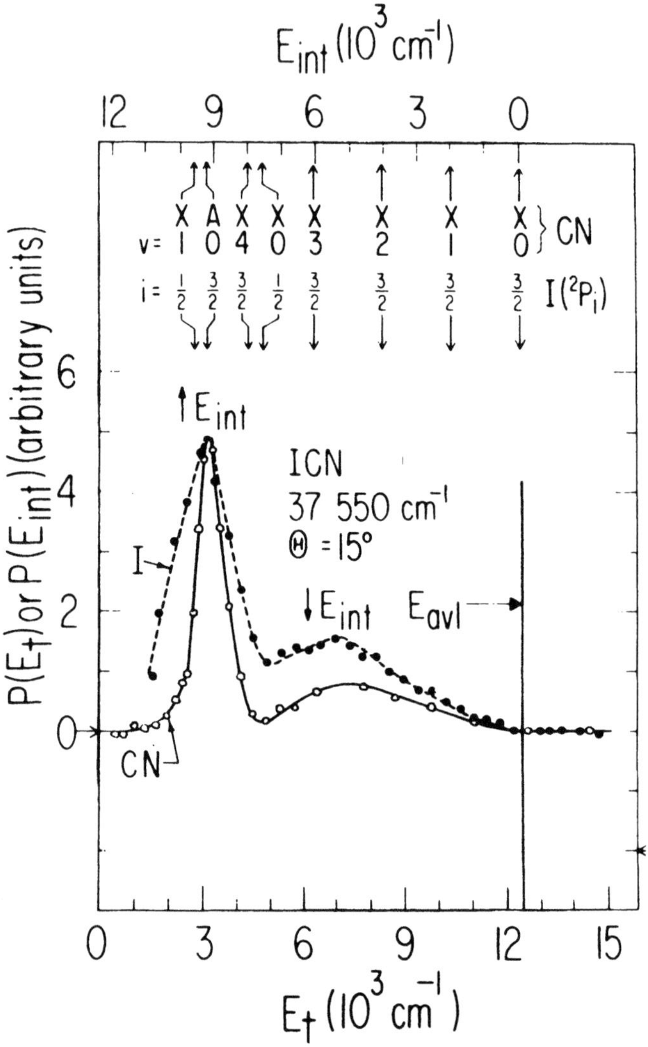

Fig. 6. Total internal and center-of-mass translational energy distributions for photofragments produced in the photodissociation of ICN at 266 nm. Arrows in the upper part of the figure represent peak positions calculated under the assumption that the I and CN fragments were formed in particular electronic and vibrational states. (Taken from [17].)

angular distribution of the resulting photofragments. In the center-of-mass coordinate system, this distribution will take the general form characteristic of a dipole radiation pattern:

$$I(\theta) = \frac{1}{4\pi} \{1 + \beta P_2(\cos\theta)\}$$

where θ is the angle between the polarization vector of the exciting beam and the direction of recoil of the separating fragments; and $P_2(\cos\theta)$ is the second-order Legendre polynomial

$$P_2(\cos\theta) = \frac{1}{2}(3\cos^2\theta - 1) .$$

The asymmetry parameter β depends upon χ, the angle between the electric-dipole transition moment $\underset{\sim}{\mu}$ and the direction of fragment recoil, through the relationship

$$\beta = c\, P_2(\cos\chi)$$

and can range in value from $\beta = +2$ ($I(\theta)$ shows a $\cos^2\theta$ distribution) to $\beta = -1$ ($I(\theta)$ shows a $\sin^2\theta$ distribution). For a diatomic molecule (except, for example, in the case of "near-threshold" dissociations, where the relative velocity of the separating fragments is sufficiently low that a significant tangential velocity contribution can arise from the rotational motion of the photoexcited molecule), fragment recoil must occur along the internuclear axis and χ is therefore restricted to the values $\chi = 0^\circ$ (a "parallel" transition) or $\chi = 90^\circ$ (when the transition moment lies perpendicular to the molecular axis). For polyatomics, however, there is no such restriction upon the direction of fragment recoil and, in general, there will be a range of χ values reflecting the distribution of excited-state geometries, and vibrational and rotational motions.

The asymmetry parameter β, through its dependence upon c, also depends upon the lifetime of the excited state. One finds $c = 2$ if the photoexcited molecules do not rotate at all prior to fragmentation; c decreases smoothly to a value of 0.5 if the photoexcited state survives for a number of rotations. Thus, the previously quoted limits to the range of possible β values are reduced by a factor of four if the rotational period is short compared to the excited-state lifetime [18] and, consequently, the observed angular distribution will be more uniform.

Fig. 5 shows the laboratory angular distribution of the CN fragments for both the $\downarrow E_{int}$ and $\uparrow E_{int}$ peaks arising in the 266 nm photodissociation of ICN. Since the relative velocities of the CN fragment are much greater than the average velocity of the ICN parents, transformation to the center-of-mass coordinate system little affects the form of the angular distributions, with both product channels exhibiting a maximum centered at $\theta \simeq 0^{\circ}$. This result implies that the photoexcited state (or states) leading to both product channels is accessed via a parallel transition, with the transition dipole moment lying predominantly along the I-CN bond in ICN, from which it follows that the excited state (or states) has $\Omega = 0$ if linear, or A' symmetry if bent. Values of the asymmetry parameters β for the two angular distributions ($\beta = 1.40$ for $\uparrow E_{int}$ and $\beta = 1.12$ for $\downarrow E_{int}$) were estimated [17] by fitting the experimental data points to the appropriate expression for $I(\theta)$ in the laboratory frame. The reduction from the theoretical value of $\beta = 2$ for a "pure" parallel process can be attributed to any of four factors: (i), parent molecular rotation; (ii), a contribution from a perpendicular transition that gives rise to the same products; (iii), a bent geometry for the photoexcited ICN; and/or (iv), lifetime of the ICN excited state. If one attributes *all* of the "lost" asymmetry to (iv), an absolute upper limit for the excited-state lifetime(s) can be estimated. This analysis yields maximum lifetimes of 0.2 ps and 0.1 ps, respectively, for the excited state or states leading to the $\downarrow E_{int}$ and $\uparrow E_{int}$ product peaks [17].

In summary, then, photofragment spectroscopy reveals that ICN photodissociation at 266 nm yields (at least) two product channels derived from one (or more) short-lived photoexcited states which, if bent, have predominantly A' symmetry or, if linear, have $\Omega = 0$. One fragment channel, which comprises ∿40% of the total dissociation yield, gives rise to I and CN fragments in their ground electronic states; the internal energy partitioning within the fragments which arise from the other 60% of dissociations cannot be unambiguously ascertained. Clearly, this information is required if we wish to gain a deeper insight into this particular photodissociation process.

Fortunately, a number of recent studies have provided much complementary information. For example, West and Berry [19], using synchrotron radiation, measured the relative quantum yield, as a function of wavelength, for formation of electronically excited $CN(A^2\Pi_i)$ fragments in the photodissociation of ICN and found *no* CN (A → X) emission following excitation at wavelengths longer than 210 nm, thus ruling out one possible attribution for the "slow" peak in the photofragment time-of-flight spectrum. Quantum yield measurements have also been used to establish the relative importance of another of the possible dissociation channels, namely,

$$ICN \rightarrow I(^2P_{1/2}) + CN(X^2\Sigma^+)$$

following excitation at various wavelengths within the ICN $\tilde{A}$-continuum [20]. By measuring the intensity of the $I(^2P_{1/2} \rightarrow {}^2P_{3/2})$ emission at 1.315 μm, following ICN photolysis at any particular wavelength, and by comparing this intensity with the $I(^2P_{1/2})$ emission intensity observed from photolysis of either iso-C_3F_7I or CF_3I (for both of which the $I(^2P_{1/2})$ quantum yields are known) under identical conditions, the quantum yield measurements can be put on an absolute scale. The product of these experimentally determined quantum yields Φ_{I*}, multiplied by the molar extinction coefficient ε_{ICN} of ICN at each wavelength, provides the absorption profile for the electronic state(s) correlating with $I(^2P_{1/2})$ + CN(X). This is shown in Fig. 7, along with the "difference" spectrum obtained by the subtraction $\varepsilon_{ICN} - \phi_{I*}\varepsilon_{ICN}$: In view of West and Berry's observation, these latter absorptions must correlate with ground electronic state I and CN fragments. The results

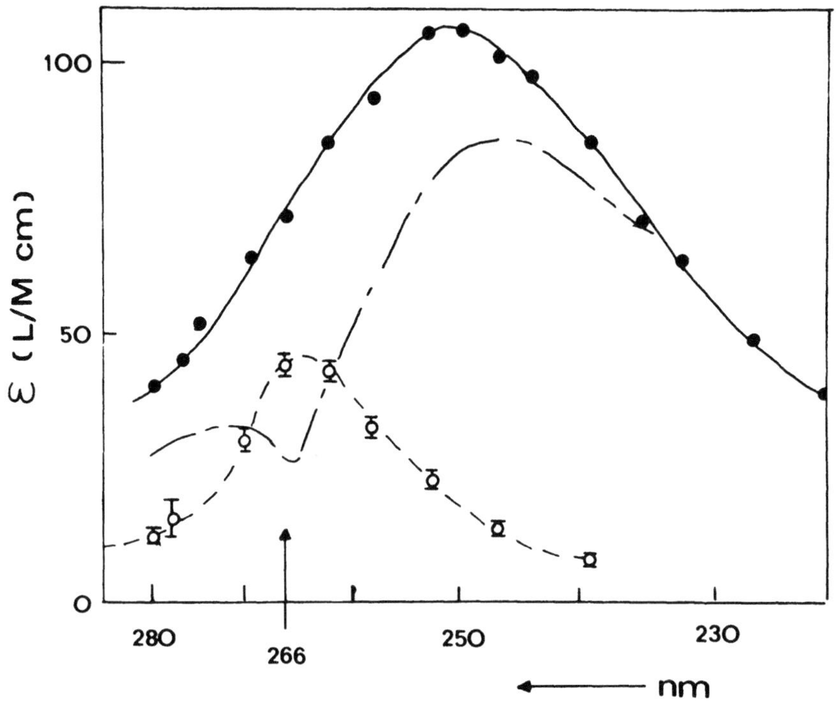

Fig. 7. Wavelength dependence of the production of $I(^2P_{1/2})$ in the photodissociation of ICN($\tilde{A}$). The solid points and curve represent the ICN absorption spectrum; open circles joined by the dashed curve show the product $\phi_{I*}\ \varepsilon_{ICN}$. Also shown is the resulting "difference" spectrum $\varepsilon_{ICN} - \phi_{I*}\ \varepsilon_{ICN}$. (Adapted from [20].)

strongly suggest, therefore, that (at least) three excited states contribute to the broad $\tilde{A}$-continuum in the ICN absorption spectrum, all of which may well be accessed by photon absorption at 266 nm. The two peaks observed in the translational photofragment spectrum, following 266 nm photodissociation of ICN, can be assigned to electronic branching in the atomic iodine product, with the channel leading to I* formation accounting for 61% of the total dissociation yield at this wavelength. Finally, the observation that the product branching ratios display such a sensitive dependence upon photolysis wavelength should sound a cautionary note in any attempt to interpret results from broadband flash photodissociation experiments.

Laser-induced fluorescence (LIF) has been used to estimate the partitioning of internal (vibrational and rotational) energy within the CN(X) fragments formed in the 266 nm photodissociation of ICN [21-23]. In these experiments, a tunable dye laser is used to selectively excite CN fragments in a specific quantum state (or states), leading to fluorescence; a plot of the resulting LIF intensity against the probing laser wavelength yields an LIF spectrum. Fig. 8 shows such a spectrum obtained following laser excitation of the CN(B $\leftarrow$ X) violet transition [22]. Nascent product state distributions can be determined by arranging for the pulsed dye laser to probe at a sufficiently short time delay after the photolysis pulse so that collisional relaxation is unimportant. Relative quantum state populations can be extracted with confidence from well-resolved spectra, such as that shown in Fig. 8, when the appropriate Franck-Condon factors and rotational line strengths are known. Analysis of this particular spectrum reveals that >95% of the CN(X) fragments formed in the collisionless 266 nm photodissociation of ICN are produced in their $v'' = 0$ level, with a markedly nonstatistical distribution of rotational state populations that peaks at very low rotational quantum numbers, $N'' \simeq 1-2$; however, the tail of the distribution extends to high rotational levels (population of the level $v'' = 0$, $N'' = 56$ can be discerned from Fig. 8) [22].

Unfortunately, in view of the previous analysis, we recognize that this rotational state distribution in the CN(X) fragment must represent the superposition of the two separate rotational state distributions resulting from the two product channels yielding $I(^2P_{3/2})$ and $I(^2P_{1/2})$, respectively. However, this does not prevent a qualitative estimate of the manner in which the total available energy is partitioned: For both of the dominant fragmentation channels, product vibrational excitation is clearly negligible; roughly 10% of the excess energy goes into CN rotation, with the balance appearing as relative translational excitation of the separating fragments [22]. This preferential partitioning of the available energy into interfragment recoil apparently occurs for ICN photodissociation throughout the $\tilde{A}$-continuum.

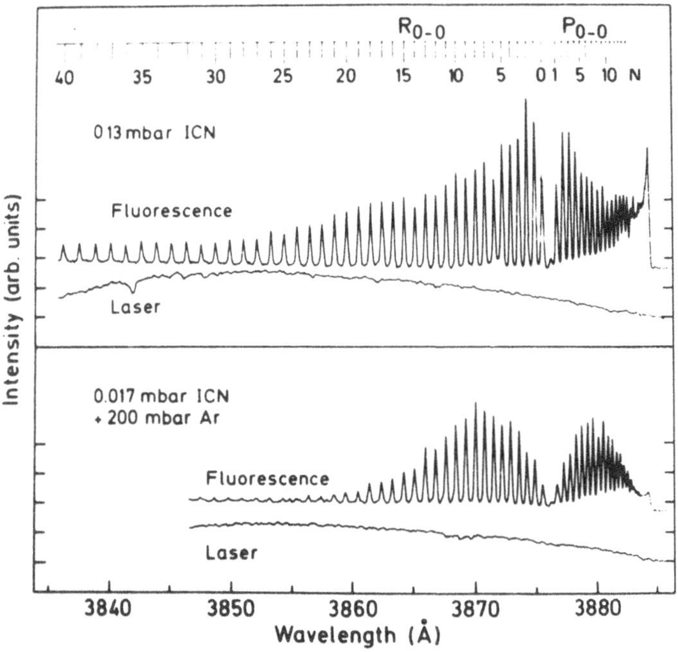

Fig. 8. Laser-induced fluorescence spectrum of CN(X) fragments produced in the photodissociation of ICN at 266 nm under collision-free conditions (upper spectrum), and collisionally relaxed by the presence of excess argon (lower spectrum). The wavelength variation of the probing dye laser intensity is plotted below each spectrum. (Taken from [22].)

LIF has been used to show that the CN(X) fragments produced in the broadband flash photolysis of ICN at λ > 220 nm exhibit a very similar overall energy partitioning amongst the various degrees of freedom [24]. Collision-free quantum state distributions for the CN(X) fragments produced in the 299.4 nm photodissociation of ICN have recently been reported [23]. From the interpretative viewpoint, excitation at this wavelength (which may be obtained as the first Stokes-shifted line in the stimulated Raman scattering of 266 nm radiation in H_2) provides two potentially significant advantages. First, the previously described quantum yield measurements [20] have been interpreted [23] as showing that ICN absorption at this wavelength is principally due to just one (admittedly uncharacterized) electronically excited state; second, the dissociation channel leading to excited iodine atom product formation is barely accessible energetically. Thus, it may be hoped that the observed product state distributions -- which, while again showing that translational fragment recoil accounts for the bulk of the available energy, also reveal a slightly higher level of vibrational excitation and more simple, Boltzmann-like rotational state distributions in the CN(X) product -- will prove more amenable to detailed theoretical analysis.

Results for the vibrational and rotational energy partitioning within the $\tilde{X}$ CN(X) fragments produced in the photodissociation of ICN in the $\tilde{A}$-continuum have proved popular with theoreticians for testing various models of the photodissociation process [25-29]. Despite the apparent wealth of experimental data for this system, the continuing inadequate characterization of the nature and geometry of the photoexcited state or states involved makes this a far from ideal test case. Discussion of the principal features of some of the models of the photodissociation process will therefore be reserved for a later section.

The following points should be stressed in conclusion. Despite the fact that considerable selectivity has been exercised in this presentation, it should be clear that the photodissociation of ICN, at 266 nm in particular, has been the subject of intensive study. For this wavelength at least, we feel confident in our identification of the primary product channels and their relative branching ratios, and we have a fair idea of the energy partitioning within these channels. However, our understanding of the dissociation process is poor, and will remain so in the absence of more precise information on the identity, geometry and, indeed, the number of photoexcited states from which these dissociation channels arise.

3. VUV PHOTODISSOCIATION OF HCN

In contrast to the extensively studied ICN photofragmentation summarized in the previous section, comparatively few investigations of the VUV photodissociation of HCN have been reported. Nevertheless, we shall see that those results which have appeared are able to provide a more complete picture of this particular photodissociation.

The first intense features in the VUV absorption spectrum of HCN lie in the 130-150 nm wavelength region. These features take the form of a progression of strong bands separated by 750 - 850 cm^{-1} (cf. Fig. 9), and have been attributed to excitation of the ν_2' bending vibrational mode in the $\tilde{C}$ electronic state [30-32]. Some of the more intense members of a second, weaker progression are apparent at shorter wavelengths; these correspond to the series $(0,v_2',1)$, in which one quantum of the HC-N stretching frequency ν_3', as well as the bending mode ν_2', have been excited.

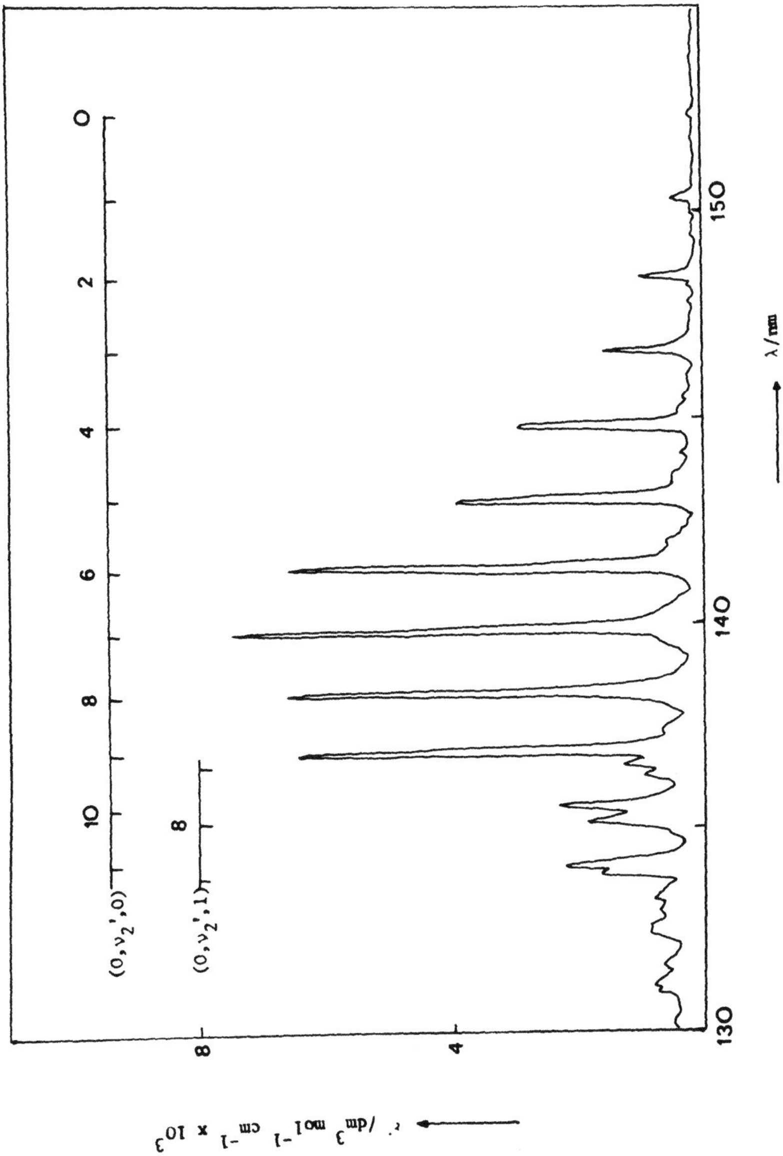

Fig. 9. Vacuum ultraviolet absorption spectrum of HCN in the 130 - 155 nm wavelength range. (Adapted from [31,32].)

Franck-Condon considerations immediately suggest that these progressions in ν_2' indicate an electronic promotion from the linear $\tilde{X}\ ^1\Sigma^+$ ground state to a *bent* excited state. However, in this instance spectroscopy can tell us far more. Despite the fact that all of the bands in the vibrational progression show some diffuseness -- indicative of predissociation in the excited state -- and that this diffuseness increases rapidly with increasing ν_2' quantum number, sufficient rotational structure for the first members of the $\tilde{C}$-$\tilde{X}$ transition has been discerned and analyzed [30] to permit identification of the symmetry (A') and equilibrium geometry ($\angle$ HCN ≃ 141°) of the photoexcited state. The ground-state electronic configuration of HCN may be represented as

$$\ldots(4\sigma)^2\ (5\sigma)^2\ (1\pi)^4 \qquad \tilde{X}\,^1\Sigma^+ \qquad (C_{\infty v})$$

or as

$$\ldots(4a')^2\ (5a')^2\ (1a'')^2\ (6a')^2 \qquad \tilde{X}\,^1A' \qquad (C_s)$$

(for a correlation, see Fig. 10). There is now reasonable consensus [8] that the $\tilde{C}$ state may be assigned to the A' component of the $^1\Pi$ state arising from the $5\sigma \rightarrow 2\pi$ (or $5a' \rightarrow 7a'$) electronic excitation; such a transition is electric-dipole allowed for both linear and bent geometries, which accounts for its strong showing in the HCN absorption spectrum.

Additional support for this analysis and assignment has been provided by studies of photofragment fluorescence excitation spectra, relative quantum yield and photofragment fluorescence polarization measurements, and by study of the isotopic species DCN. Three photodissociation channels are energetically accessible to HCN and DCN in this wavelength region:

$$\mathrm{H[D]CN}\left\{\begin{array}{lll} \rightarrow \mathrm{H[D]}\ (^2S_{1/2}) + \mathrm{CN}(X\,^2\Sigma^+) & & \lambda < 238\ \mathrm{nm} \\ \rightarrow \mathrm{H[D]}\ (^2S_{1/2}) + \mathrm{CN}(A\,^2\Pi_i) & & \lambda < 190\ \mathrm{nm} \\ \rightarrow \mathrm{H[D]}\ (^2S_{1/2}) + \mathrm{CN}(B\,^2\Sigma^+) & & \lambda < 149\ \mathrm{nm}\ . \end{array}\right.$$

The latter two fragmentation pathways yield electronically excited fragments which can be readily monitored through observation of their spontaneous fluorescence. The $CN(A^2\Pi_i)$ state fluoresces in the red and near-infrared and has a relatively long radiative lifetime (τ_{rad} ≃ 4.2 μs), while the higher energy $CN(B^2\Sigma^+)$ state radiatively decays to the ground state in the violet region of the spectrum with a lifetime of ~60 ns. *Photofragment fluores-*

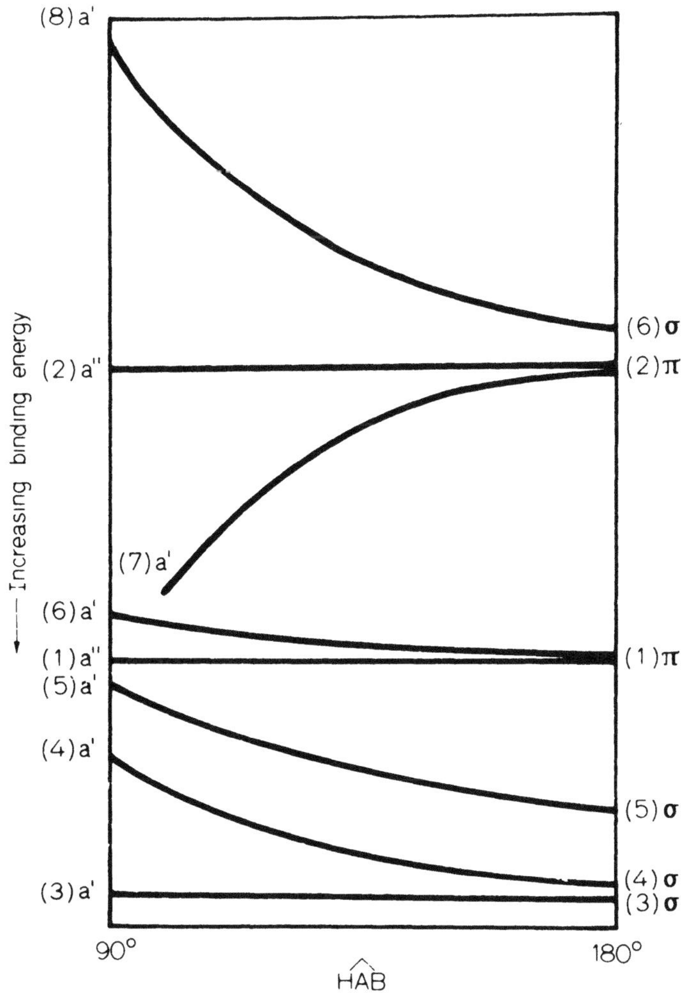

Fig. 10. Walsh diagram for an HAB molecule. (Adapted from [15].)

cence excitation spectra (PFES) have been obtained for HCN by measuring either the total CN(A→X) emission [31] or the total CN(B→X) emission [31,32] as a function of the excitation wavelength, using either synchrotron radiation or a microwave-powered rare-gas discharge as the continuously tunable source of VUV radiation. These spectra are displayed in Fig. 11. Although each spectrum provides a record of the branching into one particular dissociation channel only, comparison of the appropriate PFES with the parent molecular absorption spectrum can give a measure

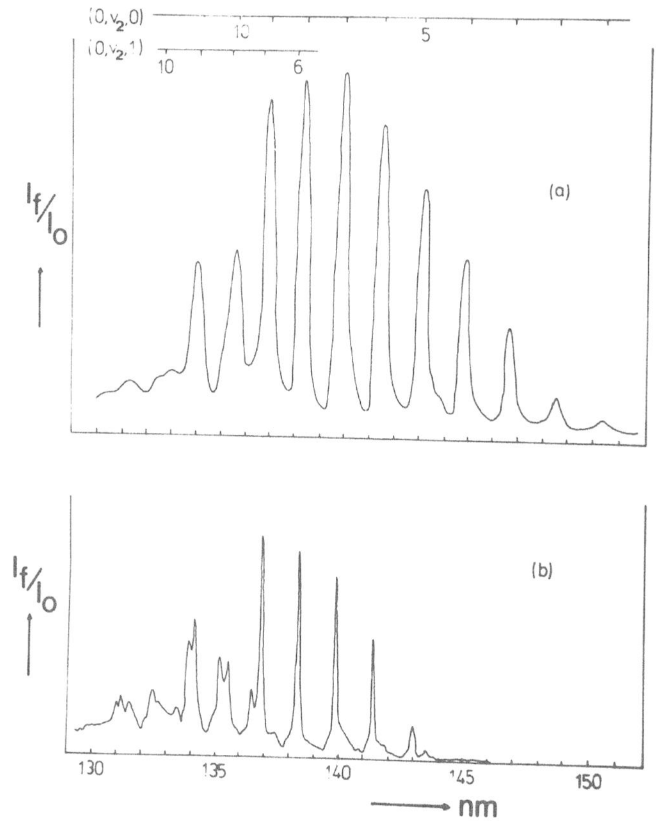

Fig. 11. Photofragment excitation spectra of (a) CN(A) and (b) CN(B) from HCN. The spectral resolutions employed were 0.8 nm and 0.04 nm, respectively, thus contributing to the very different apparent vibronic linewidths; the relative fluorescence signals are not scaled with respect to one another. (Adapted from [31,32].)

of the relative decomposition rate via that channel, as compared to the total fragmentation rate from the excited state or states prepared by photoexcitation. Such quantum yield measurements have the potential to reveal specific excited-state → product-channel correlations. In the case of HCN, the ($\tilde{C}$-$\tilde{X}$) absorption spectrum and PFES (Figs. 9 and 11) show a qualitative similarity which, on closer inspection, reveals that those dissociation channels leading to the formation of electronically excited CN products increase in importance at the shorter excitation wavelengths, but never represent much more than ∿20% of the total fragment yield.

The absorption and photofragment CN(B → X) fluorescence excitation spectra associated with the equivalent $\tilde{C}$-$\tilde{X}$ transition in DCN display a more complex appearance (cf. Fig. 12). In addition to the progressions $(0,v_2', 0)$ and $(0,v_2', 1)$ which were identified in the HCN($\tilde{C}$ → $\tilde{X}$) system, another strong series of bands can be recognized, in which each member is displaced from a corresponding band in the $(0,v_2', 0)$ sequence by a frequency appropriate to one quantum of the D-CN stretching frequency ν_1'. Just as for HCN, the quantum yield for forming CN(B), in the predissociation of DCN, gradually increases as the excitation wavelength decreases. A recent *ab initio* calculated potential energy surface for the $\tilde{C}^1A'$ state of HCN and DCN has provided a plausible interpretation for the much greater simplicity of the CN(B → X) PFES from HCN [33]. The topography of the potential energy surface is such that excitation of the ν_2' bending mode allows the ν_1' stretching motion to sample a variety of different tunnelling barriers. Near the $\tilde{C}$-state equilibrium configuration (∠ HCN ≈ 141°) the radial potential shows no barrier and, therefore, motion along the H-CN stretching coordinate, for this particular dissociation channel, will be either bound or directly dissociative depending upon whether the stretching energy is less than or greater than that of the asymptotic dissociation limit for H + CN(B). As the bending mode is excited and the H atom moves away from its equilibrium angular position, the radial potential energy curves exhibit barriers which can give rise to predissociative behavior. Thus, structure in Fig. 12 reveals predissociation from levels $(1,v_2',0)$ in DCN($\tilde{C}$), where $v_2' = 5$-12. However, the increased energy associated with one quantum of ν_1', when D is replaced by H, raises all of the corresponding excited levels in HCN($\tilde{C}$) above their respective dissociation barriers; and, direct dissociation simply generates a structureless contribution to the continuum underlying the CN(B → X) PFES of HCN in this region (cf. Fig. 11).

Thus, in the VUV photodissociation of HCN and DCN, and in marked contrast to the previously considered near-UV photofragmentation of ICN, conventional spectroscopic techniques are able to provide a fairly detailed picture of the photoselected excited state. Confirmation of the excited-state symmetry has been provided through use of *polarized photofluorescence excitation spectroscopy* [32,34,35]. The physical principles underlying this technique are essentially the same as those described previously (Section 2) for photofragment spectroscopy, which allows identification of excited-state symmetries from product angular distributions. In this case, however, the anisotropic distribution of photoexcited molecules is revealed in the polarization of the fluorescence from electronically excited fragments produced in the dissociation process.

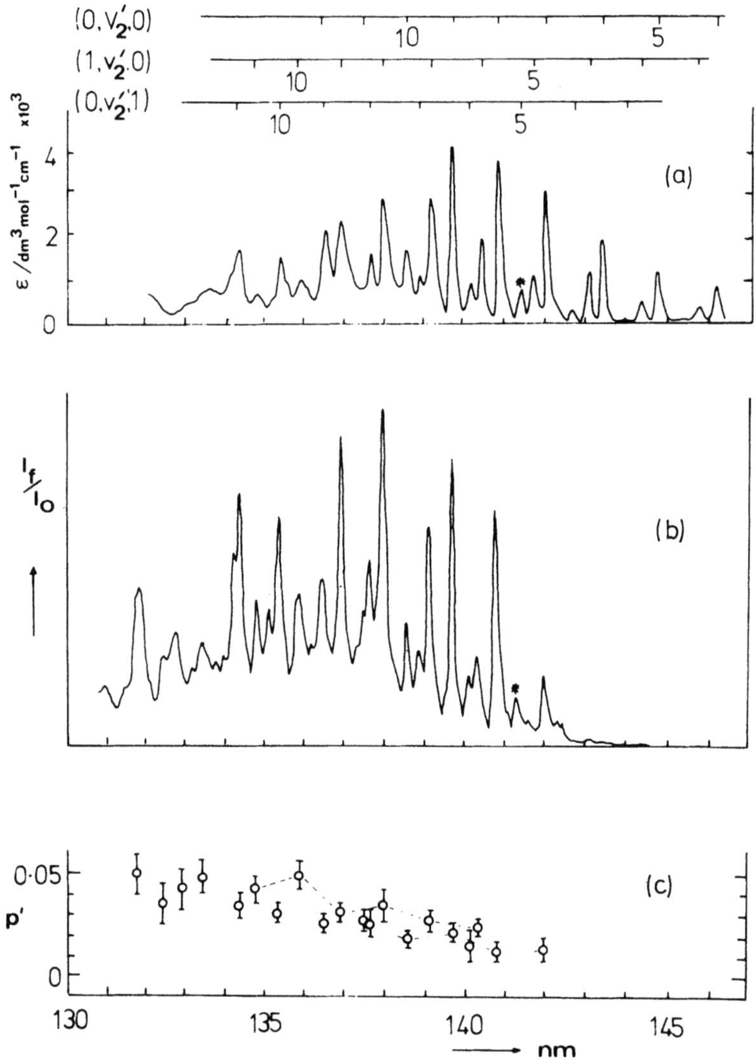

Fig. 12. (a) DCN absorption spectrum, (b) CN(B) photofragment fluorescence excitation spectrum and (c) CN ($B \rightarrow X$) fluorescence polarization measurements, as a function of vacuum ultraviolet wavelength in the range 130-150 nm. Only the one peak marked with an asterisk is significantly enhanced by the small amount of HCN impurity present in the sample. To aid the eye, the solid line in (c) joins points corresponding to the $(0,\nu_2',0)$ progression; the dashed line joins points corresponding to the $(1,\nu_2',0)$ progression. (Adapted from [32].)

By way of general illustration, consider the photodissociation of a bent triatomic molecule ABC, by a beam of linearly-polarized light, in which an electronically excited fragment is produced. As before, preferential photoselection of those molecules with transition dipole moment $\underset{\sim}{\mu}$ parallel to the direction of the electric vector $\underset{\sim}{\varepsilon}$ of the linearly-polarized light beam will occur, leading to a spatially anisotropic distribution of photoexcited ABC* molecules. Again, provided that the subsequent dissociation occurs on a timescale that is rapid in comparison to the rotational period of the photoexcited ABC*, some of this anisotropy will be transferred to the fragments. Since intramolecular motions in a bent triatomic molecule lie in a plane, the trajectory of the departing atomic fragment C will tend to be confined within the molecular plane of the photoexcited ABC*, while the electronically excited diatomic AB* will tend to be formed with its angular momentum vector $\underset{\sim}{j}$(fragment) directed perpendicular to this plane. The dissociation dynamics will be important in determining the magnitude of $\underset{\sim}{j}$(fragment). Overall angular momentum conservation requires that the separating atomic product exert a torque on the molecular fragment AB*. This torque will obviously be enhanced when photoexcitation causes a change in the equilibrium bond angle and, in these instances at least, the contribution to $\underset{\sim}{j}$(fragment) from the rotation of the parent molecule may safely be neglected. Since the fluorescence from AB* will be directed preferentially along the emission transition moment $\underset{\sim}{\mu}_{em}$, the anisotropic spatial distribution of the product diatomic rotors arising in the photofragmentation of ABC* will reveal itself as polarized fluorescence.

Two possibilities can arise when we come to consider the alternative general case, namely, when the molecule ABC has a linear ground state:

--- If the excited state ABC* prepared by photon absorption is linear as well (i.e., a linear → linear transition), the resulting photofragments will tend to separate along the ABC internuclear axis. The dynamics of such a dissociation process will cause little torque between the recoiling fragments; consequently, the rotational angular momentum of the parent molecule cannot be ignored. We shall see in more detail later that, in these circumstances, angular momentum conservation requires the mapping $\underset{\sim}{J}$(parent) → $\underset{\sim}{j}$(fragment), although this requirement must necessarily be relaxed in, for example, near-threshold dissociations, where the constraint of overall energy conservation must also be satisfied.

--- If photoexcitation leads to population of a nonlinear excited state (i.e., a linear → bent transition), the molecular transition moment must lie either in, or perpendicular to, the plane of the bent excited state ABC*. As before, linearly-polarized radiation will selectively excite those molecules for which the product $|\mu \cdot \varepsilon|^2$ is a maximum and, as a result of which, the photoexcited ABC* molecules will tend to bend into planes that are (respectively) parallel or perpendicular to the ε vector of the incident beam. Provided that the linear → bent geometry change in the photoexcitation process develops sufficient torque so that j(fragment) >> J(parent), this situation is equivalent to that described above for excitation from a nonlinear ground state. Any spatial anisotropy in the ensemble of photoexcited parent molecules ABC* will manifest itself as an anisotropic distribution of j(fragment) orientations. Thus, for either a linear → linear or a linear → bent transition, the spatial distribution of j(fragment) is anisotropic and the subsequent AB* fluorescence will be polarized. As with the photofragment angular distributions discussed earlier, measurement of the degree of polarization can provide information about the symmetry of the photoexcited state and about its lifetime with respect to dissociation.

The extent to which fluorescence is polarized is traditionally represented by the degree of polarization p defined as

$$p = \frac{I_{\parallel} - I_{\perp}}{I_{\parallel} + I_{\perp}}$$

where $I_{\parallel}$ and $I_{\perp}$ are the intensity of fluorescence measured through an analyzer aligned parallel or perpendicular, respectively, to the ε vector of the linearly-polarized photolysis beam. The degree of polarization may also be expressed in terms of the average angle γ between the electric-dipole transition moment of the parent molecule ABC (approximated by a hertzian dipole oscillator with the same directionality, μ_{abs}) and that of the electronically excited fragment AB* (again approximated by a dipole oscillator, μ_{em}) through the relationship

$$p = \frac{3\cos^2\gamma - 1}{\cos^2\gamma + 3} .$$

Clearly, it is necessary to consider the average value of $\cos^2\gamma$ in order to apply this equation; this has been done [35] for a range of possible situations in the photodissociation of triatomic molecules. Table 2 lists the values for $\cos^2\gamma$ and, hence, p determined for various cases of parallel and perpendicular photoexcitations in the parent molecule ABC [e.g., μ_{abs} parallel or perpendicular to the triatomic plane for bent → bent and linear → bent transitions or, in the case of linear → linear transitions, μ_{abs} parallel or perpendicular to J(parent)] and for μ_{em} parallel or perpendicular to j(fragment).

A number of factors can cause the experimentally observed degree of photofragment fluorescence polarization to be substantially less than the theoretical values given in Table 2. For example, in those cases where AB* radiates via a perpendicular electronic transition (e.g., $^1\Sigma$ - $^1\Pi$), Table 2 reveals that the polarized fluorescence from P and R branch transitions will show the opposite sign to that from Q branch transitions. When individual rotational lines in the emission spectrum are not resolved, the measured degree of polarization will obviously be much reduced. Similarly, it may not be possible to photoselect specifically only P and R branch, or Q branch, features in the excitation of linear → linear parent molecular transitions; the reduced degree of polarization displayed by the resulting photofragment fluorescence will reflect this reduced anisotropy in the photoexcitation stage. Obviously, the interpretation of experimentally derived degrees of polarization becomes even more complex in situations where photon absorption prepares more than one excited state. In deriving the theoretical values for p, it is assumed that the photoexcited molecule fragments directly (i.e., that it has no time to rotate before dissociating) and that the AB* fragment undergoes no collisions during its radiative lifetime; predissociations can be expected to show a reduced degree of polarization [36].

Before returning to the main theme of this section -- namely, the VUV photodissociation of HCN and DCN -- consider, as a particularly clear and illustrative example, the collisionless photodissociation of $HgBr_2$ using linearly-polarized 193 nm radiation. The visible HgBr ($B^2\Sigma^+ \rightarrow X^2\Sigma^+$) fluorescence from the resulting HgBr(B) photofragments (τ_{rad} = 23.7 ns) is found to be linearly-polarized, p = 11.9 ± 1.5% [37]. For this sixteen electron species, Walsh's rules [15] predict that the ground state is linear and that the lower-lying excited states are bent. The measured degree of polarization, combined with the knowledge that the fluorescence involves a parallel (Σ - Σ) transition, enables use of the data of Table 2 to deduce that $\underset{\sim}{\mu}_{abs}$ lies in the molecular plane of $HgBr_2$. That the observed p is so close to the theoretical value (p = 1/7) confirms the direct nature of the dissociation; assuming $D_{\infty h}$ symmetry for the ground state and C_{2v} symmetry for the excited state, the dissociative transition may be identified as $^1B_2(^1\Sigma_u^+) \leftarrow {}^1\Sigma_g^+(^1A_1)$.

The polarization of the CN(B → X) fluorescence resulting from photodissociation of HCN and DCN in the 130 - 150 nm wavelength region has been measured following excitation by an *un*polarized, parallel beam of light [32]. Under these circumstances, a reduced degree of polarization

$$p' = \frac{p}{2-p}$$

Table 2. Calculated Degrees of Polarization p of the Fluorescent Photofragments (AB)* Produced Through Photodissociation of Triatomic Parent Molecules ABC by a Beam of Linearly-Polarized Light.

Case	ABC Transition	(AB)* Transition	γ	$\cos^2\gamma$	p
		Nonlinear parent molecule or linear → bent transition			
I	μ_{abs} in (ABC)* plane	$\parallel$-type, $\mu_{em} \perp j$	45°	1/2	1/7
IIa	μ_{abs} in (ABC)* plane	$\perp$-type, $\mu_{em} \parallel j$ (Q)	90°	0	-1/3
IIb		$\mu_{em} \perp j$ (P,R)	45°	1/2	1/7
III	$\mu_{abs} \perp$ (ABC)* plane	$\parallel$-type, $\mu_{em} \perp j$	90°	0	-1/3
IVa	$\mu_{abs} \perp$ (ABC)* plane	$\perp$-type, $\mu_{em} \parallel j$ (Q)	0°	1	1/2
IVb		$\mu_{em} \perp j$ (P,R)	90°	0	-1/3

Table 2. Continued.

Case	ABC Transition	(AB)* Transition	γ	$\cos^2\gamma$	p
		Linear → linear transition			
V	$\parallel$-type, $\mu_{abs} \perp J$	$\parallel$-type, $\mu_{em} \perp j$	45°	1/2	1/7
VIa	$\parallel$-type, $\mu_{abs} \perp J$	$\perp$-type, $\mu_{em} \parallel j$ (Q)	90°	0	-1/3
VIb		$\mu_{em} \perp j$ (P,R)	45°	1/2	1/7
VIIa	$\perp$-type, $\mu_{abs} \parallel J$ (Q)	$\parallel$-type, $\mu_{em} \perp j$	90°	0	-1/3
VIIb	$\mu_{abs} \perp J$ (P,R)		45°	1/2	1/7
VIIIa	$\perp$-type, $\mu_{abs} \parallel J$ (Q)	$\perp$-type, $\mu_{em} \parallel j$ (Q)	0°	1	1/2
VIIIb	$\mu_{abs} \perp J$ (P,R)		90°	0	-1/3
IXa	$\perp$-type, $\mu_{abs} \parallel J$ (Q)	$\perp$-type, $\mu_{em} \perp j$ (P,R)	45°	1/2	1/7
IXb	$\mu_{abs} \perp J$ (P,R)				

is expected [35]. Results for DCN are displayed in Fig. 12. All of the main features of the respective PFES of HCN and DCN show positive polarization. Since the CN(B $\rightarrow$ X) transition moment is directed parallel to the internuclear axis, reference to Table 2 demonstrates that the observation of positive polarization is wholly consistent with the spectroscopic assignment of A' symmetry for the excited state of the parent molecule. That the maximum degrees of polarization observed even at the shorter wavelengths ($p' \simeq 0.04$, $p \simeq 0.08$) are less than the classically permitted maximum value ($p = 1/7$) may be attributed to some molecular rotation during the lifetime with respect to predissociation of the excited state [32]. We shall see more powerful demonstrations of the utility of polarized photofluorescence excitation spectroscopy when considering the VUV photodissociation of the cyanogen halides in the next section.

Measurements of the internal energy disposal in the fragmentation channel

$$H[D]CN + h\nu \rightarrow H[D] + CN(B^2\Sigma^+)$$

following excitation at the various monochromatic wavelengths provided by readily available atomic resonance lamps have also been used to derive some insight into the dynamics of this dissociation [38]. Armed with knowledge of the appropriate Franck-Condon factors and rotational line strengths, product vibrational and rotational state population distributions may be obtained readily through analysis of the wavelength-resolved spontaneous CN(B $\rightarrow$ X) photofragment fluorescence under collision-free conditions. For the available excitation wavelengths, energy conservation requires that the atomic H or D product be formed in its ground electronic state; thus, provided that the energetic threshold for the dissociation is known, completion of the detailed energy balance will yield the relative translational energy of the recoiling fragments.

Much theoretical effort has been devoted to modeling product-state distributions arising in photodissociations of this kind, with a view to identifying the more important features of the potential energy surface over which the dissociation proceeds. The more recent models recognize the likely importance of each of the following effects on the eventual product-state distributions:

(i) The Franck-Condon factors for the *intra*fragment, structural rearrangement of the molecular framework during transfer onto the final dissociative surface.

(ii) *Inter*fragment final-state interactions, where the impulsive recoil of the separating dissociation products affects the final translational, vibrational and rotational state distributions.

(iii) The constraints imposed by the necessity to simultaneously conserve both the total energy and angular momentum of the photodissociating system; these last constraints reveal themselves most clearly through their influence on the form of the product rotational state distributions.

Historically, models for *vibrational* energy disposal in photodissociation have varied widely in both the level of approximation introduced in formulating the problem and in the mathematical sophistication employed in solving the problem. Simons and Tasker [25] were the first to present a so-called "global" model that incorporated both intrafragment (i) and interfragment (ii) contributions to the ultimate vibrational energy disposal. For mathematical simplicicty they presumed (as have most subsequent theoretical models [26]) that linearity is retained during the fragmentation and that the reaction coordinate for the dissocciation process evolves from one of the molecule's normal modes. As applied to the predissociation of HCN, the correct normal modes are approximated by the localized H-CN and HC-N bond vibrations. An estimate for the HC-N stretching vibrational frequency in the photoexcited state can be gleaned from the absorption and photofragment fluorescence excitation spectra discussed earlier, in which a progression involving excitation of one quantum of ν_3' was identified ($\nu_3' \simeq 1800\ cm^{-1}$). Since this is significantly less than the vibrational frequency of the product diatomic CN(B) fragment ($\nu = 2164\ cm^{-1}$), the Franck-Condon factors for this substantial intramolecular rearrangement during the predissociation can be expected to make a major contribution to the eventual vibrational energy disposal. In the absence of detailed information regarding the form of the potential energy surface involved, the interfragment recoil is presumed to occur under a potential that is exponentially repulsive along the dissociation coordinate (r_{C-H}) and attractive along the C-N stretching coordinate [approximated by a simple harmonic oscillator function for CN(B)]. The importance, or otherwise, of final-state interactions upon the ultimate vibrational energy disposal is then determined principally by the steepness of the repulsive potential along the dissociation coordinate [25,28].

Fig. 13 shows that the vibrational state distributions in the CN(B) fragments produced in the VUV photodissociation of HCN and DCN at 130.4 nm, and at two slightly shorter wavelengths, can be reproduced more than adequately by the relatively simple model given above. The appeal of this approximate model is that the two adjustable parameters involved may be envisaged as having a certain, at least qualitative, physical significance. The best theoretical "fit" to each of the experimental vibrational distributions involves a considerable change in Δr_{C-N} and, hence, in

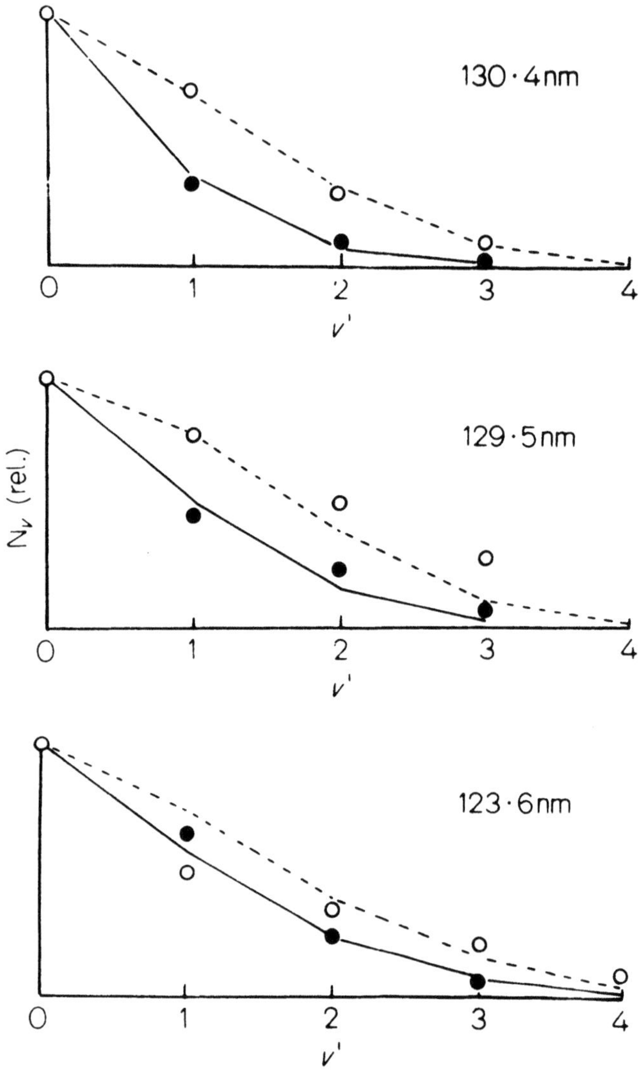

Fig. 13. Relative vibrational state populations in the $CN(B)_{v'}$ fragments produced in the collision-free photodissociation of HCN (●) and DCN (o) at three wavelengths. Modeled distributions for HCN (solid line) and DCN (dashed line) are included for comparison. (Adapted from [38].)

the C-N oscillation frequency during the radiationless transfer from the photoexited state (in accord with experimental observation for the $\tilde{C}^1A'$ state at least); in addition, a large value is required for the range parameter L in the expression $V_o \exp(-r_{C-H}/L)$ used to describe the shape of the repulsive potential along the dissociation coordinate. This latter value is consistent with a gently repulsive potential between the separating fragments, and indicates that final-state interactions make only a minor contribution to the eventual product vibrational excitation.

The more rigorous quantum mechanical models can, in principle at least, use a more correct normal mode description and the full, unseparated bound-free overlap integral [25] for the photodissociating molecule. Despite the added levels of mathematical complexity, however, any solution to the problem must remain, at best, semi-quantitative in the absence of quite precise spectroscopic data for the predissociating state. This explains the continuing appeal of the relatively simple models which, when sensibly applied, should be adequate for identifying the dominant influence upon the ultimate vibrational energy disposal in molecular photofragmentations. In the case of HCN and DCN illustrated here, geometry changes within the triatomic molecule during the predissociation largely govern the eventual product vibrational state distributions; a light H or D atom recoiling on a gently repulsive potential energy surface is unable to exert a significant impulsive contribution to the vibrating diatomic.

The comparatively minor effect of final-state interactions in this particular dissociation is further demonstrated by the form of the CN(B) photofragment's *rotational* energy disposal. Reference to Fig. 14 reveals that the rotational state distributions in the CN(B) photofragments produced in the VUV photodissociation of HCN at 147.0 nm and 141.1 nm, and of DCN at 141.1 nm, closely map the room temperature thermal distribution over rotational states in the parent HCN and DCN molecules. No detailed model for rotational energy disposal in photofragmentation is required to account for this result. These photolysis wavelengths are close to the energetic threshold for this dissociation channel and, therefore, the light, separating H and D atoms exert little impulse on the relatively heavy CN fragment as the latter recoils away. Conservation of angular momentum then requires that the rotational angular momentum of the parent molecules be carried over into the CN fragment rotation, which is precisely the behavior observed [38]. In this particular instance, it is clear that angular momentum conservation is *the* dominant influence on the rotational energy disposal. Excitation of HCN at 147.0 nm and 141.1 nm, respectively, excites the (0,3,0) and (0,6,0) bending vibronic levels of the $\tilde{C}^1A'$ state; neither the change in the excess available energy, nor the change in the level of bending mode excitation, affects the resulting fragment ro-

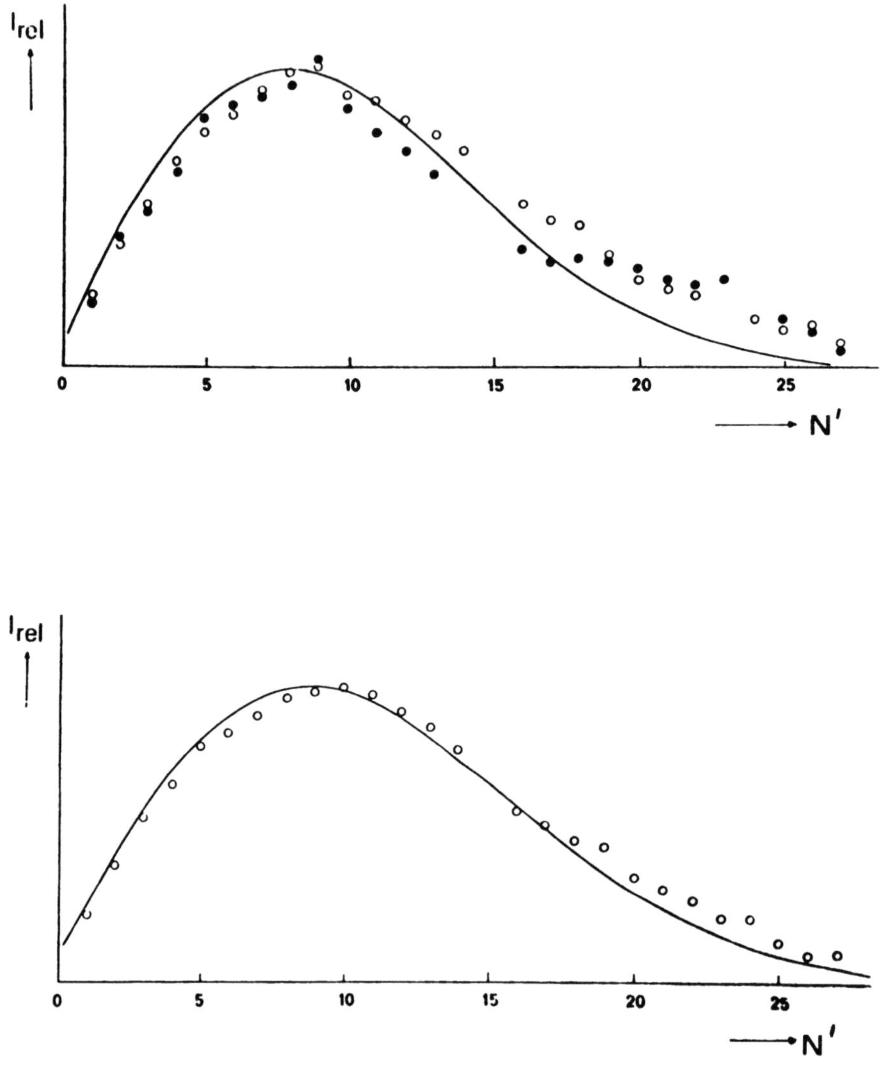

Fig. 14. Relative rotational state population distributions in the $CN(B)_{v'=0}$ fragments produced in the collision-free photodissociation of (top) HCN at 147.0 nm (●) and 141.1 nm (o), and (bottom) DCN at 141.1 nm. The solid curves show the thermal population distribution over rotational states in the ground electronic states of the respective parent molecules at 298 K. (Adapted from [38].)

tational energy partitioning. At still shorter wavelengths, however, the excess energy becomes such that a gradual increase in the level of fragment rotational excitation is observed [39].

Thus, it should be clear that the results summarized in this section provide considerable insight into the VUV photodissociation dynamics of HCN. In contrast to the previously considered near-UV dissociation of ICN, the structured absorption spectrum possessed by HCN in this wavelength region allows spectroscopy to make an important contribution to our overall understanding of this predissociative process. While it remains true that most of the detailed results relate to only one fragmentation channel -- which, in itself, represents only about 5% of the total dissociation at any wavelength -- the properties of an H or D atom, especially their light mass and their lack of low-lying excited electronic states, are such that a relatively clear picture of many of the more important dynamical features of the HCN fragmentation may be discerned.

4. VUV PHOTODISSOCIATION OF THE CYANOGEN HALIDES

As a final detailed case study, we shall consider some selected aspects of the much studied VUV photodissociation of ICN, BrCN, and CℓCN. The VUV absorption spectra of these three molecules have been reviewed in some detail [8,40] and a recent reassignment of the first two Rydberg features in the CℓCN spectrum has been offered [41]. In view of the problems outlined in our discussion of the near-UV absorption spectrum of ICN, it should come as no surprise to learn that the spectral complexity increases to shorter wavelengths. The above three cyanogen halides display qualitatively similar spectra, although the corresponding features in each spectrum shift to shorter wavelengths as the mass of the associated halogen atom decreases. A representative spectrum (for BrCN) is shown in Fig. 15. At energies above the $\tilde{A}$-continuum, each molecule reveals a second region of continuous absorption -- the $\tilde{\alpha}$-continuum. At still shorter wavelengths, relatively sharp, banded features appear superimposed upon a weak continuous absorption. In this section we shall concentrate on the photofragmentation dynamics of these three molecules following dissociation (i), from their respective $\tilde{\alpha}$-continua; and (ii), from wavelengths in the region of their first two Rydberg states, labeled $\tilde{B}$ and $\tilde{C}$.

There is now reasonable agreement that components of the intravalence $2\pi^3\pi$, $^1\Delta$ state contribute to the $\tilde{\alpha}$-continuum [8]. This conclusion has been reached through comparisions with the absorption spectra of other isovalent molecules [8,42], through analysis of photofragment fluorescence excitation spectra [19], and on the basis of correlation arguments [8]. The structured $\tilde{B}$ and

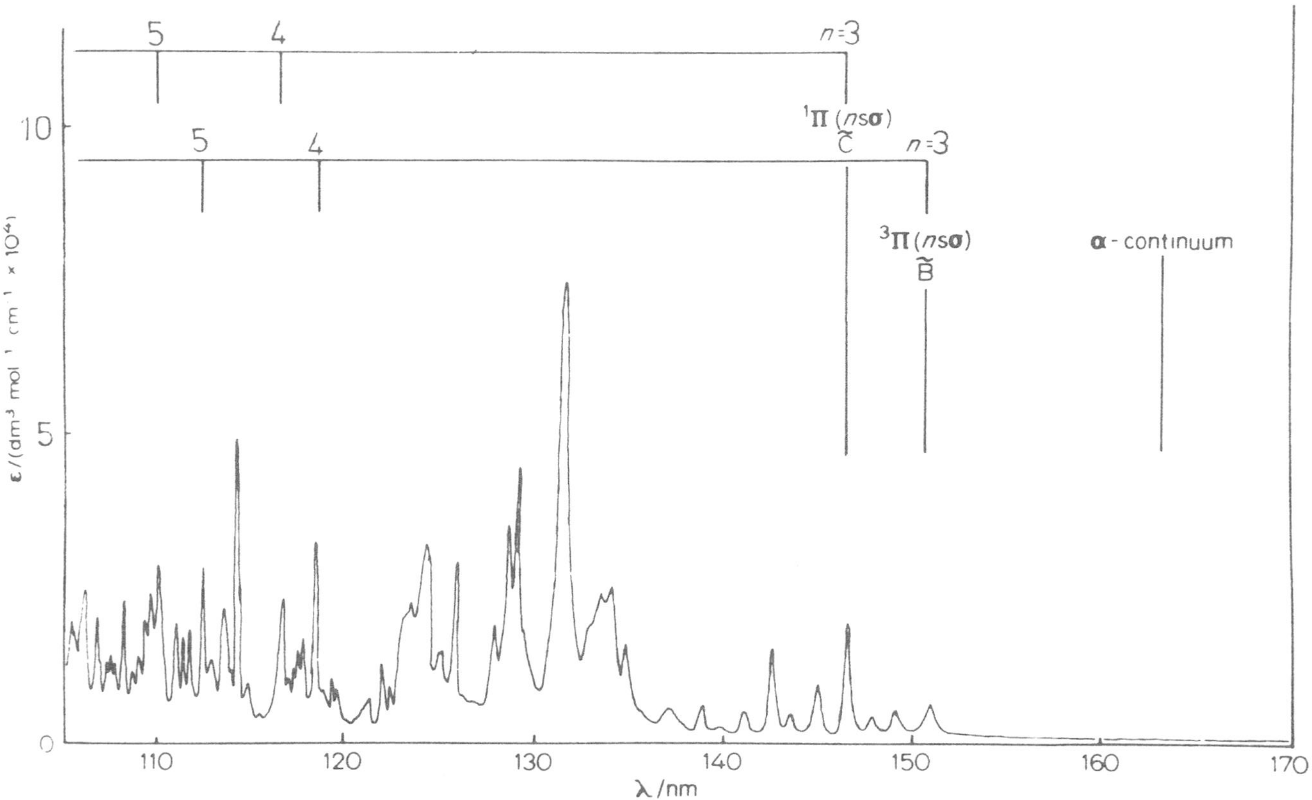

Fig. 15. Vacuum ultraviolet absorption spectrum of BrCN. (Adapted from [31].)

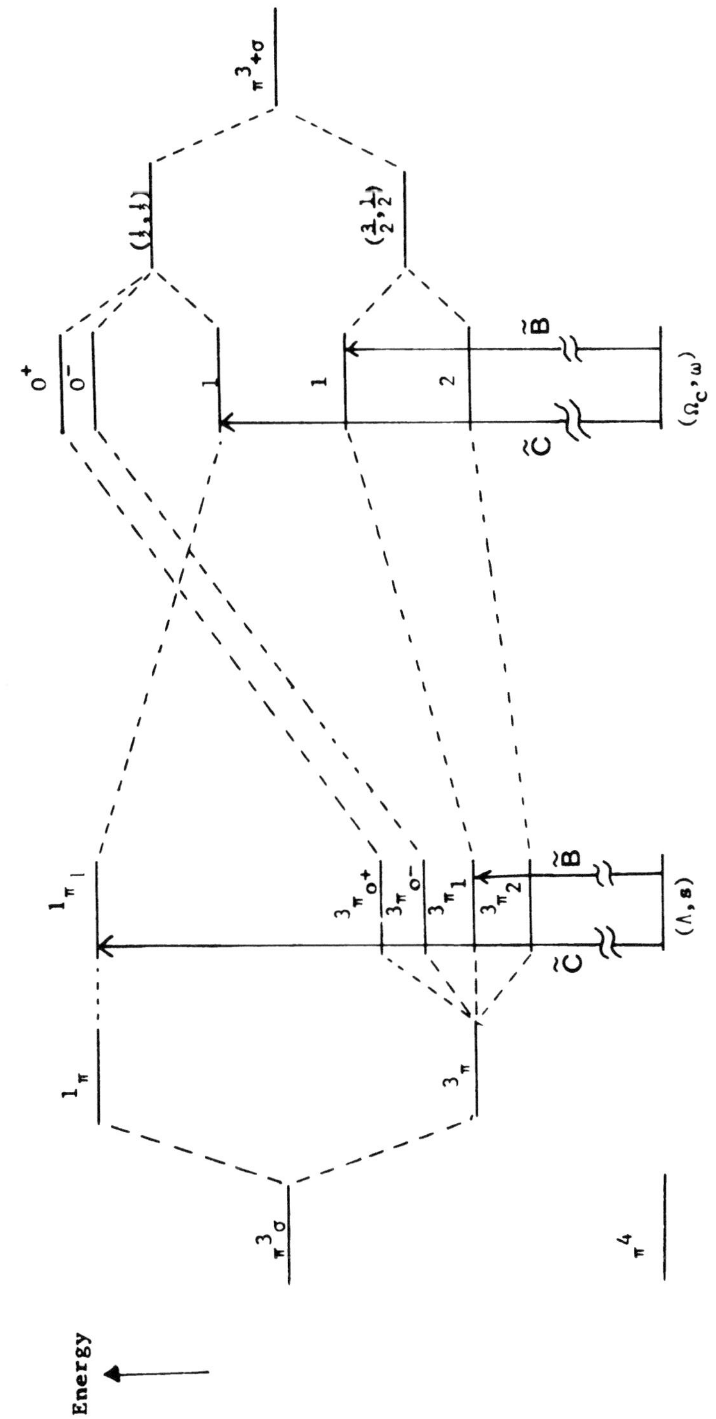

Fig. 16. Schematic correlation between (Λ,S) and (Ω_c,ω) coupling for a $\pi^3\sigma$ configuration. Energy differences are not to scale.

$\tilde{C}$ systems are assigned to the first members of the $...2\pi^3\sigma(ns)$, $^{1,3}\Pi \leftarrow ...2\pi^4, \tilde{X}^1\Sigma^+$ Rydberg series [8,40,43]; under (Ω_c, ω)-coupling, these terms correlate with the $(1/2,1/2)_1$ and $(3/2,1/2)_1$ states, respectively, as shown in Fig. 16. Quantum yield measurements, photofragment fluorescence excitation spectra and photofragment fluorescence polarization measurements for this spectral region have been reported for all three molecules [40]. Fig. 17 displays part of the CN(B → X) and CN(A → X) PFES resulting from BrCN photodissociation, along with measurements of the CN(B)

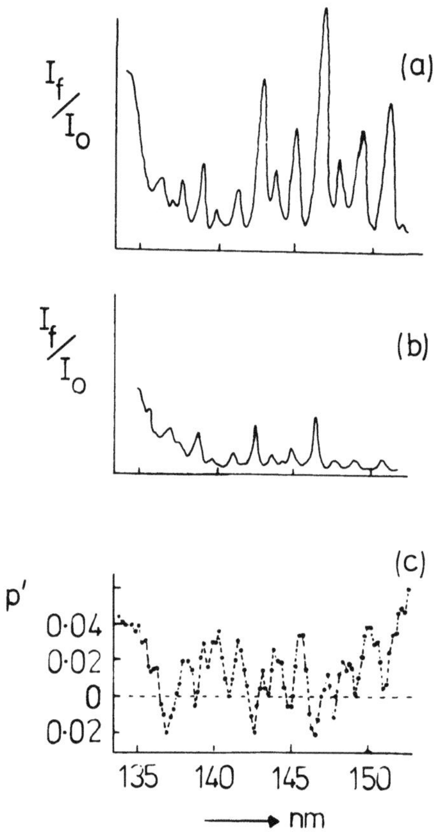

Fig. 17. Photofragment excitation spectra of (a) CN(A) and (b) CN(B) from BrCN, along with (c) polarization measurements for the total CN(B → X) fluorescence. Spectral resolution: (a), 0.35 nm; (b), 0.20 nm. (Adapted from [40,45].)

fluorescence polarization as a function of excitation wavelength. These latter data provide a dramatic illustration of the power of the polarized photofluorescence excitation technique, since it is readily apparent that the "degree of polarization" spectrum is a mirror image of the CN(B → X) PFES. This observation may be interpreted as follows. Direct dissociation from the underlying continuum generates positively-polarized CN(B → X) fluorescence, which is consistent with photoexcitation into a linear $\Sigma^+(0^+)$ or bent A' state of the parent molecule (see Table 2) since the fragment emission arises via a parallel transition. The degree of polarization is observed to fall, and even to become negative, following excitation at wavelengths resonant with the structured vibronic features associated with the predissociated Rydberg states. This, too, is consistent with the above spectroscopic assignment of these Rydberg features: a $^{1,3}\Pi(1) \leftarrow {}^1\Sigma^+(0^+)$ promotion has perpendicular character and, therefore, is expected to generate negatively-polarized CN(B → X) fluorescence, the precise magnitude of which may be reduced, of course, by parent molecular rotation during the predissociation lifetime. Thus, the actual value of p' measured at each wavelength reflects the relative contributions from the Rydberg state predissociation, and from direct dissociation from the underlying intravalence continuum, to the total CN(B) dissociation yield.

A number of detailed studies of internal energy disposal in the CN(B) photofragments produced in the VUV photodissociation of the cyanogen halides have been reported [39,44,45]. While again it should be remembered (i) that, in these molecules and at these wavelengths, the initial-state selection due to photon absorption at any particular wavelength is not as well-defined as one would ideally wish, (ii) that no definitive measure of the wavelength dependence of the branching into the two spin-orbit states of the halogen atoms in these dissociations are available and (iii) that only about 5% of the total fragmentation is ever probed by monitoring just the CN(B → X) emission, data on the collision-free rotational and vibrational state distributions in these photofragments still represent, nevertheless, some of the most complete and detailed information against which to gauge models for photodissociation dynamics.

By way of illustration, consider the wavelength-resolved CN(B → X) photofragment emission spectra displayed in Fig. 18. These spectra clearly show that the CN(B) fragment rotational state distributions which arise from the direct dissociation of ICN, BrCN and CℓCN within their respective $\tilde{\alpha}$-continuum absorptions vary markedly, despite the fact that the excitation wavelengths were selected so that, in each case, a similar total excess energy is available for partitioning within the internal and relative translational product degrees of freedom. In each case, the major fraction of the available energy must be taken up in

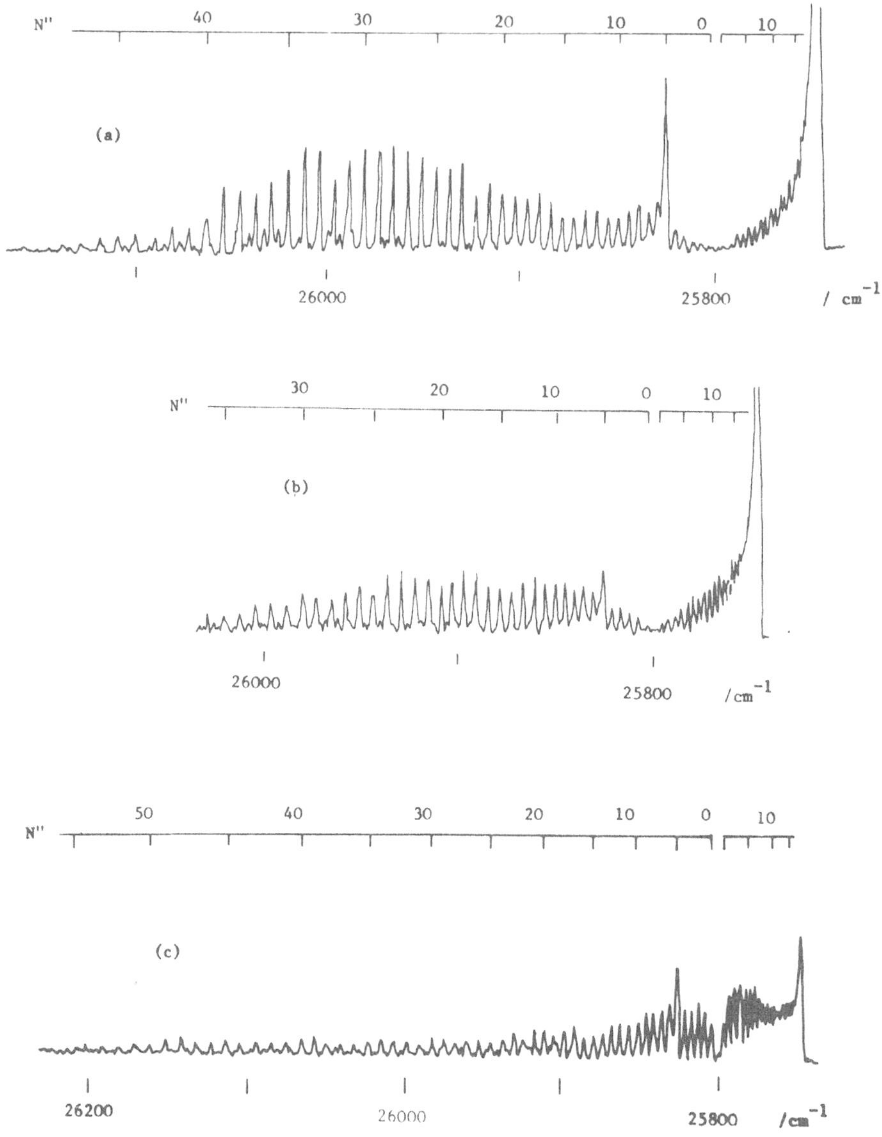

Fig. 18. Wavelength-resolved CN(B → X) fluorescence produced through photodissociation of the three cyanogen halides in their respective $\tilde{\alpha}$-continua: (a), CℓCN at 147.0 nm (E_{avl} = 7240 cm^{-1}); (b), BrCN at 158.0 nm and 163.4 nm (E_{avl} = 7110 cm^{-1} and 5020 cm^{-1}, respectively); and (c), ICN at 174.4 nm (E_{avl} = 6450 cm^{-1}). Resolution in each spectrum is ∿0.01 nm. (Adapted from [39,44].)

the form of relative translation of the separating products, since less than 10% appears as CN vibration and since a relatively constant proportion (15-20%) goes into fragment rotation. However, while the vibrational state distributions from these three dissociations show little dependence on the mass of the departing halogen atom [45], the distribution of the rotational energy amongst the various rotational states in the CN fragment does depend critically on the parent molecule. The most populated states in the $CN(B)_{v'=0}$ products from the 174.4 nm photodissociation of ICN are those with low N', although the population distribution shows a long "tail" extending to the highest energetically-accessible N' states. The maximum in the rotational state population distributions shifts progressively to higher N' in the order ICN < BrCN < CℓCN; the detailed forms of the vibrational and rotational state distributions in the $CN(B)_{v'=0}$ fragments produced through photodissociation of these three molecules at 174.4 nm, 158.0 nm and 147.0 nm, respectively, are shown in Fig. 19.

As a second illustration, Fig. 20 displays the collision-free wavelength-resolved CN(B → X) fluorescence spectra obtained following photodissociation of BrCN at 149.4 nm (an atomic nitrogen resonance line) and at 147.0 nm (a xenon emission line). Despite the fact that the available energy provided by photons at these two wavelengths differs by only ∿1000 cm^{-1}, it is clear that photodissociation at the shorter wavelength gives rise to much more highly internally-excited CN(B) fragments: Most dramatic is the observation that the fraction of the total available energy partitioned into CN(B) vibration doubles to ∿22% [39]. Transitions to certain rotational levels of the $CN(B)_{v'=0}$ state show anomalous intensity in the spectrum obtained following the 147.0 nm dissociation. A more complete analysis has revealed that these perturbed lines arise through accidental resonances with rotational levels in the v' = 10 level of the first excited $A^2\Pi_i$ state of CN [46,47]; their appearance in the fluorescence spectrum of the CN(B) fragments generated by the 147.0 nm photodissociation of BrCN indicates that the CN(A) fragments produced in this particular dissociation also carry a high level of vibrational excitation, populating vibrational states up to v' = 10 at least.

In order to account for the marked variations apparent within the above two sets of detailed observations, it is necessary to consider further the dynamics of the dissociation processes giving rise to these products. This requires a far more elaborate theoretical model than that which was previously shown to provide an adequate description for vibrational energy disposal in the photodissociation of HCN and DCN. To this end, Freed and coworkers [26,28,29,48] have developed a three-dimensional, fully quantum mechanical theory for the photodissociation of triatomic molecules that includes use of the correct normal modes of the initial bound state; lifts the constraint that dissociation neces-

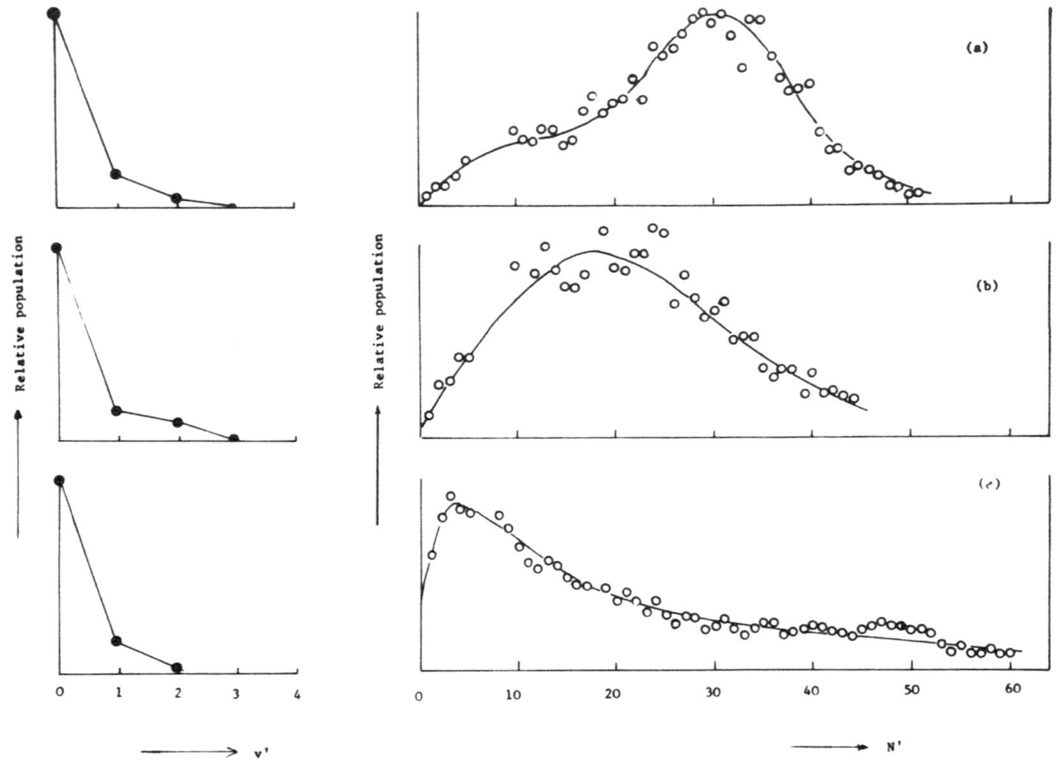

Fig. 19. Collision-free vibrational and rotational state populations in the CN(B) fragments produced through vacuum ultraviolet photodissociation of (a) CℓCN at 147.0 nm, (b) BrCN at 158.0 nm and 163.4 nm and (c) ICN at 174.4 nm. (After [39,44].)

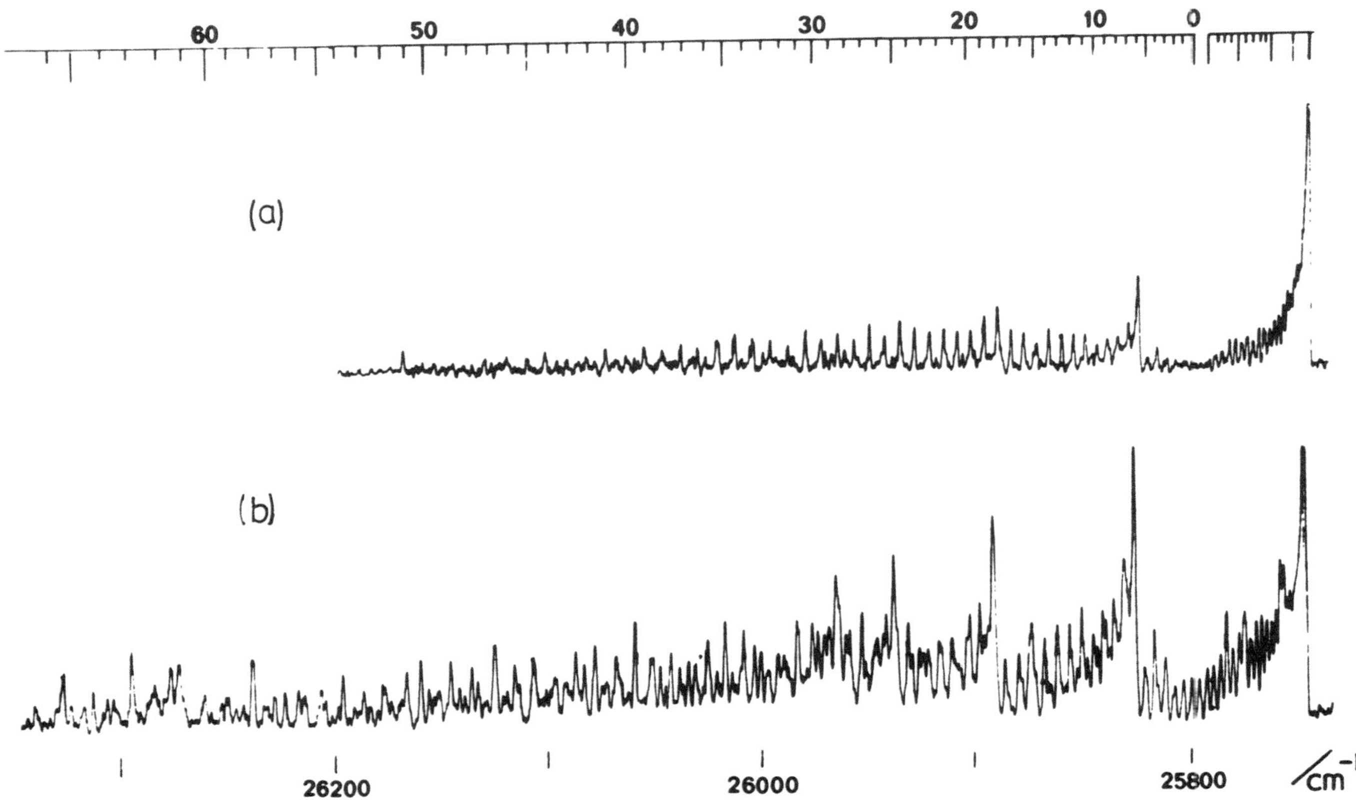

Fig. 20. Wavelength-resolved CN(B → X) fluorescence produced through photodissociation of BrCN at (a) 149.4 nm (E_{avl} = 10760 cm^{-1}) and (b) 147.0 nm (E_{avl} = 11850 cm^{-1}). (Adapted from [39].)

sarily proceeds collinearly; and properly describes both the bending vibrations in the initial bound electronic state and the rotational motions on both the initial bound and final repulsive potential energy surfaces. Thus, this model allows consideration of the influence of angular momentum, and its conservation, on the dissociation: Clearly, this is essential for any discussion of the resulting photofragment rotational distributions. In addition, this leads necessarily to multi-dimensional bound-continuum Franck-Condon integrals which, although strictly nonseparable, can be approximated by a product of one-dimensional integrals for mathematical convenience.

In the context of this discussion, we are principally interested in a physical visualization of the relative contributions made by the various components of the overall multi-dimensional overlap integral to the ultimate energy disposal. It is illustrative, therefore, to consider the multi-dimensional bound-free Franck-Condon factors as products of a one-dimensional "rotation-bending" term $|G|^2$ and a two-dimensional "vibration-translation" term $|F_{EL}|^2$: That is, the overall transition probability Γ for transfer from an initial bound state $|i\rangle$ to a repulsive state $|f\rangle$ is given by a convolution of the form

$$\Gamma_{i\to f} = |\langle f|i\rangle|^2 \propto |G|^2 |F_{EL}|^2 .$$

The G and F_{EL} integrals contain, respectively, the constraints of angular momentum conservation and energy conservation; their forms have been studied independently [26,28,29,48]. If, for example, we consider the effect of the rotation-bending term alone on the fragment rotational state distribution arising in a *strictly collinear* photodissociation of a triatomic molecule, the form of this term is such that the angular momentum initially present as parent molecular rotation $\underset{\sim}{J}$ would be efficiently channeled into fragment rotation $\underset{\sim}{j}$. This accords with intuition: The very small impact parameter associated with a collinear dissociation process cannot be expected to give rise to a significant dynamical contribution to the interfragment orbital angular momentum $\underset{\sim}{L}$. However, for the more general case in which there is, at the instant of fragmentation, some (small) displacement in the orientation of the diatomic fragment axis as compared to that of the equilibrium triatomic molecular axis in the initially bound state, Morse and Freed [29] have demonstrated that these "dynamic axis-switching" effects can substantially modify the partitioning of the parent rotational angular momentum $\underset{\sim}{J}$ into fragment rotation $\underset{\sim}{j}$ and into relative orbital angular momentum $\underset{\sim}{L}$ of the atom about the diatom. Clearly, these effects will be most noticeable when the departing atom is heavy (cf. Fig. 21). Thus, while the CN(B) fragment distributions from the photodissociation of HCN and DCN [38], described in the previous section, accurately reflect the

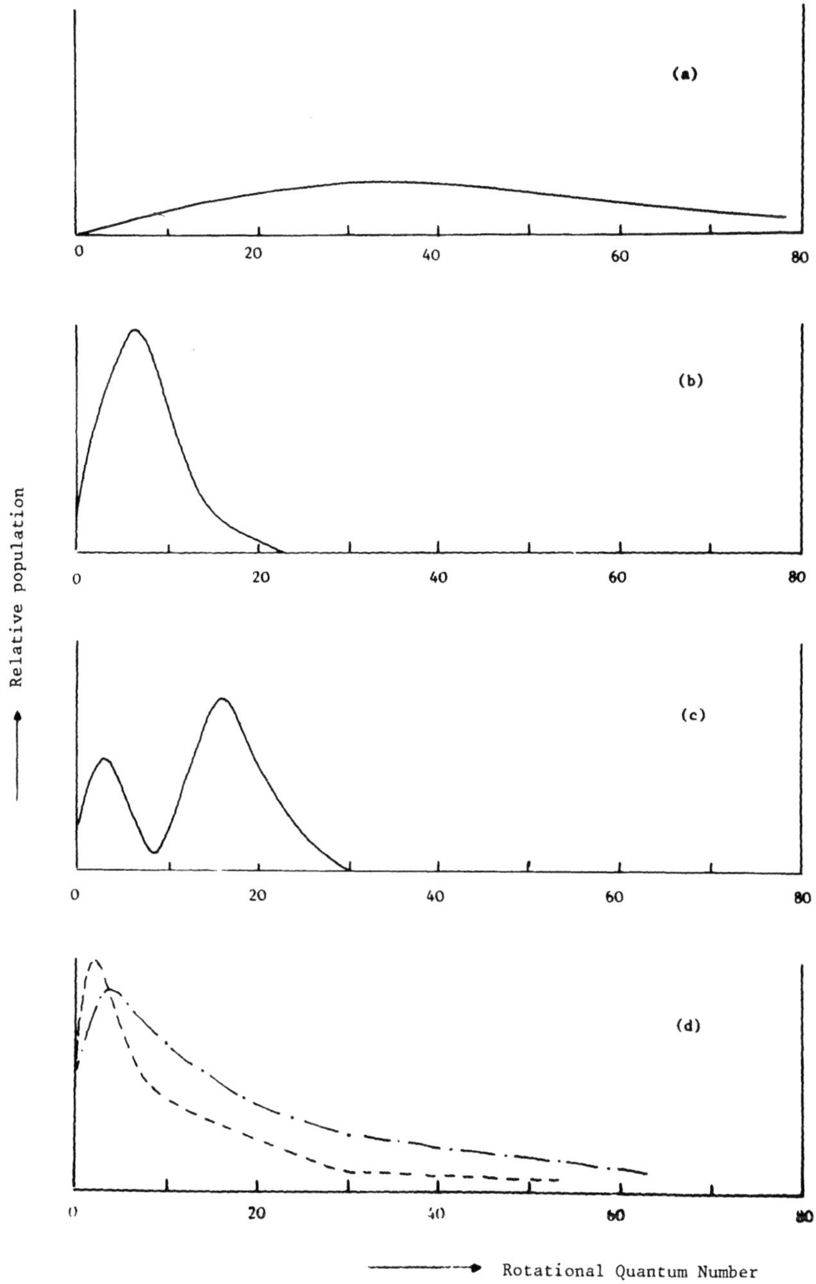

Fig. 21. Figure caption follows on next page.

Fig. 21. Calculated rotational-state population distributions in CN fragments produced in the photodissociation of ICN:

(a) As predicted by the direct mapping $\underset{\sim}{J}$(ICN) → $\underset{\sim}{j}$(CN) at 298 K.

(b) Allowing for the effects of dynamic axis switching in the photofragmentation of ICN (0,0,0). Note that for molecules such as ICN, in which the departing atom is very heavy, the ultimate distribution in $\underset{\sim}{j}$ is largely independent of the initial parent angular momentum $\underset{\sim}{J}$; as $\underset{\sim}{J}$ increases, angular momentum conservation is satisfied by a progressively more complete transfer of $\underset{\sim}{J}$ into orbital angular momentum $\underset{\sim}{L}$. (Adapted from [29].)

(c) As in (b), but for ICN in its excited bending vibrational state (0,2,0). (Adapted from [29].)

(d) For comparison, experimentally derived rotational-state population distributions for $CN(X)_{v''=0}$ fragments produced in ICN photodissociation at 266 nm (dashed line), and for $CN(B)_{v'=0}$ fragments from ICN photodissociation at 174.4 nm (dot-dashed line). (Adapted from [22,39].)

mapping $\underset{\sim}{J}$(parent) → $\underset{\sim}{j}$(fragment), the CN(B) rotational distribution produced through ICN photodissociation in its $\tilde{\alpha}$-continuum [39], and the CN(X) distributions arising in the 266 nm photodissociation of ICN from its $\tilde{A}$-continuum [21-23], peak at very low rotational quantum numbers. Further complications may arise when fragmentation occurs from excited bending vibrational modes of the parent molecule [28,29] since nodes in the bending wavefunction are carried over as undulatory structure in the fragment rotational-state distributions, as shown in Fig. 21.

Since the initial photoabsorption event is likely to select parent molecules in a range of J states, most -- if not all -- of the predicted fine structure in these rotational-state population distributions derived solely from consideration of the $|G|^2$ integrals in the multi-dimensional Franck-Condon factors may well be washed out, by the effects of thermal averaging, in any experimentally obtained spectrum. Furthermore, we must now consider the effects of the vibration-translational term $|F_{EL}|^2$, which contains the constraint of energy conservation and, therefore, depends implicitly upon $\underset{\sim}{J}$ and $\underset{\sim}{j}$ [26]. The likely influence of this term is visualized most easily in the case of a near-threshold dissociation, where the restricted available energy may reduce severely the probability of populating the higher-energy rotational states in the diatomic fragment. More generally, the role of $|F_{EL}|^2$ in determining the ultimate rotational energy disposal has been portrayed qualitatively in terms of the overlap, as shown in Fig. 22, along a dissociation coordinate Q_1 of (i), an effective one-dimensional oscillator (a theoretical construct which carries properties of both the bound triatomic initial state and the bound diatomic fragment vibration); and (ii), a repulsive potential surface [26,28]. Consider, first, the situation in which the absorbed photon carries insufficient energy E_{avl} for the translational continuum wavefunction $\psi_{cont}(Q_1)$ to overlap maximally with the effective oscillator wavefunction $\psi_{eff}(Q_1)$ [cf. Fig. 22a]. Any partitioning of the available energy into internal excitation of the molecular fragment will reduce the energy released into translation and will cause $\psi_{cont}(Q_1)$ to shift even further from the region of strong overlap. Fig. 22b demonstrates the converse situation, in which it is desirable that a substantial fraction of the energy carried by a shorter-wavelength exciting photon be channeled into fragment vibrational and rotational excitation in order to lower the translational continuum wavefunction and, thereby, maximize the bound-free overlap.

In the absence of quite detailed information about the final repulsive potential energy surface, the effect of final-state interactions in the ultimate rotational energy disposal is difficult to predict with any precision. Nevertheless, it should be clear that the preceding summary of major terms which contribute to the appropriate multi-dimensional Franck-Condon factors governing transfer from the initial bound state to the final repulsive state can provide qualitative explanations for the form of the various product energy disposals displayed in Figs. 18 and 20. For example, the dramatic change in the pattern of energy disposal in the photodissociation of BrCN, when the excitation wavelength is reduced from 149.4 nm to 147.0 nm, can be rationalized in terms of the very different character of the respective initial states. Reference to the CN(B $\rightarrow$ X) PFES and polarization data shown in Fig. 17 reveals that photon absorption at 149.4 nm excites predominantly the dissociative continuum (and, indeed, the observed energy disposal closely parallels that found following excitation at longer wavelengths in the α-continuum), while excitation at the shorter wavelength leads to significant population of the predissociated (0,0,1) vibronic level of the $\tilde{B}(^3\Pi)$ $(3/2,1/2)_1$ state, for which a reduced Br-CN bond length has been discerned [40,43]. Thus, in the simple description of Fig. 22, the disposition of the bound and free potential curves will be very different for the two situations depicted and dissimilar energy partitioning might well be expected.

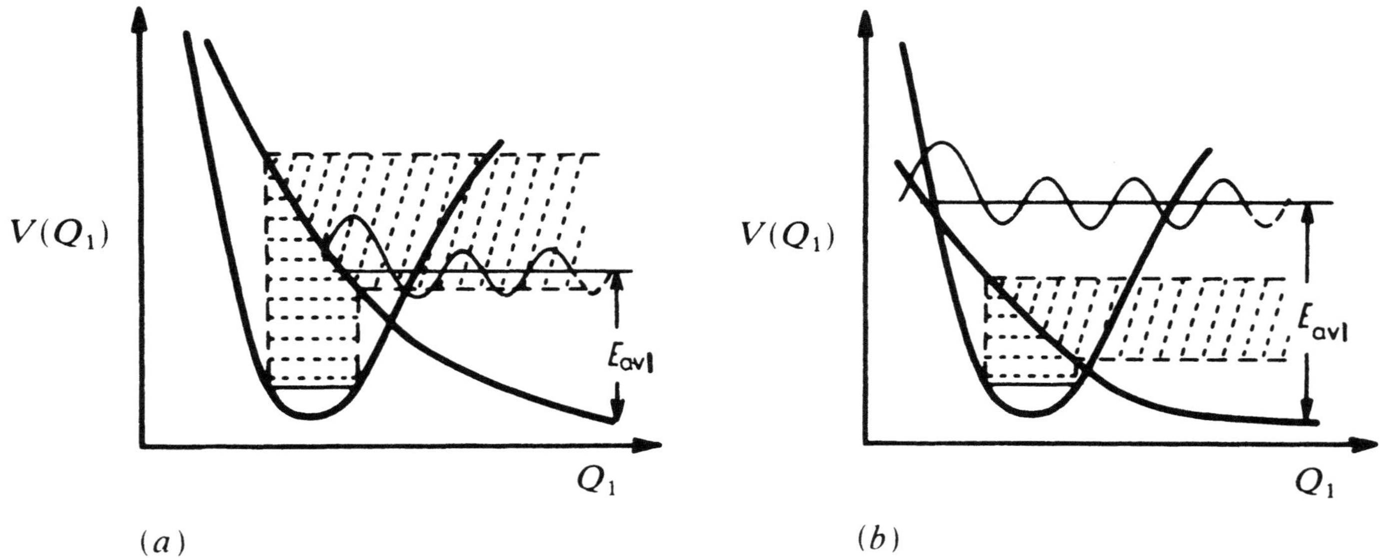

Fig. 22. Schematic representation of "bound-free" overlap in molecular photodissociation: (a), overlap favors energy disposal into translational excitation; (b), overlap favors energy disposal into internal degrees of freedom. (Adapted from [26,28].)

Several factors must be clarified before it will become possible to present an unequivocal interpretation of the various CN(B) rotational-state population distributions obtained in the direct dissociation of the cyanogen halides from their $\tilde{\alpha}$-continua [39,44,45]. Qualitatively, we have seen how these various population distributions might arise; however, any more detailed discussion must remain speculative in the absence of a more complete knowledge of (i), the photoselected state or states; (ii), their geometries; and (iii), the contribution to the overall CN(B) yield arising from photodissociation of that fraction of the thermal distribution of parent molecules in which the ν_2'' bending vibrational mode is excited.

5. CONCLUSIONS

The above three case studies have served to demonstrate many of the avenues of approach currently available in the study of molecular photodissociation processes in the vacuum ultraviolet. The most detailed information concerning the nature of the potential energy surfaces through which any given dissociation proceeds is contained in the rotational energy disposal; however, we have seen (especially in the previous section) that a very high degree of initial-state selection and characterization is necessary before it is possible to extract that information with any certainty. In this concluding section, therefore, we shall highlight two possible solutions to the problem of adequate state selection, and also outline other likely directions of advance.

As shown in Fig. 21, theory [29] predicts that, in triatomic photodissociations yielding heavy atomic fragments, the resulting population distribution over rotational states in the diatomic fragment is essentially independent of the initial parent molecular rotational angular momentum $\tilde{J}$. For lighter triatomic molecules this will not be so, and the observed photofragment rotational-state distribution will represent a superposition of different distributions over $\tilde{j}$ arising from the dissociation of parent molecules photoselected from a thermally determined range of $\tilde{J}$ states. This thermal-averaging effect can be alleviated by supersonic nozzle expansion of the parent molecule seeded in an excess of inert carrier gas (e.g., He or Ar). The use of supersonic molecular beams thus offers the possibility of studying the photodissociation process in parent molecules with very little internal energy and with $J \simeq 0$, thereby providing a far more rigorous test of the quantitative value of the current theories. Preliminary results have appeared for the 193 nm photodissociation of H_2S [49,50] and BrCN [49]. In both instances it was necessary to use LIF in order to probe the respective diatomic fragment internal-state distributions. Potential problems associated with collisional population redistribution amongst the photofragment quantum states prior to LIF probing have been demonstrated [49] and may account for the markedly different rotational energy disposals reported for the $SH(X)_{v''=0}$ fragments in the H_2S photodissociations [49,50]. More detailed and carefully controlled experiments along these lines can be anticipated in the near future.

The detail revealed in the time-of-flight spectra of photodissociation products also benefits from the greatly reduced, and more precisely defined, internal energy of molecules in a beam created by supersonic expansion. Fig. 23 shows the center-of-mass translational energy distribution of O_2 fragments produced through photodissociation of a supersonic molecular beam of O_3 at 266 nm [51]: These data clearly reveal the extent of $O(^1D)$ and $O(^3P)$ formation, as well as the fully-resolved vibrational-state populations in the $O_2(^1\Delta_g)$ product. The precise energy of the $O_2(^1\Delta_g)_{v'=0}$ peak even allows estimation of the fraction of the available energy channeled into diatomic rotational excitation. Much of the historical confusion concerning the relative impor-

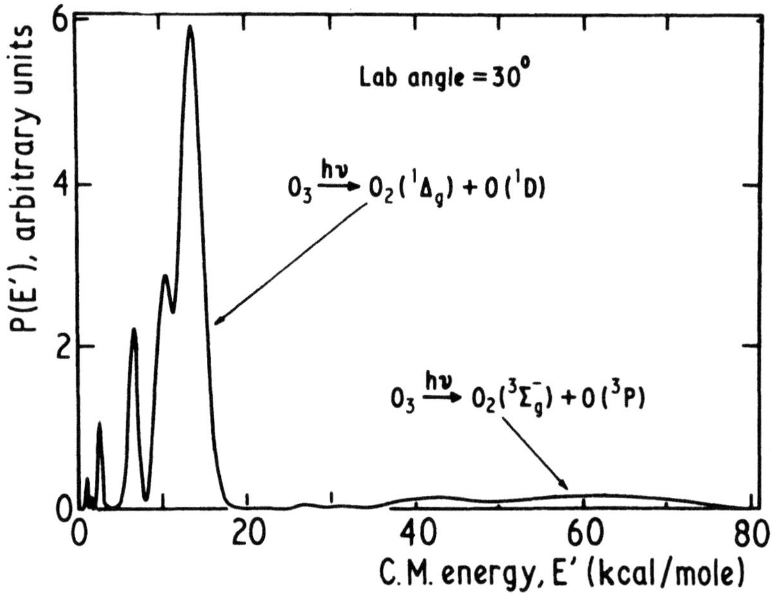

Fig. 23. Center-of-mass translational energy distributions of O_2 fragments produced in the 266 nm photodissociation of O_3, in which well-resolved peaks corresponding to the formation of $O_2(^1\Delta_g)$ in v' = 3,2,1 and 0 may be clearly discerned at the low translational energy end of the spectrum. (Adapted from [51].)

tance of the various possible product channels in the near-UV photodissociation of ICN (see Section 2) might never have arisen had energy-resolved data of this quality been obtainable in 1975.

Weakly predissociated systems provide an alternative route to the problem of well-defined state selection since they exhibit absorption spectra that may be sufficiently well-resolved to identify the initial predissociating vibrational (or even rovibrational) state. A preliminary report of the visible photofragmentation of CF_3NO via predissociation from specific vibronic levels of its first excited singlet state has appeared [52]; somewhat disappointingly, the internal state distribution of the NO fragment was found to depend only upon the available energy and not upon which vibrational mode of the parent molecule was initially excited by the photon absorption. In view of the preceding discussions of rotational energy disposal in the photofragmentation of light triatomic molecules, studies of predissociation from selected parent molecular J levels should prove more rewarding. Reference to Fig. 1 indicates that there are VUV wavelengths at which H_2O should be amenable to such experimental investigation. Similarly, very recent multiphoton ionization (MPI) spectroscopic studies of H_2S [53] and NH_3 [5,54] have identified well-defined rotational structure in a number of bands, the linewidths of which are consistent with excited-state lifetimes in the ps range (cf. Figs. 24 and 25). Further experiments of this kind are currently in progress; moreover, as intense, dedicated synchrotron radiation sources become more commonplace and readily available for VUV absorption measurements, it will be surprising if many other small molecules are not found similarly to possess Rydberg states having lifetimes with respect to predissociation that are sufficient for rotational structure to be identified spectroscopically. Indeed, rotational structure showing what appears to be instrument-limited linewidths ($<1\ cm^{-1}$) has been observed in a recent synchrotron radiation study of the $\pi_g^3 3p\pi_u\ ^3\Sigma^- - \tilde{X}^1\Sigma_g^+$ Rydberg transition in CO_2 at 110.6 nm [55].

New techniques continue to find application in studies of molecular photodissociation dynamics. Photofragment laser-induced fluorescence excited by a sufficiently narrow-band dye laser will show Doppler-limited spectral linewidths, from which translational velocity [56,57] and, in principle at least, even angular [57] distributions may be derived as functions of the internal quantum state of the product. Multiphoton ionization [58] and time-resolved measurement of the infrared chemiluminescence from vibrationally excited ground-state species [59,60] both have found use as detection methods for the primary products of photodissociation processes; both are in some respects complementary to LIF as detection techniques in that they may be applied to non-fluorescent species (e.g., the methyl radical [59-62]). Unfortun-

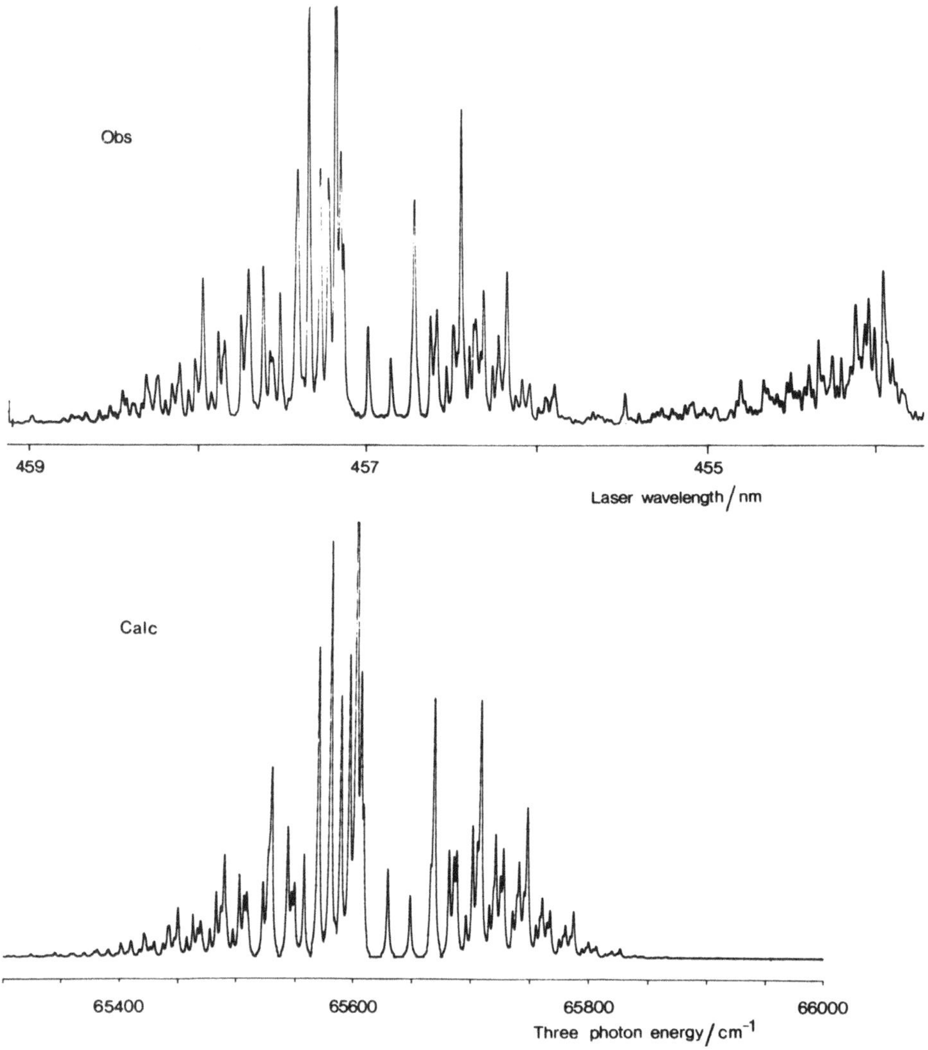

Fig. 24. (a), MPI spectrum of the $\tilde{C}'_{v'=2} \leftarrow \tilde{X}$ band of NH_3 obtained using linearly-polarized 2.5 mJ dye laser pulses, together with a computer simulation of this band assuming the entire line strength to be carried by the first-rank tensor component of the overall transition probability (selection rule: $\Delta J = 0,\pm 1$; $\Delta K = 0$) and using a linewidth of 2 cm^{-1} (FWHM) and the following rotational constants: B" = 9.944 cm^{-1}, C" = 6.196 cm^{-1}, B' = 9.93 cm^{-1}, C' = 5.35 cm^{-1} [5]. The band centered at $\sim$454 nm is due to the $\tilde{B}_{v'=7} \leftarrow \tilde{X}$ transition in NH_3. Note the weak features at 454.81 nm and 455.47 nm. (Adapted from [54].)

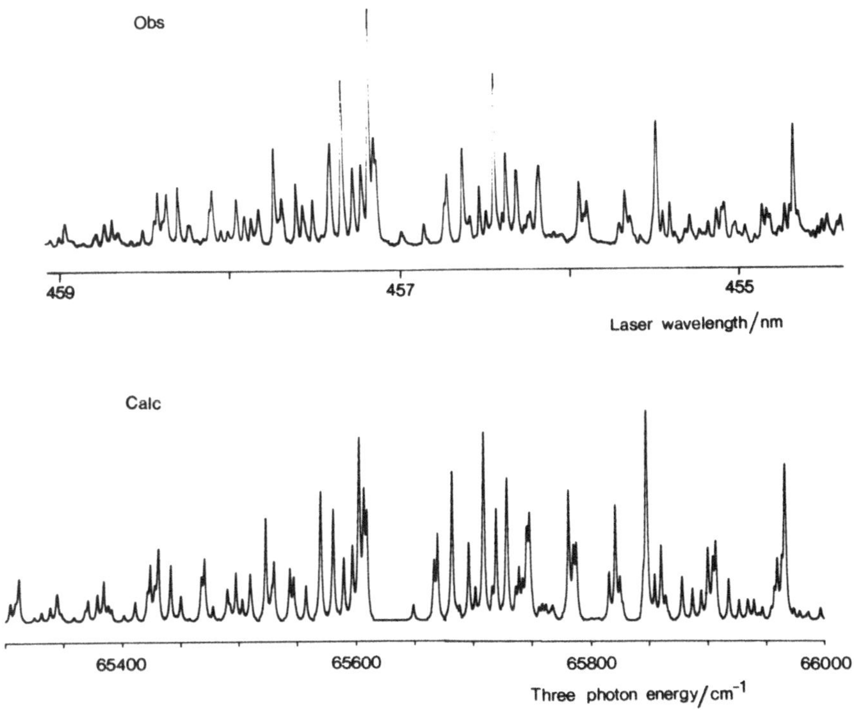

Fig. 24. (b), As in (a), except that circularly-polarized laser pulses were used. (Spectrum (b) was recorded with a four-fold increase in detection sensitivity.) Comparison of (a) and (b) reveals that these lines arise from a weak, third-rank tensor contribution to this transition (selection rule: $\Delta J = 0,\pm1,\pm2,\pm3$; $\Delta K = 0$). (Adapted from [54].)

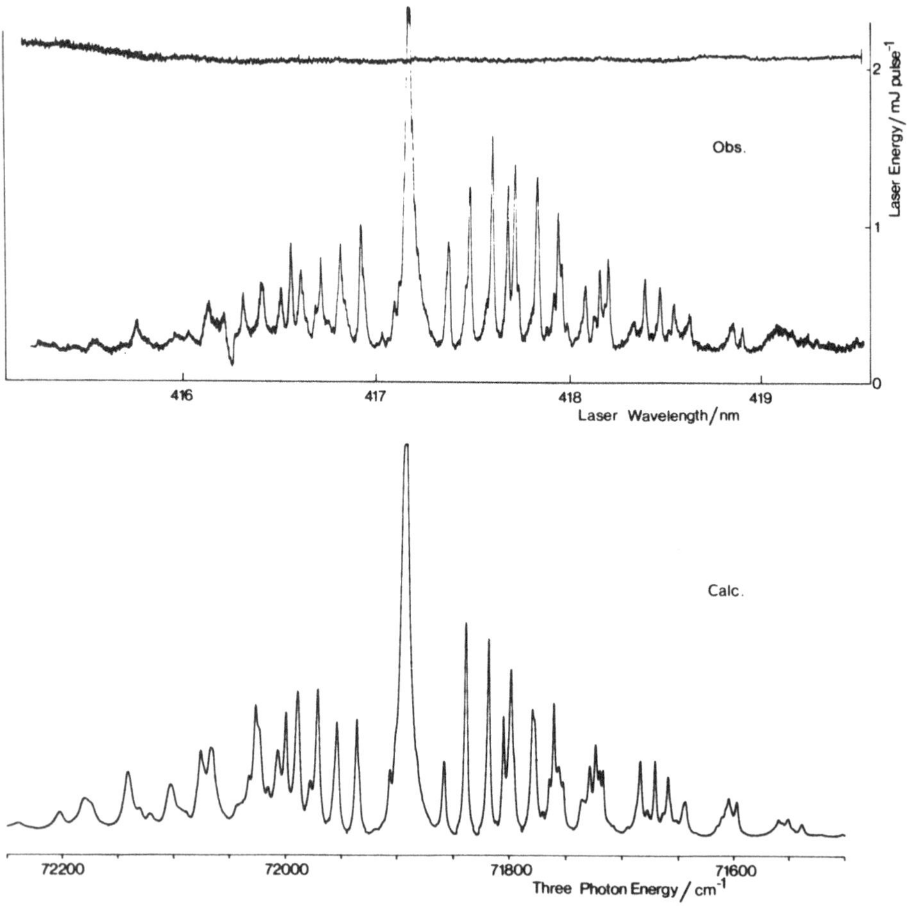

Fig. 25. Four photon ionization spectrum of H_2S using laser radiation in the 415-420 nm wavelength range, resonance enhanced by a $^1B_1$ electronic state at the three photon level, together with a simulation of this $^1B_1$ - $^1A_1$ band, in which a rotational energy dependent predissociation is presumed to be responsible for the reduced intensity and line broadening observed for the higher rotational levels. (Adapted from [53].)

ately, progress towards what must be the most important requirement of all for those studying VUV photodissociation processes -- namely, readily available, intense, narrow-bandwidth tunable sources of VUV radiation -- has been somewhat less spectacular. However, the increasing availability of pulsed synchrotron radiation, the demonstration of photochemically-useful tunable VUV intensities through third-harmonic generation of relatively narrow-band dye laser radiation in the rare gases Kr and Xe [63], and the prospects for tunable multiphoton excitation into the VUV offered by the more recent high-power excimer or Nd-YAG pumped dye laser systems should ensure maintenance of the current intense level of activity in this field.

REFERENCES

1. K. Watanabe and M. Zelikoff, J. Opt. Soc. Amer. 43, 753 (1958).
2. J. W. C. Johns, Can. J. Phys. 41, 209 (1963).
3. P. M. Johnson, Acc. Chem. Res. 13, 20 (1980).
4. P. M. Johnson and C. E. Otis, Ann. Rev. Phys. Chem. 32, 139 (1981).
5. J. H. Glownia, S. J. Riley, S. D. Colson and G. C. Nieman, J. Chem. Phys. 72, 5998 (1980); 73, 4296 (1980).
6. M. B. Robin, *Higher Excited States of Polyatomic Molecules*, Vols. 1 and 2 (Academic Press, New York, 1974).
7. H. Okabe, *Photochemistry of Small Molecules*, (Wiley, New York, 1978).
8. M. N. R. Ashfold, M. T. Macpherson and J. P. Simons, Top. Cur. Chem. 86, 1 (1979).
9. R. N. Zare, Mol. Photochem. 4, 1 (1972).
10. R. Bersohn, Israel J. Chem. 11, 675 (1973).
11. J. P. Simons, Roy. Soc. Chem. Spec. Per. Rep. 2, 58 (1977).
12. W. M. Gelbart, Ann. Rev. Phys. Chem. 28, 323 (1977).
13. S. R. Leone, Adv. Chem. Phys. 50, 255 (1982).
14. J. A. Myer and J. A. R. Samson, J. Chem. Phys. 52, 266 (1970).
15. A. D. Walsh, J. Chem. Soc. 1953, 2266.
16. K. R. Wilson, in: *Excited State Chemistry*, J. N. Pitts, Jr., ed. (Gordon & Breach, New York, 1970).
17. J. H. Ling and K. R. Wilson, J. Chem. Phys. 63, 101 (1975).
18. S. -C. Yang and R. Bersohn, J. Chem. Phys. 61, 4400 (1974).
19. G. A. West and M. J. Berry, Chem. Phys. Lett. 56, 423 (1978).
20. W. M. Pitts and A. P. Baronavski, Chem. Phys. Lett. 71, 395 (1980).
21. A. P. Baronavski and J. R. McDonald, Chem. Phys. Lett. 45, 172 (1977).
22. W. Krieger, J. Hager and J. Pfab, Chem. Phys. Lett. 85, 69 (1982).
23. A. P. Baronavski, Chem. Phys. 66, 217 (1982).
24. M. J. Sabety-Dzvonik and R. J. Cody, J. Chem. Phys. 66, 125 (1977).
25. J. P. Simons and P. W. Tasker, Mol. Phys. 26, 1267 (1973); 27, 1691 (1974).

26. Y. B. Band and K. F. Freed, in: *Excited States*, Vol. 3, E. C. Lim, ed. (Academic Press, New York, 1977), p.110.
27. J. A. Beswick and J. Jortner, Chem. Phys. 24, 1 (1977).
28. M. D. Morse, K. F. Freed and Y. B. Band, J. Chem. Phys. 70, 3604, 3620 (1979).
29. M. D. Morse and K. F. Freed, J. Chem. Phys. 74, 4395 (1981).
30. G. Herzberg, *Molecular Spectra and Molecular Structure*. III. *Electronic Spectra and Electronic Structure of Polyatomic Molecules* (Van Nostrand Reinhold, New York, 1966).
31. G. A. West, Ph.D. Thesis, (University of Wisconsin, Madison, 1975).
32. M. T. Macpherson and J. P. Simons, J. Chem. Soc. Faraday II 74, 1965 (1978).
33. D. T. Chuljian and J. Simons, J. Am. Chem. Soc. 104, 646 (1982).
34. G. A. Chamberlain and J. P. Simons, Chem. Phys. Lett. 32, 355 (1975); J. Chem. Soc. Faraday II 71, 2043 (1975).
35. M. T. Macpherson, J. P. Simons and R. N. Zare, Mol. Phys. 38, 2049 (1979).
36. G. W. Loge and R. N. Zare, Mol. Phys. 43, 1419 (1981).
37. J. Husain, J. R. Wiesenfeld and R. N. Zare, J. Chem. Phys. 72, 2479 (1980).
38. M. N. R. Ashfold, M. T. Macpherson and J. P. Simons, Chem. Phys. Lett. 55, 84 (1978).
39. M. N. R. Ashfold and J. P. Simons, J. Chem. Soc. Faraday II 74, 280 (1978).
40. M. T. Macpherson and J. P. Simons, J. Chem. Soc. Faraday II 75, 1572 (1979).
41. S. P. McGlynn, W. S. Felps and G. L. Findley, Chem. Phys. Lett. 78, 89 (1981); J. Mol. Spectrosc. 86, 71 (1981).
42. J. W. Rabalais, J. M. McDonald, V. Scherr and S. P. McGlynn, Chem. Rev. 71, 73 (1971).
43. G. W. King and A. W. Richardson, J. Mol. Spect. 21, 339, 353 (1966).
44. M. N. R. Ashfold and J. P. Simons, J. Chem. Soc. Faraday II 73, 858 (1977).
45. M. N. R. Ashfold, A. S. Georgiou, A. M. Quinton and J. P. Simons, J. Chem. Soc. Faraday II 77, 259 (1981).
46. M. N. R. Ashfold and J. P. Simons, Chem. Phys. Lett. 47, 65 (1977).
47. S. R. Long and J. P. Reilly, J. Phys. Chem. 86, 56 (1982).
48. M. D. Morse and K. F. Freed, Chem. Phys. Lett. 74, 49 (1980).
49. M. Heaven, T. A. Miller and V. E. Bondybey, Chem. Phys. Lett. 84, 1 (1981).
50. W. G. Hawkins and P. L. Houston, J. Chem. Phys. 76, 729 (1982).
51. R. K. Sparks, L. R. Carlson, K. Shobatake, M. L. Kowalczyk and Y. T. Lee, J. Chem. Phys. 72, 1401 (1980).

52. R. W. Jones, R. D. Bower and P. L. Houston, J. Chem. Phys. 76, 3339 (1982).
53. M. N. R. Ashfold and R. N. Dixon, Chem. Phys. Lett., 93, 5 (1982).
54. M. N. R. Ashfold, R. N. Dixon and R. J. Strickland, Chem. Phys. (submitted).
55. C. Cossart-Magos, S. Leach, M. Eidelsberg, F. Launay and F. Rostas, J. Chem. Soc. Faraday II 78, 1477 (1982).
56. H. Zacharias, M. Geilhaupt, K. Meier and K. H. Welge, J. Chem. Phys. 74, 218 (1981).
57. E. J. Murphy, J. H. Brophy and J. L. Kinsey, J. Chem. Phys. 74, 331 (1981).
58. G. Radhakrishnan, D. Ng and R. C. Estler, Chem. Phys. Lett. 84, 260 (1981).
59. S. L. Baughcum and S. R. Leone, J. Chem. Phys. 72, 6531 (1980).
60. H. W. Hermann and S. R. Leone, J. Chem. Phys. 76, 4759, 4766 (1982).
61. T. G. DiGiuseppe, J. W. Hudgens and M. C. Lin, Chem. Phys. Lett. 82, 267 (1981); J. Phys. Chem. 86, 36 (1982).
62. J. Danon, H. Zacharias, H. Rottke and K. H. Welge, J. Chem. Phys. 76, 2399 (1982).
63. J. C. Miller, R. N. Compton and C. D. Cooper, J. Chem. Phys. 76, 3967 (1982).
64. G. Vazquez, unpublished results.

VUV SPECTROSCOPY OF RARE-GAS VAN DER WAALS DIMERS

P. M. Dehmer and S. T. Pratt

Argonne National Laboratory
Argonne, Illinois 60439, USA

1. INTRODUCTION

It is well known that ground-state rare-gas atoms repel each other, except for van der Waals forces which lead to weak molecular binding in all of the homonuclear and heteronuclear dimers except, possibly, He_2 [1]. Removal of the outermost electron, which is antibonding in character, leads to a stable ground ionic state for all of the dimers. Several of the excited ionic states are bound as well although, in general, the dissociation energies of the excited states are much smaller than that of the ground state. The addition of an electron in a Rydberg orbital to an ion in one of the bound states may result in a bound Rydberg state; however, little is known about the stable excited states of the neutral rare-gas dimers, except for those molecular states which arise from the lowest atomic resonance states (i.e., the molecular excimer states). A notable exception is the detailed multichannel quantum defect theory analysis of the Rydberg states in He_2 presented recently by Ginter and Ginter [2].

With this in mind, we have undertaken a systematic study of the photoionization spectra of the homonuclear and heteronuclear rare-gas dimers in order to better understand the nature of the bonding in the Rydberg states and ions of these molecules. We have obtained results for Ar_2 [3,4], Kr_2 [5], Xe_2 [6], NeAr, NeKr, NeXe [7], ArKr, ArXe and KrXe [8]. Of the remaining dimer species (Ne_2 and the He/rare-gas dimers), only Ne_2 has been studied using photoionization mass spectrometry [9].

The results of the present series of experiments provide information both on the excited states of the neutral dimers and

S. P. McGlynn et al. (eds.), Photophysics and Photochemistry in the Vacuum Ultraviolet, 467–513.

on the ground and excited states of the dimer ions. Using the data obtained in these measurements, we are able to compile, for the first time, a nearly complete list of ground-state dissociation energies for the homonuclear and heteronuclear rare-gas dimer ions. Somewhat less complete results are obtained for the excited states of these species. The observed trends in binding energies provide an excellent example of those systematic changes which occur as a result of changes in atomic orbital energies, polarizability and internuclear distance; and, these trends can be explained qualitatively in terms of simple molecular orbital theory.

2. GENERAL BACKGROUND

Since the potential energy curve of a Rydberg state is built on, and therefore closely resembles, its ionic convergence limit, a knowledge of the bonding in the ground and excited states of the dimer ions is required in order to understand the bonding in the Rydberg states. A homonuclear rare-gas dimer ion X_2^+ has a bound $^2\Sigma^+_{1/2u}$ ground state and five excited states, all of which are repulsive except for a small polarization minimum at large internuclear distances (R) [1]. At small R, a molecular orbital description of the states is appropriate, with the states being labeled, in order of increasing energy, $^2\Pi_{3/2g}$, $^2\Pi_{1/2g}$, $^2\Pi_{3/2u}$, $^2\Pi_{1/2u}$ and $^2\Sigma^+_{1/2g}$. At large R, these states are labeled according to case-c coupling rules, which may result in a different ordering; however, this ordering must be consistent with the appropriate noncrossing rules. Wadt [10] has shown that, for Ar_2^+, Kr_2^+ and Xe_2^+, a crossing occurs between the $^2\Pi_{1/2g}$ and $^2\Pi_{3/2u}$ states at intermediate R. The ground and first three excited states, given above, dissociate to $X(^1S_0) + X^+(^2P^o_{3/2})$; the remaining two excited states dissociate to $X(^1S_0) + X^+(^2P^o_{1/2})$. Fig. 1 gives a schematic potential energy diagram for X_2^+, in which the dissociation energies and spin-orbit splitting are patterned after those of Xe_2^+ [11]. Following Mulliken's convention [1], the states are labeled A $^2\Sigma^+_u$, B $^2\Pi_g$, C $^2\Pi_u$ and D $^2\Sigma^+_g$. Huber and Herzberg [12] have renamed the ground state X but have not changed the notation for the excited states.

By analogy with the above, and from recent theoretical calculations by Bender and Winter [13] on the potential curves of the heteronuclear dimer ions XY^+ (where X has the greater ionization potential), it is possible to construct a schematic potential energy diagram for the heteronuclear ions. A total of six ionic states remain; however, the three lower-energy curves now dissociate to $X(^1S_0) + Y^+(^2P^o_{3/2\text{ or }1/2})$ while the three higher-energy curves dissociate to $X^+(^2P^o_{3/2\text{ or }1/2}) + Y(^1S_0)$. Representative potential energy curves for XY^+, patterned after the $ArKr^+$ curves

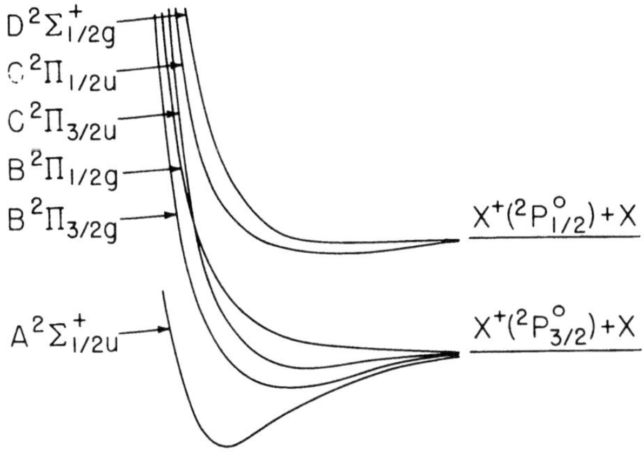

Fig. 1. Schematic potential energy curves for the homonuclear rare-gas dimer ions.

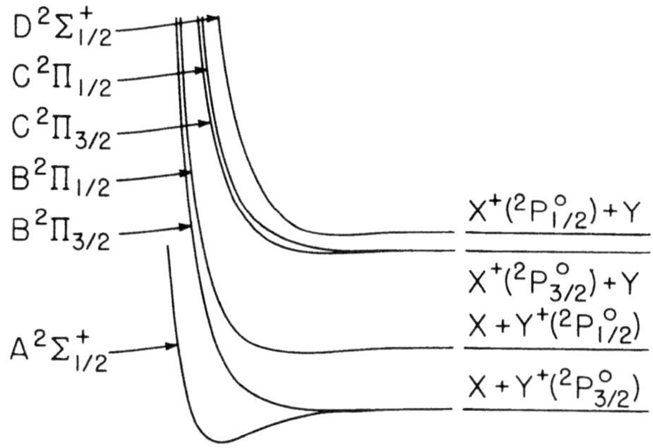

Fig. 2. Schematic potential energy curves for the heteronuclear rare-gas dimer ions.

calculated by Bender and Winter [13], are shown in Fig. 2. Again, the ground ionic state is the most strongly bound. (For HeY^+, only one state dissociates to $He^+(^2S_{1/2}) + Y(^1S_0)$ and this state is more strongly bound than the ground ionic state in $HeAr^+$ $HeKr^+$ and $HeXe^+$.) By analogy with the homonuclear dimers, we label the ground $^2\Sigma^+_{1/2}$ state A, the two $^2\Pi$ states B and C (in order of increasing energy) and the highest-energy, $^2\Sigma^+_{1/2}$ state D. (Note that Dabrowski and Herzberg [14], in their analysis of $HeNe^+$, and Dabrowski, Herzberg and Yoshino [15], in their analysis of $NeAr^+$, label the ground state X and the first two excited states A_1 and A_2 (not B $^2\Pi_{3/2}$ and B $^2\Pi_{1/2}$). The notation of Huber and Herzberg [12] for the heavier diatomics also differs from that employed here.)

The photoionization spectrum of X_2 or XY consists of contributions both from direct ionization and from autoionization of Rydberg states of the neutral molecule. Since the Franck-Condon overlap between the ground state of the neutral molecule and the ground state of the ion is poor for several tenths of a volt above the ionization threshold [11,16], the photoionization spectrum in the region of the ionization threshold consists solely of autoionization structure. At shorter wavelengths, the autoionization structure appears superimposed on the direct-ionization continuum. In the present work, we are concerned primarily with molecular autoionization features and their relationship to the atomic Rydberg states.

For the homonuclear diatomic molecules, the most intense transitions are to states that dissociate to $X^* + X(^1S_0)$, where X^* is an excited atomic state that is connected to the ground state via a dipole-allowed transition. There are five dipole-allowed atomic Rydberg series in X, three of which ($ns[3/2]^o_1$, $nd[1/2]^o_1$ and $nd[3/2]^o_1$) converge to the $^2P^o_{3/2}$ ionic limit, and two of which ($ns'[1/2]^o_1$ and $nd'[3/2]^o_1$) converge to the $^2P^o_{1/2}$ ionic limit. Each of the excited states in these series may combine with a rare-gas atom in the ground state to form both 0^+_u and 1_u dipole-allowed molecular states. In addition, there will be a much larger number of weaker 0^+_u and 1_u molecular Rydberg series resulting from transitions to states that dissociate to $X^*(np, np', nf, nf',...) + X(^1S_0)$ and $X^*(ns, ns', nd, nd')_{J\neq 1} + X(^1S_0)$. Similar considerations apply to the heteronuclear diatomic molecules.

The most significant advantage of photoionization mass spectrometry over photoabsorption, for the study of the Rydberg states of the homonuclear and heteronuclear rare-gas dimers, is the ability of the former technique to discriminate against atomic structure through mass analysis of the product ions. In photoabsorption, the molecular structure is completely masked by the much more intense atomic structure near all of the optically allowed atomic Rydberg transitions [17]. This is not a serious problem near the low-lying atomic Rydberg transitions since the atomic structure is widely spaced; however, in the region between the atomic fine-structure thresholds, the atomic structure is not discrete but, rather, is a true (albeit structured) continuum. Thus, it is impossible to observe weak dimer-structure superimposed on this very intense atomic continuum in photoabsorption studies.

The obvious -- and only -- serious disadvantage of the present technique is the inability to measure these spectra at high wavelength-resolution. At present, the best resolution used to study a van der Waals molecule in a photoionization mass spectrometry experiment is 0.07 Å for Ar_2 [4]. The spectra reported here were determined at a resolution of 0.07 to 0.45 Å. This resolution frequently is sufficient to resolve much of the structure (indeed, many features are broadened as a result of autoionization and/or predissociation) but higher resolution is necessary to resolve all of the details of various vibrational progressions.

3. EXPERIMENTAL APPARATUS

A photoionization mass spectrometer incorporating a high-intensity VUV light source was used for these studies. The apparatus consists of a helium continuum light source; a GCA/McPherson model 225M16 1-meter, near-normal-incidence monochromator which disperses and then refocuses the selected light onto the exit slit; a set of ion-extraction and focusing lenses; and a differentially-pumped quadrupole mass filter for ion detection.

The Hopfield continuum of helium is generated in a 50 cm long capillary lamp by a high-voltage pulsed dc discharge. The lamp is operated at a pressure of approximately 100 torr, with helium admitted at the anode (the end nearest the monochromator entrance slit). The gas flows through the entrance slit into the monochromator; however, an auxillary pump connected to the cathode maintains a substantial gas flow from anode to cathode, which reduces slit clogging from sputtered material passing through the slits [18]. A Velonex model 350 high-power pulse generator is used as the driver for two additional stages of high-power pulse amplification. The final output stage is a pair of Machlett 6544 triode tubes in parallel [19]. The photon flux from this lamp at the peak of the helium continuum (820 Å) is approximately 1×10^{10} photons/sec/Å.

Fig. 3 shows an overview of the ionization region, ion optics and quadrupole mass filter. In the present experiments, the ionization region is defined by a small gridded cage which is

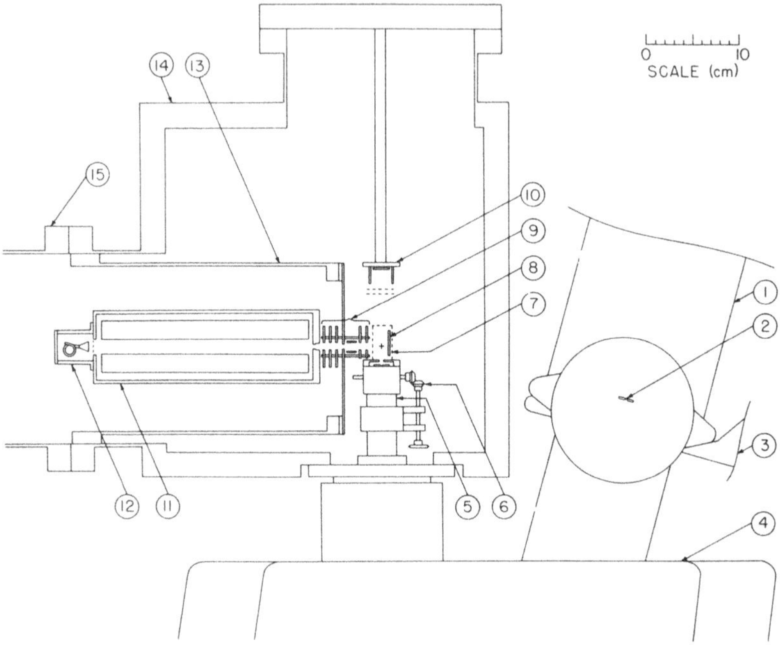

Fig. 3. Top view of the ionization and detection regions of of the photoionization mass spectrometer, showing (1) lamp housing; (2) monochromator entrance slit; (3) differential-pumping assembly for the lamp; (4) monochromator body; (5) exit-slit assembly; (6) exit-slit adjust mechanism; (7) gridded cage defining the ionization region; (8) ion-repeller plate; (9) ion-focusing lenses; (10) aluminum oxide photodiode; (11) quadrupole mass filter and housing; (12) channeltron electron multiplier and housing; (13) differential-pumping septum; (14) main vacuum chamber; and (15) detector vacuum chamber.

attached to the exit arm of the monochromator. Photoions are repelled out of this cage by a flat repeller plate and are focused onto the entrance aperture of the quadrupole mass filter (Extranuclear model 4-324-9) by the simple stack of focusing and deflection plates. In normal operation, the cage and exit arm of the monochromator are floated at a small positive potential (+15 V) and the rf shield of the quadrupole mass filter is grounded. The use of higher ion-injection energies (of up to +100 V) increases the ion-collection efficiency but both the mass resolution and the shape of the mass peak are significantly degraded.

Mass-selected ions are detected with a channeltron electron multiplier using standard pulse-counting techniques. Photons are detected with an aluminum oxide photodiode, the output of which is measured by a vibrating-reed electrometer and subsequently fed into a voltage-to-frequency converter.

Typical ion backgrounds, with either the photon beam flagged or the molecular beam off, are a few counts per minute and are due solely to multiplier dark counts, with no contribution from the high-voltage pulser. Thus, gating of the ion and photon detectors is unnecessary. The primary source of ion background during an experiment comes from ionization due to scattered light. The scattered-light background measured at wavelengths shorter than 580 Å (where there is no light from the helium continuum) is approximately 0.2% of the intensity at the peak of the continuum. This background cannot be gated out by chopping either the photon beam or the molecular beam.

The present experiments use a free supersonic molecular-beam source to produce the rare-gas dimer species. The molecular beam is parallel to, and approximately 1.8 cm from, the exit slit of the monochromator and the jet tip is positioned so that it is as close as possible to the photon beam without intersecting it. The jet is adjustable in three dimensions in order to maximize overlap of the molecular beam with the photon beam. The jet is fabricated by laser microdrilling a hole in a 25 μm thick stainless-steel disk, which is then heliarc welded onto the end of a stainless steel tube.

The jet uses no differential pumping and has the advantages of ease of fabrication and of the ability to position the tip to within 1 mm of the photon beam, resulting in very high number densities in the ionization region. Typical number densities in a skimmed molecular beam are (1-5) x 10^{12} molecules/cm^3 [(0.3-1.4) x 10^{-4} torr] [20-22], which is at least two orders of magnitude less than the number densities obtained using the present experimental arrangement. However, the high mass-throughput means that with the unskimmed jet the main vacuum chamber operates at high background pressures, occasionally as high as 1 x 10^{-4} torr. As a result, secondary processes such as associative ionization, collisional ionization and ion-molecule reactions may occur. In the present experiments, these secondary processes produce no significant interference and the very large number densities in the ionization region enable high-resolution experiments to be performed.

4. DEPENDENCE OF THE CLUSTER ION SIGNALS ON NOZZLE STAGNATION PRESSURE

The dimer and heavier cluster ion signals depend critically upon the combination of nozzle diameter and nozzle stagnation pressure. For example, Fig. 4 shows a plot of the pressure dependence of the Ar_n^+ ion signals, for n=2-6, measured at a resolution of 0.45 Å. At this wavelength, which is above the $Ar^+\ ^2P^o_{1/2}$

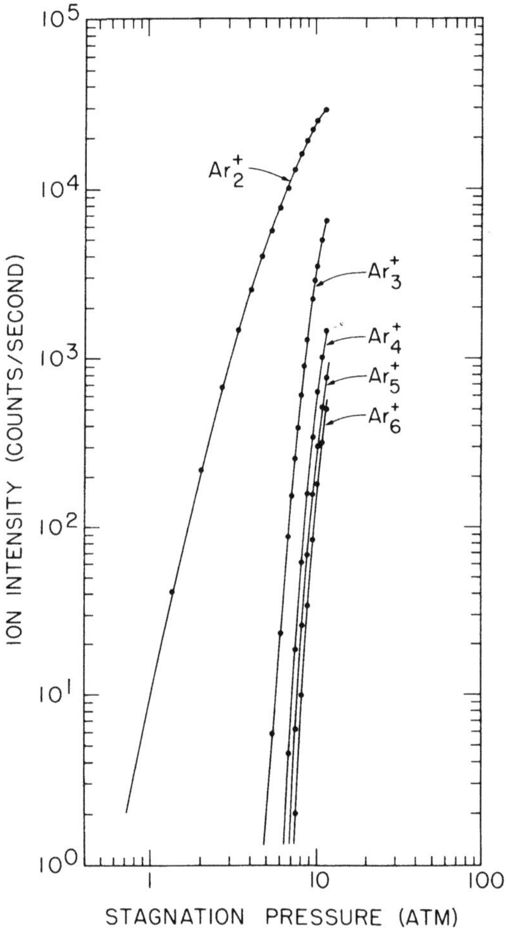

Fig. 4. Dependence of the Ar_n^+ ion signal on nozzle stagnation pressure. The data were taken at a wavelength of 772.5 Å with a resolution of 0.45 Å using pure argon expanded at room temperature through a 10 μm diameter nozzle.

ionization limit, neither the atom nor any cluster exhibits sharp structure. The ordinate in Fig. 4 shows the actual ion counting rate obtained with a 10 μm diameter orifice. Under these conditions, the monomer signal increases linearly with stagnation pressure and has a value of 5.5×10^4 counts per second at 2.0 atm. The shapes and relative intensities of the curves in Fig. 4 are insensitive to small changes (<20%) in orifice diameter; however, the exact counting rates, especially for the heavier clusters, are very sensitive to the orifice size. A small decrease in orifice diameter uniformly shifts the curves for the heavier clusters to higher stagnation pressures. For a large change in orifice diameter (>50%), the shapes and relative intensities of the curves are also altered. It has been observed in previous studies [23,24], and may be seen in Fig. 4, that the dimer can be prepared in the beam over a wide range of pressures where the presence of larger clusters is not detectable. Preparation of a beam containing only monomer, dimer and trimer is much more difficult, and perhaps impossible, since the tetramer and larger clusters are observed at a stagnation pressure only slightly higher than that at which the trimer is first observed.

There have been several earlier studies on condensation in supersonic beams of Ar using electron-impact ionization and varying the nozzle temperature, orifice diameter and stagnation pressure [23-27]. In a previous electron-impact ionization study [24], which used large orifices (>100 μm) and a skimmed jet, the dimer-ion signal was observed to rise to a maximum, to decrease and then to rise again with increasing stagnation pressure. It was postulated that the decrease in dimer signal is due to condensation to form larger clusters, while the second rise is due to fragmentation of larger clusters into dimer. The present photoionization results with an unskimmed jet are consistent with the previous experiments. In all of the results reported here, the nozzle stagnation pressure was adjusted so as to produce a beam that contained primarily monomer and dimer in order to minimize contributions to the dimer-ion signal from fragmentation.

5. APPEARANCE POTENTIALS OF THE DIMER IONS AND DISSOCIATION ENERGIES OF THE DIMER-ION GROUND STATES

An accurate value of the adiabatic ionization potential (IP) of a rare-gas dimer is required to calculate the dissociation energy (D_0) of the corresponding dimer ion since

$$D_0(XY^+) = IP(Y) + D_0(XY) - IP(XY) \quad . \qquad (1)$$

A fundamental problem for rare-gas dimers is that, owing to the large change in geometry that occurs upon ionization, the Franck-

Condon factor for ionization to the vibrational state is extremely small. For example, the Franck-Condon factor for the argon dimer is negligible up to approximately 0.8 eV above the adiabatic ionization potential [16]. Ionization is observed in the threshold region but is due solely to autoionization. Thus, the adiabatic ionization potential will be observed only if there is a high density of autoionizing molecular Rydberg states in the threshold region.

The study of the appearance-potential pressure dependence for a rare-gas dimer helps determine whether or not the adiabatic ionization potential is actually observed. A comparison of the appearance potential obtained with a low-pressure beam (containing only monomer and dimer) and with a high-pressure beam (containing a mixture of heavy clusters) is very instructive. A shift of the appearance potential to longer wavelengths with increasing nozzle stagnation pressure is evidence that the appearance potential is not the adiabatic ionization potential. A null shift of the appearance potential suggests that the adiabatic ionization potential is observed.

Fig. 5 shows the Ar_2 threshold region taken at a wavelength resolution of 0.45 Å and at nozzle stagnation pressures of 3.06 and 10.18 atm. In both runs, the threshold region shows a weak onset which increases only slowly as a function of excess energy. This gradual rise makes the determination of the appearance potential extremely difficult and the effects of collisional ionization must be considered carefully. Collisional ionization occurs when neutral dimers are excited to high-Rydberg states and gain the energy needed to ionize via collisions. Evidence against significant interference from collisional ionization comes from an examination of the amount of collisional ionization in the photoionization cross-section of atomic argon taken under similar experimental conditions. At approximately 1.0 Å below the Ar^+ $^2P^o_{3/2}$ threshold, the intensity of collisional ionization is less than 1% of the intensity of the continuum ionization. Collisional ionization is expected to be even weaker in molecules since molecular Rydberg states may decay by predissociation, a process which is faster than any of its competitors -- fluorescence, associative ionization or collisional ionization. The latter three decay processes are the only ones available to an excited atom. Collisional dissociation of the excited van der Waals molecule may also compete with collisional ionization. Finally, collisional ionization will be structured, reflecting the absorption cross-section [28], rather than monotonically increasing.

Using the guideline that at 1 Å above the appearance potential the intensity of collisional ionization is 1% of the intensity at the band maximum, and taking into account the effect of

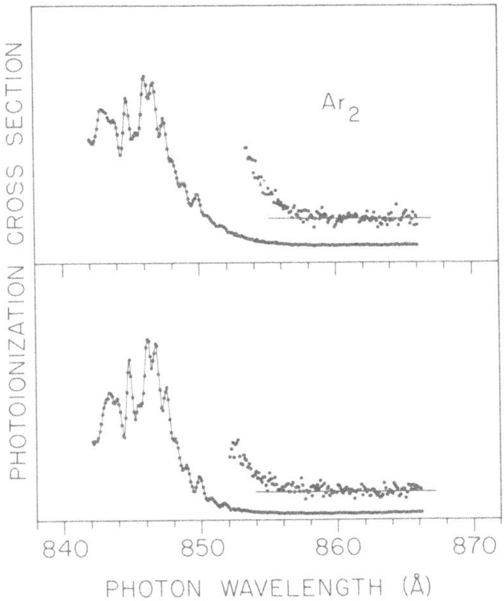

Fig. 5. Relative photoionization cross-section for Ar_2^+ near threshold taken at a wavelength resolution of 0.45 Å. The Ar_2 was produced in a supersonic molecular-beam expansion of pure argon at room temperature using a 12 μm diameter jet and nozzle stagnation pressure of 3.06 atm (lower frame) and 10.16 atm (upper frame). The region from 852-866 Å in each frame is expanded by a factor of ten; the baseline of the expanded portions has been shifted to keep the figure compact.

instrumental resolution [29], the appearance potential of Ar_2^+ is determined to be 855.0 ± 1.5 Å for both sets of data shown in Fig. 5. Although the threshold region of the high-pressure run rises noticeably more steeply than that of the low-pressure run as a result of fragmentation, the appearance potential does not shift.

The Kr_2^+ threshold is shown in Fig. 6 and, like the Ar_2^+ threshold, it does not shift with nozzle stagnation pressure. A sharp step is observed at threshold and there is no evidence of tailing (i.e., of collisional ionization). The steplike behavior of the Kr_2^+ cross-section at threshold is qualitatively different from that of Ar_2^+, which shows a gradual onset. The absence of collisional ionization in Kr_2^+ suggests that the tailing in

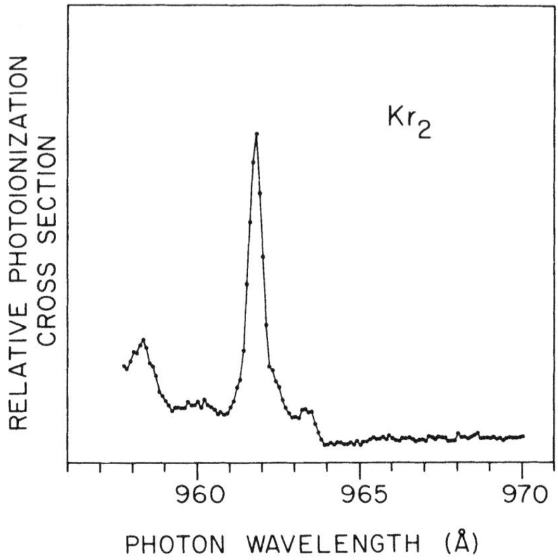

Fig. 6. Relative photoionization cross-section for Kr_2 taken at a wavelength resolution of 0.45 Å in the region near the ionization threshold.

the Ar_2^+ cross-section is real and not due to collisional ionization. This, of course, would shift the Ar_2^+ appearance potential to slightly longer wavelengths.

Table 1 summarizes the best values of the appearance potential of the rare-gas dimer ions and the values of the ionization potentials of the free atoms; Table 2 summarizes the values of the ground-state dissociation energies of the neutral molecules; and Table 3 summarizes the values of the ground-state dissoçiation energies of the ions calculated using Eq. (1) and the data from Tables 1 and 2.

Several trends are immediately apparent from the data in Table 3. The dissociation energies of the homonuclear dimer ions are much larger than those of the heteronuclear dimer ions. Furthermore, the dissociation energy of a heteronuclear dimer ion XY^+ decreases as the mass of Y increases for constant X, and as the mass of X decreases for constant Y. These trends can be explained qualitatively using simple molecular orbital theory. Briefly, it can be shown that the bond dissociation energy for a heteronuclear dimer ion is approximately given by [30]

$$D_0 \simeq \frac{H_{ab}^2}{H_{aa} - H_{bb}} \quad . \tag{2}$$

Table 1. Ionization Potentials of the Rare-Gas Dimers (eV)

	He	Ne	Ar	Kr	Xe
He					
Ne		20.21 ± 0.05[a]	15.685 ± 0.004[b]	13.950 ± 0.003[b]	12.094 ± 0.004[b]
Ar			14.50 ± 0.03[c]	13.484 ± 0.015[d]	11.968 ± 0.012[d]
Kr				12.866 ± 0.003[e]	11.763 ± 0.011[d]
Xe					11.127 ± 0.010[f]
Atomic IP	24.587[g]	21.565[g]	15.760[g,h]	14.000[g,i]	12.130[g]

[a] Ref. [9].
[b] Ref. [7].
[c] P. M. Dehmer and S. T. Pratt, J. Chem. Phys. 76, 843 (1982).
[d] Ref. [8].
[e] Ref. [5].
[f] C. Y. Ng, D. J. Trevor, B. H. Mahan and Y. T. Lee, J. Chem. Phys. 65, 4327 (1976).
[g] Ref. [35].
[h] K. Yoshino, J. Opt. Soc. Am. 60, 1220 (1970).
[i] K. Yoshino and Y. Tanaka, J. Opt. Soc. Am. 69, 159 (1978).

Table 2. Dissociation Energies of the Rare-Gas Dimers (cm^{-1})[a]

	He	Ne	Ar	Kr	Xe
He	---	1.9	8.2	7.0	8.3
Ne		17.3[b]	35.9	38.7	39.1
Ar			84.0[c]	100.6	115.3
Kr				126.8[d]	148.3
Xe					185.2[e]

[a]For all of the heteronuclear diatomic molecules, the tabulated values are from M. V. Bobetic and J. A. Barker, J. Chem. Phys. 64, 2367 (1976).
[b]D. E. Freeman, K. Yoshino and Y. Tanaka, J. Chem. Phys. 67, 3462 (1977).
[c]R. J. LeRoy, J. Chem. Phys. 57, 573 (1972).
[d]Y. Tanaka, K. Yoshino and D. E. Freeman, J. Chem. Phys. 59, 5160 (1973).
[e]D. E. Freeman, K. Yoshino and Y. Tanaka, J. Chem. Phys. 61, 4880 (1974).

Here H_{ij} are the usual Hamiltonian integrals $\int \phi_i H \phi_j dr$, where ϕ_i and ϕ_j are the atomic orbitals and where the integration is performed over all space. The denominator of Eq. (2) is usually taken to be the difference in ionization potentials of the free atoms. This makes use of Koopmans' theorem [31], which approximates the atomic orbital energy by the ionization potential corresponding to the removal of an electron from that orbital. Eq. (2) further assumes that the effect, on the atomic energies, of the presence of the other atom is small. Finally, Eq. (2) predicts that the binding energy increases as the difference in ionization energies of the atomic rare-gas partners decreases, which is in agreement with observation in both the rows (constant X) and the columns (constant Y) of Table 3.

If one assumes that the charge on a heteronuclear dimer ion resides primarily on one of the atoms, then the rows in Table 3 are composed of dimers having the same neutral partner throughout while the columns are composed of dimers having the same charged partner throughout. In addition to changes in dissociation energy caused by the difference in orbital energies of the free atoms, one must also consider those changes resulting from a

Table 3. Ground-State Dissociation Energies of the Rare-Gas Dimer Ions (eV)

	He	Ne	Ar	Kr	Xe
He	2.365 ± 0.006[a]	0.692 ± 0.037[b]	0.026 ± 0.002[c]		
Ne		1.36 ± 0.05[d]	0.079 ± 0.004[e]	0.055 ± 0.003[e]	0.041 ± 0.004[e]
Ar			1.269 ± 0.017[f]	0.528 ± 0.015[g]	0.176 ± 0.012[g]
Kr				1.150 ± 0.004[h]	0.385 ± 0.011[g]
Xe					1.03 ± 0.01[i]

[a] Ref. [12], p. 298.
[b] Ref. [14].
[c] Ref. [15].
[d] Ref. [9].
[e] Ref. [7].
[f] P. M. Dehmer and S. T. Pratt, J. Chem. Phys. 76, 843 (1982).
[g] Ref. [8].
[h] Ref. [5].
[i] C. Y. Ng, D. J. Trevor, B. H. Mahan and Y. T. Lee, J. Chem. Phys. 65, 4327 (1976).

variation in the polarizability of the neutral partner. A further effect, although one which is sometimes difficult to observe because of stronger competing effects, results from the change in R_e throughout the rows and columns.

For systems in which chemical effects are small, a substantial part of the binding should be due to the ion-induced dipole interaction, which is given by $-\alpha e^2/2R^4$, where α is the atomic polarizability. In the sequence $NeXe^+$, $ArXe^+$, $KrXe^+$, the increase in the ion-induced dipole energy from $NeXe^+$ to $KrXe^+$, which results from the increase in polarizability, is more than a factor of 6 [32]; the decrease in the ion-induced dipole energy caused by the increase in R_e, however, is only about a factor of 2 [33]. We conclude that the increase in dissociation energy in going from $NeXe^+$ to $KrXe^+$, or from $NeKr^+$ to $ArKr^+$, is due to increased polarizability of the neutral partner as well as to increased chemical effects predicted by the difference in orbital energies of the free atoms.

The numerator in Eq. (2) [i.e., the "interaction energy"] is often assumed to be proportional to the overlap integral S_{ab} ($\equiv \int \phi_a \phi_b dr$) [34]. This overlap is expected to increase as R_e decreases, an expectation which is in agreement with the observed trend in the rows of Table 3 and is the same as the trend predicted by the difference in ionization potentials of the free atoms. In the columns of Table 3, even though R_e decreases in going from $KrXe^+$ to $NeXe^+$, the dissociation energy decreases because of the overwhelming effects of (i), the increased difference in orbital energies of the free atoms; and (ii), the rapid decrease in polarizability of the neutral partner.

The dissociation energies of the ground states of the homonuclear dimer ions follow the same trend as that observed for the heteronuclear rare-gas dimer ions, namely, $D_0(Ar_2^+) > D_0(Kr_2^+) > D_0(Xe_2^+)$; however, the explanation is more complicated. The decrease in dissociation energy in going from Ar_2^+ to Xe_2^+ depends upon a number of factors and is ultimately the result of spin-orbit interaction. Recent calculations [10] show that, in the absence of spin-orbit coupling, the dissociation energies show only a very slight decrease with increasing reduced mass.

6. MOLECULAR RYDBERG STRUCTURE BETWEEN THE ATOMIC FINE-STRUCTURE THRESHOLDS: DISSOCIATION ENERGIES OF $X_2^+(C\ ^2\Pi_{1/2u})$ AND $XY^+(B\ ^2\Pi_{1/2}$ and $D\ ^2\Sigma^+_{1/2})$

The molecular spectra of the homonuclear and heteronuclear rare-gas dimers show strong similarities to the corresponding atomic spectra in the energy region between the atomic $^2P^o_{3/2}$ and $^2P^o_{1/2}$ fine-structure thresholds. In the homonuclear dimers, molecular Rydberg series are observed that converge to the C $^2\Pi_{1/2u}$ and D $^2\Sigma^+_{1/2g}$ states of the dimer ion (see Fig. 1); however, the D $^2\Sigma^+_{1/2g}$ state is either very weakly bound or purely repulsive, and molecular Rydberg states converging to it are not seen in photoionization. In the heteronuclear dimers, the molecular Rydberg series converge to either the B $^2\Pi_{1/2}$ or D $^2\Sigma^+_{1/2}$ state of the ion, depending upon whether the energy region corresponds to the region between the Y^+ or the X^+ fine-structure thresholds (see Fig. 2). This is a particularly simple region of the spectrum for most of the heteronuclear rare-gas dimers since all of the molecular Rydberg states converge to a single ionic limit. In addition, much of the molecular structure retains the characteristic Beutler-Fano profiles observed in the atomic spectra.

Figs. 7 through 9 summarize the results in the following format: The uppermost plot in each figure shows the atomic spectrum in the region between the $^2P^o_{3/2}$ and $^2P^o_{1/2}$ fine-structure thresholds. Figs. 7, 8 and 9 present the results for Ar, Kr and Xe, respectively. The additional plots in each figure show the homonuclear and heteronuclear dimer spectra in the same energy region as the corresponding atomic spectrum. (Note that, in each of the figures, one of the heteronuclear dimer spectra is missing, either because it is too weak to be observed on the direct-ionization background or because it is masked by Rydberg structure associated with other series.) Fig. 9 shows the spectra of Xe, NeXe, ArXe and Xe_2 and is a striking example of the progression from a spectrum exhibiting a strong resemblance to the atomic spectrum (NeXe), to one exhibiting only a modest resemblance to the atomic spectrum (Xe_2). Similar results are obtained for the Ar series (Fig. 7) and the Kr series (Fig. 8).

The effective principal quantum number n* of each molecular state was calculated from

$$n^* = [R/(E_\infty - E_{obs})]^{1/2} \quad (3)$$

where R is the Rydberg constant; E_{obs} is the observed excitation energy; and E_∞ is the vertical ionization potential of the ionic convergence limit. In all cases, peak positions were taken to be the band maxima. Because of the pronounced Beutler-Fano profiles in some of the bands, band maxima can be red-shifted from their respective centerlines. Care was taken to choose the atomic and molecular peak positions in a consistent manner. For the homonuclear dimers, the vertical ionization potential of the C $^2\Pi_{1/2u}$ state, which is obtained from high-resolution photoelectron spectra [11,16], is used as the series limit. Since no photoelectron spectra have been reported for the heteronuclear dimers, an initial estimate of the vertical ionization potential of the ionic

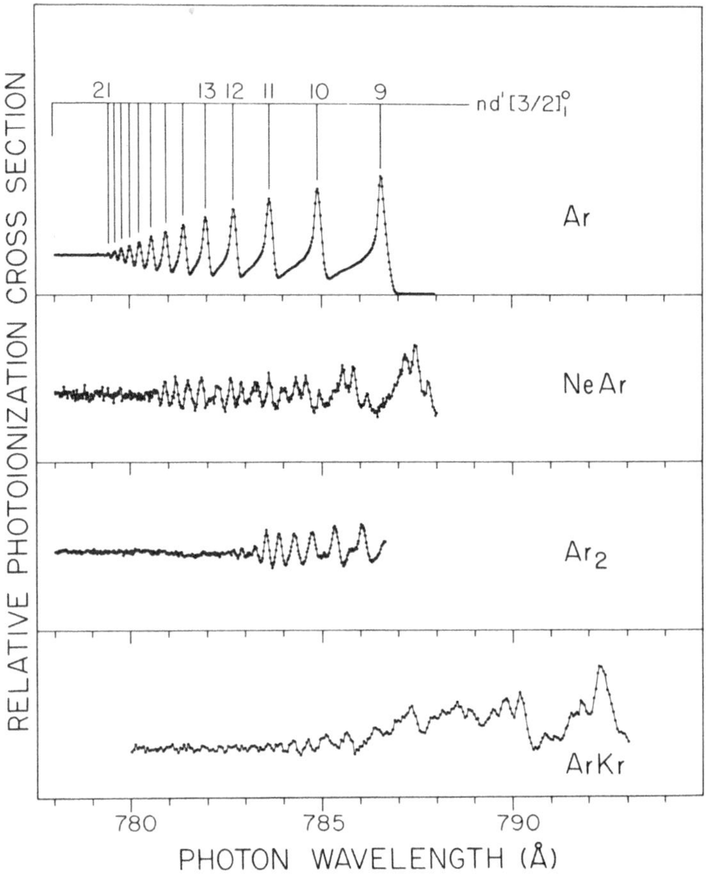

Fig. 7. Relative photoionization cross-sections for Ar, NeAr, Ar_2 and ArKr in the energy region between the Ar fine-structure thresholds. The wavelength resolution is 0.28 Å for the ArKr spectrum and 0.15 Å for the remaining spectra.

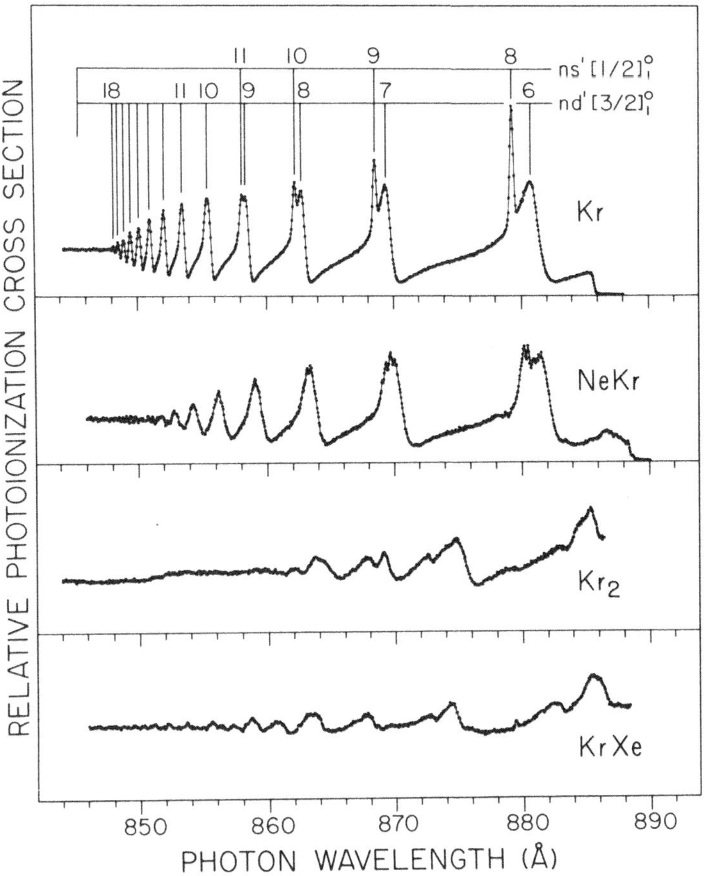

Fig. 8. Relative photoionization cross-sections for Kr, NeKr, Kr_2 and KrXe in the energy region between the Kr fine-structure thresholds. The wavelength resolution is 0.28 Å for all spectra.

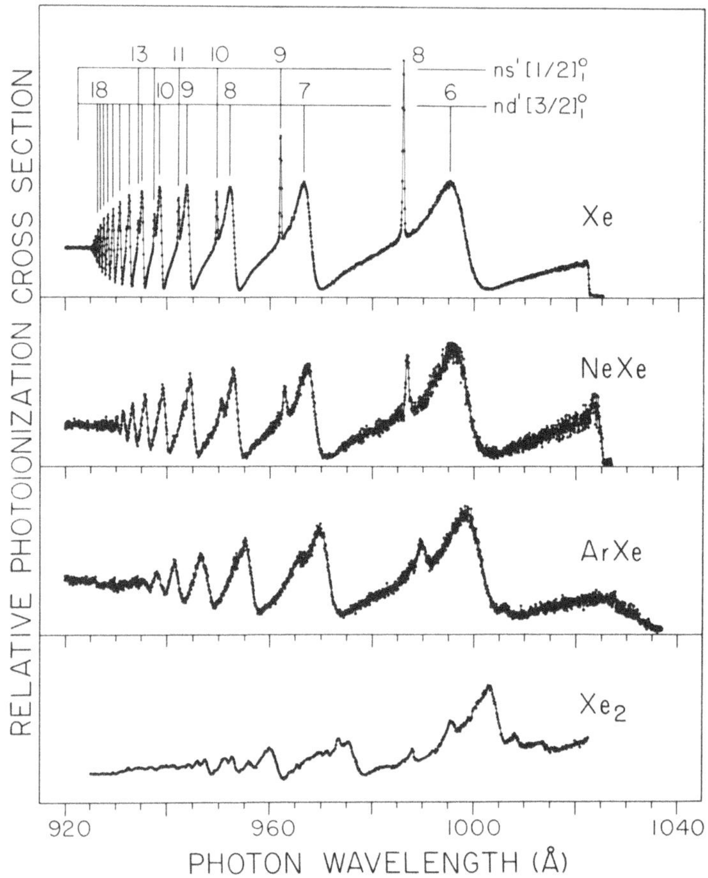

Fig. 9. Relative photoionization cross-sections for Xe, NeXe, ArXe and Xe_2 in the energy region between the Xe fine-structure thresholds. The wavelength resolution is 0.45 Å for Xe_2 and 0.28 Å for the remaining systems.

convergence limit was obtained by shifting the atomic ionization potential by the same amount that the molecular peaks are shifted from the corresponding atomic peaks. The value of E_∞ was varied until the most nearly constant set of values for the non-integral part of n* was obtained; and the final value of E_∞ was taken to be the vertical ionization potential of the ionic state. Since there is only one ionic state to which these series may converge, the ns' and nd' series must both be fit by the same value of E_∞.

The dissociation energy of a specific Rydberg state is given by

$$D_0(XY^*) = E(X^* \text{ or } Y^*) + D_0(XY) - E_{obs}(XY^*) + \Delta \tag{4}$$

where E(X* or Y*) is the excitation energy of the atomic Rydberg state to which XY* dissociates; $D_0(XY)$ is the dissociation energy of the neutral ground state; and Δ is the difference in energy between the vertical and adiabatic excitation energies (i.e., the difference in energy between the transition to the maximum of the Franck-Condon envelope and the transition to the v'=0 level of the Rydberg state). Because the potential energy curves of highly-excited Rydberg states are similar to the potential energy curve of the ionic convergence limit, Δ was estimated, for the homonuclear dimers, to be the difference between the vertical and adiabatic ionization potentials, with both of the latter quantities being obtained from photoelectron spectra. No values of the adiabatic ionization potentials exist for the B $^2\Pi_{1/2}$ and D $^2\Sigma^+_{1/2}$ states of heteronuclear dimers; hence, only values of the quantity $(D_0-\Delta)$ are reported. For both the homonuclear [11,16] and the heteronuclear dimers, Δ is expected to be a small fraction of the dissociation energy.

Tables 4 and 5 give representative results for a homonuclear diatomic molecule (Ar_2) and a heteronuclear diatomic molecule (ArXe), respectively. Included in these tables are the molecular peak positions in Å and eV; the value of n* calculated from Eq. (3); the energetics of the assumed dissociation limit, including the value of n* for the atomic transition corresponding to the dissociation limit; and the value of $(D_0-\Delta)$ for each assigned molecular transition and for the series limit (i.e., the ionic state). Table 6 summarizes the values of D_0 and $(D_0-\Delta)$ for the excited states of all of the rare-gas dimers and includes values determined from a variety of spectroscopic and molecular-beam scattering measurements.

At large internuclear separation, the potentials of the excited states of the rare-gas dimer ions are described by dispersion forces, with the dominant contribution arising from the ion-induced dipole term given by $-\alpha e^2/2R^4$. This potential has been evaluated at the equilibrium internuclear separation of each neutral rare-gas dimer: The results are compared with experiment in Table 7. With respect to the homonuclear dimers, agreement is fair for Ar_2 but becomes progressively worse for Kr_2 and Xe_2 This variation is best understood by remembering that spin-orbit coupling mixes the A $^2\Sigma^+_{1/2u}$ and C $^2\Pi_{1/2u}$ ionic states [1]. This mixing, which increases from Ar_2 to Xe_2, results in a decrease in dissociation energy of the relatively strongly-bound A $^2\Sigma^+_{1/2u}$ state, with a concomitant increase in dissociation energy of the

Table 4. Bands of Ar_2 Between the Ar^+ $^2P^o_{3/2}$ and $^2P^o_{1/2}$ Thresholds

Observed molecular bands			Energetics of the atomic dissociation limit[a]				
λ(Å)	E(eV)	n*	Assumed Dissociation Limit	λ(Å)	E(eV)	n*	$(D_0-\Delta)$ (meV)[b,c]
786.048	15.7732	11.80	12d'	782.74	15.8399	11.82	77.0
785.329	15.7877	12.78	13d'	782.02	15.8545	12.82	77.2
784.736	15.7996	13.81	14d'	781.45	15.8660	13.82	76.9
784.278	15.8088	14.79	15d'	780.99	15.8753	14.82	76.9
783.884	15.8168	15.84	16d'	780.62	15.8829	15.82	76.5
783.550	15.8235	16.93	17d'	780.31	15.8892	16.8	76.1
783.251	15.8296	18.12	18d'	780.05	15.8944	17.8	75.3
783.080	15.8330	18.93	19d'	779.84	15.8989	18.8	76.3
782.900	15.8367	19.90	20d'	779.65	15.9026	19.8	76.4
782.705	15.8406	21.16	21d'	779.49	15.9059	20.8	76.8
781.206[d]	15.8710	∞		777.957	15.9373	∞	76.7

[a]The assignments and positions of the atomic Rydberg series are from K. Yoshino, J. Opt. Soc. Am. 60, 1220 (1970).

[b]Calculated using $D_0(Ar_2)$ = 10.4 meV from R. J. LeRoy, J. Chem. Phys. 5, 573 (1972).

[c]$\Delta \simeq 15 \pm 10$ meV.

[d]The series limit is taken as the vertical ionization potential of the C $^2\Pi_{1/2u}$ state from Ref.[16].

Table 5. Bands of ArXe Between the $Xe^{+}\ ^2P^o_{3/2}$ and $^2P^o_{1/2}$ Thresholds

Observed molecular bands			Energetics of the atomic dissociation limit				
λ(Å)	E(eV)	n^*	Assumed Dissociation Limit	λ(Å)	E(eV)	n^*	$(D_0-\Delta)^a$ (meV)
998.47	12.4175	3.74	6d'	995.47	12.4549	3.72	52
989.79	12.5264	3.97	8s'	986.07	12.5737	3.97	62
969.78	12.7849	4.74	7d'	966.64	12.8264	4.72	56
965.82	12.8373	4.96	9s'	962.01	12.8881	4.93	65
955.18	12.9803	5.76	8d'	952.24	13.0204	5.72	54
953.17	13.0077	5.96	10s'	949.56	13.0571	5.99	64
946.62	13.0977	6.81	9d'	943.84	13.1363	6.73	53
941.55	13.1682	7.82	10d'	938.67	13.2086	7.73	55
938.21	13.2151	8.80	11d'	935.14	13.2585	8.74	58
935.57	13.2524	9.92	12d'	932.64	13.2940	9.77	56
925.90±0.2	13.3908	∞		922.75	13.4365	∞	60±3

[a] Calculated using D_0(ArXe) = 14.3 meV from M. V. Bobetic and J. A. Barker, J. Chem. Phys. 64, 2367 (1976).

Table 6. Excited-State Dissociation Energies of the Rare-Gas Dimer Ions (eV)[a]

	He	Ne	Ar
He	b	$D_0(A_2\ ^2\Pi_{1/2}) = 0.0350 \pm 0.001$[c] $D_0(B\ ^2\Sigma^+) = 0.0363 \pm 0.0001$[c]	$D_0(A_2\ ^2\Pi_{1/2}) = 0.015 \pm 0.002$[d] $D_0(B\ ^2\Sigma^+) = 0.156 \pm 0.002$[d]
Ne			$(D_0-\Delta)(B\ ^2\Pi_{1/2}) = 0.022 \pm 0.002$[f]
Ar			$D_0(B\ ^2\Pi_{3/2g}) = 0.14 \pm 0.02$[g] $D_0(C\ ^2\Pi_{1/2u}) = 0.10 \pm 0.02$[g] $(D_0-\Delta)(C\ ^2\Pi_{1/2u}) = 0.077 \pm 0.002$[g,h]

[a]Depending on availability, the vertical $(D_0-\Delta)$ and/or adiabatic (D_0) dissociation energies are given. The spectroscopic labels are as shown in Figs. 1 and 2, except for HeY. For these molecules, the $A_1\ ^2\Pi_{3/2}$ and $A_2\ ^2\Pi_{1/2}$ excited electronic states dissociate to $He(^1S_0) + Y^+(^2P^o_{1/2})$ and the $B\ ^2\Sigma^+$ excited electronic state dissociates to $He^+(^2S_{1/2}) + Y(^1S_0)$.

[b]The single excited state of He_2^+ is repulsive except for a very small van der Waals minimum at 5.3 Å. See ref. [12], p. 298.

[c]Ref. [14].

[d]Ref. [15].

[e]H. P. Weise, Ber. Bunsenges. Phys. Chem. 77, 578 (1973).

Table 6. Continued.

	Kr	Xe
He	$D_0(B\ ^2\Sigma^+) = 0.22$[e]	$D_0(B\ ^2\Sigma^+) = 0.28$[e]
Ne	$(D_0-\Delta)(B\ ^2\Pi_{1/2}) = 0.018 \pm 0.004$[f]	$(D_0-\Delta)(B\ ^2\Pi_{1/2}) = 0.014 \pm 0.002$[f]
Ar	$(D_0-\Delta)(D\ ^2\Sigma^+_{1/2}) = 0.087 \pm 0.004$[i]	$(D_0-\Delta)(B\ ^2\Pi_{1/2}) = 0.060 \pm 0.003$[i]
Kr	$D_0(B\ ^2\Pi_{3/2g}) = 0.16 \pm 0.02$[g] $D_0(C\ ^2\Pi_{1/2u}) = 0.14 \pm 0.02$[g] $(D_0-\Delta)(C\ ^2\Pi_{1/2u}) = 0.113 \pm 0.002$[g,h]	$(D_0-\Delta)(B\ ^2\Pi_{1/2}) = 0.14 \pm 0.02$[i] $(D_0-\Delta)(D\ ^2\Sigma^+_{1/2}) = 0.106 \pm 0.003$[i]
Xe		$D_0(B\ ^2\Pi_{3/2g}) = 0.18 \pm 0.02$[g] $D_0(C\ ^2\Pi_{1/2u}) = 0.19 \pm 0.02$[g] $(D_0-\Delta)(C\ ^2\Pi_{1/2u}) = 0.145 \pm 0.002$[g,h]

[f] Ref. [7].
[g] Ref. [16].
[h] Ref. [6].
[i] Ref. [8].

Table 7. Evaluation of the Ion-Induced Dipole Potential at R_e of the Neutral Dimers

Molecule	R_e(Å)	α(Å^3)[a]	$V = \frac{-e^2\alpha}{2R_e^4}$ (meV)	Ionic Convergence Limit	$(D_0-\Delta)$ (meV)
NeAr	3.43[b]	0.395	21	B $^2\Pi_{1/2}$	22 ± 2[c]
NeKr	3.58[b]	0.395	17	B $^2\Pi_{1/2}$	18 ± 4[c]
NeXe	3.74[b]	0.395	14	B $^2\Pi_{1/2}$	14 ± 2[c]
Ar_2	3.76[d]	1.642	59	C $^2\Pi_{1/2u}$	76.7 ± 0.7[e]
ArKr	3.88[f]	2.480	79	D $^2\Sigma^+_{1/2}$	87 ± 4[g]
ArXe	4.06[f]	1.642	44	B $^2\Pi_{1/2}$	60 ± 3[g]
Kr_2	4.01[h]	2.480	69	C $^2\Pi_{1/2u}$	112 ± 1.5[e]
KrXe	4.18[f]	2.480	59	B $^2\Pi_{1/2}$	140 ± 20[g]
KrXe	4.18[f]	4.044	95	D $^2\Sigma^+_{1/2}$	106 ± 3[g]
Xe_2	4.36[h]	4.044	81	C $^2\Pi_{1/2u}$	145 ± 1.5[e]

[a]A. Dalgarno and A. E. Kingston, Proc. Roy. Soc. (London) A259, 424 (1960); Proc. Phys. Soc. (London) 78, 607 (1961).

[b]C. Y. Ng, Y. T. Lee and J. A. Barker, J. Chem. Phys. 61, 1996 (1974).

[c]Ref. [7].

[d]R. J. LeRoy, J. Chem. Phys. 57, 573 (1972).

[e]Ref. [6].

[f]D. E. Freeman, K. Yoshino and Y. Tanaka, J. Chem. Phys. 67, 3462 (1977).

[g]Ref. [8].

[h]J. A. Barker, R. O. Watts, J. K. Lee, T. P. Schafer and Y. T. Lee, J. Chem. Phys. 61, 3081 (1974).

C $^2\Pi_{1/2u}$ state. The increase in binding energy of the C $^2\Pi_{1/2u}$ state is considerably greater than that predicted by the ion-induced dipole potential. Hence, while this potential appears to be a reasonable approximation for the C $^2\Pi_{1/2u}$ state of Ar_2^+ and the Rydberg states converging to it (at R = R_e of Ar_2), the increased binding in Kr_2 and Xe_2 caused by spin-orbit mixing makes the ion-induced dipole potential a poor approximation for these latter molecules. The agreement of Table 7 is considerably better for the heteronuclear dimers, especially for those containing Ne; in general, however, the ion-induced dipole approximation is expected to fail at the internuclear distance corresponding to R_e of the ground state. And, this failure may even extend to weakly-bound van der Waals molecules, for which R_e can be quite large.

7. MOLECULAR RYDBERG STRUCTURE IN THE SPECTRA OF NeY NEAR THE Ne 3s ATOMIC RESONANCE

In principle, the energy regime near the lowest-lying Rydberg state (i.e., the resonance transition) of a rare-gas atom will provide the best opportunity for studying molecular states that are well-separated from adjacent levels. The resonance levels of the rare-gas atoms are below the ionization potential of the homonuclear dimers and, therefore, molecular states arising from these dissociation limits cannot be observed by photoionization mass spectrometry; however, this is not the case for the heteronuclear Ne/rare-gas dimers. The Ne*3s resonance levels are above the molecular ionization potential of these dimers; thus, molecular states in this energy region can be studied using photoionization techniques. Only five dipole-allowed molecular Rydberg states arise from the combination of a rare-gas atom, in the $^1S_0$ ground state, and a neon atom in one of the four atomic states arising from the configuration $1s^22s^22p^53s$ (i.e., $3s[3/2]_1^o$, $3s[3/2]_2^o$, $3s'[1/2]_1^o$ and $3s'[1/2]_0^o$). The atomic states with J=1 are the familiar dipole-allowed neon resonance lines and those with J=0 or 2 are the dipole-forbidden (metastable) states employed in neon scattering experiments.

Fig. 10 shows the relative photoionization cross-sections of NeAr, NeKr and NeXe in the wavelength region from 730 to 750 Å. The same NeAr spectrum, together with the NeAr spectrum obtained with an effusive rather than a supersonic molecular-beam source, is shown in Fig. 11. The four atomic states derived from the excited-state neon configuration $1s^22s^22p^53s$ are [35] $3s[3/2]_2^o$ at 746.025 Å, $3s[3/2]_1^o$ at 743.7195 Å, $3s'[1/2]_0^o$ at 741.726 Å and $3s'[1/2]_1^o$ at 735.8962 Å. There are no other excited states of Ne, or of Ar, Kr or Xe, in this wavelength region.

All three molecular spectra show a narrow band, at approxi-

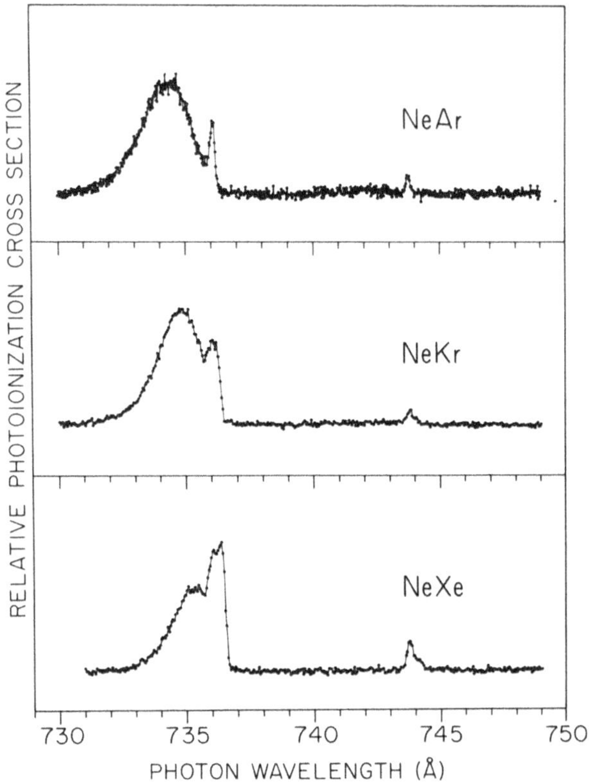

Fig. 10. Relative photoionization cross-sections for NeAr, NeKr and NeXe in the region of the 3s and 3s' resonance transitions. The wavelength resolution is 0.15 Å for NeAr and 0.28 Å for NeKr and NeXe.

mately the position of the $3s'[1/2]_1^o$ transition, and a broader band blue-shifted from this level. The position of the dip between the two bands occurs at the energy of the molecular dissociation-limit (which is blue-shifted from the position of the Ne* $3s'[1/2]_1^o$ transition by an amount equal to the dissociation energy of the NeY ground state). The narrower feature is attributed to a transition to bound levels of a molecular state dissociating to Ne*($3s'[1/2]_1^o$) + Y($^1S_0$). This band is characterized by a sharp onset, red-shifted from the position of the resonance transition, and by a width that is broader than the spectrometer resolution. Both the peak width and the magnitude of the red-shift increase from NeAr to NeXe, thus giving an indication of the binding in this state. As the pressure in the ionization

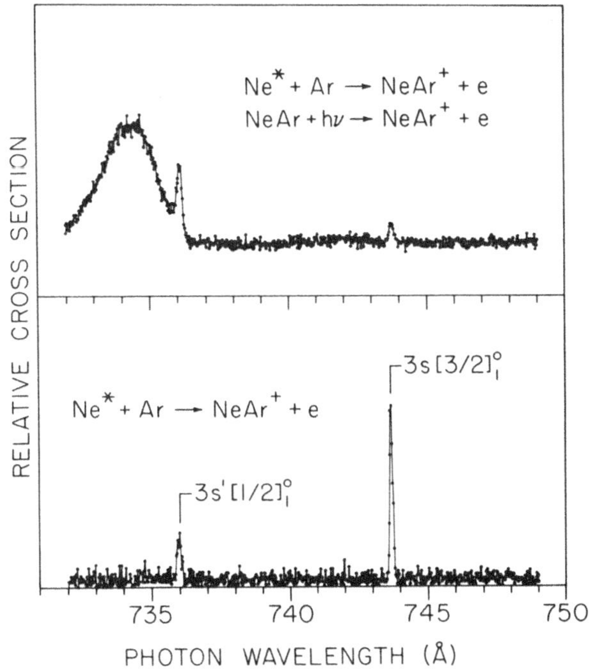

Fig. 11. Relative photoionization cross-sections for NeAr obtained with a supersonic molecular-beam source (upper frame) and an effusive molecular-beam source (lower frame) at a wavelength resolution of 0.15 Å. The dimer concentration in the spectra obtained with the effusive source is negligible and $NeAr^+$ is produced solely by associative ionization.

region is increased, the relative intensity of the narrow band decreases slightly with respect to that of the broad band.

The broad peak which is blue-shifted from the Ne* $3s'[1/2]_1^o$ atomic resonance position in all three Ne/rare-gas dimer spectra is also attributed to a transition to a molecular state dissociating into $Ne^*(3s'[1/2]_1^o) + Y(^1S_0)$. This peak results from excitation to the repulsive wall of the NeY* potential curve, followed by electronic autoionization to give $NeY^+ + e$. This process can be likened to an associative-ionization half-reaction in which, at a given wavelength, the NeY* collision complex is prepared with a fixed relative translational energy determined by the position of excitation on the repulsive wall. Because the rotational energy of the NeY dimer is not altered significantly by the absorption of a photon, the collision complex is formed with approximately thermal (20 K) rotational energy, which corresponds to a relatively small impact-parameter collision.

If one assumes that the molecular absorption oscillator strengths for transitions to states dissociating into $Ne^*(3s[3/2]_1^o$ or $3s'[1/2]_1^o) + Y(^1S_0)$ are proportional to the corresponding atomic oscillator strengths, then molecular structure should be observed in the NeY photoionization cross-sections near the $3s[3/2]_1^o$ level, and should be approximately a factor of 15 weaker than the structure near the $3s'[1/2]_1^o$ level [36]. An examination of the data shown in Fig. 10 reveals a weak broad peak in the NeAr spectrum, to the blue of the $3s[3/2]_1^o$ transition, possessing a width approximately equal to that of the molecular peak near the $3s'[1/2]_1^o$ transition. No similar features were observed in the NeKr or the NeXe spectra; however, small variations in transition probability, Franck-Condon factors or autoionization efficiency could make such features undetectable. No structure is observed in any of the spectra near the Ne metastable states, a result expected from the forbidden nature of these atomic transitions.

The NeKr and NeXe associative-ionization spectra obtained with an effusive molecular-beam source are similar to the spectrum shown in Fig. 11 for NeAr. The background pressure in the ionization chamber during the effusive-beam experiments was identical to that during the corresponding supersonic-beam experiments; thus, the intensity of the associative-ionization structure should be similar in both spectra. In all cases, the intensity of the associative-ionization peak at the position of the Ne* $3s[3/2]_1^o$ resonance transition was nearly equal to the intensity of the corresponding peak in the spectrum obtained using a supersonic molecular-beam source, suggesting that this structure is due in large part to associative ionization. Since associative ionization is about a factor of 4 weaker for the $3s'[1/2]_1^o$ state than for the $3s[3/2]_1^o$ state, it contributes negligibly to the molecular structure observed near the $3s'[1/2]_1^o$ transition in spectra measured with the supersonic molecular-beam source.

The two features observed in the Ne/rare-gas photoionization spectra near the Ne* $3s'[1/2]_1^o$ resonance line must result from transitions to one, or possibly both, of the dipole-allowed molecular states arising from the $Ne^*(3s'[1/2]_1^o) + Y(^1S_0)$ dissociation limit. These states have 0^+ and 1 symmetry. It is possible to hypothesize a mechanism to explain the observed bimodal spectral structure, one that involves transitions to a single excited state. According to this hypothesis, a single state of the Ne*/rare-gas excimer is populated in photoabsorption. It is well known that the absorption cross-section must be continuous at the dissociation limit [37]. We postulate that the dip in the photoionization cross-section at the dissociation limit results from the difference in decay rates between the bound and continuum vibrational levels. The bound levels may decay via fluorescence, autoionization or predissociation; the continuum levels may decay via direct dissociation as well as the above mechanisms, and direct dissociation is expected to rapidly depopulate the molecular states. Thus, the ionization cross-section may well show a discontinuity at the dissociation limit, with that part of the cross-section corresponding to transitions to bound vibrational levels being relatively more intense. This is indeed what is observed. The pressure dependence of the cross-section is also consistent with this mechanism; namely, the long-lived bound vibrational levels are more pressure sensitive than the short-lived continuum levels.

We have performed calculations to determine whether or not the shape of the photoionization cross-section above the molecular dissociation limit can be reproduced using the potential energy curves obtained from scattering data. The calculations assume that the energy dependence of the absorption cross-section is proportional to the Franck-Condon factor and that the autoionization rate is proportional to the associative-ionization cross-section at the appropriate relative translational energy [38]. This further assumes that the electronic matrix element is invariant with energy and that the interaction between the autoionizing state and the direct-ionization continuum does not affect the line shape of the discrete state (i.e., the Fano q parameter is large) [39].

For each Ne/rare-gas system, the ground-state potential energy curve was taken from the work of Ng, *et al.* [40] and the upper-state NeY* potential energy curve was the IAMMSV (ion-atom Morse-Morse spline van der Waals) potential determined by Gregor and Siska [41]. Fig. 12 shows an example of these potential energy curves for NeAr. The vibrational temperature of the dimers formed in the expansion was assumed to be 20 K; however, this value is not known accurately and the effect of varying it was examined.

The Franck-Condon factors were determined using the delta function approximation [42], which has been employed in a number of calculations involving bound-free transitions. In this approximation, the upper-state continuum wavefunction is set equal to a delta function at the classical turning point and the Franck-Condon distribution is obtained by reflecting the square of the ground-state wavefunction off of the upper-state potential curve. Briefly, the steps in the calculation were as follows:

--- The ground-state vibrational wavefunctions $\psi_i(R)$ were calculated for each initial vibrational level i.
--- The absorption cross-section for each vibrational level (i.e., the Franck-Condon distribution) was determined

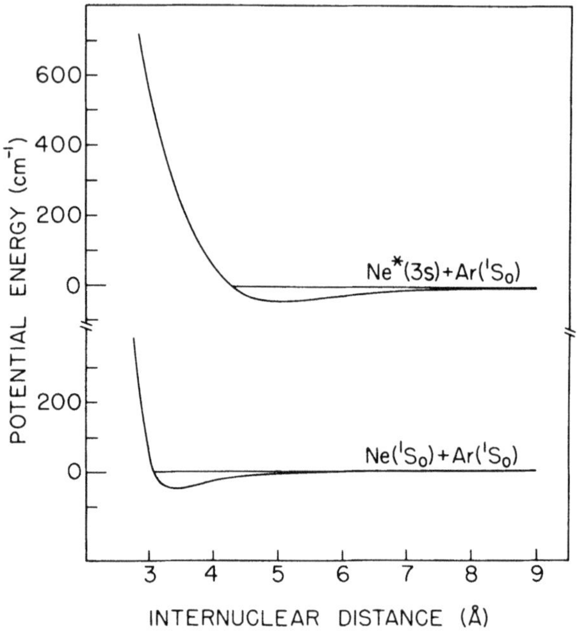

Fig. 12. Example of the potential energy curves for NeY and NeY* that were used in the model calculations described in the text.

by reflecting $\psi_i^2(R)$ from the upper-state potential energy curve to give $\psi_i^2(\Delta E)$, where ΔE is the transition energy.

--- The weighted cross-sections were calculated by multiplying the cross-section for each vibrational level by the appropriate Boltzman factor f_i at 20 K to give $f_i\psi_i^2(\Delta E)$.

--- The relative photoionization cross-section for a given initial vibrational level was calculated by multiplying each weighted cross-section by the (energy-dependent) associative-ionization cross-section to give $\sigma a_i(\Delta E)f_i\psi_i^2(\Delta E)$.

--- The total photoionization cross-section was calculated by summing the contributions from the individual vibrational levels.

The relative ionization cross-section was then normalized to the experimental data by equating the maxima of the theoretical and experimental distributions. Because the experimental associative-ionization cross-sections are limited to translational energies greater than 0.01 eV, the calculated cross-sections start slightly to the blue of the position of the Ne* $3s'[1/2]_1^o$ transition.

The results are shown in Fig. 13. The agreement with experiment is very good for NeXe but the width of the theoretical distribution is smaller than that of the experimental distribution for both NeKr and NeAr. Increasing the assumed temperature of the beam increases the width of the theoretical distribution only slightly since the dominant effect of a higher temperature is to increase the contributions of large R values in the calcu-

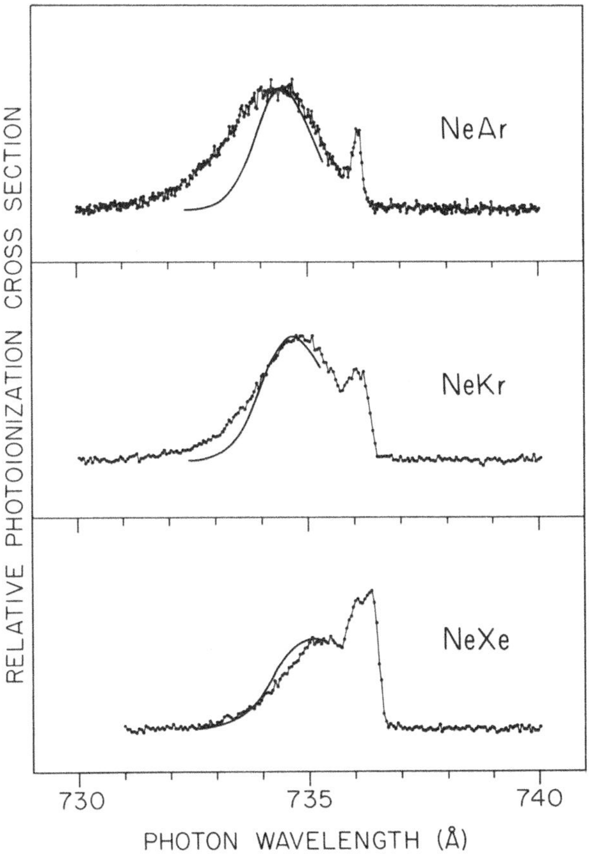

Fig. 13. Experimental and calculated (solid line) relative photoionization cross-sections near the neon $3s'[1/2]_1^o$ resonance transition.

lations of the Franck-Condon factor, which leads to transitions to the bound portion of the excited state.

The shape of the theoretical distribution is extremely sensitive to the slope of the repulsive wall of the excited state: Increasing this slope dramatically improves the fit to the NeAr data. We expect that this simple model can be improved and extended by more accurate theoretical work and, when coupled with data from scattering experiments, may lead to more accurate potentials for these neon/rare-gas excimers.

8. MOLECULAR RYDBERG STRUCTURE IN THE SPECTRA OF ArKr AND ArXe NEAR THE Ar 5s AND 3d ATOMIC RESONANCES

The energy region near the first resonance transition in Ar (the 4s levels) is not accessible in photoionization since it is energetically below the ionization thresholds of ArKr and ArXe; however, the energy region near the second transition in Ar does lie above the ionization thresholds of these heteronuclear dimers. Unfortunately, the atomic Ar 5s energy levels are nearly degenerate with the 3d levels; thus, the photoionization cross-sections for ArKr and ArXe near these atomic levels are expected to be much more complex than those observed, near the Ne 3s levels, for the heteronuclear dimers containing Ne.

Fig. 14 shows the relative photoionization cross-sections for ArKr and ArXe between 860 and 890 Å determined at a wavelength resolution of 0.28 Å. Going to still higher resolution (0.15 Å) resolves no new features, indicating that the spectral features are broadened by autoionization and/or predissociation. This wavelength region contains five dipole-allowed states of atomic Ar based on the 5s and 3d configurations. These states are indicated in Fig. 14. Autoionizing molecular states of 0^+ and 1 symmetry, derived from the combination Ar*(5s or 3d)$_{J=1}$ + Xe($^1S_0$), are expected to account for much of the prominent structure in the ArXe spectrum. The situation is more complex for ArKr owing to the presence of the atomic Kr*(ns' and nd') Rydberg levels; however, the strong similarity of the ArKr and ArXe spectra in this wavelength region indicates that the molecular states dissociating to Ar*(5s or 3d) + Kr($^1S_0$) dominate the ArKr spectrum. This is in accord with the fact that the atomic transition probabilities are greater to the low-lying states of Ar than to the higher-lying states of Kr.

There is a very approximate correlation of the structure in ArKr and ArXe spectra, with the structure in the ArXe spectrum being red-shifted by 1-2 Å from the corresponding structure in the ArKr spectrum. The detailed correlation with the atomic Ar spectrum is not obvious and the spectra are clearly more complex

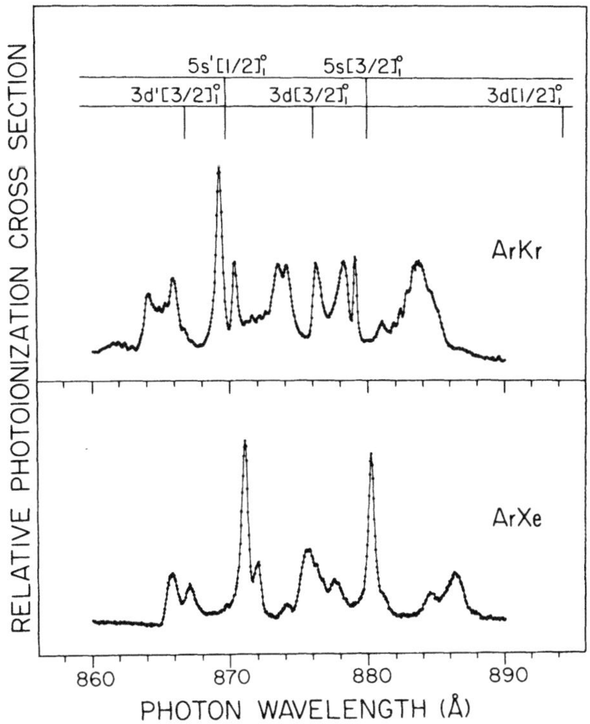

Fig. 14. Relative photoionization cross-sections for ArKr and ArXe in the region of the Ar* 5s and 5d transitions, taken at a wavelength resolution of 0.28 Å.

than the simple red- or blue-shifted bands observed in the heteronuclear dimers containing Ne. It is possible, as was suggested by Ng, *et al.* [43], that the bands that are blue-shifted from the dipole-allowed atomic Ar levels result from transitions to molecular states with potential humps. However, at least two other explanations may also account for these features. First, as in the case of the Ne/rare-gas dimers, the blue-shifted bands may result from excitation to the repulsive wall of ArY* followed by rapid autoionization to form ArY^+. Second, the bands may result from transitions to molecular states that dissociate to one of the many dipole-forbidden atomic Ar levels that are in this wavelength region. It is doubtful that these spectra will be understood completely without the help of theoretical calculations.

9. MOLECULAR RYDBERG STRUCTURE IN THE SPECTRA OF X_2 AND XY NEAR THE HIGH-LYING ATOMIC RYDBERG STATES

The structure in the molecular spectra near the high-lying atomic Rydberg states is extremely complex and is only partially amenable to analysis. For example, the spectra of Ar_2, ArKr and ArXe, determined at a resolution of 0.15 Å, in the region 800 to 850 Å are shown in Figs. 15a and 15b. The features in the photoionization cross-sections of ArKr and ArXe in this wavelength region result from transitions to states that dissociate to Ar*(ns or nd) + Y($^1S_0$), where n≥6 for s states and n≥4 for d

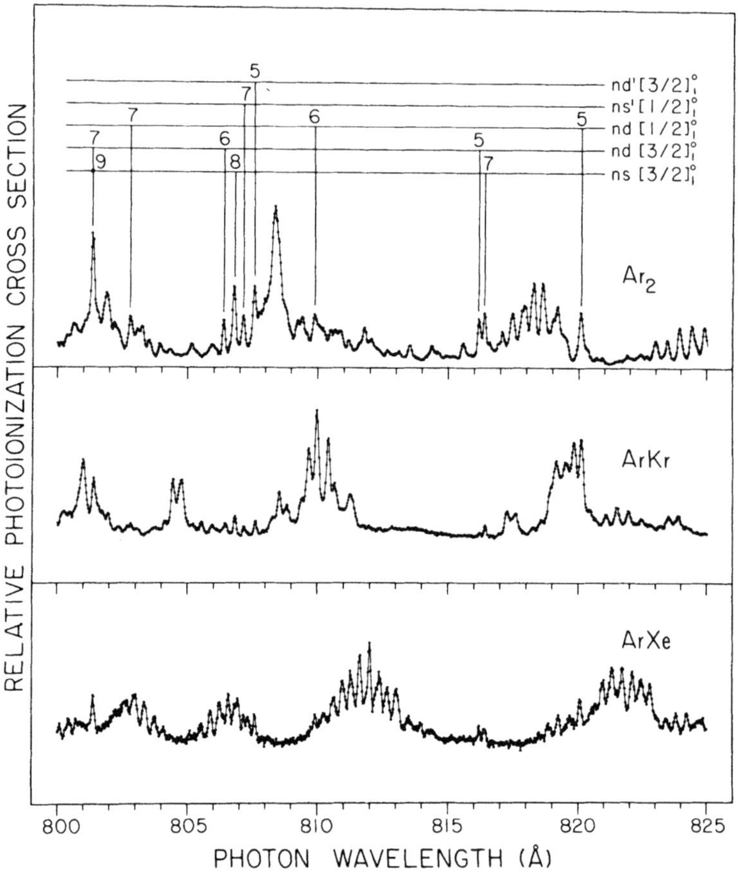

Fig. 15a. Relative photoionization cross-sections for Ar_2, ArKr and ArXe in the wavelength region from 800 to 825 Å, taken at a wavelength resolution of 0.15 Å.

states. The atomic Ar Rydberg series are indicated in the figures: It is immediately obvious that there is little correlation between the atomic and the molecular Rydberg series in this wavelength region. (It should be remembered that the weak, sharp peaks occuring at the exact positions of the atomic Ar transitions are due to the associative-ionization reaction $Ar^* + Y \rightarrow ArY^+ + e$).

There are similarities in the spectra of ArKr and ArXe, although the ArXe bands are red-shifted by approximately 1-1/2 to 2 Å from the corresponding bands in ArKr and show more discrete

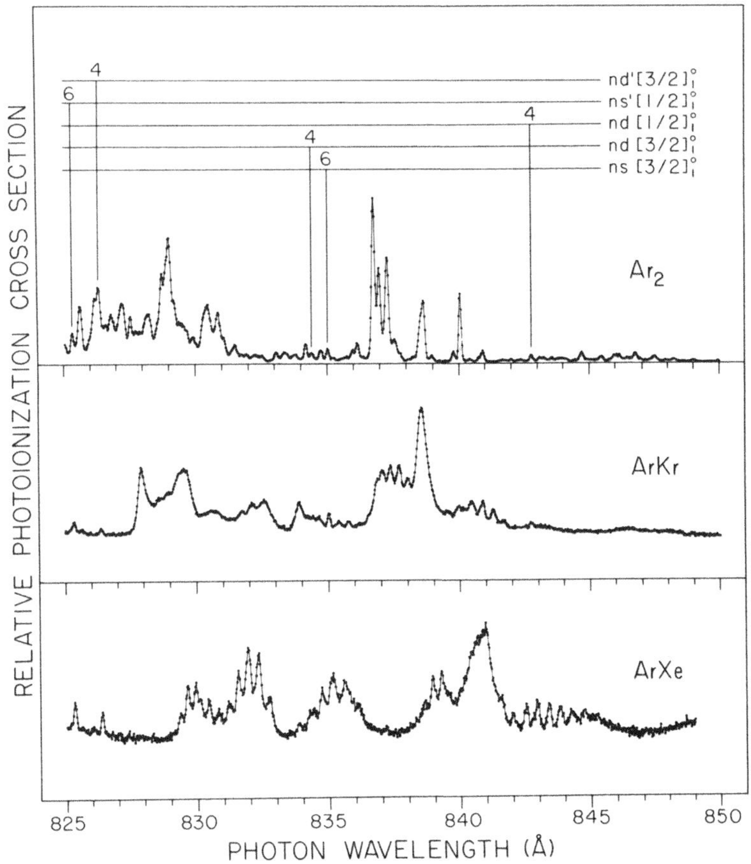

Fig. 15b. Relative photoionization cross-sections for Ar_2, ArKr and ArXe in the wavelength region from 825 to 850 Å, taken at a wavelength resolution of 0.15 Å.

structure than do the ArKr bands. The correlation of features in ArKr and ArXe with those in Ar_2 is less obvious. This probably results from the differences in the molecular-ion convergence limits for the homonuclear and heteronuclear dimers and the consequent differences in the Rydberg structure.

In general, a detailed analysis of the bands is not possible using spectra obtained at 0.15 Å resolution. We have determined the energy spacings for all peaks within (what appear to be) the various progressions; however, the energy differences are erratic and there is considerable overlapping of bands. The fact that excitation occurs at relatively large internuclear distances, corresponding to R_e of the neutral dimers, may be partially responsible for the complicated spectra. Vertical transitions are made to states in a region where configuration mixing is often extremely important [44] and can lead to complicated potential energy curves.

Many of these bands have vibrational structure with spacings of 50-70 cm^{-1}, which is consistent with the magnitudes of the dissociation energies of the excited states of the ions reported in Table 6. Using this observation, we have performed a simple analysis of the regularly spaced features in the Ar_2 spectrum in the wavelength region from 822 to 826 Å. The prominent peaks are listed in Table 8 along with the assignments and energy differences. The energy spacings between peaks are similar to the vibrational spacings in the B $^2\Pi_{3/2g}$ state of the ion, suggesting that this structure corresponds to vibrational members of a Rydberg state converging to this limit. In order to quantify this comparison, Franck-Condon factors were calculated for the X $^1\Sigma_g^+$, $v''=0 \rightarrow$ B $^2\Pi_{3/2g}, v'$ ionizing transition. For this calculation, a Morse potential was used for the X $^1\Sigma_g^+$ ground state, with R_e = 3.76 Å, ω_e = 30.68 cm^{-1} and D_e = 98.70 cm^{-1} [4]. A numerical potential was used for the B $^2\Pi_{3/2g}$ state, based on the shape of the curve calculated by Wadt [10], but was scaled to give the experimental dissociation energy of 0.14 eV [16]. The vibrational energy levels, values of $\Delta G_{v'+1/2}$ and Franck-Condon factors from X $^1\Sigma_g^+$, $v''=0$, are given in Table 9. Near the maximum in the Franck-Condon distribution, the values of $\Delta G_{v'+1/2}$ range from 65 to 100 cm^{-1} and encompass seven vibrational levels, which is in reasonable agreement with observation. The most probable dissociation limit for this state is the $5d[1/2]_1^o$ atomic level.

Part of the reason for the complexity of these spectra may be seen by considering the spectrum of Ar_2 in the wavelength region from 836 to 842 Å. The structure in this region is red-shifted by several angstroms from the $6s[3/2]_1^o$ atomic transition. It is of interest to compare the molecular structure in this region with that near the $4s[3/2]_1^o$ resonance at 1066.66 Å, and that

Table 8. Bands of Ar_2 Between 822 and 826 Å [a]

		Tentative Assignments			
λ(Å)	ν (cm^{-1})	v'	v''	$\Delta G_{v''+1/2}$ (cm^{-1})	$\Delta G_{v'+1/2}$ (cm^{-1})
822.762	121541.8	v	0		44.3
823.062	121497.5	v-1	0		68.9
823.529	121428.6	v-2	0		70.8
824.010	121357.8	v-3	0		71.6
824.496	121286.2	v-4	0	25.1	71.7
824.667	121261.1		1		
824.984	121214.5	v-5	0		104.5
825.696	121110.0	v-6	0		82.4
826.258	121027.6	v-7	0		

[a] The wavelength scale is accurate to ±0.02 Å; differences are accurate to ±0.005 Å.

Table 9. Energy Levels and Franck-Condon Factors for the Transition $Ar_2\ X^1\Sigma_g^+$, v"=0 → $Ar_2^+\ B^2\Pi_{3/2g}$, v'

v'	$E_{v'}$ (cm^{-1})	$\Delta G_{v'+1/2}$ (cm^{-1})	FCF
0	0		0.0007
1	111.4	111.4	0.0058
2	218.5	107.2	0.0241
3	320.6	102.1	0.0659
4	416.9	96.2	0.1300
5	506.8	89.9	0.1927
6	589.9	83.1	0.2171
7	666.1	76.2	0.1836
8	735.4	69.3	0.1143
9	797.4	62.0	0.0502
10	851.4	54.0	0.0139
11	896.2	44.8	0.0016

near the $5s[3/2]_1^o$ resonance at 879.95 Å [35]. A strong band system containing 30 bands appears in the region 1065-1073 Å (Tanaka and Yoshino Band System II) [17] and transitions are observed to seven vibrational levels of the upper state with $\Delta G_{v'+1/2}$ values ranging from 78 to 137 cm^{-1}. A similar, although weaker, band system appears in the region 884-892 Å (Tanaka and Yoshino Band System VIII) [17] and shows long v' progressions with individual bands degraded toward short wavelengths. The $\Delta G_{v'+1/2}$ values range from 60 to 128 cm^{-1}. The structure in the wavelength region to the red of the $6s[3/2]_1^o$ atomic transition is significantly different from that just described for the $4s[3/2]_1^o$ and $5s[3/2]_1^o$ atomic transitions. Fig. 15b shows that, in the photoionization spectrum, no regular extended vibrational progression is observed. (The absorption and ionization spectra are very similar throughout most of this region; in other words, the photoionization spectrum does not discriminate against absorption features that predissociate.) Thus, the molecular Rydberg states in the region near the $6s[3/2]_1^o$ atomic resonance must be significantly different from their counterparts at longer wavelengths. One obvious complication is that the atomic $6s[3/2]_1^o$ and $4d[3/2]_1^o$ energy levels are nearly coincident and, as a result, the molecular states arising from these dissociation limits will be subject to mutual perturbations. Moreover, in addition to these two dipole-allowed energy levels, there exists a large number of dipole-forbidden atomic energy levels which can combine with a ground-state Ar atom to form 0_u^+ and 1_u molecular states. While transitions to these molecular states are expected to be weaker than those to states dissociating to a J=1 atomic level, they will undoubtedly complicate the spectrum through perturbations and avoided crossings.

Fig. 16 shows the data in this region taken at a resolution of 0.07 Å. All of the possible atomic dissociation limits are shown, with the optically allowed atomic energy levels indicated by heavy lines. The 0_u^+ and 1_u molecular states arising from the combination of the excited-state atom with a ground-state atom are also shown. The atomic transitions are shifted in energy by an amount equal to $D_0(Ar_2)$, which places the atomic and molecular transitions on the same energy scale, assuming that all of the molecular transitions originate in v''=0 of the ground state. There are fifteen 0_u^+ and 1_u molecular states in the energy region shown in Fig. 16, resulting in a very large number of possible interactions. It is thus not surprising that few regularities are observed in the spectrum.

A second example of the complexity of the dimer spectra is shown in the region from 925 to 940 Å in the spectrum of the heteronuclear dimer KrXe. This region, which is shown in Fig. 17, exhibits a long series of weak peaks superimposed on the direct-ionization continuum. The wavelength positions and energy separations between peaks are given in Table 10. This region contains the convergence limit of the molecular Rydberg series that dissociate to Kr($^1S_0$) + Xe(ns' or nd'). The analogous series in ArXe converge to a limit at 925.90 Å while those in Xe_2 converge to a limit at 931.24 Å. However, the spacings of the observed peaks in KrXe do not indicate a convergent series and it is therefore more likely that these peaks are part of a vibrational progression associated with a state that dissociates to Kr* + Xe($^1S_0$). The most likely candidates for the Kr* state are the $8s[3/2]_1^o$ level at 922.73 Å and the $6d[3/2]_1^o$ level at 923.71 Å [35]; however, the spacings of the peaks listed in Table 10 are irregular and no reasonable Birge-Sponer plot results. The origin of the intense peak at 928.5 Å is not known but it is possible that the weaker peaks gain sufficient intensity to be observed through interaction with this state.

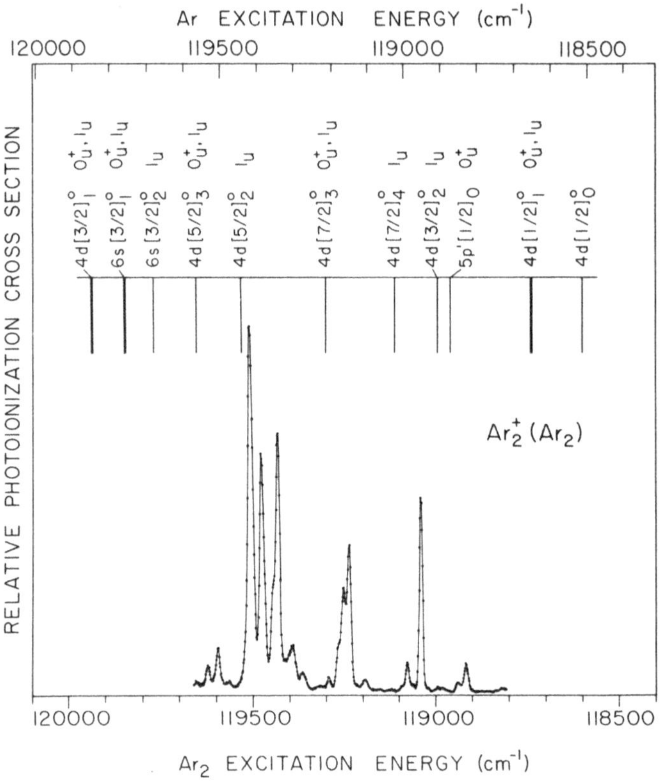

Fig. 16. Relative photoionization cross-section for Ar_2 taken at a wavelength resolution of 0.07 Å in the region from 118,800 to 119,650 cm^{-1}. All of the atomic Ar energy levels are shown in the figure, with those that are optically allowed indicated by the heavy lines. The atomic transitions are shifted in energy with respect to the molecular transitions by an amount equal to D_0 of the Ar_2 ground state. This places the atomic and molecular transitions on the same energy scale assuming that all of the molecular transitions originate in v''=0 of the ground state.

Table 10. Bands of KrXe Between 925 and 940 Å.

λ (Å)	ν (cm^{-1})	Δν (cm^{-1})
937.484	106668.5	
		57.1
936.982	106725.6	
		54.0
936.508	106779.6	
		48.3
936.085	106827.9	
		53.3
935.618	106881.2	
		51.0
935.172	106932.2	
		70.8
934.553	107003.0	
		65.9
933.978	107068.9	
		60.3
933.451	107129.2	
		55.2
932.972	107184.4	
		56.2
932.483	107240.6	
		55.5
932.000	107296.1	
		57.1
931.505	107353.2	
		50.2
931.069	107403.4	
		49.3
930.642	107452.7	
		50.1
930.208	107502.8	
		43.3
929.834	107546.1	

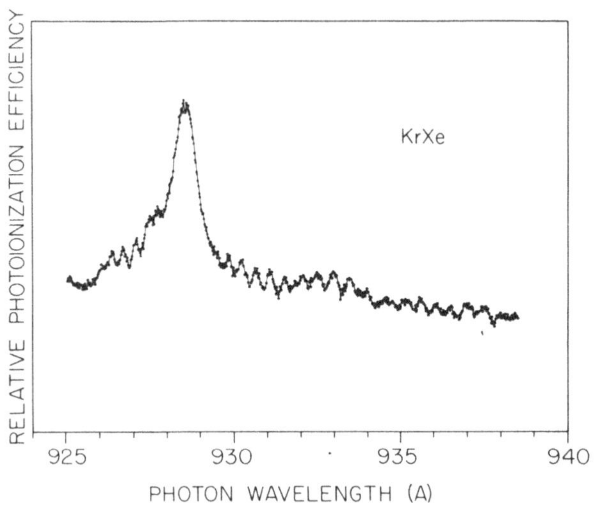

Fig. 17. Relative photoionization cross-section for KrXe in the wavelength region from 925 to 940 Å, taken at a wavelength resolution of 0.15 Å.

10. SUMMARY AND CONCLUSIONS

We have described the photoionization cross-sections for homonuclear and heteronuclear rare-gas dimers in the wavelength region from the molecular ionization potential to the $^2P^o_{1/2}$ ionization limit of the lighter atom. In the spectral region between the atomic fine-structure thresholds, a simple analysis based on the value of the effective principal quantum number n* and on the observed similarity between the atomic and molecular structure shows that most of the structure can be readily interpreted in terms of Rydberg series converging to an excited state of the ion. Vertical binding energies are obtained for both the Rydberg states and the ionic convergence limits.

In comparing the atomic and molecular spectra in the region between the atomic fine-structure thresholds, one cannot help but note the striking similarities among them; and, it is tempting to generalize this observation and conclude that molecular spectra of van der Waals molecules are merely weakly-perturbed atomic spectra. However, there are significant differences in the atomic and molecular spectra in regions other than that between the atomic fine-structure thresholds. Long vibrational progressions are observed in the spectra determined at a resolution of

0.15 Å, indicating that it is an oversimplification to characterize the molecular spectra as perturbed atomic spectra.

Ginter, *et al.* [45] have cautioned against the use of low-resolution spectra (i.e., rotationally-unresolved), such as those presented here, for the determination of structural assignments since the probability of guessing a correct solution is very low using an analysis based solely on quantum defects. We have found that higher-resolution data certainly would be necessary to analyze the complex vibrational progressions observed in Ar_2, ArKr and ArXe between 800 and 850 Å (Figs. 15a and 15b). In the region of the Beutler-Fano profiles, the quantum defect analysis is more reliable, however, because of the additional information contained in the lineshapes of the bands. This rather fortunate situation allows the determination of the (vertical) dissociation energies of the Rydberg states and of the excited states of the ions, quantities which cannot be obtained from a conventional absorption spectrum. Furthermore, the adiabatic ionization potential (and, hence, the dissociation energy of the ground state of the ion) can be determined from the present experiments, but not from the absorption spectra, since the molecular Rydberg series converging to the ionization limit have not been analyzed.

The present data thus serve as an effective complement to absorption spectra obtained using high-resolution instruments (i.e., those having resolving powers greater than 100,000) and the technique of photoionization mass spectrometry is shown to be a useful method for obtaining certain spectral and thermodynamic information on van der Waals molecules, information not readily obtainable using conventional absorption spectroscopy.

ACKNOWLEDGMENT

This work was supported by the U. S. Department of Energy and the Office of Naval Research.

REFERENCES

1. R. S. Mulliken, J. Chem. Phys. 52, 5170 (1970).
2. D. S. Ginter and M. L. Ginter, J. Mol. Spectrosc. 82, 152 (1980) and references therein.
3. P. M. Dehmer and E. D. Poliakoff, Chem. Phys. Lett. 77, 326 (1981).
4. P. M. Dehmer, J. Chem. Phys. 76, 1263 (1982).
5. S. T. Pratt and P. M. Dehmer, Chem. Phys. Lett. 87, 533 (1982).
6. P. M. Dehmer and S. T. Pratt, J. Chem. Phys. 75, 5265 (1981).

7. S. T. Pratt and P. M. Dehmer, J. Chem. Phys. 76, 3433 (1982).
8. P. M. Dehmer and S. T. Pratt, J. Chem. Phys. 77, 4804 (1982).
9. D. J. Trevor, Ph.D. thesis, University of California, Berkeley, 1980.
10. W. R. Wadt, J. Chem. Phys. 68, 402 (1978).
11. P. M. Dehmer and J. L. Dehmer, J. Chem. Phys. 68, 3462 (1978).
12. K. P. Huber and G. Herzberg, *Molecular Spectra and Molecular Structure IV. Constants of Diatomic Molecules* (Van Nostrand Reinhold Company, New York, 1979).
13. C. F. Bender and N. W. Winter, Appl. Phys. Lett. 33, 29 (1978).
14. I. Dabrowski and G. Herzberg, J. Mol. Spectrosc. 73, 183 (1978).
15. I. Dabrowski, G. Herzberg and K. Yoshino, J. Mol. Spectrosc. 89, 491 (1981).
16. P. M. Dehmer and J. L. Dehmer, J. Chem. Phys. 69, 125 (1978).).
17. See for example Y. Tanaka and K. Yoshino, J. Chem. Phys. 53, 2012 (1970).
18. W. A. Chupka and J. Berkowitz, J. Chem. Phys. 51, 4244 (1969).
19. W. A. Chupka, P. M. Dehmer and W. T. Jivery, J. Chem. Phys. 63, 3929 (1975).
20. C. Y. Ng, D. J. Trevor, B. H. Mahan and Y. T. Lee, J. Chem. Phys. 66, 446 (1977).
21. W. M. Trott, N. C. Blais and E. A. Walters, J. Chem. Phys. 69, 3150 (1978).
22. Y. Ono, S. H. Linn, H. F. Prest, M. E. Gress and C. Y. Ng, J. Chem. Phys. 73, 2523 (1980).
23. A. van Deursen, A. van Lumig and J. Reuss, Int. J. Mass Spectrom. Ion Phys. 18, 129 (1975).
24. (a) D. Golomb, R. E. Good and R. F. Brown, J. Chem. Phys. 52, 1545 (1970); (b) D. Golomb, R. E. Good, A. B. Bailey, M. R. Busby and R. Dawbarn, J. Chem. Phys. 57, 3844 (1972).
25. (a) T. A. Milne and F. T. Greene, J. Chem. Phys. 47, 4095 (1967); (b) T. A. Milne, A. E. Vandergrift and F. T. Greene, J. Chem. Phys. 52, 1552 (1970).
26. R. E. Leckenby, E. J. Robbins and P. A. Trevalios, Proc. R. Soc. (London) Ser. A280, 409 (1964).
27. N. Lee and J. B. Fenn, Rev. Sci. Instrum. 49, 1269 (1978).
28. K. Radler and J. Berkowitz, J. Chem. Phys. 70, 221 (1979).
29. P. M. Guyon and J. Berkowitz, J. Chem. Phys. 54, 1814 (1971).
30. J. N. Murrell, S. F. A. Kettle and J. M. Tedder, *The Chemical Bond* (Wiley, New York, 1978), pp. 82-85, 91, 168.
31. J. H. D. Eland, *Photoelectron Spectroscopy* (Butterworth, London, 1974).

32. A. Dalgarno and A. E. Kingston, Proc. R. Soc. (London) Ser. A259, 424 (1961).
33. This assumes that $R_e(XXe^+) \approx 1/2\ R_e(X_2) + [R_e(Xe_2{}^+) - 1/2\ R_e(Xe_2)]$. While this is admittedly a very rough approximation, it does permit an estimate of the ratio of $R_e{}^4$.
34. M. Wolfsberg and L. Helmholz, J. Chem. Phys. 20, 837 (1952).
35. C. E. Moore, Natl. Bur. Stand. Circ. 467, Vol. III (1949).
36. W. L. Wiese, M. W. Smith and B. M. Glennon, *Atomic Transition Probabilities*, Vol. I, Natl. Stand. Ref. Data Ser., Natl. Bur. Stand. 4, 1966.
37. (a) A. C. Allison and A. Dalgarno, J. Chem. Phys. 55, 4342 (1971); (b) A. L. Smith, J. Chem. Phys. 55, 4344 (1971).
38. (a) R. H. Neynaber and G. D. Magnusen, Phys. Rev. A11, 865 (1975); (b) R. H. Neynaber and G. D. Magnusen, Phys. Rev. A14, 961 (1976); (c) R. H. Neynaber and S. Y. Tang, J. Chem. Phys. 70, 4272 (1979).
39. (a) U. Fano, Phys. Rev. 124, 1866 (1961); (b) U. Fano and J. W. Cooper, Phys. Rev. 137A, 1364 (1965).
40. C. Y. Ng, Y. T. Lee and J. A. Barker, J. Chem. Phys. 61, 1996 (1974).
41. R. W. Gregor and P. E. Siska, J. Chem. Phys. 74, 1078 (1981).
42. G. Herzberg, *Molecular Spectra and Molecular Structure I. Spectra of Diatomic Molecules* (Van Nostrand, Princeton, 1950), pp. 393-394.
43. C. Y. Ng, P. W. Tiedemann, B. H. Mahan and Y. T. Lee, J. Chem. Phys. 66, 5737 (1977).
44. R. S. Mulliken, J. Am. Chem. Soc. 88, 1849 (1966).
45. M. L. Ginter, D. S. Ginter and C. M. Brown, Appl. Opt. 19, 4015 (1980).

EXCITONS AND ENERGY TRANSFER IN INSULATORS

J. Jortner[a], E. E. Koch[b] and N. Schwentner[c]

[a]Department of Chemistry
Tel Aviv University
69978 Tel Aviv, Israel

[b]Hamburger Synchrotronstrahlungslabor HASYLAB
Deutsches Elektronen-Synchrotron DESY
2000 Hamburg 52, West Germany

[c]Institüt für Atom- und Festkörperphysik
Freie Universität Berlin
1000 Berlin, West Germany

1. PROLOGUE

We concern ourselves, in this paper, with the electronic level structure and excited-state nonradiative relaxation phenomena in molecular solids [1] in the energy range 6-50 eV. Fig. 1 provides an overview of the energy regions for the basic excitations and radiationless decay channels in molecular solids. Phonons, and vibrational and librational excitations, mark the lower energy limit at 0.01 eV. The elementary excitations at higher energies involve intervalence and low-intravalence electronic excitations which can be described in terms of the tight-binding Frenkel scheme [2-4], as well as Wannier exciton [5] and impurity states. At even higher energies, one encounters interband transitions and impurity ionization states. Dynamical processes in electronically excited states [6] involve a variety of intrastate and interstate relaxation phenomena, such as exciton trapping via vibrational relaxation; electronic relaxation of exciton states manifested by interconversion of electronic energy to photons; electronic energy transfer; autoionization of metastable excitons located above the band gap; electron-hole recombination; and charge-transfer processes. In what follows, we dwell on electronic excitations in insulators in the energy range 6-50 eV, focus-

S. P. McGlynn et al. (eds.), Photophysics and Photochemistry in the Vacuum Ultraviolet, 515–577.

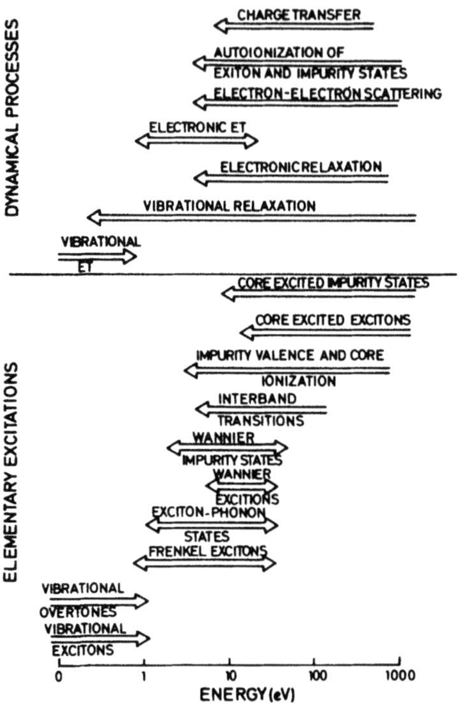

Fig. 1. Energy regions for excited-state photophysical and chemical processes in molecular crystals [after J. Jortner and S. Leach, J. Chim. Phys. 77, 7 (1980)].

ing attention on the following features:

--- The nature of the electronic excitations in molecular solids, liquids and fluids.

--- The relation between the electronic level structure and the geometrical structure of the condensed phase, emphasizing the effects of structural and compositional disorder on the electronic excitations.

--- The parentage problem. The inter-relationship between the electronic-vibrational excitations of the constituents, and the excitations in the condensed phase.

--- Nonradiative relaxation of electronic excitations via intrastate nuclear deformation and interstate electronic relaxation.

--- The influence of the state of aggregation on nonradiative relaxation processes.

--- Electronic energy transfer phenomena, which pertain to basic issues of coherence and localization.

The present discussion is selective rather than exhaustive. We have provided a recent extensive review [1] of electronic excitations in condensed rare gases. Our major goal here is to focus on the relations and correlations between the molecular approach and the solid-state approach for the elucidation of the excited-state energetics and dynamics of electronic excitations in insulators.

2. FRENKEL EXCITONS

These electronic excitations in pure molecular crystals have a definite parentage in an intravalence molecular (or atomic) excitation of the constituents [2-5]. Typical examples involve exciton states originating from $\pi \rightarrow \pi^*$ transitions in organic crystals [3,4], with which we shall not be concerned, and the lowest electronic excitations in simple molecular crystals [7,11] (e.g., solid CO and N_2), which are located in the VUV range and which are of considerable interest for the present discussion. The description of the intermolecular interactions is considerably simplified since the intravalence excitations are characterized by a minor expansion of the excited-state charge distribution, whereupon intermolecular overlap corrections are negligible. Accordingly, the tight-binding approximation is applicable for the description of exciton states [2-5]. The crystal Hamiltonian is

$$H = \sum_n \sum_\alpha H_{n\alpha} + \frac{1}{2} \sum_n \sum_m \sum_\alpha \sum_\beta V_{n\alpha,m\beta} \tag{1}$$

where $H_{n\alpha}$ is the single-molecule Hamiltonian for a molecule located at the site α ($\alpha=1,\ldots,\sigma$) in the nth unit cell ($n=1,\ldots,N$), while $V_{n\alpha,m\beta}$ denotes the intermolecular interaction. The single-molecule wavefunctions $\phi^0_{n\alpha}$ and $\phi^f_{n\alpha}$ for the ground electronic state (0) and for the electronically excited state (f), respectively, can be used to construct the excitation-transfer integrals

$$M^f_{n\alpha,m\beta} = \int (A\phi^0_{n\alpha}\, \phi^f_{m\beta})\, V_{n\alpha,m\beta}\, \phi^f_{n\alpha}\, \phi^0_{m\beta}\, d\tau \tag{2}$$

where A is the intermolecular electron antisymmetrizer [4]. Following the conventional tight-binding approach, the lattice sums for the f excited state are

$$L^f_{\alpha\beta}(\underset{\sim}{k}) = \sum_n e^{i\underset{\sim}{k}\cdot\underset{\sim}{n}} M^f_{n\alpha,m\beta} \tag{3}$$

where $\underset{\sim}{k}$ is the lattice quasi-momentum. The $\sigma \times \sigma$ energy matrix is

$$H_{\alpha\beta}(\underset{\sim}{k}) = \left[\varepsilon^f + D^f + L^f_{\alpha\alpha}(\underset{\sim}{k})\right] \delta_{\alpha\beta} + L^f_{\alpha\beta}(\underset{\sim}{k})(1 - \delta_{\alpha\beta}) \tag{4}$$

where ε^f is the isolated-molecule excitation energy, while D^f is the environmental shift term consisting of electrostatic and dispersive interactions. Diagonalization of this energy matrix results in the energies $E_j(\underset{\sim}{k})$ $[j=1,\ldots,\sigma]$ which determine the exciton dispersion curves and yield the exciton density of states. As is well known [2-5], the momentum conservation rule implies that the exciton states amenable to optical excitation from the ground state of the crystal are characterized by $\underset{\sim}{k} = \underset{\sim}{Q}$, where $\underset{\sim}{Q}$ is the wavevector of exciting light (i.e., $\|\underset{\sim}{Q}\| = 2\pi/\lambda \simeq 0$). The $\underset{\sim}{k} \simeq 0$ optically-accessible exciton states for a crystal containing two molecules per unit cell ($\sigma = 2$) are characterized by the energies

$$\begin{aligned} E_1 &= \varepsilon^f + D^f + L^f_{11}(\underset{\sim}{k} = 0) + L^f_{12}(\underset{\sim}{k} = 0) \\ E_2 &= \varepsilon^f + D^f + L^f_{11}(\underset{\sim}{k} = 0) - L^f_{12}(\underset{\sim}{k} = 0) \ . \end{aligned} \tag{5}$$

The following physical implications of the tight-binding approach to exciton states are noteworthy.

Davydov splitting. The optical spectra exhibit the celebrated Davydov splitting, which for a crystal with $\sigma = 2$ is given by $\Delta E_{DAVYDOV} = 2L_{12}(\underset{\sim}{k} = 0)$, reflecting the coupling between translationally inequivalent molecules.

Weak exciton-vibron coupling [3,4]. When the Davydov splitting is low relative to the intramolecular vibrational frequency, the excitation-transfer integrals and the lattice sums are modified by the nuclear vibrational overlap Franck-Condon factors $\langle FC \rangle$, which are given by $L_{\alpha\beta}(\underset{\sim}{k}=0) = L^f_{\alpha\beta}(\underset{\sim}{k}=0)\langle FC \rangle$, where $L^f_{\alpha\beta}$ is the electronic contribution [i.e., Eq. (3)]. This effective reduction of the electronic contribution amounts to a polaron effect [6]. In this limit, the Davydov splitting is proportional to the intensity of each vibronic component.

Strong exciton-vibron coupling [3,4]. When the Davydov splitting exceeds the intramolecular frequency, the excitation transfer occurs at a frozen nuclear configuration. The lattice sums are determined by the "bare" excitation-transfer integrals -- that is,

$L_{\alpha\beta}(\underset{\sim}{k}=0) = L^{f}_{\alpha\beta}(\underset{\sim}{k}=0)$.

Solvent shifts. The energetic shift of the center of gravity of the manifold of Davydov components relative to the excitation energy of the bare molecule is given by [2] $\Delta E_S = D^f + L_{\alpha\alpha}(\underset{\sim}{k}=0)$, and is determined by the environmental shift term as well as by the resonance interaction with translationally equivalent molecules. In the weak-coupling limit, the contribution of $L_{\alpha\alpha}$ to each ΔE_S is determined by the relative intensity of the particular vibronic component.

Crystal-field mixing. The first-order scheme outlined above requires extension beyond a single electronically excited configuration. Off-resonance interactions, induced by the crystal Hamiltonian of Eq. (1), with high electronically excited states can readily be incorporated in the tight-binding scheme by extending the matrix of Eq. (4) to include several electronic states [2-4]. Crystal-field mixing will result in gross modifications of the absolute intensities, as well as of the relative intensities of the Davydov components. The most dramatic effect of crystal-field mixing pertains to the appearance of the electronic origin of crystal excitons, which have a parentage in symmetry-forbidden molecular transitions. These electronic excitations become allowed, in principle, in the crystal, the Hamiltonian of which is of lower symmetry, and are induced by the crystal Hamiltonian stealing intensity from higher-energy allowed first-order exciton states.

Longitudinal and transverse excitons. When the exciton band structure is determined by dipole-dipole interactions, the dipole lattice sums [Eq. (3)] are piecewise discontinuous at $\underset{\sim}{k} \to 0$ [12]. For an exciton state which has its parentage in a triply-degenerate p atomic state, two types of $\underset{\sim}{k} \simeq 0$ exciton states will be exhibited [5,13], namely, longitudinal excitons with $\underset{\sim}{\mu} || \underset{\sim}{k}$, and transverse excitons with $\underset{\sim}{\mu} \perp \underset{\sim}{k}$, where $\underset{\sim}{\mu}$ is the transition moment. The longitudinal-transverse splitting is $4\pi\rho|\mu|^2$, where ρ is the molecular density. For optical excitations in an ideal crystal with $\underset{\sim}{k} = \underset{\sim}{0}$, only the transverse excitons are allowed [5,13]; however, exciton scattering at imperfections will result in the appearance of the longitudinal exciton branch in the optical spectrum.

Cooperative electronic-vibrational excitons [14-16]. Up to this point, we have been concerned with electronic-vibrational excitations located on a single molecule, which in solid-state language are referred to as "one-particle" excitations. One can construct "two-particle" electronic-vibrational excitations, where the electronic excitation and the vibrational excitation reside on different molecules. In view of the frequency difference between the ground and electronically excited states, the one-particle and two-particle excitations will appear at different energies. The two-particle excitation is optically forbidden in zero order, but excitonic interactions result in mixing between the one-particle and two-particle states. Since the total momentum $\tilde{k}_t$ of a two-particle state is a superposition of the exciton momentum $\tilde{k}$ and the vibron momentum $\tilde{k}_v$ (i.e., $\tilde{k}_t = \tilde{k} + \tilde{k}_v$), the conservation rule for optical excitation ($\tilde{k}_t = \tilde{0}$) results in excitation with $\tilde{k} = -\tilde{k}_v$, whereupon two-particle exciton states with all $\tilde{k}$ values will appear in the optical crystal spectrum, thus providing central information on the exciton density of states [15,16].

The experimental implications of Frenkel excitons for the elucidation of VUV spectra of simple molecular crystals below the conduction band are of considerable interest [7-11]. Figure 2 shows the absorption spectrum of crystalline CO, in the energy range 7.5-10.0 eV, which exhibits electronic-vibrational Frenkel excitons, originating from the $X^1\Sigma^+ \rightarrow A^1\Pi$ allowed transition of the CO molecule. The vibrational spacings considerably exceed the intermolecular interactions, so that the exciton states correspond to the weak-coupling situation. The vibronic excitons with v = 1,2 and 3 exhibit a definite splitting which may tentatively be attributed to Davydov splitting in the crystal containing two molecules per unit cell [7]. The appearance of the Davydov splitting for the most intense vibronic components of the excitonic transition is borne out by the theoretical description of the weak exciton-vibron coupling situations. It should be noted, however, that the double hump structure exhibited for these intense vibronic components may originate from phonon side bands [17], and that further studies are required to resolve the issue of whether excitonic splitting is indeed exhibited in the first singlet exciton state of solid CO. The N_2 molecule is isoelectronic with CO and the corresponding first spin-allowed molecular excitation is the $X^1\Sigma_g^+ \rightarrow a^1\Pi_g$ transition, which is symmetry forbidden in N_2. Crystal-field mixing makes this transition partly allowed in solid N_2, as is apparent from Fig. 3. No Davydov splitting seems to be exhibited for the vibronic components of this excitonic transition. Similarly, the second spin-allowed excitonic transition of solid N_2 in the energy range 9-12 eV (cf. Fig. 3), which has its parentage in the symmetry-forbidden $X^1\Sigma_g^+ \rightarrow w^1\Delta_u$ molecular excitation, is induced in the crystal by crystal-field mixing. The third Frenkel exciton state of solid N_2, which originates from the symmetry-allowed $X^1\Sigma_g \rightarrow b^1\Pi_u$ molecular transition, appears in the N_2 crystal in the energy range 12-14 eV (cf. Fig. 4). This excitonic transition is characterized by quite large dipole-dipole coupling, which contributes to the intermolecular excitation-transfer integrals. The intense vibronic components exhibit the Davydov splitting (cf. Fig. 4), as is expected for the weak-coupling situation. So far, we have been concerned with one-particle excitations in solid CO and N_2. The two-particle excitations in these crystals are characterized by

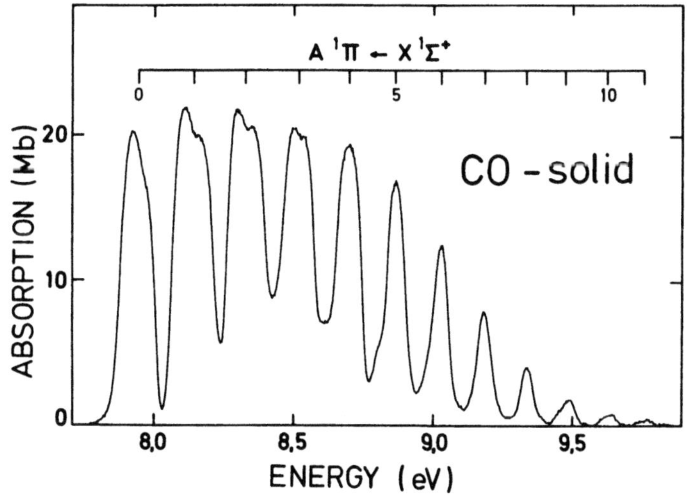

Fig. 2. Absorption spectrum of solid CO in the energy range 7.5-10 eV (from Gürtler [11]).

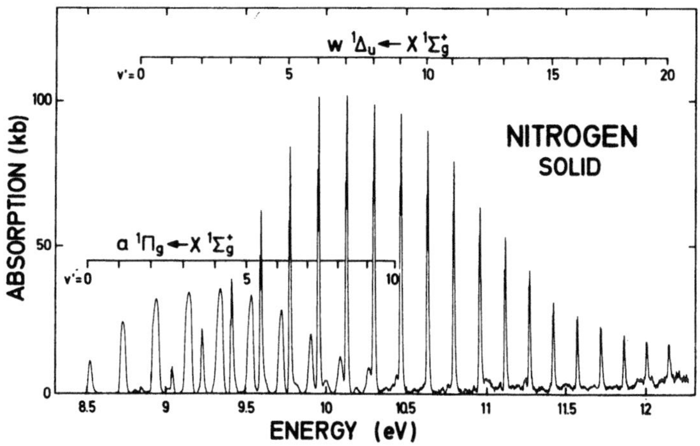

Fig. 3. Absorption spectrum of solid N_2 in the energy range 8.0-12.5 eV (from Gürtler [11]).

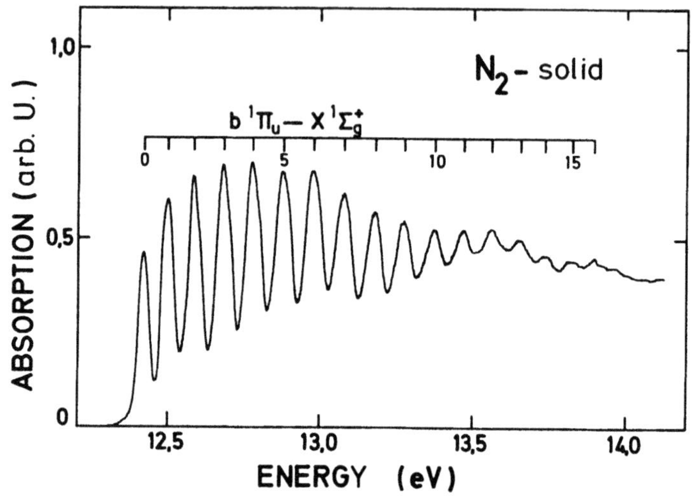

Fig. 4. Absorption spectrum of solid N_2 in the energy range 12.0-14.0 eV (from Gürtler [9,11]).

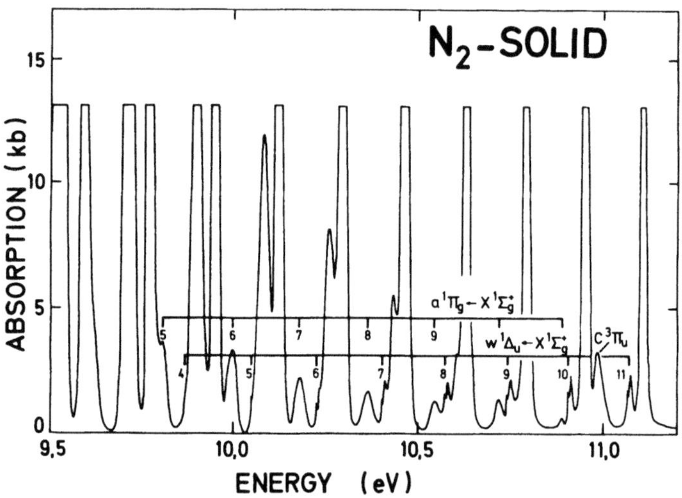

Fig. 5. Cooperative electronic-vibrational excitations in the $X^1\Sigma_g^+ \rightarrow w^1\Delta_u$ transition of solid N_2 (after Gürtler [11]).

two features. First, they are well-separated in energy (e.g., the frequency change between the $X^1\Sigma_g$ and $a^1\Pi_g$ states of N_2 is ∿500 cm^{-1}). Second, in view of the large energy spacing between the two-particle excitation and the corresponding one-particle state, the intensity of the former is low. Fig. 5 shows the weak two-particle excitations in solid N_2, which originate from the $X^1\Sigma_g^+ \rightarrow a^1\Pi_g$ and $X^1\Sigma_g^+ \rightarrow w^1\Delta_u$ molecular transitions. A careful comparison between the lineshapes of some conventional one-particle, and the corresponding two-particle excitations, as shown in Fig. 6, reveals that the sideband is broader in the latter case, thus exhibiting the effect of excitation of all of the states in the exciton band for the two-particle cooperative excitations.

The one-particle and two-particle Frenkel excitations in solid CO and N_2 have a definite parentage in electronic excitations of a single molecule, and constitute stable crystal excitations below the interband transition to the conduction band. An interesting example of stable exciton states was observed by Lau, Fock

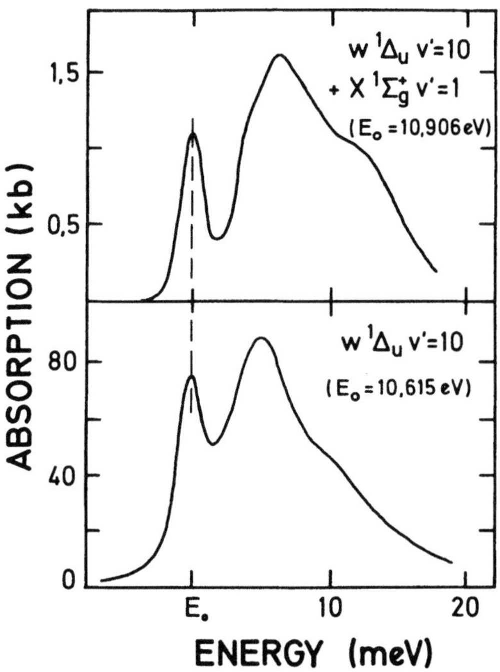

Fig. 6. Lineshapes of one-particle (lower spectrum) and two-particle (upper spectrum) electronic-vibrational excitations in solid N_2 (after Gürtler [11]).

and Koch [18,19] for solid CO_2, in the energy range 8-13 eV. Fig. 7 shows the absorption spectrum, together with the photoelectron yield curves, for solid CO_2 in the energy range 8-30 eV. From the comparison between absorption and photoelectron yield curves, it is apparent that the absorption spectrum above 15 eV corresponds to interband transitions. The structure in the absorption spectrum below the onset of interband transitions at 13 eV corresponds to stable exciton states. The sharp peak at 11 eV in solid CO_2 constitutes a stable exciton. Searching for the molecular parentage of this state, we assign it to a Frenkel-type exciton which has its parentage in the excited molecular states of CO_2 [20]. One of these, the $^3\Sigma_u$ state, which is located at 11.2 eV in the isolated molecule, has recently been shown to be essentially a Rydberg state with a substantial (~30%) intravalence character [20]. Thus, the 11 eV exciton state in solid CO_2 should be considered as an intermediate exciton, which can be envisaged as a Frenkel exciton with substantial non-orthogonality corrections [21,22]. An alternative description of this exciton state, which has its parentage in a molecular Rydberg state, rests on a modified Wannier scheme [21,22]. And, this brings us to consider Wannier states in molecular crystals.

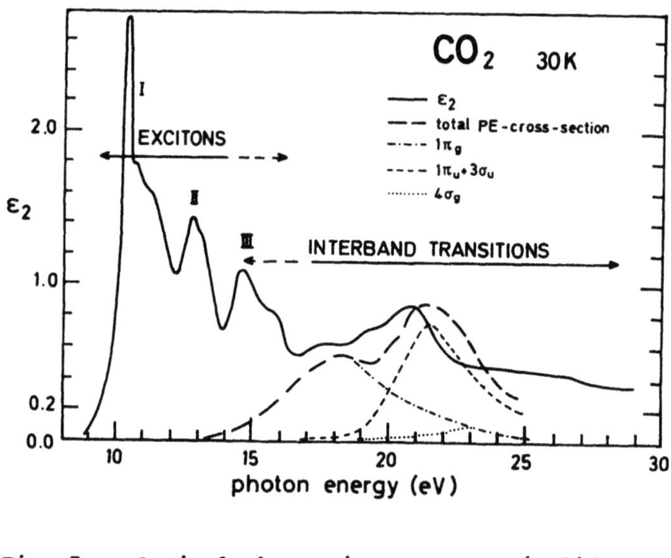

Fig. 7. Optical absorption spectrum (solid curve) and photoelectron cross-sections for solid CO_2 in the energy ranges 9-30 eV (from Fock, Lau and Koch [18,19]).

3. WANNIER EXCITONS

The traditional description of Wannier excitons [5] considers bound states in the band gap, resulting from the electron-hole attractive Coulomb interaction, which do not have parentage in the electronic excitations of the constituents (cf. Fig. 8). The traditional treatment of Wannier states rests on the following four approaches.

The conventional effective mass approximation (EMA) [23]. This approach considers a single pair of valence and conduction bands, with the effect of higher transitions being subsumed in the dielectric constant, which serves to screen the electron-hole Coulomb interaction. Next, and most important, the EMA rests on the cardinal assumption that the spatial extent of the electronic wavefunction is large relative to the lattice constant r_c. In a crystal, where the conduction band is parabolic, the EMA results in a hydrogenic wave equation for the envelope function F(r) and for the Wannier states:

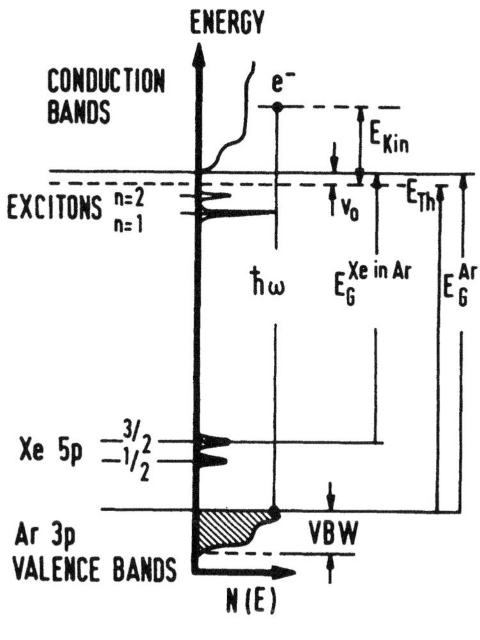

Fig. 8. Schematic of the energy levels of excitons and atomic impurity states in a solid rare-gas matrix, depicted for the case of Xe in Ar.

$$\left[-\frac{\hbar^2}{2\mu}\nabla^2 - \frac{e^2}{\varepsilon r}\right] F(r) = EF(r) \tag{6}$$

where ε is the (static) dielectric constant and μ is the exciton effective mass (i.e., $\mu^{-1} = m_e^{-1} + m_h^{-1}$). Eq. (6) results in a hydrogenic series of Wannier excitations, which are characterized by the principal quantum number $n = 1,2,...$ (and angular momentum quantum number ℓ) and which have energies, relative to the bottom of the conduction band, given by

$$E_n = -\mu e^2/2\varepsilon^2\hbar^2 n^2 \tag{7}$$

with the radii of these states being given by

$$r_n = n^2\hbar^2\varepsilon/\mu e^2 . \tag{8}$$

The radius of the lowest (n=1) exciton state is $r^{n=1} = \hbar^2\varepsilon/\mu e^2$; thus, the validity condition for the applicability of the EMA is $r^{n=1} \gg r_c$. An interesting class of simple molecular crystals, where a search for Wannier states can be conducted, involves rare-gas solids (RGS). As is apparent from Table 1, for RGS $r^{n=1} < r_c$ and the EMA is expected to be inapplicable; therefore, this theoretical scheme predicts that in these molecular crystals, which are characterized by low dielectric constants, Wannier excitons will not be exhibited. Contrary to these theoretical considerations, which rest on a conventional solid-state approach, experimental studies pioneered by Schnepp and Dressler [24] and by Baldini [25] have established that RGS exhibit a rich spectrum of Wannier excitons [26-28]. From the spectra portrayed in Figs. 9 and 10, it is apparent that the low-lying electronic

Table 1. Radii of the n=1 Excitons for Ne, Ar, Kr and Xe in Comparison with the Nearest-Neighbor Distances r_c in the fcc-Lattice.

	Ne	Ar	Kr	Xe
$r^{n=1}$ (Å)[a]		1.8	2.5	3.2
r_c (Å)[b]	3.156	3.755	3.992	4.335

[a] Calculated with Eq. (8).
[b] G. K. Horton, Am. J. Phys. 36, 93 (1968).

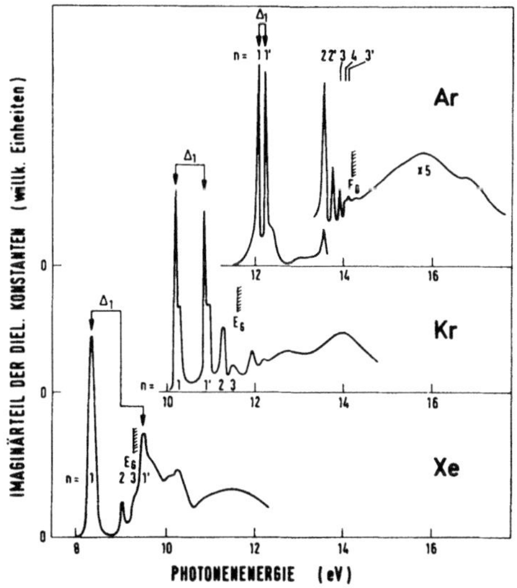

Fig. 9. The spectra of the imaginary part of the dielectric constant, ε_2, for solid Ar, Kr and Xe in the region of the valence excitons. The spin-orbit splitting in the n=1 excitons is Δ_1; the assignments of the higher members of the exciton series and the band gap E_G are indicated.

excitations of solid Ne, Ar, Kr and Xe reveal a long series of Wannier excitons with principal quantum numbers up to n=5. The general features of these electronic spectra of RGS are:

--- The appearance of Wannier states with n = 2-5. These "new" states have no apparent atomic parentage.

--- The energies of these "new" states obey the hydrogenic relation, Eq. (7), with n = 2,3,... .

--- The "new" states exhibit spin-orbit splitting.

--- The lowest electronic excitation is accompanied by a second intense transition separated from it by the spin-orbit splitting ΔE_{SO}. The two intense transitions are labelled n=1 and n'=1 in Figs. 9 and 10. These two intermediate excitons can be described in terms of two alternative pictures: (i), n=1 Wannier states subjected to large central-cell corrections; and (ii), Frenkel excitons with parentage in the $^1S_0 \rightarrow {}^3P_1$ and $^1S_0 \rightarrow {}^1P_1$ atomic excitations modified by large non-orthogonality corrections.

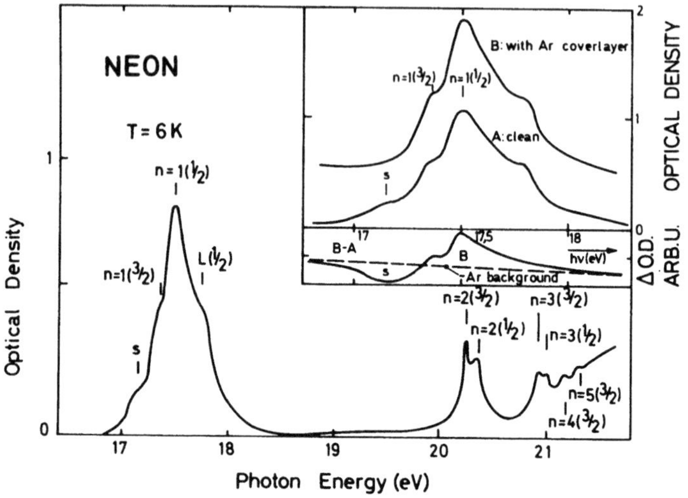

Fig. 10. Optical density of solid Ne at 6 K in the exciton range of the spectrum. The main features can be grouped into two series, split by spin-orbit interactions, which converge to the band gap. In the inset, results of a surface-coverage experiment for solid Ne in the range of the n=1 exciton are shown. Curve A is the optical density of a clean Ne sample. Upon evaporation of a thin Ar coverlayer, the structure denoted by S disappears, whereas the band shape of the remaining peak remains unchanged. In the lower part of the inset, the difference spectrum B-A is shown and the smooth and structureless Ar background is indicated by the broken line (from Saile and Koch [27]).

The phenomenology of the energetics of deep Wannier excitons in RGS (cf. Fig. 8) rests on the hydrogenic level structures

$$E_n = E_G - \frac{B}{n^2} \tag{9}$$

where E_G is the gap energy, while B represents the exciton binding energy

$$B = \mu e^4/2\varepsilon^2\hbar^2 \; . \quad (10)$$

These exciton states involve the excitations of an electron from the top of the valence band to a bound state below the bottom of the conduction band, as shown in Fig. 8. Some of the features of the valence and conduction bands in RGS are summarized in Fig. 11.

The energy relation of Eq. (9) applies well to states with n > 1, resulting in estimates of E_G and B for RGS given in Table 2. The values of B, together with the experimental data for the dielectric constant ε, result in the evaluation of the exciton

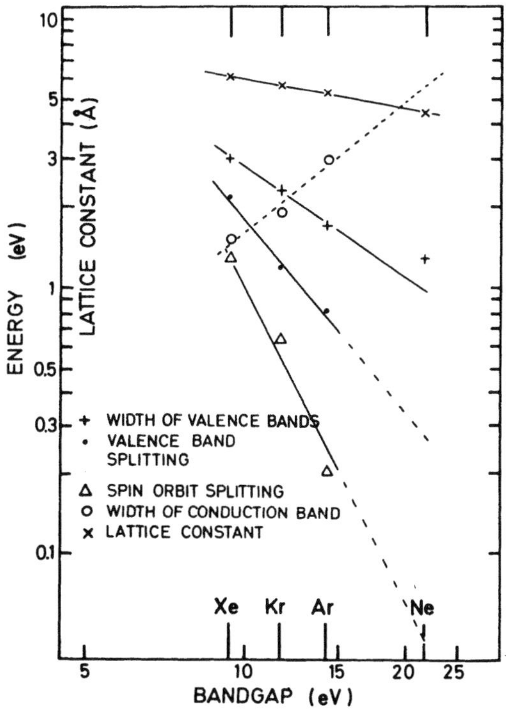

Fig. 11. The plot of characteristic parameters of the valence and conduction bands (+, width of valence bands; •, valence band splitting; Δ, spin-orbit splitting; o, width of conduction band; x, lattice constant) on a linear band-gap scale shows a continuous change of these parameters from Xe to Ne [after N. Schwentner, Ph.D. Thesis (Univ. of München, 1974)].

Table 2. Energetic Parameters for Exciton States in Pure RGS

Solid	E (eV)	B (eV)	ΔE_{SO} (eV)	Δ (eV)
Ne	21.58	4.98	0.09	0.86
Ar	14.16	2.36	0.18	0.26
Kr	11.61	1.53	0.69	0.09
Xe	9.33	1.02	1.32	0.06

effective mass μ, according to Eq. (10). On the other hand, Eq. (9) fails miserably for the lowest n=1 exciton states in RGS. Adhering to the Wannier model, we can assert that in this case the central-cell correction [29-31]

$$\Delta = E^{n=1}(\text{expt}) - E^{n=1} = E^{n=1}(\text{expt}) - (E_G - B) \qquad (11)$$

is substantial.

This success of the phenomenological description raises the intriguing question of why, in spite of the failure of EMA, does the Wannier model work for RGS. In this context, a unified approach to deep Wannier excitons is desirable. And, several theoretical models have been advanced for this purpose.

Pseudopotential theory. Phillips [31] has considered contributions to central-cell corrections for the energies of deep Wannier excitons, which originate from the following sources: (i), spatial dispersion of the effective mass; (ii), spatial dispersion of the dielectric constant; and (iii), core non-orthogonality corrections. This approach suggests the use of a model potential [29-31] which is Coulombic at large distances, and constant for small distances, from the positive ion core.

The integral equation approach. Andreoni, *et al.* [32,33] have adopted a Green's function approach with a local perturbation potential to account for the energies of Wannier states.

The quantum defect method (QDM). This approach [34-36] is of considerable interest in the context of the present discussion since it bridges the gap between atomic (molecular) excitations and solid-state large-radius exciton states for all values of n. Starting from the one-electron picture for deep Wannier excitons,

$$\left[-\frac{\hbar^2}{2\mu}\nabla^2 + U(r)\right]\psi(r) = E\psi(r) \quad , \tag{12}$$

the asymptotic form of the potential is

$$U(r) \xrightarrow{r\to\infty} -\frac{e^2}{\varepsilon r} \ . \tag{13}$$

The energy levels can be written in the form

$$E_n = E_G - \frac{B}{(n+\delta_n)^2} \tag{14}$$

where n is the principal quantum number, while δ_n is the quantum defect, which depends on n. Table 3 provides a comparison between experimental transition energies for bulk excitons in solid neon, including derived parameters [27], and theoretical calculations for Wannier excitons in terms of the QDM. It is interesting to note that the δ_n values exhibit a very weak dependence on n. A heuristic approach to the QDM rests on the use of a model potential $V_n(r)$, which is different for each principal quantum number n, and which has the explicit form [36]

$$\begin{aligned} V_n(r) &= -C_n \ ; \ r < \rho_n \\ V_n(r) &= -\frac{e^2}{\varepsilon r} \ ; \ r > \rho_n \ . \end{aligned} \tag{15}$$

Thus, $V_n(r)$ is constant for small distances and Coulombic for large r. Advancement of this model potential is in the spirit of the pseudopotential theory, mentioned above. The Schrödinger equation is now

$$\left[-\frac{\hbar^2}{2\mu}\nabla^2 + V_n(r)\right]\psi_n = (E_G - E_n)\,\psi_n \tag{16}$$

with the boundary condition $\psi_n(r) \xrightarrow{r\to\infty} 0$, which makes it possible to solve this equation by equating the logarithmic derivatives of ψ_n at $r = \rho_n$. The input data for the solution of the QDM Eqs. (15,16) for RGS are of two types: (i) Atomic parameters. The energies for the excitations of the rare-gas atoms can be used for the calculation of the potential parameters C_n and ρ_n.

Table 3. Comparison of Experimental Transition Energies, and Other Parameters, for Bulk Excitons in Solid Neon with Theoretical Calculations Based on QDM Fits.
[$B_1 = E_G - E(n = 1)$ are the experimental binding energies of the n = 1 excitons, and μ is the effective mass. All energies are in eV.]

	Experiment[a]		Theory			
J =	3/2	1/2	3/2[a]	1/2[b]	3/2[c]	1/2[c]
n = 1	17.36	17.50	17.75	17.85	17.58	17.79
2	20.25	20.36	20.32	20.42	20.24	20.35
3	20.94	21.02	20.94	21.04	20.91	21.03
4	21.19	21.29	21.19	21.29	21.19	21.31
5	21.32		21.31	21.41	21.33	21.45
E_G	21.58	21.62	21.55	21.65	21.61	21.73
B_1	4.22	4.12				
Δ_{so}	0.09		0.10		0.12	
δ	0.28	0.24	0.35	0.35	∿0.5	∿0.5
μ	0.8	0.7	0.8		0.97	

[a] Ref. [27].

[b] Ref. [36]; gap energies are obtained by a fit of the experimental data of [27], and Δ_{so} is the atomic value. From these gap energies we subtracted the theoretical exciton binding energies. The values for δ vary only little with n for the theoretical QDM results. For more detailes see [27].

[c] L. Resca, R. Resta and S. Rodriguez, Sol. State Comm. 26, 849 (1978), and [34].

(ii) Solid-state parameters μ, ε and E_G. The QDM equations are then solved for the quantum defects δ_n, the energies E_n [Eq. (14)] and the relative intensities $I_n \propto (n + \delta_n)^{-3}$. The practical advantages of QDM can be summarized as follows:

--- This scheme is applicable for valence excitations [34] and for core excitations [35] in RGS, as well as for atomic impurity states.

--- The energies of deep Wannier exciton states, including the problematic n=1 state, are properly accounted for.

The practical disadvantages of this scheme are:

--- Proliferation of parameters.

--- Relative intensities of the Wannier states are not properly accounted for [28].

In spite of these technical shortcomings, QDM is of considerable interest from the viewpoint of general methodology because of the following reasons:

--- It establishes the atomic parentage of Wannier states with high n.

--- It is capable of dealing, in principle, with the transition from the Frenkel (atomic) to the Wannier (solid-state) picture.

Before concluding the discussion of deep Wannier exciton states in RGS, it is appropriate to point out that, with the increase of experimental sophistication of sample preparations and of spectroscopic techniques, new types of electronic excitations in films of solid rare gases were observed (Figs. 10 and 12). These fall into two categories:

--- Surface Wannier exciton states with n=1 and n=2 were observed in thin films [28] (Figs. 10 and 12).

--- The appearance of transverse-longitudinal exciton splitting was recorded [1,26,28]. While, in an ideal solid, in optical absorption only transverse excitons appear, scattering of excitons by the surface and by imperfections gives a finite intensity to longitudinal excitons. The transverse-longitudinal exciton splittings in RGS, summarized in Table 4, yield some information concerning the strength of resonance coupling in RGS.

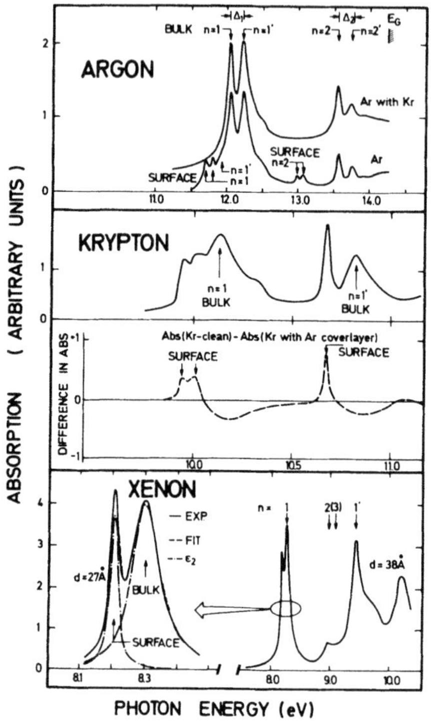

Fig. 12. High-resolution absorption spectra for solid Ar, Kr and Xe in the range of the valence excitons. Volume and surface excitons are observed for all three samples. For Ar and Kr, the results of surface-coverage experiments are also shown. For Xe, the experimentally determined spectrum in the range of the n=1 surface and volume exciton is displayed on an expanded scale, together with a line-shape analysis using two Lorentzians [after V. Saile, Ph.D. Thesis, (Univ. of München, 1976)].

4. DEEP WANNIER ATOMIC IMPURITY STATES

Rare-gas guest atoms in a solid rare-gas host provide the best known example of deep impurity states in insulators. Absorption spectra of these diluent rare-gas alloys were studied by Baldini and Knox [37], Baldini [38], Gedanken, *et al.* [39] and Pudewill, *et al.* [40], among others. The central result of these

Table 4. Longitudinal-Transverse Splitting (eV) for Bulk Excitons in RGS[a]

	Ne		Ar		Kr		Xe	
n = 1	3/2	1/2	3/2	1/2	3/2	1/2	3/2	1/2
Δ^{opt}_{LO-TO}	---	0.27	---	0.26	0.12	0.09	0.06	---

[a]V. Saile and E. E. Koch, Phys. Rev. B20, 784 (1974), and unpublished results.

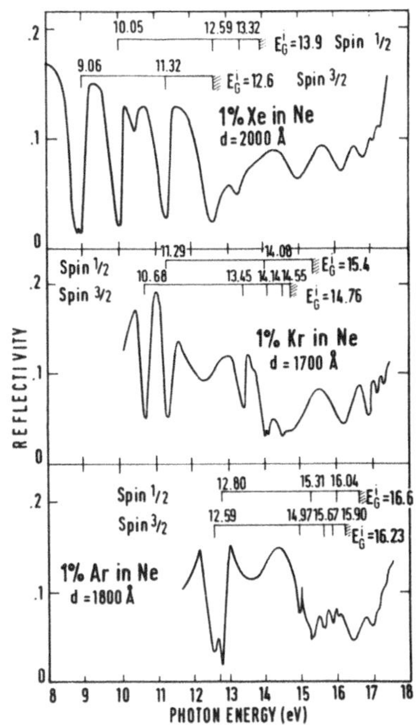

Fig. 13. Reflectance spectra of Ne films doped with 1% Xe, d = 2000 Å (upper panel); 1% Kr, d = 1700 Å (middle panel); and 1% Ar, d = 1800 Å (lower panel) [d ≡ film thickness] evaporated onto a gold substrate. The spin-orbit split impurity exciton series are indicated on top of each spectrum (from Pudewill, *et al.* [40]).

spectroscopic studies is the appearance of a long Wannier series containing lines up to n=5, as shown in Fig. 13. In each dilute alloy, two distinct series are exhibited that are split by the spin-orbit interaction and converge to the bottom of the conduction band of the host matrix (cf. Fig. 8). The energy levels of these deep Wannier impurity states, with $n \geq 2$, obey the hydrogenic equation

$$E_n^i = E_G^i - \frac{B^i}{n^2} \tag{17}$$

where E_G^i is the impurity gap (i.e., the impurity ionization potential in the solid), while

$$B^i = \frac{\mu_e e^4}{2\varepsilon^2 \hbar^2} \tag{18}$$

is the exciton binding energy, which is analogous to Eq. (10), with the exciton effective mass being replaced by the electron mass, μ_e, in the conduction band. As is apparent from Fig. 14, the Wannier formula, Eq. (17), is extremely well-obeyed for deep impurity states in solid Ne. The values of B_i obtained for different impurities are independent of the nature of the impurity and are close to the value of B for pure Ne, since $\mu_e \simeq \mu$. On the other hand, the energy gaps derived from Fig. 14 depend on the nature of the impurity, as is expected. The values of E_G^i and B^i obtained from optical absorption spectroscopy are summarized in Table 5. Supplementary information on E_G^i (and E_G) can be obtained from the threshold for internal photoconductivity. Such data are not yet available for RGS, but for liquid rare gases this information has been obtained [41]. Novel information on the energetics of both exciton and impurity states originate from the threshold for photoemission E_{Th} and E_{Th}^i (cf. Fig. 8), which are given by

$$E_{TH} = E_G - V_0 \tag{19}$$

$$E_{Th}^i = E_G^i - V_0 \tag{20}$$

where V_0 is the energy of an electron at the bottom of the conduction band relative to the vacuum level. Figure 15 demonstrates the determination of the E_{Th}^i values for impurity states in RGS. These experimental data for photoemission thresholds are summarized in Table 5.

We are now in a position to provide a quantitative discussion of impurity ionization processes in solids, liquids and

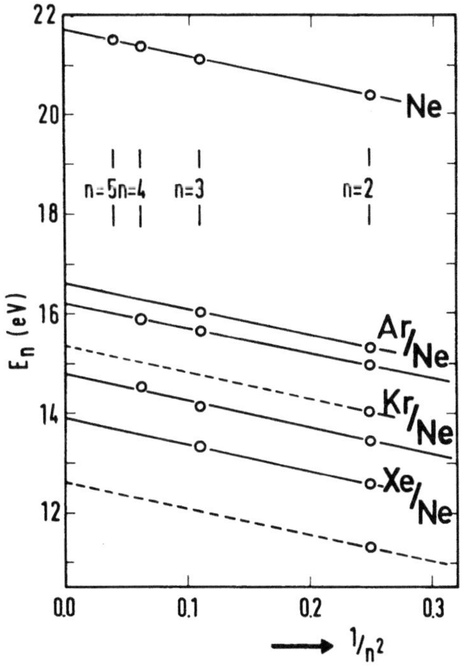

Fig. 14. Excitation energies of excitons in pure Ne (E_n) and of the impurity states of Ar, Kr and Xe in a Ne matrix ($E_n^{\,i}$), plotted versus $1/n^2$ according to the Wannier model (from Pudewill, *et al.* [40]).

Table 5. Energetic Parameters for Wannier Impurity States in RGS

Guest	Host	E_G (eV)	B (eV)	ΔE_{SO} (eV)	Δ_i (eV)	E^i_{TH} (eV)
Ar	Ne	16.1	6.93	0.34	1.39	15.05
Kr	Ne	14.6	6.93	0.61	1.03	13.48
Xe	Ne	12.6	6.93	1.27	1.53	11.60
Kr	Ar	12.5	2.36		0.5	12.2
Xe	Ar	10.54	1.53	1.08	1.0	10.23
Xe	Kr	10.1	1.02		0.4	10.30

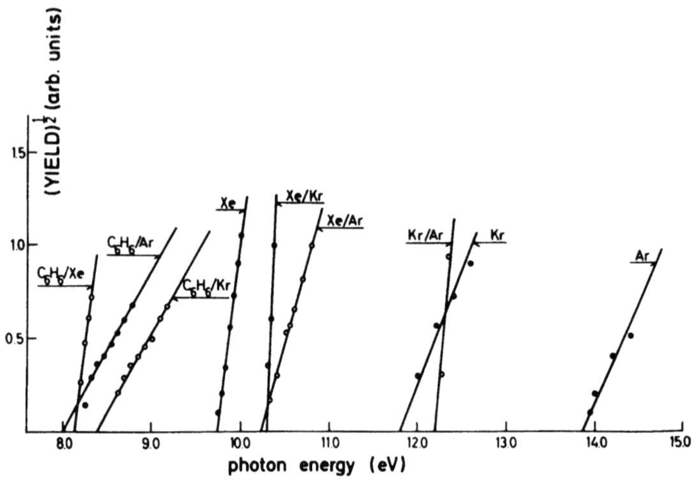

Fig. 15. Photoemission onsets from pure rare gases and from impurity states in RGS (from [57]).

dense fluids. The impurity ionization potential in a condensed phase is given by [42]

$$E_G^i = I_{GAS} + P_+^i + V_0 \tag{21}$$

where I_{GAS} is the gas-phase ionization potential, while P_+^i represents the medium polarization energy induced by the impurity. From Eqs. (20) and (21), we immediately infer that the threshold for photoemission from an impurity state is

$$E_{Th}^i = I_{GAS} + P_+^i \quad . \tag{22}$$

For doped solid rare gases, the combination of E_G^i data from absorption spectroscopy with photoemission yield data resulted in the determination of P_+^i values and, more important, the establishment of V_0 values for RGS (Table 6). Extensive data are available on V_0 values for the liquid rare gases [1,43]. A theoretical description advanced by Springett, Jortner and Cohen [44] considers two major cumulative contributions to V_0, namely,

$$V_0 = T + U_p \tag{23}$$

originating from short-range repulsive interactions T, which are subsumed in a kinetic energy term, and a background polarization

Table 6. Experimental V_0 Scale for Solid and Liquid Rare Gases Obtained from Optical and Photo-Emission Experiments (cf. Table 3) Compared to the Calculation of V_0 Based on the SPRINGETT, JORTNER, COHEN (SJC) Theory.
[A superscript i denotes values for the impurity states. $V_0 = E_G - E_{TH} = U_p + T$. U_p = background polarization potential; T = kinetic energy term.][a,b]

System		Reference	Experiment V_0(eV)	Theory[a,b] T (eV)	Theory[a,b] U_p (eV)	Theory[a,b] V_0 (eV)
Liquid ^{4}He	4.2 K	c	+ 1.05	1.66	- 0.3	+ 1.28
Solid Ne	4 K	d	+ 1.3	2.44	- 1.85	+ 0.60
Liquid Ne	25 K	e	+ 0.67 ± 0.05	1.91	- 1.46	+ 0.45
Ar in solid Ne		d	+ 1.1			
Kr in solid Ne		d	+ 1.1			
Xe in solid Ne		d	+ 1.0			
Solid Ar	6 K	d	+ 0.4	3.46	- 3.54	- 0.08
Liquid Ar	87 K	e,f	- 0.2 ± 0.02	2.53	- 2.73	- 0.20
Kr in solid Ar		d	+ 0.3			
Xe in solid Ar		d	+ 0.3			
Solid Kr		d	- 0.3	3.84	- 4.05	- 0.21
Liquid Kr	123 K	e	- 0.45 ± 0.05	2.73	- 3.11	- 0.38
Xe in solid Kr		d	- 0.2			
Solid Xe			- 0.4	4.12	- 4.64	- 0.52
Liquid Xe	165 K	e	- 0.61 ± 0.05	2.85	- 3.46	- 0.61

[a] Ref. [44]; [b] Ref. [43]; [c] W. I. Sommer, Phys. Rev. Lett. 12, 271 (1964); [d] Ref. [1]; [e] W. K. Tanchert, Ph.D. Thesis (Freie Universität Berlin, 1975) and W. K. Tanchert and W. F. Schmidt, Z. Naturforsch. 30a, 1085 (1975); [f] J. Leckner, et al., Phys. Rev. 156, 351 (1972).

energy U_p (Table 6). From this analysis of the V_0 scale, it is apparent that a uniform description is provided for the energetics of an electron in the conduction band, both in a solid and in a liquid. Accordingly, we are led to the important conclusion that the notion of the conduction band is applicable to liquids and other structurally disordered dense fluids, so that there is no need to invoke the notion of translational symmetry for the description of this basic universal property of electronic states of condensed phases [45].

5. EXCITONS AND IMPURITY STATES IN RARE-GAS FLUIDS

The applicability of the universal concept of the conduction band to liquid rare gases raises the interesting question as to whether or not solid-state type excitations, such as excitons and impurity states, are amenable to experimental observation in structurally disordered materials. Obviously, Frenkel-type states, which have a unique parentage in the states of the constituents, will definitely be observed. What is more interesting to inquire is whether or not Wannier exciton and impurity states "exist" in dense fluids. On the basis of theoretical arguments, it was conjectured [46] that large-radius Wannier excitons are amenable to experimental observation in a dense fluid, where the electron-medium interaction is sufficiently weak. Liquid rare gases constitute a medium where the electron-atom interaction is moderately weak, ensuring that the linewidth of Wannier states is not excessive. Some experimental work has established that n=2 and n=3 Wannier excitons [47,48], and n=2 Wannier impurity states [49,50], are amenable to experimental observation in liquid rare gases.

In order to establish the nature of Wannier states in simple dense fluids, it is interesting to obtain a description of the electronic excitations of a guest atom or molecule in a host fluid, where the fluid density is continuously varied over a broad density range [50-52]. In this context, three types of electronic transitions of the guest atom should be considered, involving (i), intravalence excitations; (ii), low extravalence excitations which are accompanied by a change $\Delta n_a=1$ in the principal quantum number n_a; and (iii), high extravalence excitations, characterized by $\Delta n_a=2$. Figs. 16 and 17 show the absorption spectra of Xe-Ar, spanning a broad density range of the Ar host, from the atomic limit through the liquid to the solid. In the energy range 9.5-10.8 eV, the following five transitions can be identified:

Transition I to 6s'[1/2]J=1 [$5p^5$ ($^2P_{1/2}$) 6s] at 9.57 eV (1296 Å).

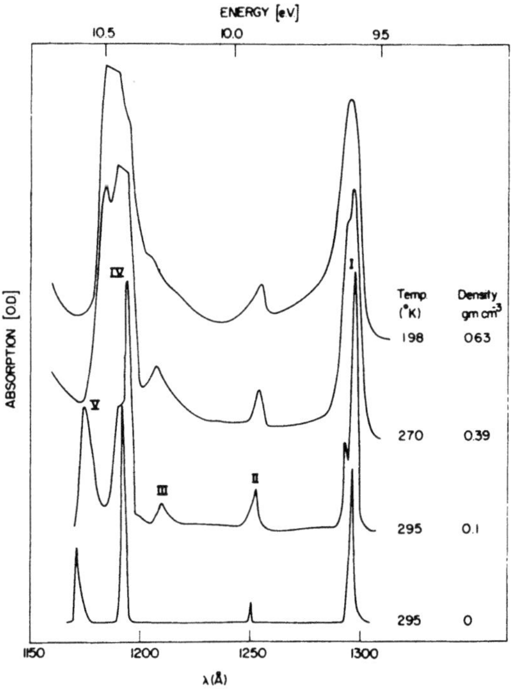

Fig. 16. The absorption spectra of Xe in fluid Ar in the low and intermediate Ar density range. The absorption curves are horizontally shifted, all being displayed on the same optical density (OD) scale (from Messing, *et al.* [50]).

Transition II to 5d[1/2]J=1 $[5p^5\ (^2P_{3/2})\ 5d]$ at 9.92 eV (1250 Å).

Transition III to 5d[5/2]J=3 $[5p^5(^2P_{3/2})\ 5d]$ at 10.25 eV (1210 Å), which is symmetry forbidden in the atomic limit.

Transition IV to 5d[3/2]J=1 $[5p^5\ (^2P_{3/2})\ 5d]$ at 10.40 eV (1192 Å).

Transition V to 7s[3/2]J=1 $[5p^5\ (^2P_{3/2})\ 7s]$ at 10.60 eV (1170 Å).

Transitions II - IV are due to intravalence excitations. Transition I corresponds to a low extravalence excitation, with Δn_a=1, of a Rydberg series which starts with the first resonance line 6s

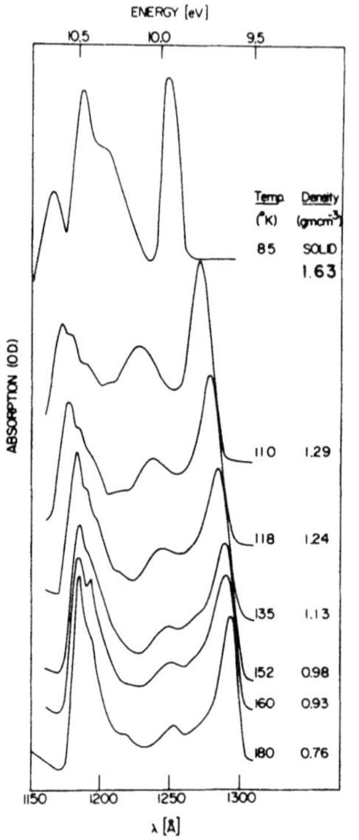

Fig. 17. The absorption spectra of Xe in fluid Ar at high Ar density and in solid Ar. The Xe concentration is roughly 1 ppm. The absorption curves are horizontally shifted, all being displayed on the same optical density (OD) scale (from Messing, *et al.* [50]).

[3/2]J=1 $[5p^5(^2P_{3/2})\ 6s]$ at 8.44 eV (1470 Å). Transition V constitutes a high extravalence excitation, with Δn_a=2. The following medium effects have been observed at low and moderate densities (cf. Fig. 16):

--- Satellite bands. In the density range 0.1-0.4 g cm^{-3}, a single blue satellite band accompanies each of Transitions I and IV. The intensity at these blue satellites increases with increasing density until, at Ar density ≈0.6 g cm^{-3}, they merge together with the corresponding resonance line into a single band.

--- Symmetry-breaking effects. Transition III, to a J=3 state, is symmetry forbidden in the atomic limit, as is evident from Fig. 16, but allowedness is induced by the host fluid.

--- Medium-induced shifts of the lowest $\Delta n_a=1$ extravalence excitations. These shifts are summarized in Fig. 18. The Xe resonance lines 6s[3/2] (not shown in Figs. 16-18) and 6s'[1/2] (Transition I) are weakly red-shifted at low densities, and exhibit a large blue shift at higher densities. A similar behavior is observed for Transition IV. These features have been explained in terms of a strongly repulsive Xe-Ar interaction potential in the excited state [50-52].

--- Medium-induced shifts of the higher $\Delta n_a=2$ extravalence

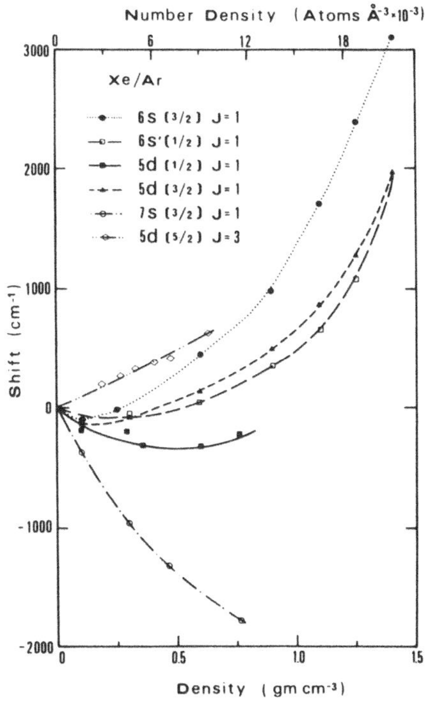

Fig. 18. The density dependence of the energies of the maxima of the absorption bands relative to the unperturbed atomic transition energy for the absorption lines of Xe in fluid Ar observed in the spectral range 8.5 eV (≈1460 Å) to 10.8 eV (1150 Å); see Figs. 16 and 17. The spectral lines are designated by the conventional atomic notation (from Messing, *et al.* [50]).

excitation. The spectral shift of Transition V (to the 7s[3/2] state) drastically differs from those revealed by the 6s[3/2] and 6s'[1/2] transitions. While the latter low extravalence excitations exhibit predominantly a large spectral blue shift (cf. Fig. 18), the former "high" extravalence transition reveals an appreciable spectral red shift, even at high densities. In the high density range (cf. Fig. 17), the first transition at ∿9.7 eV (1265-1290 Å) was assigned by Messing, *et al.* [50] to an excitation of the 6s'[1/2] state (Transition I), and the third main transition at ∿10.5 eV (1170-1190 Å) to the 5d[3/2] state (Transition IV). These transitions exhibit large blue shifts (cf. Fig. 17) at high densities. The most interesting feature of the spectra in Fig. 17 is the absorption band at 9.9-10.2 eV. This second band, centered at 10.12 eV (1225 Å) in the high-density fluid, was assigned by Messing, *et al.* [50] to an n=2 Wannier state on the basis of the following arguments. First, this transition is close in energy to the n=2 Wannier impurity state in solid Xe/Ar alloys. Second, the intensity of this absorption in the Xe/Ar system at high (> 7×10^{-3} atoms $Å^{-3}$) densities is anomalously enhanced, indicating the appearance of a new absorption band at these densities. This behavior is unique to the Xe/Ar system and was not observed for Xe/Ne and Xe/He in the same range of atomic densities.

The assignment of the n=2 [3/2] Wannier impurity state of Xe in Ar fluid makes it possible to establish a correlation between atomic states and Wannier excitations. Fig. 19 provides a correlation diagram, advanced by Messing, *et al.* [50], which relates the high extravalence atomic excitations and "high" n=2 states. From this analysis of the spectroscopy of impurity states in a dense fluid, the following conclusions emerge:

--- There exists a possible parentage of high n=2 Wannier states in high Δn_a=2 extravalence atomic excitations of the impurity constituent. This calls for an extension of the QDM approach to impurity states in fluids.

--- Both s and d atomic impurity states were observed in the dense fluid.

--- Metastable impurity states (e.g., $^2P_{3/2}(5d)J=1$ excitation of Xe), were observed above the onset of the conduction band [31]. The absorption lineshapes for such autoionizing states should exhibit Fano resonances [53]. However, these interesting effects may be masked by phonon broadening and inhomogeneous broadening phenomena.

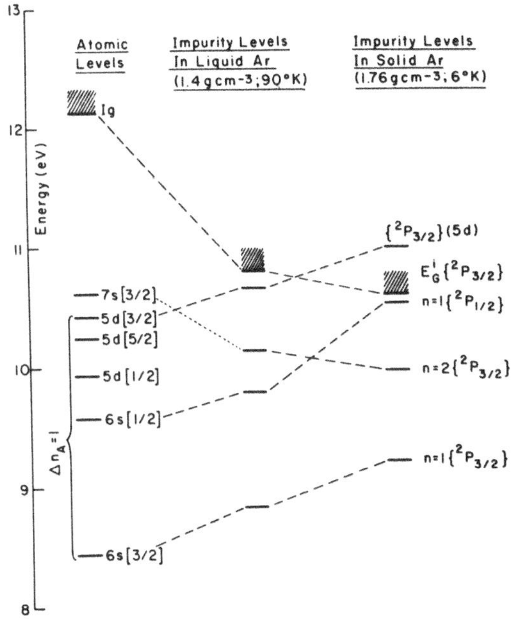

Fig. 19. Energy correlation diagram of the unperturbed atomic states of Xe, of Xe impurity states in liquid Ar and of Xe impurity states in solid Ar. The correlation of the 7s[3/2] atomic state is rather tentative, as denoted by the dashed line (from Messing, *et al.* [50]).

6. MOLECULAR WANNIER IMPURITY STATES IN RARE-GAS SOLIDS

The observation of intermediate and Wannier-type atomic impurity states in RGS and in fluids implies that analogous molecular excitations should be amenable to experimental observation. There has been a lively controversy as to whether molecular Rydberg states "exist" in solid matrices [22,54-56]. This question should be paraphrased to ask: "What are the relevant matrix effects which will prohibit the spectroscopic detection of molecular extravalence excitations in simple solids" [22]. Extensive experimental work [22,39], some of which is presented in Figs. 20 and 21, has established that intermediate-type and higher Wannier-type molecular excitations are observed in RGS. The following features of molecular Rydberg states in RGS should be emphasized:

--- The first molecular extravalence excitation corresponds to an intermediate case, which can be described either in terms of the first molecular Rydberg state subjected to large non-ortho-

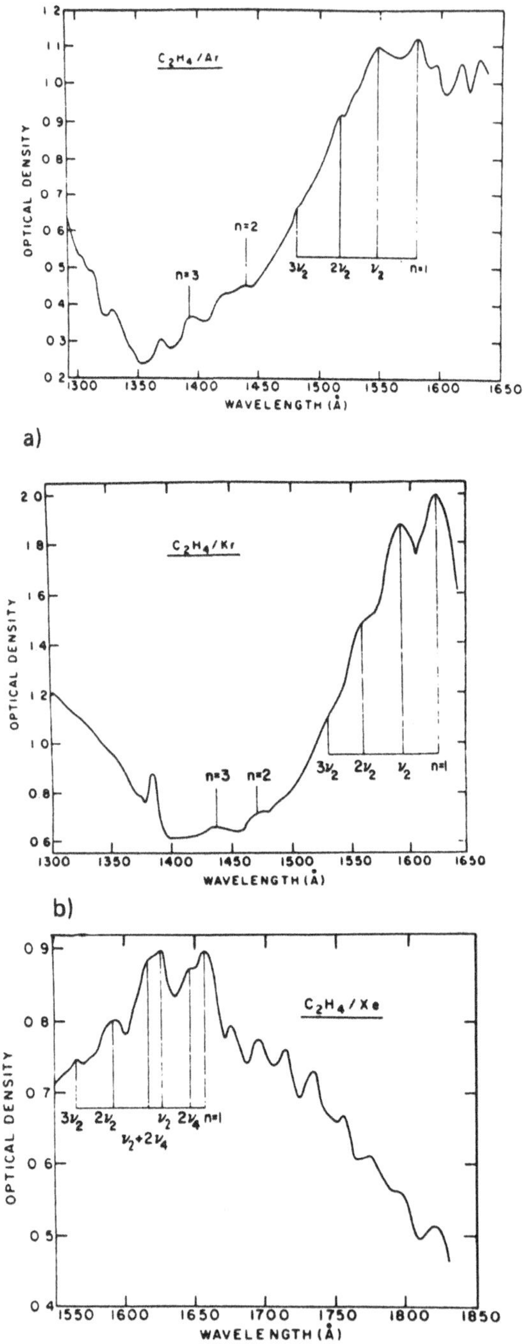

Fig. 20. Extravalence electronic excitations of ethylene in RGS (from Gedanken, *et al.* [39]).

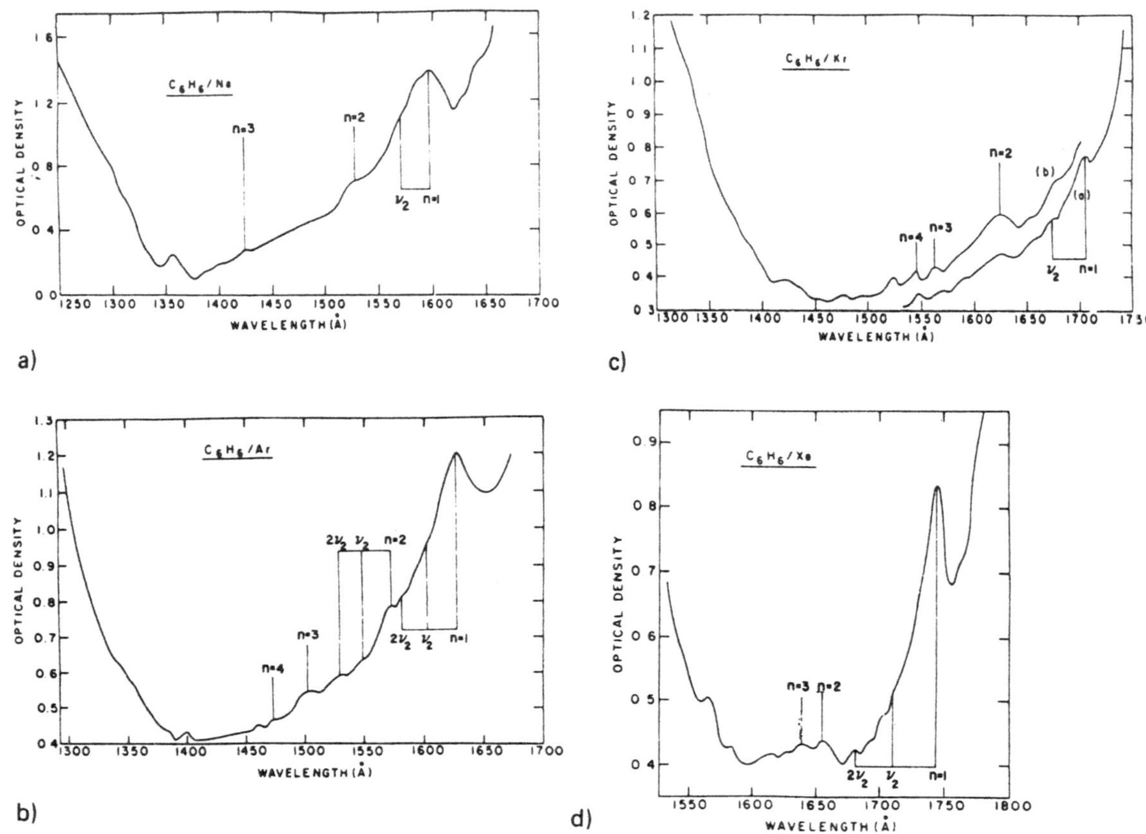

Fig. 21. Extravalence electronic excitations of benzene in insulators (from Gedanken, *et al.* [39]).

gonality corrections or, alternatively, in terms of an n=1 Wannier state with large central-cell corrections. The former description implies large blue spectral shifts of the lowest molecular Rydberg state in the matrix (cf. Table 7). Furthermore, extensive line broadening is exhibited (cf. Figs. 20 and 21 and Table 8), originating from phonon broadening and also from some inhomogeneous broadening effects. We note in passing that the diagnostic criterion advanced by Robin [54,55] for the identification of molecular Rydberg states should be modified from "Molecular Rydberg states disappear in a condensed phase" to "The lowest molecular Rydberg transition is considerably blue-shifted in a condensed phase".

--- Higher Rydberg states, corresponding to an n=2 Wannier state of an electron bound to a positive ion in an RGS, have been observed [22,39,56].

--- Vibrational structure for all n molecular intermediate and Wannier-type excitations has been observed [22,39,56]. This new spectroscopic feature of molecular extravalence excitations in a dense medium corresponds to vibrational excitations of the positive ion core.

--- Overlap of the molecular Wannier-type excitations with other broad structureless transitions, which correspond to dissociative states and which complicate the observation of molecular extravalence excitations in condensed phases.

--- The spectroscopic identification of high molecular Wannier states provides a direct method for the determination of the molecular ionization potential in the solid, via Eqs. (17) and (21), and the threshold E^i_{Th} for photoemission, via Eq. (22). The E^i_{Th} values determined directly by photoemission [57,58] from molecular impurity states in RGS are in good agreement with the spectroscopic data [22,39]. The consistency obtained between the spectroscopic and photoemission data provides strong support for the identification of molecular Wannier states in RGS.

Up to this point, we have been concerned with molecular Wannier impurity states in RGS. It is interesting to inquire whether or not Wannier exciton states can be observed in pure molecular crystals. The absorption spectrum of the acetylene crystal [59], shown in Fig. 22, reveals two transitions: (i) The intravalence $\pi \rightarrow \pi^*$ transition in the range 6.8-7.8 eV. The vibrational structure of this transition shows an extensive progression of the $2\nu_4$ twisting vibration (472 cm^{-1} in the gas phase). This transition reveals a modest medium-dependent spectral shift, being red-shifted by 0.1 eV for the crystal relative to C_2H_2 in solid Ne (cf. Fig. 22). (ii) An intense crystal transition originating at 8.2 eV, which shows the following characteristics: A

Table 7. Spectral Shifts (cm^{-1}) of the n = 1 Transitions of Molecular Impurities in Solids Relative to the Electronic Origin of the Lowest Rydberg.

Impurity \ Matrix	Xe	Kr	Ar	Ne	CF_4	N_2
CH_3I	--	5095	6480	8950	12880	--
C_6H_6	1400	2920	5640	6800	--	5520
C_2H_4	2850	4160	5720	6330	--	--
C_2H_2	--	4210	6650	6910	8700	--

Table 8. Linebroadening (cm^{-1}) of the n = 1 Transitions of Molecular Impurities in Solids.

Impurity \ Matrix	Xe	Kr	Ar	Ne	CF_4	N_2
C_2H_4	620	1010	--	--	--	--
C_6H_6	570	1070	1240	2760	--	3580
C_2H_2	--	345	760	1590	1250	--
CH_3I	--	550	760	1140	2320	--

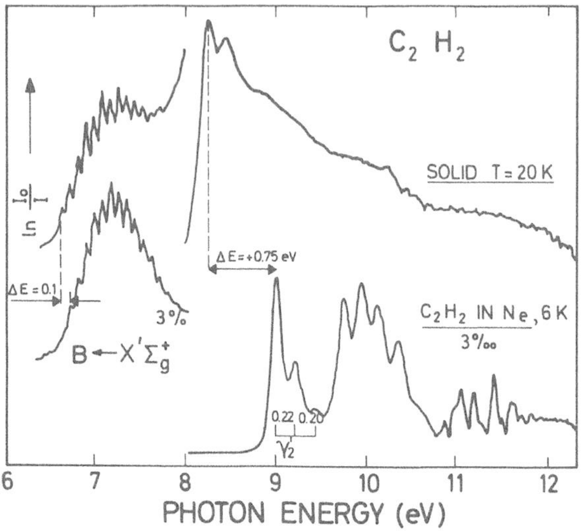

Fig. 22. The absorption spectra of solid acetylene and of acetylene in solid Ne (from E. E. Koch and J. H. Fock [59]).

short vibrational progression of 1300 cm^{-1}, as is the case for the Rydberg state of acetylene; and, a large spectral shift (0.75 eV) in the crystal relative to acetylene in Ne (Fig. 22). This transition is assigned as the n=1 intermediate Wannier excitation in crystalline acetylene, the parentage of which can be traced to the first Rydberg state of the acetylene molecule.

A novel experimental approach to the study of molecular Rydberg states in dense media is provided by spectroscopic studies of van der Waals (vdW) molecules [60] and clusters [61]. Let us consider a vdW molecule M·R, consisting of a diatomic or polyatomic molecule M bound to a rare-gas atom R -- for example, NO·Ar [62] or benzene·Ar. A Rydberg excitation of M will exert a short range M-R repulsive interaction, whereupon the potential curve will be repulsive, resulting in a dissociative continuum and manifesting direct M-R photodissociation. Increasing the size of the vdW complex by adding more R atoms will result in an MR_n large cluster. The Rydberg states of M in a large cluster will not lead to direct dissociation but rather to nuclear relaxation around the excited state, resulting in the expansion of the cluster around the extravalence molecular excitations. The spectroscopy of Rydberg states of vdW molecules and clusters will provide direct information on medium effects on molecular Rydberg excitations, as explored from the microscopic point of view.

7. EXCITED-STATE DYNAMICS IN RARE-GAS SOLIDS

We have concerned ourselves, so far, with the electronic structure of pure and doped insulators, which provides the essential input data for the elucidation of excited-state dynamics. Two general classes of relaxation processes in electronically excited states can be considered.

Intrastate processes. The system is produced in a nonequilibrium nuclear configuration, so that the balance of interatomic (or intramolecular) interactions is destroyed. Nuclear relaxation occurs within a single electronically excited state. The following phenomena correspond to this category:

--- Exciton trapping by nuclear deformation in pure RGS and in rare-gas liquids [1,22,63-67].

--- Lattice relaxation around excited impurity states in RGS [68-70].

--- "Bubble" formation around electronically excited states in liquids [1].

--- Medium-induced vibrational relaxation of $R_2{}^*$ electronically-vibrationally excited diatomic molecules in RGS and in liquid rare gases [22].

Interstate processes. The relaxation process is accompanied by a change in the electronic state, involving partial or complete interconversion of electronic energy into phonons. Typical examples are:

--- Electron-hole recombination [6].

--- Electronic relaxation of exciton states in RGS [22].

--- Electronic relaxation of impurity states in RGS and in liquid rare gases [22,68-70].

--- Electronic energy transfer by "free" excitons in RGS [22,57,71-74].

--- Electronic energy transfer between localized states in solid and in liquid insulators [1,22,75-77].

We now proceed to discuss several examples for intrastate and interstate relaxation processes, emphasizing the role of molecular energetics and relaxation in condensed-phase dynamics.

8. SELF-TRAPPING OF EXCITONS IN RARE-GAS SOLIDS

The emission spectra of RGS [63-67], summarized in Fig. 23, provide interesting information concerning exciton trapping by nuclear deformation. The prominent emission bands from pure solid Ar, Kr and Xe are characterized by the following features:

--- Broad luminescence bands are observed, exhibiting large (1-2 eV) Stokes shifts relative to the lowest n=1 ($^2P_{3/2}$) excitonic absorption.

--- The lineshapes of these emission bands are nearly Gaussian.

--- Similar emission spectra are exhibited in the solid, in the liquid and in the cold gas.

These luminescence bands are due to [63] emission from the electronically-excited, vibrationally-relaxed, homonuclear rare-gas molecule R_2* [78]. From the schematic potential surfaces of diatomic rare-gas molecules, portrayed in Fig. 24, we note that the ground state is characterized by a shallow van der Waals potential. In the lowest excited $^3\Sigma_u$ and $^1\Sigma_u$ states, the binding energy is high ($\sim$1-2 eV). Electronic excitation results in a considerable decrease in the equilibrium separation of R_2*, which is reduced by about 30% relative to the ground state. Emission from the ground vibrational level of R_2* will give rise to R_2 in its strongly repulsive part of the potential, so that the two R atoms will fly apart. The large Stokes shift in emission, and the broad, Gaussian emission lineshapes in solid Ar, Kr and Xe, can readily be rationalized on the basis of this molecular picture. Furthermore, the practical invariance of the emission spectra with respect to the state of aggregation of the rare gas is also consistent with this molecular description of the emitting center in these RGS.

The emission of solid Ne differs drastically from the emission spectra of the heavier RGS. The prominent emission band of solid Ne is relatively narrow, peaking at 16.7 eV (cf. Fig. 23), and is Stokes shifted by 0.7 eV to lower energies relative to the n=1 ($^2P_{3/2}$) exciton states [66,67]. This emission originates from a medium-relaxed atomic state of Ne [66,67]. Medium relaxation results in dilation around the electronic excitation, so that perturbations on the atomic excitation are minor. Indeed, the energy of the prominent emission band in solid Ne is close to the low atomic excitations. In some experiments, a splitting of this "atomic-type" emission into three components was observed, which was correlated with the $^1P_1$, $^3P_1$ and $^3P_0$ states. In addition to the prominent emission band, solid Ne exhibits another broad emission band, located towards lower energies, which is as-

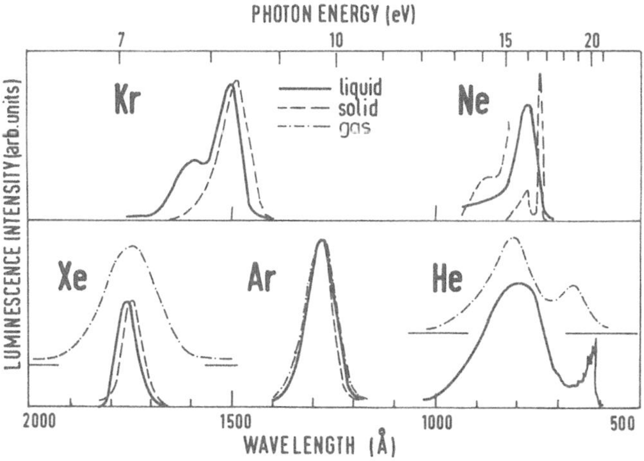

Fig. 23. Emission spectra of solid, liquid and gas-phase rare gases.

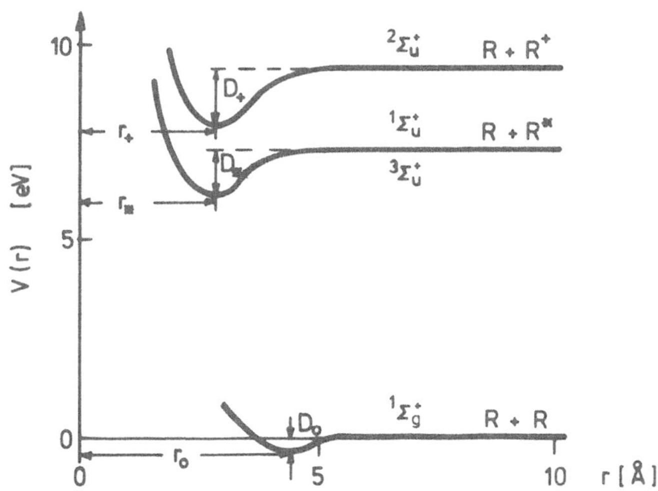

Fig. 24. Potential curves for a diatomic rare-gas molecule.

signed to the decay of the vibrationally-unrelaxed Ne_2* diatomic molecule.

To summarize the phenomenology of the emission spectroscopy of RGS, we would like to point out that four distinct emission bands were observed, which will now be listed in the order of increasing energy.

Emission from a free exciton state. The quantum yield for this emission, namely,

$$Y = \tau_t / \tau_R \tag{24}$$

is given in terms of the ratio between the exciton trapping time τ_t and the pure radiative lifetime τ_R of the free exciton. For pure solid Xe, the experimental data indicate that [65] $Y \simeq 10^{-3}$. Since $\tau_R \simeq 10^{-9}$ s, we estimate $\tau_t \approx 10^{-12}$ s, indicating the occurrence of efficient exciton trapping on the ps timescale.

Emission from a localized atomic excitation. This relatively narrow emission band is somewhat red-shifted relative to the exciton state, indicating the reduction of repulsive matrix interactions due to lattice dilation. This emission band is prominent in solid Ne.

Emission from a vibrationally-hot diatomic R_2^* *molecule.* In this case exciton trapping via the formation of R_2* took place; however, vibrational relaxation is slow on the timescale of the radiative decay of the diatomic molecule. The secondary emission band in solid Ne is due to this emission.

Emission from vibrationally-relaxed R_2^* *excited molecules.* The prominent emission bands of solid Ar, Kr and Xe correspond to this category.

In the language of solid-state physics, the emission from an R_2* center corresponds to two-center trapping, while the emission from an atomic state is attributed to one-center trapping. Two mechanisms for exciton two-center and one-center trapping in RGS were advanced:

The molecular approach [22]. Two-center localization in solid Ar, Kr and Xe originates from the formation of R_2* at high vibrational states, followed by efficient vibrational relaxation of the diatomic molecule. In view of the low vibrational frequencies of Ar_2*, Kr_2* and Xe_2*, which are comparable to the Debye frequency $\hbar\omega_D$ of the solid, vibrational relaxation is very efficient on the timescale of the radiative decay. In solid Ne, vibrational relaxation is slow due to the high vibrational frequency of Ne_2*

relative to the Debye frequency. One can then consider a local picture for exciton trapping, the assets of which are:

--- Predominance of the molecular point of view. The description of exciton trapping is universal for solids, liquids and gases.

--- The weak ($Y \simeq 10^{-3}$) free-exciton emission from solid Xe can be rationalized in terms of a low branching ratio from these excited states which evade vibrational relaxation.

--- The absence of relaxed molecular emission from solid Ne is explained.

From the point of view of general methodology, the molecular model disregards excitonic and solid-state effects, which are emphasized by the alternative approach to the problem.

The solid-state continuum model [79,80]. Exciton localization in RGS is of considerable methodological interest, since in these solids short-range coupling with acoustic phonons prevails, while long-range coupling to polar modes does not exist. Toyozawa [79] and Rashba [80] considered the transition from the free to the localized state of the exciton, which occurs when the lattice relaxation energy E_{LR} is comparable to the exciton bandwidth b. The localization process due to lattice distortion can be visualized as the splitting of a bound state from the exciton band. When E_{LR} is sufficiently large, the localized state is characterized by a minimum on the adiabatic potential surface, which is separated from the free exciton state by a potential barrier H_m. On the basis of an isotropic continuum, where the lattice distortion corresponds to an elastic dilation, the barrier height is given by $H_m = 4b^3/27E_{LR}^2$. According to the continuum model, two conditions have to be satisfied for the observation of self-trapping in a solid. First, the energetic criterion $E_{LR} \simeq b$ has to be satisfied. Second, a dynamic criterion $H_m \lesssim \hbar\omega_D$ has to be obeyed to ensure nuclear tunnelling from the free exciton to the self-trapped state. A schematic energetic scheme appropriate for self-trapping is given in Fig. 25. The continuum model is not specific concerning the microscopic aspects of self-trapping. Two stable configurations of the localized state are of particular interest (cf. Fig. 26): (i) The one center, atomic trapped exciton. Here, the nearest neighbors displace symmetrically around the excitation. (ii) The two-center trapped excitations. Here, the two neighboring atoms displace axially. The relative energetic stability of the one-center and two-center excitations can be assessed, in principle, on the basis of an excitonic model [81] (cf. Fig. 26) which considers the change in the diagonal site energies U' and the change in the off-diagonal energies V' accompanying lattice displacements. Large diagonal defor-

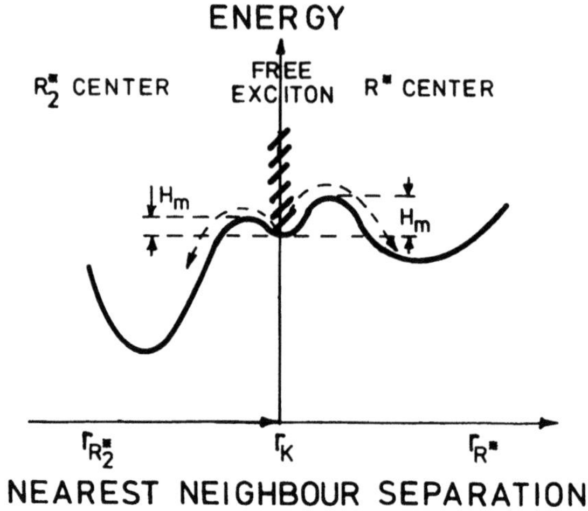

Fig. 25. Schematic potential curves and lattice distortions describing the localization of free excitons in atomic and excimer centers [after I. Ya. Fugol, Adv. in Physics 27, 1 (1978)].

mations result in one-center trapping, while large off-diagonal deformations give two-center trapping. The continuum model modified to incorporate the microscopic nature of exciton trapping (Fig. 26) in RGS, leads to the following predictions:

--- The two-center trapped exciton R_2* is energetically favored over the free exciton for all RGS.

--- The two-center trapped exciton is energetically favored over the one-center trapped state for all RGS. This prediction contradicts the experimental data for solid Ne.

--- In solid Ne and Ar, the barrier height H_m is low relative to $\hbar\omega_D$ and very efficient trapping occurs.

--- In solid Xe, the barrier height for two-center localization is large, so that the free state should be observed. The low ($Y \simeq 10^{-3}$) quantum yield for emission from free excitons in solid Xe is blamed on trapping and imperfections.

It appears that at present it is rather difficult to assess the relative merits of the molecular and the solid-state models on the basis of the confrontation of predictions with the experimental facts. The molecular model has a slight edge, since it correctly predicts one-center localization in solid Ne, indicating a

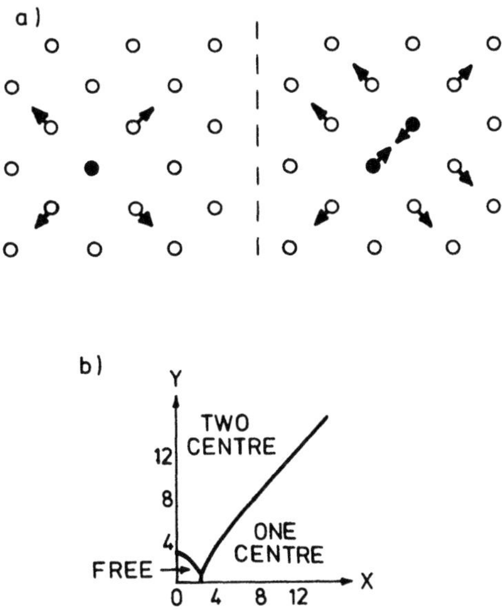

Fig. 26. (a) Schematic lattice rearrangement for one-center self-trapping (left) and two-center self-trapping (right) of excitons in RGS (after Song [81] and Jortner [22]).
(b) Map of zones of stable states of self-trapper excitons in the plane $X = U'/\sqrt{\alpha|V|}$, $Y = V'/\sqrt{\alpha|V|}$ (from Song [81]).

slight superiority of chemical concepts over the physical model.

Further support for the "chemical" description of self-trapped excitons in RGS stems from recent experimental information concerning transient absorption studies [82] and time-resolved studies [83]. Transient absorption studies [82] of the self-trapped exciton in solid Ar and Kr have established that R_2^* corresponds to the $^3\Sigma_u$ state of the homonuclear rare-gas diatomic molecule. These studies provide further support for the molecular description of the self-trapped exciton in heavy RGS. Time-resolved studies of the emission from solid Ar, Kr and Xe (Figs. 27 and 28) have established the occurrence of multiple decay. In solid Xe (Fig. 27), the decay is characterized by both a short (ns) and a long (μs) temperature-dependent component. Furthermore, the spectral distribution of the short and long components is different (cf. Fig. 28). The short component was tentatively attributed to the radiative decay of the $^1\Sigma_u^+$ molecular state,

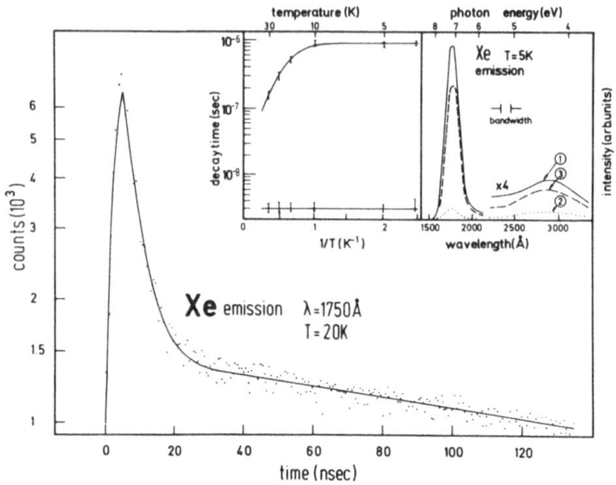

Fig. 27. Decay curves for solid Xe. Left inset, temperature dependence of the long and short components; right inset, emission bands for different times after deposition of the sample (from Hahn, *et al.* [83]).

while the long component(s) are due to the decay of the $^3\Sigma_u^+$ molecular state, which may be subject to crystal-field splitting. These time-resolved data again demonstrate the prominence of the molecular point of view for the description of the self-trapped exciton in heavy RGS. It should, however, be emphasized that both the transient absorption and the time-resolved studies interrogate the self-trapped exciton on a "long" (ns and longer) timescale and do not provide information on the trapping process, which occurs on the ps timescale. The molecular and the solid-state models for exciton trapping should not be considered as orthogonal, but rather complementary and supplementary, so that the synthesis of chemical and physical concepts will provide a general picture of the dynamics of exciton localization in simple insulators.

9. LATTICE RELAXATION AROUND IMPURITY STATES

The one-center localization of the exciton in pure solid Ne involves symmetric radial displacement of the neighboring atoms away from the electronically excited state. An analogous phenomenon is ubiquitous for electronically excited atomic impurity states in RGS. In these systems, short-range repulsive interac-

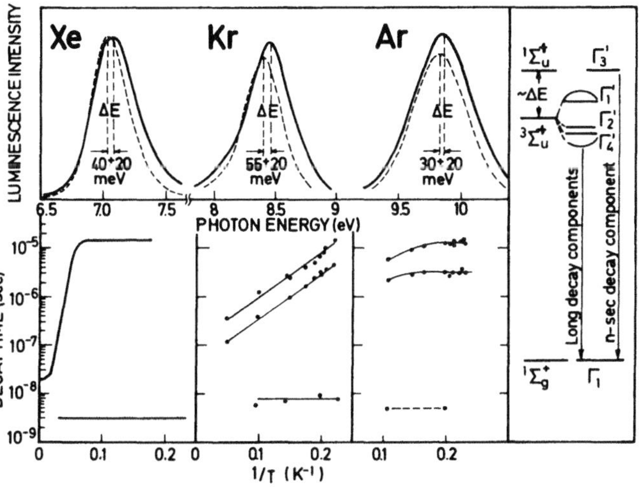

Fig. 28. Results of time-resolved luminescence experiments for the main luminescence bands of solid Xe, Kr and Ar. Full lines in the intensity curves (upper panel) correspond to the component with short decay times (10^{-9} s region), dashed lines correspond to long decay times. Emission bands and decay curves for Kr and Ar were obtained by Hahn, *et al.* [83] [from G. Zimmerer, J. Luminescence 18/19, 875 (1979)].

tions exerted by the excited atom on its neighbors result in an appreciable increase of the average separation between the electronically excited atom and its nearest host atoms. The electronically-vibrationally excited state, populated by a vertical transition (cf. Fig. 29), will relax subsequently by phonon emission to a new equilibrium configuration [6]. The configurational diagrams for the ground state and electronically excited state of an impurity in an RGS, as portrayed in Fig. 29, correspond to the strong electron-phonon coupling limit, which is characterized by a large horizontal displacement of the minima of the potential surfaces [6]. The absorption lineshape is expected to be Gaussian with a linewidth [6]

$$H = H_0 \left(\coth \frac{\hbar\langle\omega\rangle}{2kT}\right)^{1/2} \tag{25}$$

where $\langle\omega\rangle$ is a characteristic phonon frequency. These expecta-

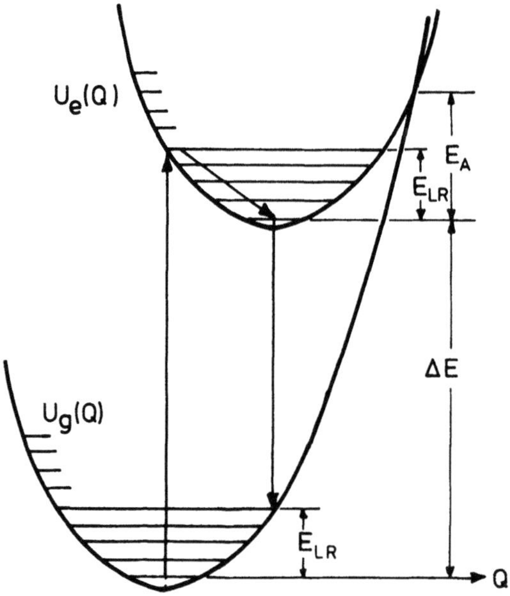

Fig. 29. A configurational diagram for an impurity center.

tions are borne out by the absorption spectra of Xe impurity in a 1 cm thick Ar crystal [84] (Figs. 30 and 31). The Gaussian lineshape and the temperature dependence of the linewidth clearly demonstrate that the system corresponds to the strong-coupling situation. The large lattice relaxation around impurity states has a profound effect on the energetics of the emission from impurity states. The large configurational change around the impurity center will drastically reduce the short-range repulsive interactions, which in turn contribute to a large blue spectral shift in absorption relative to the atomic excitation. Accordingly, one expects a large Stokes shift between absorption and emission. The older experimental data [68,69] for the emission of Xe in solid and in liquid Ar (Fig. 32) excited by x-rays, and the newer experimental data [70] for the emission of Ar, Kr and Xe in solid Ne excited by synchrotron radiation(Table 9), demonstrate the existence of such a large Stokes shift. Furthermore, the large symmetric lattice dilation around the impurity state results in appreciable reduction of the impurity-medium interaction, so that the spectral shift of the emission relative to the atomic excitation is small (cf. Table 9). To provide a physically transparent demonstration of the effect of medium dilation around the excited

Table 9. Energies of Emission Bands (E) and Comparison with the Corresponding States in Absorption (A) and in the Free Atom (G) [Energies in eV.]

		Emission Bands in Ne Matrix (E)			Atomic Levels (G)			Absorption Bands in Ne Matrix (A)		(A)-(E) Stokes Shift	(G)-(E) Matrix Shift
			8.43[b]	8.34;8.47[c]	6s	$1s_5$	8.31				- 0.16
Xe/Ne	I	8.61[a]		8.61		$1s_4$	8.44	n = 1	9.06	0.45	- 0.17
			9.56		6s'	$1s_3$	9.45				- 0.11
	II	9.80		9.77		$1s_2$	9.57	n = 1'	10.05	0.25	- 0.20
	III	10.78			5d	$3d_2$	10.40	n = 2	11.32	0.54	- 0.38
			10.03	10.02	5s	$1s_5$	9.91				- 0.11
Kr/Ne	I	10.10		10.16		$1s_4$	10.03	n = 1	10.6	0.5	- 0.13
					5s'	$1s_3$	10.51				
	II	10.78				$1s_2$	10.64	n = 1'	11.22	0.44	- 0.14
	III	12.12			4d	$3d_5$	12.04	n = 2	13.35	1.23	- 0.08
Ar/Ne			11.61	11.63	4s	$1s_5$	11.54				- 0.07
	I	11.70		11.71		$1s_4$	11.62	n = 1	12.51	0.81	- 0.09
				11.81	4s'	$1s_3$	11.72				- 0.09
	II	11.92		11.92		$1s_2$	11.83	n = 1'	12.74	0.82	- 0.09
	III	13.93			3d	$3d_5$	13.86	n = 2	14.82	0.89	- 0.07

[a] Ref. [70].
[b] Ref. [69].
[c] E. Schuberth and M. Creuzburg, Phys. Stat. Solidi B90, 189 (1978).

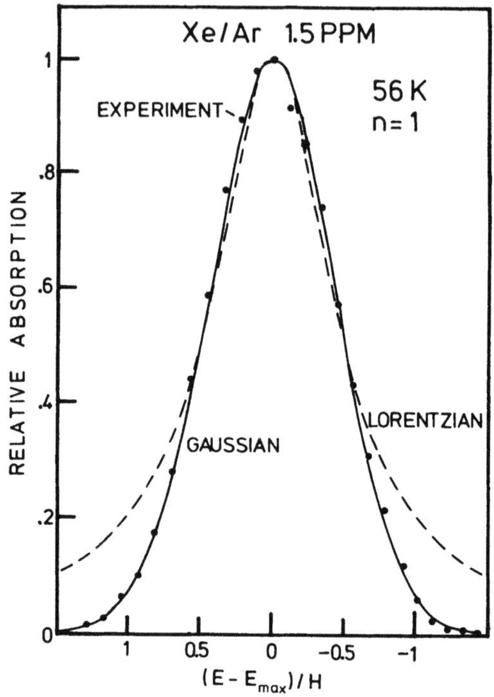

Fig. 30. Experimental absorption lineshape (points) of the n=1 exciton of Xe guest atoms in an Ar crystal at 56 K; and, numerical fits with a Gaussian (solid line) and Lorentzian (dashed line) lineshape. The center of the line E_{max} corresponds to 9.156 eV and the halfwidth (FWHM) H is 119 meV (after Ophir [84]).

impurity state, we present in Fig. 33 the excitation (absorption) spectrum and the emission spectrum of Xe in Ne [70]. The absorption and emission bands for the 6s[3/2], 6s[1/2] and 7s[3/2] atomic states are connected by a Messing plot (see Section 5) of the energies versus the density and are plotted in Fig. 33 to give a linear scale for the nearest-neighbor separation [1,70]. The emission spectra correspond to medium perturbations exerted at an average Xe-Ne separation of ∿4.5 Å, which exceeds by 30% the ground-state Xe-Ne spacing of 3.1 Å. Figure 33 demonstrates nicely the formation of a "microscopic cavity" [7,68] around the electronically excited impurity state.

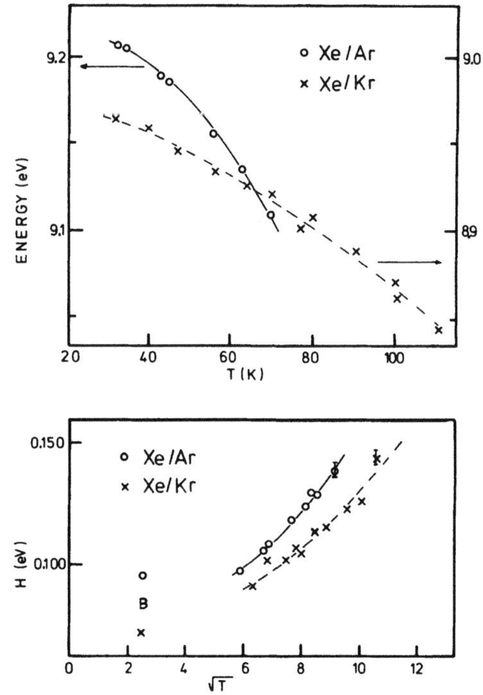

Fig. 31. Temperature dependence of the position of the n=1 excitons of Xe guest atoms in Ar crystals (circles) and Xe guest atoms in Kr crystals (crosses) (upper part), and of the corresponding halfwidth H (lower part). The result of Baldini (1965) is shown by B. The solid (Ar matrix) and dashed (Kr matrix) lines correspond to fits with Eq. (25) (lower part); $E_{max} = E_{max}(0) + \hbar\Delta\omega \coth\left(\frac{\hbar\langle\omega\rangle}{2kT}\right)$ (upper part) (after Ophir [84]).

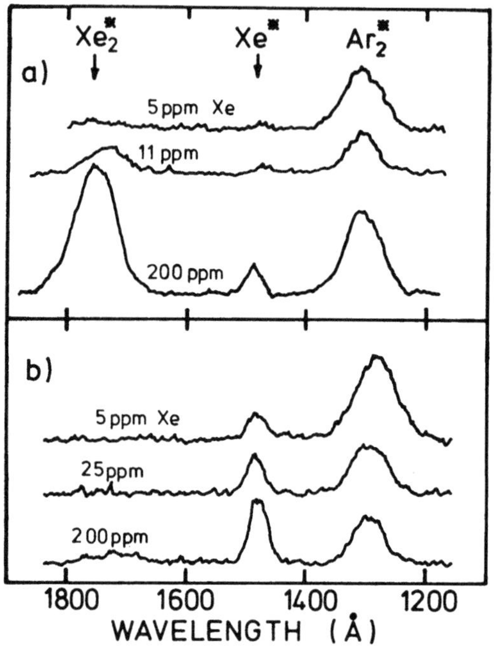

Fig. 32. Emission spectra of (a) xenon-doped liquid argon at 87 K and (b) xenon-doped solid argon at 80 K (after Cheshnovsky, *et al.* [68]).

10. ELECTRONIC RELAXATION OF IMPURITY STATES

Fig. 33 clearly demonstrates that prominent emission of Xe impurity in solid Ne occurs from the three excited states 6s [3/2], 6s[1/2] and 7s[3/2]. This observation is interesting and quite exceptional. In a variety of molecular excited states [85], and in impurity centers in solids [6], effective interstate electronic relaxation between highly-excited states results in efficient nonradiative decay, rendering these high states ineffective in emission, so that emission is observed only from the lowest electronically excited state. The Xe/Ne system, the interstate dynamics of which is summarized in Fig. 34, as well as other atomic impurity states in RGS, provide examples for rather unique, inefficient electronic relaxation between highly-excited states, which is slow on the timescale of the radiative decay [70]. The cardinal question is: Which states emit? This question can be answered by invoking the energy gap law for electronic relaxation [85]. The nonradiative rate W can be expressed in the form

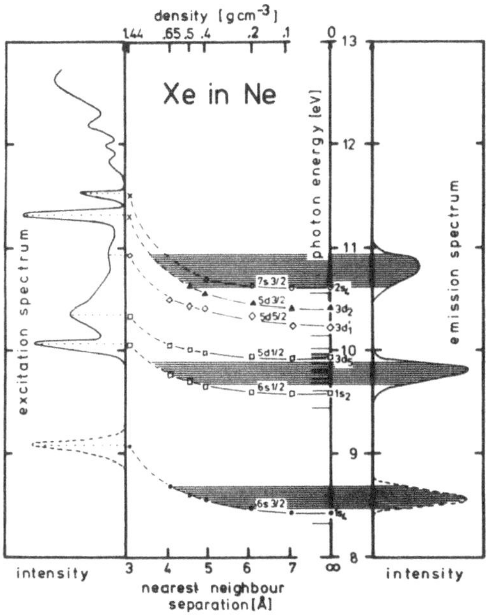

Fig. 33. Dependence of the energy position of the lowest states of Xe atoms on the density of surrounding Ne atoms. The density is expressed by the mean nearest-neighbor separation between a Xe atom and surrounding Ne atoms. The Ne density ranges from zero through the liquid phase to the solid phase. For solid Ne (density 1.44 g cm^{-3}; 3.13 Å intermolecular separation), the maxima of the excitation spectra (left-hand side) have been included. The density around relaxed Xe atoms follows from a comparison with the emission spectra (right-hand side; horizontal lines) (from Hahn and Schwentner [70]).

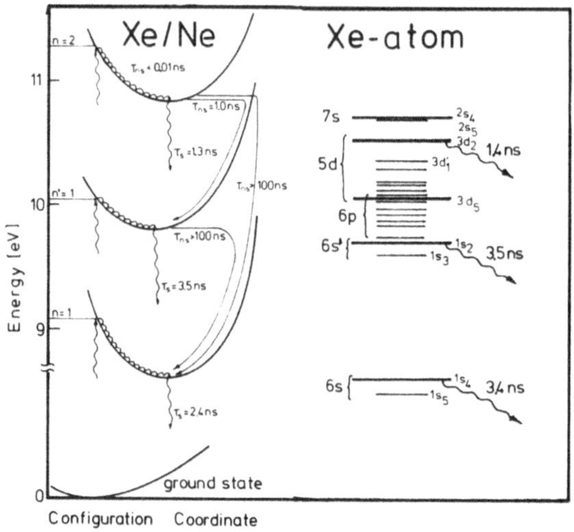

Fig. 34. Left part: experimental radiative (⁓⁓→) and nonradiative (——→) electronic relaxation pathways and time constants (in ns) for Xe in Ne in a configuration coordinate diagram. The excitation channels n=1, n'=1 and n=2, and the matrix relaxation (⁓⁓→), are included. Right part: atomic energy levels for Xe with some lifetimes for radiative transitions to the ground state.

$$W = \frac{A}{(2\pi\Delta E/\hbar\langle\omega\rangle)^{\frac{1}{2}}} \exp(-G) \exp(-\gamma\Delta E/\hbar\langle\omega\rangle) \qquad (26)$$

where A ($\simeq 10^{12} - 10^{14}$ s^{-1}) is a pre-exponential factor; G is the strength of electron-phonon coupling; γ is a numerical parameter on the order of unity; $\langle\omega\rangle$ is a characteristic phonon frequency; and ΔE is the electronic energy gap. The exponential decrease of W with increasing ΔE makes electronic relaxation ineffective for large energy gaps. This expectation is in accord with the experimental data of Fig. 34, where large energy gaps prohibit electronic relaxation between high excited states.

11. ELECTRONIC ENERGY TRANSFER IN RARE-GAS SOLIDS

Studies of electronic energy transfer in pure RGS are of interest to establish the features of exciton migration on an ultrashort timescale prior to exciton trapping. We have shown in Section 8 that the lifetimes of "free" excitons are $\tau \simeq 10^{-12}$ s. These ps energy transfer phenomena are expected to provide interesting information on the nature of exciton dynamics, which may correspond to one of the following categories:

Coherent exciton transport. The exciton lattice scattering is weak, so that the exciton mean free path Λ considerably exceeds the lattice spacing r_c. The exciton dynamics is essentially determined by its group velocity, so that the displacement of the exciton at time t is proportional to t.

Incoherent exciton transport. When exciton scattering by phonons, disorder, etc. is strong, the exciton mean free path is comparable to the lattice spacing, $\Lambda \simeq r_c$. Exciton transport is now characterized by a diffusion coefficient D, while the diffusion length ℓ is given by

$$\ell = (D\tau)^{1/2} \tag{27}$$

where $\tau(\simeq 10^{-12}$ s) is the "free" exciton lifetime.

To explore electronic energy transfer by "free" excitons, one must use sensitive methods to interrogate dynamic processes on the ps timescale. Photoemission and fluorescence measurements provide adequate tools. Three methods were advanced for the study of the diffusion lengths of "free" excitons:

Exciton-induced impurity ionization [57,58,86]. For dilute impurity states, intrinsic photoemission occurs following excitation of the host exciton states, which in turn transfer energy to the impurity, resulting in Penning-type ionization of the impurity (cf. Fig. 35). Only "free" excitons can induce this ionization process, since the energy of R_2* trapped excitons is so low that these trapped excitons are inactive. The interrogation of impurity ionization by "free" excitons is conducted by the measurement of the photoemission yield curves, or of electron energy distribution curves at several photon energies [86]. Fig. 36 shows a typical example for the emission of benzene in Xe, which originates from this mechanism. From the dependence of the photoemission yield on the excitation wavelength and impurity, the exciton diffusion length is obtained.

Extrinsic electronic emission via energy transfer to a metal substrate [71-73]. This is a "classical" energy-transfer experiment

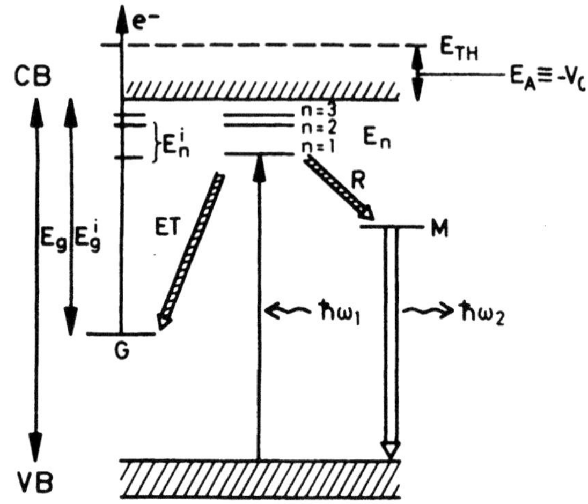

Fig. 35. Schematic energy-level scheme for impurity photoemission from doped solid rare gases.

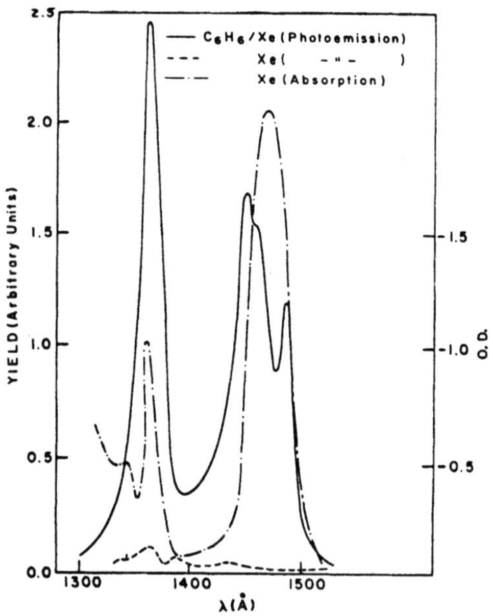

Fig. 36. Photoemission of benzene impurity (1%) in solid Xe (from [58]).

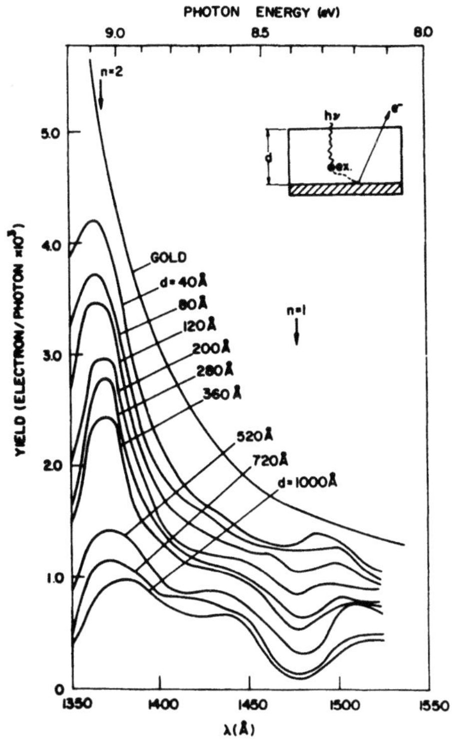

Fig. 37. Experimental photoemission yields $Y = i/I_0$ (where I_0 is the incident light intensity) from solid Xe deposited on the gold emitter electrode for various values of the Xe film thickness. The yield from the gold electrode (which is mounted on the LiF substrate) is also shown for comparison. The positions of the $n=1(^2P_{3/2})$ and $n=2(^2P_{3/2})$ excitons of solid Xe are marked by arrows. The upper inset portrays schematically the extrinsic photoemission mechanism via "free" exciton diffusion to the gold followed by electron ejection from the electrode (from [71]).

involving excitons (Fig. 37). Excitons are excited in an RGS below the direct photoemission threshold. These excitons diffuse to a metal substrate, which is used as an electrode, resulting in photoemission from the electrode (Fig. 37). By monitoring the dependence of the photoelectron yield on the film thickness (Fig. 38) and the wavelength (Fig. 39), the exciton diffusion length can be inferred.

Energy transfer to contaminated boundary surface of a rare-gas film [74,76,77]. This process reduces the quantum yield for luminescence. These luminescence quenching experiments provide information on the exciton diffusion length.

In Table 10, we have summarized experimental data for the diffusion of "free" excitons in RGS. From these results, it is apparent that:

--- The diffusion lengths considerably exceed the lattice constant.

--- The diffusion length in solid Ne is considerably higher than that in the heavier RGS, presumably reflecting a longer lifetime for the free exciton, which is subject to one-center locali-

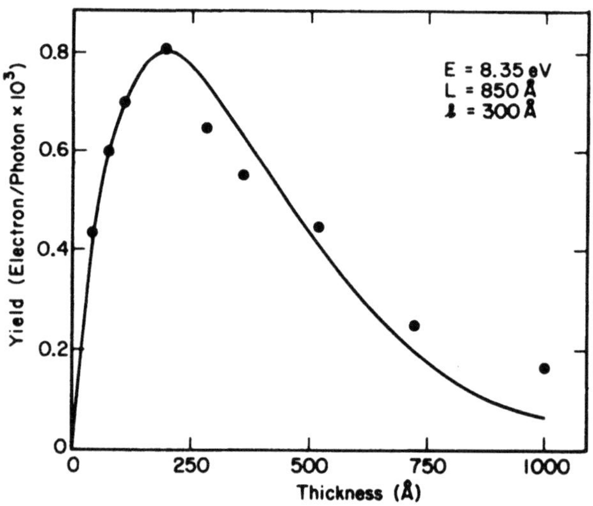

Fig. 38. The thickness dependence of the photoemission yield from Xe films on a gold substrate. L is the electron escape length for solid Xe (from [71]).

Table 10. Diffusion Lengths (Å) for Energy Transfer to Guest Atoms and to Boundaries.

		Photoelectron Emission		Luminescence	
	Excitons	Guest Atom	Metal Electrode	Guest Atom	Crystal Boundary
Ne	n = 1,1',2	2500 ± 500[a]	observed[a]		
Ar	n = 1,1' n = 2,2',3	120[b]	observed[b]		50 100 50[g]
Kr	n = 1 n = 1' n = 2	observed[b]	10 - 100 150 - 250[d] 150 - 350	300[f]	200 - 250[f] 200[g,h]
Xe	n = 1,2	170[c]	300[e]	25,260[f]	150 - 1000[i], 500[f]

[a]Ref. [72].

[b]Ref. [57].

[c]Ref. [58].

[d]N. Schwentner, G. Martens and H. W. Rudolf, Phys. Stat. Solidi 106B, 183 (1981).

[e]Ref. [71].

[f]Ref. [77].

[g]Ref. [76].

[h]Ch. Ackermann, R. Brodmann, U. Hahn, A. Suzuki and G. Zimmerer, Phys. Stat. Solidi 74B, 579 (1976).

[i]Gerick, Ph.D. Thesis (University of Hamburg, 1976).

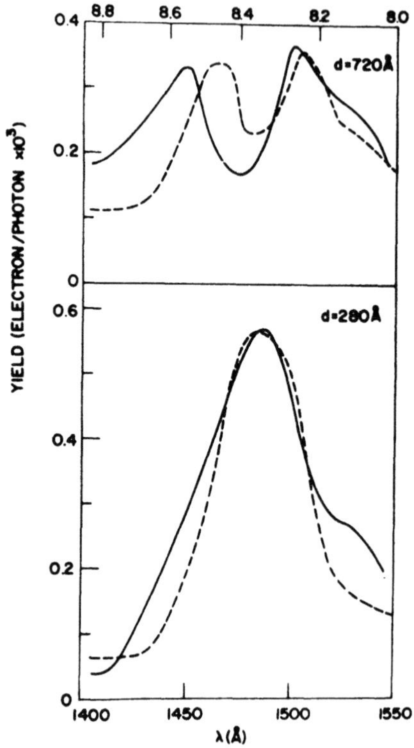

Fig. 39. The energy dependence of extrinsic photoemission from solid Xe on a gold substrate at two values of the film thickness. Solid lines correspond to experimental data, while dashed lines are calculated with ℓ = 300 Å (from [71]).

zation in solid Ne.

--- The diffusion coefficients of free excitons in Ar, Kr and Ne are $D \simeq 1\ cm^2\ s^{-1}$. Such diffusion coefficients correspond to incoherent diffusive motion.

--- In RGS studies, to date, strong scattering of excitons by phonons and by crystal imperfections dominate the exciton dynamics.

12. PERSPECTIVES

The present overview of the energetics and dynamics of electronic excitations in large-gap insulators demonstrates the interplay between molecular and solid-state concepts. Solid-state theoretical concepts were invoked for the description of electronic excitations of pure solids and of impurity states. Nevertheless, the tight-binding approach, which rests on the "molecular" properties of the constituents, provides a useful starting point. Furthermore, the quantum defect method provides a bridge between high solid-state excitation and the highly-excited atomic (or molecular) states of the constitutents. The molecular approach is also extremely useful for the elucidation of some intrastate relaxation phenomena, for example, the nature of self-trapped excitons in pure materials, as well as the energetics of emission from excited impurity states. Last but not least, the molecular local approach is extremely relevant for the understanding of the energetics and dynamics in structurally disordered insulators, namely, fluids and liquids. The merging between the molecular and solid-state points of view is of considerable interest. While our current understanding of the electronic structure of the ground and excited states of insulators, such as RGS, is fairly complete, further information will be required for the elucidation of excited-state dynamics. Most of the experiments we would like to suggest are by no means easy. However, they appear to us to be worthwhile performing, and we suggest the following methods:

--- Determination of the homogeneous linewidths for excitons and impurity states by "hole burning" and laser fluorescence line narrowing in the VUV.

--- Probing the retention of phase memory by the interrogation of coherent optical effects -- that is, free-induction decay or photon echoes in the VUV.

--- Studies of exciton diffusion in "good" well-annealed crystals of insulators to eliminate the effects of structural disorder.

--- Studies of electronic energy transfer of high-energy excitations in an impurity band of binary large-gap insulators. Some novel information regarding the problem of Anderson localization [87,88] and dispersive transport [89] of excitons will emerge from such studies.

REFERENCES

1. N. Schwentner, E. E. Koch and J. Jortner, *Electronic Excitations in Condensed Rare Gases* (Springer Tracts in Modern Physics, in press).
2. A. S. Davydov, *Theory of Molecular Excitons* (Plenum Press, New York, 1971).
3. D. P. Craig and S. H. Whalmsly, *Excitons in Molecular Crystals* (Benjamin, New York, 1968).
4. S. A. Rice and J. Jortner, in: *Physics and Chemistry of the Organic Solid State*, F. Fox, M. Labes and A. Weissberger, eds. (Interscience Publishers, New York, 1967), Vol. III, Chap. 4.
5. R. S. Knox, *Theory of Excitons* (Academic Press, New York, 1963).
6. R. Kubo and H. Kamimura, *Dynamical Processes in Solid State Optics* (Benjamin, New York, 1967).
7. M. Brith and O. Schnepp, Mol. Phys. 9, 473 (1965).
8. R. Scharber and S. E. Webber, J. Chem. Phys. 55, 3977 (1971).
9. E. Boursey, V. Chandrasekharan, P. Gürtler, E. E. Koch, P. Kunsch and V. Saile, Phys. Rev. Lett. 41, 1516 (1978).
10. P. Gürtler and E. E. Koch, Chem. Phys. 49, 305 (1980).
11. P. Gürtler, Ph.D. Thesis (University of Hamburg, 1979).
12. M. H. Cohen and F. Keffer, Phys. Rev. 99, 1128 (1955).
13. W. R. Heller,and A. Marcus, Phys. Rev. 84, 809 (1951).
14. A. S. Davydov and E. N. Myasnikov, Phys. Stat. Solidi 20, 153 (1967).
15. E. I. Rashba, Sov. Phys.-JETP 27, 292 (1968).
16. J. Klafter and J. Jortner, Chem Phys. 47, 25 (1980).
17. K. K. Rebane, *Impurity Spectra of Solids* (Plenum Press, New York, 1970).
18. H. J. Lau, Diplom. Thesis (University of Hamburg, 1982).
19. J. H. Fock, H. J. Lau and E. E. Koch, in press.
20. C. Cossart-Magos and S. Leach, J. Chem. Soc. Farad. Trans. 78, 1477 (1982). See also C. Friedh, L. Åsbrink and E. Lindholm, Chem Phys. Lett. 27, 169 (1978).
21. S. Webber, S. A. Rice and J. Jortner, J. Chem. Phys. 42, 1907 (1965).
22. J. Jortner, in: *Vacuum Ultraviolet Radiation Physics*, E. E. Koch, R. Haensel and C. Kunz, eds. (Pergamon Vieweg, 1974), p. 263.
23. W. Kohn, *Solid State Physics* (Academic Press, New York, 1958), Vol. 5, p. 258.
24. O. Schnepp and K. Dressler, J. Chem Phys. 33, 49 (1960).
25. G. Baldini, Phys. Rev. 128, 1562 (1962).
26. B. Sonntag, in: *Rare Gas Solids*, M. K. Klein and J. A. Venables, eds. (Academic Press, New York, 1977), Vol. II, p. 1021.
27. V. Saile and E. E. Koch, Phys. Rev. B20, 784 (1979).

28. V. Saile and E. E. Koch, Phys. Rev. B21, 4892 (1980).
29. J. Hermanson and J. C. Phillips, Phys. Rev. 150, 652 (1967).
30. J. Hermanson, Phys. Rev. 150, 660 (1967).
31. J. C. Phillips, *Solid State Physics* (Academic Press, New York, 1966), Vol. 18, p. 55.
32. W. Andreoni, M. Altarelli and F. Bassani, Phys. Rev. B11, 2352 (1975).
33. W. Andreoni, W. Perrot and F. Bassani, Phys. Rev. B14, 3589 (1976).
34. L. Resca, R. Resta and S. Rodriquez, Phys. Rev. B18, 696 (1978); B18, 702 (1978).
35. L. Resca and S. Rodriguez, Phys. Rev. B17, 3334 (1978).
36. R. Resta, Phys. Stat. Solidi 86, 627 (1978).
37. G. Baldini and R. S. Knox, Phys. Rev. Lett. 11, 127 (1963).
38. G. Baldini, Phys. Rev. 137A, 508 (1965).
39. A. Gedanken, B. Raz and J. Jortner, J. Chem. Phys. 58, 1178 (1973).
40. D. Pudewill, F. J. Himpsel, V. Saile, N. Schwentner, M. Skibowski and E. E. Koch, Phys. Stat. Solidi 74, 485 (1976).
41. U. Asaf and I. T. Steinberger, Phys. Rev. B10, 4464 (1974).
42. B. Raz and J. Jortner, Chem. Phys. Lett. 4, 511 (1970).
43. J. Jortner and A. Gaathon, Can. J. Chem. 55, 1801 (1977).
44. B. Springett, J. Jortner, and M. H. Cohen, J. Chem. Phys. 48, 2720 (1968).
45. N. F. Mott and E. A. Davis, *Electronic Processes in Non-Crystalline Solids* (Oxford University Press, Oxford, 1971).
46. S. A. Rice and J. Jortner, J. Chem. Phys. 44, 4470 (1966).
47. D. Beaglehole, Phys. Rev. Lett. 15, 551 (1965).
48. I. T. Steinberger and U. Asaf, Phys. Rev. B8, 914 (1973).
49. B. Raz and J. Jortner, Proc. Roy. Soc. (London) A317, 113 (1970).
50. I. Messing, B. Raz and J. Jortner, Chem. Phys. 23, 23 (1977).
51. I. Messing, B. Raz and J. Jortner, J. Chem. Phys. 66, 2239 (1977).
52. I. Messing, B. Raz and J. Jortner, J. Chem. Phys. 66, 4577 (1977).
53. U. Fano, Phys. Rev. 124, 1866 (1961).
54. M. Robin, H. Basch, N. A. Kuebler, B. E. Kaplan and J. Meinwald, J. Chem. Phys. 48, 5037 (1968).
55. M. Robin and N. A. Kuebler, J. Mol. Spect. 33, 274 (1970).
56. B. Katz, M. Brith, B. Scharf and J. Jortner, J. Chem. Phys. 50, 5195 (1969).
57. Z. Ophir, B. Raz, J. Jortner, V. Saile, N. Schwentner, E. E. Koch, M. Skibowski and W. Steinmann, J. Chem. Phys. 62, 650 (1975).

58. Z. Ophir, B. Raz and J. Jortner, Phys. Rev. Lett. 33, 415 (1974).
59. E. E. Koch and J. H. Fock, unpublished results.
60. D. H. Levy, Adv. Chem. Phys. 47, 323 (1981).
61. A. Amirav, U. Even and J. Jortner, J. Phys. Chem. (in press).
62. P. R. R. Langridge-Smith, E. Carrasquillo and D. H. Levy, J. Chem. Phys. 74, 6513 (1981).
63. J. Jortner, L. Meyer, S. A. Rice and E. G. Wilson, J. Chem. Phys. 42, 4250 (1965).
64. R. Heumüler, Ph.D. Thesis (University of Regensburg, 1978).
65. F. Colletti and A. M. Bonnot, Chem. Phys. Lett. 52, 92 (1978).
66. R. E. Packard, F. Reif and C. M Surko, Phys. Rev. Lett. 25, 1435 (1970).
67. A. Gedanken, B. Raz and J. Jortner, J. Chem. Phys. 53, 1630 (1973).
68. O. Cheshnovsky, B. Raz and J. Jortner, J. Chem. Phys. 57, 4628 (1972).
69. A. Gedanken, B. Raz and J. Jortner, J. Chem. Phys. 59, 5741 (1973).
70. U. Hahn and N. Schwentner, Chem. Phys. 48, 53 (1980); J. Luminescence 18/19, 23 (1979).
71. Z. Ophir, N. Schwentner, B. Raz, M. Skibowski and J. Jortner, J. Chem. Phys. 63, 1072 (1975).
72. D. Pudewill, F. Himpsel, V. Saile, M. Skibowski, N. Schwentner, E. E. Koch, and J. Jortner, J. Chem. Phys. 65, 5226 (1976).
73. N. Schwentner, Appl. Opt. 19, 4104 (1980).
74. N. Nagasawa and T. Namba, Optics Comm. 11, 152 (1974).
75. O. Cheshnovsky, B. Raz and J. Jortner, J. Chem. Phys. 59, 5554 (1973).
76. Ch. Ackerman, R. Brodman, G. Tolkiehn, G. Zimmerer, R. Haensel and U. Hahn, J. Luminescence 12/13, 315 (1976).
77. Ch. Ackerman, Thesis, University of Hamburg (1976).
78. R. S. Mulliken, J. Chem. Phys. 52, 5170 (1970); Rad. Res. 59, 357 (1974).
79. Y. Toyozawa, in: *Vacuum Ultraviolet Radiation Physics*, E. E. Koch, R. Haensel and C. Kunz, eds. (Pergamon Vieweg, 1974), p. 317.
80. E. I. Rashba, Izv. USSR Ser. Fiz. 40, 535 (1976).
81. K. S. Song, J. Phys. Soc. Japan 26, 1131 (1969).
82. T. Suemoto and J. Kanzaki, J. Phys. Soc. Japan 46, 1554 (1979).
83. U. Hahn, N. Schwentner, and G. Zimmerer, Optics Com. 21, 237 (1977).
84. Z. Ophir, M.Sc. Thesis, Tel-Aviv University (1970).
85. R. Englman, and J. Jortner, Mol. Phys. 18, 145 (1970).
86. N. Schwentner and E. E. Koch, Phys. Rev. B 14, 4687 (1976).
87. P. W. Anderson, Phys. Rev. 109, 1492 (1958).

88. S. K. Lyo, Phys. Rev. B3, 3331 (1973).
89. K. Godzik and J. Jortner, J. Chem. Phys. 72, 4471 (1980).

REGULAR AND IRREGULAR MOTION IN CLASSICAL AND QUANTUM SYSTEMS

Marko Robnik

Astronomische Institute
Universität Bonn
Auf dem Hügel 71
D-5300 Bonn, West Germany

1. INTRODUCTION: ORDER AND CHAOS

It is now widely known that simple, nonlinear, deterministic dynamical systems can exhibit chaotic motion, which is unpredictable even in principle [1-4]. By *simple*, we mean that the system has only a few degrees of freedom N, say N = 2. By *deterministic*, we mean that the laws of motion are perfectly known locally; they are given by some velocity vector field in phase space. Time evolution of the system is governed by some ordinary differential equation, so that the future and the past are completely determined by the initial state of the system. The motion in time is obtained by finding the integral curves of the vector field (i.e., by solving the differential equation). This can always be done (at least numerically) for small times; and yet, the motion can still be unpredictable for large times. The reason for such a chaotic motion lies in the instability of trajectories, which can separate exponentially as time goes on. The motion thus displays a sensitive dependence on initial conditions. In such a case, the number of accurately calculated *digits* will decrease in proportion to the time elapsed from the initial condition, so that after a finite time our prediction will be completely wrong, no matter how good the measurement of the initial state was. An exponential instability can result in a type of motion which is as random as the outcome of tossing a coin. In fact, there is a whole hierarchy of chaotic systems, ranging from Bernoulli systems (most chaotic), through mixing and ergodic systems, down to almost-integrable (KAM) systems. An ergodic (or more chaotic) system will pass arbitrarily close to any point on the energy surface in phase space.

S. P. McGlynn et al. (eds.), Photophysics and Photochemistry in the Vacuum Ultraviolet, 579–629.

Regular motion can arise if the trajectory is confined to some smooth submanifold of phase space, and if the trajectory is periodic or quasi-periodic there. This can be the case when the system has some global, smooth integrals of the motion. The invariant submanifolds are then level sets of such constants of the motion.

The best example of ordered motion is that of the planets around the Sun. Indeed, this motion is so regular that we can use it as a natural clock. Chaotic motion, on the other hand, is responsible for the approach to equilibrium, and for the ergodic properties of many-body systems treated by statistical mechanics and thermodynamics. In fact, the chaotic, ergodic nature of motion in these latter systems was postulated by the fathers of statistical mechanics [5]. For a long time, it was believed that chaotic motion is a property of systems with many degrees of freedom, and that systems with only a few degrees of freedom behave regularly and predictably. However, in the study of the three-body problem connected with the stability of the solar system, a breakthrough was provided by the works of Poincaré, Birkhoff, Kolmogorov, Arnold, Moser, Smale, Feigenbaum, and many others, and it became more and more evident that even simple systems, possessing only a few degrees of freedom, can be chaotic.

The modern theory of nonlinear dynamical systems provides a bridge between the analytical mechanics of few-body systems and the statistical mechanics of many-body systems. This theory has wide applications in almost every branch of science, since most natural processes are modeled by differential equations (e.g., chemical reactions, evolution of populations in biology, nonlinear oscillations in electrical and mechanical engineering, etc.). In physics, applications are found not only in celestial mechanics, but also in the theory of turbulence [6,7], plasma physics [8,9], molecular physics [10,11], atomic physics [12-16], nuclear physics [17,18], and in particle physics (e.g., beam-beam interactions in intersecting storage rings [19], where millions of dollars are spent to control chaos, by keeping the particle beams collimated.)

We know that our universe is neither completely chaotic nor everywhere ordered. In fact, it seems that both chaotic and ordered structures are present on all scales. Examples of ordered structures extend from the scale of galaxy clusters to that of the interstellar medium (where strongly nonequilibrium processes such as masers can be observed), the planetary systems and life on Earth [20]. Indeed, this last example is clear proof that nonlinear interactions, in addition to inducing chaos, can also produce ordered and self-organized structures (synergetics) [21] -- that is, systems far from thermal equilibrium. These ordered structures can persist for a very long time, provided that driving forces and energy sources are present. On the other hand, we know of systems which are isolated and which approach thermodynamic equilibrium as a result of dynamical instabilities. Therefore, chaos and order are equally important in nature. It may be that the universe is, was or will be mostly chaotic, somehow; however, as intelligent beings we naturally are more interested in ordered structures since these are used to make life stable. Nonetheless, to better understand the nature of order, one must understand the origin of chaos.

In what follows, we consider only Hamiltonian systems. Both regular and chaotic behavior in classical systems are reviewed first, and then the consequences of stochasticity in quantum mechanics are explored.

2. INTEGRABLE SYSTEMS

The Kepler problem, the harmonic oscillator and a few other systems (such as various kinds of Toda lattices [22]) are examples of integrable Hamiltonian systems. The motion of these systems is regular and simple, thus enabling us to integrate the equations of motion explicitly. Until quite recently, almost all textbooks on classical mechanics were filled only with such nicely calculable examples. However, we must emphasize the fact that integrable systems are the exception rather than the rule: Indeed, the measure of the set of integrable systems in the space of all possible Hamiltonian systems is zero. (In other words, if we randomly choose a system realized in nature, the probability is one that the system chosen is nonintegrable.) Nevertheless, integrable systems are important for any understanding of the generic properties of Hamiltonian systems since many of the more general properties appear in integrable systems when a perturbation is applied. This perturbation approach leads, both historically and heuristically, to the present day theory of dynamical systems. (We return to the perturbation problem in the next section.)

Consider a system given by the Hamiltonian $H = H(q,p)$, where $\tilde{q} = (q_1,\ldots,q_N)$, $\tilde{p} = (p_1,\ldots,p_N)$ are, respectively, canonically conjugate coordinates and momenta. (The dimension of the configuration space will always be denoted by N.) The motion in the 2N-dimensional phase space M is governed by Hamilton's equations:

$$\frac{d\tilde{q}}{dt} = \frac{\partial H}{\partial \tilde{p}}, \quad \frac{d\tilde{p}}{dt} = -\frac{\partial H}{\partial \tilde{q}} . \tag{1}$$

A Hamiltonian system is *integrable* if there exist N *global*, *analytic*, *single-valued* and functionally *independent* integrals of the motion, $F_i = F_i(\tilde{q},\tilde{p})$, $1 \leq i \leq N$ (one of them, say F_1, is the

Hamiltonian itself), which are in *involution*, (i.e., the Poisson bracket

$$\{F_i, F_j\} = \frac{\partial F_i}{\partial \underset{\sim}{q}} \frac{\partial F_j}{\partial \underset{\sim}{p}} - \frac{\partial F_i}{\partial \underset{\sim}{p}} \frac{\partial F_j}{\partial \underset{\sim}{q}}$$

vanishes pairwise.)

Note that each level set defined by the equation $F_i(\underset{\sim}{q}, \underset{\sim}{p}) = f_i$ = constant, $1 \leq i \leq N$, is a smooth (2N-1)-dimensional manifold. Since the F_i's are independent, their intersection is an N-dimensional smooth manifold. And, because the integrals of the motion are global, (i.e., defined everywhere on M), the phase space is decomposed into a family of such invariant N-dimensional surfaces, to which the motion is confined. Two important questions follow: (i) What is the topology of such N-dimensional invariant surfaces? (ii) How do the orbits lie on these surfaces?

The answer to the first question is provided by a theorem from topology, and has been made explicit by Arnold [23]: If the N-dimensional invariant surface mentioned above is compact and connected, then it has the topology of an N-dimensional torus; we call it the *invariant torus*. The topological reason for this is that a compact, connected manifold must be an N-dimensional torus, if there are N independent and commutative vector fields defined on it. But such a set of vector fields can be immediately constructed from the integrals of the motion F_i: Namely, take

$$\underset{\sim}{v}_k = \left(\frac{\partial F_k}{\partial \underset{\sim}{p}} , - \frac{\partial F_k}{\partial \underset{\sim}{q}} \right) , \quad k = 1, \ldots, N .$$

Each $\underset{\sim}{v}_k$ is tangent to each surface $F_m(\underset{\sim}{q}, \underset{\sim}{p}) = f_m$ = constant,

$$\underset{\sim}{v}_k \cdot \left(\frac{\partial F_m}{\partial \underset{\sim}{q}} , \frac{\partial F_m}{\partial \underset{\sim}{p}} \right) = \{F_m, F_k\} = 0 \quad ,$$

and they satisfy the required conditions. The proof can be found in [23].

The motion of an integrable system, for any initial conditions, is thus confined to an N-dimensional torus. It is then convenient to use the topologically natural coordinate system, namely, action-angle variables. The actions $\underset{\sim}{I}$ label a given torus, while the angles $\underset{\sim}{\Theta}$ specify the position of a phase point on the torus. In addition, we require that $(\underset{\sim}{I}, \underset{\sim}{\Theta})$ be canonically conjugate momenta and coordinates, respectively. One can construct these variables from the integral

$$S(\underset{\sim}{q},\underset{\sim}{f}) = \int_{\underset{\sim}{q}_0}^{\underset{\sim}{q}} \underset{\sim}{p}(\underset{\sim}{q}',\underset{\sim}{f})d\underset{\sim}{q}' \quad , \tag{2}$$

where $\underset{\sim}{f} = (f_1,\ldots,f_N)$ specifies the N-dimensional torus. First, note that this integral is independent of the path chosen on $\underset{\sim}{f}$, which is a consequence of the Lagrangian property

$$\frac{\partial p_i}{\partial q_j} - \frac{\partial p_j}{\partial q_i} = 0 \ , \quad i,j = 1,\ldots,N \ .$$

Thus, the integral of Eq. (2), taken along a closed path which can be continuously deformed into a point, vanishes. However, on a torus $\underset{\sim}{f}$ we have paths γ_i, $1 \leq i \leq N$, which cannot be shrunk to a point because of the topology of the torus (cf. Fig. 1). We call these paths *irreducible cycles*. Since the integral

$$\Delta S = 2\pi I_i = \oint_{\gamma_i} \underset{\sim}{p}(\underset{\sim}{q}',\underset{\sim}{f})d\underset{\sim}{q}' \tag{3}$$

is the same for all paths continuously deformable into γ_i (i.e., for all paths homotopic to γ_i), it is a natural invariant of the torus. There are N independent irreducible cycles on an N-dimensional torus, so we can use the actions I_i, $1 \leq i \leq N$, defined by Eq. (3) as a label of the torus. In so doing, we can express $\underset{\sim}{f}$ in Eq. (3) in terms of $\underset{\sim}{I}$, namely $\underset{\sim}{f} = \underset{\sim}{f}(\underset{\sim}{I})$, and after inserting this into Eq. (2) we obtain the generating function

$$S(\underset{\sim}{q},\underset{\sim}{I}) = \int_{\underset{\sim}{q}_0}^{\underset{\sim}{q}} \underset{\sim}{p}(\underset{\sim}{q}',\underset{\sim}{I})d\underset{\sim}{q}' \tag{4}$$

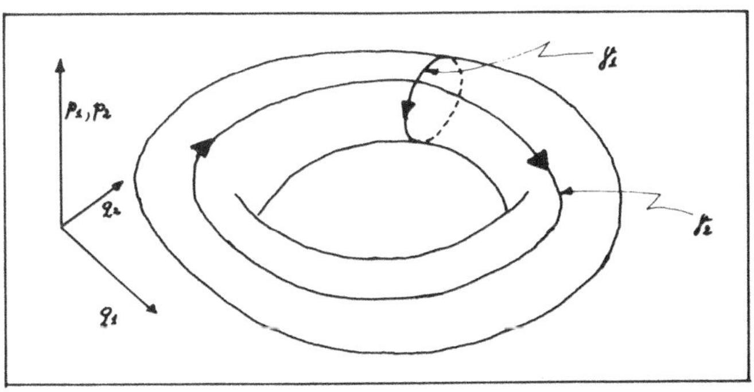

Fig. 1. Two independent irreducible cycles γ_1 and γ_2 on a two-dimensional torus.

as a function of the old coordinates $\tilde{q}$ and the new momenta $\tilde{I}$ (the actions). The standard theory of canonical transformations [23, 24] then yields the angles

$$\tilde{\Theta} = \frac{\partial S}{\partial \tilde{I}} \tag{5}$$

which are the coordinates canonically conjugate to the momenta $\tilde{I}$.

In these new variables, the Hamiltonian is solely a function of the actions,

$$H = H(\tilde{I}) \ , \tag{6}$$

since the angles are cyclic variables by construction. We call these variables *angles* because Θ_i changes by 2π along an irreducible cycle γ_i, while the Θ_j, $j \neq i$, do not change:

$$(\Delta\Theta_i)_{\gamma_j} = \Delta_{\gamma_j} \frac{\partial S}{\partial I_i} = \frac{\partial}{\partial I_i}(\Delta_{\gamma_j} S) = 2\pi\delta_{ij} \ , \tag{7}$$

where the last equality follows from Eq. (3).

Hamilton's equations are generated by Eq. (6):

$$\dot{\tilde{I}} = -\frac{\partial H}{\partial \tilde{\Theta}} = 0 \ , \quad \dot{\tilde{\Theta}} = \frac{\partial H}{\partial \tilde{I}} =: \tilde{\omega}(\tilde{I}) \ . \tag{8}$$

These equations can be immediately integrated to give

$$\tilde{I} = \text{constant} \ , \quad \tilde{\Theta} = \tilde{\omega}(\tilde{I})t + \Theta_0 \ , \tag{9}$$

which shows that the actions are constants of the motion, while the angles rotate uniformly with angular velocity $\omega(\tilde{I})$ depending upon the actions [i.e., $\tilde{\omega}$ can vary from torus to torus (non-isochronicity)].

When the old momenta $\tilde{p}$ and coordinates $\tilde{q}$ are expressed in terms of $(\tilde{I},\tilde{\Theta})$, we find that they are periodic functions of the angles, with Fourier coefficients depending upon the actions. That is,

$$\tilde{q} = \sum_{\tilde{m}} \tilde{q}_{\tilde{m}}(\tilde{I}) e^{i\tilde{m}\cdot\tilde{\Theta}} \ , \quad \tilde{p} = \sum_{\tilde{m}} \tilde{p}_{\tilde{m}}(\tilde{I}) e^{i\tilde{m}\cdot\tilde{\Theta}} \tag{10}$$

where $\tilde{m} = (m_1,\ldots,m_N)$ is a vector of integers, called the lattice vector. Thus, the motion on an invariant torus will, in general, be quasi-periodic. It can be periodic on *resonant tori*, for which the frequencies $\underset{\sim}{\omega}(\underset{\sim}{I})$ are incidentally rationally connected (i.e., $\underset{\sim}{k}\cdot\underset{\sim}{\omega} = 0$ for some vector $\underset{\sim}{k} = (k_1,\ldots,k_N) \neq 0$ of integers). We thus have the answer to the second question posed above: The orbits wind quasi-periodically on the torus (cf. Fig. 2), and can be periodic (closed) on a resonant torus.

EXAMPLES:

<u>1-dimensional harmonic oscillator (Fig. 3):</u> $H = \frac{\omega}{2}(q^2 + p^2)$ and $I = (1/2\pi)\oint dq\ [(2H/\omega)-q^2]^{\frac{1}{2}} = H/\omega$. Thus, $H = I\omega$ and $\partial H/\partial I = \omega =$ constant. The oscillation frequency is independent of both the

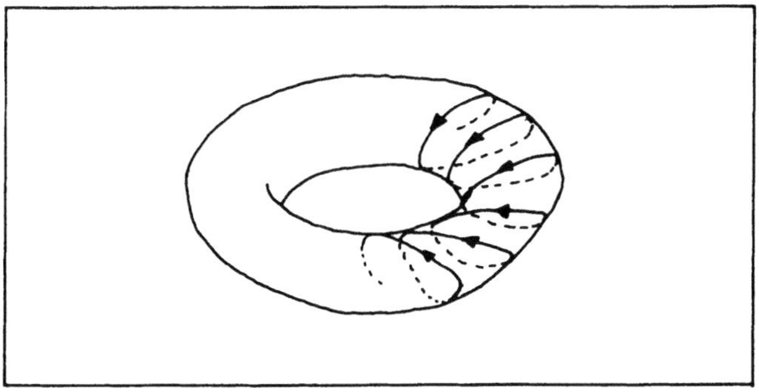

Fig. 2. The orbits wind quasi-periodically on a non-resonant torus, and are closed on a resonant torus.

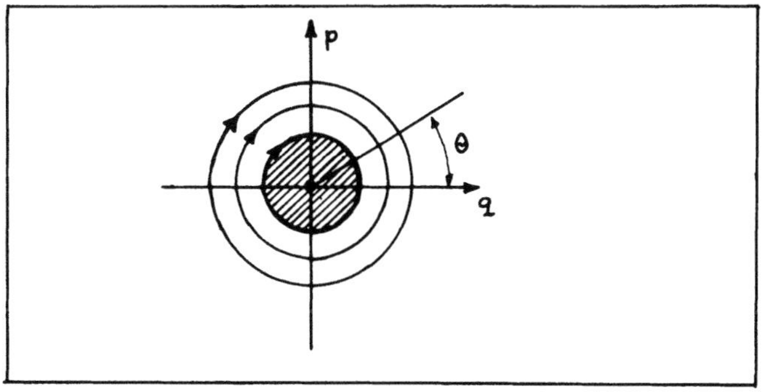

Fig. 3. The invariant tori of the one-dimensional harmonic oscillator are circles of constant energy. The action of a torus equals the area/2π of a circular disk enclosed by the torus.

action and energy (isochronicity). The invariant torus is the circle of constant energy, with the area inside this circle being equal to the action I. The angle Θ rotates uniformly with period $2\pi/\omega$.

2-dimensional box (Fig. 4): $H = (p_x^2 + p_y^2)/2m$, with a,b, being the sides of the box. Also,

$$I_x = \frac{1}{2\pi}\oint p_x dx = \frac{p_x a}{\pi}\ ,\quad I_y = \frac{1}{2\pi}\oint p_y dy = \frac{p_y b}{\pi}$$

$$H = \frac{\pi^2}{2m}\left(\frac{I_x^2}{a^2} + \frac{I_y^2}{b^2}\right)$$

$$\omega_x = \frac{\pi^2}{ma^2} I_x\ ,\quad \omega_y = \frac{\pi^2}{mb^2} I_y \qquad \text{(non-isochronicity)}$$

The motion on a torus can be represented as shown in Fig. 4. When $\omega_x/\omega_y = -k/\ell$, where k and ℓ are integers, the orbits are periodic; after ℓ cycles of Θ_y have been completed, k cycles of Θ_x will have been completed too, and the motion returns to the original position.

3-dimensional Kepler problem: In spherical coordinates (r,θ,ϕ) the Hamiltonian is

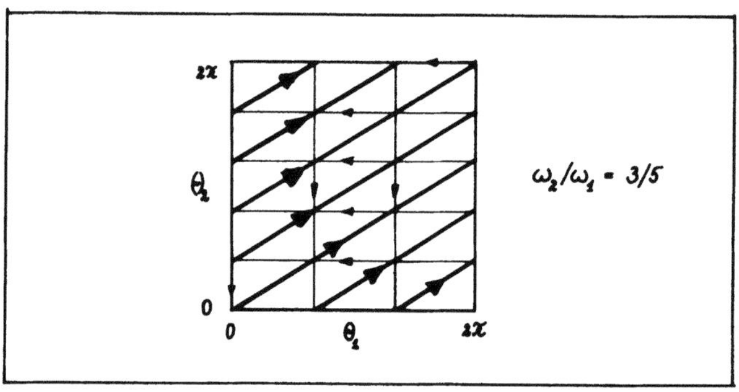

Fig. 4. A two-dimensional torus may be represented by a $2\pi \times 2\pi$ square, with opposite sides being identified. The motion described by the angles θ_1, θ_2 is then a simple uniform motion, modulo 2π. A closed orbit corresponding to $\omega_1/\omega_2 = 3/5$ is shown. In the nonresonant case, the orbit will densely fill the entire torus.

$$H = \frac{1}{2}\left(p_r^2 + \frac{p_\theta^2}{r^2} + \frac{p_\phi^2}{r^2 \sin^2\theta}\right) - \frac{\alpha}{r} \quad ; \quad \alpha = \text{constant}$$

We know the following:

$$p_\phi = L_z = \text{constant}$$

$$p_\theta^2 + \frac{p_\phi^2}{\sin^2\theta} = L^2 = (\text{angular momentum})^2 = \text{constant}$$

$$\frac{1}{2}\left(p_r^2 + \frac{L^2}{r^2}\right) - \frac{\alpha}{r} = E = \text{energy} = \text{constant}$$

Hence,

$$I_\phi = \frac{1}{2\pi}\int_0^{2\pi} p_\phi d\phi = p_\phi$$

$$I_\theta = \frac{1}{2\pi}\oint p_\theta d\theta = L - I_\phi$$

$$I_r = \frac{1}{2\pi}\oint p_r dr = \frac{\alpha}{(-2E)^{1/2}} - L$$

and, therefore,

$$H = -\frac{\alpha^2}{2(I_r + I_\theta + I_\phi)^2}$$

The frequencies $\omega_r = \omega_\theta = \omega_\phi \equiv \omega_{Kepl}$ are given by

$$\omega_{Kepl} = \frac{\partial H}{\partial I_r} = + \frac{\alpha^2}{(I_r + I_\theta + I_\phi)^3} = \frac{(-2E)^{3/2}}{\alpha} ,$$

and depend only on the energy E. Thus, each torus is resonant and all orbits are periodic, as is well known! This periodicity is connected with the super-symmetry of the Kepler (Coulomb) problem, where in addition to conservation of the angular momentum vector, the Runge-Lenz vector is also conserved. In fact, as a result of this degeneracy, the invariant tori are not uniquely determined: There are several different families of invariant tori into which the Kepler orbits can be embedded [23].

3. PERTURBATION OF INTEGRABLE SYSTEMS AND THE KAM-THEOREM

As mentioned above, integrable systems are isolated points

in the space of all Hamiltonians. This means that in a sufficiently small neighborhood of an integrable Hamiltonian $H_o(\underset{\sim}{I})$, there is no other integrable one. How can we know this?

Consider a perturbation of H_o, such that the perturbed system is given by

$$H = H_o(\underset{\sim}{I}) + \varepsilon H_1(\underset{\sim}{I},\underset{\sim}{\Theta}) \ , \tag{11}$$

where ε is a (small) perturbation parameter. $\underset{\sim}{I}$ and $\underset{\sim}{\Theta}$ are still good canonical variables but, as a result of the perturbation, they are no longer action-angle variables. There is no *a priori* reason to believe that the system is still integrable, with invariant tori filling the whole of phase space. Indeed, the question of what happens to the invariant tori of an integrable system under a perturbation has been a long standing and most important problem of classical mechanics.

Some people, such as Landau, thought that all systems are integrable but that we lack the tools to find the integrals of the motion. Others, such as Fermi, believed that integrable systems are exceptional (which is true) and become ergodic [25] when slightly perturbed (which is not true).

The integrability of H would be proved if action-angle variables could be found through some canonical transformation. This was attempted, in the framework of classical perturbation theory, by a series expansion. However, considerable difficulties arise because of the presence of small divisors [26] and, in its original form, the convergence of the perturbation series has never been shown. But later, Kolmogorov, Arnold and Moser (KAM-Theory) invented a method of accelerated, quadratic convergence, which is based on the Newton method in function space. They were able to prove the convergence of this perturbation scheme, with a highly nontrivial proviso: The series converges sufficiently far from resonant (i.e., rational) tori, but convergence is not guaranteed near rational tori. Thus, the perturbed system is *locally* integrable on some (complicated) subset of phase space and, therefore, is neither integrable nor ergodic. We state now the important

KAM-Theorem: All tori of the integrable system $H_o(\underset{\sim}{I})$ which are outside the resonant gaps survive a perturbation $H_o \to H = H_o + \varepsilon H_1$. The resonant gaps are centered around the rational tori, with the width of each gap being determined by the inequality [26-28]

$$\left| \sum_{j=1}^{N} k_j \omega_j \right| \geq K(\varepsilon) \left(\sum_{j=1}^{N} |k_j| \right)^{-a} \ , \tag{12}$$

where $K(\varepsilon)$ is some constant which vanishes as $\varepsilon \to 0$, k_j are integers, $\omega_j = \partial H_o/\partial I_j$ are the frequencies, and a is a constant greater than (N-1).

The volume of each resonant gap is of the order of ε, which is also true for the sum of all gaps (since this series is convergent). This observation is the basis for the condensed versions of the KAM-Theorem, such as: "most tori survive" or "all sufficiently irrational tori survive", albeit they can be distorted. (Later we shall see an example showing what is meant by "sufficiently irrational".)

One may ask whether convergence might be proved inside the resonant gaps as well, thereby restoring the complete (global) integrability. For a *general* perturbation the answer is no: First, because the KAM method approaches the resonances as closely as possible (in the sense of continued fractions; see the beautiful review by Berry [28]); second, because examples of the destruction of tori within the resonant gaps are known. However, there also exist examples of the persistence of resonant tori under some special perturbation (the first such example was given by Henon and Heiles [29].) Since the KAM-Theorem deals with an arbitrary perturbation ($H_1(I,\Theta)$ may be any smooth function), it cannot and does not say anything about the motion within the resonant gaps. It is very difficult to predict what happens in the resonant gaps since the motion there depends upon the details of the perturbation. However, there are some general results on the structure of the gaps which will be considered later.

EXAMPLE (N = 2):

Consider a two-dimensional integrable system. The invariant tori are two-dimensional and they can bound regions of the three-dimensional energy surface. (In order to bound regions of a (2N-1)-dimensional energy surface, tori must have 2N-2 dimensions, which is possible, for invariant tori, only for N = 2. This has important consequences, as explained below.) Let $H_o = H_o(I_1,I_2)$, where I_1 and I_2 are the actions. For the frequencies

$$\omega_i = \omega_i(I_1,I_2) = \frac{\partial H_o}{\partial I_i}, \quad i = 1,2 \quad , \tag{13}$$

we assume non-isochronicity: That is,

$$\det \frac{\partial^2 H_o}{\partial I_i \partial I_j} = \frac{\partial\omega_1}{\partial I_1}\frac{\partial\omega_2}{\partial I_2} - \frac{\partial\omega_1}{\partial I_2}\frac{\partial\omega_2}{\partial I_1} \neq 0 \quad , \tag{14}$$

so that the frequencies vary continuously and are functionally independent (i.e., the frequencies are nondegenerate). Consequently, the ratio

$$\nu := \frac{\omega_1}{\omega_2} = \nu(I_1, I_2) \tag{15}$$

is a nonconstant differentiable function of the actions. For the resonant tori, ν is rational ($\nu = r/s$, r and s being integers without a common factor). Without loss of generality, we assume $0 \leq \nu \leq 1$.

For a general, sufficiently smooth perturbation H_1, with $H_0 \rightarrow H = H_0 + \varepsilon H_1$, the KAM-Theorem predicts that the existence of invariant tori whenever [28]

$$|\nu - \frac{r}{s}| \geq K(\varepsilon) s^{-5/2} \tag{16}$$

We can visualize the resonant gaps as shown in Fig. 5. The resonant tori are given by ν = 1, 1/2, (1/3, 2/3), (1/4, 3/4), (1/5, 2/5, 3/5, 4/5), ..., and the half-widths $\Delta\nu$ of the corresponding resonant gaps are equal to

$$\Delta\nu = K(\varepsilon),\ 2^{-5/2}K(\varepsilon),\ 3^{-5/2}K(\varepsilon),\ 4^{-5/2}K(\varepsilon),\ 5^{-5/2}K(\varepsilon),\ \ldots \quad .$$

In general, the number of resonant gaps of order s is less than s; and, since each of these gaps possesses the half-width $s^{-5/2}K(\varepsilon)$, the total length L of all gaps satisfies the inequality

$$L < \sum_{s=1}^{\infty} 2\ s\ s^{-5/2}K(\varepsilon) = 2\ K(\varepsilon) \sum_{s=1}^{\infty} s^{-3/2} \simeq 5.224\ K(\varepsilon)$$

Thus, L is finite and of the order of the perturbation parameter ε (since $K \rightarrow 0$ as $\varepsilon \rightarrow 0$). Hence, in spite of the fact that resonant tori are dense in phase space, the measure of resonant gaps

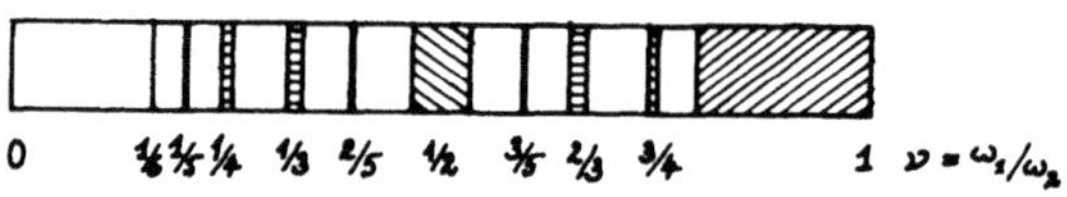

Fig. 5. The resonant gaps of a two-dimensional system; ν is the frequency ratio. Although the rational numbers are dense in the interval $0 \leq \nu \leq 1$, the widths of the gaps around them decrease so rapidly that their total length is finite and of the order of the perturbation parameter ε. For a perturbed system, the invariant tori exist in phase space with the exception of a complicated subset, the measure of which is of the order of ε.

in which tori can be destroyed is of order ε. Now, since the system is assumed nondegenerate [cf. Eqs. (14,15)], the smallness of $\Delta\nu$ implies the smallness of ΔI_1, ΔI_2. Thus, "most tori survive" (cf. Fig. 6).

In this two-dimensional example, the resonant gaps are bounded between KAM invariant tori, so that even if the motion inside the gap is chaotic, it is confined between the exact tori. The chaotic regions are thus strictly localized as long as invariant tori exist. In higher dimensions ($N > 2$), the invariant tori cannot be a boundary of a region of the energy surface and, therefore, resonant gaps can form a connected lattice. Although the measure of these gaps is of the order of ε, a phase point can slowly wander through such a lattice (called an *Arnold web*) and a significant departure from the initial torus can eventually occur. For $N > 2$, then, a universal instability arises from an arbitrarily small perturbation; this is termed *Arnold diffusion* [2].

4. AREA-PRESERVING MAPPINGS

One important concept for the analysis of Hamiltonian systems -- for example, in the resonant gaps of a KAM-system -- is the Poincaré mapping of a *Surface-of-Section* (SOS). An SOS is some surface Σ in phase space. The Poincaré mapping maps Σ onto itself, and is generated by the phase flow (cf. Fig. 7). For the case in which the energy is the only integral of the motion that we know, an SOS should be (2N-2)-dimensional. And, if each orbit of a given energy passes through the SOS, the Poincaré mapping will fully describe the stability of the motion.

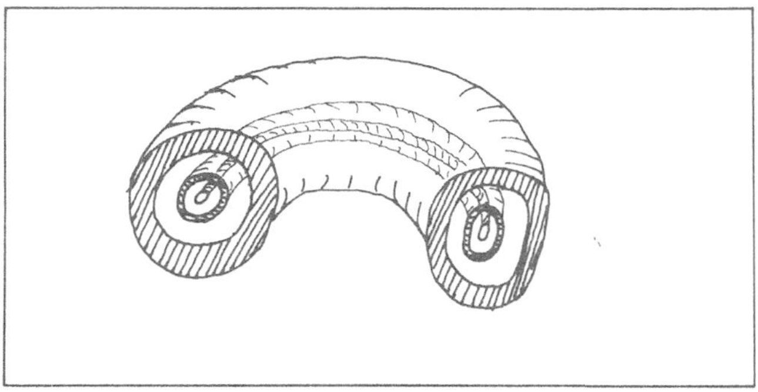

Fig. 6. If the Hamiltonian system is nondegenerate (i.e., $\det\left|\frac{\partial \underset{\sim}{\omega}}{\partial \underset{\sim}{I}}\right| = \det \frac{\partial^2 H}{\partial I_i \partial I_j} \neq 0$) the smallness of the resonant gaps in the frequency scale (cf. Fig. 5) implies that the gaps in phase space have a measure of the order of the perturbation parameter.

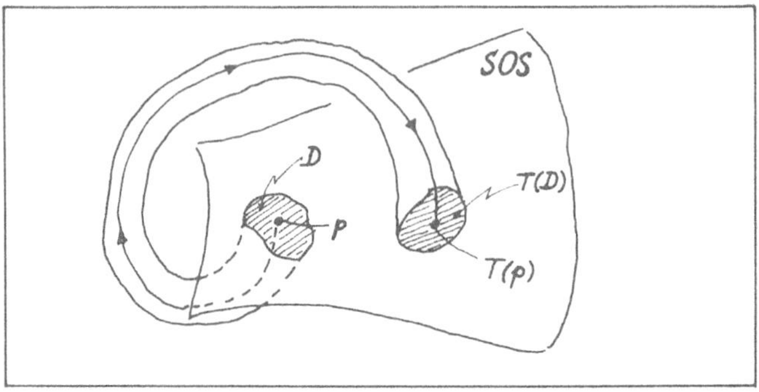

Fig. 7. The Poincaré mapping T of a Surface-of-Section (SOS) is generated by the phase flow: A point p on the SOS is left to move according to Hamilton's equations until it returns to the SOS at another point, defined as T(p). T is an area-preserving mapping (i.e., area (D) = area (T(D)) for a region D in the SOS).

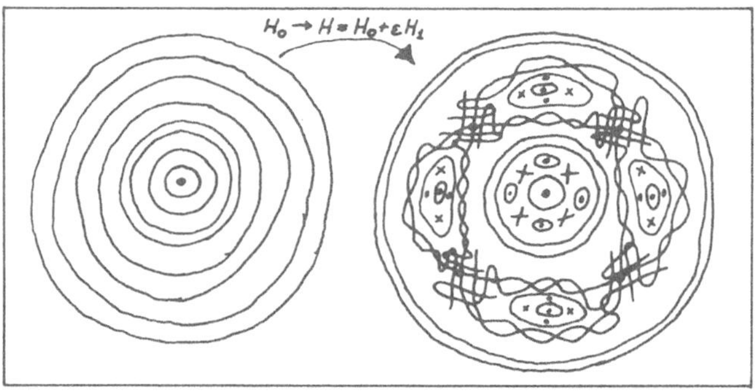

Fig. 8. The destruction of resonant tori when a perturbation H_1 is applied. An even number of periodic orbits survives a perturbation according to the Poincaré-Birkhoff Theorem. Half of these orbits are hyperbolic and are surrounded by chaotic regions. The stable closed orbits are surrounded by a family of stable orbits, and appear in the SOS as elliptic islands. Each of these islands has a similar structure, so that we find an infinite hierarchy of smaller and smaller almost self-similar elliptic islands.

An SOS is most useful when it is two dimensional since we can then employ graphical representations. In this case, a periodic orbit will appear in the SOS as a fixed point or, more generally, as a chain of n periodic points termed an n-*cycle*. Each point of an n-cycle is a fixed point of the nth iterate of the Poincaré mapping. Further, if there are any irrational tori at a given energy, they will occur in the SOS as invariant curves (the points of these curves are not fixed points for any finite iterate, however). Finally, if there exists a resonant torus on which all orbits are periodic, it appears in the SOS as an invariant curve consisting of periodic points. The Poincaré mappings have the important property that they are area-preserving, which is a consequence of the existence of integral invariants. Many features of the motion within the resonant gaps of a KAM-system can be understood by means of general area-preserving mappings. A typical picture of the manner in which the SOS of an integrable system changes under a small perturbation is shown in Fig. 8.

In order to understand the chaotic motion within the resonant gaps, we now consider a general area-preserving mapping T. T is, in general, nonlinear and, therefore, difficult to study. However, we are mainly interested in the stability of fixed points which correspond to periodic orbits in phase space. A *fixed point* x_0 has the property that $T(x_0) = x_0$. x_0 is a *stable* fixed point if for each neighborhood U of x_0 there exists a subneighborhood $V \subset U$, containing x_0, such that all iterates $T^k(V)$ lie in U, k = 1,2,3, ... (cf. Fig. 9). (If x_0 is stable under the T-mapping, the corresponding periodic orbit in phase space is also stable [23].) Without loss of generality, we can choose x_0 as the coordinate origin. The linearization around x_0 shows that the linear part L of T is a 2x2 unimodular matrix with constant real coefficients. The eigenvalues λ of L are determined by

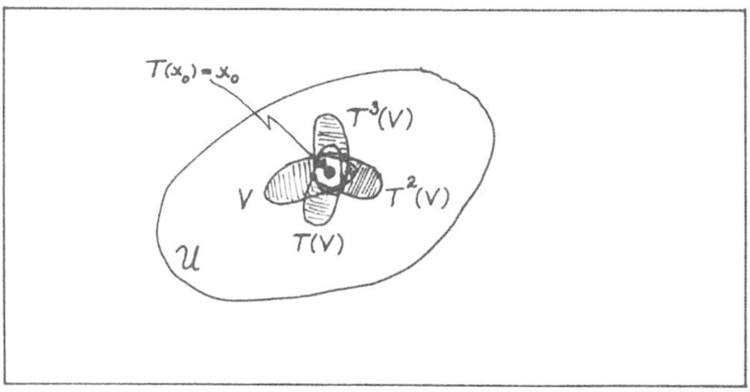

Fig. 9. Stability of a fixed point $x_0 = T(x_0)$ of an area-preserving mapping T.

$$\lambda^2 - \lambda \mathrm{Tr}(L) + 1 = 0 \tag{17}$$

These eigenvalues are either complex conjugates λ, λ^* on the unit circle, or reciprocals λ, $1/\lambda$ on the real axis. In the first case, L describes elliptic rotation and x_0 is called an *elliptic* fixed point of T. In the second case, L describes the hyperbolic rotation on the same or interchanging branches, depending upon whether the eigenvalues are positive or negative, respectively (cf. Fig. 10); the fixed point is then called *ordinary hyperbolic* or *inversion hyperbolic*, respectively.

According to the Moser Twist Theorem [27], elliptic fixed points are stable (and nearby orbits separate linearly with time) except in cases of low-order resonances (i.e., those cases in which the rotation angle of L is $2\pi/m$, $m = 1,2,3,4$), where the linear stability analysis is not sufficient. Hyperbolic fixed points, on the other hand, are always unstable, as shown by Hadamard [27]. Near a hyperbolic fixed point, the orbit will soon leave the neighborhood in which a linear description is valid; moreover, nearby orbits separate exponentially.

In the resonant gaps of Fig. 8, the elliptic islands surrounding the stable periodic points are shown. Hyperbolic points also appear, and in the vicinity of these points complicated behavior can be observed, as explained below. Clearly, the resonant gap was born out of a resonant torus of the unperturbed integrable system. Such a resonant torus appears in the SOS as an invariant curve, with each of its points being a fixed point of some iterate $T = P^n$ of the Poincaré mapping P. The resonant torus is thus represented by a curve consisting of fixed points of T.

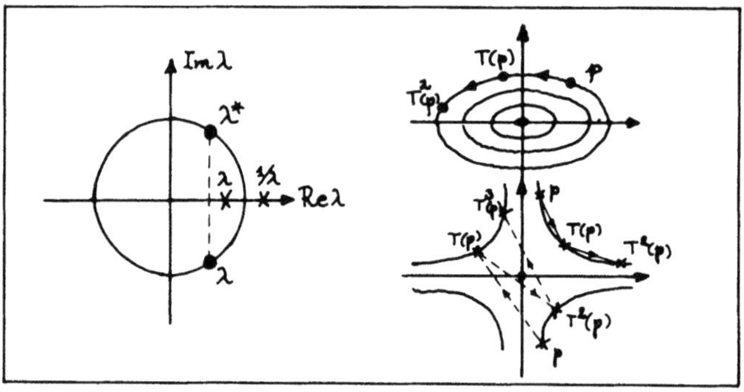

Fig. 10. The spectrum of a linear area-preserving mapping (left side) and the neighborhood of an elliptic, and of an hyperbolic fixed point (right side; top and bottom, respectively).

When the perturbation is applied, the resonant torus will be destroyed, but not all of the fixed points on it will disappear. The *Poincaré-Birkhoff Theorem* [27] guarantees that an even number of these fixed points will remain: Half of them are elliptic and half of them are hyperbolic. Because the proof of this theorem is so elegant, we give it here.

Consider a neighborhood of a curve C consisting of fixed points of T. (It is the image of the unperturbed resonant torus on the SOS). When the perturbation is applied, $H_0 \to H_0 + \varepsilon H_1$, the Poincaré mapping P, as well as T, will slightly change. The perturbed mapping is denoted by T_ε. By the KAM-Theorem, there exists two curves, C_- and C_+, in a neighborhood of C which are invariant with respect to T_ε (cf. Fig. 11). Let C_- be such that T_ε rotates points on it anticlockwise, while the action of T_ε on C_+ is opposite. Take a fixed point O inside the region bounded by C_- (at least one fixed point exists there) and consider points on a ray R emanating from O. The points p_-, p_+ of the intersections $R \cap C_-$, $R \cap C_+$ will move in opposite directions under the mapping T_ε. But, since T_ε is a smooth mapping, there must be a point p_0 on R that is not rotated, but only displaced along R (i.e., $T_\varepsilon(p_0) \in R$). If T_ε is assumed to be nondegenerate, the rotation velocity will be a strictly monotonic function of the position R, so that on each ray R one and only one such point p_0 exists. The collection of all such points is a curve K, which is not invariant under T_ε (i.e., $T_\varepsilon(K) \neq K$), but has the special property that its points are displaced along the rays but not rotated. Consider now the image $T_\varepsilon(K)$. Clearly, $T_\varepsilon(K)$ can lie neither inside nor outside K since T_ε is an area-preserving mapping. Hence, $T_\varepsilon(K)$ must intersect K. If it does so, it must do

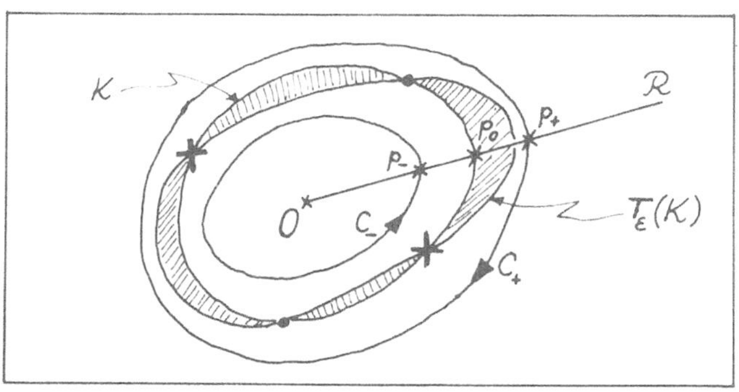

Fig. 11. The Poincaré-Birkhoff Theorem guarantees that an even number of periodic points survives a perturbation $T \to T_\varepsilon$. These are the points of intersection of the curves K and $T_\varepsilon(K)$.

it an even number of times. The intersection points $K \cap T_\varepsilon(K)$ are fixed points of T: They neither rotate nor move along the ray. This proves that an even number of fixed points remain after the perturbation. To show that half of them must be elliptic and half of them hyperbolic, we mention that the Poincaré index [30,31] of the annulus between the curves C_- and C_+ is zero. And, since this index is a conserved quantity, it must be zero after the perturbation as well. Finally, since the Poincaré index is additive, and equals +1,-1 for elliptic and ordinary hyperbolic fixed points, respectively, the number of these fixed points must be the same.

We now understand how the elliptic islands and the hyperbolic points can occur in a resonant gap. The infinite hierarchy shown in Fig. 8 is explained by the observation that all of the above considerations are valid for any iterate of the Poincaré mapping. It remains, however, for us to understand the motion in the *nonlinear* region around a hyperbolic point, which is connected with the chaotic motion. We shall concentrate on the asymptotes of the hyperbolae and the motion of them (cf. Fig. 12), which is described by the linearization discussed above. We call these asymptotes incoming or outgoing *strands*, according to whether L acts as a contraction or an expansion, respectively. In fact, these linear parts can be extended into the nonlinear region of T and are then called stable and unstable manifolds [32], respectively. The most important thing to be considered is how these manifolds intersect each other. A smooth joining of them is of course exceptional, as is immediately clear from the geometry of Fig. 12. In the case when they do join smoothly,

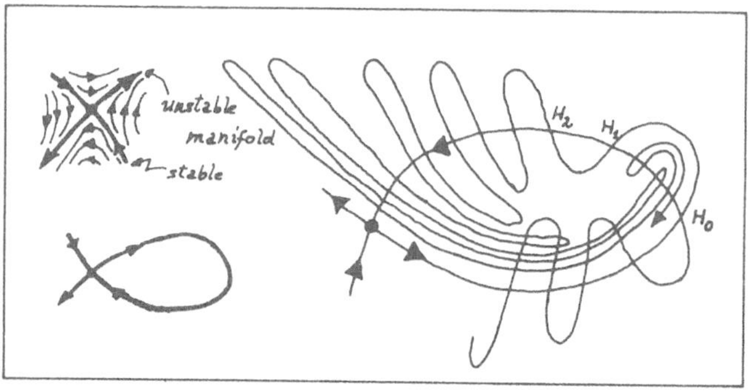

Fig. 12. If the stable and unstable manifolds of a hyperbolic point meet transversally at a homoclinic point H_0, then there exists an infinity of homoclinic points $H_1, H_2, \ldots$. No smooth integral of the motion exists in homoclinic oscillations; this is a prototype of chaotic motion.

nothing special happens. However, in almost all cases they will meet transversally at a point, called a *homoclinic* point, which is not a (hyperbolic) fixed point; and if they do so, they will do it infinitely many times. Moreover, the area-preserving property implies that the resulting oscillations will become larger and larger, so that a very complicated motion, called homoclinic oscillation, arises. One can see immediately that this has nothing in common with regular, integrable motion. Indeed, it has been rigorously shown that no smooth integral of the motion exists for this type of motion [33,34]: This is a prototype of chaotic motion. We see that *hyperbolicity* + *transversality* imply chaotic motion. We also understand, from qualitative geometrical considerations, that such a situation will occur in almost all cases; thus, it is a generic property. Moreover, the resultant motion is *structurally stable* since a small perturbation will not change the two basic properties, hyperbolicity and transverality.

5. CHAOTIC SYSTEMS

We can now understand the nature of the extreme example provided by completely unstable systems: These systems must be full of hyperbolic orbits which are surrounded by homoclinic points.

A standard example which illustrates what happens to some geometric object when it is mapped by an "everywhere hyperbolic" mapping is the so-called Arnold's cat. This is a linear mapping on a 2-torus = square [0,1] x [0,1], with opposite sides being identified. It is defined by the matrix

$$T = \begin{pmatrix} 1 & 1 \\ 1 & 2 \end{pmatrix} .$$

This mapping is full of hyperbolic fixed points, as can easily be verified. Now, take a cat as an interesting geometric object and look at what happens to it when T is applied. In Fig. 13, we show CAT, T(CAT) and T^2(CAT). Obviously, T^2(CAT) is already hardly recognizable as a CAT!

In physics, and in particular in statistical physics, ergodic systems are important. There are several equivalent definitions of an ergodic system [5,23,35]: (i) Each invariant set has either measure zero or measure one; the measure in question is the normalized Liouville measure, and this property is termed metrical transitivity. (ii) The microcanonical distribution is the only invariant distribution. (iii) The time average of each integrable function is almost everywhere equal to the phase average. It can be most clearly seen from (ii) that an ergodic Hamiltonian system can have no integrals of the motion except for the Hamiltonian itself (Jeans Theorem!). Roughly speaking, in an ergodic Hamiltonian system, almost every point of the (2N-1)-

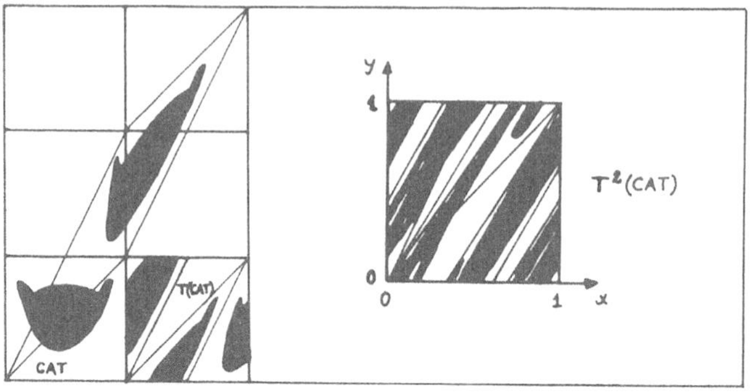

Fig. 13. Arnold's cat is a linear mapping T on a torus. It is full of hyperbolic fixed points. Can the second iterate T^2(CAT) be identified as a CAT?

dimensional energy surface will visit each neighborhood of each point on this surface as time goes on. An ergodic system must be full of hyperbolic periodic orbits. The measure of these orbits is zero, however, so their existence does not contradict property (i). We could say, actually, that the hyperbolic periodic orbits span the flow of an ergodic Hamiltonian system. Are there any Hamiltonian systems rigorously known to be ergodic? One example is the Sinai billiard, discussed below. The anisotropic Kepler problem [36] is most probably also ergodic.

In the framework of statistical mechanics, ergodicity is important for justifying the use of the microcanonical ensemble. Clearly, if a system is less chaotic than ergodic, there exist invariant distributions different from the uniform distribution on the energy surface. Ergodicity provides a necessary and sufficient condition for the uniqueness of the microcanonical ensemble. It does not guarantee the approach to equilibrium, however. To see this, we note that one-dimensional Hamiltonian systems are integrable and ergodic at the same time. Consider the harmonic oscillator (cf. Fig. 3). The motion is simply a continuous rotation of the energy circle. Therefore, only a uniform distribution on the energy "surface" is invariant. But, consider a non-uniform distribution: It is neither invariant, nor does it approach a uniform distribution, but periodically repeats itself in time.

For the approach to equilibrium, a stronger property is needed, namely, *mixing*. A system is mixing, if its phase flow Φ_t (Φ_t is a mapping which carries a given phase point along its orbit to the position occupied at time t) has the following property:

$$\lim_{t\to\infty} \mu[(\Phi_t A) \cap B] = \mu(A) \cdot \mu(B) \quad , \tag{18}$$

where μ is the Liouville measure, and A and B are two arbitrary sets on the energy surface. Mixing means that points which are initially in A distribute themselves uniformly over the entire energy surface in the course of time. Therefore, in mixing systems it is difficult, or even impossible, to distinguish points which were in a given set A in a sufficiently remote past. What happens in a mixing system can be illustrated by mixing milk in a cup of coffee: Although the volumes of both fluids are conserved, the drink will quickly be perfectly mixed on some sufficiently coarsened scale (cf. Fig. 14). A mixing system has the property that the correlation functions decay in time, whereby the system eventually "forgets" the initial conditions.

An example of a mixing system is the motion of a point particle in a hard-sphere gas, as shown by Sinai [35]. To see the dynamical instability which arises here, consider the particle colliding with a ball of radius R. From the geometry given in Fig. 15, it is clear that a small change $\delta\phi_0$ in the initial velocity direction will be rapidly amplified in subsequent collisions, because of convexity. Thus,

$$\delta\phi_n = (1 + \frac{2\ell}{R})\ \delta\phi_{n-1} \quad ,$$

where $\delta\phi_n$ is the change of the velocity direction after the nth collision, and ℓ is the mean distance between the balls. We find therefore,

$$\delta\phi_n = (1 + \frac{2\ell}{R})^n\ \delta\phi_o$$

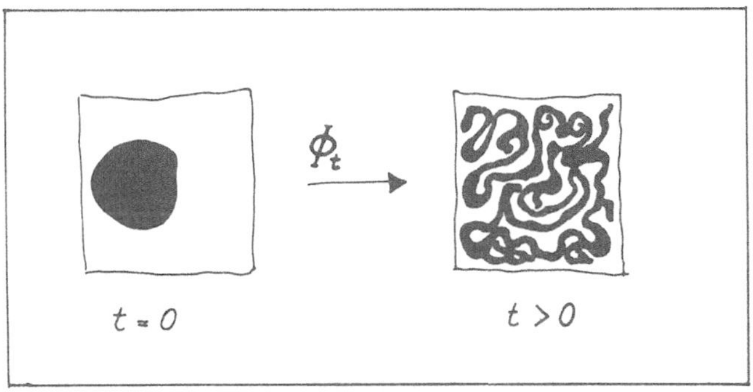

Fig. 14. Mixing in phase space.

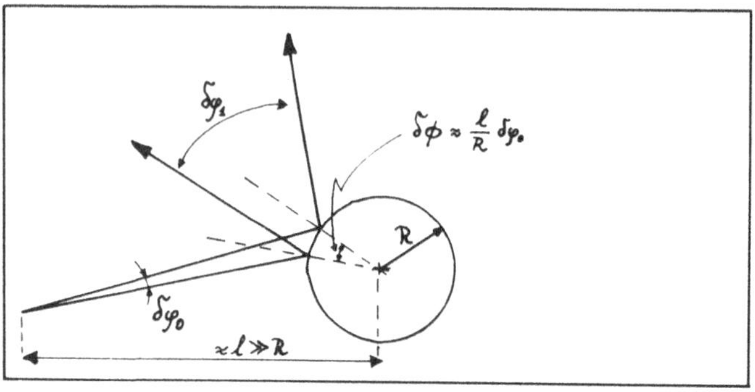

Fig. 15. Collision of a point particle with a ball of a rigid-sphere gas. A small change in the initial velocity direction increases exponentially with the number of collisions.

Since the total distance traveled by the ball is $s \simeq n\ell$, in the limit $n \to \infty$,

$$\delta\phi_{s=n\ell} \simeq \delta\phi_0 \exp\frac{2s}{R} = \delta\phi_0 \exp\frac{2n\ell}{R} \tag{19}$$

Thus, a small initial perturbation of the velocity direction increases exponentially with the number of collisions. This instability is very dramatic, as the following example [28] shows. Let us model air molecules by rigid spheres, with $R \simeq 1$ Å and with ℓ determined at room temperature and pressure. If, at the limit of the observable universe, there is a single electron acting as a *gravitational* perturber of our rigid-sphere system, we find that after some ten collisions $\delta\phi$ will be of the order of 360^0, so that all information on initial conditions is irretrievably lost. Everyone doing molecular-dynamics calculations on a computer knows very well that the computations are not reversible. The instability just mentioned amplifies the round-off errors exponentially, so that the number of correctly given digits decreases in proportion to the number of collisions.

The motion along individual orbits in such strongly chaotic systems is thus unpredictable. In this case, the point structure and individual trajectories in phase space have little meaning and, instead, we consider bundles of orbits. Two neighboring orbits, separated by a distance D(0) at t=0, will separate exponentially in time (for times which are not too large -- that is, as long as their separation is much smaller than the typical dimension of the confining energy surface). Thus,

$$D(t) = e^{\lambda t} D(0) \quad . \tag{20}$$

The coefficient λ is called the *Lyapunov exponent*. Since λ depends, of course, on the position in phase space, it is a local quantity; and, the instability expressed in Eq. (20) is sometimes referred to as local instability [4]. Although the Liouville Theorem says that the phase volume is conserved on the finest scale, in mixing systems it is appropriate to define a coarsened phase volume which can change. Then, as a result of the local instability [Eq. (20)], a small coarsened volume element Γ will spread exponentially with time:

$$\text{Vol}\ (\Gamma(t)) = e^{\lambda t}\ \text{Vol}\ (\Gamma(0)) \quad . \tag{21}$$

A global, integral property of a system is described by the phase average of the Lyapunov exponent,

$$h = \int \lambda \ d\mu = \langle\lambda\rangle \quad , \tag{22}$$

which is termed K-*entropy*. From the definition, it follows that a volume element will increase in the mean according to Eq. (21), with λ replaced by h. In other words,

$$h = \frac{d}{dt} \ln \text{Vol}\ (\Gamma(t)) \quad , \tag{23}$$

so that the K-entropy is the rate of change of entropy, if this latter quantity is defined in the usual manner.

Systems for which the K-entropy is positive ($h > 0$) are called K-systems. These have the important property that any two systems with the same h have statistically equivalent dynamics in phase space [4]. Such systems are physically very similar.

Now, the curious thing is that there are systems in which *chaotic*, mixing regions are *coexistent with* islands of stable, *regular* motion in phase space. For such systems $h > 0$, so they are K-systems, but not mixing or ergodic in a rigorous sense. (To be precise, such systems can be decomposed into several invariant components, some of which can be ergodic or mixing in the reduced phase space.) What is perhaps even more surprising serves as the title of the next section.

6. A GENERIC SYSTEM IS NEITHER INTEGRABLE NOR ERGODIC, BUT SHOWS A STOCHASTIC TRANSITION

Consider a Hamiltonian system

$$H = H(\underset{\sim}{q},\underset{\sim}{p}) = \frac{1}{2}\underset{\sim}{p}^2 + V(\underset{\sim}{q}) \quad , \tag{24}$$

where the potential energy $V(\underset{\sim}{q})$ is assumed to have a minimum at $\underset{\sim}{q}_o$ and a Taylor expansion around $\underset{\sim}{q}_o$ which begins with harmonic terms. We shall assume $\underset{\sim}{q}_o = 0$, and $E_{min} = 0$. Suppose that the energy $E = H(\underset{\sim}{q},\underset{\sim}{p})$ is the only integral of the motion that we know. Are there any others?

First of all, it should be noted that there is no general method for the construction of these integrals, if any exist. But, it is also difficult to prove that integrals do not exist since, for such a proof, details of the dynamics must be known. For example, if we know that a hyperbolic point with homoclinic tangencies exists, the Smale horseshoe map can be embedded into the homoclinic motion [32] and this would indeed prove the non-integrability. This is difficult and we usually make only numerical tests, which are of course not rigorous proofs. However, when a system shows chaotic behavior on the finest scale available by present-day computers (27 digits, say), then such a system is chaotic for all practical applications; if it is integrable in a rigorous sense, integrability has theoretical, but not practical, significance.

Numerical tests of integrability make use of the Poincaré mappings of the SOS. As discussed in Section 4, these mappings are most convenient for systems with N=2, which we now consider. Let us take the surface-of-section defined by the (q_1,p_1)-plane in phase space. The intersection of the SOS with the energy surface is assumed to be a bounded connected region for each energy E. Then, for each E we produce a Poincaré mapping of the SOS by choosing a point on it, and numerically integrating the equations of motion until the orbit returns to SOS. By repeating this procedure, we can plot a sequence of iterated maps of the given initial point. If all iterates lie on a smooth curve, an integral of the motion (independent of H and in involution with H) exists for that initial condition. In other words, an invariant curve in the SOS corresponds to an invariant torus in phase space. And, everything which has been said about the correspondence between orbits in phase space and Poincaré mappings (cf. Section 4) applies.

In Fig. 16, we show schematically three Poincaré maps at energies $E_1 < E_2 < E_3$. The generic case revealed in numerical calculations is as follows: At sufficiently low energies, near the potential minimum, the motion is regular. Invariant tori exist almost everywhere on the energy surface. However, the structure of these tori changes with increasing energy: Bifurcations of periodic orbits occur, resulting in new families of tori and new separatrices. At higher energies, the resonant tori are seen to be destroyed as a result of such bifurcations, and the first signs of chaotic motion in the resonant gaps appear. With increasing energy, tori are destroyed very rapidly and, eventual-

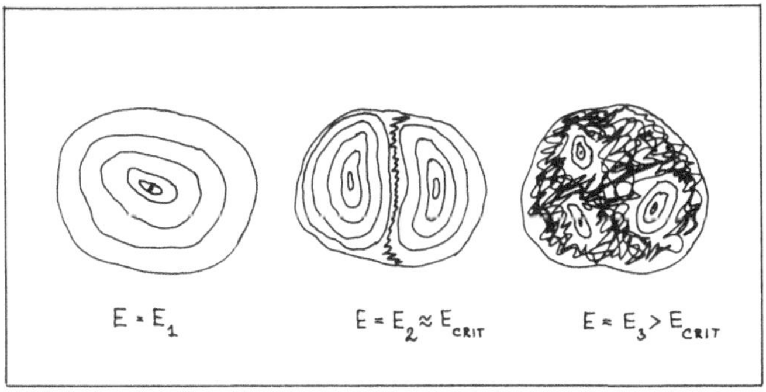

Fig. 16. The invariant curves and their destruction on the SOS of a typical Hamiltonian system. At low energy E_1 the motion appears integrable, but at the critical energy E_{crit} a stochastic transition occurs. At higher energies, $E > E_{crit}$, the system is chaotic, but islets of stability may exist.

ly, the "last" large-scale torus will be destroyed. The motion then becomes largely chaotic, but there may be some islands of stable, regular motion. The size of these islands diminishes as E increases, so that at high energies the motion can become completely chaotic, and perhaps even ergodic or mixing. An example concerning the hydrogen atom in a strong magnetic field can be found in [12].

A typical Hamiltonian system is thus a K-system, with h=0 for energies E smaller than some critical energy E_{crit}; however, h increases almost discontinuously (probably exponentially fast) as E increases above E_{crit}, so that $h > 0$ for $E > E_{crit}$. This phenomenon is termed a *stochastic transition*, and the condition for determining E_{crit}, at which this happens, is the *stochasticity criterion*. There are several approximate methods of predicting the stochastic transition [4,31,47], for example, Chirikov's criterion of resonance overlap [2], the renormalization approach by Escande & Doveil [40] and negative curvature considerations. For some special systems, such as the Henon-Heiles potential, it is possible to obtain a few rigorous results on periodic orbits and a proof of nonintegrability for $E > E_{crit}$ has been found [42]. In a recent work on period-doubling bifurcations in area-preserving mappings, a new perspective on the transition to chaos has been given [31]. In good analogy to the bifurcation properties of one-dimensional dissipative mappings [43], it was found that, as a parameter p of an area-preserving mapping varies, a stable elliptic periodic point may become unstable at the parameter value $p = p_n$, and may give birth to two new periodic orbits of twice the original period. With parameter variation, we thereby follow up a whole tree of *period-doubling bifurcations*. The parameter intervals between two subsequent bifurcations converge geometrically to

$$\lim_{n\to\infty} \frac{p_{n+1} - p_n}{p_n - p_{n-1}} = 1/\delta \quad ,$$

where $\delta = 8.721097200\ldots$ is a universal rate. At the accumulation point, we reach a mapping which has inifinitely many hyperbolic orbits, and which displays chaotic behavior (at least locally). The surprising property of such bifurcations is their universality, an explanation for which has been found in the discovery of a universal mapping which acts as an attractor in function space. Note, finally, that the stochastic transition thus observed is rather abrupt, since $1/\delta$ is so small.

7. THE BIRKHOFF-GUSTAVSON NORMAL FORM AND INVARIANT TORI BELOW THE CRITICAL ENERGY

When we detect (by a numerical experiment) the existence of tori below the critical energy E_{crit}, we would like to find at least approximate analytical expressions for them. In other words, we wish to find the action-angle variables at least approximately. This is possible by using the Birkhoff-Gustavson algorithm [23,37], which was invented for the multi-dimensional systems introduced in Eq. (24). For simplicity, we consider the case N = 2. (The general case is given in Gastavson's work [37].) We assume that the Taylor expansion of Eq. (24) is known:

$$H = \frac{\omega_1}{2}(p_1^2 + q_1^2) + \frac{\omega_2}{2}(p_2^2 + q_2^2) + \sum_{j=3}^{s} V^{(j)}(q_1,q_2) + \text{h.o.t.} \,, \qquad (25)$$

where ω_1, ω_2 are the oscillation frequencies at $E = E_{min} = 0$, while $V^{(j)}(q_1,q_2)$ are monomials of order j. It can be shown that a canonical transformation exists, with generating function

$$G^{(s)} = \underset{\sim}{q} \cdot \underset{\sim}{P} + W^{(s)}(\underset{\sim}{q},\underset{\sim}{P}) \,, \qquad (26)$$

where $\underset{\sim}{P}$ are new momenta, $\underset{\sim}{Q} = \partial G/\partial P$ are new coordinates, and $W^{(s)}(\underset{\sim}{q},\underset{\sim}{P})$ is a polynomial of degree s, such that the new Hamiltonian

$$\Gamma^{(s)}(Q,P) = \frac{\omega_1}{2}(Q_1^2 + P_1^2) + \frac{\omega_2}{2}(Q_2^2 + P_2^2) + \sum_{j=3}^{s} g^{(j)}(\underset{\sim}{Q},\underset{\sim}{P}) + \text{h.o.t.} \qquad (27)$$

is in *normal form*: That is, each monomial $g^{(j)}(Q,P)$ of order $j \leq s$ commutes with the *harmonic part* of $\Gamma^{(s)}$. In other words,

$$Dg^{(j)} = 0 \quad , \tag{28}$$

where

$$D := \omega_1(Q_1 \frac{\partial}{\partial P_1} - P_1 \frac{\partial}{\partial Q_1}) + \omega_2(Q_2 \frac{\partial}{\partial P_2} - P_2 \frac{\partial}{\partial Q_2}) \quad . \tag{29}$$

Moreover, in the nonresonant case (ω_1 and ω_2 incommensurable),

$$I_1 := \frac{1}{2}(Q_1^2 + P_1^2) \ , \quad I_2 := \frac{1}{2}(Q_2^2 + P_2^2) \tag{30}$$

are integrals of the motion when expressed as functions of the old variables (q,p), and $\Gamma^{(s)} = \Gamma^{(s)}(I_1,I_2)$ depends only on the actions I_1, I_2. In the resonant case (ω_1 and ω_2 rationally connected),

$$\tau = \omega_1 I_1 + \omega_2 I_2$$

is an integral of the motion, independent of H and in involution with H, so that again the system is formally integrable.

Note that a normal form is obtained here only up to order s; by series expansion, however, it can be expanded arbitrarily ($s \to \infty$). Moreover, this formal series diverges (i.e., its convergence radius vanishes). Clearly this must be so, since we know rigorously that nonintegrable systems exist. In such a case, the convergence of the series would lead to a contradiction.

The technical difficulty which causes the divergence of the series is the existence of small divisors. Nevertheless, at sufficiently small energies, the truncated Birkhoff-Gustavson series yields a good approximation for the invariant tori. Indeed, it is useful even for energies E just below E_{crit}. An example worked out for the Henon-Heiles system can be found in the original paper of Gustavson [37]. While the applicability of the Birkhoff-Gustavson normal form in classical mechanics is limited to K-systems at energies $E < E_{crit}$, it seems to be of considerable importance in quantum mechanics, as we shall see later.

To calculate the normal form to a sufficiently high order, say s = 8, is an almost impossible task: For N = 2, this already involves polynomials with 495 terms, and requires 495 x 495 matrices to be inverted. To our knowledge, all calculations reported so far in the literature have been mostly numerical, even when computers were used. However, we have calculated algebraically the eigth-order normal form for N = 2 Hamiltonians using the

REDUCE programming language [38]; this now enables an easy and direct applicability of the normal form.

EXAMPLE:

$$H = \frac{\omega_1}{2}(q_1^2 + p_1^2) + \frac{\omega_2}{2}(q_2^2 + p_2^2) + a\, q_1 q_2^2 + b\, q_1^2 q_2^2 \quad . \tag{31}$$

This example is calculable by hand up to fourth order. In the nonresonant case, the normal Hamiltonian is a function of the actions given by Eq. (30):

$$\Gamma^{(4)} = \omega_1 I_1 + \omega_2 I_2 + I_1 I_2 \left(b + \frac{2a^2 \omega_2}{\omega_1^2 - 4\omega_2^2} \right) - \frac{a^2 I_2^2}{4\omega_1} \frac{3\omega_1^2 - 8\omega_2^2}{\omega_1^2 - 4\omega_2^2} \quad . \tag{32}$$

A recent discovery [39] reveals dramatically that this purely classical algorithm has important applications in quantum-mechanical perturbation calculations of energy spectra. We shall return to this problem, and to the example of Eq. (32), in subsequent sections, after first reviewing the nature of stochasticity in quantum mechanics.

8. DO QUANTUM SYSTEMS EXHIBIT CHAOTIC BEHAVIOR?

While the chaotic behavior of classical Hamiltonian systems is well understood, stochasticity in quantum mechanics is still an open problem [3,4]. Let us repeat the rhetorical question which serves as the title for this section: When a Hamiltonian $H = H(\underset{\sim}{q},\underset{\sim}{p})$ is known to be chaotic classically, does the corresponding quantum system display any features which might, in some sense, be called chaotic? The answer is yes, which might have been guessed from the correspondence principle, or from the limit $\hbar \to 0$.

In quantum mechanics, we are faced with two basic problems: (i), the stationary problem of determining the spectra of the complete set of (mutually commuting) observables, in particular the energy spectrum; and (ii), the nonstationary problem of the time evolution of a state (wavefunction). Concerning the nonstationary problem, we note the following difficulties. Classically, the evolution operator (≡ Liouville operator) of a bound system can possess either a discrete or a continuous spectrum. In the former case one finds quasi-periodic (regular) motion of invariant tori, while in the latter case the motion is chaotic, perhaps even ergodic or mixing. In quantum mechanics, however, the spectrum of the evolution operator (Hamiltonian operator) of a *bound* system is always discrete and, therefore, corresponds to regular motion. On these grounds, one would expect that no chaos is possible in quantum mechanics. The resolution of this paradox has been explained as follows [44]: For sufficiently small times (smaller than a quasi-period), wavefunctions do behave chaotically, but for a sufficiently long time, quasi-periodicity will be observed. In other words, if the quasi-period T is sufficiently long, chaotic behavior occurs for times $t \lesssim T$. We shall ignore the nonstationary problem in what follows, but interested readers may appreciate the reviews by Berry [3] and Zaslavsky [4], and the references given therein. As far as the stationary problem is concerned, we shall discuss mainly the energy spectrum, with all other aspects being relegated to the final section of this paper.

Chaotic features in energy spectra have been known for a long time. For example, in nuclear physics it is well known that the spectra of complex nuclei can be statistical in character, in the sense that individual energy levels may have no meaning, while the statistical properties of the energy-level distribution does. Moreover, there exist examples of spectra which are regular at low energies, with individual lines resolved, but which become irregular and random at high energies (cf. Figs. 17,18). Such behavior corresponds to the stochastic transition in a generic classical Hamiltonian system. The statistical theory of irregular nuclear spectra was initiated in the early works of Dyson, Porter, Wigner and others. However, at that time it was not clear what gives rise to the stochastic properties. In the works of Percival [45], Pomphrey [46] and others, it has been pointed out that this stochasticity is connected with the instability of energy levels. Although the rigorous theory of spectral stability is just at its beginning [47], there exists a lot

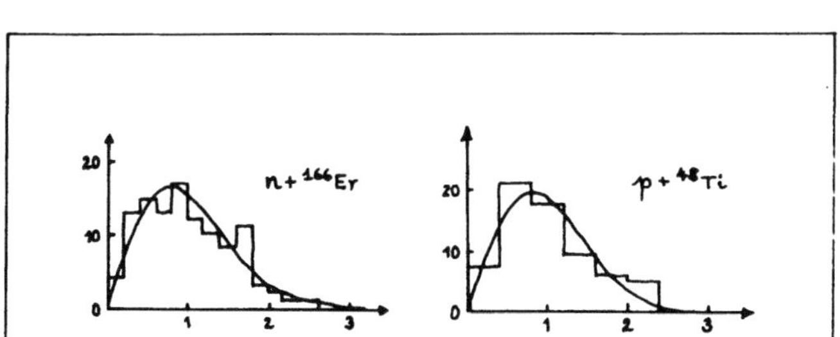

Fig. 17. Random nuclear spectra adapted from [18]. The ordinate represents the number of cases in which the spacing S, in units of mean level separation, lies between S and S + dS.

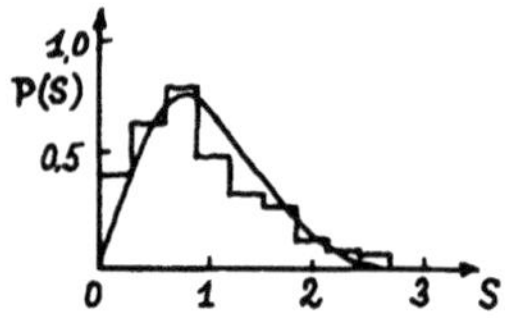

Fig. 18. Histogram showing the distribution of atomic level spacings S (normalized to the mean level separation) for a group of elements in the osmium region (Hf, Ta, W, Re, Os and Ir). The spectral randomness is a consequence of strong spin-orbit interactions. [Plotted according to the data in C. E. Porter, N. Rosenzweig, Ann. Acad. Sci. Fennicae Ser. A 6, No. 44 (1960)].

of evidence that energy levels corresponding to the chaotic regions of the classical phase space are unstable, while those corresponding to the motion on invariant tori are stable. We call these energy levels irregular and regular, respectively. Similar observations have been made by Noid [48], Marcus [49] and other workers [50] in the study of molecular spectra. Moreover, such phenomena are expected in atomic physics [12] as well. We thus have enough observational facts to motivate our theoretical study of the chaotic behavior of quantum systems. But, someone may ask whether or not we really can expect new physics? Recall, however, that the fact that the Schrödinger equation is in principle solvable for any system helps just as little as the similar fact concerning the Hamilton-Jacobi equation of classical mechanics. What we need are more explicit quantization conditions, since only then can we see which phenomena may occur and which of them are generic (i.e., most typical).

9. INTEGRABLE SYSTEMS AND TORI QUANTIZATION

In those rare cases in which a Hamiltonian $H(\underset{\sim}{q},\underset{\sim}{p})$ is integrable, we have the semiclassical quantization conditions. The theory for separable systems was started by Bohr and Sommerfeld in the years of the early quantum theory. Einstein [51] pointed out that this original formulation is not invariant under canonical transformations. He argued, correctly, that in the integrable systems we have (classical) invariant tori; the actions, defined as integrals along irreducible cycles, are canonically invariant, and he thus suggested that they must be quantized. He also saw that integrability need not be the rule, and that chaotic systems with no integrals might exist -- in which case no quantization conditions could be given.

Maslov [52] has studied the problem of the *short-wave* approximation [23] (also called the *eikonal*, or *ray*, or *geometrical* approximation). This is a broad problem related to wave-particle duality. Quite generally, the short-wave approximation reduces the linear wave equation (acoustic waves, electromagnetic waves, de Broglie waves) to a nonlinear equation for the phase, which is the Hamilton-Jacobi equation of classical mechanics. As is well known, the "ray" approximation of wave optics meets difficulties because of amplitude singularities at the caustics. However, Maslov pointed out that these singularities are just projection singularities and can be removed if the waves are studied in phase space, so that Hamilton's equations replace the Hamilton-Jacobi equation. His calculation [52] essentially reproduced Einstein's result, but now the presence of caustics has been taken into account: They affect the result only insofar as the phase of a wave changes by $\pi/2$ upon crossing a caustic. The Maslov quantization conditions read, therefore,

$$I_j = (n_j + \alpha_j/4)\hbar \tag{33}$$

where α_j is the number of caustics encountered in configuration space upon traversing the jth irreducible cycle of the torus, $1 \leq j \leq N$. α_j is called the *Maslov index* and the semiclassical quantization condition [Eq. (33)] is referred to as *tori quantization*.

We now give an elementary derivation [3,28]. The short-wave approximation describes the wave in terms of a wavefunction

$$\psi(\underset{\sim}{q},t) = A(\underset{\sim}{q},t)\, e^{\frac{i}{\hbar} B(\underset{\sim}{q},t)} \tag{34}$$

where the amplitude A is a slowly-varying function, while the phase B changes rapidly. This ansatz can be inserted into the linear wave equation, which then yields a relation between A and B, and the Hamilton-Jacobi equation for B. The phase B is thus the classical action:

$$B(\underset{\sim}{q},t) = S(\underset{\sim}{q}) - Et \quad . \tag{35}$$

The action for a torus is given by Eq. (4). We see, therefore, that this correspondence is nothing but the de Broglie principle that the wavenumber is equal to (momentum)/$\hbar$. Now, instead of solving the rigorous equations, we use a second principle which says that the probability density $|\psi|^2$ is equal to the density of classical trajectories. Therefore, since the orbits are uniformly distributed on a torus,

$$|\psi|^2 = |A|^2 = \det\left(\frac{\partial \underset{\sim}{\Theta}}{\partial \underset{\sim}{q}}\right), \tag{36}$$

where $\underset{\sim}{\Theta}$ are angles on the torus (cf. Fig. 19). This is just a geometric projection of a piece of a torus onto configuration space. As is apparent from Fig. 19, this projection will result in an infinite density whenever the projection ray is perpendicular to the normal of a surface element (caustics). For the one-dimensional case shown in Fig. 19, there are two caustics, also called classical turning points. The Maslov index α_1 equals two for this case.

Now, from Eq. (5) we have

$$|A|^2 = \det \frac{\partial^2 S}{\partial q_i \partial I_j} \tag{37}$$

which implies that

$$\psi(\underset{\sim}{q},t) = \left|\frac{\partial^2 S}{\partial q_i \partial I_j}\right|^{\frac{1}{2}} \exp\left\{\frac{i}{\hbar}\int_{\underset{\sim}{q}_0}^{\underset{\sim}{q}} \underset{\sim}{p}(\underset{\sim}{q}',\underset{\sim}{I})\,d\underset{\sim}{q}' - \frac{i}{\hbar}Et\right\} . \tag{38}$$

This wavefunction is stationary. And, it will be single-valued when the change ΔS of the action [cf. Eq. (4)] along an arbitrary closed path on the torus (with the phase changes $\pi/2$ on the caustics being subtracted) is equal to an integral multiple of $2\pi\hbar$. This condition is satisfied if it is satisfied for each irreducible cycle γ_j [53]. Hence,

$$(\Delta S)_{\gamma_j} - \alpha_j \frac{\pi}{2}\hbar = n_j\, 2\pi\hbar \quad ,$$

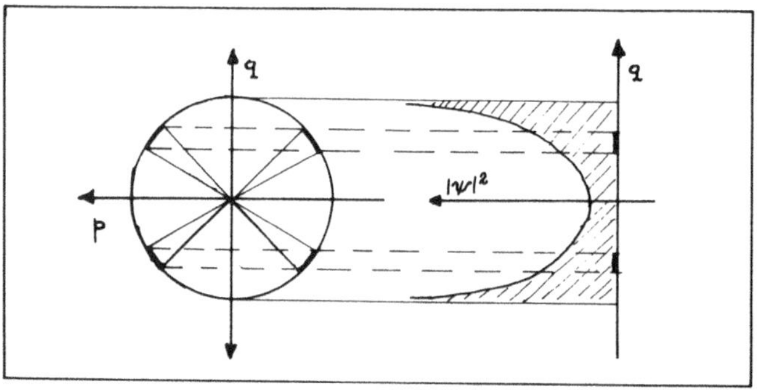

Fig. 19. Caustics at the boundary of a classically allowed region are a projection effect.

and with the definition $(\Delta S)_{\gamma_j} = 2\pi I_j$ [cf. Eq. (3)] we get Maslov's result [Eq. (33)]. We thus have N quantization conditions which determine the quantized value of the actions $I_{\underset{\sim}{n}} = (\underset{\sim}{n} + \underset{\sim}{\alpha}/4)\hbar$, with $\underset{\sim}{n}$ and $\underset{\sim}{\alpha}$ being lattice vectors. The energy E is given by [cf. Eq. (6)]

$$E_{\underset{\sim}{n}} = H(I_{\underset{\sim}{n}}) \quad , \tag{39}$$

which is the semiclassical energy spectrum; $\underset{\sim}{n} = (n_1,\ldots,n_N)$ are the quantum numbers.

The caustics are a projection effect, as noted above. The Maslov index α_j therefore depends upon how a torus lies in phase space. For example, when the projection of the torus from Fig. 1 onto the configuration space is an annulus, then $\alpha_1 = 2$, $\alpha_2 = 0$ (cf. Fig. 20).

Note that, in the derivation of Eqs. (38,39), it is essential that a torus exists. If the system is nonintegrable, there is no invariant surface in phase space with the Lagrangian property; the integral in Eq. (2) then depends upon the path of integration and, hence, does not define a single-valued function. The association between an invariant ergodic component of the phase space (on which the uniform distribution of orbits exists) and the wavefunction is still possible for a nonintegrable system [3]. However, in this case it is impossible to obtain quantization conditions. Thus, a different quantization method is needed for chaotic systems.

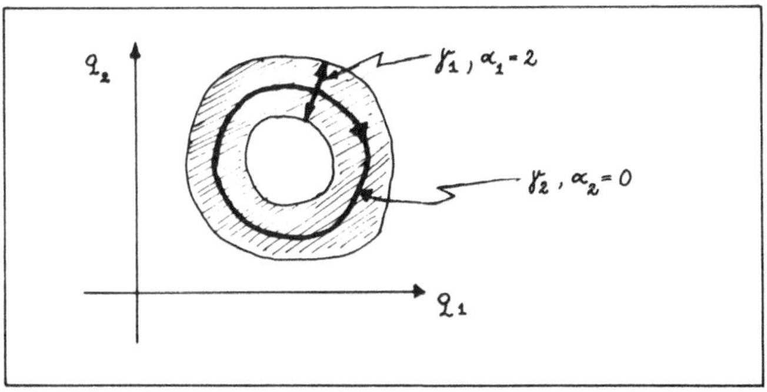

Fig. 20. If the projection of the torus of Fig. 1 onto the configuration space is an annulus, the Maslov indices α_1, α_2 of two irreducible cycles γ_1, γ_2 equal 2 and 0, respectively.

EXAMPLES:

For the examples of Section 2 we obtain (incidentally exact) spectra:

$$E_n = (n + 1/2)\hbar\omega \quad ,$$

$$E_{n_x n_y} = \frac{\hbar^2\pi^2}{2m}\left(\frac{n_x^2}{a^2} + \frac{n_y^2}{b^2}\right),$$

and

$$E_{n_r n_\theta n_\phi} = -\frac{\alpha^2/\hbar^2}{2(n_r + n_\theta + n_\phi + 1)^2} ,$$

where all quantum numbers are integers greater than or equal to zero.

10. QUANTIZATION OF CHAOTIC SYSTEMS

So far, no quantization conditions as explicit as tori quantization -- Eqs. (33,39) -- are known for irregular, chaotic systems. One might suggest quantizing the action of each periodic orbit. However, this procedure is wrong since it gives too many energy levels: A counter-example is the motion in a rectangle [54]. In this case, there is no unique relation between individual closed orbits and individual energy levels. However, a collective relation exists: Namely, a sum over all closed orbits determines the energy spectrum in the semiclassical approximation (small but finite $\hbar$).

The above result was obtained using Feynman path integrals, as we shall see shortly. It is not surprising that the search for correct quantization conditions is directed towards the application of path integrals, since these integrals represent the most intimate link between classical and quantum mechanics. The method was initiated by Gutzwiller [55], Balian and Bloch [56] and Zaslavsky [4,57].

Consider the kernel $K(\tilde{q}_a, \tilde{q}_b, t)$ for a Schrödinger equation having energy eigenvalues E_j and eigenfunctions $\psi_j(\tilde{q})$. This kernel is given by the expansion

$$K(\tilde{q}_a, \tilde{q}_b, t) = \sum_j \psi_j(\tilde{q}_a)\, \psi_j^*(\tilde{q}_b)\, e^{-iE_j t/\hbar} \quad . \tag{40}$$

Its Fourier transform, a Green function, equals

$$G(q_a, q_b, t) = \frac{1}{2\pi\hbar} \int dt\, e^{iEt/\hbar} K(q_a, q_b, t)$$

$$= \sum_j \psi_j(q_a)\, \psi_j^*(q_b)\, \delta(E - E_j) \quad , \tag{41}$$

where the $\delta(E - E_j)$ are Dirac delta functions. Putting $q_a = q_b = q$ and integrating over q yields the *response* d(E),

$$d(E) = \int d^N q G(q, q, E) = \sum_j \delta(E - E_j) \quad , \tag{42}$$

which is simply the sum of the delta spikes at the energy eigenvalues. Hence, knowing the kernel [Eq. (40)], we know the spectrum [Eq. (42)]. But, Feynman [58] has shown that the kernel can be given exactly by the sum of all paths going from q_a at t = 0 to q_b at time t: That is,

$$K(q_a, q_b, t) = \int e^{iS[a,b]/\hbar}\, Dq(t) \quad , \tag{43}$$

where

$$S[a,b] := \int_a^b L(q, \dot{q}, t)\, dt \tag{44}$$

is the action of a path between a = (q_a, t = 0) and b = (q_b, t).

In the semiclassical limit of small but finite $\hbar$, the contributions to Eq. (43) will display rapid oscillations, so that only the stationary ones ($\delta S[a,b] = 0$) will be non-negligible. But, the action [Eq. (44)] is stationary precisely if the path a → b is a classical trajectory (cf. Fig. 21). Further, the identification $q_a = q_b$ in passing from Eq. (41) to Eq. (42) requires that

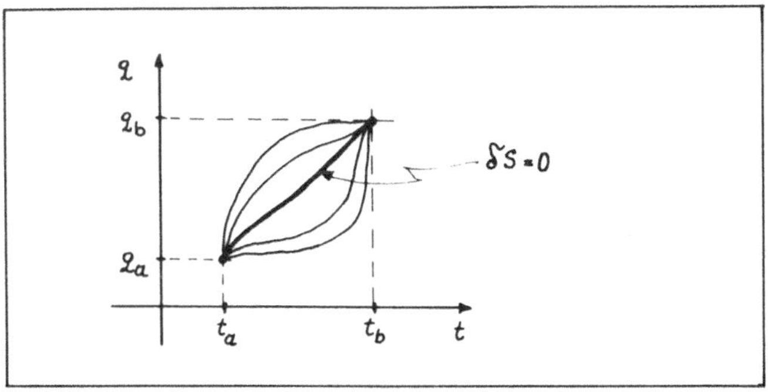

Fig. 21. The stationary path is the classical orbit.

only those classical paths which are recurrent in configuration space should be taken. These paths must also be stationary with respect to the endpoints, that is,

$$\left.\frac{\partial S}{\partial \underset{\sim}{q}_a}\right|_{\underset{\sim}{q}} + \left.\frac{\partial S}{\partial \underset{\sim}{q}_b}\right|_{\underset{\sim}{q}} = \underset{\sim}{p}_a - \underset{\sim}{p}_b = 0 \quad , \tag{45}$$

since only then is their contribution to the integral of Eq. (42) significant. Thus, we see that the momenta $\underset{\sim}{p}_a$ and $\underset{\sim}{p}_b$ must be equal too. We conclude, then, the following: In the semiclassical limit, only the closed (i.e., periodic) classical paths contribute to the response d(E). In this sense, the semiclassical spectrum is spanned by the periodic orbits.

This result seems paradoxical since the periodic orbits have as a rule, measure zero in phase space, no matter whether a system is regular and irregular. In an integrable system, the periodic orbits lie on a resonant torus, so that according to the response formula [Eq. (42)] the semiclassical energy spectrum is spanned by resonant tori. On the other hand, the result of the preceding section is that tori labeled by the quantized *actions* generate the spectrum, which is *not* equivalent to the resonant condition: In other words, $\underset{\sim}{I} = \underset{\sim}{I}_{\underset{\sim}{n}}$ does *not* imply

$$\underset{\sim}{k} \cdot \underset{\sim}{\omega}(\underset{\sim}{I}_{\underset{\sim}{n}}) = \underset{\sim}{k} \cdot \left.\frac{\partial H}{\partial \underset{\sim}{I}}\right|_{\underset{\sim}{I}_{\underset{\sim}{n}}} = 0$$

for some nonvanishing lattice vector $\underset{\sim}{k}$. To resolve the paradox, we first mention that the resonant tori are dense in phase space, although their measure is zero (just as the set of rational numbers, which has measure zero, is dense in the space of real numbers). Therefore, in an arbitrarily small neighborhood of an irrational torus, there is always a resonant torus. This fact indicates that the two quantization methods may be equivalent.

Berry and Tabor [59] have shown, indeed, that the response

$$d(E) = \sum_{\underset{\sim}{n}} \delta(E - H(\underset{\sim}{I}_{\underset{\sim}{n}})) \quad , \tag{46}$$

with the eigenvalues given by Eq. (39), namely $E_{\underset{\sim}{n}} = H(\underset{\sim}{I}_{\underset{\sim}{n}})$, results in a sum involving only the closed classical orbits, thereby resolving the "paradox". We show only how the resonance conditions are obtained.

Let us transform Eq. (46) into a series of integrals, using the Poisson sum formula

$$d(E) = \hbar^{-N} \sum_{\underset{\sim}{m}} e^{-\frac{i\pi}{2} \underset{\sim}{\alpha} \cdot \underset{\sim}{m}} \int_{+} d^N \underset{\sim}{I}\ \delta(E - H(\underset{\sim}{I}))\ e^{(2\pi i/\hbar)\underset{\sim}{m} \cdot \underset{\sim}{I}}, \tag{47}$$

where the sum is taken over all lattice vectors $\underset{\sim}{m}$; $\underset{\sim}{\alpha}$ are the Maslov indices from $\underset{\sim}{I}_n = (\underset{\sim}{n} + \underset{\sim}{\alpha}/4)\hbar$; and the integration extends over positive values of the integration variables $\underset{\sim}{I}$. (Clearly, if summation and integration are interchanged, we get precisely Eq. (46).) The oscillatory integrals

$$d_m := \int d^N \underset{\sim}{I}\ \delta(E - H(\underset{\sim}{I}))\ e^{(2\pi i/\hbar)\underset{\sim}{m} \cdot \underset{\sim}{I}} \tag{48}$$

can be evaluated as follows. The energy surface given by $E = H(\underset{\sim}{I})$ in action space is (N-1)-dimensional. The volume element $d^N\underset{\sim}{I}\,\delta(E - H(\underset{\sim}{I})) =: dV$ around a piece of that surface equals

$$dV = d\sigma \left|\frac{\partial H}{\partial \underset{\sim}{I}}\right|_{H(\underset{\sim}{I}) = E} ,$$

where $d\sigma = du_1 du_2 \ldots du_{N-1}$ is the surface element in an orthogonal coordinate system $(u_1, \ldots, u_{N-1}) =: \underset{\sim}{u}$. Since $\underset{\sim}{\omega} = \partial H/\partial \underset{\sim}{I}$, we obtain

$$d_m = \int d\sigma \frac{\exp\ [(2\pi i/\hbar)\underset{\sim}{m} \cdot \underset{\sim}{I}(\underset{\sim}{u})]}{\left|\underset{\sim}{\omega}(\underset{\sim}{u})\right|_{H(\underset{\sim}{I}(\underset{\sim}{u})) = E}} . \tag{49}$$

In the limit of small but finite $\hbar$, the intergrand in Eq. (49) will be a rapidly oscillating function. Therefore, only the stationary contributions should be taken. These are determined by

$$\underset{\sim}{m} \cdot \frac{\partial \underset{\sim}{I}}{\partial u_i} = 0 \ , \ i = 1,2,\ldots N-1 \quad . \tag{50}$$

On the other hand, one has

$$\frac{\partial H}{\partial u_i} = \frac{\partial H}{\partial \underset{\sim}{I}} \cdot \frac{\partial \underset{\sim}{I}}{\partial u_i} = \underset{\sim}{\omega}(\underset{\sim}{I}) \cdot \frac{\partial \underset{\sim}{I}}{\partial u_i} = 0 \quad , \tag{51}$$

since $H = H(\underset{\sim}{I}(\underset{\sim}{u})) = E = \text{constant}$. The condition of Eq. (50) shows that $\partial \underset{\sim}{I}/\partial u_i$ must be a vector of the reciprocal lattice, orthogonal to $\underset{\sim}{\omega}(\underset{\sim}{I})$. Therefore, $\underset{\sim}{\omega}(\underset{\sim}{I})$ must be proportional to a lattice vector. That is,

$$\underset{\sim}{\omega}(\underset{\sim}{I}) \propto \underset{\sim}{m} \quad , \tag{52}$$

which is a condition for commensurability. We conclude, then, that only resonant tori contribute to the response [Eq. (46)].

We now give the explicit sum formula resulting from the semiclassical Feynman propagator [Eq. (43)]. The derivation can be found in the original papers by Gutzwillwer [55] and Balian and Bloch [56]. The appropriate Green-function theory has been reviewed by deWitt-Morette, *et al.* [60]. The response d(E), which is the distribution function of energy levels, is a sum of a smooth part $\bar{d}(E)$, related to the mean separation of levels, and oscillatory corrections $d_{osc}(E)$:

$$d(E) = \bar{d}(E) + d_{osc}(E) \quad . \tag{53}$$

The mean level density $\bar{d}(E)$ is given by the Thomas-Fermi rule that each quantum state "occupies" a phase space volume h^N, so that

$$\bar{d}(E) \propto \frac{1}{h^N} \int d^N \underset{\sim}{q} d^N \underset{\sim}{p} \; \delta(E - H(\underset{\sim}{q},\underset{\sim}{p})) \quad . \tag{54}$$

Thus, the mean level separation $1/\bar{d}(E)$ is proportional to $\hbar^N$. The oscillatory part $d_{osc}(E)$ contains contributions for which the energy wavelength of oscillation is smaller and smaller. Summing all contributions produces the delta spikes of Eq. (40). The result is

$$d_{osc}(E) = \sum_j \sum_{p=1}^{\infty} \frac{A_{j,p}(E)}{1 + \ell_j/2} \sin\left(\frac{pS_j(E)}{h} + p\beta_j\right) . \tag{55}$$

This includes (p-fold) repetitions of a simple closed orbit, labeled by j, with action

$$S_j(E) = \oint \underset{\sim}{p}_j \cdot d\underset{\sim}{q} \quad . \tag{56}$$

The constants β_j, which depend upon the focusing of orbits in the vicinity of closed orbits, appear in the phase. The wavelength ΔE of oscillation in energy scale is given by

$$\frac{p}{h} \frac{dS_j}{dE} \Delta E = 2\pi \quad ,$$

and, because $T_j = dS_j/dE$ is the period of the jth simple closed orbit, we have

$$\Delta E = \frac{h}{pT_j} \quad . \tag{57}$$

This is of order $\hbar$ (and *not* $\hbar^N$, which would be much smaller as $\hbar \to 0$). The longer the orbit (larger p), the smaller ΔE.

The amplitudes $A_{j,p}(E)$ depend upon whether the jth orbit is isolated or nonisolated. The integer ℓ_j is the number of parameters for the family of orbits to which the jth path belongs; it can vary between zero (isolated orbit) to N-1 (for closed orbits on an invariant torus).

The stability properties of orbits determine the behavior of $A_{j,p}$ as a function of repetitions. $A_{j,p}$ oscillates with p in the case of a stable isolated orbit, and decays exponentially if the orbit is unstable. For stable orbits on tori, $A_{j,p}$ decreases as $p^{-(N-1)/2}$.

Berry [54] has shown that to see delta functions emerging from the oscillatory contributions of Eq. (53), one must include in the summation all trajectories having periods shorter than $T_{max} \propto \hbar^{-(N-1)}$. As $\hbar \to 0$, T_{max} increases and the number of operations goes as $\hbar^{-N(N-1)}$ for integrable systems, and even faster -- exp(+ constant/$\hbar$) -- for chaotic systems, which have many more closed orbits, all being unstable. This means that the response formula [Eqs. (53-55)] is not a practical way of calculating individual levels, but merely gives the distribution of levels.

However, this is not the only limitation. As explained earlier, chaotic systems (e.g., ergodic or even mixing) are full of (isolated, hyperbolic) closed orbits. To find all of these orbits requires a detailed solution of the dynamics, which is an almost impossible task.

Zaslavsky [4,57] points out that K-systems possessing the same K-entropy h(E) have (statistically) similar dynamics. Moreover, because of local instability, there is no sense in studying individual classical trajectories but, rather, one should consider a bundle of orbits instead. He emphasizes that the task of summation is greatly facilitated if one considers *almost* closed orbits, rather than exactly closed orbits. This procedure should be satisfactory for a determination of the statistical properties of classical dynamics, and for determining the semiclassical energy spectrum. We understand that this need not be a contradiction to the assertion "only closed orbits contribute to d(E)", since we have already seen, by way of example, that irrational tori (on which the orbits are quasi-periodic) can be approximated by rational tori as accurately as required.

Berry [61], however, insists on taking only closed orbits, but this is not necessary. To see this, we return to the stationarity condition [Eq. (45)] at the endpoints a,b of a recurrent ($q_a = q_b$) classical orbit. This condition can be weakened since all that is required is that the variation δS of the action [Eq. (44)] must be smaller than $\hbar$; that is,

$$\delta S = \left(\frac{\partial S}{\partial g_a}\bigg|_{\tilde{q}} + \frac{\partial S}{\partial g_b}\bigg|_{\tilde{q}} \right) \cdot \delta \tilde{q} = \delta \tilde{p} \cdot \delta \tilde{q} \leq \hbar \quad , \tag{58}$$

so that

$$|\delta S| \leq \hbar \tag{59}$$

if

$$|\delta \tilde{p}| \; |\delta \tilde{q}| \leq \hbar \quad . \tag{60}$$

If a classical orbit emanates from a 2N-dimensional cube of volume ε in phase space M and returns to it, the condition of stationarity given by Eq. (60) will be fulfilled if

$$\varepsilon \leq \hbar^N \quad . \tag{61}$$

Approximately closed orbits are therefore allowed and they still give constructive interference in Eq. (41). (Note that the conditions of Eq. (60) and Eq. (61) are complementary to the Heisenberg and Thomas-Fermi inequalities, respectively.)

11. STATISTICS OF ENERGY LEVELS

What does the spectrum of a chaotic system look like? Since a chaotic system has no continuous symmetries, this problem is essentially a study of the spectra of general matrices (i.e., those with no symmetry). A result of von Neumann and Wigner [62] is that, typically, the spectrum will be *non*degenerate. They also studied the question of whether or not one can obtain a degeneracy by varying a parameter. The answer is no: At least *two* parameters of the Hamiltonian must be varied in order to reach a degeneracy. The reason, which is simple, comes from perturbation theory: In order for a degeneracy to occur, two conditions must be satisfied, which is always possible when two parameters can be varied, but which can occur only accidentally if only one parameter can be varied. Consider a two parameter Hamiltonian, $H = H(\tilde{q},\tilde{p}; A,B)$. The energy levels $E = E(A,B)$ can be represented in the three-dimensional space (E,A,B). Suppose that at $A = A^*$, $B = B^*$ there exists a degeneracy $E_1 = E_2 = E^*$ of the levels $E_1 = E_1(A,B)$ and $E_2 = E_2(A,B)$. In a neighborhood of $(E^*,A^*,B^*) = P^*$, the level surfaces E_1 and E_2 are two cones, as will be proved shortly. When one parameter is varied (i.e., when we move along a curve in the parameter space), it is clear from the geometry (cf. Fig. 22) that the degeneracy point P^* will be missed with probability one, although a curve can pass very close to it. In this case, the levels along the curve display a repulsion, as shown in Fig. 22.

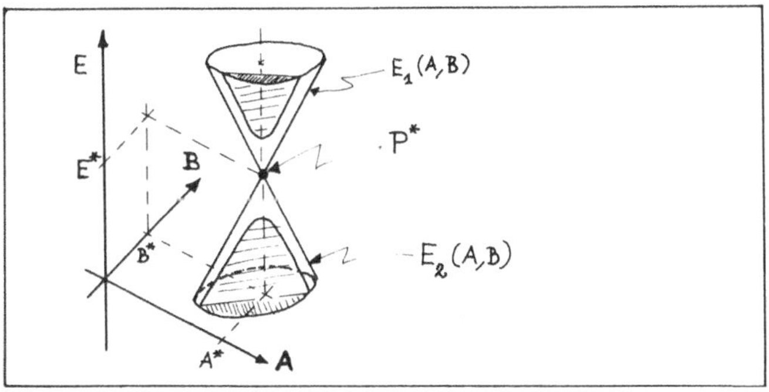

Fig. 22. The double cone of energy levels near a degeneracy point P* = (E*, A*, B*) for a Hamiltonian depending on two parameters A and B.

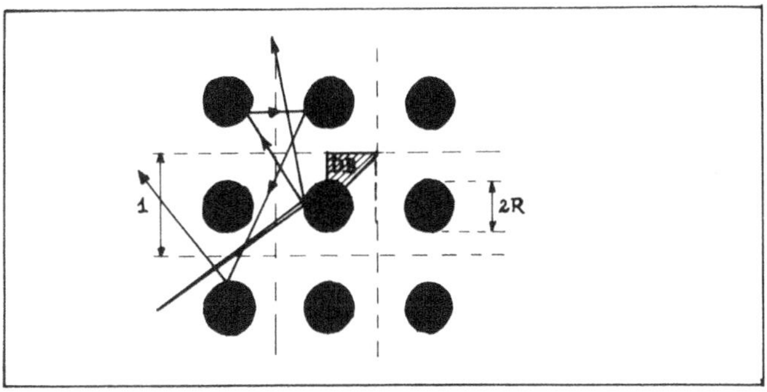

Fig. 23. The Sinai billiard: rigid disks of radius R placed at the sites of a square lattice with lattice constant equal to one. Two nearby trajectories of a point particle are shown. The problem is to find the eigenfunctions (with periodic boundary conditions) on a unit cell. The desymmetrized billiard (DB) is the shaded area.

The spectrum of a one-parameter family of ergodic systems should therefore possess no degeneracies, but should exhibit avoided crossings instead. Indeed, Berry [61] has quantized the Sinai billiard (cf. Fig. 23), known to be classically ergodic, by using the Korringa-Kohn-Rostoker (KKR) determinant of solid-state physics. The "exactly" calculated spectrum of the desymmetrized billiard, as a function of the disk radius R, shows no degeneracies (cf. Fig. 24).

However, some avoided crossing between levels are so close that it is almost impossible to decide whether a repulsion or a crossing occurs. There is a nice theorem, given by Arnold [23] and discussed by Berry [61], which treats this difficulty: The signs of the wavefunctions ψ_1 and ψ_2, which correspond to the upper and lower cones E_1 and E_2, respectively, change over a closed path C in parameter space if C encircles the degeneracy point (A^*,B^*), and do not change otherwise. Thus, we have the surprising result that the single-valuedness of wavefunctions, with respect to coordinates, does *not* imply that they are single-valued functions of the parameters. In fact, they are two sheeted.

We now prove this "small" theorem. Let u and v be two degenerate eigenfunctions of $H^* = H(A^*,B^*)$:

$$H^*u = E^*u \quad , \quad H^*v = E^*v \quad . \tag{62}$$

For convenience, we put $A^* = B^* = 0$. Then, the Hamiltonian $H(A,B)$ near the degeneracy point P^* is given by

$$H(A,B) = H^* + H_1(A,B) \quad , \tag{63}$$

where the perturbation term is a linear function of A and B, namely,

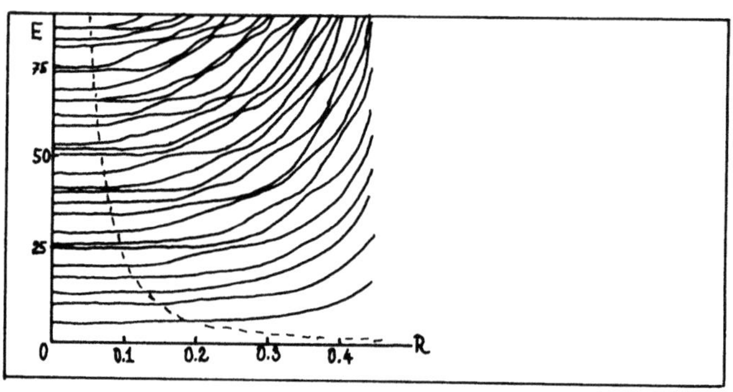

Fig. 24. Energy levels as a function of the disk radius R for the Sinai billiard (cf. Fig. 23). The dashed curve is the boundary of the region (small energies E and small radius R) where perturbation theory is applicable. (Adapted from [61].)

$$H_1(A,B) = \left.\frac{\partial H}{\partial A}\right|_{p*} A + \left.\frac{\partial H}{\partial B}\right|_{p*} B =: h_1 A + h_2 B \quad . \tag{64}$$

Let us write ψ as a linear combination of u and v,

$$\psi = u \cos \chi - v \sin \chi \quad , \tag{65}$$

where the parameter χ depends upon A and B, and is determined by degenerate perturbation theory. Solving the secular equation yields

$$\Delta E = E(A,B) - E^* = \tfrac{1}{2}\{\langle u|H_1|u\rangle + \langle v|H_1|v\rangle \pm$$
$$\pm[(\langle u|H_1|u\rangle - \langle v|H_1|v\rangle)^2 + 4(\langle u|H_1|v\rangle)^2]^{\frac{1}{2}}\} \quad . \tag{66}$$

(Here, we see that in order to produce a degeneracy, two conditions must be satisfied, namely $\langle u|H_1|u\rangle = \langle v|H_1|v\rangle$ and $\langle u|H_1|v\rangle = 0$, as was claimed above.) But according to Eq. (64), all matrix elements in Eq. (66) are linear combinations of A and B. Therefore, $\Delta E = \Delta E(A,B)$ is indeed the equation of a double cone. The value of the parameter χ is given by

$$\tan 2\chi = \frac{2\langle u|H_1|v\rangle}{\langle v|H_1|v\rangle - \langle u|H_1|u\rangle} \quad , \tag{67}$$

so that for each A and B there are *two* solutions for χ, which differ by $\pi/2$.

Let us define

$$A = r \cos \phi \quad , \quad B = r \sin \phi \tag{68}$$

and investigate how ψ changes on a circle of constant r, the path being parametrized by ϕ. From Eq. (64) and Eqs. (67,68), we find

$$\tan 2\chi(\phi) = \frac{2(\langle u|h_1|v\rangle + \tan \phi \, \langle u|h_2|v\rangle)}{\langle v|h_1|v\rangle - \langle u|h_1|u\rangle + \tan \phi(\langle v|h_2|v\rangle - \langle u|h_2|u\rangle)} \tag{69}$$

which is independent of the radius. If ϕ changes by π (half-circuit), χ changes by $\pi/2$ and the wavefunctions ψ_1 and ψ_2 are interchanged. During a complete cycle C, ϕ changes by 2π and χ by π; thus, the eigenfunction of Eq. (65) changes sign, as asserted.

After this brief discussion of spectral fine-structure, we now consider the distribution of levels. Let P(S)dS be the probability that the separation of two nearby levels is between S and

S + dS. [S is measured in units of the mean level spacing $1/\bar{d}(E)$.] The behavior of P(S) for small S (i.e., $S \to 0$) is related to the existence of level repulsions. On this smallest scale, the energy levels behave as described by the sections of the level cone in the two-parameter space (A,B). Moreover, for a chaotic system the double cones will be randomly and uniformly distributed in (E,A,B), and P(S) will then be given by the probability of cutting a cone near a degeneracy point, with $E_1 - E_2 = S$. This is a problem in geometric probability theory, and the result is

$$P(S) = \text{constant} \cdot S \quad , \tag{70}$$

as $S \to 0$. The probability of degeneracy (S = 0) vanishes since avoided crossings are typical of chaotic systems. For large S, the dependence is Gaussian [63]:

$$P(S) = \text{constant} \cdot \exp\,(-\ \text{constant} \cdot S^2) \quad . \tag{71}$$

In Fig. 25, the distribution of level spacings S between the eigenvalues for the Sinai billiard, for $0.20 \leq R \leq 0.44$, is shown. This distribution seems to obey the limiting laws of Eqs. (70, 71). Similar distributions can be observed in Figs. (17,18). However, for a precise test of theoretical predictions, more numerical or experimental data are needed.

For an integrable system with invariant tori, the energy levels are Poisson-distributed [64],

$$P(S) \propto e^{-S} \quad , \tag{72}$$

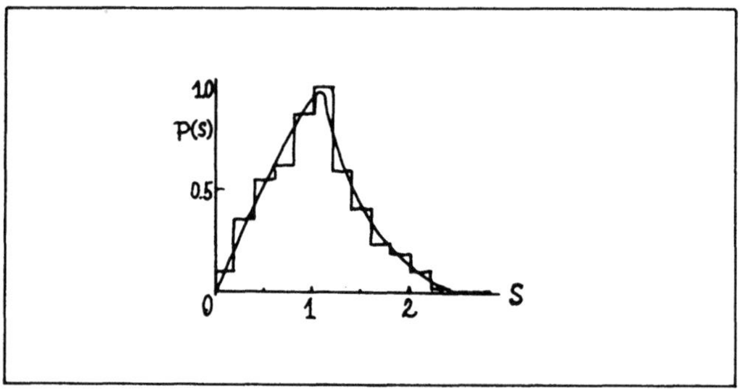

Fig. 25. The level-spacings distribution P(S) for the energy levels of the Sinai billiard. S is measured in units of the mean level separation.

so that $P(S) \to$ constant as $S \to 0$, which shows that degeneracies are allowed for an integrable system and, *a fortiori*, for a separable one.

We see, then, that the consequence of tori destruction is the drastic change from the level satistics of Eq. (72) to the distribution laws given by Eqs. (70,71). The analysis of the level-spacings distribution is thus a sensitive test of whether a system is regular or chaotic. The converse is useful too: Knowing the properties of the corresponding classical system enables one to predict P(S).

Zaslavsky [4,57] has derived a different law for chaotic systems, namely,

$$P(S) \propto S^{\text{constant}/h(E)} \quad , \tag{73}$$

as $S \to 0$, where h(E) is the K-entropy around the levels considered. Because h(E) generically increases with increasing energy, $[\text{constant}/h(E)] - 1$ may become negative, thus predicting a singularity in the derivative of P(S) [i.e., $P'(S) \to \infty$ as $S \to 0$]. This method is based on statistical considerations of almost-closed orbits. However, because of the crudeness of numerically calculated histograms, it is difficult to distinguish between Eq. (70) and Eq. (73).

12. QUANTIZING A GENERIC SYSTEM

As explained in Section 6, a generic smooth Hamiltonian of the type given in Eq. (24) will display a stochastic transition at the critical energy $E = E_{crit}$. However, for $E < E_{crit}$ the energy surface in phase space is densely filled with invariant tori. These tori can be constructed in an approximate way by applying the Birkhoff-Gustavson normal form (cf. Section 7), and then quantized according to the results of Section 9. Examples of such calculations in nuclear, atomic and molecular physics can be found in [17], [14] and [65], respectively. These calculations usually are extremely accurate, with typical errors being 0.1% or even 0.01%. Indeed, one cannot ignore the empirical fact that there appears to be an integrable system behind the spectrum of every Hamiltonian [67], even a chaotic one. This seems paradoxical, but we shall explain below that this need not contradict the results of the preceding section, where we claimed that the level statistics changes drastically when an integrable system becomes chaotic.

We first note that "we cannot hear the shape of a drum" [4, 57]. In other words, it is impossible to reconstruct the form of a resonator when only the spectrum is known. In one dimension, we have the analogy that the discrete spectrum is not sufficient for a determination of the potential of a Hamiltonian. It is a result of inverse scattering theory [66] that this inverse problem has a unique solution only if the scattering amplitudes are known too. Thus, the same bound spectrum can be generated by any system belonging to a rather large family. Similar results are assumed to hold in multi-dimensional cases. We are left, then, with the following question: Does an integrable system exist in such an equivalence class of Hamiltonians? I expect that the answer is yes. Moreover, such a system is given to the required accuracy by a truncated Birkhoff-Gustavson normal form of sufficiently high degree. A full account of this is given elsewhere [39], and we shall present only an example.

Note that the normal form of a nonresonant Hamiltonian is a type of decomposition into harmonic oscillators [cf. Eq. (30)]. The spectra of these oscillators are known exactly, namely $I_i = (n_i + 1/2)\hbar$. And, the spectrum of the normal form $\Gamma^{(s)}$ is simply a function of the harmonic spectra. In the example provided by Eqs. (31,32), there are then two quantum numbers, $n_1, n_2 = 0,1,2,\ldots$, and

$$E^{(4)}_{n_1n_2} = (n_1+\tfrac{1}{2})\hbar\omega_1 + (n_2+\tfrac{1}{2})\hbar\omega_2 + \hbar^2\left\{(n_1+\tfrac{1}{2})(n_2+\tfrac{1}{2})\left(b + \frac{2a^2\omega_2}{\omega_1^2 - 4\omega_2^2}\right) - (n+\tfrac{1}{2})^2\,\frac{a^2}{4\omega_1}\,\frac{3\omega_1^2 - 8\omega_2^2}{\omega_1^2 - 4\omega_2^2}\right\} + \text{h.o.t.} \tag{74}$$

This is actually very similar to tori quantization, although there may be some differences [39]. But now comes a most surprising result: The purely quantum-mechanical perturbation series up to second order, starting with the harmonic part of the Hamiltonian as the unperturbed system, yields precisely the expansion of Eq. (74), apart from an additive constant.

Let $\psi_{n_1n_2}$ be the eigenfunctions of the unperturbed (harmonic) operator H_o and let

$$H_1 = axy^2 + bx^2y^2 \tag{75}$$

be the perturbation. We denote the eigenvalues of H_o by $E^{(0)}_{n_1n_2}$. Then, the energy eigenvalues of the perturbed system $H = H_o + H_1$, up to second order, are given by

$$E_{n_1n_2} = E^{(0)}_{n_1n_2} + \langle n_1n_2|bx^2y^2|n_1n_2\rangle + \sum_{\substack{n_1'n_2'=0\\(n_1',n_2')\neq(n_1,n_2)}}^{\infty} \frac{|\langle n_1n_2|axy^2|n_1'n_2'\rangle|^2}{E^{(0)}_{n_1n_2} - E^{(0)}_{n_1'n_2'}} + \text{h.o.t.} \tag{76}$$

The calculation is straightforward, but lengthy, and the details are given in [39]. The result is identical to Eq. (72), apart from the additive constant

$$E_{n_1n_2}(\text{Birkhoff-Gustavson}) - E_{n_1n_2}(\text{perturbation})$$
$$= \frac{3a^2}{16}\,\frac{\omega_1}{\omega_1^2 - 4\omega_2^2}$$

which is indeed independent of the quantum numbers n_1, n_2.

A generalization of the hypothesis discussed above may be formulated as follows: The operators corresponding to the terms of the Birkhoff-Gustavson normal form are diagonal in the basis of harmonic functions; the spectrum of the original system is the same as that of the Birkhoff-Gustavson normal form, except for an additive constant.

This has important technical implications since the classical normal form can be calculated much more easily than can the perturbation series, which involves infinite matrices. Moreover, using the REDUCE programming language, this calculation need by performed only once for all times [38].

Now, recall that the Birkhoff-Gustavson normal form almost always diverges, which is connected to the destruction of integrals of the motion and of invariant tori. On the other hand, the quantum-mechanical perturbation series diverges too, because the operator H_1 is unbounded. This quantum divergence is related to the instability of levels: Below the critical energy E_{crit} the divergence is slow, resulting in stable levels; however, the divergence is expected to become very strong for $E \geq E_{crit}$, thus yielding unstable levels.

One can imagine the enormous advantage of such an approach by noting that the vibrational spectrum of a complicated molecule could then be obtained from explicit formulae for the spectrum as a function of the potential.

13. CONCLUSIONS AND DISCUSSION

Although progress has been made in our understanding of chaotic behavior in quantum systems, it is at present still unclear what the quantum mechanical analogy of the KAM Theorem is. In a recent work, Korsch [68] has shown that classical and quantum mechanical integrability are in precise correspondence (at least for sufficiently smooth systems): If there are N classical integrals of the motion in involution, then there are N commuting observables of the quantum system, and vice versa. To show this he uses the Weyl mapping between the classical phase-space functions and the quantum operator algebra. However, while local integrability is well-defined in classical mechanics -- when a perturbation is applied most tori still exist -- it has no meaning in quantum mechanics. Therefore, under a small perturbation all quantum integrals of the motion (observables) will be destroyed, except for the Hamiltonian itself. Therefore, this cannot be a way towards a quantum KAM Theorem. However, as indicated in Section 8 and explained in Section 12, a quantum analogy has been obtained only insofar as stability and instability of levels correspond to classical regular and chaotic regions in phase space, respectively. This is, in fact, the suggestion made by Percival [45]. But, a quantum analogy of the KAM Theorem must be formulated in a purely quantum language, with no reference to classical mechanics except where rigorous relations exist. This may be possible by studying the properties of quantum perturbation series [39] and the general problem of spectral stability [47].

ACKNOWLEDGMENT

The author thanks Sonderforschungsbereich Radioastronomie (SFB 131) for financial support. Sincere thanks for stimulating discussions are due to H. J. Korsch, W. P. Reinhardt and, particularly, to E. Schrüfer for his kind support and many helpful comments. Finally, I should like to thank G. M. Zaslavsky for comments, and M. Nepveu for his interest in this work and for "linguistics". This work was not supported by any military agency.

REFERENCES

1. R. H. G. Helleman, in: *Fundamental Problems in Statistical Mechanics*, Vol. 5, E. G. D. Cohen, ed. (North-Holland, Amsterdam, 1980).
2. B. V. Chirikov, Phys. Rep. 52, 263 (1979).
3. M. V. Berry, in: *Proceedings of the July 1981 'Les Houches' Summer School on Chaotic Behaviour of Deterministic Systems*, R. H. G. Helleman and G. Iooss, eds. (North-Holland, Amsterdam, 1982).
4. G. M. Zaslavsky, Phys. Rep. 80, 158 (1981).
5. L. J. Lebowitz and O. Penrose, Physics Today, 23 (Feb., 1973).
6. M. I. Rabinovich, Sov. Phys. Usp. 21, 443 (1978).
7. A. Libchaber and J. Maurer, J. Physique Lett. 39, L-369 (1978).
8. For applications to magnetic confinement in fusion and plasma physics see: *Intrinsic Stochasticity in Plasmas*, G. Laval and D. Gresillon, eds. (Les Editions de Physique Publ., Orsay, 1980).

9. F. Doveil, Phys. Rev. Lett. 46, 532 (1981).
10. J. B. Delos and R. T. Swimm, Chem. Phys. Lett. 47, 76 (1977).
11. R. A. Marcus, Ann. N.Y. Acad. Sci. 357, 169 (1980).
12. M. Robnik, J. Phys. A 14, 3195 (1981).
13. M. Robnik, in: *Proceedings of the Colloque International CNRS on Atomic and Molecular Physics near Ionisation Thresholds* (Aussois, France, 1982), J. de Physique Colloque C2, Supplement 11 (1982).
14. W. P. Reinhardt and D. Farrelly, in: *Proceedings of the Colloque International CNRS on Atomic and Molecular Physics near Ionization Thresholds* (Aussois, France, 1982); J. de Physique Colloque C2, Supplement 11 (1982).
15. A. R. Edmonds and R. A. Pullen, Preprints Imperial College, London (1980).
16. D. Richards, J. Phys. B., submitted.
17. R. D. Williams and S. E. Koonin, "Semiclassical Quantization of the Shell Model," Preprint Caltech (1982).
18. H. A. Weidenmüller, Phys. Blätter 38, 176 (1982).
19. See the contributions in: *Nonlinear Dynamics and the Beam-Beam Interaction*, M. Month and J. C. Herrera, eds. (AIP Conf. Proc. 57, 1979).
20. *Life in the Universe*, J. Billingham, ed. (MIT Press, Cambridge, 1981).
21. H. Haken, *Synergetics* (Springer, Berlin, 1978).
22. M. A. Olshanetsky and A. M. Perelomov, Phys. Rep. 71, 313 (1981).
23. V. I. Arnold, *Mathematical Methods of Classical Mechanics* (Springer, New York, 1980).
24. H. Goldstein, *Classical Mechanics* (Addison-Wesley, Reading, (1950).
25. For the moment "ergodic" means just that there are no integrals of the motion except for the energy. Later, we give the precise definitions.
26. R. J. Whiteman, Rep. Prog. Phys. 40, 1033 (1977).
27. V. I. Arnold and A. Avez, *Ergodic Problems of Classical Mechanics* (Benjamin, New York, 1968).
28. M. V. Berry, in: AIP Conf. Proc. 46, S. Jorna, ed. (AIP, New York, 1978).
29. M. Henon and C. Heiles, Astron. J. 69, 73 (1964).
30. V. I. Arnold, *Ordinary Differential Equations* (MIT Press, Cambridge, 1980).
31. J. M. Greene, R. S. MacKay, F. Vivaldi and M. J. Feigenbaum, Physica 3D, 468 (1981).
32. J. Guckenheimer, Lect. in Appl. Math. 17, 187 (1979).
33. C. L. Siegel and J. K. Moser, *Lectures on Celestial Mechanics* (Springer, New York, 1971).
34. J. K. Moser, *Stable and Random Motions in Dynamical Systems* (Princeton Univ. Press, Princeton, 1973).

35. Ya. G. Sinai, *Introduction to Ergodic Theory* (Princeton Univ. Press, Princeton, 1976).
36. M. C. Gutzwiller, Phys. Rev. Lett. 45, 150 (1980).
37. F. G. Gustavson, Astron. J. 71, 670 (1966).
38. E. Schrüfer and M. Robnik, in preparation.
39. M. Robnik, Preprint Bonn (1982).
40. D. F. Escande and F. Doveil, J. Stat. Phys. 26, 257 (1981).
41. J. M. Greene, J. Math. Phys. 20, 1183 (1979).
42. R. C. Churchill, G. Pecelli and D. L. Rod, in: Lecture Notes in Physics 93 (Springer, Berlin, 1978).
43. P. Collet and J. -P. Eckmann, *Iterated Maps on the Interval as Dynamical Systems* (Birkhäuser, Basel, 1980).
44. B. V. Chirikov, F. M. Izrailev and D. L. Shepelyansky, "Transient Stochasticity in Quantum Mechanics," Preprint 80-210 Novosibirsk (1980).
45. I. C. Percival, J. Phys. B 6, L229 (1973).
46. N. Pomphrey, J. Phys. B 7, 1909 (1974).
47. E. Vock and W. Hunziker, "Stability of Schrödinger Eigenvalue Problems," Preprint Zürich (1982).
48. D. W. Noid, M. L. Koszykowski and R. A. Marcus, J. Chem. Phys. 67, 404 (1977).
49. See [11], and references therein.
50. R. Ramaswamy and R. A. Marcus, J. Chem. Phys. 74, 1379 (1981); J. Chem. Phys. 74, 1385 (1981).
51. A. Einstein, Verh. d. Deutsch. Phys. Ges. 19, 82 (1917).
52. V. P. Maslov and M. V. Fedoryuk, *Quasiclassical Approximation for Equations of Quantum Mechanics* (Nauka, Moscow, 1976); in Russian.
53. The rule is that the phase of a wave decreases by $\pi/2$ upon crossing a caustic with the classical allowed region on the right; similarly, it increases by $\pi/2$ in the case of opposite orientation. For an irreducible cycle, the orientation is the same for all caustics encountered during such a closed path.
54. M. V. Berry, Preprint Utrecht (1981).
55. M. C. Gutzwiller, J. Math. Phys. 12, 343 (1971); in: *Path Integrals*, J. T. Devreese, ed. (Plenum, New York, 1978), p. 163.
56. R. Balian and C. Bloch, Ann. Phys. (N.Y.) 69, 76 (1972).
57. G. M. Zaslavsky, Sov. Phys. Usp. 22, 788 (1979).
58. R. P. Feynman and A. R. Hibbs, *Quantum Mechanics and Path Integrals* (McGraw-Hill, New York, 1965).
59. M. V. Berry and M. Tabor, Proc. Roy. Soc. London A 349, 101 (1976).
60. C. DeWitt-Morette, A. Maheshwari and B. Nelson, Phys. Rep. 50, 255 (1979).
61. M. V. Berry, Ann. Phys. (N.Y.) 131, 163 (1981).
62. J. von Neumann and E. Wigner, Phys. Z. 30, 467 (1929).
63. G. M. Zaslavsky and N. N. Filonenko, Sov. Phys. JETP 38, 317 (1974).

64. M. V. Berry and M. Tabor, Proc. Roy. Soc. London A 356, 375 (1977).
65. R. T. Swimm and J. B. Delos, J. Chem. Phys. 71, 1706 (1979).
66. I. M. Gelfand and B. M. Levitan, Amer. Math. Soc. Transl. 1, 253 (1955); F. Calogero and A. Degasperis, Preprint Rome (1978).
67. W. P. Reinhardt, private discussion; I should like to thank him for bringing this fact to my attention.
68. H. J. Korsch, Phys. Lett. 90A, 113 (1982).

NEW TRENDS IN ATOMIC DIAMAGNETISM

J. C. Gay

Laboratoire de Spectroscopie Hertzienne*
L'Ecole Normale Superieure
4 Place Jussieu
Paris 75230 Cedex 05, France

1. INTRODUCTION

The problem of an atom in a magnetic field is concerned fundamentally with diamagnetism and, basically, with what happens to an electron when subjected to the joint action of the Coulomb and diamagnetic forces. Overall, this remains an unsolved problem; however, recent experimental and theoretical advances in atomic physics have renewed interest in the phenomenon.

Interest in this problem arises, first of all, from its fundamental simplicity, since it serves as a prototype for a wide variety of questions. It is a nonseparable problem in R(3), with two forces of different symmetries and of comparable strengths acting on the electron. Major questions are then those of the stability of the classical trajectories, of convenient semi-classical quantization and of the existence of adiabatic invariants. The conceptual importance of the problem follows from the fact that the two limiting cases -- that is, the Coulomb and Landau (oscillator-like) problems, associated with the early Bohr-Sommerfeld and Landau quantizations, respectively -- are the only two exactly separable [in R(3)] systems known in classical or quantum physics.

Second, all of the questions connected with atomic diamagnetism have wide applications in various domains of physics such as, for example, in solid-state physics of excitonic systems [1,2], where studies were developed long ago, although without any success so far as a general solution of the problem is concerned.

*Associated with C.N.R.S. (LA18)

S. P. McGlynn et al. (eds.), Photophysics and Photochemistry in the Vacuum Ultraviolet, 631–705.

Astrophysical implications [3] are also of importance, but the range of field strengths relevant to pulsars and white dwarfs is quite different from that encountered in solid-state physics and far outside practical limitations under laboratory conditions. Plasma physics applications may also turn out to be of extreme importance, although they have not yet been developed: Firstly, for tokamaks, since the magnetic-field action completely modifies basic processes such as electronic recombination, ionization and the energy balance; second, it is now well established that self-generation of ultra-intense magnetic fields occurs due to various kinds of magnetohydrodynamical processes in the plasma with, as a consequence, a complete alteration of the local atomic properties of the surroundings.

Historically, the study of the interaction between atoms (or atomic vapors) and a magnetic field has been a source of constant interest since 1850. This was long before the advent of quantum mechanics, which happened rather in consequence of this early interest in the field actions. Studies by investigators from Faraday to Lorentz and Zeeman permitted a "settling up" of the theory. However, these early works were concerned with only one aspect of the magnetic interaction in atoms, namely paramagnetism, which is indeed the dominant feature for low-lying atomic states. The second aspect of the magnetic interaction, namely that associated with diamagnetism and with Faraday's law, was not obviously revealed in the atomic spectra since it is usually a small perturbation for low-lying atomic states.

The importance of the diamagnetic interaction for weakly-bound electronic systems was first established by Landau [4], who derived the quantum mechanical spectrum of "free" particles in a magnetic field. This spectrum is basically oscillator like. The first experimental attempt directed toward the effects of diamagnetism in atomic spectra was performed by Jenkins and Segre, in 1939, on Rydberg series of sodium atoms [5]; they studied the region where the diamagnetic interaction is still small in comparison to the Coulomb binding energy. On the other hand, the existence of a Landau regime in the exciton spectrum, which is nearly Coulombic in zero-field, was demonstrated in the 1950's. But, the story of diamagnetism really began in 1969: Garton and Tomkins [6] first demonstrated for barium the existence near threshold of a new signature in the atomic spectrum, later called the "strong field mixing regime" by A. R. P. Rau [7], which was neither the Coulombic nor the Landau regime. Basically, this new regime was characterized by the existence of broad modulations, with a spacing of about 1.5 $\hbar\omega_c$, which extend far beyond the ionization limit into the continuum.

What appears to be extremely important in the above result is the evidence that an atomic spectrum still presents some organization in a region where the Coulomb and magnetic forces acting on the electron are of the same magnitude. Since the symmetry properties of these forces differ deeply, intuitive arguments would be in favor of a featureless, but certainly complicated, spectrum. But, *for all atoms* this turns out to be far from the truth: Since 1969, experiments have provided striking evidence that, for all atoms, the general behavior is the same, which means that the intermediate regime at threshold is as basic as the Coulomb and Landau regimes.

In Sections 2 and 3, we recall the classical and quantum properties of the Coulomb and Landau problems, respectively, which will then be used in the following sections. General considerations of an atom in a magnetic field follow in Section 4. Next, we come back to a classical analysis of the phenomenon (Section 5), which is presently a source of numerous discoveries and advances; the nature of classical trajectories is discussed and consequences are drawn from the discovery of an adiabatic invariant in the Coulomb limit. Section 6 is devoted to the quantum approach to the spectrum in the Coulomb limit. Then, in the absence of any other approach, semi-classical methods for dealing with the strong mixing and Landau regimes are discussed (Section 7). In Section 8, some important aspects of experiments are stressed, aspects which have enabled recent progress in our understanding of the problem. (This seems the rule since all of the basic spectral features have been discovered experimentally before being understood theoretically!) Finally, we conclude in Section 9 with a few prospects, some of them already under experimental realization.

2. THE COULOMB SPECTRUM

In this Section, we recall some well-known results on the pure Coulomb spectrum of atomic hydrogen. These results, obviously, have wide generality whatever the atomic species. But, in addition, purely Coulombic potentials lead to fundamental degeneracies in the atomic spectrum, which are associated with the underlying group structure [8,9].

2.1 Classical Mechanics of the Coulomb Problem

The expression of the Coulomb potential is

$$V_C = - q^2/4\pi\varepsilon_o r = - e^2/r \tag{1}$$

and the equation of motion for the electron in a Coulomb field is

$$m\ddot{\underset{\sim}{r}} = m\underset{\sim}{\gamma} = \underset{\sim}{F}_C = - e^2\underset{\sim}{r}/r^3 \tag{2}$$

The Hamiltonian for the problem is readily written as

$$H = \frac{p^2}{2m} + V_C(r) \quad , \tag{3}$$

while the angular momentum $\underset{\sim}{L}$ is defined by

$$\underset{\sim}{L} = \underset{\sim}{r} \times \underset{\sim}{p} \quad . \tag{4}$$

It is straightforward to verify that H and $\underset{\sim}{L}$ are constants of the motion which, for $\underset{\sim}{L}$, means that

$$\frac{d}{dt} \underset{\sim}{L} = \underset{\sim}{0} \quad . \tag{5}$$

Eq. (5) expresses nothing but the conservation of $\underset{\sim}{L}$, for any motion in a central potential, due to the rotational invariance of the system around any axis in space. In addition to this well-known result, however, there is another constant of the motion which is *specific* to the Coulomb potential and is called the *Laplace-Runge-Lenz Vector* $\underset{\sim}{A}$ [10]:

$$\underset{\sim}{A} = \underset{\sim}{p} \times \underset{\sim}{L} - me^2 \frac{\underset{\sim}{r}}{r} \quad . \tag{6}$$

Using Eq. (2), it is straightforward to show that

$$\frac{d}{dt} \underset{\sim}{A} = \underset{\sim}{0} \quad . \tag{7}$$

Obviously, $\underset{\sim}{L}$ and $\underset{\sim}{A}$ fulfill the condition

$$\underset{\sim}{L} \cdot \underset{\sim}{A} = 0 \tag{8}$$

while the energy can be expressed as a function of L^2 and A^2:

$$A^2 = 2mHL^2 + m^2e^4 = 2mEL^2 + m^2e^4 \quad . \tag{9}$$

The trajectories are easily deduced from the scalar product $\underset{\sim}{r} \cdot \underset{\sim}{A}$; one finds

$$\underset{\sim}{r} \cdot \underset{\sim}{A} = L^2 - me^2 r \quad .$$

Hence,

$$\frac{1}{r} = \frac{me^2}{L^2} \left(1 + \frac{A}{me^2} \cos\phi\right) \quad , \tag{10}$$

where ϕ is the angle between $\underset{\sim}{r}$ and $\underset{\sim}{A}$. These trajectories are the well-known *Kepler* ellipses, shown in Fig. 1: The Runge-Lenz vector is directed along the major axis of the ellipse; the focus is at the nucleus; and the excentricity of the ellipse is given by

$$\xi = A/me^2 \quad . \tag{11}$$

From Eq. (9), it is clear that the classical motion is highly degenerate since the trajectory and its energy are completely defined once $\underset{\sim}{L}$ and $\underset{\sim}{A}$ are given. The existence of the extra constant of the motion $\underset{\sim}{A}$ is responsible for this feature and for the *closed character* of the trajectories.

2.2 The Coulomb Quantum Spectrum

The Hamiltonian is given by Eq. (3). We take advantage of the existence of the two constants of the motion for the classical problem in deriving the Coulomb quantum spectrum. We first introduce the following quantum expression of the Runge-Lenz vector:

$$\underset{\sim}{A} = \frac{1}{2}\,(\underset{\sim}{p} \times \underset{\sim}{L} - \underset{\sim}{L} \times \underset{\sim}{p}) - me^2\,\frac{\underset{\sim}{r}}{r} \quad . \tag{12}$$

The commutation relations for H, $\underset{\sim}{L}$ and $\underset{\sim}{A}$ are as follows [8,11]:

$$[H,L_i] = [H,A_i] = 0$$

$$[L_j,L_k] = i\hbar\,\varepsilon_{jk\ell}\,L_\ell \;;\; [A_j,A_k] = -\,i\hbar\,\varepsilon_{jk\ell}\,L_\ell(2mH) \tag{13}$$

$$[L_j,A_k] = i\hbar\,\varepsilon_{jk\ell}\,A_\ell$$

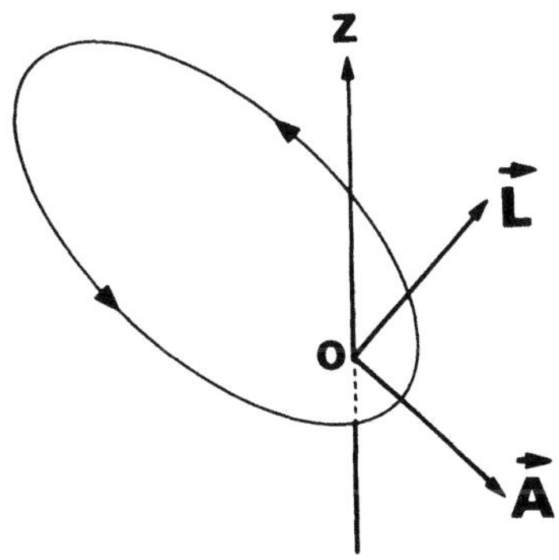

Fig. 1. Zero-field Coulomb ellipse ($\underset{\sim}{L}$ ≡ angular momentum; $\underset{\sim}{A}$ ≡ Runge-Lenz vector).

with

$$\underset{\sim}{L} \cdot \underset{\sim}{A} = \underset{\sim}{A} \cdot \underset{\sim}{L} = 0 \quad \text{and} \quad A^2 = 2mH(L^2 + \hbar^2) + m^2e^4 \quad . \tag{14}$$

Since $\underset{\sim}{L}$ and $\underset{\sim}{A}$ commute with H, Eq. (13) suggests consideration of a subspace belonging to the eigenvalue E (E < 0). In this subspace, we introduce the operator $\underset{\sim}{a}$:

$$\underset{\sim}{a} = (-\ 2mH)^{-1/2}\ \underset{\sim}{A} \quad . \tag{15}$$

From the new sets of commutation relations, it becomes clear that $\underset{\sim}{J}_+ = \frac{1}{2}(\underset{\sim}{L} + \underset{\sim}{a})$ and $\underset{\sim}{J}_- = \frac{1}{2}(\underset{\sim}{L} - \underset{\sim}{a})$ build up two commuting sets of operators, each one satisfying the commutation relations of ordinary angular momentum; hence, their eigenvalues are readily derived:

$$J_+^2 = \hbar^2 j_+(j_+ + 1) \quad \text{and} \quad J_-^2 = \hbar^2 j_-(j_- + 1) \quad . \tag{16}$$

According to Eq. (14), $\underset{\sim}{L}\cdot\underset{\sim}{A} = \underset{\sim}{A}\cdot\underset{\sim}{L} = 0$, so that $j_+ = j_- = j$. Then, from Eq. (14) it is straightforward to show that

$$a^2 + L^2 + \hbar^2 = (2j + 1)^2\hbar^2 = -\ me^4/2E \quad . \tag{17}$$

Writing $n = 2j + 1$, one recovers the usual Coulomb spectrum since j is allowed to take on every half-integer value:

$$E = -\ me^4/2\hbar^2\ (2j + 1)^2 = -\ me^4/2\hbar^2 n^2 \quad . \tag{18}$$

The possible values of L for each n follow from the definitions of $\underset{\sim}{J}_+$ and $\underset{\sim}{J}_-$: These values are $\ell = 0,1,2,\ldots,2j$. Thus, the degeneracy of each n level is

$$\sum_{\ell=0}^{2j} (2\ell + 1) = (2j + 1)^2 = n^2 \tag{19}$$

2.3 Basis Set in Spherical Coordinates

Due to the fundamental degeneracy of the problem or, equivalently, due to the existence of several independent sets of operators commuting with the Hamiltonian, there are several possible choices of a basis of eigenvectors.

The first, well-known basis is associated with the eigenvectors of $\{H,L^2,L_z\}$. The angular part of the wavefunction is then the usual spherical harmonics $Y_\ell^M(\theta,\phi)$, while the solution of the radial equation is a Laguerre polynomial. One finds (using atomic units) [11-13]

$$<r\theta\phi|n\ell M> = R_{n\ell}(r)\, Y_{\ell M}(\theta,\phi)$$

with

$$R_{n\ell}(r) = \frac{2}{n^{\ell+2}(2\ell+1)!}\left[\frac{(n+1)!}{(n-\ell-1)!}\right]^{1/2}(2r)^{\ell} \cdot e^{-r/n}\, F(-n+\ell+1,\ 2\ell+2,\ 2r/n) \ . \tag{20}$$

In Eq. (20), F is the confluent hypergeometric function and

$$n = n_r + \ell + 1 \tag{21}$$

where n_r is the radial Coulomb quantum number.

2.4 Basis Set in Parabolic Coordinates

The set of wavefunctions which are eigenstates of $\{H, A_z, L_z\}$ is also of importance. This set is associated with the separability of the Schrödinger equation in parabolic coordinates which are defined by

$$\xi = r + z \quad \text{and} \quad \eta = r - z \ . \tag{22}$$

These coordinates turn out to be extremely useful for studying the Stark effect in hydrogenic species [13]. The expression for the wavefunction (in atomic units) is

$$<\xi\eta\phi|n_1 n_2 M> = (\pi^{1/2} n^2)^{-1}\, e^{iM\phi}\, F_{n_1 M}(\xi/n)\, F_{n_2 M}(\eta/n)$$

where

$$F_{pM}(x) = \frac{1}{|M|!}\left[\frac{(p+|M|)!}{p!}\right]^{1/2} F(-p,\ |M|+1,\ x)\, e^{-x/2} x^{|M|/2} \tag{23}$$

with

$$n = n_1 + n_2 + |M| + 1 \ . \tag{24}$$

In this basis, the z component of the Runge-Lenz vector is diagonal:

$$A_z = (n_2 - n_1)/n \tag{25}$$

while L^2 and parity are no longer defined since they do not commute with A_z.

2.5 Group Theoretical Meaning

L^2 and L_z are constants of the motion; this allows us to infer that the Coulomb problem possesses O(3) symmetry in coordinate space. L_z and A_z are also constants of the motion, which means that the problem has another O(3) symmetry independent of the previous one. The group O(3) ⊗ O(3) is O(4). And, this O(4) symmetry explains the specific degeneracy of the Coulomb problem.

The O(4) symmetry is much more evident using the Fock method [9] in momentum space $\{\underset{\sim}{p}\}$. Moreover, it is especially clear, using the Fock approach, that passage from the spherical-harmonic basis to the parabolic basis is straightforwardly performed through a rotation, with the expansion coefficients being given by a Clebsch-Gordan series. Within a given n manifold, of course, it is possible to define other basis sets which may be more appropriate for the study of specialized questions. Details can be found in [9].

3. THE LANDAU SPECTRUM

We now recall some well-known classical and quantum mechanical results on the motion of a free charged particle in a uniform magnetic field, a problem which is fully as basic as the Coulomb problem [12]. This is the second (and last) example of a physical system for which exact quantum and classical solutions exist, since it is a separable problem in R(3). Indeed, this separability obtains because the Landau problem is almost equivalent to that of the two-dimensional harmonic oscillator.

3.1 Classical Mechanics of the Landau Problem

The equation of motion for a free charged particle in a magnetic field is (in units such that c = 1)

$$m\ddot{\underset{\sim}{r}} \equiv m\underset{\sim}{\gamma} = \underset{\sim}{F}_L = q\,\underset{\sim}{v} \times \underset{\sim}{B} \quad . \tag{26}$$

This leads, after one integration, to

$$\underset{\sim}{\Pi} = m\underset{\sim}{v} = q\,\underset{\sim}{r} \times \underset{\sim}{B} + \underset{\sim}{\Pi}_0 \tag{27}$$

where $\underset{\sim}{\Pi}_0$ is a constant vector and $\underset{\sim}{\Pi}$ is the momentum of the velocity. After projection of Eq. (27) onto the field axis and onto the plane perpendicular to the field, respectively, one finds, with obvious notations ($\underset{\sim}{\rho}$ is the radius vector in cylindrical coordinates)

$$\Pi_{\parallel} = mv_{\parallel} = \Pi_{o\parallel}$$
$$\underset{\sim}{\Pi}_{\perp} = m\underset{\sim}{v}_{\perp} = q\,\underset{\sim}{\rho} \times \underset{\sim}{B} + \underset{\sim}{\Pi}_{o\perp} \quad . \tag{28}$$

Since $\underset{\sim}{\Pi}_{\parallel}$ is a constant, the motion along $\underset{\sim}{B}$ is uniform. Writing $\underset{\sim}{\Pi}_{o\perp} = -q\,\underset{\sim}{r}_o \times \underset{\sim}{B}$, the transverse part of Eq. (28) becomes

$$\underset{\sim}{\Pi}_{\perp} = q(\underset{\sim}{\rho} - \underset{\sim}{r}_o)\times \underset{\sim}{B} \quad . \tag{29}$$

Thus, the motion in the plane perpendicular to the field is a uniform circular motion, with the following characteristics: The angular velocity is the cyclotron frequency ω_c

$$\omega_c = -\,qB/m \tag{30}$$

and the position of the center of the circle is

$$\underset{\sim}{r}_o = \underset{\sim}{\rho} + \frac{\underset{\sim}{\Pi}_{\perp} \times \underset{\sim}{B}}{qB^2} \tag{31}$$

which is a constant of the motion. These conclusions are summarized in Fig. 2.

An additional constant of the motion can be derived from Eq. (28), using the angular momentum $\underset{\sim}{\ell}$ of $\underset{\sim}{\Pi} = m\underset{\sim}{v}$ taken at the origin. One finds

$$\frac{d}{dt}\underset{\sim}{\ell} = \frac{d}{dt}(\underset{\sim}{r} \times m\underset{\sim}{v}) = \underset{\sim}{r} \times (q\,\underset{\sim}{v} \times \underset{\sim}{B}) \tag{32}$$

and, projecting this equation onto the z-axis parallel to the field direction, one obtains

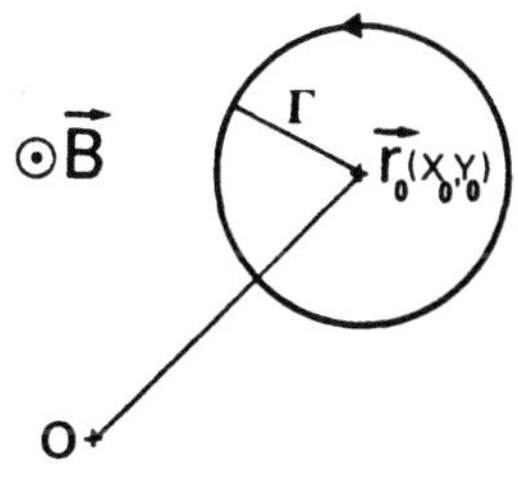

Fig. 2. Classical Landau trajectory in the plane perpendicular to the magnetic field ($\underset{\sim}{r}_o$ and Γ are constants of the motion).

$$\frac{d}{dt} \ell_z = - q(\underset{\sim}{\rho} \cdot \underset{\sim}{v}_\perp) B = - q(\underset{\sim}{\rho} \cdot \dot{\underset{\sim}{\rho}}) B \quad .$$

In other words, the quantity

$$M_z = \ell_z + \frac{1}{2} qB\rho^2 \tag{33}$$

is a constant of the classical motion. Obviously $\underset{\sim}{\ell}$, the angular momentum of $m\underset{\sim}{v}$, is no longer a constant of the motion in the presence of an external $\underset{\sim}{B}$ field, in the laboratory frame. Performing a transformation to a Larmor frame rotating at the constant angular frequency $\frac{1}{2}\omega_c$ shows that M_z coincides with the momentum of the impulsion in this new frame and, therefore, has an obvious physical meaning. Moreover, in the Larmor frame the equations of motion are simply

$$m\underset{\sim}{\gamma}_L = - (q^2B^2/4m)\underset{\sim}{\rho} \tag{34}$$

which makes clear the equivalence of this problem with that of the two-dimensional symmetric harmonic oscillator. Thus, trajectories in the Larmor frame are ellipses centered at the origin.

Expressing the energy of the system in the laboratory frame is straightforward:

$$H = \frac{1}{2} mv^2 = \Pi^2/2m = (\Pi_\perp^2/2m) + (\Pi_\parallel^2/2m) = E_\perp + E_\parallel \quad , \tag{35}$$

where $E_\perp$ and $E_\parallel$ are, respectively, the energies associated with the transverse and longitudinal motions which, from Eq. (28), are conserved independently. From Eqs. (29) and (35), one deduces immediately the relation between the radius Γ of the circle and the transverse part of the energy:

$$\Gamma^2 = 2E_\perp/m\omega_c^2 \quad . \tag{36}$$

To sum up, the classical trajectories, as shown in Fig. 2, are circular helices described at the cyclotron frequency ω_c. The constants of the classical motion are the coordinates $\underset{\sim}{r}_\rho(x_o,y_o)$ of the center of the circle; the radius Γ of the circle, which is proportional to the transverse part of the energy; and M_z, defined through Eq. (33). One can derive another expression for M_z, namely,

$$M_z = \frac{1}{2} qB(r_o^2 - \Gamma^2) \quad . \tag{37}$$

Eq. (36) has an obvious physical consequence. It shows that the classical two-dimensional problem, in the plane perpendicular to the field, is highly degenerate since the set of trajectories with fixed Γ but with different values of $\underset{\sim}{r}_0$ will have the same energy. Indeed, in the Larmor frame, the problem was shown (see above) to be equivalent to that of the two-dimensional symmetric harmonic oscillator, the high degeneracies of which are well-known [12]. This allows us to explain why the trajectories are closed.

3.2 The Landau Quantum Spectrum

The main differences between the classical and quantum derivations of the Landau spectrum are associated with the use of the Hamiltonian formalism for the latter. The use of this formalism requires that one choose a gauge. And, even though the physical results must be gauge-independent, the mathematical expressions of the physical constants of the motion in terms of the canonical variables will depend upon the choice of gauge [11,12].

The Hamiltonian of a free charged particle in a magnetic field is

$$H = \Pi^2/2m \tag{38}$$

with

$$\underset{\sim}{\Pi} = m\underset{\sim}{v} = \underset{\sim}{p} - q\underset{\sim}{A} \tag{39}$$

where $\underset{\sim}{A}$ is the vector potential ($\underset{\sim}{\nabla} \times \underset{\sim}{A} = \underset{\sim}{B}$). (We use the same symbol for the vector potential and the Runge-Lenz vector. Confusion will be avoided by the context.) We assume that $\underset{\sim}{B}$ is parallel to the z-axis. The $\underset{\sim}{p}$ and $\underset{\sim}{r}$ canonical operators fulfill the standard commutation relations, namely,

$$[p_i,p_j] = 0 \ , \quad [x_j,p_k] = i\hbar\, \delta_{jk} \ , \tag{40}$$

while the $\underset{\sim}{\Pi}$ (velocity) operators do not obey the canonical relations since

$$\begin{aligned} [\Pi_x,\Pi_z] &= [\Pi_y,\Pi_z] = 0 \\ [\Pi_x,\Pi_y] &= i\hbar\, qB = -\, i\hbar\, m\omega_c \end{aligned} \tag{41}$$

and

$$[x_j,\Pi_k] = i\hbar\, \delta_{jk} \ . \tag{42}$$

And, although the $\underset{\sim}{p}$ operator fulfills the canonical relations, the velocity operator is the one which has physical meaning.

The angular momentum $\underset{\sim}{L}$ is defined from $\underset{\sim}{p}$ as

$$\underset{\sim}{L} = \underset{\sim}{r} \times \underset{\sim}{p} \qquad (43)$$

while the angular momentum $\underset{\sim}{\ell}$ of the velocity operator is defined as

$$\underset{\sim}{\ell} = \underset{\sim}{r} \times \underset{\sim}{\Pi} \quad . \qquad (44)$$

The $\underset{\sim}{L}$ operator obviously fulfills the usual canonical commutation relations of an angular momentum, namely,

$$\begin{aligned} [L_j, x_k] &= i\hbar\, \varepsilon_{jk\ell}\, x_\ell \\ [L_j, p_k] &= i\hbar\, \varepsilon_{jk\ell}\, p_\ell \\ [L_j, L_k] &= i\hbar\, \varepsilon_{jk\ell}\, L_\ell \end{aligned} \qquad (45)$$

while $\underset{\sim}{\ell}$ does not fulfill these relations and is not an angular momentum since

$$[\ell_j, \ell_k] = i\hbar\, \varepsilon_{jk\ell}(\ell_\ell + q\, x_\ell\, \underset{\sim}{r} \cdot \underset{\sim}{B}) \ , \qquad (46)$$

although it does possess physical meaning.

The constants of the motion can be deduced straightforwardly from their classical analogs. The Hamiltonian in Eq. (38) can be divided into two parts:

$$H = H_\perp + H_\parallel$$

where

$$H_\perp = \Pi_\perp^2/2m = (\Pi_x^{\ 2} + \Pi_y^{\ 2})/2m \qquad (47)$$

and where $H_\parallel$ and $H_\perp$ commute. The motion parallel to the $\underset{\sim}{B}$ field is a free motion; thus

$$H_\parallel = \Pi_z^{\ 2}/2m = \frac{1}{2}\, m v_z^{\ 2} \quad . \qquad (48)$$

Obviously, $H_\parallel$ possesses a continuous spectrum. For the transverse motion, the operators corresponding to the constants of the motion in the classical problem are:

$$x_o = x + (\Pi_y/qB)$$

$$y_o = y - (\Pi_x/qB) \tag{49}$$

$$r_o^2 = x_o^2 + y_o^2$$

where (x_o,y_o) defines the position of the center of the classical trajectory; and

$$M_z = \ell_z + \frac{qB}{2}\rho^2 = \frac{1}{2}\ qB(r_o^2 - \Gamma^2) \quad , \tag{50}$$

where Γ, which defines the radius of the classical circle, is given by

$$\Gamma^2 = (x - x_o)^2 + (y - y_o)^2 = 2H_\perp/m\omega_c^2 \quad . \tag{51}$$

One can verify easily that these operators commute with H:

$$[H,x_o] = [H,y_o] = [H,r_o^2] = [H,M_z] = [H,\Gamma^2] = 0 \quad .$$

But, in contrast to what is clearly evident in classical mechanics, the operators x_o and y_o do not commute:

$$[x_o,y_o] = \frac{i\hbar}{qB} \equiv i\ a_c^2 \tag{52}$$

where a_c, termed the cyclotron radius, is the unit of length in the problem. One should also note that

$$[M_z,x_o] \neq 0 \ , \ [M_z,y_o] \neq 0 \ , \ [r_o^2,x_o] \neq 0 \quad . \tag{53}$$

It is important to stress that $[H,\ell_z] \neq 0$ and, usually, $[H,L_z] \neq 0$ -- that is, the z projection of the angular momentum is usually not a constant of the motion. Indeed, L_z has no physical meaning and its expression depends upon the choice of gauge. Thus, it is likely that L_z will not be a constant of the motion for an arbitrary choice of the gauge.

One can derive the quantum spectrum of the transverse Hamiltonian without any choice of gauge. From the expression for $H_\perp$, and from the commutation relations obeyed by Π_x and Π_y,

$$[H,\Pi_x] = i\hbar\ \omega_c\ \Pi_y \ , \ [H,\Pi_y] = -\ i\hbar\ \omega_c\ \Pi_x \quad , \tag{54}$$

one can interpret $H_\perp$ as the Hamiltonian of a one-dimensional harmonic oscillator in which Π_x and Π_y play the role of momentum and coordinate, respectively. Writing

$$u = \Pi_y/(-\hbar qB)^{1/2}\ ,\quad v = \Pi_x/(-\hbar qB)^{1/2} \tag{55}$$

one finds $[u,v] = i$ and

$$H_\perp = \frac{1}{2}\ \hbar\,\omega_c(u^2 + v^2) \tag{56}$$

which is really the Hamiltonian of a one-dimensional harmonic oscillator. Rewriting Eq. (56) in terms of the usual ladder operators

$$b = 2^{-1/2}(u + iv) \text{ and } b^\dagger = 2^{-1/2}(u - iv) \tag{57}$$

gives

$$H_\perp = \hbar\,\omega_c(b^\dagger b + 1/2)\quad . \tag{58}$$

Then, the Landau spectrum is given by

$$E = \hbar\,\omega_c(N + 1/2) + (\Pi_z{}^2/2m)\quad . \tag{59}$$

The discrete part of the Landau spectrum is equivalent to a one-dimensional harmonic oscillator spectrum while the continua are associated with the free motion along $\underset{\sim}{B}$. Such a spectrum is shown in Fig. 3. Nevertheless, this does not mean that the equivalence is complete. The expression of b in Eq. (57) shows that the degeneracy of each discrete state is infinite, as expected from the classical behavior. In contrast to what happens for a true one-dimensional oscillator, Eq. (57) depends upon the two

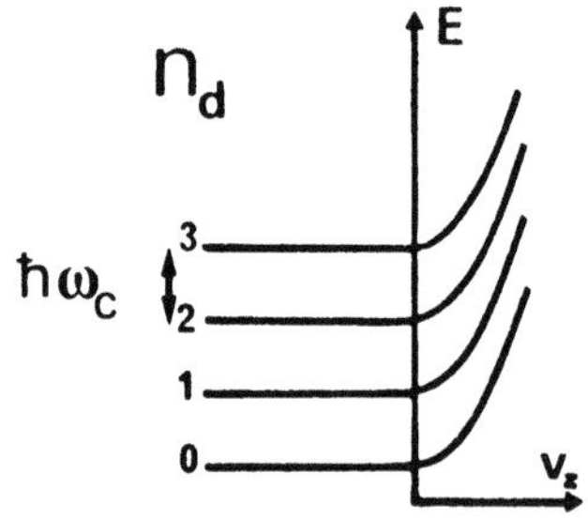

Fig. 3. Landau spectrum of a charged particle.

coordinates (x,y), which means that the ground state of our equivalent harmonic oscillator has infinite degeneracy. This also holds true for the excited states, which are proportional to $(b^{\dagger})^N|0\rangle$.

3.3 Choice of Gauge

One needs to choose a particular gauge in order to derive an expression for the wavefunctions. As mentioned above, the physical results will not depend upon this choice of gauge but the mathematical expressions for the various operators will.

The two gauges which are most widely used in this problem are the Landau [13] gauge and the so-called symmetric gauge. The vector potential $\underset{\sim}{A}$ in the Landau gauge is

$$A_x = -By \quad , \quad A_y = A_z = 0 \quad . \tag{60}$$

The use of this gauge is of interest in problems exhibiting some translational invariance, such as problems in solid-state physics applications. The second possible choice of a gauge, which is more convenient for atomic physics applications, is the so-called symmetric gauge:

$$\underset{\sim}{A} = -\frac{1}{2}\,\underset{\sim}{r} \times \underset{\sim}{B} \quad . \tag{61}$$

This gauge exhibits explicitly the rotational invariance along $\underset{\sim}{B}$.

Aside from a trivial phase factor, the eigenfunctions of the problem in the two gauges are the same, although the mathematical expressions of the constants of the motion $(x_o, y_o, r_o^2, M_z, H_\perp)$, as functions of the canonical operators $(\underset{\sim}{p}, \underset{\sim}{L})$, vary from one gauge to the other.

Some striking examples of the above features are as follows. Using the Landau gauge, it is straightforward to show that the expression for y_o is simply

$$y_o = -p_x/qB \quad , \tag{62}$$

which means that p_x is here a constant of the motion. The problem, then, is readily reduced to a one-dimensional harmonic oscillator along the y axis, with the equilibrium position of the oscillator shifted to y_o [13]. The associated wavefunctions are those of the one-dimensional harmonic oscillator, shifted along the y-axis, and are proportional to Hermite polynomials. Translational invariance along the x-axis is clear from the fact that, y_o being defined, x_o cannot be defined [cf. Eq. (52)]. The fundamental degeneracy of the problem is described here through the continuous index y_o. The wavefunctions are associated with the set of observables $\{H_\perp, H_\parallel, y_o\}$. And, neither L_z nor M_z are defined in this representation.

Using the symmetric gauge, on the other hand, one immediately sees that the situation is of interest for atomic physics applications. From Eqs. (39,43,44,50,61), one finds

$$L_z = \ell_z + \frac{1}{2} qB \rho^2 = M_z \quad . \tag{63}$$

That is, the z component of the angular momentum in the symmetric gauge coincides with the constant of the motion M_z. The rotational invariance around the z-axis is explicit here. The wavefunctions associated with this $\{H_\perp, H_\parallel, L_z\}$-representation are labeled with the integer index $M = L_z/\hbar$ associated with the quantized projection of the angular momentum. This provides us with a discrete description of the fundamental degeneracy. From Eq. (50), it is clear that r_o^2 is defined in this representation and that $(r_o^2 - \Gamma^2)$ is quantized, rather than r_o^2 (the square of the distance to the center of the classical circle) [11].

A schematic representation of these two main types of wavefunction is given in Fig. 4.

3.4 Fock Representation of the Landau Spectrum

We hereafter adopt the symmetric gauge formulation in order to stress the rotational invariance of the problem [12]. The expression for the transverse part of the Hamiltonian is

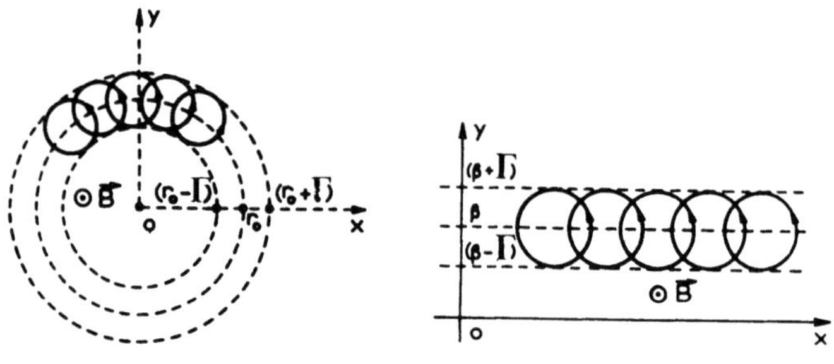

Fig. 4. Schematic "semi-classical" representation of the wavefunctions which possess rotational and translational symmetries in the Landau problem.

$$H_{\perp} = (p_x + \frac{1}{2} qBy)^2/2m + (p_y - \frac{1}{2} qBx)^2/2m \tag{64}$$

and the four basic constants of the motion are

$$\Pi_z = mv_z = p_z$$

$$x_o = \frac{1}{2} x + (p_y/qB) \ , \ y_o = \frac{1}{2} y - (p_x/qB) \tag{65}$$

$$M_z = L_z = \frac{1}{2} qB(r_o^2 - \Gamma^2) \ .$$

From Eqs. (49,51,64,65), one finds

$$H_{\perp} = \frac{1}{2m} (p_x^2 + p_y^2) + \frac{\omega_c}{2} L_z + \frac{m\omega_c^2}{8} (x^2 + y^2) \ . \tag{66}$$

Except for the paramagnetic term $\frac{1}{2}\omega_c L_z$, this is the Hamiltonian of a two-dimensional symmetric harmonic oscillator. The introduction of the usual ladder operators $(a_x, a_x^\dagger)$ and $(a_y, a_y^\dagger)$, associated with annihilation and creation of excitation along the x and y axes, respectively, is a source of difficulty since these operators obviously do not commute with L_z. It is far better to use the operators associated with annihilation and creation of left- $(a_+, a_+^\dagger)$ and right- $(a_-, a_-^\dagger)$ handed circular excitation in the system:

$$a_+ = 2^{-1/2} (a_x + ia_y) \ , \quad a_- = 2^{-1/2} (a_x - ia_y) \ . \tag{67}$$

These operators fulfill the usual commutation rules:

$$[a_i, a_k^\dagger] = \delta_{ik} \ . \tag{68}$$

The expressions of the various constants of the motion are, then,

$$H_{\perp} = (a_-^\dagger a_- + 1/2)\hbar\omega_c \ , \ L_z = (a_-^\dagger a_- - a_+^\dagger a_+)\hbar \tag{69}$$

$$x_o = (\hbar/2m\omega_c)^{1/2}(a_+^\dagger + a_+) \ , \ y_o = (\hbar/2m\omega_c)^{1/2} i(a_+^\dagger - a_+) \tag{70}$$

and

$$r_o^2 = x_o^2 + y_o^2 = a_c^2(a_+^\dagger a_+ + a_+ a_+^\dagger) \tag{71a}$$

$$\Gamma_o^{\,2} = (x - x_o)^2 + (y - y_o)^2 = a_c^{\,2}(a_-^\dagger a_- + a_- a_-^\dagger) \quad . \tag{71b}$$

Introducing the notation n_-, n_+ for the number of right- and left-handed circular excitations in the system, respectively, the transverse part of the energy becomes [12]

$$E_\perp = (n_- + 1/2)\,\hbar\,\omega_c \quad . \tag{72}$$

Thus, for the electron, the energy of the system does not depend on n_+ (the number of left-handed circular excitations in the system), which makes clear the infinite degeneracy of each n_- Landau manifold. The angular-momentum projection is given by

$$L_z = (n_- - n_+)\,\hbar = M\hbar \quad . \tag{73}$$

That is, in a given n_- manifold, the degeneracy can be labeled with the integer index n_+ varying from 0 to $+\infty$; or, equivalently M varies between n_- and $-\infty$. A straightforward choice for the wavefunction is

$$|n_- n_+\rangle = (n_-!\; n_+!)^{-1/2}\;(a_-^\dagger)^{n_-}(a_+^\dagger)^{n_+}|00\rangle \quad . \tag{74}$$

In cylindrical coordinates, $\langle\rho\theta\phi|n_- n_+\rangle$ can be expressed as a generalized Laguerre polynomial [11].

The above choice of wavefunction is associated with the $\{H, H_\perp, H_\parallel, L_z\}$-representation. The values of r_o^2 and Γ^2 are then defined as

$$\Gamma^2 = (2n_- + 1)a_c^{\,2} \;, \quad r_o^{\,2} = (2n_+ + 1)a_c^{\,2} \quad . \tag{75}$$

In other words, coming back to the classical picture, the radius Γ of the circle is $(2n_- + 1)^{\frac{1}{2}}\,a_c$, which is constant in a given Landau band. The degeneracy of the Landau band is associated with the distance $r_o = (2n_+ + 1)^{\frac{1}{2}}\,a_c$ between the origin and the center of the circle. From Eq. (74), it is clear that any choice of basis is possible within a given n_- manifold provided that one has

$$|\Psi\rangle = (a_-^\dagger)^{n_-}|0\rangle|f\rangle$$

with $|f\rangle$ a linear combination of $|n_+\rangle$ states only. For example, one can build, in a given manifold, quasi-classical states in the sense of Glauber [12], out of left-handed circular excitations for which the fluctuations $\Delta x_o \Delta y_o$ take the minimum value $2^{-\frac{1}{2}}a_c$, while the energy is still defined. These states will have the general form

$$|n_-\alpha_+\rangle = (n_-!)^{-1/2}(a_-^\dagger)^{n_-}\exp(\alpha_+a_+^\dagger - \alpha_+^* a_+)|00\rangle \quad . \tag{76}$$

Of course, L_z will not be diagonal in such a basis.

In addition, the other usual set of wavefunctions for the $\{H,y_o\}$-representation can be obtained by choosing another basis for the left-handed circular excitations. One obvious choice is that associated with the wavefunctions of the one-dimensional left-handed harmonic oscillator in momentum space:

$$|p_x(= - qBy_o)\rangle = \sum_{n_+} \psi_{n_+}^*(p_x)|n_+\rangle$$

where the $\psi_{n_+}^*$ are the usual wavefunctions of the one-dimensional harmonic oscillator. The complete wavefunction is then

$$|n_-p_x(= - qBy_o)\rangle = \sum_{n_+} \psi_{n_+}^*(p_x)|n_-n_+\rangle \quad . \tag{77}$$

Of course, this is nothing more than the wavefunction which we discussed in Section 3.3. L_z and r_o^2 are no longer defined in this representation and the wavefunctions exhibit a clear translational invariance.

Finally, our last remark about the pure Landau problem is connected with the building of semi-classical wave packets. Writing $D(\alpha)$ for the Glauber operator:

$$D(\alpha) = \exp(\alpha a^\dagger - \alpha^* a) \tag{78}$$

one finds

$$|\alpha_-\alpha_+\rangle = D(\alpha_-)D(\alpha_+)|00\rangle \quad . \tag{79}$$

It is possible to show that, with

$$\alpha_- = \frac{\Gamma}{2^{1/2}a_c} e^{i\phi_-} \quad \text{and} \quad \alpha_+ = \frac{r_o}{2^{1/2}a_c} e^{i\phi_+} \quad ,$$

the state $|\alpha_-\alpha_+\rangle$ represents the motion of a particle on the circle (center r_o, radius Γ) which is followed with the minimum possible fluctuation [12].

4. THE ATOM IN A MAGNETIC FIELD

We consider the simplest atomic situation, namely, the two-particle problem, in a magnetic field. In contrast to what happens in the purely Coulombic situation, the two (interacting) particle problem in a magnetic field does not separate in the center-of-mass frame [14-16].

4.1 Classical Mechanics of the Two-Particle Problem in a Magnetic Field

Writing (q_i, m_i), $i = 1,2$, for the charges and masses of the two particles, the classical equations of motion are

$$
\begin{aligned}
m_1 \ddot{\underset{\sim}{r}}_1 &= q_1 \dot{\underset{\sim}{r}}_1 \times \underset{\sim}{B} + (q_1 q_2/4\pi\varepsilon_o) \frac{\underset{\sim}{r}_1 - \underset{\sim}{r}_2}{|\underset{\sim}{r}_1 - \underset{\sim}{r}_2|^3} \\
m_2 \ddot{\underset{\sim}{r}}_2 &= q_2 \dot{\underset{\sim}{r}}_2 \times \underset{\sim}{B} + (q_1 q_2/4\pi\varepsilon_o) \frac{\underset{\sim}{r}_2 - \underset{\sim}{r}_1}{|\underset{\sim}{r}_1 - \underset{\sim}{r}_2|^3}
\end{aligned} \tag{80}
$$

We will write these equations in terms of the center-of-mass variables, namely,

$$
\begin{aligned}
\underset{\sim}{R} &= \sum_i m_i \underset{\sim}{r}_i / M \ , \quad M = m_1 + m_2 \\
\underset{\sim}{r} &= \underset{\sim}{r}_1 - \underset{\sim}{r}_2 \quad .
\end{aligned} \tag{81}
$$

Combining Eqs. (80) and (81), one finds

$$
(m_1 + m_2)\ddot{\underset{\sim}{R}} = (q_1 \dot{\underset{\sim}{r}}_1 + q_2 \dot{\underset{\sim}{r}}_2) \times \underset{\sim}{B}
$$

One integration, with $q_1 = - q_2 = q$, gives

$$
(m_1 + m_2)\dot{\underset{\sim}{R}} - q\, \underset{\sim}{r} \times \underset{\sim}{B} = \underset{\sim}{C} = \text{constant} \quad . \tag{82}
$$

Eq. (82) implies that the motion of the reduced particle and the motion of the center of mass are always coupled in the presence of the magnetic field, for the two particles interacting through the Coulomb field [14].

4.2 Quantum Approach for the Two-Particle Problem in a Magnetic Field

The Hamiltonian of the system in the presence of the $\underset{\sim}{B}$ field is

$$H = (\Pi_1^{\,2}/2m_1) + (\Pi_2^{\,2}/2m_2) - (q^2/4\pi\varepsilon_o)\,|\underset{\sim}{r}_1 - \underset{\sim}{r}_2|^{-1} \tag{83}$$

with

$$\underset{\sim}{\Pi}_i = \underset{\sim}{p}_i - q_i\underset{\sim}{A}(\underset{\sim}{r}_i) \quad . \tag{84}$$

It is straightforward to show that the operator

$$\underset{\sim}{C} = \underset{\sim}{\Pi} - q\,\underset{\sim}{r} \times \underset{\sim}{B} \quad , \tag{85}$$

which is associated with Eq. (82) from classical mechanics, commutes with the Hamiltonian, and that

$$[C_i, C_j] = 0 \quad . \tag{86}$$

Then, although the problem is not separable (i.e., the reduced-particle motion and the center-of-mass motion are not decoupled), the existence of the operator $\underset{\sim}{C}$ allows a quasi-separability [14, 17]. Choosing the symmetric gauge, one finds

$$\underset{\sim}{C} = \underset{\sim}{P} - \frac{1}{2}\,q\,\underset{\sim}{r} \times \underset{\sim}{B} \tag{87}$$

where $\underset{\sim}{P}$ is the momentum of the center of mass in this gauge (i.e., $\underset{\sim}{P} = -i\hbar\nabla_{\underset{\sim}{R}}$).

The reduction of the two-particle problem now follows. The eigenfunctions of Eq. (87) are [15]:

$$\Psi_{\underset{\sim}{C}}(\underset{\sim}{R},\underset{\sim}{r}) = \exp\left[\frac{i}{\hbar}\,(\underset{\sim}{C} + \frac{q}{2}\,\underset{\sim}{r} \times \underset{\sim}{B})\cdot\underset{\sim}{R}\right]\,\psi_{\underset{\sim}{C}}(r) \quad . \tag{88}$$

Substitution of Eq. (88) into Eq. (83) yields the expression of the reduced Hamiltonian H_C:

$$H_C = \frac{C^2}{2m} - \frac{q}{M}(\underset{\sim}{C}\times\underset{\sim}{B})\cdot\underset{\sim}{r} + \frac{p^2}{2\mu} - \frac{q}{2}\left(\frac{1}{m_1} - \frac{1}{m_2}\right)\underset{\sim}{B}\cdot\underset{\sim}{L} + \frac{q^2}{8\mu}(\underset{\sim}{r}\times\underset{\sim}{B})^2 - \frac{q^2}{4\pi\varepsilon_o}r^{-1} \tag{89}$$

where $\underset{\sim}{p}$ and $\underset{\sim}{L}$ are the canonical momentum and angular momentum operators, respectively, for the reduced particle of mass $\mu = m_1m_2/(m_1 + m_2)$; and where $\underset{\sim}{C}$ is the eigenvalue of the operator of Eq. (87).

4.3 Physical Meaning of the Reduced Two-Particle Hamiltonian

The fact that the problem is not separable in the center-of-mass frame follows from the $\underset{\sim}{C}$ dependence of H_C. Clearly, $\underset{\sim}{C}$ is not the eigenvalue of the center-of-mass momentum operator and is responsible for a coupling between the center-of-mass motion and the reduced-particle motion. The interpretation of the various terms in H_C is as follows. The first term represents the translational energy of the system. The second term couples the relative and translational motions; the quantity $(\mathbf{C}/M) \times \mathbf{B}$ is the motional electric field seen in the rest frame of the reduced particle and is perpendicular to $\mathbf{B}$. The other field-dependent terms are, respectively, the paramagnetic interaction (in which the gyromagnetic ratio is altered) and the diamagnetic term (which is proportional to B^2).

In summary, it is clear that the center-of-mass motion and the reduced-particle motion are no longer independent in the presence of a magnetic field. Specifically, the individual energies are no longer independently conserved, which means that the problem is nonseparable, although it is soluble [16,17]. Complete solution of the problem of a one-electron atom in a magnetic field really requires that we solve an infinite set of one-particle problems in crossed electric and magnetic fields. To do so is presently outside practical possibilities. Nevertheless, the approximation -- which we make in the next section -- of infinite proton mass is justified for most situations in atomic physics. Of course, this approximation is no longer correct if one is concerned with positronium!

4.4 One-Particle Hamiltonian in a Magnetic Field

Assuming $(\mathbf{C}/M) \simeq 0$ and $m \equiv m_1 \ll m_2$ in Eq. (89), one finds

$$\begin{aligned} H &= \frac{p^2}{2m} - \frac{qB}{2} L_z + \frac{q^2B^2}{8m} (x^2 + y^2) - \frac{e^2}{r} \\ &= \frac{p^2}{2m} + \frac{\omega_c}{2} L_z + \frac{m\omega_c^2}{2} (x^2 + y^2) - \frac{e^2}{r} \quad , \end{aligned} \tag{90}$$

which is the expression for the Hamiltonian in the symmetric gauge.

By setting $B = 0$ in Eq. (90), one recovers the Coulomb Hamiltonian, while setting $e = 0$ yields the Landau problem [cf. Eq. (66)]. There are two field-dependent terms in the Hamiltonian. The first one, which is linear in B, is the well-known paramagnetic term associated with the Zeeman effect; it is essentially constant throughout the spectrum and does not depend upon the degree of excitation of the electron's motion. The second term is the diamagnetic interaction, which depends upon the square of B; also, it strongly depends upon the degree of excitation of the electron's motion and, in some situations, it can completely overwhelm the Coulomb interaction. This is especially true for high-lying Rydberg states, for which the Coulomb interaction becomes progressively weaker, and for continuum states.

4.5 Constants of the Motion

From the analysis of Sections 2 and 3, it is clear that the only constants of the motion in this problem are L_z and parity. (More generally, whatever the choice of gauge, the operator M_z, defined in Eq. (50), is a constant of the motion.)

Since the diamagnetic interaction breaks the spherical symmetry of the atomic problem, L^2 is no longer a constant of the motion. The same conclusion is valid for the Runge-Lenz vector $\mathbf{A}$, which no longer commutes with the Hamiltonian (and especially does not commute with the diamagnetic interaction). Thus, the diamagnetic interaction is responsible for a breakdown of the super-symmetry of the Coulomb problem. Moreover, all of these conclusions remain valid for the constants of the motion which were deduced for the Landau problem: None among them, save L_z and parity, are constants of the motion in the present situation since they do not commute with the Coulomb interaction. In fact, one can verify that the Hamiltonian of Eq. (90) is not separable in any of the thirteen sets of natural coordinates for R(3). This means that there will not exist any additional constant of the motion aside from L_z and parity. A major question is thus laid out: What are the various aspects of the atomic spectrum in a magnetic field? How do the Coulomb energy levels branch to the Landau levels when either the field or the energy is varied?

4.6 Phenomenology of the Various Magnetic Regimes

Associated with the various terms in Eq. (90), there are special characteristics and alterations of the atomic spectra and approximate sets of constants of the motion [18].

For low-lying excited states, the diamagnetic term is negligible. The dominant term, then, is the Zeeman term (proportional to B), which is small in comparison to the Coulomb interaction. These regimes, which are well-understood, are associated basically with a removal of the L_z degeneracy of the spectrum.

The more fundamental aspects of the problem are those connected with diamagnetism. The diamagnetic interaction, denoted H_D, depends tremendously on the degree of excitation of the electron's motion and is of the order of (a_o is the Bohr radius)

$$H_D \simeq (q^2B^2/8m)\; n^4\; a_o^{\,2} \quad . \tag{91}$$

Thus, H_D totally dominates the Zeeman contribution in highly-excited Rydberg states and, in some circumstances, dominates the electrostatic contribution to the binding energy as well (especially for continuum states).

Four characteristic regimes associated with diamagnetism have been recognized in atomic spectra, although each one is really an expression of the same fundamental feature [19]. The *inter-ℓ mixing regime* occurs for $2R/n^3 \gg H_D \simeq (e^2B^2/8m)n^4a_o^2$. In this case, L^2 is no longer a constant of the motion but the Coulomb radial quantum number is still defined. Energies behave as B^2, with some deviations associated with quantum defects in nonhydrogenic situations. In the *inter-*n *mixing regime*, for which $2R/n^3 \simeq H_D$, the atomic spectrum is completely altered through a merging of the various adjacent hydrogenic manifolds. The third regime, which is the most important one, has been termed the *strong field mixing regime* [7]; it takes place when the electrostatic and magnetic contributions to the energy of the electron are comparable, and $2R/n^3 \simeq \hbar\omega_c$. Finally, when the magnetic force completely overwhelms the Coulomb force, far into the continuum, the *Landau regime* is reached. The way of passing from the Coulomb regime to the Landau regime in atomic spectra is discussed in the following Sections.

5. CLASSICAL MECHANICS OF DIAMAGNETISM

The difficulties associated with finding the quantum spectrum of a nonseparable problem are basically of the same nature as those encountered in studying the behavior of the classical trajectories. In addition, classical studies are of importance for setting up the proper semi-classical quantization of a problem (see Section 7). For the problem of atomic diamagnetism, these studies have allowed recent theoretical progress culminating in the discovery of an adiabatic invariant in the Coulomb limit. Moreover, classical studies allow us to show that the strong field mixing regime is somewhat analogous to the classical resonance region, or Hill's region, which exists in the problem of coupled harmonic oscillators.

5.1 Classical Equations of Motion

Assuming the proton to have infinite mass, the motion of the electron is described by

$$m\ddot{\underset{\sim}{r}} = q\, \underset{\sim}{v} \times \underset{\sim}{B} + \underset{\sim}{F}_C \quad . \tag{92}$$

The Lorentz force, which depends linearly on the B field, recovers the two classes of magnetic interactions, namely, paramagnetism and diamagnetism. Changing to a Larmor frame rotating at a constant angular velocity $\frac{1}{2}\omega_c$ around the $\underset{\sim}{B}$-axis allows us to get rid of any effects of the paramagnetic interaction. One finds, in the Larmor frame,

$$m\ddot{\underset{\sim}{r}}_L = \underset{\sim}{F}_C - (q^2B^2/4m)\underset{\sim}{\rho} \quad , \tag{93}$$

which means that the Lorentz force just cancels the Coriolis term in the transformation ($\underset{\sim}{\rho}$, again, is the radius vector in cylindrical coordinates). The second term in Eq. (93) is the diamagnetic force, which we first saw in Eq. (34). If one neglects this term one recovers the usual equation of motion for the Coulomb problem in the Larmor frame. As expected, this shows that the trajectories in the Larmor frame are the field-free ones (i.e., ellipses with the focus at the origin, a well-known result of the Larmor theorem). On the other hand, if one neglects F_C in Eq. (93), the trajectories are the Landau ones in the Larmor frame (i.e., ellipses with the center at the origin). Between these two limits, *where other closed trajectories exist*, no general solution of the problem is known.

The diamagnetic force is a two-dimensional, attractive harmonic force with cylindrical symmetry. It derives from the harmonic potential $H_D = (q^2B^2/8m)\rho^2$ which follows from the Hamiltonian formulation in the symmetric gauge [(see, for example, Eq. (90)].

5.2 Symmetries and Coupling Constants

From a classical point of view, the problem of the one-electron atom in a magnetic field is equivalent, in the Larmor frame, to that of a particle experiencing the joint actions of a central Coulomb force, and of an axial harmonic force associated with the diamagnetic interaction. The axial character of the diamagnetic force is responsible for a breakdown of the super-symmetry of the field-free problem, whatever the magnetic field strength. The conclusions are the same from the Landau approach in Eq. (93). Indeed, none of the thirteen sets of natural coordinates for R(3) [17] permits a separation of Eq. (93). No clear evidence of any inner symmetry of the motion exists in the intermediate regime, where both forces act on the electron.

Another key point in Eq. (93) follows from the evaluation of the relative strengths of the diamagnetic ($\underset{\sim}{F}_D$) and Coulomb ($\underset{\sim}{F}_C$) forces. Using atomic units, the coupling parameter ξ can be written as

$$\xi = F_D/F_C = \frac{1}{4}\,(a_o/a_c)^4\,\rho r^2 \quad , \tag{94}$$

where a_c is the cyclotron radius defined in Eq. (52). ρ and r are dimensionless parameters of the order of atomic dimensions: That is, they depend upon the degree of excitation of the electron's motion. One then introduces the coupling constant γ,

$$\gamma = (a_o/a_c)^2 = \hbar\omega_c/2R = B/B_c \quad , \tag{95}$$

which measures the relative strength of the magnetic and electrostatic interactions. B_c is the critical value of the magnetic field, for which the Larmor frequency equals the Rydberg constant

$$B_c = 2.35 \times 10^5 \text{ T} \quad . \tag{96}$$

Under laboratory conditions, B is, at maximum, of the order of 25 T in DC operation; that is, γ is, at maximum, of the order of 10^{-4}, which is fairly small, as expected. Thus, one concludes that the diamagnetic interaction is always negligible in comparison to the Coulomb interaction, provided that the degree of excitation of the electron's motion is weak (i.e., provided that the sizes of the orbits are a few atomic units). However, these conclusions are no longer correct if one deals with high-lying atomic states, since the coupling parameter ξ depends not only on γ but also on the degree of excitation of the electron's motion, through ρr^2:

$$\xi = \frac{1}{4} \gamma^2 \rho r^2 \quad . \tag{97}$$

In contrast, the paramagnetic interaction, which is associated with rotation to the Larmor frame, is always of the order of $\frac{1}{2}\hbar\omega_c$ and constant, regardless of the degree of motional excitation. In addition, this interaction never breaks the spherical symmetry of the problem.

An order of magnitude of the coupling constant in Eq. (97) can be obtained by assuming that the degree of excitation of the electron's motion is described with an index n, such that r or ρ are of the order of n^2. Then, ξ takes the form [19]

$$\xi = F_D/F_C = \frac{1}{4} (\gamma n^3)^2 \quad . \tag{98}$$

This means that the coupling parameter in the problemm is γn^3, instead of γ. Although γ is fairly small under normal atomic-physics conditions, the ratio of the two forces can take every value, which means that the diamagnetic force may completely overwhelm the Coulomb force.

5.3 Phenomenology of the Various Regimes

Three types of motion follow from Eq. (98) [7]:

--- If $\gamma n^3 \ll 1$, the diamagnetic force is negligible in comparison to the Coulomb force and the motion of the electron is quasi-Coulombic. Such a regime does not exist if $B > B_c$ (i.e., $\gamma > 1$).

--- If $\gamma n^3 \gg 1$, the diamagnetic force dominates the Coulomb force and the motion of the electron is mainly of the Landau type, in the plane perpendicular to the field, while the Coulomb force still has some binding action along the $\tilde{B}$ field.

--- If $\gamma n^3 \simeq 1$, the diamagnetic and Coulomb forces are of comparable strengths but have far different symmetry properties. This is the strong field mixing, or quasi-Landau, regime [7,19, 20]. No perturbative approach for such a regime, starting from the Coulomb or Landau limit, can be valid. The very different symmetries of the forces and the lack of separability suggest that the motion of the electron in such a regime might be completely featureless and disorganized. But, surprisingly, experiments have shown that, for all atoms, this is far from the case [3,6]. Thus, in between the two regimes of Coulomb and Landau, where the trajectories are closed and periodic, a third regime exists in which the trajectories may have some regularity, although intuitive arguments suggest exactly the opposite.

One further remark is that the classical problem obviously depends upon two parameters, namely, the energy of the electron and the magnetic field strength. The analysis of the motion can be performed either at constant field, while varying the energy, or at constant energy, while varying the field. The somewhat equivalent role of E and B can be rendered much more explicit by using an R(4) formulation of the problem (see Section 5.6). Several ways of investigating the behavior of the classical trajectories are then possible, by playing with the initial conditions in Eq. (92). The increase of the energy at fixed field will result in a more important role for the diamagnetic force, while the increase of the field at fixed energy will result in a decrease of the diamagnetic contribution. The region where $\gamma n^3 \simeq 1$ is associated with a transition curve in the (E,B) plot.

5.4 Phase-Space Trajectories

Phase-space studies of the motion of the particle are extremely useful for the exhibition of the inner symmetries and peculiarities of the problem. And, since the two limiting cases are quite interesting because of their highly degenerate character, it is natural to try to gain some insight into what happens in between these limits.

Trajectory studies can be performed in cylindrical coordinates [21]. Since L_z is a constant of the motion, the angle ϕ is a cyclic coordinate and, therefore, the relevant dimension of phase space is four: $(\rho, \dot{\rho}, z, \dot{z})$. Some aspects of these trajectories are shown in Fig. 5, which represents two-dimensional projections of the four-dimensional motion (at constant energy). Fig. 5a is a plot in the (ρ,z) plane for low field values, where the diamagnetic term is almost negligible. The projection is a well-defined curve which is stable for more than 100 times the period; in fact this is nothing more than the projection of a Kepler ellipse. Other features, which appear in a $(\dot{\rho},\dot{z})$ plot, are shown in Fig. 5b. Fig. 5c represents a (ρ,z) projection for the same set of initial conditions, but for a far greater value of the magnetic field. In this latter case, the perturbation of the Coulombic trajectory is no longer negligible and distortions of the Kepler ellipse appear. A complete interpretation of the evolution can be achieved using the analysis of Section 5.7. For the sake of illustrating some other aspects of the classical behavior, Fig. 6 presents some of the two-dimensional projections obtained at several values of the energy, but at a fixed magnetic field strength. The initial conditions were chosen such that the excitation of the motion along $\underset{\sim}{B}$ is weak, with the initial value of ρ at $t = 0$ being increased successively. Trajectories are expected to lie close to the $z = 0$ plane. In Fig. 6a-d, $(\rho,\dot{\rho})$ plots are shown for increasing values of the diamagnetic contribu-

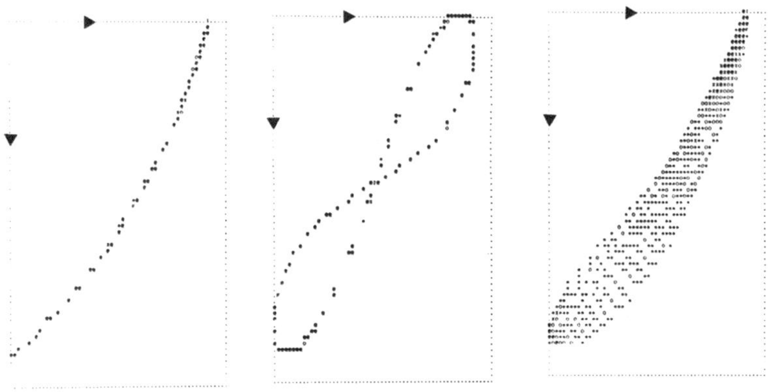

Fig. 5. Two-dimensional projections of the phase-space motion for initial conditions $(\rho, z, \dot{\rho}, \dot{z}) = (0.816, 0.577, 0, 0)$ [atomic units] and $M = 1$. The trajectory is followed for a duration of approximately 100 times the period. (a) and (b) are, respectively, (ρ, z) and $(\dot{\rho}, \dot{z})$ plots for $B = 100$ kG, $E = -0.5$ au; (c) is a (ρ, z) plot for $B = 10^5$ kG, $E = -0.4572$ au. (Taken from [21].)

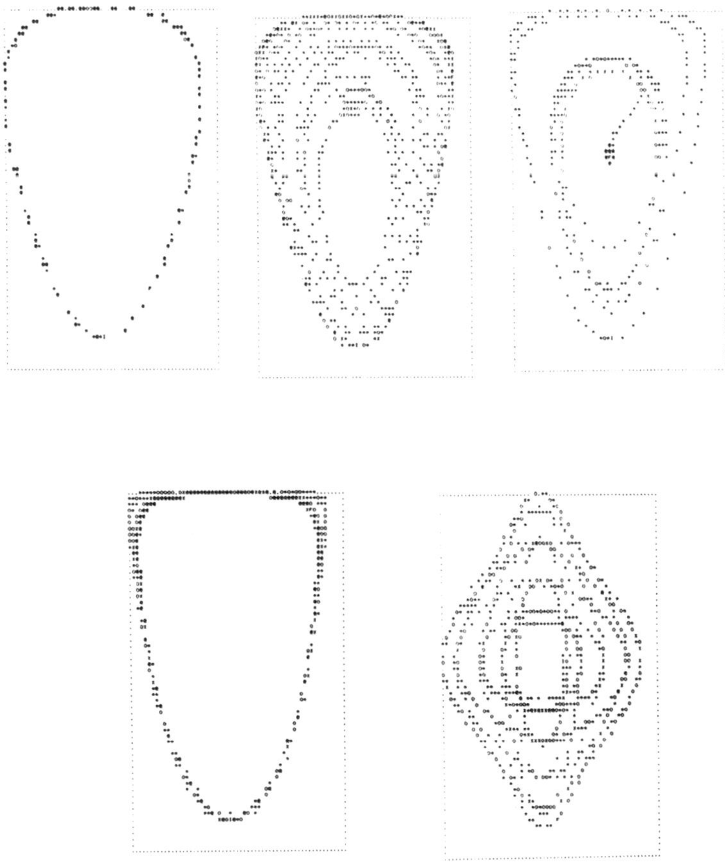

Fig. 6. Two-dimensional projections of the phase-space motion for B = 1.175 x 10^5 T (γ = 0.5), M = 1 and initial conditions of the general form (ρ, z, $\dot{\rho}$, $\dot{z}$) = (ρ, 0, 0, 10^{-2}) [atomic units]. The initial value of ρ is successively varied so as to increase the relative diamagnetic contribution to the total energy. (a) - (d) are (ρ, $\dot{\rho}$) plots for: ρ = 2, E = 10^{-4} (a); ρ = 2.8, E = + 0.40 (b); ρ = 4, E = + 1.06 (c); and ρ = 10, E = + 6.56 (d). (e) is a (z, $\dot{z}$) plot for E = 10^{-4}, (ρ, z, $\dot{\rho}$, $\dot{z}$) = (2, 0, 0, 10^{-2}). (Taken from [21].)

tion to the energy. In Fig. 6a, the curves are still well-defined and quasi-periodic. In Fig. 6b, however, a complete destabilization of the trajectory occurs due to the onset of the strong mixing regime, since the diamagnetic contribution to the energy is now extremely important. In Fig. 6c, one sees an example of ionization occuring in the z-direction. If the diamagnetic part of the energy is increased still further, one recovers a very simple situation in which the trace is a closed curve followed for more than 100 times the period of the motion (cf. Fig. 6d). For these very positive values of the energy, the motion is once again quasi-periodic as one enters the Landau regime [21]. In Fig. 6e one can see that in the strong field mixing regime there are still some regularities in $(z,\dot{z})$ plots, although they are completely different in nature from those seen in the highly degenerate Landau and Coulomb regimes [22]. The purpose of Figs. 5 and 6 is illustrative: These figures show that the passing from the Coulomb to the Landau regime fundamentally modifies the behavior of the system, which becomes more or less chaotic although some regular features still remain. The initial conditions which were adopted in the strong field mixing regime are the worst ones in the sense that they are very close to a hyperbolic point, as found in the linear stability analysis of the system [21].

5.5 Poincaré Surfaces-of-Section

[A fuller discussion of this topic will be found in the paper by M. Robnik in this volume.]

The most general trajectory in phase space lies on a three-dimensional surface. In contrast, the Coulomb and Landau limits are associated with one-dimensional surfaces (i.e., closed curves). Finding an extra constant of the motion in a problem permits a reduction of the dimension of the surface by one. For exploring such a possibility, one is led to show that, for a set of trajectories having the same energy, two-dimensional sections of phase space do not exhibit a random distribution of points. On the contrary, the distribution must be organized in curves which are sections of a torus. A Poincaré surface-of-section is the intersection of the four-dimensional phase-space trajectory $(\rho(t), \dot{\rho}(t), z(t), \dot{z}(t))$ with the plane $z = 0$, as projected onto the $(\dot{\rho},\rho)$ plane when $\dot{z} > 0$. If the section is a closed curve, an extra constant of the motion will exist (in addition to the energy), which serves to confine the four-dimensional motion to the surface of a torus.

Several attempts at this problem have been made recently [23-26] and there is now evidence that approximate constants of the motion do exist. Under some conditions, the features of the surface-of-section follow the coarse classification of Section 5.3.

For energies and field values such that $E < E_c \simeq -0.5(\gamma)^{2/3}$ [24], where E_c is a critical energy, the electronic motion is regular, as shown in Fig. 7, and the Poincaré section gives evidence of the existence of an approximate constant of the motion. The trajectories lie on a torus in phase space, the sections of which are seen in the plot of Fig. 7. These trajectories are quasi-periodic ones and obviously are associated with two very different classes of symmetries. This facet, which will be interpreted in Section 5.7, is connected with the existence of an adiabatic invariant in the Coulomb limit. One must remark that the critical energy fulfills

$$(\gamma/E_c^{3/2})^{2/3} \simeq (n_c^3 B)^{2/3} \simeq (n_c^3 \gamma)^{2/3} \simeq \xi^{1/3} \simeq 1 \tag{99}$$

and is associated with the onset of the strong mixing regime, as expected.

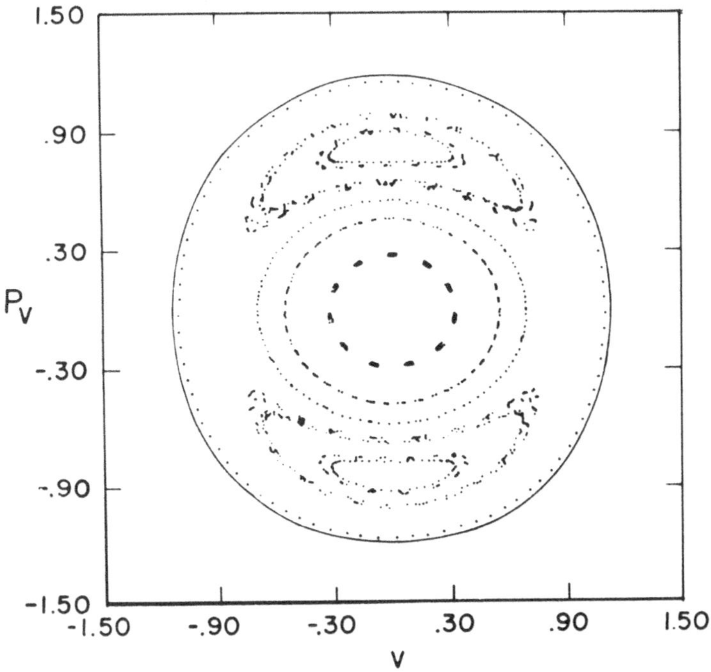

Fig. 7. Poincaré surface-of-section [26] for E = - 1.0 au. The (p_v, v) coordinates are analogous to the ones defined in Eq. (102) of the text. Several different trajectories are depicted in this plot. (Courtesy of W. P. Reinhardt.)

If one crosses the critical line of Eq. (99), irregular features appear in the motion of the electron, as depicted in Fig. 8. These irregularities show that some of the trajectories at a given energy are no longer quasi-periodic. This is the onset of classical chaos in the system. A Poincaré section for such trajectories is shown in Fig. 9. Clearly, the generic point almost fills the whole volume of phase space bounded by the caustics. Nevertheless, this does not preclude the existence of remnant structures of tori, or the coexistence of quasi-periodic orbits with regions where classical chaos takes place. This is clear in Fig. 8, where the two types of behavior can be seen [26].

For higher values of energy and field, when

$$E > \frac{1}{2} \gamma L_z = E_I \quad , \tag{100}$$

where E_I is the ionization energy, it is likely that one will recover the more regular behavior associated with the onset of the Landau regime. As in the Coulomb limit, adiabatic invariants

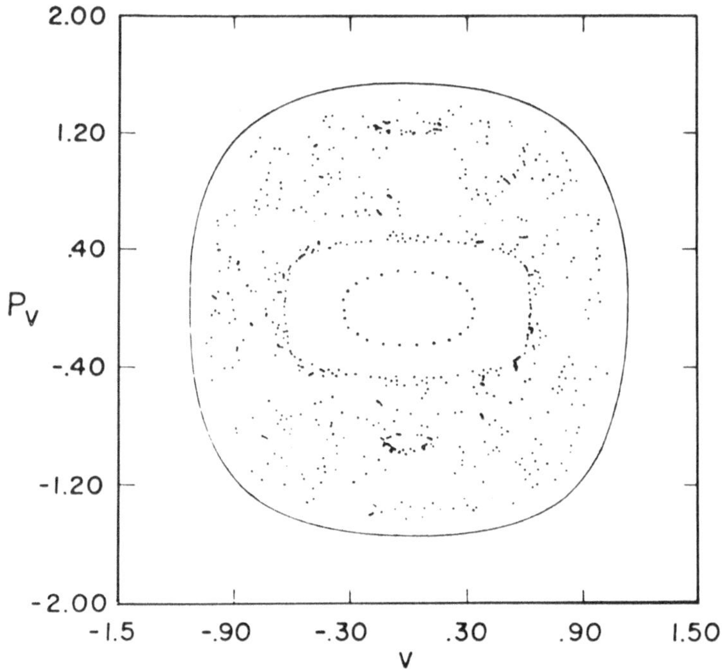

Fig. 8. Poincaré surface-of-section [26] for E = − 0.35 au, showing partial destruction of the regular features of Fig. 7. (Courtesy of W. P. Reinhardt.)

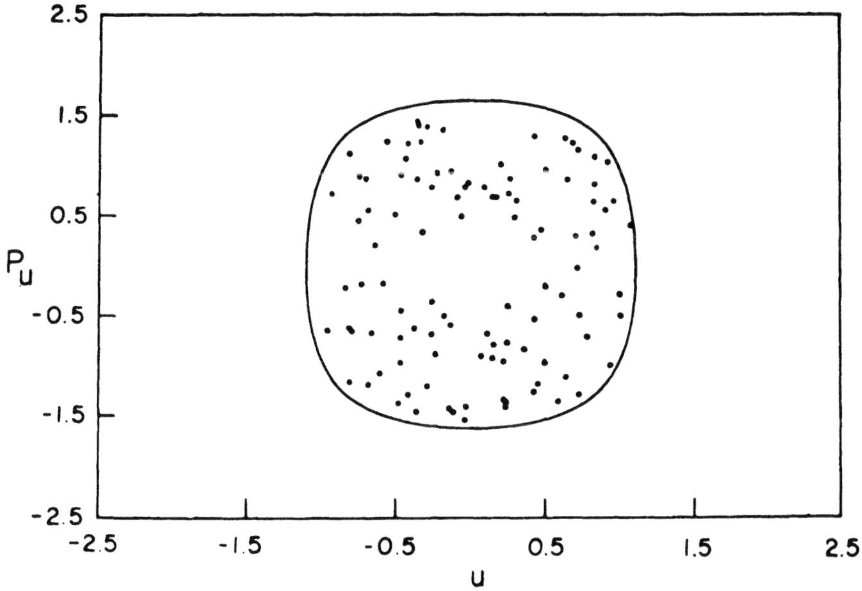

Fig. 9. Poincaré surface-of-section [26] for E = - 0.275 au, for a chaotic classical trajectory. (Courtesy of W. P. Reinhardt)

will exist, although they have not been found so far. Such invariants will allow us to explain the transition to the pure Landau situation.

To conclude, these Poincaré sections facilitate analysis of the evolution of trajectories in between the Coulomb and Landau limits, where highly degenerate motion takes place. For example, consider the high degeneracies associated with the two limiting cases. These degeneracies permit the electron to choose from among the set of classical ellipses at constant energy (either in the Coulomb or in the Landau limit) a trajectory which is still an ellipse, but which has parameters slowly evolving with the perturbation (either the diamagnetic one or the Coulomb one). In between, in the strong mixing regime, the existence of a torus is not forbidden, but a large part of the motion is irregular. In some sense, this region is quite similar to the so-called classical resonance region, or Hill's region, in the problem of coupled harmonic oscillators [27,28].

5.6 The Strong Mixing Regime as a Classical Resonance Region

Well-known examples of classical systems exhibiting instabilities and classical chaos are those dealing with coupled harmonic oscillators. Since the equation of motion usually reduces to the Whittaker-Hill form, the instability regions are called Hill's, or classical resonance, regions [28]. This is also true for many phenomena associated with parametric resonances. Our intent, here, is to show that the strong mixing regime of diamagnetism shares some common characteristics with these interesting questions of modern physics [29-31].

How can the present problem be put into the form of two harmonic oscillators interacting through anharmonic corrections? This can be obtained using a convenient set of coordinates in R(4) [9,17]. The Hamiltonian for the Coulomb problem is then transformed into the one for a four-dimensional isotropic harmonic oscillator, namely,

$$\tilde{H} = \underset{\sim}{p}^2 - E_o \underset{\sim}{r}^2 \quad , \tag{101}$$

where $\underset{\sim}{p}$ and $\underset{\sim}{r}$ are the four-dimensional impulsion and coordinate, respectively; and where E_o is the Coulombic energy which plays the role of $(-\omega^2)$, with ω being the oscillator frequency. The Coulomb problem is equivalent to solving the harmonic oscillator problem in R(4), but with constant energy [9]. A particular choice of the coordinates in R(4) is of interest in dealing with the diamagnetic interaction [17], namely,

$$\mu = (r + z)^{1/2} = \xi^{1/2} \quad \text{and} \quad \nu = (r - z)^{1/2} = \eta^{1/2} \quad , \tag{102}$$

where ξ and η are the usual parabolic coordinates (cf. Section 2.4). This coordinate system permits us to reduce the four-dimensional oscillator problem to two two-dimensional, uncoupled harmonic oscillators in perpendicular planes [17]. The expression of the diamagnetic interaction is then

$$H_D = \frac{\gamma^2}{4} \mu^2 \nu^2 \quad . \tag{103}$$

A further straightforward reduction of the problem can be achieved by writing the equations in terms of a new parameter τ, defined such that $(dt/d\tau)\alpha = (\rho^2 + z^2)^{1/2}$. If we neglect the rotational term in the equations of motion, the new form of the Hamiltonian, in (μ,ν) coordinates, is

$$\tilde{H} = p_\mu^{\;2} + p_\nu^{\;2} - E(\mu^2 + \nu^2) + \frac{\gamma^2}{4} \mu^2 \nu^2 (\mu^2 + \nu^2) \quad , \tag{104}$$

which shows the equivalence to the problem of two harmonic oscillators, possessing the same frequency, coupled through anharmonic corrections. Those solutions of Eq. (104) which are relevant to the physical problem for the energy E are the ones for which $\tilde{H}$ =

4. These results have been obtained in a slightly different form using the Levi-Civita regularization procedure [32], by several authors [23,26].

An interesting feature is apparent from the form of the potential, which is shown in Fig. 10 with the rotational energy included. The $\mu = \nu$ line corresponds to motion in the plane z = 0 in R(3), and is associated with a ridge where the potential energy is maximum and varies as $\gamma^2\mu^6$. Two valleys lie on either side of the ridge. Fano [33,34] has pointed out the importance of motion near this ridge for the dynamics and understanding of the quantum mechanical problem. Writing E = 0 in Eq. (104) permits an exact solution of the equations of motion. One then finds the exact condition

$$n\,\gamma^{1/3} = 1.16 \quad , \tag{105}$$

which is associated with the strong mixing regime [17,21].

Studies of the stability of the motion along the ridge can be made after changing to a new set of coordinates, namely:

$$u = 2^{-1/2}(\mu + \nu) \quad \text{and} \quad v = 2^{-1/2}(\mu - \nu) \quad .$$

The motion along the ridge is then associated with $v \simeq \varepsilon \simeq 0$. The u(t) solution of the equations of motion is then periodic,

$$u(t) \sim u_o \cos(\alpha t) \quad ,$$

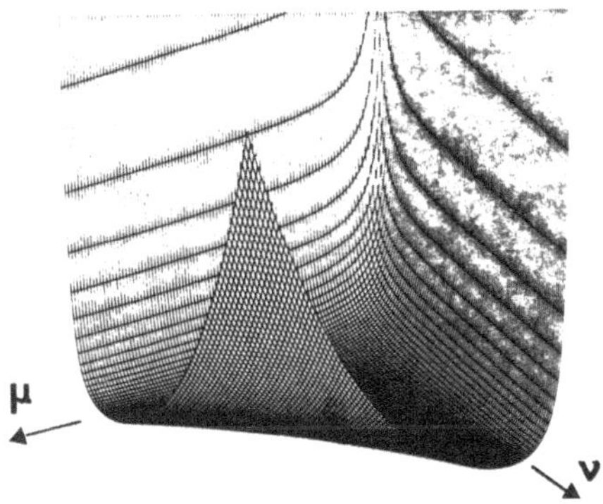

Fig. 10. Effective potential (including rotational energy) associated with Eq. (104) of the text. Motions in the z = 0 plane in real space are associated with motions along the ridge $\mu = \nu$.

and, substituting this solution into the above equation for v, we find that v(t) obeys ($s = \alpha t$)

$$\frac{d^2v}{ds^2} + (A + B\cos s + C\cos 2s) = 0 \quad . \qquad (106)$$

Eq. (106) is of Whittaker-Hill type and the regions of instability are well-known [28] and bounded, with some curves depending on (A,B,C) as well as the energy E. One can show that for energy values greater than $E_c \simeq -0.40$, the system enters an unstable region, which explains the coexistence of chaotic and regular motion in the plot of Fig. 8 for E = - 0.35. The motion along the ridge is then not necessarily stable. Such an analysis (which has obvious limitations) shows that the strong field mixing phenomenon is partly governed by the same kind of equations as those for the coupled-oscillators problem.

The fact that classical chaos appears in the strong mixing regime does not necessarily exclude the existence of an approximate dynamical symmetry. The true physical problem concerns the sizes of the regions of classical chaos, as compared to $\hbar^4$. From preliminary results by W. P. Reinhardt [26], it seems that there is still a possibility of constructing an invariant which would be valid in this intermediate regime.

5.7 Adiabatic Invariant in the Coulomb Limit

An explanation of the regular behavior observed below the transition energy, and described in Sections 5.3 and 5.5, has become possible with the recent discovery of an adiabatic invariant valid in the Coulomb limit [35]. The search for this invariant was, previously, not really unsuccessful [21,24], but no mathematical self-consistent formulation was found. This prior work [21, 24] was really a by-product of an earlier (1976) paper [36] dealing with group theory. The basic idea is that the problem is separable in impulsion space on the R(4) Fock hypersphere (in elliptico-cylindrical coordinates) [35]. This does not at all imply any separability in coordinate space, since such separability really does not exist. Nevertheless, this result is of importance for the classification of the quantum spectrum. But, first, it allows us to derive the expression for an adiabatic invariant in the Coulomb limit, which is

$$\Lambda = 4A^2 - 5A_z^2 \quad , \qquad (107)$$

where $\underset{\sim}{A}$ is the Runge-Lenz vector, defined in Eq. (6). Λ is an adiabatic invariant and not a constant of the motion. Λ permits us to describe how the zero-field Kepler ellipses are perturbed once the magnetic field is applied and thereby breaks the supersymmetry of the Coulomb problem. Classical trajectories are still ellipses, at any time, but their parameters are slowly evolving under the field action. How this occurs can be investigated using classical perturbation theory, by averaging the classical equation of motion on the unperturbed Coulombic ellipse. Then, it is possible to show that

$$\frac{d}{dt} \langle\Lambda\rangle = 0 \tag{108}$$

up to fourth order, in the B field. The average value of ρ^2 (proportional to the diamagnetic interaction) on an unperturbed ellipse is:

$$\langle\rho^2\rangle = -\frac{1}{2E^{(o)}}\left[4A^2 - 5A_z^{\ 2} + L_z^{\ 2} - \frac{1}{E^{(o)}}\right] , \tag{109}$$

where $E^{(o)}$ is the unperturbed energy which fulfills Eq. (9): That is, in atomic units,

$$A^2 - 1 = E^{(o)}L^2 \quad \text{with} \quad \underset{\sim}{A} \cdot \underset{\sim}{L} = 0 \quad . \tag{110}$$

The perturbed energy E is then

$$E = E^{(o)} - \frac{\gamma^2}{8E^{(o)}}\left[-\frac{1}{E^{(o)}} + L_z^{\ 2} - \frac{\Lambda}{E^{(o)}}\right] \tag{111}$$

or, writing $E^{(o)} = -1/n^2$,

$$E = E_o + \langle H_D\rangle = -\frac{1}{n^2} + \frac{\gamma^2}{8} n^2 (n^2 + M^2 + n^2\Lambda) \quad . \tag{112}$$

Practically, this means that the adiabatic turning on of the diamagnetic perturbation, on a hydrogenic orbit with a specified $(\underset{\sim}{A},\underset{\sim}{L})$, will induce a slow evolution of the parameters (excentricity, directions of the axes) of the trajectory, such that Λ is conserved to the fourth order in B. Of course, such an analysis is valid provided that the period of the evolution of the parameters of the ellipse is much larger than the period of the motion of the electron on the ellipse: That is,

$$\gamma n^3 << 1 \quad . \tag{113}$$

Direct numerical simulations show that Λ is no longer constant if $\gamma n^3 \simeq 1$ [21].

The existence of the adiabatic invariant allows us to gain some qualitative ideas concerning the evolution of the classical ellipse, as well as to derive some very important results about the symmetries of the trajectories in the Coulomb limit. From Eqs. (107-112), one easily finds (in atomic units; n and M not necessarily integers here!)

$$0 \leq A^2 \leq n^2-M^2, \ - n^2 \leq \Lambda \leq 4(n^2-M^2) \text{ and } 0 \leq \langle\rho^2\rangle \leq \frac{5}{2} n^4 \ . \tag{114}$$

Thus, the highest value of Λ corresponds to the highest value of $\langle\rho^2\rangle$ and of the energy. The evolution and symmetries of the orbits will be completely different depending upon the sign of Λ. During the slow deformation of the ellipse, the extremity of the Runge-Lenz vector $\underset{\sim}{A}$ moves on the surface defined by

$$4 A_x^2 + 4 A_y^2 - A_z^2 = \Lambda = \text{constant.} \tag{115}$$

This surface is a one-sheet hyperboloid if $\Lambda > 0$ and a two-sheet hyperboloid if $\Lambda < 0$. The orbits will then belong to two different classes. This is shown in Fig. 11a,b.

When $\Lambda < 0$ the $\underset{\sim}{A}$ vector, which is parallel to the major axis of the ellipse and fulfills Eq. (11) [i.e., the excentricity is $\xi = A/n$], is on the average directed along the $\underset{\sim}{B}$-axis, with one side of the z = 0 plane being perpendicular to the field. The ellipse, which is highly excentric if $\Lambda \simeq - n^2$, will have a stretching and vibrational motion in the $\underset{\sim}{B}$-direction. The same

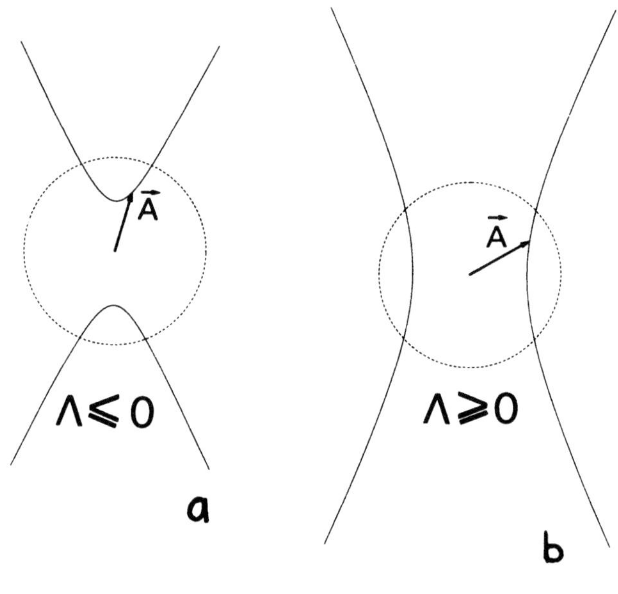

Fig. 11. Trajectories of the Runge-Lenz vector $\underset{\sim}{A}$ associated with negative (a) and positive (b) values of the adiabatic invariant, in the Coulomb limit. The dotted curve corresponds to an energy condition.

type of motion will take place on the other side of the z = 0 plane. However, coupling between these two types of trajectories can take place in classical mechanics.

On the other hand, if $\Lambda > 0$ the motion is completely different, as can be seen from Fig. 11b. The $\tilde{A}$ vector is allowed to move on the one-sheet hyperboloid on each side of the z = 0 plane. Thus, this set of trajectories will have some rotational symmetry around the $\tilde{B}$-direction. If Λ is close to its maximum value, $\tilde{A}$ will be confined in the z = 0 plane. The very excentric motion of the electron will then take place in the plane perpendicular to the field.

This evidence for the existence of two extreme types of symmetry of the classical trajectories in the Coulomb limit -- roughly associated with libration and rotation -- is an important result. Those orbits having mainly a rotational symmetry are associated with motions in the plane $z \simeq 0$, or motion along the ridge $\mu = \nu$. This appears to have important implications for tackling the strong field mixing regime. On the other hand, it is clear that Λ is no longer an invariant in this regime.

5.8 Adiabatic Invariant in the Landau Limit

The search for an adiabatic invariant in the Coulomb limit -- and its discovery -- is probably a major step forward in our understanding of the diamagnetic problem. Of course, another major step would be to have an explicit analytical form for that adiabatic invariant which is likely to exist in the other limit, namely the Landau limit. This turns out to be complicated, if one uses classical perturbation theory [21]. Nevertheless, one can get a coarse idea about the structure of such an invariant from the various constants of the motion (r_o^2, x_o, y_o) in the Landau limit (cf. Section 3).

5.9 Other Aspects of the Strong Field Mixing Regime

We will end these considerations of classical mechanics by giving some equivalent formulations of the strong field mixing regime [19,20].

From the equality of the diamagnetic and Coulomb forces, one can deduce that the strong mixing regime occurs when

$$\gamma n^3 \simeq 1 \quad . \tag{116}$$

Since the Coulomb contribution to the energy is of the order of $(-1/n^2)$, while the diamagnetic contribution is of the order of $n\gamma$ (in atomic units), when Eq. (116) is fulfilled it follows that

$$E \simeq E_o + E_D \simeq 0 \quad . \tag{117}$$

Thus, the regime occurs near the zero-field ionization threshold.

Now, if we consider atomic sizes, the orbit radius is of the order of n^2a_o in the Coulomb limit, and of the order of $n^{\frac{1}{2}} a_c$ in the Landau limit. When Eq. (116) is fulfilled, $n^2a_o \simeq n^{\frac{1}{2}} a_c$, which means that the dimension of the orbit is the same when evaluated in the two approaches. Eq. (116) also implies that the classical frequencies associated with the Coulomb and Landau problems are equal.

In summary, one can say that the strong field mixing regime occurs when the two forces are of the same order of magnitude, or when the classical frequencies of the equivalent oscillations are equal. This occurs near zero energy of the electron, close to the zero-field ionization limit. The sizes of the atomic orbits are then condensed in comparison to the zero-field situation, due to the extra binding action of the diamagnetic force. A more rigorous formulation can be derived from generalized Bohr models [3,37].

6. QUANTUM MECHANICAL APPROACH

Our understanding of the quantized version of the diamagnetic spectrum is far from complete. This is, of course, associated with the paucity of constants of the motion (i.e., L_z and parity) which commute with the Hamiltonian of Eq. (90). In spite of the fact that the two limiting situations, Landau and Coulomb, are well-understood and have conceptual importance, it is quite likely that no further attempts at this problem, after those of Jenkins and Segre [5] and Schiff and Snyder [18] in 1939, would have been made if experiments had not shown that, *for all atoms*, atomic spectra still present very simple characteristics in the intermediate regime, near and above threshold. Such simplicity in the quasi-Landau spectrum measured by Garton and Tomkins [6], in a situation where all intuitive arguments suggest the existence of strong complexity, reveals the basic conceptual importance of the problem [7].

As far as quantum aspects are concerned, only a moderate amount of progress has occurred since 1939 [37]. Because of the necessity of dealing with continuum states, the problem is really a tremendous one. In comparison, the Stark problem, which shares some common characteristics, can be solved easily since the Hamiltonian is separable in parabolic coordinates because of the existence of an exact dynamical symmetry associated with the Runge-Lenz vector. Since the existence of such symmetry is still an elusive question in the present problem, quantum approaches are rather undeveloped. Near and above threshold, no attempts have ever been made, except in the framework of WKB theories (see Section 7). Finally, quantum attempts have been developed essentially for dealing with the inter-ℓ and inter-n mixing regimes. Indeed, these approaches have provided us with a deeper understanding of the physics of these regimes, which were first studied by Schiff and Snyder [18]. However, this is in consequence of the recent discovery of an adiabatic invariant in the Coulomb limit, as discussed in Section 5.

Hereafter, we will use atomic units. The one-particle Hamiltonian of Eq. (90) then takes the form

$$H = p^2 + \gamma L_z + \frac{\gamma^2}{4} \rho^2 - \frac{2}{r} \tag{118}$$

where γ is defined in Eq. (95). The diamagnetic Hamiltonian H_D is

$$H_D = \frac{\gamma^2}{4} \rho^2 \quad . \tag{119}$$

6.1 Coulomb Inter-ℓ Mixing Regime

In this region, the diamagnetic term is small in comparison to the zero-field Hamiltonian. Nevertheless, the role of this term cannot be neglected since it breaks the super-symmetry of the Coulomb problem once the magnetic field is turned on. Obviously, this symmetry breaking is associated with noncommutation of the diamagnetic Hamiltonian H_D with both A_z and L^2. Neither ℓ nor $(n_2 - n_1)$ is a good quantum number; however, the principal quantum number n is still defined. As a first step, the diamagnetic interaction can be viewed as a perturbation acting on the degenerate hydrogenic manifold. Since the diamagnetic term is of the order of $\gamma^2 n^4$, this is valid provided that the perturbation is small when compared to the spacing $2/n^3$ of adjacent manifolds: That is,

$$\gamma n^{7/2} \ll 1 \quad . \tag{120}$$

Since L_z and parity are constants of the motion, the diagonalization of H_D can be performed separately on each subspace with fixed M and parity. From Eq. (119), we see that diagonalizing H_D is equivalent to finding the eigenstates and eigenvalues of ρ^2 in the hydrogenic manifold. And, this can be done in one of the numerous bases of the Coulomb problem, defined in Section 2.

The eigenstates which diagonalize H_D will be denoted $|n\,K\,M\,\Pi\rangle$, where K is a label replacing the ℓ or $(n_2 - n_1)$ value of the usual basis; they are field independent and are solutions of the field-free Coulomb problem, although not of the usual type. Writing

$$H = H_o + H_D \tag{121}$$

one has

$$H_o|n\ K\ M\ \Pi> = (-\frac{1}{n^2} + \gamma M)|n\ K\ M\ \Pi>$$

and

$$H_D|n\ K\ M\ \Pi> = E^{nK}\ (M,\ \Pi)|n\ K\ M\ \Pi>$$

Obviously, at this order of perturbation theory, E^{nK} (the diamagnetic correction to the energy) varies as B^2, and can be expressed in the general form

$$E^{nK}\ (M,\ \Pi) = \frac{\gamma^2 n^2}{4}\ (n^2 + M^2 + 1 + n^2 \Lambda_K) \tag{122}$$

where Λ_K is the eigenvalue of the quantum operator associated with the adiabatic invariant defined in Eq. (107).

The $|n\ K\ M\ \Pi>$ are eigenstates of ρ^2 in the hydrogenic manifold. The label K varies from 1 to the maximumm possible value, which is of the order of n/2 since these eigenstates have definite parity. They can be expressed as

$$|n\ K\ M\ \Pi> = \sum_{\ell} C_{\ell}^{nK}|n\ \ell\ M> = \sum_{n_1 n_2} D_{n_1 n_2}^{nK}|n_1\ n_2\ M> \tag{123}$$

in the spherical $\{nL^2L_z\}$ or parabolic $\{nA_zL_z\}$ representation. Typical variations of the energies E^{nK}, and the structures of the diamagnetic manifold, are displayed in Fig. 12. Under the present conventions, the K = 1 state is the one for which the energy increases fastest; that is, from Eq. (122), Λ_K is maximal. States associated with maximum values of K, on the other hand, exhibit the smallest increase in energy.

6.2 Properties of the Eigenfunctions in the Inter-ℓ Mixing Regime

The $|n\ K\ M\ \Pi>$ are common eigenvectors of H_o, and of the projection of ρ^2, on a given hydrogenic manifold. Calling $\{|\tilde{n}\ K\ M\ \Pi>\}$ the basis of the eigenvectors of Eq. (118), we find

$$|\tilde{n}\ K\ M\ \Pi> \rightarrow |n\ K\ M\ \Pi>\ ,\ \text{when}\ \ \gamma^2 n^7 << 1\ \ , \tag{124}$$

which shows that the $|n\ K\ M\ \Pi\rangle$ are the appropriate eigenvectors in the Coulomb limit. Moreover, these are the only eigenvectors which respect the symmetry in the manifold, which is broken once the field is turned on [38,39].

The $|n\ K\ M\ \Pi\rangle$ can be characterized by means of their spatial electronic density and nodal surface distributions, both of which obviously have cylindrical symmetry around $\tilde{B}$. Such a (ρ,z) plot of nodal surfaces is shown in Fig. 13, for $\tilde{n} = 17$, $M = \pm 3$ and odd parity. The lines represent the ensemble of points where the $\langle\rho\phi|nKM\rangle$ wavefunction is zero. Obviously, these nodal surfaces

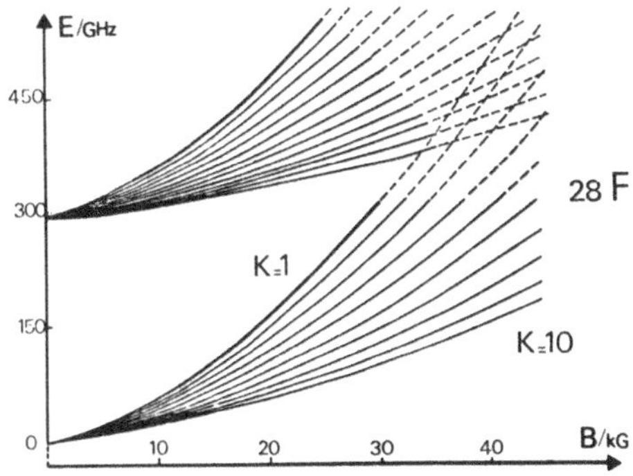

Fig. 12. Diamagnetic structure of the hydrogenic manifold in the inter-ℓ mixing regime.

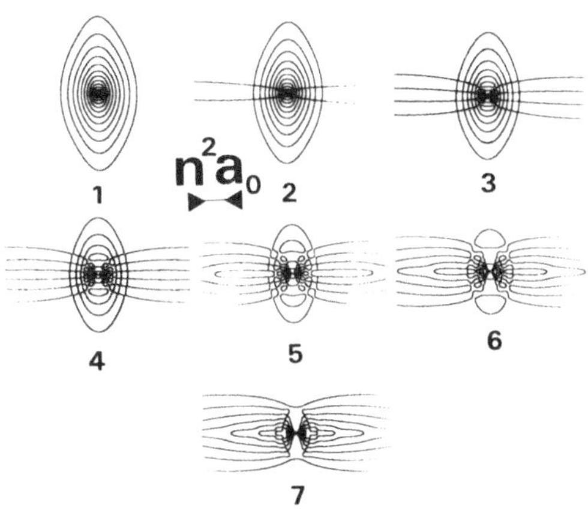

Fig. 13. Plot of the nodal surfaces of the $|nKM\Pi\rangle$ wavefunctions ($n = 17$, $M = 3$, odd parity) in the inter-ℓ mixing regime. (Taken from [39].)

are very different from those of the usual spherical basis [cf. Eq. (20)], which are an ensemble of intersecting spheres and cones (a result associated with the separability of the angular and radial equations, and with the spherical symmetry of the zero-field problem). The eigenstates of the restriction of ρ^2 in the manifold cannot have these properties, except for a rotational invariance around $\underset{\sim}{B}$. From Fig. 13, one can see how the nodal surfaces are built up from a merging of cones and spheres of the usual spherical basis -- with, as a result, completely new characteristic patterns and symmetries. As shown in enlarged form, in Fig. 14, the nodal lines never cross, which means that no separ-

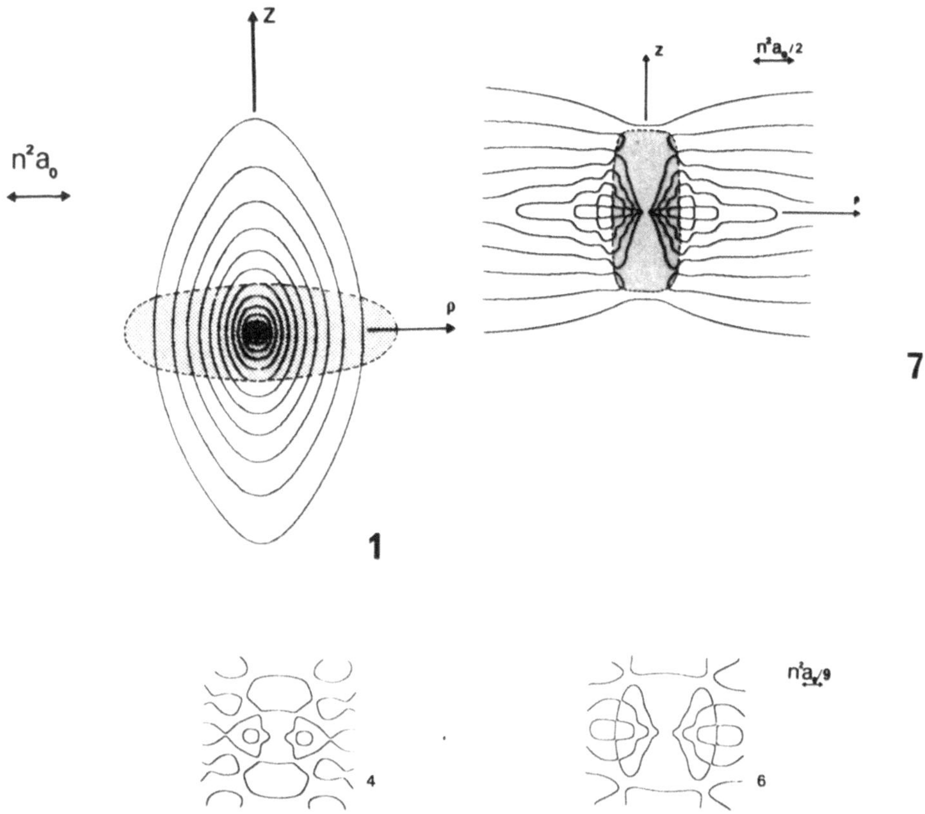

Fig. 14. Details of the nodal surfaces of Fig. 13. The shaded area corresponds to absolute values of the wavefunctions (K = 1 and K = 7) greater than 3 x 10^{-5}. The size of the spatial anti-crossings is of the order of $3a_0$ [39].

ability in coordinate space is allowed. However, the spatial anti-crossings are extremely small (of the order of a_0), while the extension of the wavefunction is of the order of $n^2 a_0$ [39].

The second important feature connected with these nodal surfaces is that the drawings near the core are much more complicated than the one far from the core. Inside, the details suggest an analogy with what happens in bifurcation processes, such as, for example, in systems of two coupled harmonic oscillators [27]. Outside, the patterns are much more regular and can be associated with vibrational patterns along $\underset{\sim}{B}$ for the maximum K values.

The major feature in Fig. 13 is the evidence of the very different symmetries of the wavefunctions, as K changes. From K = 1 to K = 7, the symmetry evolves from rotation to vibration along $\underset{\sim}{B}$. The diamagnetic structure of the hydrogenic manifold looks like the rovibrational bands of molecules. In Fig. 14, we have plotted the isoelectronic lines for which $|\Psi|$ is greater than 3×10^{-5}. Obviously, the K = 1 density distribution looks like that of a prolate spherical top, with a maximum concentration in a disk around the z = 0 plane, perpendicular to the field. The extension along the field direction is small in comparison to that along the ρ-axis. Indeed, such a state can be shown to possess an overall O(3) rotational symmetry [40], although not of the usual type. The electronic density distribution for the K = 7 state, at the bottom of the diamagnetic band, completely differs from the K = 1 state. The extension along $\underset{\sim}{B}$ is now several times greater than that transverse to $\underset{\sim}{B}$. Indeed, this looks like a vibrational state in the $\underset{\sim}{B}$-direction. Of course, it is not surprising that K = 1 and K = 7 states have very different transverse extensions; indeed, these extensions are proportional to $\langle\rho^2\rangle_K$, which is maximal for K = 1 states in the manifold.

Such a rovibrational structure of the diamagnetic manifold can be justified using more powerful arguments from group theory (see Section 6.5). In Fig. 15, we give a plot of the electronic densities of the $|n\ K\ M, \Pi = \pm 1\rangle$ states for the two parity values. (The label k used in this figure is slightly different from the K introduced previously; k now denotes each state in the (n,M) manifold [38], regardless of the parity.) These plots show that our previous conclusions are roughly valid but, certainly, not in detail.

Another interesting feature is that, in following the parity along the z-axis, Π_z, the states will either have or not have a node in the z = 0 plane. In the plot of nodal surfaces given in Fig. 13, the parity is odd, with M = 3. Thus, the parity along the z-axis is even since

$$\Pi_z = \Pi(-)^M \quad . \tag{125}$$

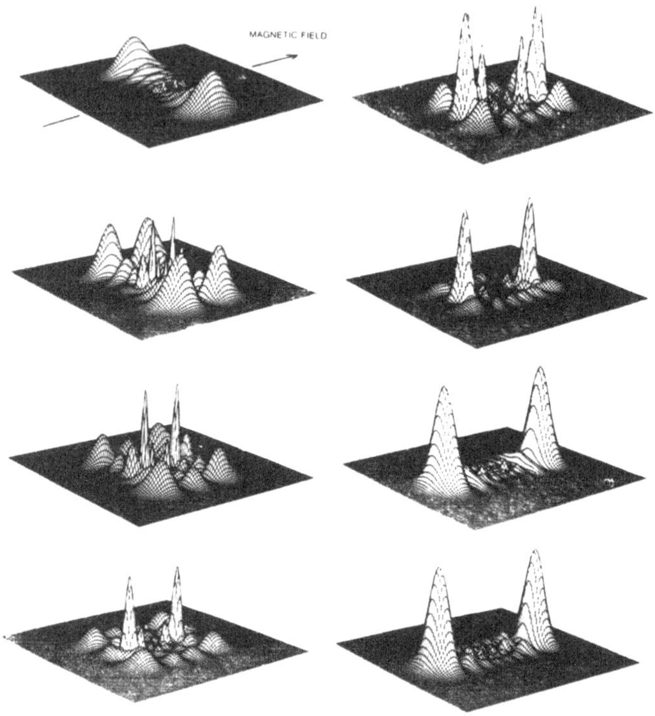

Fig. 15. Plot of the electronic densities of the wavefunctions for n = 8, M = 0 and the two different parity values: Π_z even (left) and odd (right). Only one of these two types of states can be optically excited with polarized radiation. (Courtesy of M. L. Zimmerman.)

And, therefore, the electronic density in the z = 0 plane is non-zero. Of course, for the other set of states belonging to the same manifold, with the opposite parity value, there is a node in the z = 0 plane. This can be seen on the right part of the plot in Fig. 15. This feature is extremely important. The electronic density distribution in Fig. 15, for states with small K values and opposite parities, are completely different. Consequently, their diamagnetic energies, proportional to $\langle\rho^2\rangle$, will differ. Such states with rotational symmetry at the top of the diamagnetic band are not likely to have the same properties and energies. On the other hand, as can be seen from Fig. 15, states with maximum K values are strongly peaked along the z-axis, whatever the parity, and present the same general vibrational symmetry. The fact that there is, or is not, a node in the z = 0 plane is not important here since the wavefunction is almost zero in this region. This means that the states at the bottom of the diamagnetic band will have almost the same diamagnetic energy and will be more or less degenerate depending upon the parity. This result was derived previously by C. W. Clark [41].

All of the above results become evident when we refer to the classical analysis of Section 5.7. States with minimum values of K, at the top of the diamagnetic band, are associated with maximum values both of $\langle \rho^2 \rangle$ and of the adiabatic invariant Λ, in accord with Eq. (114). These states are associated with classical trajectories in which A^2 is maximal and A belongs to the one-sheet hyperboloid of Fig. 11. Thus, it is not surprising that these states are concentrated near the z = 0 plane and exhibit an approximate O(3) symmetry. States at the bottom of the band are associated with the minimum values of Λ and ρ^2, and with trajectories in which the Runge-Lenz vector $\underset{\sim}{A}$ lies on the two-sheet hyperboloid of Fig. 11. The symmetries of these states will be roughly of vibrational type. Quantum mechanically speaking, $\underset{\sim}{A}$ can tunnel from one sheet to the other. This explains why states with different parities at the bottom of the band will be almost degenerate.

Finally, one should note that the states with minimum values of K are associated with motion near the z = 0 plane, or motion along the ridge $\mu = \nu$ in Fig. 10. Depending upon the values of K and Π_z, these states will be concentrated near the ridge, or will have nodes on the ridge [33] and in the direction transverse to the ridge [42]. However, those states which are confined in the potential valleys of Fig. 10, far from the ridge, will have an approximate $O(2) \otimes O(2)$ symmetry associated with vibration, and only small couplings through the barrier. (This is another way of explaining the degeneracy at the bottom of the diamagnetic bands.) The very special patterns of the nodal surfaces of these states may possibly be a remnant of the chaotic behavior of the classical trajectories observed under strong field mixing conditions. The pattern of nodal surfaces of K = 1 states, associated with motion along the ridge, is obviously much more regular.

The last important question is that of the properties of the above eigenstates under optical excitation. The intensity of the various lines are proportional to $|\langle i|\underset{\sim}{\varepsilon}\cdot\underset{\sim}{r}|n\ K\ M\ \Pi\rangle|^2$, where $|i\rangle$ is the lower state of the transition (usually unperturbed by the diamagnetic interaction and proportional to the spherical harmonic function), and where $\underset{\sim}{\varepsilon}$ is the radiation polarization vector. From Eq. (123), all of the $|n\ K\ M\ \Pi\rangle$ states with appropriate values of parity and M will be excited. However, depending upon the parity Π_z, the $|n\ K\ M\ \Pi\rangle$ may have a node in the z = 0 plane. Then, the efficiency of optical excitation will vary with the polarization $\underset{\sim}{\varepsilon}$. This is quite clearly shown in Fig. 16, where a

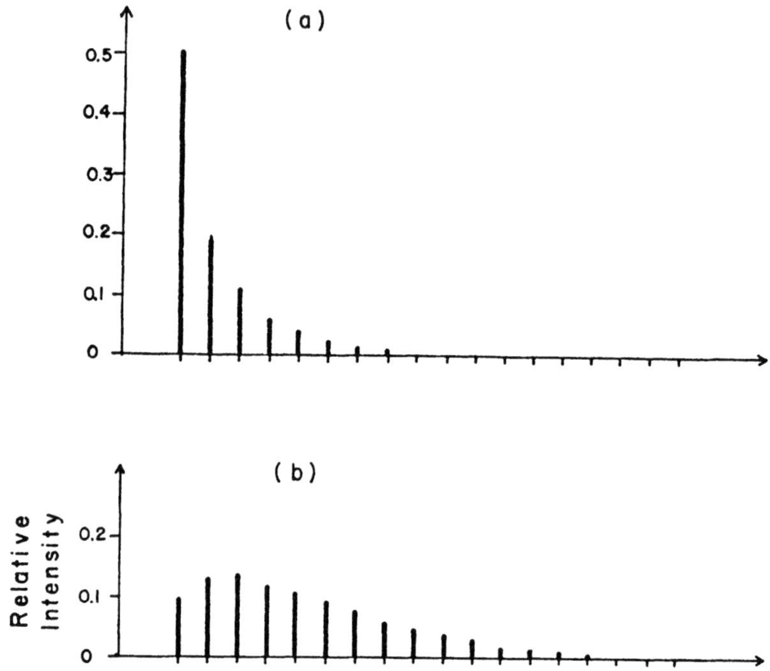

Fig. 16. Efficiency of optical excitation of the various K components in the manifold for Π_z even (a) and odd (b). (Courtesy of J. C. Castro-Neto.)

plot of line intensities is given for even and odd parity Π_z. For Π_z even, the efficiency of optical excitation of K = 1 states is very important, and decreases with the value of K. This means that the overlap of the $|n\ K\ M\ \Pi\rangle$ state with the spherical harmonics of lowest ℓ_0 value in the manifold is maximal, while the overlap between $|n\ \ell_0\ M\rangle$ and the other K states is smaller as K decreases. Then, in Eq. (129) the $C^{nK=1}_{\ell=\ell_0}$ coefficient is maximal for Π_z even. On the other hand, if the z-parity is odd, the $|n\ K{=}1\rangle$ states have a node in the z = 0 plane, which means that the overlap with the lowest-order spherical harmonics $|n\ \ell_0\rangle$ will be weaker. The optical absorption strength will be more sensitive to the spatial extension of the wavefunction and, in comparison to the previous situation, must be smaller since the ground state is peaked around the origin, where $\langle\rho\phi|n\ K{=}1\rangle$ does have a minimum at z = 0. In fact, as shown in Fig. 16, the K = 1 state no longer has the greatest probability of optical excitation, and the distribution of oscillator strengths is spread over the various K values. These features turn out to be of extreme importance for the interpretation of experimental data, and for an understanding of the quasi-Landau phenomenon. Some of these results, however, could have been derived from the classical analysis of Section 5.7.

6.3 Inter-n Mixing Regime

As shown in Fig. 12, if the field or the energy is increased, one encounters the inter-n mixing regime, in which the diamagnetic manifolds merge. This occurs when

$$\gamma^2 n^7 > 1 \quad , \quad \gamma n^3 << 1 \quad . \tag{126}$$

The Coulomb principal quantum number n is no longer a good quantum number, but the diamagnetic interaction is not yet dominant over the Coulomb interaction. The eigenstates $|\tilde{n}\ K\ M\ \Pi\rangle$ are not known; however, they can be obtained, with limited accuracy, through the diagonalization of the Hamiltonian in an $\{|n\ K\ M\ \Pi\rangle\}$ basis, involving as many manifolds as possible. That is,

$$|\tilde{n}\ K\ M\ \Pi\rangle = \sum_{n_o K_o} \alpha_{n_o K_o}^{\tilde{n}K} |n_o\ K_o\ M\ \Pi\rangle \quad , \tag{127}$$

where the α coefficients now depend upon the magnetic field strength. No closed form of Eq. (127) has ever been obtained, which is a consequence of the R(3) nonseparability of the Hamiltonian. The only way of getting the eigenvectors in the inter-n mixing regime is through numerical calculation. But, since H_D couples any state with, in particular, continuum states, there are practical limitations!

One can, nevertheless, develop some intuitive arguments which help in understanding this regime. First of all, the diamagnetic contribution to the energy will no longer behave as B^2. This is the main result of the interaction of a given n manifold with the other ones: Since the repelling action of those states above n is stronger, the increase in energy will become smaller than B^2. This is to be expected since, in the Landau limit, the energy must increase linearly with B. Second, important information can be drawn from the fact that the states in the inter-ℓ mixing regime have very different limiting symmetries. Numerical diagonalizations showed, in 1980 [38,39], that the $|n, K=1\rangle$ states and the $|n+1, K_{max}\rangle$ states interact only weakly through H_D. As shown in Fig. 17, the sizes of the anti-crossings between these states are exponentially small, varying as e^{-2n}. On the other hand, one can show that the states $|n, K=1\rangle$ and $|n+1, K=1\rangle$ interact strongly and, in consequence, exhibit a smaller rate of increase of energy with B. Such features were considered to be a clue to the existence of a hidden, approximate dynamical symmetry in the problem [38]. But, of course, this is now quite well-understood from the symmetry arguments developed in Sections 5.7

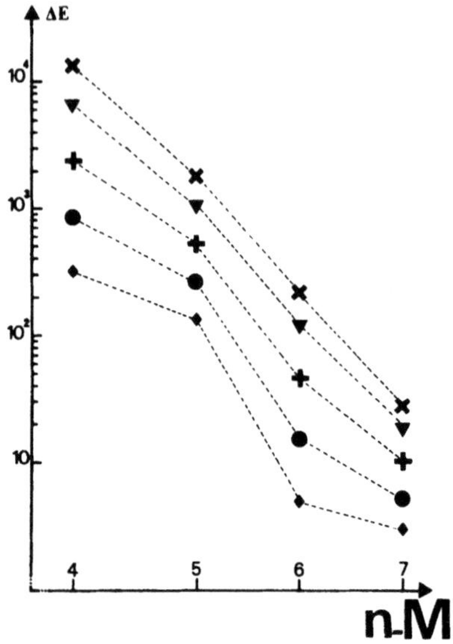

Fig. 17. Size of the anti-crossing (in GHz) between $(n, K = 1)$ and $(n+1, K_{max})$ for the lowest energy levels, in logarithmic coordinates, as a function of $(n - M)$. The size varies almost as e^{-2n}. Each broken curve corresponds to a fixed M value which is, from the upper curve to the lower one, $M = 0$ to $M = 4$. For $M = 4$, even Π_z, the first anti-crossing between $(n, K = 1)$ and $(n+1, K_{max} = 2)$ occurs for $n = 8$ at a magnetic field strength of 588 T, while for $M = 4$, Π_z even, $n = 12$ it is at 144 T.

and 6.2, which demonstrated the existence of an adiabatic invariant in the Coulomb limit. The approximate O(3) and O(2) ⊗ O(2) symmetries of the $|n, K=1\rangle$ and $|n, K_{max}\rangle$ states are quite different, thus ensuring the weakness of anti-crossings between these states. The exponential size of the anti-crossing follows from the fact that the spatial overlap of the wavefunctions occurs in classically forbidden regions, where the wavefunctions behave as e^{-n}. This is also to be expected from the picture of motion along the ridge [33,34]: The two types of motion will take place, respectively, on the ridge or in the double valleys and, therefore, the overlap will be small.

One can conclude that the $|n\ K\ M\ \Pi\rangle$ states will interact essentially with those other states belonging to the same symmetry class. This supplies us with a coarse model in which the spectrum is composed of a set of several weakly interacting series [43]. Each series is identified through {K M Π}, where the label K is associated with the Λ_K eigenvalue of

$$\Lambda = 4\,A^2 - 5\,A_z^{\;2}$$

in the low-field limit. Such a view is in agreement with the analysis of Section 6.4, as well as with experimental results.

Given the lack of any real separability in coordinate space, but taking advantage of our previous remarks, it is likely that some clever choice of basis will simplify numerical calculations. Of course, suitable bases are not expected to be the same for K_{min} and K_{max} states because of the different symmetries of these states. For those states associated with small K values, which possess approximate O(3) symmetry, the use of a rotated parabolic basis gives a convenient representation [40,44]. Indeed, the off-diagonal elements of H_D can be rendered small, given an appropriate choice of the rotation (which is not of the same type as the usual rotation in real space). This is intimately associated with the possibility of using several independent sets of generators for the Lie group involved in the R(4) approach to the Coulomb problem [9].

The above remarks can be motivated by the following argument, which is based upon the well-known Fermi replacement

$$\underset{\sim}{r} \rightarrow \frac{3}{2}\, n\, \underset{\sim}{A} \quad . \tag{128}$$

in a given energy shell. Making this replacement in H_D gives

$$H_D = \frac{\gamma^2}{4}\rho^2 = \frac{9\gamma^2}{16}\,(A_x^{\;2} + A_y^{\;2}) \tag{129}$$

which is, however, *incorrect*. From the previous sections, we have seen that H_D must be proportional to $(4A^2 - 5A_z^2)$. This latter result, which is correct, follows from taking proper account of the symmetries of H_D in a given energy shell, or from a convenient generalization of the Wigner-Eckardt theorem. But, although Eq. (129) is wrong, it turns out to coincide with the exact *limiting* expression of H_D at the top of the diamagnetic band, when $A_z \ll A_x, A_y$. To show this, consider the operator $\underset{\sim}{\lambda} \equiv (A_x, A_y, L_z)$, which possesses all of the properties of an angular momentum (see the results of Section 2.2). Then, from Eq. (129), a *limiting* expression of the restriction of H_D to the manifold is proportional to $(\lambda^2 - \lambda_z^{\;2})$. From Eqs. (14-16), which define $\underset{\sim}{J}_+$ and J_-, an obvious choice of representation for the Coulomb problem is $\{J_+^2\; J_-^2\; J_{+z}\; J_{-z}\}$, where the two angular momenta are not coupled. This is associated with the well-known set of eigenfunctions in parabolic coordinates (provided with some convenient symmetrization). In a given manifold, the choice of the coupled representation $\{J_+^2\ J_-^2\ \lambda^2\ \lambda_z\}$ diagonalizes one of the limiting forms of the diamagnetic Hamiltonian. The resulting wavefunctions are (up to phase factors) deduced from the parabolic basis through a rotation. This is not a rotation in real space since it involves $\tilde{A}$; however, the coefficients of the expansion are basically the Clebsch-Gordan coefficients. These approximate wavefunctions would certainly be extremely useful for tackling the inter-n mixing regime. In addition, since H_D is proportional to $(\lambda(\lambda+1) - M^2)$ in this representation, the analogy to states of the oblate spherical top becomes clear. And, from the definition of $\tilde{\lambda}$, one can also conclude that the O(3) state symmetry is not the usual spatial type [40] since, basically, it involves $\tilde{A}$ instead of $\tilde{L}$.

For states associated with large K values, at the bottom of the diamagnetic band, the limiting form of H_D given in Eq. (129) is obviously wrong. In this case, the conditions are $A_z \gg A_x, A_y$ and H_D is simply proportional to A_z^2. The symmetries of the states are approximately $O(2) \otimes O(2)$ and an approximate set of constants of the motion is $\{A_z^2, L_z\}$. This means, as first pointed out by C. W. Clark [41], that the use of the parabolic basis is appropriate for dealing with such states in the inter-n mixing regime.

6.4 The Sturmian Approach to the Inter-n and Strong Mixing Regimes

So far, the only accurate quantum study of the inter-n and strong mixing regimes employed a Sturmian representation of the Coulomb spectrum [42]. The Sturmians form a complete set of discrete, nonorthogonal functions which allow us to represent the hydrogenic spectrum, and to take continuum states [45] into account through a discrete representation. Expansion convergence requires a large basis of Sturmian functions and the use of a Cray computer. Some of these numerical simulations [42,43] are shown in Fig. 18, for atomic hydrogen. They are limited to negative energies because of accuracy and convergence requirements. The results, however, partly confirm the intuitive arguments given in Sections 6.2 and 6.3; they also are in agreement with experiment (see Section 8).

From Fig. 18, one can see that the atomic spectrum is far different for states with different z-parity. This was explained in Section 6.2 for inter-ℓ mixing conditions, and that discussion remains valid for inter-n mixing conditions and also near the zero-field threshold. The main result is that when the z-parity is even (i.e., $(-)^{\ell+M} = +1$), the K = 1 lines dominate in the inter-n mixing regime. In contrast, regular features are hardly

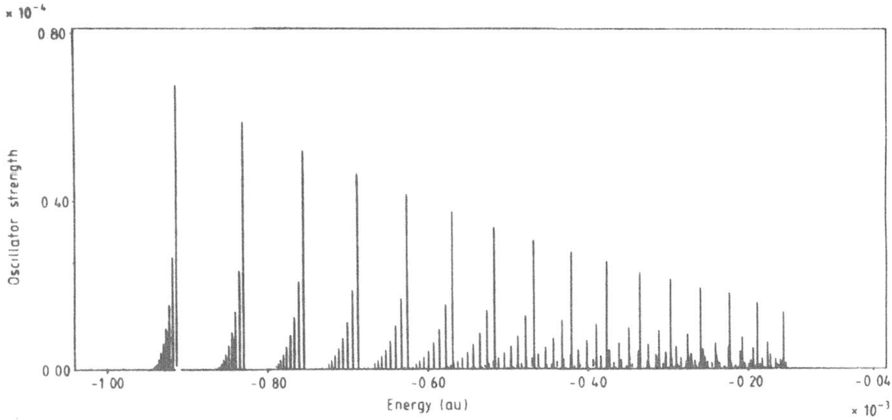

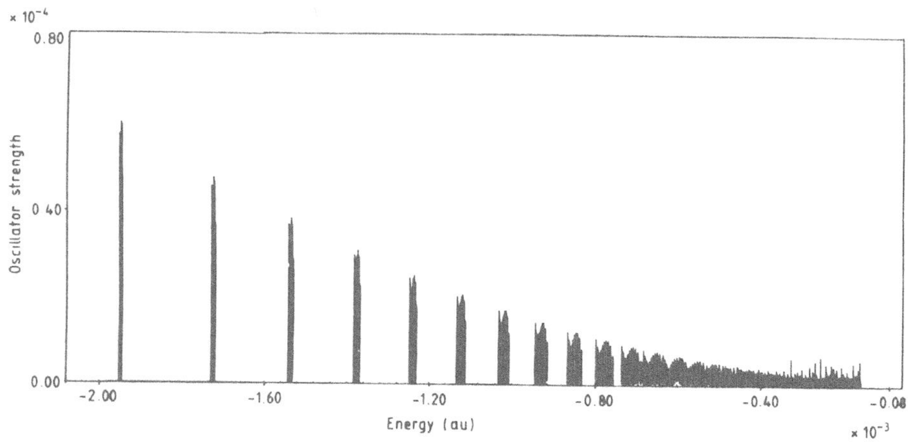

Fig. 18. Numerical Sturmian simulations of the spectrum for atomic hydrogen for B = 47 kG, for even (top) and odd (bottom) values of Π_z. (Courtesy of C. W. Clark.)

distinguishable for odd z-parity, once the various manifolds begin to merge. Moreover, in the plot for even z-parity, the secondary lines, associated with the K = 2,... states, do not strongly interact with the other ones. This confirms our previous view of noninteracting, or only weakly interacting, series of lines associated with the various K states as precursors of the quasi-Landau spectrum in the low-field inter-ℓ mixing regime. In addition -- and this will be discussed in Section 7 -- the positions of the various lines agree fairly well with the predictions of semi-classical theories, in which one assumes a very weakly excited motion along B. These various series, for several K values, are shown in Fig. 19.

6.5 Group Theoretical Meaning

The fact that the wavefunctions have very nice symmetry properties in the inter-ℓ mixing regime can also be deduced from a group theoretical analysis, in the {p}-representation, of the Fock hypersphere [40]. This allows us to deduce the form of the adiabatic invariant [Eq. (107)] and to find the wavefunctions in the inter-ℓ regime. Such an analysis also points out the O(3) symmetry of the states at the top of the band (in strong analogy

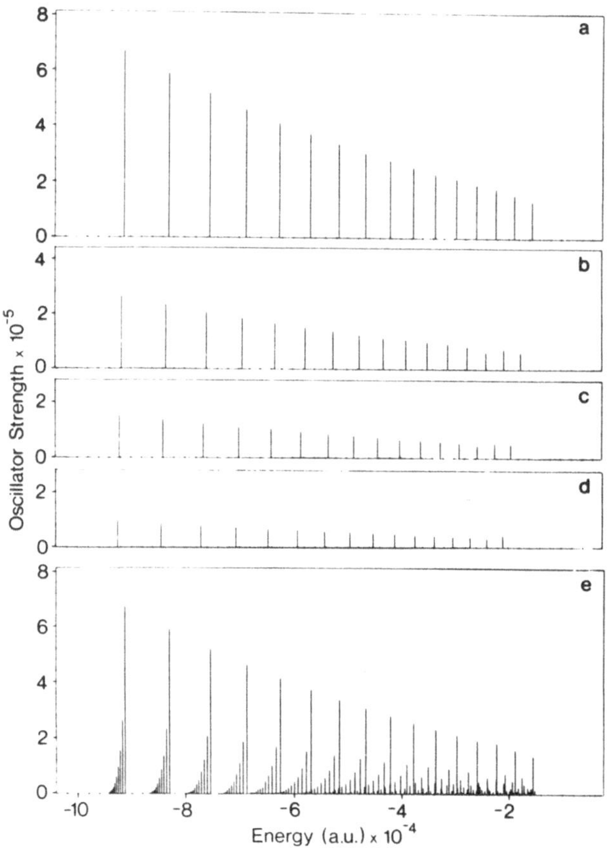

Fig. 19. Components associated with K = 1 (a), 2 (b), 3 (c) and 4 (d) lines in the even parity spectrum (e) of Fig. 18. (Courtesy of C. W. Clark.)

to those of an oblate spherical top), while the states at the bottom of the band lie in a double potential well and have an approximate $O(2) \otimes O(2)$ vibrational symmetry. These latter states are weakly coupled through tunneling, which explains the quasi-degeneracy of these states under the parity operation. (The diamagnetic band looks like the rovibrational structure of a molecule.) This approach also confirms the relationship between the present problem and that of the classification of doubly-excited states in atoms. This was first pointed out by Fano [46] on the basis of the ridge behavior shared by the two problems [40].

Of course, the fact that the problem separates in momentum space in R(4), in elliptico-cylindrical coordinates [9,35,40], does not mean that it separates in real space, since the former separability occurs on a given energy shell of R(4). In other words, this means that

$$\Lambda = 4\, A^2 - 5\, A_z^{\;2}$$

commutes with the Hamiltonian only approximately.

6.6 Strong Mixing and Landau Regimes

The previous considerations, some of them from very recent work, sum up the state-of-the-art as far as quantum approaches to diamagnetism are concerned. Because of the lack of separability, and the involvement of continuum states, no quantum approach has been developed near or above threshold. Continuum states obviously play an important role. Once the field is applied, regardless of its strength, a complete discretization of the continuum occurs, leading to the appearance of resonances rather than bound states, and recalling the possibility of ionization along the B-field direction. Actually, no quantum description of these states exists and their properties are unknown. The nature of the correlation diagram between the Coulomb and Landau limits is, therefore, still an open question.

6.7 Nonhydrogenic Atoms

The first effect of non-Coulombic corrections to the potential will be to break the super-symmetry of the Coulomb problem. This means that the $|n\ell\rangle$ states will no longer be degenerate in zero-field. For low field values, the diamagnetic interaction will be negligible in comparison to the quantum defects associated with neighboring levels. A first regime will exist in which the only role of the diamagnetic interaction will be to increase the energy of each level by the amount

$$\Delta E_D = \langle n\ell M | H_D | n\ell M \rangle$$

$$= \frac{q^2B^2}{8m} a_o^2 \frac{n^2[5n^2+1 - 3\ell(\ell+1)][\ell^2+\ell-1+M^2\ell]}{(2\ell-1)(2\ell+3)} , \qquad (130)$$

which is derived from conventional first-order perturbation theory. In addition, because of the mixing of wavefunctions, several weaker lines will appear.

In a second regime, the diamagnetic term will become dominant over the non-Coulombic contributions and the patterns will look more or less like the one seen in the purely hydrogenic case, at least as far as the energies are concerned. The main difference is that the oscillator strengths associated with the various K components will be extremely sensitive to the short-range corrections to the potential. And, this means that the intensity patterns will be extremely different from the ones shown in Figs. 16 and 18. It is likely that, in the absence of the dominant features associated with the K = 1 lines of the purely hydrogenic case, the interpretation of experimental data will be difficult.

6.8 Ultra-High Field Limits

If $B > B_c$ or $\gamma > 1$, the quantum problem is almost completely solved. This is the domain of ultra-high fields [37], or of solid-state physics of excitons [1]. The wavefunctions are then of Landau type as far as the transverse part is concerned, while the longitudinal part is the solution to a one-dimensional Coulomb problem [47,48]. Such an approach is valid since $\gamma \gg 1$ permits us to make the adiabatic approximation for the z motion. Several interesting questions are still unsolved but we will not discuss these high-field aspects here.

7. SEMI-CLASSICAL METHODS

In the absence of any quantum theoretical approach to the strong mixing regime, the only way of making predictions is through semi-classical theories. A semi-classical approach seems quite appropriate, however, since we are concerned with highly-excited states. Indeed, such attempts have been made earlier and have resulted in important advances in the understanding of the various mechanisms operative in the strong mixing regime [49-51]. Unfortunately, there remains a major difficulty which is, as expected, associated with the nonseparability of the Hamiltonian.

7.1 Semi-Classical Quantization of Nonseparable Problems

The usual Bohr-Sommerfeld quantization conditions apply only to one-dimensional problems, and have the form (γ_i is the Maslov index; for further details, see the paper by M. Robnik in this volume)

$$\oint p_i \, dq_i = (n_i + \gamma_i)h \quad . \tag{131}$$

These conditions, therefore, cannot be used for a nonseparable problem. Eq. (131) is of no use for the diamagnetic problem except in the Landau and Coulomb limits; in particular, it does not apply in the region between these two limits. The right way to perform a semi-classical quantization of the diamagnetic problem is to use EBK (Einstein-Brillouin-Keller) quantization, which was first suggested by Einstein [26,53]:

$$\oint_{C_i} \underset{\sim}{p} \cdot d\underset{\sim}{q} = (n_i + \gamma_i)h \quad . \tag{132}$$

The paths C_i are not classical trajectories but, rather, topologically-distinct fundamental paths on the surface of an invariant torus in phase space, while the n_i are the quantum numbers associated with these independent paths. Tori quantization, however, requires a study of the classical motion. And, this is one of the reasons why classical studies, such as those described in Section 5, are of importance for our present problem. Actually, there is no real proof of the survival of a torus in the intermediate regime, where $n^3\gamma \simeq 1$. Thus, in this regime Eq. (132) may not apply. However, more general methods, based on Feynmann path integral techniques, do not require the survival of tori as a condition for quantization [31,54]. Nevertheless, even though these methods are the only rigorous ones available, they have not yet been applied to the diamagnetic problem because of their evident complexity. The conclusion, then, is clear: There are, at present, no correct semi-classical predictions available. This does not mean, however, that the various approximate semi-classical theories which have been developed are not of interest.

7.2 Approximate Semi-Classical Quantization of the Transverse Motion

As shown in Section 6, the dominant lines in the diamagnetic spectrum are those associated with wavefunctions concentrated near the $z = 0$ plane and (at least) having a low degree of excitation along $\underset{\sim}{B}$. In addition, the main reason for the nonseparability of the Hamiltonian in cylindrical coordinates is the coupling between the z and ρ motions. Making the approximation $z = 0$ in the Hamiltonian allows us to deal with a one-dimensional problem in ρ. Bohr-Sommerfeld quantization of such a problem is then possible; one finds (in atomic units)

$$\int_{\rho_1}^{\rho_2} \left[E_\perp - \frac{1}{4}\gamma^2\rho^2 + \frac{2}{\rho} - \frac{(|M| + 1/2)^2}{\rho^2}\right]^{\frac{1}{2}} d\rho = (n_r + \alpha)\pi \tag{133}$$

where (ρ_1,ρ_2) are the classical turning points, while the expression for the rotational term has been choosen in order to respect the low-field diamagnetic behavior of the K = 1 line. In this limit, n_r and the Coulomb quantum number n are related by

$$n = n_r + |M| + 1$$

if we assume that $\alpha = 1/2$ in Eq. (133).

Obviously, the above approach suffers from various shortcomings and difficulties. For example, it does not account for the effects of the motion along $\underset{\sim}{B}$, which will be responsible for the presence of secondary lines and other important features such as autoionization! Furthermore, it does not reproduce correctly either the Coulomb limit[$\alpha = 1/2$ in Eq. (133)] or the Landau limit [$\alpha = 1/4$ in Eq. (133)]. This means, at least, that α depends smoothly on both the field and the energy. Thus, Eq. (133) cannot afford exact predictions for the energy levels because of its two-dimensional approximation of what is, in fact, a three-dimensional problem.

Nevertheless, Eq. (133) is of interest provided that one does not require absolute predictions. Indeed, predictions from Eq. (133) are in good overall agreement with both the experimental results and the numerical simulations of Clark and Taylor for the K = 1 lines [43]. Even if one cannot trust Eq. (133) for absolute determinations of line positions, the predictions of line spacings are correct and in good agreement with quantum calculations of the Coulomb inter-n and inter-ℓ mixing regimes. Therefore, Eq. (133) fairly well describes the general behavior of the K = 1 lines, which are those associated with wavefunctions concentrated near the z = 0 plane.

From Eq. (133), one easily derives the major features of the strong mixing regime: That is, near the zero energy threshold, one finds

$$n_r + 1/2 = 1.16\ (B_c/B)^{1/3} \quad , \tag{134}$$

which is the exact form of the quantization law $n^3\gamma \simeq 1$. Moreover, the expression for the spacing of the resonances at threshold is [50,51]

$$dE/dn_r = \frac{3}{2}\hbar\omega_c \quad , \tag{135}$$

which is the spacing found experimentally by Garton and Tomkins [6,55]. This explains why Eq. (133), or other modified forms, are still widely used today. Finally, recent developments have shown that Eq. (133) can be put in a closed analytical form using elliptic integrals [49], and that it applies to many two-dimensional problems.

7.3 Exact Semi-Classical Quantization of the Three-Dimensional Motion in the Coulomb Limit

The recent discovery of the adiabatic invariant Λ permits an exact semi-classical treatment of the three-dimensional problem in the Coulomb limit. Since this problem is approximately separable, one can use either the Bohr-Sommerfeld conditions, or quantization on invariant tori via Eq. (132); one finds [35]

$$\int_{\theta_1}^{\theta_2} L_\perp(\theta)\,d\theta = \pi(K + 1/2) \quad , \tag{136}$$

where θ is the angle between $\underset{\sim}{B}$ and the Runge-Lenz vector $\underset{\sim}{A}$, and where $L_\perp$ is the component of the angular momentum perpendicular to the $(\underset{\sim}{A}, \underset{\sim}{B})$ plane. The explicit expression for $L_\perp^2$ is

$$L_\perp^2(\theta) = n^2\left(1 + \frac{\Lambda}{1 - 5\sin^2\theta}\right) - \frac{M^2}{\sin^2\theta} \quad , \tag{137}$$

where n and M are associated with quantization of the energy and L_z, respectively. From Eq. (136) one gets the quantized eigenvalues Λ_K of Λ and, finally, Eq. (112) for the energy. Such a procedure is rigorous in the Coulomb limit since the three integrals of the motion which are in involution (H_o, Λ, L_z) do define a torus in phase space. This is then equivalent to the topological, semi-classical EBK quantization [26]. As a result, the predictions are in excellent agreement with the quantum calculations for *all* the K states in the manifold. And, of course, the method does not suffer from the various shortcomings described in Section 7.2.

7.4 Three-Dimensional Semi-Classical Approach in the Landau Limit

Finally, we must stress the fact that a very early attempt at the problem which serves as the title of this section was made in 1966 by Monozon and Zhilich [56,57]. Although the model was developed under conditions where $\gamma \gg 1$, it seems to have a fairly wider validity. The basic idea is to use the adiabatic approximation in the Landau limit and a relation similar to Eq. (133), in which the classical turning points ρ_1 and ρ_2 now depend on the variable z. This allows one to deduce proper adiabatic potentials for the z motion. Furthermore, semi-classical quantization of the three-dimensional motion gives two limiting formulae for the energy spectrum, namely:

--- For excited motion along z, with quantum numbers n,

$$E = \gamma(2n_r + |M| + M + 1) - (n + \delta_n)^{-2} \quad . \tag{138}$$

--- For a low degree of excitation along $\underset{\sim}{B}$,

$$E = \gamma(2n_r + 1)\ [1 + 2(2\gamma)^{-1/2}(2n_r + 1)^{-3/2} + (2\gamma)^{-1/4}(n + 1/2)(2n_r + 1)^{-7/4}] \quad . \tag{139}$$

Eq. (138) is the well-known [37] expression, valid when $\gamma >> 1$, in which the motion along $\underset{\sim}{B}$ is described through a one-dimensional Coulomb-type potential. Eq. (139), however, is much more interesting for our present purposes since it is a three-dimensional approach to the spectrum when the degree of excitation of the longitudinal motion is weak. Surprisingly, one finds from Eq. (139) that

$$\frac{dE}{dn_r} = \frac{3}{2}\ \hbar\omega_c \quad \text{and} \quad n_r^{\,3}B \simeq B_c \quad , \quad \text{for } E \simeq 0 \quad , \tag{140}$$

while the secondary structures associated with the logitudinal motion fulfill

$$\frac{dE}{dn} = \frac{2^{-1/2}}{2}\ \hbar\omega_c \quad , \quad \text{for } E \simeq 0 \quad . \tag{141}$$

This means that between two successive principal lines (n_r, n=0) and (n_r + 1, n=0), three secondary lines associated with (n_r, n = 1,2,3) will appear. Since the frequencies are not commensurate and, therefore, will slightly evolve with E and B, the spectrum will begin to exhibit chaotic features as a local reinforcement of lines occurs. This is not far from the picture found by Clark and Taylor (cf. Fig. 19). From Eq. (139), quasi-degeneracies will occur for those lines associated with (n_r, n), (n_r - 1, n + 4),... .

In this model, the K = 1 lines are associated with (n_r, n=0) states (with weak excitation along $\underset{\sim}{B}$), while the other K lines are associated with n = 1,2,... degrees of excitation along $\underset{\sim}{B}$. Finally, Eq. (139) suggests that the quasi-Landau spectrum at threshold is band-structured [56], with the bottom of the bands being associated with K = 1 states.

In the end, Eq. (139) implies that a rigorous semi-classical quantization is possible starting from the Landau limit: One need only discover the analytic form of an adiabatic invariant in the Landau regime. This, indeed, would be another major piece of the puzzle.

8. EXPERIMENTAL STUDIES IN ATOMIC PHYSICS

Experiment has played a major role in the evolution of our understanding of the diamagnetic problem. The story begins in 1969, with the discovery [6] of the quasi-Landau resonances, which possess 1.5 $\hbar\omega_c$ spacings near threshold. At that time, these resonances were thought to be smooth modulations of the continuum oscillator strength. Further experimental work, however, has shown the universal character of this phenomenon for all atoms. These early experiments used optical absorption techniques, with classical sources and high-resolution spectrometers. Since 1977, however, with the advent of pulsed and CW dye lasers, numerous kinds of new experiments have been developed [58-61]. Some of these experiments, employing very high resolution techniques, have achieved major advances such as, for example, experimentally demonstrating that the quasi-Landau phenomenon affects well-defined series of discrete lines, from the Coulomb to the Landau regimes. The results of these experiments have provided strong motivation for the development of new theoretical methods such as those discussed in Section 7. In this section, we review the more striking features of these experiments. But, before doing so, we first stress those characteristics of the ideal experiment which, assuredly, will be developed in the near future.

8.1 The Ideal Experiment: CW Dye Laser Excitation of Hydrogen In Atomic Beams

The ideal experiment must be concerned with the search for new ideas for solving the problem or, at least, stimulating theory; therefore, it is clear that only experiments which are as simple as possible should be considered. Atomic hydrogen is thus the only possibility -- which does not mean that the effects of diamagnetism on nonhydrogenic species are not of interest! In the absence of any short-range corrections to the Coulomb potential, however, one can be sure that no extra effects of configuration interaction will hinder the study of the diamagnetic perturbation.

Moreover, experiments employing atomic beams are highly desirable since this will allow a total control of motional Stark-field effects. As shown in Eq. (89) of Section 4.3, the center-of-mass motion is not separable, which will cause some alterations of the reduced-particle motion, especially through the interaction of the reduced particle with the motional Stark field. Indeed, the presence of crossed ($\underset{\sim}{E}$, $\underset{\sim}{B}$) fields can produce a complete alteration of the quasi-Landau spectrum (see Section 9). It seems necessary, therefore, to reduce the impact of motional Stark effects on the spectrum by choosing an arrangement in which the velocity of the atom in the beam is parallel to the magnetic field. This ensures, from Eq. (89), that the motional Stark term is negligible, provided that the beam is sufficiently well-collimated. Another requirement is that the beam should be as monokinetic as possible, because of the $C^2/2M$ term in Eq. (89). From Eqs. (87,89) it is clear that any change in the overall motion of the two-particle system will affect the relative motion. Since this problem, which is equivalent to the one-particle problem in crossed ($\underset{\sim}{E}$, $\underset{\sim}{B}$) fields, is not solved, it is impossible to draw any final conclusions about the effects of a non-monokinetic beam on the spectrum. However, these effects are likely to be small in comparison to the residual Stark broadening associated with imperfect collimation. Finally, low velocity beams or traps (for the neutral particle) seem an appropriate choice for observing the spectrum under good conditions.

Beam experiments permit accurate determinations of Rydberg state parameters. However, the production of Rydberg states with $n > 80$ is hardly possible or, at least, requires huge resolving powers, which does not favor the requirement of optical purity for the excitation.

Spectral purity and tunability of the optical source are obviously important factors for resolving all of the details of the quasi-Landau spectrum (e.g., accidental coincidence of lines or weak anti-crossings), or for dealing with high ($n > 100$) Rydberg states. For $n = 140$, for example, the quasi-Landau regime occurs for field values of 150 Gauss, when $n^3B \simeq B_c$! It is highly desirable, for various reasons, to study the spectrum under such conditions. And, this turns out to be possible [55], but under gas-phase conditions, using single-mode CW dye laser excitation. In such an experiment, of course, motional Stark-field effects will exist, thus requiring the choice of heavy atoms and conditions of low temperature and low field.

Beam experiments on high Rydberg states of hydrogen, using pure CW dye laser excitation, have not yet been achieved, but some progress is likely to occur in the near future. Until now, beam experiments [62] have been done on alkali atoms using pulsed dye laser excitation. Usually, this does not allow total control of the width and lineshape of the source and, therefore, of the intensity features of the experimental data. On the other hand, gas-phase experiments (also on alkali atoms), combined with the use of a single-mode CW dye laser possessing a well-controlled linewidth [63], are likely to suffer from a lack of control of the motional Stark-field effects.

8.2 Actual Experiments

Many experimental attempts have been made in the past few years using either conventional light sources or dye laser excitation, cells or beam apparatuses, and various kinds of atomic species and excitation schemes [59-64]. In some of these experiments, the choice of atoms which depart strongly from hydrogenic behavior hindered any possible detailed interpretation of the spectra, especially of the intensities. In the next section, we discuss two recent experiments which showed that the quasi-Landau spectrum has a discrete nature, and that the dominant lines in this spectrum are associated with the $K = 1$ lines in the Coulomb limit, as discussed in Sections 6 and 7. These studies were performed in rather complementary ways, by either scanning the energy or scanning the magnetic field.

The first experiment uses beam techniques and pulsed dye laser (GHz linewidths) excitation of sodium nD Rydberg states. These attempts have been limited to the bound spectrum below threshold. The second experiment uses highly-selective single-mode CW dye laser excitation of cesium nF Rydberg series, which are highly hydrogenic ($\delta \simeq 0.033$). The pumping scheme is the so-called hybrid two-photon process [17], which is quite efficient. Production of states with $n \lesssim 162$ has been achieved [55]. These experiments were performed at constant laser frequency while scanning the B field, which permits, in particular, a check on the n^3B quantization law obeyed at threshold, and the nB law obeyed in the Landau regime.

8.3 The Discrete Nature of the Quasi-Landau Spectrum

These experiments first showed that the strong field mixing phenomenon affects series of well-defined sharp lines from the Coulomb limit to the Landau limit. The dominant series is associated with $K = 1$ states (cf. Sections 6 and 7) in the low-field Coulomb inter-ℓ mixing regime, and its components obey the approximate predictions of Eq. (133). The spacing of the components at threshold is $1.5\ \hbar\omega_c$, while the quantization law at constant energy is $nB^{1/3} = 1.16\ B_c^{1/3}$. In addition, other series of discrete lines obey approximately the same quantization law, and are associated with the $K = 2,3,...$ states of the Coulomb limit.

The quasi-Landau spectrum for variable energy

The results of the beam experiment on sodium atoms are shown in Fig. 20. Approximate quantum calculations are shown in comparison, and provide a rough assignment of the lines. The intensity

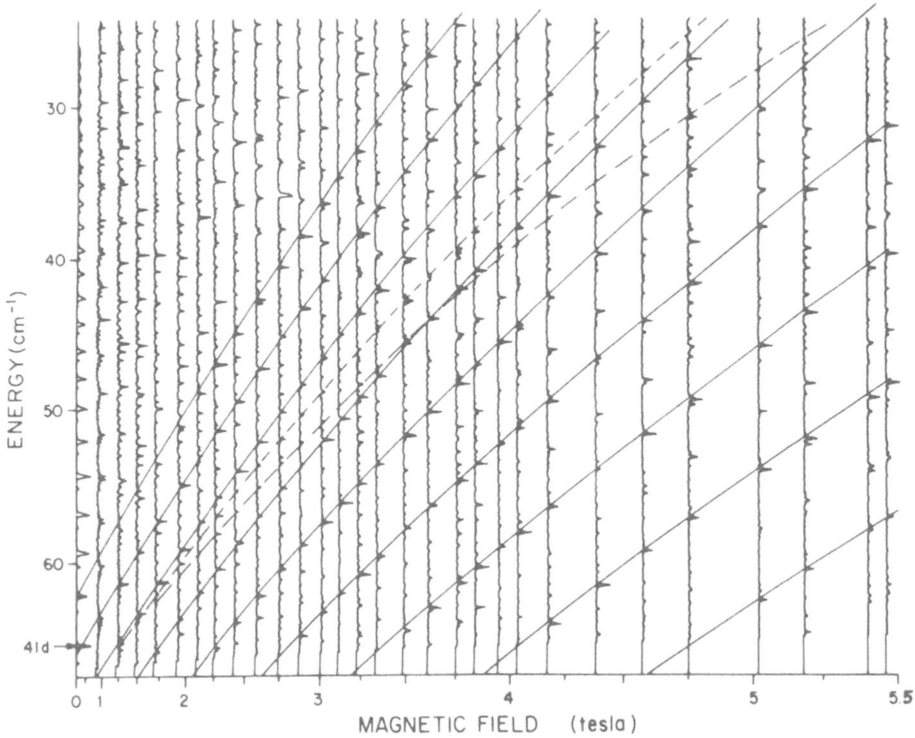

Fig. 20. Results of the beam experiment on sodium atoms showing the evolution of the components of the K = 1 series as a function of B^2. (Courtesy of M. L. Zimmerman.)

of the lines is not relevant because of fluctuations in intensity and lineshape of the pulsed dye laser output. All of these results were obtained below threshold.

Fig. 21 shows absorption spectra, measured under cell conditions [61], for strontium atoms near and above threshold. These data demonstrate that the phenomena below and above threshold are in continuity, except for the lineshapes.

Landau spectrum for variable magnetic field

A typical spectrum in the diamagnetic region, as a function

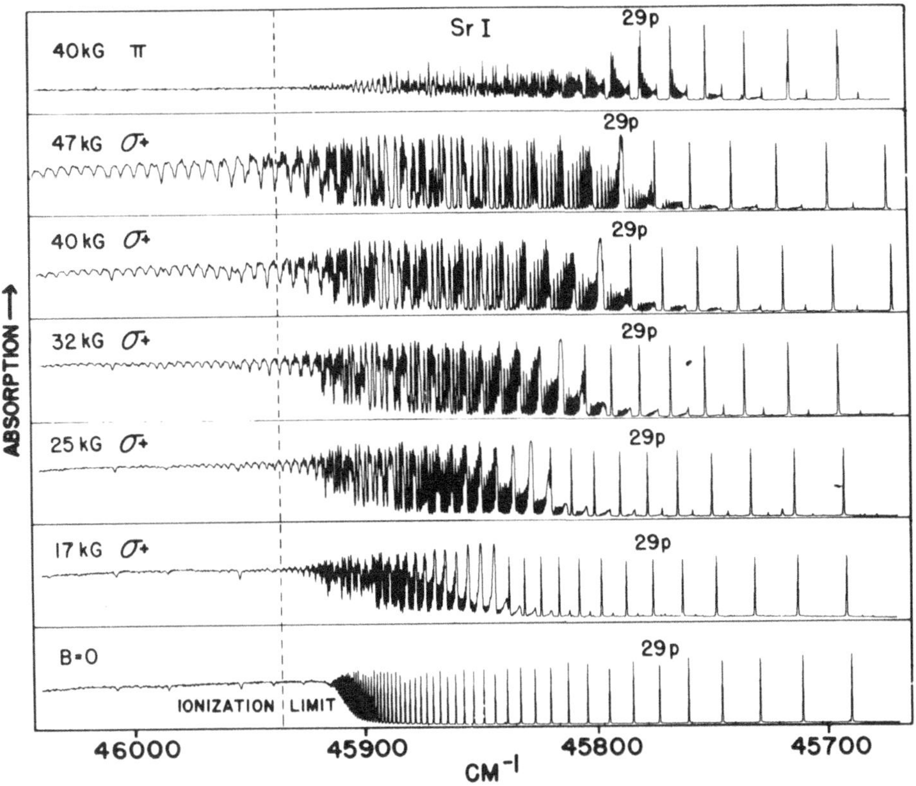

Fig. 21. Complete pattern of the diamagnetic regimes, as seen in gas-phase absorption spectra of strontium, plotted as a function of the frequency of the classical excitation source. (Courtesy of F. S. Tomkins.)

of laser frequency, is shown in Fig. 22, for the n = 40, M = + 3 states of cesium. The various components of the diamagnetic manifold are displayed, and all are in good agreement with theoretical calculations [39]. The fact that the K = 1 line does not have the maximum intensity is associated with the small perturbation due to the quantum defect ($\delta \simeq 0.033$). At higher field values, this spurious effect will disappear. A complementary aspect of the diamagnetic pattern is shown in Fig. 23, where the magnetic field is scanned at constant laser frequency. Another aspect of the spectrum below threshold is shown in Fig. 24, where (successively) the inter-ℓ and inter-n mixing regimes, and the onset of the strong mixing regime are displayed at constant laser frequency.

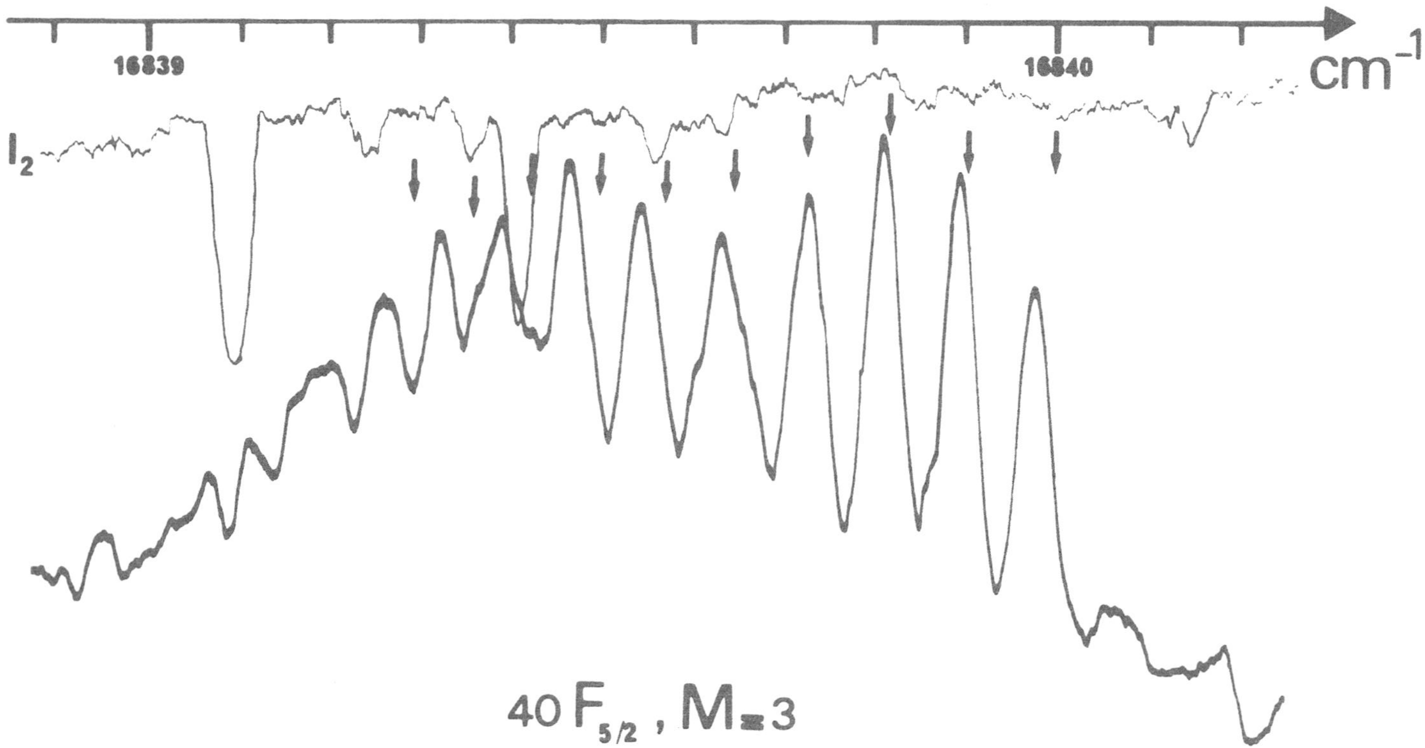

Fig. 22. Structure of the n = 40, M = 3 diamagnetic manifold of cesium (B ≃ 6 kG), plotted as a function of the CW dye laser frequency. The top trace is the molecular iodine spectrum.

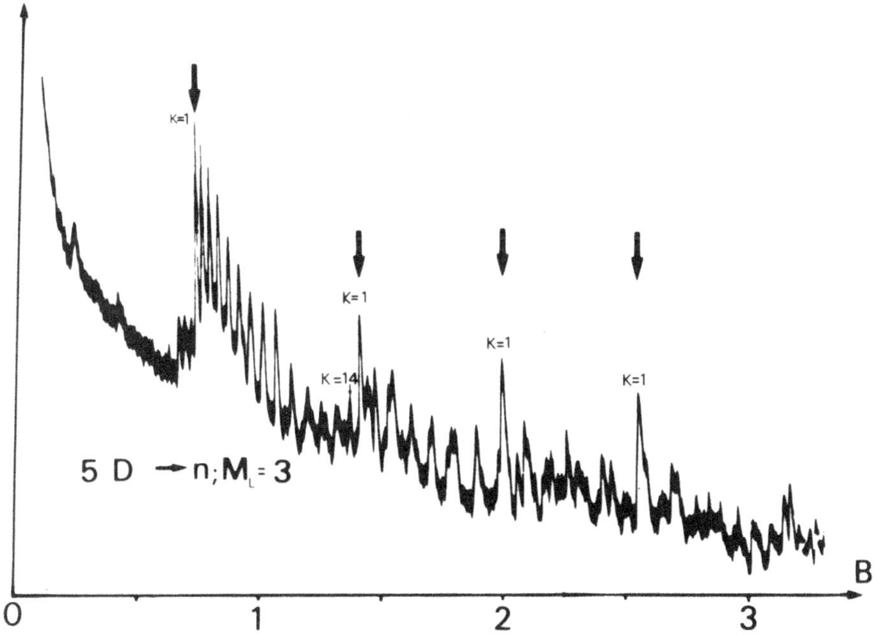

Fig. 23. Diamagnetic patterns of the n = 50,49,... states of cesium, plotted as a function of B for a fixed laser frequency.

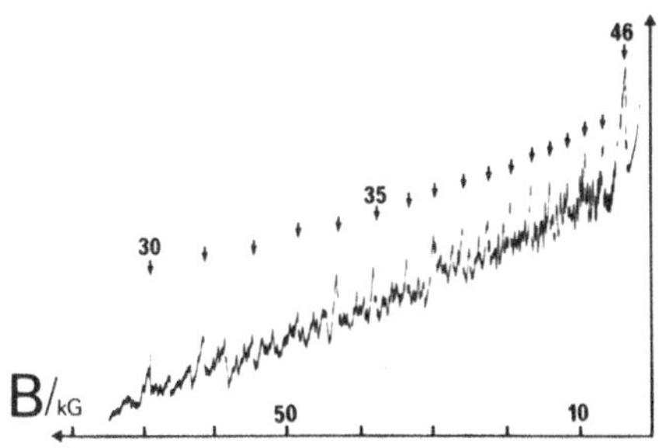

Fig. 24. Absorption spectrum of cesium as a function of magnetic field, for a constant laser frequency (fixed around the n = 50 Coulomb state), showing the various diamagnetic regimes.

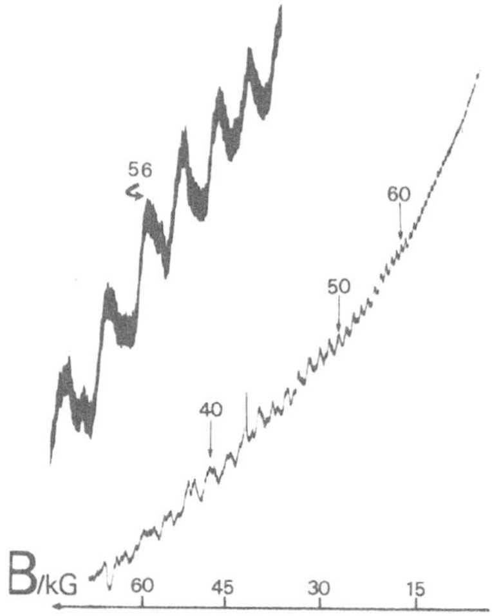

Fig. 25. Absorption spectrum of cesium as a function of the magnetic field, for a fixed laser frequency. Electronic energy E = + 10 cm^{-1} above threshold.

Fig. 25 is a scan in magnetic field, for energies E = +10 cm^{-1} above threshold. These states are quasi-Landau states. Similar spectral studies have been performed for constant energies ranging from 120 cm^{-1} below the ionization threshold (n = 30 Coulomb states) to 120 cm^{-1} above threshold, for magnetic field strengths between 0 and 8 Tesla. The results for the dominant (K = 1) series are plotted in Fig. 26, and compared to the approximate predictions from Eq. (133). Especially at threshold (E = 0), the quantization law $n_r{}^3B$ = constant is obeyed exactly. The average energy spacing at threshold is 1.52 $\hbar\omega_c$. In Fig. 27, the radial quantum number is plotted against 1/B. For the higher energy values, the experimental points clearly lie on a straight line, thus indicating that the Landau limit of the spectrum has almost been reached. The spacing of the resonances, for E ≃ 120 cm^{-1}, B ≃ 25 kG and n_r ≃ 90, is 1.1 $\hbar\omega_c$. For this series of dominant lines, which are discrete and associated with well-defined states in the Coulomb limit, one passes, by varying the field and energy, from a Coulomb to a Landau signature in the spectrum. The n^3B = constant quantization law, obeyed at threshold, is the basic feature which demarcates these two limits. Indeed, the experimental results in the bound-state region, as well as in the resonance region above threshold, converge at high fields toward this new law. Of course, other series associated with the other K states are likely to produce the same kind of pattern. But, this has

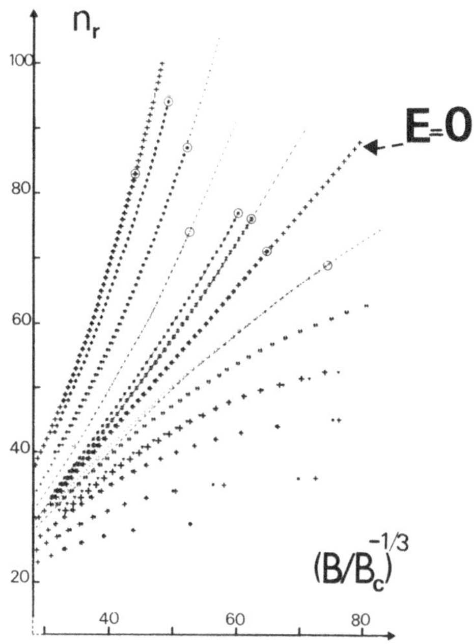

Fig. 26. Experimental and theoretical plots of the radial quantum number n_r versus $(B/B_c)^{-1/3}$, at various energies, for the dominant discrete series of the spectrum. Theoretical points (crossed or broken curves) are from semi-classical calculations. Experimental points are for energies 121.5 (top line), 99.5, 65.7, 37.3, 17.2, 11.5, 0, - 10.9, - 20, - 29.3, - 43.6, and - 66.6 cm^{-1} (lower line), respectively, from the ionization limit. At the threshold the curve is a straight line. For positive and negative energies of the electron, the results curve in opposite directions towards the Landau and Coulomb regions, respectively. (Taken from [17], [55].)

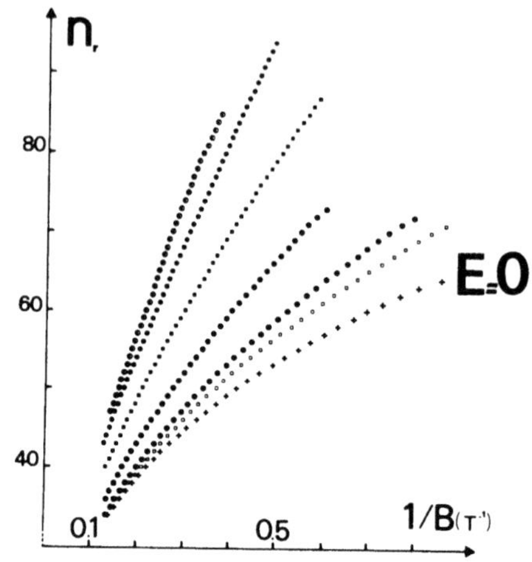

Fig. 27. Plot of n_r as a function of 1/B for the resonance above threshold, showing that the Landau limit is almost reached for E ≃ 100 cm^{-1}. (See the caption to Fig. 26.)

not yet been demonstrated, either theoretically or experimentally.

9. PROSPECTS AND CONCLUSIONS

Among the various problems connected with atomic diamagnetism, there is one which has attracted considerable attention, namely, the crossed electric and magnetic field problem [65-67]. The Hamiltonian in this case is that of Eq. (89), supplemented by the $q\mathbf{r}\cdot\mathbf{E}$ electric dipolar interaction. For this problem, there are no longer any constants of the motion. Nevertheless, experiments have indicated that the atomic spectrum can still be organized, exhibiting a set of resonances with an energy spacing of $\frac{1}{2}\hbar\omega_c$ [59] near threshold.

Several spectral regimes can be distinguished. In the first one, where we neglect the diamagnetic interaction, only the linear Stark effect and the paramagnetic term are present [68]. Under such conditions, it may happen that the atomic spectrum for highly hydrogenic species will still present some regular features, which are neither the Zeeman ones nor the Stark ones. This is connected with the existence of an adiabatic invariant [69] in this intermediate region. Denoting the frequency associated with the linear Stark effect as $\boldsymbol{\omega}_S$, where

$$\underset{\sim}{\omega}_S = -\frac{3}{2} \cdot n \cdot \underset{\sim}{E} \text{ (a.u.)} , \tag{142}$$

and using classical perturbation theory, the equations of evolution for the classical ellipses are found to be

$$\begin{aligned} \frac{d}{dt} \langle \underset{\sim}{J}_+ \rangle &= \left(\frac{1}{2} \underset{\sim}{\omega}_c + \underset{\sim}{\omega}_S \right) \times \underset{\sim}{J}_+ = \underset{\sim}{\omega}_+ \times \underset{\sim}{J}_+ \\ \frac{d}{dt} \langle \underset{\sim}{J}_- \rangle &= \left(\frac{1}{2} \underset{\sim}{\omega}_c - \underset{\sim}{\omega}_S \right) \times \underset{\sim}{J}_- = \underset{\sim}{\omega}_- \times \underset{\sim}{J}_- \end{aligned} \tag{143}$$

where $\underset{\sim}{J}_+$ and $\underset{\sim}{J}_-$ are defined in Eqs. (14-16). Eq. (143) implies that $\underset{\sim}{J}_+$ and $\underset{\sim}{J}_-$ still precess independently, but along the two axes $\underset{\sim}{\omega}_+$ and $\underset{\sim}{\omega}_-$, respectively. Independent quantization can be performed along these two axes, with the energy then being

$$E = -(1/n^2) + \frac{1}{2} \underset{\sim}{\omega}_c \cdot \underset{\sim}{L} + q \underset{\sim}{r} \cdot \underset{\sim}{E} \tag{144a}$$

or

$$E = -(1/n^2) + \underset{\sim}{J}_+ \cdot \underset{\sim}{\omega}_+ + \underset{\sim}{J}_- \cdot \underset{\sim}{\omega}_- . \tag{144b}$$

That is,

$$E = -(1/n^2) + (m_+\omega_+ + m_-\omega_-) . \tag{145}$$

In the general case, then, the mixed Stark-Zeeman patterns of the manifold will be quite complicated. If $\underset{\sim}{E}$ and $\underset{\sim}{B}$ are crossed, however, one finds

$$\begin{aligned} \omega_+^2 &= \omega_-^2 \equiv \omega^2 = \left(\frac{1}{4} \omega_c^2 + \omega_S^2 \right) , \\ E &= (-1/n^2) + (m_+ + m_-)\omega . \end{aligned} \tag{146}$$

Eq. (146) indicates the existence of a new regular pattern which is neither the paramagnetic nor the Stark pattern. This follows from the existence of an adiabatic invariant when the diamagnetic term is weak.

The second class of regimes is associated with the mixed effect of the diamagnetic, Coulomb and dipolar electric interactions. In some cases, as shown in Fig. 28, a non-Coulombic valley may exist at large electron-proton separation distances. In such a one-dimensional valley, resonances will exist with an energy spacing of $\frac{1}{2} \hbar\omega_c$ [51,65,67,68]. This regime is intermediate between the quasi-Landau and Stark-resonance [51] situations.

Other types of studies and developments are connected with

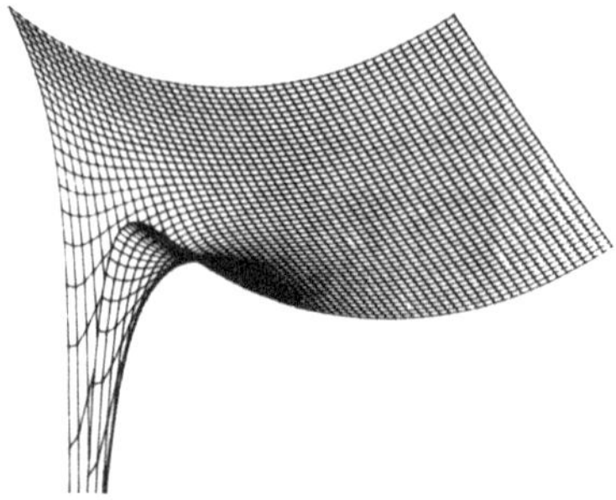

Fig. 28. Plot of the approximate effective potential for the crossed ($\underset{\sim}{E}$, $\underset{\sim}{B}$) field configuration, showing the non-Coulombic valley at large electron-proton separation distances.

the physical properties of quasi-Landau resonances. Indeed, the collisional and radiative properties of these resonances are likely to be quite interesting. The tunable character of the Landau ladder is certainly of interest for studies concerning superradiance or detection of radiation in the 100 GHz range. The partial discretization of atomic continua and the local increase of oscillator strengths will completely modify the mechanisms of electronic capture or electronic collisions. Such features are already seen in cyclotron resonance experiments on cold plasmas. Finally, it may turn out that diamagnetism will be as fine a tool as the Zeeman effect, but for MQDT analyses of high-Rydberg series and autoionizing resonances. The main direction of interest for this latter area is not yet clear. What is clear, however, is that the effects of the diamagnetic interaction on the core will be weak, while the continuum channels will be discretized into Landau resonances. It is not unlikely that some important modifications of the properties of autoionizing resonances will be achieved in this way.

Of course, a major breakthrough would be to understand completely the basic problem of diamagnetism for the hydrogen atom. And, a large step in this direction would be provided by an approach to the spectrum in the resonance region above threshold, including approximate wavefunctions. Some pieces of this puzzle have been put in the right place recently, using one of the various possible methods. Other pieces, however, are still missing.

ACKNOWLEDGMENTS

Part of this work was done in collaboration with Dominique

Delande, to whom the author is very grateful; he also kindly supplied some of the figures presented here. This work was supported in part by NATO Grant No. 107.82. Several stimulating discussions, over the past few years, with C. W. Clark, D. Delande, U. Fano, W. R. S. Garton, K. T. Lu, R. F. O'Connell, A. R. P. Rau, M. Robnik, K. T. Taylor, F. S. Tomkins, W. P. Reinhardt and M. L. Zimmerman are gratefully acknowledged.

REFERENCES

1. H. Hasegawa, in: *Physics of Solids in Intense Magnetic Fields*, E. D. Haidemenakis, ed. (Plenum, New York, 1969).
2. J. G. Mavroides, in: *Optical Properties of Solids*, F. Abeles, ed. (North Holland, Amsterdam, 1972).
3. R. H. Garstang, Rep. Prog. Phys. 40, 105 (1977).
4. L. D. Landau, Z. Phys. 64, 629 (1930).
5. F. A. Jenkins and E. Segre, Phys. Rev. 55, 52 (1939).
6. W. R. S. Garton and F. S. Tomkins, Astrophys. J. 158, 839 (1969).
7. A. R. P. Rau, Phys. Rev. A16, 613 (1977).
8. See, for example, M. Bander and C. Itzykson, Rev. Mod. Phys. 38, 330 (1966).
9. B. R. Judd, *Angular Momentum Theory for Diatomic Molecules* (Academic Press, New York, 1975).
10. L. D. Landau and E. M. Lifschitz, *Mécanique* (Mir, Moscow, 1966).
11. A. Durand, *Mécanique Quantique* (Dunod, Paris, 1970).
12. C. Cohen-Tannoudji, B. Diu and F. Laloe, *Mécanique Quantique* (Hermann, Paris, 1973).
13. L. D. Landau and E. M. Lifschitz, *Mécanique Quantique* (Mir, Moscow, 1966).
14. W. E. Lamb, Phys. Rev. 85, 259 (1952).
15. L. P. Gorkov and I. E. Dzyaloshinskii, JETP 26, 449 (1968).
16. B. P. Carter, J. Math. Phys. 10, 788 (1968).
17. D. Delande, Thèse 3e Cycle Paris (1981).
18. L. I. Schiff and H. Snyder, Phys. Rev. 55, 59 (1939).
19. J. C. Gay, Comm. At. Mol. Phys. 9, 97 (1980).
20. A. R. P. Rau, Comm. At. Mol. Phys. 10, 19 (1980).
21. D. Delande and J. C. Gay, unpublished.
22. A. R. Edmonds, J. Phys. (Paris Colloque C4) 31, 71 (1970).
23. A. R. Edmonds, reprints, Imperial College, London.
24. M. Robnik, J. Phys. A14, 3195 (1981).
25. M. Robnik, in: *Proceedings of the Colloque International CNRS* (Aussois, France, 1982), J. de Physique (France), Colloque C2, Supplement 11 (1982).
26. W. P. Reinhardt and D. Farrely, in: *Proceedings of the Colloque International CNRS* (Aussois, France, 1982), J. de Physique (France), Colloque C2, Supplement 11 (1982).

27. E. J. Heller, E. B. Stechel and M. J. Davis, J. Chem. Phys. 73, 4720 (1980).
28. W. Magnus and S. Winkler, *Hill's Equation*(Wiley, New York, 1966).
29. I. C. Percival, Adv. Chem. Phys. 36, 1 (1977).
30. M. P. Strand and W. P. Reinhardt, J. Chem. Phys. 70, 3812 (1979).
31. R. Balian and C. Bloch, Ann. Phys. (NY) 85, 514 (1974).
32. V. Szebehely, *Theory of Orbits* (Academic Press, New York, 1967).
33. U. Fano, J. Phys. B13, L519 (1980).
34. U. Fano, Phys. Rev. A22, 2660 (1980).
35. E. A. Solovev, JETP Lett. 34, 265 (1981).
36. E. G. Kalnins, W. Miller and P. Winternitz, Siam J. Appl. Math. 30, 630 (1976).
37. See, for example, J. C. Gay, in: *Progress in Atomic Spectroscopy*, Part C, Hanle and Kleinpoppen, eds. (Plenum, New York, to be published).
38. M. L. Zimmermann, M. M. Kash and D. Kleppner, Phys. Rev. Lett. 45, 1092 (1980).
39. D. Delande and J. C. Gay, Phys. Lett. 82A, 393 (1981).
40. D. R. Herrick, Phys. Rev. A26, 323 (1982).
41. C. W. Clark, Phys. Rev. A24, 605 (1981).
42. C. W. Clark and K. T. Taylor, J. Phys. B15, 1175 (1982).
43. C. W. Clark and K. T. Taylor, J. Phys. B13, L737 (1980).
44. J. J. Labarthe, J. Phys. B14, L467 (1981). A correct derivation of Eq. (129) is given in D. Delande and J. C. Gay, Comm. At. Mol. Phys., to be published.
45. A. R. Edmonds, J. Phys. B6, 1603 (1973).
46. U. Fano, in: Coll. Int. CNRS 283 (CNRS, Paris, 1977).
47. G. Wunner, H. Herold and G. Ruder, Phys. Lett. 88A, 344 (1982).
48. J. Virtamo, J. Phys. B9, 751 (1976).
49. R. F. O'Connell, Phys. Lett. 60A, 481 (1977). New developments can be found in J. A. C. Gallas and R. F. O'Connell, J. Phys. B15, L309, L593 (1982) and the paper by R. F. O'Connell in this volume.
50. A. F. Starace, J. Phys. B6, 585 (1973).
51. A. R. P. Rau, J. Phys. B12, L183 (1979).
52. V. P. Maslov, *Théorie des Perturbations et Methodes Asymptotiques* (Dunod, Paris, 1972).
53. A. Einstein, Verh. Deutsch. Phys. Ges. Berlin 19, 82 (1917).
54. M. C. Gutzwiller, Phys. Rev. Lett. 45, 150 (1980).
55. J. C. Gay, D. Delande and F. Biraben, J. Phys. B13, L720 (1980).
56. A. G. Zhilich and B. S. Monozon, Sov. Phys. Sem. 1, 563 (1967).
57. A. G. Zhilich and B. S. Monozon, Sov. Phys. Solid State 8, 2846 (1967).

58. M. L. Zimmerman, J. C. Castro and D. Kleppner, Phys. Rev. Lett. 40, 1083 (1978).
59. K. T. Lu, F. S. Tomkins, H. M. Crosswhite and H. Crosswhite, Phys. Rev. Lett. 41, 1034 (1978).
60. N. P. Economou, R. R. Freeman and P. F. Liao, Phys. Rev. A18, 2506 (1979).
61. K. T. Lu, F. S. Tomkins and W. R. S. Garton, Proc. Roy. Soc. A 364, 421 (1978).
62. J. C. Castro, M. L. Zimmerman, R. G. Hulet and D. Kleppner, Phys. Rev. Lett. 15, 1780 (1980).
63. D. Delande and J. C. Gay, Phys. Lett. 82A, 399 (1981).
64. D. Delande, C. Chardonnet and J. C. Gay, Optics Comm. 42, 25 (1982).
65. H. M. Crosswhite, K. T. Lu, U. Fano and A. R. P. Rau, Phys. Rev. Lett. 42, 963 (1979).
66. G. P. Drukarev and B. S. Monozon, JETP 34, 509 (1979).
67. J. C. Gay, L. R. Pendrill and B. Cagnac, Phys. Lett. 72A, 315 (1979).
68. A. R. P. Rau and K. T. Lu, Phys. Rev. A21, 1957 (1980).
69. T. P. Grozdanov and E. A. Solovev, J. Phys. B15, 1195 (1982).

EVIDENCE FOR MOTIONAL ELECTRIC FIELD EFFECTS IN THE ABSORPTION SPECTRUM OF LITHIUM VAPOR IN A MAGNETIC FIELD

H. M. Crosswhite, F. S. Tomkins and H. Crosswhite

Chemistry Division
Argonne National Laboratory
Argonne, Illinois 60439, USA

W. R. S. Garton

The Blackett Laboratory
Imperial College of Science amd Technology
Prince Consort Road
London SW7 2BZ, England

1. INTRODUCTION

Since the discovery, by Garton and Tomkins in 1969 [1], of magnetically-induced periodic fluctuations in the vicinity of the ionization limit in the Ba I absorption spectrum, there has been a growing interest in high-energy spectroscopic phenomena. Developments in fundamental theory, particularly by the Fano school, have opened up new perceptions of the importance of such phenomena to basic physical and chemical processes. Magnetic effects, of course, constitute only a small part of high-energy spectroscopic phenomena; however, a brief review of some of the experimental results, and of the interpretation of these results, is trenchant in defining the place of magnetic effects in the overall high-energy context. Such, indeed, is the purpose of this paper.

An extension of the Ba I experimental results was given recently by Garton, *et al.* [2] and a much more general review of atomic diamagnetism has been given by J. C. Gay [previous paper in this volume]. As an illustration of some of the spectroscopic effects described by Gay, we present below some details of our analysis of the diamagnetic spectrum of Li I, showing par-

S. P. McGlynn et al. (eds.), Photophysics and Photochemistry in the Vacuum Ultraviolet, 707–720.

ity mixing arising from a motionally-induced electric field crossed with an applied magnetic field [3,4], as well as our analysis of the more conventional ℓ- and m-mixing phenomena.

The absorption spectrum of lithium vapor, in the presence of a 50 kG magnetic field produced with a superconducting solenoid, was photographed in the second order of the Argonne 30 ft Paschen-Runge spectrograph. The light entering the spectrograph was polarized in such a way as to select preferentially only those transitions from the 2s $^2S_{1/2}$ ground state to states for which the magnetic quantum number m_ℓ changed by +1 (σ^+ spectrum) or -1 (σ^- spectrum). Since the spin-orbit interaction is negligibly small for all of the levels studied here, the analysis given below is entirely at the Paschen-Bach limit; and, since the selection rule on m_s prevents any spin flip, the spin magnetic energy remains a constant for all transitions. Therefore, although the ground state splits into two sublevels (separated by $\mu_0 B$ cm^{-1}, where μ_0 is the electron Bohr magneton and B is the magnetic flux), all upper states are modified similarly, with the result that two spectroscopically indistinguishable absorption spectra arise from the magnetically-split ground state. In what follows, therefore, we ignore spin effects altogether.

Although this paper is primarily a presentation of experimental results, it is instructive to include interpretations based upon a simple quantum mechanical model, given in an (n, ℓ, m_ℓ) basis, which includes paramagnetic and diamagnetic terms, as well as a term representing the induced electric field generated by the component of electronic motion perpendicular to the magnetic axis. The field-free energies are expressed by a Rydberg formula with a constant quantum defect $\delta = 0.05$. For our purposes, this quantum defect is quite adequate to label all important experimental features for $n < 25$. Finally, radial contributions to the matrix elements are calculated by a modified Bates-Damgaard method [5].

2. RESULTS AND DISCUSSION

An overall view of the Li I absorption spectrum is given in Fig. 1. The strong absorption features, as shown below, can be explained by a more or less straightforward application of standard theory; of equal interest, however, are several weaker features, barely visible in Fig. 1, which require more detailed attention. For the purposes of this discussion, we divide the spectrum into the following six regions, with each region possessing distinct characteristics as listed in Table 1:

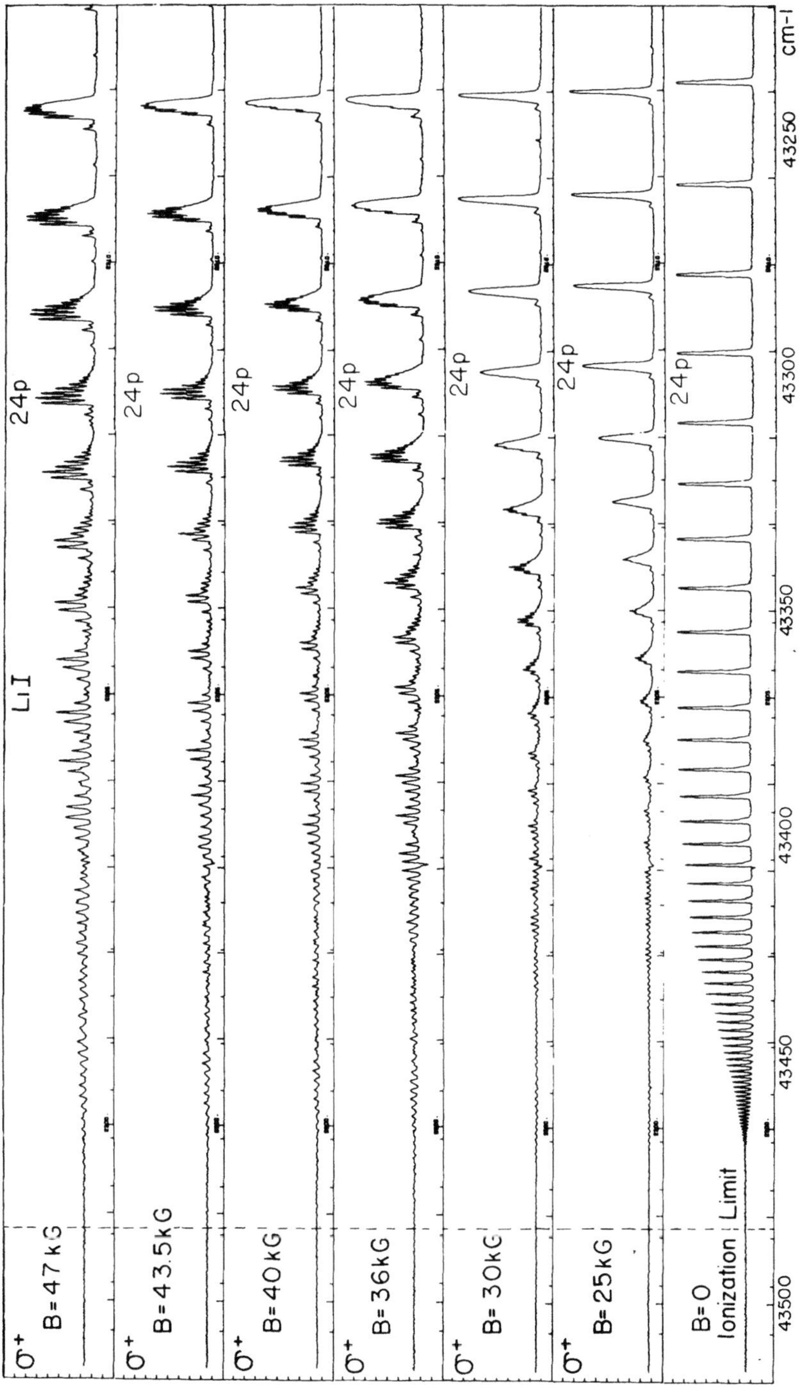

Fig. 1. Polarized (σ^+) absorption spectra of Rydberg series in Li I for various magnetic fields. This spectrum is taken from [3].

Table 1: Rydberg Spectral Regions for Li I (B = 48 kG)

Region		Range
I.	Linear Zeeman effect	$n \leq 8$
II.	Quadratic Zeeman effect; forbidden lines (m-mixing)	$8 < n < 15$
III.	ℓ-mixing (intensity redistribution)	$15 < n < 28$
IV.	n-mixing	$n \geq 28$
V.	Transition to quasi-Landau	$35 < n < 42$
VI.	Quasi-Landau resonances	$n > 42$

Region 1 (linear Zeeman region). For very low quantum numbers, the dominant contribution to the Hamiltonian is given by the paramagnetic energy $g\mu_0 B$. Since we assume that g = 1, we expect, for $\sigma^+(\sigma^-)$ polarized light, a single level at +1(-1) Lorentz units (LU) from the unperturbed level position. And, we can use these splittings as internal checks on magnetic field strengths. For the fields used here ($\leq$ 48 kG), these considerations hold up to n = 8.

Region II (quadratic Zeeman region). This region is superficially similar to Region I, except that the diamagnetic term in the Hamiltonian has become significant. Diamagnetic contributions to levels with $m_\ell = \pm 1$ are identical (and positive). Differences in σ^+ and σ^- spectra can still be used to verify magnetic field measurements, although the lines are no longer centered on the field-free positions. For B = 48 kG, Region II extends approximately over the range n = 9-13. And, it is in this region that evidence for parity-forbidden transitions first appears. Fig. 2 shows densitometer tracings of n = 9, 11 and 13, for fields of 41.2 and 47.8 kG. (In these exposures, a rather high lithium vapor pressure (3.0 torr) was used.) To the high-energy side of the very heavily over-absorbed np transition, three satellites appear: One exhibits very little field dependence; one moves parallel to the parent np line; and one moves approximately twice this latter amount. Extrapolation back to zero field indicates that all of these satellites come from a common point, corresponding to zero quantum defect, and that the np lines (from exposures taken at lower vapor pressures) extrapolate to positions corresponding to $\delta \simeq 0.046$.

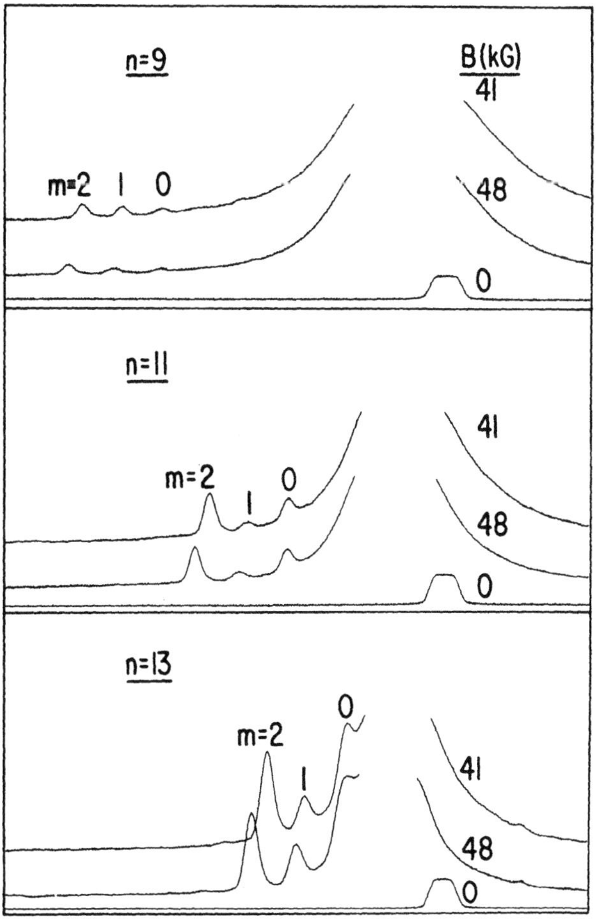

Fig. 2. Magnetically-induced σ^+-polarized satellites near 9p, 11p and 13p absorption lines of Li I, at B fields of 0, 41 and 48 kG.

Only the middle one of the three satellites was expected: We identify it as a forbidden transition to a predominantly nf (m = 1) level which, however, has been mixed slightly with np by the lower symmetry resulting from the diamagnetic term in the Hamiltonian. From the differential shifts of the other two satellites, at 41 and 48 kG, we assign magnetic quantum numbers of m = 0 and m = 2 to these features (cf. Fig. 2); and, the intensities of these satellites indicate a mixing with np comparable to that for the nf (m = 1) level. Diagonalization of the Hamiltonian for the B fields of Fig. 2 gives eigenvalues which confirm our interpretation, if even-parity states are included in the basis set. Intensity estimates, assuming proportionality of the intensity to the square of the (ℓ = 1, m_ℓ = 1) content of each particular eigenvector, are in qualitative agreement with the data of Fig. 2. Thus, we classify the outer satellites as predominantly nd (m = 0) and nd (m = 2).

Fig. 3 shows the transition to 14p for various magnetic

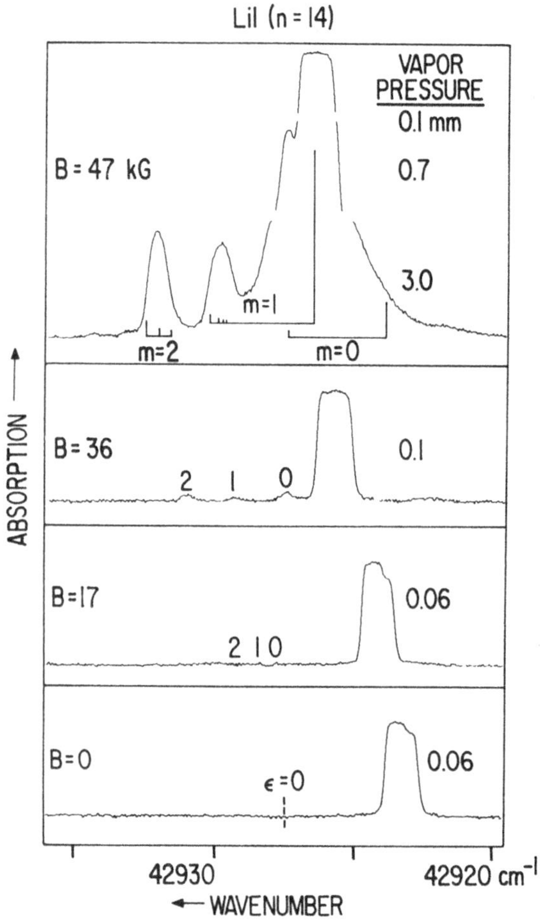

Fig. 3. σ^+ absorption spectrum near 14p for various magnetic fields and lithium vapor pressures.

fields, the top frame being a composite of densitometer traces taken for three different lithium vapor pressures at the highest B field. Here the 14p(1) transition is still dominant, but at the highest field it nearly overlaps the 14d(0), which has now increased considerably in strength. The 14f(1) is still the weakest of the satellites. Vertical lines are drawn at the positions calculated by diagonalizing the Hamiltonian for this n value, with lengths proportional to the square of the 14p(1) component. The calculations show the beginnings of ℓ-mixing, indicated by the incipient breakup of 14d(2) and 14f(1), but the experimental conditions are not adequate to resolve the components. Finally, the ground-state hyperfine structure is responsible for the indications of structure in the two lower traces of Fig. 3.

Region III (ℓ-mixing region). Beginning at n = 16, the nf (1) satellite shows first a broadening and then a splitting into a number of separate lines. At high enough field and/or n, the group of satellites becomes indistinguishably mixed with np, with the latter exhibiting a much reduced intensity. Fig. 4 shows this progression for n = 16, 18 and 20, with B = 48 kG. The nd(0) is no longer observed as a distinct feature and the nd(2) has broken up into several components. At this point, the orbital quantum number has little validity as a label for the levels, as is verified by diagonalization of the Hamiltonian for these n values.

Transitions to ns states also begin to appear in this region. Quantum defects in the neighborhood of 0.4 for these transitions are consistent with values found for known states of lower n.

For n = 21, as shown in Fig. 5, ℓ-mixing phenomena resulting from the diamagnetic term, and m_ℓ-mixing phenomena resulting from the motional Stark effect, are both well-developed. The line strength of the transition to the original 21p(1) level is now nearly equally shared with initially forbidden transitions to 21f(1), 21h(1), ..., 21y(1), with all of the odd-parity states being consistent with n = 21. The m = 2 group (also originally forbidden) shows both the motional Stark mixing of 21p(1) with 21d(2) and the further redistribution over 21d(2), 21g(2), ..., as well as a beginning of the transfer of 21d(2) to 21f(3), ... Again, the vertical bars are drawn according to calculational results. In Fig. 5, the value for the Bohr magneton at B = 47.8 kG is indicated for comparison, as is the 21p(1) field-free position (43237.3 cm^{-1}). The theoretical implications of this figure are discussed in more detail in [4].

Region IV (n-mixing region). Up to the neighborhood of n = 30, it is still possible to distinguish separate groupings of

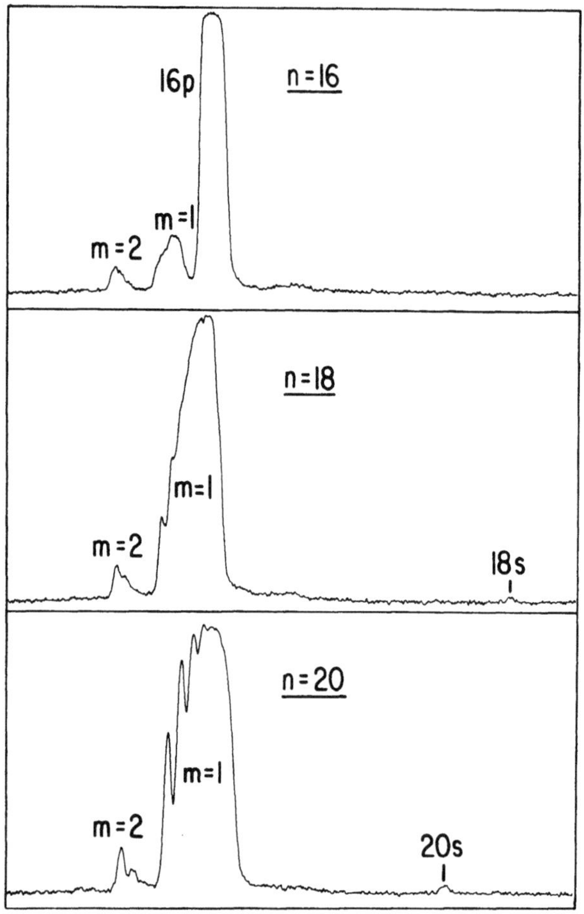

Fig. 4. σ^+ absorption near 16p, 18p and 20p for a magnetic field of 48 kG and lithium vapor pressure of 0.1 mm.

lines, although it is obvious by now that n also must be losing its significance as a quantum number. This region is marked not only by a blending of groups, indicating a mixing of states with different principal quantum numbers, but also by a gradual change of character in the intensity distribution within a group. At lower n, where the groups are well-separated, the intensities are more or less uniformly distributed over the whole manifold, with the higher-energy members tending to be a little stronger and a little better separated. As n increases, the spacing also increases; and, when this spacing reaches the

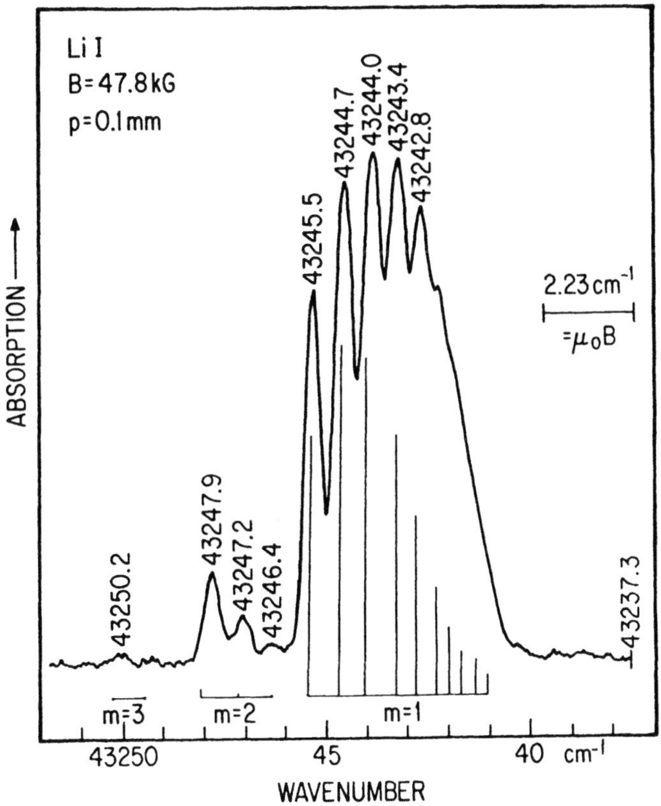

Fig. 5. Magnetically-induced structure of 21p σ^+ absorption line at 48 kG and 0.1 mm lithium pressure.

neighborhood of μ_0B, we find that the intensity is now concentrated in only a few of the leading members, with the trailing members being lost. Fig. 6 shows this progression as a function of field strength. Note that the Stark-induced ns (m = 0) lines, which are weak and easily recognized as interlopers between the dominant parent lines, become indistinguishable in character from the m_ℓ = 1 lines at the highest fields. Note also a growing tendency for the line strength to be concentrated into the leading members of each m_ℓ-group as the spacing of the group components approaches μ_0B, the separation between groups.

Region V (sub-Landau resonance region). Lines in this region appear to be undergoing a metamorphosis, as shown in Fig. 7. The spectral description in terms of atomic (n,ℓ, m_ℓ) manifolds appears to be breaking down, but the quasi-Landau condi-

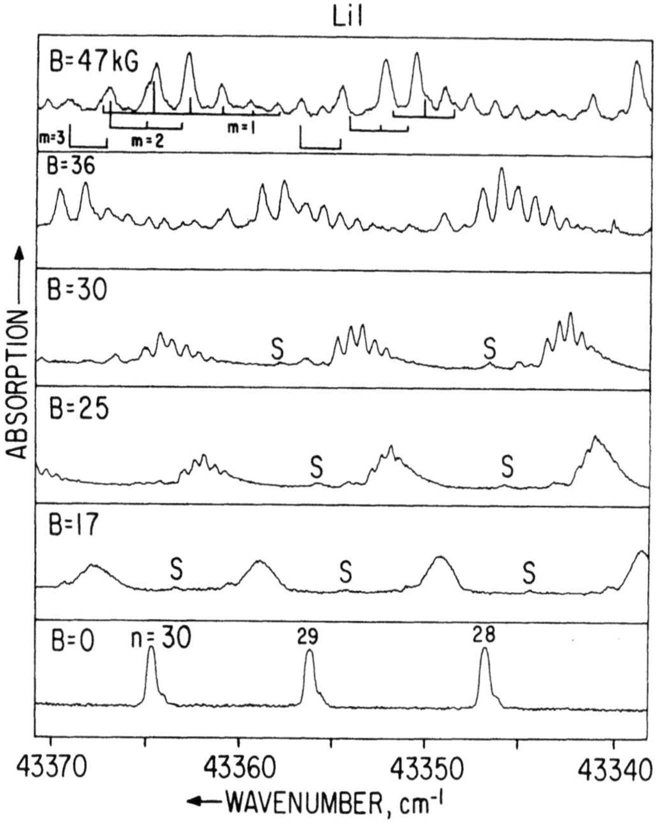

Fig. 6. σ^+ magnetic spectrum near 28p for various fields and lithium pressure of 0.1 mm.

tions have not yet been fully established. Note the persistent recurrance of the spacing $\mu_0 B$, which appears to result from two sources: (i), the separation between the leading components of the different m-groups, which agrees almost exactly with the nominal value $\mu_0 B$; and (ii), the spacing between components of a group, which is only approximately equal to $\mu_0 B$. The latter source is the one noted by Edmonds [6] in his discussion of the original work by Garton and Tomkins [1].

For the calculations shown in Fig. 7, it was necessary to expand the basis set to include four n values. Eigenvectors for a typical component are given in Table 2.

Fig. 8 shows the situation for several fields at higher n,

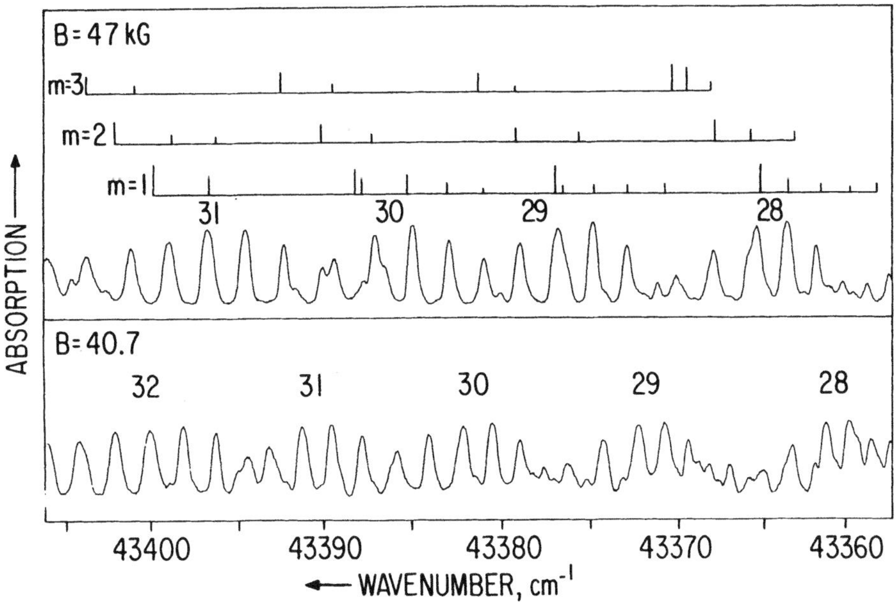

Fig. 7. σ^+ magnetic spectrum near 30p for fields of 41 and 47 kG and pressure of 0.1 mm.

as the ionization limit is approached. There is now a pronounced overlapping of groups but, by careful comparison of spectra at different fields, it is possible to trace a given line through all of these states and to associate it with a specific

Table 2. Eigenvector for the 43387.6 cm^{-1} Computed Level ($n \simeq 30$), $m = 1$ component; $B = 47$ kG

ℓ =	p	f	h	k	o	a	c
n = 29	0.32	-0.27	0.22	-0.15	---	---	---
30	0.37	-0.22	0.21	-0.21	---	---	---
31	-0.20	0.36	-0.24	---	0.16	-0.13	0.12
32	---	-0.14	0.13	---	---	---	---

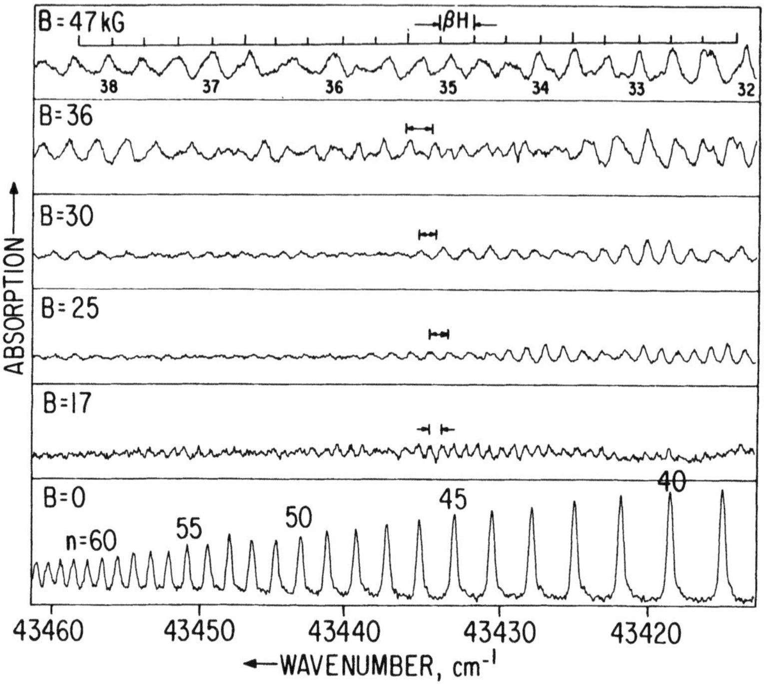

Fig. 8. Variations with magnetic field of σ^+ absorption spectrum slightly below the ionization limit.

low-field quantum state. The high-field labels (cf. Fig. 8), however, should only be regarded as convenient sequence numbers.

Region VI (quasi-Landau region). Fig. 9 shows the resonances spanning the field-free ionization threshold. A regular sequence of lines (or blends of lines of different origins), separated by $\mu_0 B$, has now been established. For a given m, the spacing is $3\mu_0 B$, a value identical to that found for all previously reported cases and to that expected from theory. Each m-sequence may be thought of as a separate quasi-Landau series, displaced from the next one by $\mu_0 B$. The indexing corresponds to those quantum numbers that each feature assumes when the field approaches zero.

M. Robnik, in his discussion of the transition to chaos in quantum mechanical systems [M. Robnik, this volume], has commented that, in such a case as that presented by the strong n- and ℓ-mixing regime, the appearance of unexpected regularities may suggest a hidden symmetry. If this speculation is valid, it

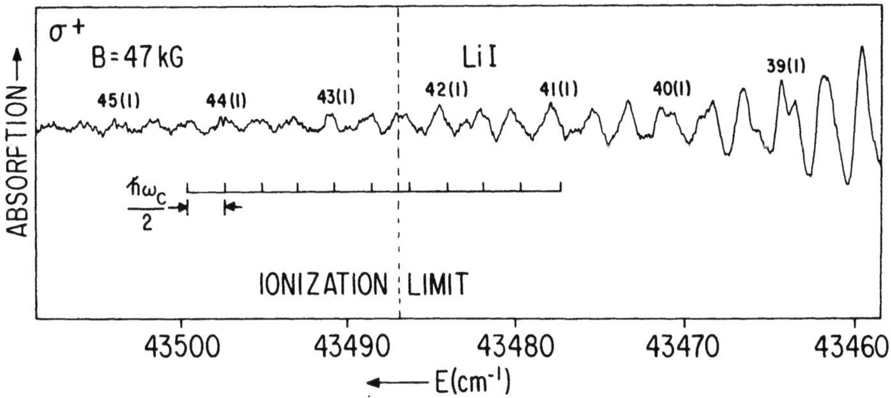

Fig. 9. Quasi-Landau and sub-Landau resonances in the vicinity of the ionization limit. The Bohr magneton scale $\mu_0 B = \frac{1}{2}\hbar\omega_c$ is drawn below.

implies that there may exist a more convenient basis set for describing the above experimental observations. Some regularity does indeed appear in Region V (cf. Fig. 7), in both the experimental and calculated spectra. And, if a new basis which reproduces this regularity can be found, spectral analysis of Region V will require only a more thorough study of the boundary conditions on the conventional Hamiltonian itself, rather than an extension of the fundamental formalism.

ACKNOWLEDGMENT

This work was performed under the auspices of the Office of Basic Energy Sciences, Division of Chemical Sciences, U. S. Department of Energy, under contract W-31-109-Eng-38.

REFERENCES

1. W. R. S. Garton and F. S. Tomkins, Astrophys. J. 158, 839 (1969).
2. W. R. S. Garton, F. S. Tomkins and H. M. Crosswhite, Proc. R. Soc. London A 373, 189 (1980).
3. K. T. Lu, F. S. Tomkins, H. M. Crosswhite and H. Crosswhite, Phys. Rev. Lett. 41, 1034 (1978).

4. H. Crosswhite, U. Fano, K. T. Lu and A. R. P. Rau, Phys. Rev. Lett. 42, 963 (1979).
5. W. R. Bates and A. Damgaard, Philos. Trans. R. Soc. London, Ser. A 242, 101 (1949).
6. A. R. Edmonds, J. Phys. (Paris) 31, C4 (1970).

ON THE SPECTRUM OF $V(\rho) = -\frac{qt}{\rho} + \frac{rt}{\rho^2} + st\rho + t\rho^2$. PHYSICAL IMPLICATIONS FOR A VARIETY OF PROBLEMS

J. A. C. Gallas*

Max-Planck Institut für Quantenoptik
D-8046 Garching bei München
West Germany

R. F. O'Connell

Department of Physics and Astronomy
Louisiana State University
Baton Rouge, Louisiana 70803, USA

1. INTRODUCTION

In recent years there has been much interest in obtaining eigenvalues for relatively complicated potentials, particularly for those situations where perturbation theory is not applicable. Motivation for this interest is not hard to discern: On the one hand, there are the various strong-field problems of condensed-matter-, atomic- and astro-physics [1-7]; and, on the other hand, there are the extensive investigations of the charmonium spectrum [8-10]. Many of these studies have been concerned with the one-dimensional potential

$$V(\rho) = -\frac{qt}{\rho} + \frac{rt}{\rho^2} + st\rho + t\rho^2 \tag{1}$$

where $r \geq 0$ and $q,t>0$. For simplicity, we may also take $s>0$, since this covers many of the cases of interest.

*Permanent address: Departamento de Fisica
Universidade Federal de Santa Catarina
88000 Florianópolis, SC, Brasil

S. P. McGlynn et al. (eds.), Photophysics and Photochemistry in the Vacuum Ultraviolet, 721–728.

Table 1 provides a listing of those physical problems encompassed by the potential of Eq. (1). In particular, this potential is precisely that one considered in many investigations of the charmonium energy spectrum [8-10]. When r=s=0, the quasi-Landau spectrum with magnetic quantum number m=0 results [2]; whereas, if only s=0, the more general quasi-Landau spectrum (m≠0) results [3]. If the ρ^2 term is omitted, the potential has the same form as that for the well-known Stark problem [7]. And, finally, this potential is also appropriate for treating the case of electrons at the surface of liquid helium [1,5] in the presence of crossed electric and magnetic fields [6].

In Section 2 we obtain a WKB solution of the eigenvalue problem resulting from Eq. (1); and, in Section 3 we present a brief discussion of some applications.

2. SOLUTION OF THE EIGENVALUE PROBLEM

Our purpose here is to extend a previously employed technique [4] in order to solve the spectral problem for Eq. (1). We start from the well-known WKB result for the energy E:

$$(2M)^{\frac{1}{2}} \int_{\rho_1}^{\rho_2} [E - V(\rho)]^{\frac{1}{2}} \, d\rho = (n + \tfrac{1}{2}) \pi\hbar \tag{2}$$

where M is a particle mass (usually that of the electron), and where ρ_1 and ρ_2 are the WKB turning points. Thus, from

Table 1. A List of Physical Problems Encompassed by the Potential $V(\rho) = -\frac{qt}{\rho} + \frac{rt}{\rho^2} + st\rho + t\rho^2$.

Physical Problems	Non-Contributing Terms in V(ρ)
Quasi-Landau (m=0)	r=s=0
Quasi-Landau (m≠0)	s=0
Stark Effect	ρ^2 term is omitted
Electrons at the Surface of Liquid Helium in an Electric Field	ρ^2 term is omitted
Electrons at the Surface of Liquid Helium in Crossed Electric and Magnetic Fields	---
Charmonium	---

Eqs. (1,2) we may write

$$I = \int_{\rho_1}^{\rho_2} \frac{d\rho}{\rho}(-\rho^4 - s\rho^3 + p\rho^2 + q\rho - r)^{\frac{1}{2}} = (n+\tfrac{1}{2})\frac{\pi\hbar}{(2Mt)^{\frac{1}{2}}} \tag{3}$$

where $p \equiv E/t$, and where $\rho_1 < \rho_2$ are the two real and non-negative roots of

$$-\rho^4 - s\rho^3 + p\rho^2 + q\rho - r \equiv (\rho-\rho_1)(\rho_2-\rho)(\rho^2 - 2X\rho + Y) = 0. \tag{4}$$

It is convenient to define a generic integral

$$I_i = \int_{\rho_1}^{\rho_2} \rho^i(-\rho^4 - s\rho^3 + p\rho^2 + q\rho - r)^{-\frac{1}{2}} d\rho \tag{5}$$

which allows us to rewrite Eq. (3) as

$$I = -I_3 - sI_2 + pI_1 + qI_0 - rI_{-1}. \tag{6}$$

One may also verify, using integration by parts, that [4]

$$I = I_3 + \frac{s}{2}I_2 + \frac{1}{2}qI_0 - rI_{-1}. \tag{7}$$

Combining Eqs. (6,7), we obtain

$$I = -\frac{s}{4}I_2 + \frac{1}{2}pI_1 + \frac{3}{4}qI_0 - rI_{-1}. \tag{8}$$

A very useful observation can now be made: Namely, that the *structure* of the polynomial occurring in Eq. (4) is the same as that for the s=0 case [4], with the only difference being the numerical values of the four roots ρ_1, ρ_2, and $X \pm (X^2-Y)^{\frac{1}{2}}$. Thus, the integrals I_1, I_0 and I_{-1} are exactly the same as described previously [4] and depend only upon K and Π (the complete elliptic integrals of the first and third kinds, respectively) [10]. We refer to [4] for explicit results.

The only new integral we must evaluate is I_2; and, we find that this integral depends not only upon K and Π but also upon E (the complete elliptic integral of the second kind). As in all problems of this nature [2-4], there are two cases to consider, namely $E > V_c$ and $E \leq V_c$, where V_c is the relative minimum of $V(\rho)$ along the negative ρ-axis. As before [4], we use the results of Byrd and Friedman [11] to obtain:

(i) if $E > V_c$

$$I_2 = g\left\{d^2 K(k) + 2d(\rho_2 - d)\,\Pi(\alpha^2,k) + \frac{(\rho_2-d)^2}{2(\alpha^2-1)(k^2-\alpha^2)} \times\left[\alpha^2 E(k) + (k^2-\alpha^2)K(k) + (2\alpha^2k^2 + 2\alpha^2 - \alpha^4 - 3k^2)\,\Pi(\alpha^2,k)\right]\right\} \tag{9}$$

(ii) if $E \le V_c$

$$I_2 = g\frac{(\rho_1 A-\rho_2 B)^2}{(A-B)^2}\Bigg\{2K(k) + \frac{2(\rho_2-\rho_1)(A+B)}{(\rho_1 A+\rho_2 B)}\Pi(\alpha_1^2,k) - \frac{2AB(A+B)^2}{(\rho_1 A+\rho_2 B)^2} \times\left[\frac{(A-B)^2-2(\rho_1-\rho_2)^2}{4AB}\Pi(\alpha_1^2,k) + \left(\frac{\rho_1-\rho_2}{A+B}\right)^2 K(k) - \left(\frac{A-B}{A+B}\right)^2 E(k)\right]\Bigg\} \quad (10)$$

The various symbols introduced in Eqs. (9,10) are defined in [4], with the exception of α:

$$\alpha^2 = (\rho_1-\rho_2)/(\rho_1-d) \quad . \quad (11)$$

To summarize, we have analyzed a relatively complicated four term potential, using WKB techniques, and have succeeded in obtaining an analytic solution for the eigenvalues in terms of elliptic functions, of which there are only three kinds. We restricted ourselves to s>0 (which is the actual case for many applications and, in particular, for those discussed below) in order to avoid a double-well potential and the associated tunneling and energy shifts [12,13]. In the future, we will return to the s<0 situation in order to consider potentials similar to those depicted in Fig. 2 of [12]. This will require an extension of the work of [12], which was restricted to a harmonic double-well potential, by inclusion of the more general potential. For now, we turn to a brief discussion of a variety of physical problems to which our results are applicable.

3. APPLICATIONS

3.1 Quasi-Landau Resonances

A problem which has attracted enormous interest in recent years is that of the motion of a spinless electron in combined Coulomb and magnetic fields. It is relevant to investigations in such diverse areas as condensed-matter-, atomic- and astrophysics; and, it is fascinating *per se*, from a theoretical point of view, because it is an outstanding example of a nonseparable problem. If the Schrödinger equation for this combined-field problem is written in cylindrical coordinates, it can be shown that the problem is reducible to two-dimensional motion in ρ-z space, where $\rho^2 = x^2 + y^2$. However, if we set z=0, motion in a one-dimensional potential results [2]:

$$V(\rho) = \frac{\hbar^2}{2M}\frac{T}{\rho^2} - \frac{e^2}{\rho} + \frac{1}{8}M\omega^2\rho^2 \quad (12)$$

where ω = eB/Mc is the cyclotron frequency; B is the magnetic field strength; and T is a known function of the magnetic quantum number m.

The z=0 approximation has been shown to give good agreement with the experimental results of Garton and Tomkins [14]. These authors [14] studied the absorption spectrum of barium in a magnetic field of 24 kG, for n values as high as 75, and found broad resonances spaced by approximately 1.5 $\hbar\omega$ near E=0. These so-called quasi-Landau resonances were observed only in the σ spectrum ($\Delta m=\pm 1$); and, since the σ lines result from states which are essentially localized in the x-y plane, we obtain a better appreciation of why the z=0 approximation works so well.

It is clear that Eq. (12) is a special case of the general potential given in Eq. (1) and corresponds to taking s=0. Since we have already discussed the solution to this problem at length, we refer to our previous work [2-4] for details.

3.2 Stark Effect

This is a classic problem which has recently returned to the forefront of interest because of the plethora of experimental investigations of highly-excited states (Rydberg states) brought about by the availability of laser techniques. For many of these investigations, perturbation theory is no longer adequate because of the relative weakness of the Coulomb field, vis-à-vis the electric field, for high-Rydberg states. It is well-known that the Hamiltonian for this problem can be separated in parabolic coordinates, resulting in two Schrödinger radial-like equations in the now familiar ξ = r+z and η = r-z coordinates. The corresponding potentials are special cases of Eq. (1) [the ρ^2 term is missing] and a detailed WKB solution, in terms of the elliptic integrals discussed above, has recently been presented [7].

3.3 Electrons at the Surface of Liquid Helium

Electrons outside a free surface of liquid helium are trapped in an image potential, which is essentially one-dimensional Coulombic [1,5,6]. In addition, there is a 1 eV potential barrier at the surface which prevents electron penetration. As a result, the electrons are trapped in a one-dimensional well, giving rise to quantized states in a direction (z say) perpendicular to the surface. The potential is given by

$$V(z) = -Ze^2/z \tag{13}$$

where $Z = \frac{1}{4}(\varepsilon-1)/(\varepsilon+1) = 6.951 \times 10^{-3}$. If we now assume that the barrier is infinite at the origin, we are led to a "hydrogenic" spectrum for electrons trapped at the surface. Some interesting characteristics of this spectrum have been discussed by Grimes, *et al.* [15]. It was found to be convenient experimentally to apply an electric field F to the system [15] since this configuration permitted investigations to be carried out at a fixed frequency, while the energy separations were varied by means of the electric field. However, the presence of an electric field gives rise to an extra term in the potential, namely eFz. We have already presented a WKB solution to this problem [5], the potential being the same as that of Eq. (1) with the ρ^2 term omitted; the corresponding z^{-2} term which arises in this problem comes from the Langer-type correction to the potential [4,16], a correction which is necessary if one wishes to use the WKB method, derived for the one-dimensional range $(-\infty,+\infty)$, in problems where the range is restricted to $(0,+\infty)$.

Zipfel, Brown and Grimes [6] next applied a magnetic field along the surface ($B=B_y$ say). Following these authors [6], if one chooses a gauge $A = (zB,0,0)$, the Hamiltonian can be written as

$$H = \frac{1}{2M}\left(p_x - \frac{e}{c}Bz\right)^2 + \frac{1}{2M}\left(p_y^{\;2} + p_z^{\;2}\right) - \frac{Ze^2}{z} + eFz \ . \quad (14)$$

Since p_x and p_y commute with H, it is clear that they are both constants of the motion. Moreover, the motion in the z-direction is governed by the potential

$$V(z) = (-\omega p_x + eF)z + \tfrac{1}{2} M\omega^2 z^2 - \frac{Ze^2}{z} \quad (15)$$

where, again, $\omega = eB/Mc$, and where p_x now refers to the eigenvalue rather than to the operator.

Finally, since $0<z<\infty$, the Langer correction contributes a z^{-2} term to Eq. (15); hence, the potential which results is similar to that of Eq. (1), provided F is large enough (or else p_x and/or B are small enough) that the coefficient of z in Eq. (15) is positive. In essence, then, the problem is solved and a detailed exposition of the numerical predictions will be presented in another venue.

3.4 Charmonium

The suggestion, by Gell-Mann and Zweig in 1964, that quarks are the fundamental building blocks of elementary particles has had a profound influence on high-energy physics research. The original proposal was a three-quark model (up, down and strange quarks) but, shortly thereafter, Bjorken and Glashow suggested the existence of a fourth quark (called "charm"). Then, in 1974, a revolution occured in high-energy physics following the discovery of the famous ψ/J particle. It is now well-established that this particle is a particular bound state of the charmed quark and its anti-particle. Such a bound state, the net charm quantum number of which is zero, is called "charmonium". The ψ/J particle is simply the 1^3S_1 state of the charmonium spectrum, a spectrum which, at least as far as nomenclature is concerned, can be compared to the positronium spectrum. Our knowledge of the interquark forces is still somewhat phenomenological but it is generally accepted that the potential is of the form

$$V(r) = -\frac{\beta}{r} + \mu r + \lambda r^2 + \cdots \quad . \tag{16}$$

The force between quarks is propagated by gluons, in analogy to the photon propagation of electromagnetic forces, but here the situation is more complicated in that one must deal with a non-Abelian gauge theory (known as quantum chromodynamics, QCD). However, in the non-relativistic limit it is known that a Coulomb-type force arises at small distances, which rationalizes the first term of Eq. (16), where β is a measure of the gluon force (observations imply that $\beta \simeq 0.25$). The other terms in Eq. (16) represent the "confining potential" which prevents break-up of the system; these terms are introduced, without any basic theoretical underpinning, to understand the observed level spacings. Perhaps it should be emphasized that the potential given in Eq. (16) is inserted into the three-dimensional Schrödinger equation; then, the usual transformation $R(r) \to \xi(r) = r\,R(r)$ provides a one-dimensional equation for the modified radial wavefunction $\xi(r)$, with the potential also being modified by the addition of a centrifugal term $\hbar^2\,\ell(\ell+1)/2Mr^2$. This latter term is further modified by the Langer correction to give $\hbar^2\,(\ell+1/2)^2/2Mr^2$. Such a term, when added to the V(r) appearing in Eq. (16), results in, once more, a potential which is a particular example of Eq. (1).

4. SUMMARY

We have used WKB techniques to obtain analytic solutions, in terms of elliptic functions, for the eigenvalues of the potential given in Eq. (1). In addition, we have shown that this solution provides a *general framework* which can be used for the consideration of a large class of problems, the so-called *strong-field* problems: That is, problems in which more than one field makes comparable contributions to the basic forces and energies. Also, for some of these strong-field problems we have already demonstrated in detail the accuracy of our results, by comparison with experiment and other theoretical investigations [2-5].

ACKNOWLEDGMENT

JACG acknowledges support from the UFSC-Florianopolis/Brasil and DAAD/Germany. RFO'C acknowledges support from the Department of Energy, Division of Material Sciences, contract DE-AS05-79ER10459; he would also like to thank the Max-Planck Institute of Quantum Optics for hospitality during the summer of 1982, when this work was completed.

REFERENCES

1. R. F. O'Connell, Phys. Lett. 60A, 481 (1977); J. A. C. Gallas, E. Gerck and R. F. O'Connell, Phys. Rev. Lett. 50, 324 (1983).
2. J. A. C. Gallas and R. F. O'Connell, J. Phys. B: At. Mol. Phys. 15, L75 (1982).
3. J. A. C. Gallas and R. F. O'Connell, J. Phys. B: At. Mol. Phys. 15, L309 (1982).
4. J. A. C. Gallas and R. F. O'Connell, J. Phys. B: At. Mol. Phys., 15, L593 (1982).
5. R. F. O'Connell, in: *Proceedings of the Colloque International C.N.R.S.* (Aussois, France, 1982), J. de Physique (France), Colloque C2, Supplement 11, C2-81 (1982).
6. C. L. Zipfel, T. R. Brown and C. C. Grimes, Phys. Rev. Lett. 37, 1760 (1976).
7. J. A. C. Gallas, H. Walther and E. Werner, Phys. Rev. A26, 1775 (1982); Phys. Rev. Lett. 49, 867 (1982).
8. C. Quigg and J. L. Rosner, Phys. Reports 56C, 167 (1979); J. Killingbeck, J. Phys. A: Math. Gen. 13, L393 (1980).
9. D. P. Datta and S. Mukherjee, J. Phys. A: Math. Gen. 13, L221 (1980).
10. J. Killingbeck and S. Galicia, J. Phys. A: Math. Gen. 13, 3419 (1980); G. P. Flessas, J. Phys. A: Math. Gen. 15, L1 (1982).
11. B. P. Byrd and M. D. Friedman, *Handbook of Elliptic Integrals for Engineers and Scientists*, 2nd rev. ed. (Springer-Verlag, Berlin, 1975); in particular, see p. 124 and also Eq. (336.02), p. 201.
12. J. H. Weiner, J. Chem. Phys. 69, 4743 (1978).
13. J. H. Weiner and S. T. Tse, J. Chem. Phys. 74, 2419 (1981).
14. W. R. S. Garton and F. S. Tomkins, Astrophys. J. 158, 839 (1969).
15. C. C. Grimes and T. R. Brown, Phys. Rev. Lett. 32, 280 (1974); C. C. Grimes, T. R. Brown, M. L. Burns and C. L. Zipfel, Phys. Rev. B 13, 140 (1976).
16. R. E. Langer, Phys. Rev. 51, 669 (1937).

PERTURBATION SPECTROSCOPY

John D. Scott

Department of Chemistry
University of Montana
Missoula, Montana 59812, USA

1. INTRODUCTION

Perturbation Spectroscopy, for the purposes of this paper, is defined as the absorption spectroscopy of samples which are exposed to an externally applied perturbation. The experiments to be discussed are (i), magnetic circular dichroism (MCD); (ii), electric linear dichroism (ELD) and electrochromism; and (iii), pressure-effect spectroscopy.

Two problems which arise when interpreting effects of perturbations on spectra are:

--- Deduction of observables from theories of microscopic structure: That is, determination of a model ideally structured to make certain molecular parameters (properties) addressable by experiment.

--- Determination of the mode of spectral analysis appropriate for a practical realization of these properties.

The model may be more (or less) well-structured; those for MCD and electric field effects are very clear, whereas that for pressure-effect studies is less clear, a result which follows from the complex nature of the latter perturbation.

In the present paper we give a generalized model for the analysis of perturbation-induced spectral effects. The philosophy underlying this approach is based on the assumption of two effects produced by the perturbation:

S. P. McGlynn et al. (eds.), Photophysics and Photochemistry in the Vacuum Ultraviolet, 729–763.

--- An energy shift of absorption; the absorption band retains its shape, but is dispersed according to a distribution of the ground-state molecules in the perturbing field.

--- A change in intensity caused by (i), a change in the absorber coordinate distribution; and (ii), a change in the transition probability, induced by a perturbation of the absorber wavefunction.

Measurable quantities of interest include the magnitude and, possibly, coordinate distribution of the perturbation; $A^P(\tilde{\nu})$ and $A^O(\tilde{\nu})$, the absorption spectra with and without the perturbation applied, respectively; and/or $A_1(\tilde{\nu})$ and $A_2(\tilde{\nu})$, the absorption spectra with the perturbation applied, but with two different polarizations of the excitation radiation. Two coordinate systems must be considered. The energy shift $D(\underset{\sim}{Q})$ of the transition is a function of the absorber/perturber coordinate Q; $D(Q)$ must be realized in terms of $\Omega(\underset{\sim}{Q})$, the molecular distribution in Q. The excitation mechanism is a function of the laboratory frame $\underset{\sim}{q}$; the absorber distribution in $\underset{\sim}{q}$ is $\omega(\underset{\sim}{q})$. For the magnetic and electric perturbations, $\{\underset{\sim}{Q}\}$ and $\{\underset{\sim}{q}\}$ are identical sets since the perturbation is fixed in the laboratory frame; for pressure-effect studies, however, $\underset{\sim}{Q}$ and $\underset{\sim}{q}$ are independent. An additional important quantity of interest is the absorption probability. This can be considered, and expressed, relative to the zero-perturbation probability and will be given as a product of two functions, $\Gamma(\underset{\sim}{Q})$ and $\gamma(\underset{\sim}{q})$. $\gamma(\underset{\sim}{q})$ arises from an orientation effect different from that at zero perturbation, whereas $\Gamma(\underset{\sim}{Q})$ arises because of wavefunction perturbation.

Formulation of the above effects involves the expansion of $A^P(\underset{\sim}{\nu})$ in a Taylor series about $A^O(\underset{\sim}{\nu})$:

$$A^P(\tilde{\nu}) = \tilde{\nu} \sum_{i=0} a_i \, d^i[A^O(\tilde{\nu})/\tilde{\nu}]/d\tilde{\nu}^i \qquad (1)$$

where the a_i's are parametrically dependent on the perturbation, temperature and radiation polarization, and are expressed as

$$a_i = \frac{(-1)^i}{i!} \; \frac{\iint \Omega(Q)\,\omega(q)\,D^i(Q)\,\Gamma(Q)\,\gamma(q)\,dv_q\,dv_Q}{\iint \Omega(Q)\,\omega(q)\,dv_q\,dv_Q} \qquad (2)$$

The terms in the integrands depend upon the particular experiment and are determined from theory. The experimental $A^P(\tilde{\nu})$ is analyzed in terms of $A^O(\tilde{\nu})$ and a set of empirical expansion coefficients is found; these are equated with respective theoretical a_i's and determination of certain molecular properties is then made. How well theory is contacted is determined by knowledge of the theoretical forms of the a_i's.

In actual practice the perturbed spectrum is often measured as a difference spectrum, $\Delta A(\tilde{\nu})$, which also can be expressed as an expansion:

$$\Delta A(\tilde{\nu}) = \tilde{\nu} \sum_{i=0} c_i \, d^i[A^o(\tilde{\nu})/\tilde{\nu}]/d\tilde{\nu}^i. \quad (3)$$

For a *chromic* effect,

$$\Delta A(\tilde{\nu}) \equiv A^P(\tilde{\nu}) - A^o(\tilde{\nu}) \quad (4)$$

and for a *dichroic* effect,

$$\Delta A(\tilde{\nu}) \equiv A_1^P(\tilde{\nu}) - A_2^P(\tilde{\nu}) \; . \quad (5)$$

The derivations of the c_i's in Eq. (3) from the a_i's in Eq. (1) should be obvious for both definitions of $\Delta A(\tilde{\nu})$.

The following three sections provide descriptions of the specific perturbations.

1.1 Magnetic Circular Dichroism

MCD represents circular dichroism induced by a magnetic field $\underset{\sim}{B}$ applied parallel to the radiation propagation vector. In applying Eq. (5), polarization "1" is taken to be left-circular polarization and "2" to be right-circular polarization. Stephens [1] has written a comprehensive review of MCD. Application of MCD to the study of Rydberg transitions [2] is relatively recent; the wealth of information available from such studies certainly warrants further investigations, both experimental and theoretical. The technical aspects of MCD measurement in the VUV have been discussed in the literature [3].

For modest field strengths (B < 5T) the intensity of $\Delta A(\tilde{\nu})$ is linear in B and the expansion need be carried through only the first derivative if the molecule is C_{3v}, D_{3h}, D_{2d} or higher symmetry; if the symmetry is lower, then only c_0 is non-zero. The molecules considered in this paper all have closed-shell ground states; in addition, we neglect any rotational magnetic moment contribution. Thus, the coefficients are

$$c_0 = bB \quad (6)$$

where b results from the field-induced intensity (the integrated effect of $\Gamma(\underset{\sim}{Q})$), and

$$c_1 = \underset{\sim}{\mu}_m \cdot \underset{\sim}{m}_k B/m_k \quad (7)$$

where μ_m is the electronic magnetic moment of the excited state created by absorbance of right ($\underset{\sim}{m}_2 = \hbar \underset{\sim}{e}_3$) or left ($\underset{\sim}{m}_1 = -\hbar \underset{\sim}{e}_3$) circularly-polarized radiation, and where $\underset{\sim}{e}_3$ is a unit vector in the molecular figure axis. It should be noted that this simple form is applicable only over spectral regions corresponding to transitions for which there is no mixing of electronic states, and for which the vibrational and rotational parts are Franck-Condon allowed (or are made allowed by identical coupling mechanisms). The reason for presenting c_1 in the form of Eq. (7) will become apparent below (Section 2.2).

1.2 Electric Field-Effect Spectroscopy

This study can be made as either a "chromic" (electrochromism) or a "linear-dichroic" (ELD) investigation. The underlying mechanism is the electro-optical Kerr effect [4]. The theoretical form [5] of $A^P(\tilde{\nu})$ has been known for about twenty years and application to Rydberg states was first reported in the middle of the last decade [6]. Technical aspects of spectral measurements in the VUV have been summarized in the literature [3].

An electric field $\underset{\sim}{E}$ is applied perpendicular to the propagation vector of a beam of monochromatic radiation, plane polarized at angle β with respect to $\underset{\sim}{E}$. For modest field strengths ($E <$ 100 kV/cm), $\Delta A(\tilde{\nu})$ is quadratic in E and expansion through the second derivative is sufficient. For electrochromism, the coefficients are

$$c_i = \ln(10)\ [c_i' - (1 - 3\cos^2\beta)c_i'']\ E^2 \qquad (8)$$

where c_i' and c_i'' are given in terms of molecular parameters in Table 1. For ELD, the coefficients are

$$c_i = \ln(10)\ 3(\cos^2\beta_1 - \cos^2\beta_2)\ c_i''\ E^2 \qquad (9)$$

The techniques employed in the actual measurement of these coefficients have been discussed in the literature [3,6]. The values given in Table 1 are for molecules which possess principal symmetry axes (i.e., C_{2v} or higher symmetry); the symbols (//) and ($\perp$) signify transition-moment orientation parallel and perpendicular to the figure axis, respectively.

Electric field-effect spectroscopy has been used to measure permanent electric-dipole moment changes and to determine excited-state electronic symmetries for Rydberg states. It is hoped that, in the near future, temperature-dependence studies will make accurate determinations of other properties possible.

Table 1. Analytic Expressions for c_i' and c_i''.

	c_i'	$c_i''(//)$	$c_i''(\perp)$
i=0	$\frac{1}{3}\left[\frac{R^{(1)}\mu_e}{kT} + S^{(1)}\right]$	$\frac{1}{30}\left[\frac{2\mu_e^2}{k^2T^2} + \frac{3(\alpha_{ii} - \bar{\alpha})}{kT} - 2\left(\frac{R^{(1)}\mu_e}{kT} - S^{(1)}\right) + 3\left(\frac{R^{(2)}\mu_e}{kT} - S^{(2)}\right)\right]$	$c_1''(//) - \frac{\mu_e^2}{10k^2T^2}$
i=1	$\frac{1}{hc}\left[\frac{\mu_e\Delta\mu_e}{3kT} + \frac{\Delta\bar{\alpha}}{2} + \frac{R^{(1)}\Delta\mu_e}{3}\right]$	$\frac{1}{15hc}\left[\frac{2\mu_e\Delta\mu_e}{kT} + \frac{3}{2}(\Delta\alpha_{ii} - \Delta\bar{\alpha}) + \frac{1}{2}(2R^{(1)} - 3R^{(2)})\Delta\mu_e\right]$	$c_2''(//) - \frac{\mu_e\Delta\mu_e}{5kThc}$
i=2	$\frac{(\Delta\mu_e)^2}{6h^2c^2}$	$\frac{(\Delta\mu_e)^2}{15h^2c^2}$	$-\frac{(\Delta\mu_e)^2}{30h^2c^2}$

$\boldsymbol{\mu}_e$ and $\boldsymbol{\mu}_e'$ are ground- and excited-state permanent electric-dipole moments, respectively; $\mu_e = |\boldsymbol{\mu}_e|$; $\Delta\mu_e \equiv (\boldsymbol{\mu}_e' - \boldsymbol{\mu}_e)\cdot\boldsymbol{\mu}_e/|\boldsymbol{\mu}_e|$; $\bar{\alpha}$ and $\bar{\alpha}'$ are ground- and excited-state mean electric polarizabilities, respectively; $\Delta\bar{\alpha} = \bar{\alpha}' - \bar{\alpha}$; α_{ii} and α_{ii}' are elements of the ground- and excited-state polarizability tensors, respectively, i being the Cartesian, molecule-fixed axis in which the transition moment lies; $\Delta\alpha_{ii} = \alpha_{ii}' - \alpha_{ii}$; $\boldsymbol{R}^{(1)}$ and $\boldsymbol{R}^{(2)}$ are vectors which originate in a 1st-order treatment of intensities; $S^{(1)}$ and $S^{(2)}$ are scalars which arise from a 2nd-order treatment of intensities; $R^{(1)} \equiv \boldsymbol{R}^{(1)}\cdot\boldsymbol{\mu}_e/|\boldsymbol{\mu}_e|$; $R^{(2)} \equiv \boldsymbol{R}^{(2)}\cdot\boldsymbol{\mu}_e/|\boldsymbol{\mu}_e|$. (See Ref. [5] for derivations and analytical forms of $S^{(1)}$, $S^{(2)}$, $\boldsymbol{R}^{(1)}$ and $\boldsymbol{R}^{(2)}$.)

1.3 Pressure-Effect Spectroscopy

This *chromic* effect can be studied by measuring the absorption spectrum of a gas sample (the "absorber") in the presence of a nonabsorbing "perturber" gas. Analysis of $A^P(\tilde{\nu})$, in the context of Eq. (1), then gives rise to the various coefficients a_i. In order to extract particular molecular properties from these coefficients, however, a theoretical model must be chosen. And, since the interaction mechanisms for this perturbation are not as simple as were those for the external fields discussed in the previous two sections, the choice of a theoretical model is not as clearly defined. The following discussion presents a model that we are currently developing for use in pressure-effect spectroscopy.

The approximations and assumptions involved in the model are as follows:

--- We assume that the chromic effect results from an energy shift only; no wavefunction perturbation is considered (i.e., $\Gamma(\underset{\sim}{Q}) = 1$).

--- Only pairwise absorber/perturber interactions are considered.

--- The only absorber/perturber coordinate considered is the intermolecular separation r (i.e., $\underset{\sim}{Q} = r$).

--- The interaction potentials for the ground (unprimed) and excited (primed) absorber states are of van der Waals type:

$$\begin{aligned} U(r) &= 2D_e[-(r_e/r)^6 + \tfrac{1}{2}(r_e/r)^{12}] \\ U'(r) &= 2D_e'[-(r_e'/r)^6 + \tfrac{1}{2}(r_e'/r)^{12}] \end{aligned} \tag{10}$$

where D_e and D_e' are the respective potential-well depths, while r_e and r_e' are the respective positions of minimum potential.

--- Each perturber molecule is assigned the ideal-gas volume

$$v = kT/P \tag{11}$$

where k = Boltzmann constant, T = absolute temperature and P = pressure. We assume that each volume is a sphere of radius

$$R = (3kT/4\pi P)^{1/3} \tag{12}$$

and that R is the upper integration limit of Eq. (2) [see Eq. (14), below].

--- We assume a Boltzmann distribution of the ground-state absorber/perturber orientation. In addition, we assume a continuum energy distribution for both ground and excited states.

The energy shift D(r) is given by

$$D(r) = U'(r) - U(r) \tag{13}$$

By defining

$$D_P^i \equiv \int_0^R D^i(r)\ \exp[-U(r)/kT]\ r^2\ dr \Big/ \int_0^R \exp[-U(r)/kT]\ r^3\ dr \tag{14}$$

Eq. (2) can be written

$$a_i = (-1)^i D_p^i / i! \tag{15}$$

The dependence of the a_i's on the perturbation (pressure), the interaction parameters (D_e, D_e', r_e and r_e') and temperature is apparent. The first three D_P^i 's are

$$D_P^0 = 1 \tag{16}$$

$$\begin{aligned} D_P^1 = \{&2[D_e - D_e'(r_e'/r_e)^6]\ F_{-1}(d,r_e/R) \\ &-[D_e - D_e'(r_e'^{12}/r_e^{12}]\ F_1(d,r_e/R)\}/F_{-3}(d,r_e/R) \end{aligned} \tag{17}$$

and

$$\begin{aligned} D_P^2 = \{&4[D_e - D_e'(r_e'/r_e)^6]^2\ F_1(d,r_e/R) \\ &-4[D_e - D_e'(r_e'/r_e)^6][D_e - D_e'(r_e'/r_e)^{12}]\ F_3(d,r_e/R) \\ &+[D_e - D_e'(r_e'/r_e)^{12}]^2\ F_5(d,r_e/R)\}/F_{-3}(d,r_e/R) \end{aligned} \tag{18}$$

where

$$d \equiv D_e/kT \tag{19}$$

and

$$F_n(d,r_e/R) \equiv \int_{(r_e/R)^6}^{\infty} \exp[-d(x-1)^2]\ x^{n/2}\ dx \quad . \tag{20}$$

Eqs. (15-18) provide an ansatz for analyzing the experimentally determined coefficients to obtain molecular properties, namely, the excited- and ground-state D_e and r_e.

The possibility of problems arising as a result of the various approximations and assumptions invoked above must be considered:

--- If analysis gives a value of $a_0 \neq 1$, the neglect of wavefunction perturbation is invalid. One may expect that deviations from $a_0 = 1$ will be the most sensitive measure of the effects of neglecting wavefunction perturbation; thus, the inadequacy of this approximation can be assessed through knowledge of a_0.

--- The appearance of satellite bands would imply (i), breakdown of the continuum-distribution assumption; and/or (ii), that the set of zero-order absorber wavefunctions does not form a basis for representation of the absorber/perturber pair.

--- Divergence, with increasing pressure, from expected values of the coefficients would indicate (i), inadequacy of the spherical model; and/or (ii), the need to consider absorber/perturber interactions of higher order than pairwise.

All of these problems may become important, especially for higher Rydberg states and at higher pressures.

The study of pressure effect on band shape is, of course, not new. A classic work by Margenau and Watson [7] appeared in 1936, and included a discussion of low-energy shifts of higher Rydbergs in the alkali metals. Robin and Kuebler [8], in 1970, dealt specifically with pressure effects on Rydberg transitions. Other models have appeared which invoked different interaction potentials, namely, an harmonic potential [9] and a Morse potential [10]. The above approach is unique in that it provides analysis of experimental data along those lines common to other chromic measurements. And, since this form of analysis involves all points within a band, as opposed to considering only a few points, one may expect that this approach will be a more sensitive measure of perturbation effects than, for example, a simple determination of changes in intensity maxima, band widths and energies of band maxima.

2. PERTURBATION THEORY AND MCD SPECTROSCOPY

2.1 Theory

Since the phenomena underlying MCD spectroscopy are most easily understood and explained in terms of a conserved angular momentum, we choose first-order perturbation approaches based on expansions in a linear molecule basis. Consider the linear molecules A-C and A-B-C, which belong to the $C_{\infty v}$ point group ($D_{\infty h}$, if A = C), with nuclei fixed in the Z-axis. For an n-electron molecule, the electron positions can be given in cylindrical coordinates ρ_j, z_j and ϕ_j ($j = 1,\ldots,n$). By defining $\phi_j' \equiv \phi_j - \phi_1$, the potential energy can be made independent of ϕ_1. The result is an electronic wavefunction

$$\psi_{e\pm} = F(\rho_j, z_j, \phi_j'; j = 1,\ldots,n)\, e^{\pm i\Lambda\phi_1} \tag{21}$$

where $\psi_{e-} = \psi_{e+}^*$ for states with $\Lambda \neq 0$. The conserved orbital angular momentum $\underset{\sim}{\Lambda}$ for ψ_{e+} and ψ_{e-}, respectively, is

$$\underset{\sim}{\Lambda}_+ = \Lambda\hbar \underset{\sim}{e}_3 \quad \text{and} \quad \underset{\sim}{\Lambda}_- = -\Lambda\hbar \underset{\sim}{e}_3 \tag{22}$$

where $\underset{\sim}{e}_3$ is a unit vector in the Z-axis.

For our purposes, the most important element of the wavefunction is the ϕ_1-dependent factor:

$$\Phi_\pm(\phi_1) = e^{\pm i\Lambda\phi_1} \quad . \tag{23}$$

It is in this function that the angular momentum information resides, and it is this function that determines how a particlar perturbation Hamiltonian operator manifests itself. In what follows, several perturbation operators are considered and the results which are expected for MCD spectroscopy are presented. The models invoked are more heuristic than computational; however, they are very useful in interpreting MCD data.

2.1a Molecules of Symmetry Lower than $C_{\infty v}$ or $D_{\infty h}$

The most important effect of lowering the molecular symmetry is to introduce a dependence of the potential energy on ϕ_1, thereby destroying the quantization of axial angular momentum. Within the context of our perturbation format, this is accomplished by mixing states of different Λ.

As an example of this technique, consider a C_{mv} or D_{mh} (m = integer) molecule as a "perturbed" linear molecule. The approximate perturbation Hamiltonian we use is

$$\hat{H}'_m = \sum_{j=0}^{\infty} \sum_{k=1}^{n} a_j(\rho_k, z_k) \cos jm\phi_k$$

$$= \tfrac{1}{4} \sum_{j=0}^{\infty} \sum_{k=1}^{n} a_j(\rho_k, z_k)[e^{ijm\phi'_k} e^{ijm\phi_1} + e^{-ijm\phi'_k} e^{-ijm\phi_1}]$$

$$= \sum_{j=0}^{\infty} [v_j e^{ijm\phi_1} + v_j^* e^{-ijm\phi_1}] \tag{24}$$

where $v_j = v_j(\rho_k, z_k, \phi_k'; k = 1,\ldots,n)$. While the v_j terms determine the magnitude of the perturbation effect, the primary symmetry constraints reside in the $\exp(\pm ijm\phi_1)$ terms. These effects will manifest themselves by (i), changing the energy barycenter of the states; (ii), lifting the degeneracy of some states for which $\Lambda \neq 0$; and (iii), mixing states having different Λ's. The barycenter change V_0 is

$$V_0 = \int \psi^* v_0 \psi d\tau \tag{25}$$

Since there is no experimental quantity from which to derive an empirical V_0, we will not dwell on this term. Effects (ii) and (iii), however, can be addressed experimentally and are discussed below.

First-order energy correction:

Application of first-order degenerate perturbation theory to the degenerate pairs of zero-order states is standard. Solution of the 2x2 secular determinant

$$\begin{vmatrix} H_{++}-E' & H_{+-} \\ H_{-+} & H_{--}-E' \end{vmatrix} = 0 \tag{26}$$

for states with $\Lambda \neq 0$ gives E'. The matrix elements are

$$H_{++} = \int \psi_{e+}^* \hat{H}'_m \psi_{e+} d\tau = V_0 \tag{27a}$$

$$H_{--} = V_0 \tag{27b}$$

$$H_{+-} = \int \psi_{e-}^{*}\hat{H}_m'\psi_{e+}d\tau$$

$$= \sum_{j=0}^{\infty} \int F^2 v_j \; e^{2i\Lambda\phi_1} e^{-ijm\phi_1} d\tau$$

$$= \sum_{j=0}^{\infty} v_j \int \exp[i(2\Lambda - jm)\phi_1]d\tau_1 \qquad (27c)$$

and

$$H_{-+} = H_{+-} \quad . \qquad (27d)$$

From Eqs. (26,27c) one sees that when $2\Lambda/m$ is an integer, there will be symmetry-forced degeneracy lifting; otherwise, $H_{+-} = 0$. The energy correction, through first order, and the corresponding proper zero-order wavefunctions are

$$E_S' = V_0 + V_p, \quad \psi_S = (1/2)^{\frac{1}{2}}[\psi_{e+} + \psi_{e-}] \qquad (28a)$$

$$E_A' = V_0 - V_p, \quad \psi_A = (1/2)^{\frac{1}{2}}[\psi_{e+} - \psi_{e-}] \qquad (28b)$$

when $p(=2\Lambda/m)$ is an integer. If p is noninteger, the degeneracy is not lifted and the wavefunctions remain ψ_{e+} and ψ_{e-}. This is, of course, exactly the same result as that recorded in tables giving resolution of linear-molecule species into those of lower symmetry molecules [11].

First-order wavefunction correction:

While MCD will yield information regarding degeneracy lifting [$c_1 = 0$ in Eq. (7)], it will also give information regarding states that remain degenerate. We now proceed to investigate the symmetry properties of integrands containing two different state functions. Consider a "parent" state 1 which mixes with a "contaminant" state 2 via the perturbation $\hat{H}_m'$. The wavefunction expanded in the linear-molecule basis set is

$$\psi = \psi_1^0 + c_2^1\psi_2^0 + \ldots \qquad (29)$$

where

$$C_2^1 = \frac{\int \psi_2^{0*} \hat{H}_m' \psi_1^0 d\tau}{E_1^0 - E_2^0} \tag{30}$$

$$= \left\{ \sum_{j=0} \left[\int F_2^{0*} v_j F_1^0 d\tau' \int e^{\pm i\Lambda_2\phi_1} e^{ijm\phi_1} e^{\pm i\Lambda_1\phi_1} d\phi_1 \right.\right.$$

$$\left.\left. + \int F_2^{0*} v_j^* F_1^0 d\tau' \int e^{\pm i\Lambda_2\phi_1} e^{-ijm\phi_1} e^{\pm i\Lambda_1\phi_1} d\phi_1 \right] \right\} \Big/ (E_1^0 - E_2^0) \; .$$

($d\tau'$ is the differential volume element excluding $d\phi_1$.)

The coefficient C_2^1 is nonzero only when at least one of the following conditions is satisfied:

$$\begin{aligned} &\textit{condition 1} \qquad \Lambda_1 + \Lambda_2 = jm \\ &\textit{condition 2} \qquad \Lambda_1 - \Lambda_2 = \pm\, jm \end{aligned} \tag{31}$$

where j can be any integer. For *condition 1*, states 2 and 1 combine such that $\underset{\sim}{\Lambda}_2$ and $\underset{\sim}{\Lambda}_1$ are antiparallel; for *condition 2*, $\underset{\sim}{\Lambda}_2$ and $\underset{\sim}{\Lambda}_1$ are parallel. In interpreting MCD spectra this alignment feature will be shown to be important.

Tables 2 and 3 contain summaries of the effects of $\hat{H}_3'$ and $\hat{H}_6'$ perturbations in "creating" C_{3v} (or D_{3h}) and C_{6v} (or D_{6h}) molecules.

2.1b Vibronic Coupling

The coupling of vibrational and electronic motions can be treated by first-order degenerate perturbation theory to yield results that are immediately applicable to MCD spectra. We present this for linear triatomics first (Renner-Teller effect), followed by nonlinear polyatomics (addition of Jahn-Teller effects).

Linear molecules:

Of the three normal vibrational modes of a $C_{\infty v}$ triatomic molecule, two are totally symmetric (σ^+) and the remaining one transforms as π. For a $D_{\infty h}$ triatomic the symmetry species are σ_g^+, σ_u^+ and π_u. A transition from a ground electronic state, having no π vibration excited, to an excited electronic state to which a singly-excited π vibration is coupled is nominally forbid-

Table 2. Effect of the Perturbation Associated with m=3

Parent Λ	Contaminant State Λ* k=0	k=1	k=2	Symmetry Species in C_{3v}
1	1	(-)2,4	(-)5,7	E
2	2	(-)1,5	(-)4,8	E
3†	3	0,6	(-)3,9	$A_1(\psi_s), A_2(\psi_a)$
4	4	1,7	(-)2,10	E
5	5	2,8	(-)1,11	E
6†	6	3,9	0,12	$A_1(\psi_s), A_2(\psi_a)$

*The sign indicates the relative directions of the vector $\underset{\sim}{\Lambda}$ in the mixing and parent states.

†Levels for which degeneracy is removed.

Table 3. Effect of the Perturbation Associated with m=6

Parent Λ	Contaminant State Λ* k=0	k=1	k=2	Symmetry Species in C_{6v}
1	1	(-)5,7	(-)11,13	E_1
2	2	(-)4,8	(-)10,14	E_2
3†	3	(-)3,9	(-)9,15	$B_1(\psi_s), B_2(\psi_a)$
4	4	(-)2,10	(-)8,16	E_2
5	5	(-)1,11	(-)7,17	E_1
6†	6	0,12	(-)6,18	$A_1(\psi_s), A_2(\psi_a)$
7	7	1,13	(-)5,19	E_1

*,†See footnotes for Table 2.

den by the Franck-Condon principle. The occurrence of such a transition must involve vibronic interaction.

The potential energy introduced by the nuclear motions associated with a π vibration is considered here as a static perturbation. The nuclear positions are given in cylindrical coordinates; considering center-of-mass and rotational coordinates as separable, and employing two equations of constraint, the potential energy may be expressed in terms of electronic coordinates and two nuclear coordinates, ρ_N and ϕ_N. The form of the perturbation Hamiltonian is

$$H'_{C_{\infty v}} = \sum_{j=0} \sum_{k=0} \Big\{ a_{jk}(\rho_\ell, z_\ell, \phi'_\ell; \ell = 1,\ldots,n)\, \rho_N^j\, e^{ik\phi_1} e^{-ik\phi_N} + a^*_{jk}(\rho_\ell, z_\ell, \phi'_\ell; \ell = 1,\ldots,n)\, \rho_N^j\, e^{-ik\phi_1} e^{ik\phi_N} \Big\} \ . \quad (32)$$

The zero-order wavefunction is

$$\psi^0_{ev} = \psi^0_e \psi^0_v$$

where ψ_e is the electronic wavefunction [see Eq. (21)] and ψ_v is [12]

$$\psi_v \equiv \psi_{v\pm} = f_{n',\ell}(\rho_N)\, e^{\pm i\ell\phi_N} \quad (33)$$

where n' is the number of excited quanta and

$$\ell = n', n'-2, \ldots, 1 \text{ or } 0 \ . \quad (34)$$

The function $f_{n',\ell}(\rho_N)$, while numerically important, contains no symmetry restrictions in $C_{\infty v}$ vibronic perturbation and, therefore, we do not consider it.

When the electronic and vibrational states are each doubly degenerate, the wavefunction is

$$\psi_{ev} = \psi_{e\pm}\psi_{v\pm} \ . \quad (35)$$

All four combinations are possible; they are denoted

$$\psi_1 = \psi_{e+}\psi_{v+} \qquad \psi_3 = \psi_{e-}\psi_{v+}$$
$$\psi_2 = \psi_{e+}\psi_{v-} \qquad \psi_4 = \psi_{e-}\psi_{v-} \ . \quad (36)$$

The perturbation energy is found by solving the 4x4 determinant

$$|H'_{tu} - \delta_{tu}E'| = 0 \tag{37}$$

where

$$H'_{tu} = \int \psi_u^* \hat{H}'_{C_{\infty v}} \psi_t d\tau \ . \tag{38}$$

The only nonzero off-diagonal elements are H'_{23} and H'_{32} where

$$H'_{23} = H'_{32} \ . \tag{39}$$

These are nonzero only if $\ell = \Lambda$; otherwise, second-order perturbation theory must be employed to account for the degeneracy splitting of states 2 and 3 away from 1 and 4.

Information regarding a property of interest to us, namely, the presence of a nonzero electric-dipole transition moment for a nominally forbidden transition, lies in the perturbation wavefunctions. The perturbation Hamiltonian must mix a Franck-Condon allowed "contaminant" into a forbidden "parent". *Allowed*, from the standpoint of our basis functions, means that

$$\begin{aligned} &\Delta\Lambda = 0, \pm 1; \qquad \Sigma^+ \not\leftrightarrow \Sigma^- \\ &\Delta\ell = 0. \end{aligned} \tag{40}$$

Consider a Σ^+, non-vibrationally-excited ground state as an example. The Franck-Condon allowed, electric-dipole allowed excited state must have

$$\begin{aligned} &\Lambda = 0 \text{ or } 1 \qquad (\Sigma^+ \text{ or } \Pi) \\ &\ell = 0. \end{aligned} \tag{41}$$

If the parent state ψ_I is forbidden, and a contaminant state ψ_{II} is allowed, the perturbed state function may be written

$$\psi = \psi_I + C_{II}\psi_{II} + \ldots \tag{42}$$

where

$$C_{II} = \frac{\int \psi_{II}^* \hat{H}'_{C_{\infty v}} \psi_I d\tau}{E_I^0 - E_{II}^0} \ . \tag{43}$$

Considering the forms of $\hat{H}'_{C\infty v}$ and the zero-order wavefunctions, the possibilities of $C_{II} \neq 0$ (by symmetry), where ψ_{II} is allowed, are listed for a few forbidden transitions in Table 4. It should be noted that in some instances $\tilde{\Lambda}_I$ and $\tilde{\Lambda}_{II}$ are parallel, while in others they are antiparallel.

Nonlinear polyatomics:

We assume that the zero-order functions used in previous sections provide an adequate representation of the electronic states of a C_{mv} (or D_{mh}) molecule. We also assume that the wavefunctions of the doubly-degenerate vibrations can be given in the form of Eq. (33), with the coordinate ϕ_N being defined as

$$\phi_N = \cos^{-1} \xi_a/(\xi_a^2 + \xi_b^2)^{\frac{1}{2}} \tag{44}$$

where ξ_a and ξ_b are the normal coordinates of a doubly-degenerate vibration.

The vibronic perturbation Hamiltonian for a C_{mv} molecule differs somewhat from that of the linear case:

$$\hat{H}_{C_{mv}} = \sum_{j=0} \sum_{k=0} \sum_{n=0} a_{jkn}\rho_N^j \cos nm\phi_1 \cos k(\phi_1 - \phi_N)$$

$$= \tfrac{1}{2} \sum_{j=0} \sum_{k=0} a_{jk0}\rho_N^j \left[e^{ik\phi_1} e^{-ik\phi_N} + e^{-ik\phi_1} e^{ik\phi_N}\right]$$

$$+ \tfrac{1}{4} \sum_{j=0} \sum_{k=0} \sum_{n=1} a_{jkn}\rho_N^j \left\{\left[e^{i(k+nm)\phi_1} e^{-ik\phi_N} + e^{-i(k+mn)\phi_1} e^{ik\phi_N}\right]\right.$$

$$\left. + \left[e^{i(k-nm)\phi_1} e^{-ik\phi_N} + e^{-i(k-mn)\phi_1} e^{ik\phi_N}\right]\right\} \quad . \tag{45}$$

The Renner-Teller terms from Eq. (32) are recognizable. The additions are the Jahn-Teller terms, which increase the possibilities for inducing allowedness into vibronically forbidden transitions. Table 5 is a compilation of the intensity-conferring aspects of vibronic coupling in a C_{3v} molecule. Note again the parallel/antiparallel alignments of parent and contaminant $\tilde{\Lambda}$ vectors.

Table 4. Vibronic Coupling in $C_{\infty v}$

Electronic State		Vibrational State		Allowedness*		Symmetry of Allowed Electronic Contaminant Which is Franck-Condon Allowed†
Λ	Symmetry Species	ℓ	Symmetry Species	Electric Dipole	Franck-Condon	
0	Σ^+	1	π	a	f	Π
1	Π	1	π	a	f	Σ^+
		2	δ	a	f	$(-)\Pi$
2	Δ	1	π	f	f	Π
		2	δ	f	f	Σ^+

*The allowedness refers to that of the non-interacting electronic/vibrational state. "a" means allowed; "f" means forbidden.

†(-) indicates antiparallel orientation of parent and contaminant state Λ.

Table 5. Vibronic Coupling in C_{3v} *

Electronic State			Vibrational State			Allowedness		Symmetry of Allowed Electronic Contaminant Which is Franck-Condon Allowed
Λ	Symmetry Species $C_{\infty v}$	Symmetry Species C_{3v}	ℓ	Symmetry Species $C_{\infty v}$	Symmetry Species C_{3v}	Electric Dipole	Franck-Condon	
0	Σ^+	A_1	1	π	e	a	f	Π
1	Π	E	1	π	e	a	f	$\Sigma^+, (-)\Pi$
			2	δ	e	a	f	$\Sigma^+, (-)\Pi$
2	Δ	E	1	π	e	f	f	Σ^+, Π
			2	δ	e	f	f	Σ^+, Π

*See footnotes of Table 4.

2.2 Experiment: Practical Realization of Perturbation Models

2.2a Manifestations of "State Mixing"

This discussion is limited to (i), allowed transitions; and (ii), forbidden transitions made observable through single perturbations. And, our emphasis is placed on the sign and magnitude of c_1. As indicated in Eq. (7), the sign of c_1 is determined by the relative alignment of $\underset{\sim}{\mu}_m$ and $\underset{\sim}{m}_k$ -- parallel for positive c_1 and antiparallel for negative c_1. An example is the electric-dipole allowed transition $^1\Sigma^+ \rightarrow {}^1\Pi$ of a $C_{\infty v}$ molecule. In the $^1\Pi$ state, $\underset{\sim}{\Lambda}$ and $\underset{\sim}{\mu}_m$ are antiparallel; therefore, $\underset{\sim}{m}_k$ and $\underset{\sim}{\mu}_m$ are also antiparallel. The value of c_1 is

$$c_1 = - \mu_m B \tag{46}$$

The sign of c_1, for a nominally forbidden transition, is determined by the intensity-conferring mechanism. The following examples serve as illustrations of this.

C_{3v} perturbation and a $^1\Sigma^+ \rightarrow {}^1\Delta$ transition:

This transition is electric-dipole forbidden in $C_{\infty v}$ molecules; however, it becomes an allowed $^1A_1 \rightarrow {}^1E$ transition in a C_{3v} molecule. From Table 2 one sees that the allowedness is gained through mixing with $^1\Pi$ states. The mixing is such that the angular momenta of the parent and $^1\Pi$ contaminant are antiparallel:

$$\begin{aligned} \psi_+(C_{3v}) &= \psi_+{}^0(\Lambda=2) + a\psi_-{}^0(\Lambda=1) \\ \psi_-(C_{3v}) &= \psi_-{}^0(\Lambda=2) + a\psi_+{}^0(\Lambda=1). \end{aligned} \tag{47}$$

For small perturbations $\underset{\sim}{\mu}_m$ is determined principally by $\psi^0(\Lambda=2)$; however, the transition is allowed through the $\psi^0(\Lambda=1)$ state. Since $\underset{\sim}{\mu}_m$ and $\underset{\sim}{\Lambda}$ of the $\psi^0(\Lambda=2)$ parent state are antiparallel, the coupling of the allowed contaminant predicts a positive c_1.

Vibronic coupling in a $C_{\infty v}$ molecule:

Consider first the transition $^1\Sigma(\ell=0) \rightarrow {}^1\Pi(\ell=1)$, where the vibrational quantum number ℓ was discussed in Section (2.1b). This is a Franck-Condon forbidden transition; it may gain intensity (cf. Table 4) by the mixing of a $^1\Sigma^+(\ell=0)$ with a $^1\Pi(\ell=1)$ state. The degeneracy of a pair of the states is removed [see Eq. (36)] and transition to the $^1\Sigma^+$ component is allowed. The c_1 for such a transition is zero.

Consider next the transition $^1\Sigma^+(\ell=0) \rightarrow {}^1\Delta(\ell=1)$, which is both electric-dipole and Franck-Condon forbidden. From Table 4 one sees that intensity may be gained through mixing of a $^1\Pi$ $(\ell=0)$ with the $^1\Delta(\ell=1)$ state. In the $^1\Delta$ electronic state, Λ and $\underset{\sim}{\mu}_m$ are antiparallel. The $^1\Pi$ contaminant mixes in a parallel fashion with the $^1\Delta$ parent giving an antiparallel $\underset{\sim}{\mu}_m$ and $\underset{\sim}{m}_k$; a negative c_1 results.

Vibronic coupling in a C_{3v} *molecule:*

The transition $^1\Sigma^+(\ell=0) \rightarrow {}^1\Pi(\ell=1)$ becomes the Franck-Condon forbidden $^1A_1a_1 \rightarrow {}^1Ee$ transition in a C_{3v} molecule. From Table 5 one finds that there are two allowed mechanisms arising from vibronic interactions. Two of the degenerate zero-order states lose degeneracy; one gains allowedness by mixing with $^1\Sigma^+(\ell=0)$ states, resulting in $c_1=0$. Two remain degenerate and intensity is conferred by their mixing with $^1\Pi(\ell=0)$ states. This is an antiparallel mixing and results in a positive c_1. If the energy splitting is not resolvable, the MCD band will lie somewhere between two extremes. At the one extreme the $^1\Pi(\ell=0)$ transition has no intensity and $c_1=0$; at the other extreme the $^1\Sigma^+$ $(\ell=0)$ transition has no intensity and $c_1 = \underset{\sim}{\mu}_m B$.

Similar phenomena occur for the transition $^1\Sigma^+(\ell=0) \rightarrow {}^1\Delta(\ell=1)$, which becomes $^1A_1a_1 \rightarrow {}^1Ee$ in a C_{3v} molecule. In this case, as shown in Table 5, the transition can gain intensity through mixing of the excited parent state with either a $^1\Pi(\ell=0)$ or $^1\Sigma^+(\ell=0)$ state. Thus, two components become allowed and either one or both may be observed. The mixing with the $^1\Pi(\ell=0)$ is parallel, producing a negative c_1. As discussed in the foregoing paragraph, the MCD spectrum may lie between the two extremes, with $-\underset{\sim}{\mu}_m B \leq c_1 \leq 0$.

2.2b Spectral Interpretation

To exemplify applications of the above perturbation models, we now turn to experimental results for several methyliodide transitions. (A thorough discussion of these bands has been given previously [2,13].) The band system of interest, shown in Figs. 1-6, contains the first s Rydberg transitions; the linear-molecule analogues are $^1\Sigma^+ \rightarrow {}^3\Pi$ and $^1\Sigma^+ \rightarrow {}^1\Pi$. This is a strong spin-orbit coupling regime, as evidenced by equal intensities in the $^1\Pi$ and $^3\Pi_1$ components; both have equal amounts of $^1\Pi$ character and thus are equally spin-allowed from the $^1\Sigma^+$ state. The MCD spectra of these two origins [$^3\Pi_1$ is denoted "2" in Fig.1; $^1\Pi$ is denoted "4" in Fig. 4] demonstrate negative c_1 values and analysis indicates the ratio of 1:0.82 for the magnetic moments of states 2 and 4, respectively. The negative sign of c_1 is indicative of the antiparallel alignment of the angular momentum and $\underset{\sim}{\mu}_m$ in the two states.

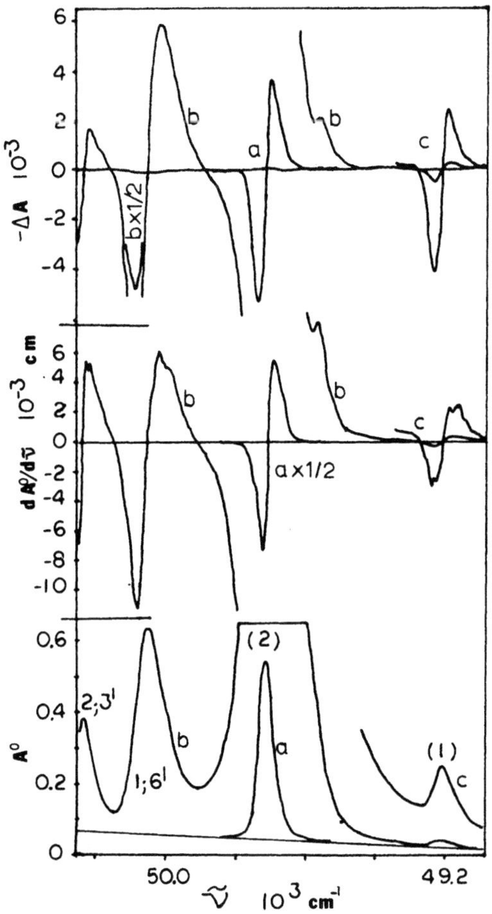

Fig. 1. Absorption $dA^{o}/d\tilde{\nu}$ and $-\Delta A$ spectra of gas-phase methyliodide. T = 300 K; sample pressures a=0.02, b= 1.0 and c=10.0 torr; magnet current = 43.0 amps and magnet constant = 5 Tesla/45 amps. Assignment notation $\alpha;\beta_i^j$: α(=1-4) designates origin; β (=1-6) designates normal mode (ν_1-ν_6); and i,j are the number of excited quanta in normal coordinate ν_β of the ground and excited electronic states, respectively.

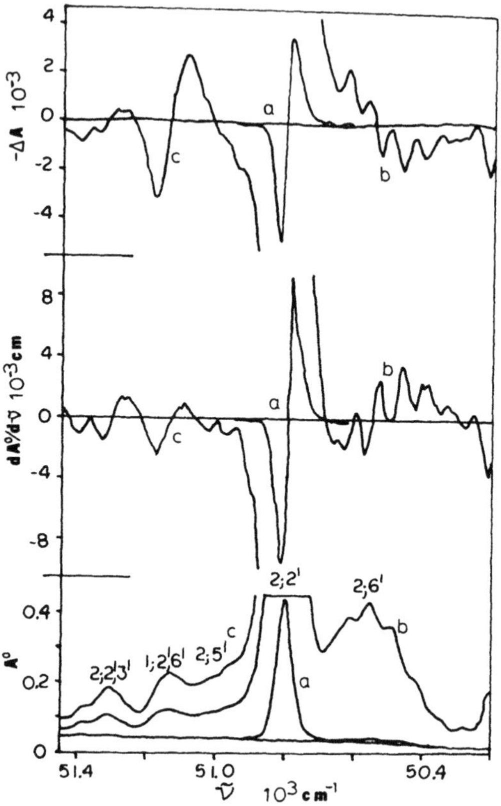

Fig. 2. Absorption, $dA^0/d\tilde{\nu}$ and $-\Delta A$ spectra of gas-phase methyliodide. Sample pressures: a=0.05, b=3.0 and c=6.5 torr; magnet current = 43.0 amps. See Fig. 1 caption for other information.

Origin 1 in Fig. 1 corresponds to the $^3\Pi_2$ component. Spin-orbit coupling mixes the $^3\Pi_2$ state with a spin-allowed $^1\Delta$ state, which remains forbidden by electric-dipole selection rules. Among the possible mechanisms for inducing allowedness is the C_{3v} symmetry of the molecule; this could couple a $^1\Pi$ state with the $^1\Delta$ state. As discussed in earlier sections, this would be an anti-parallel mixing resulting in a positive c_1. The MCD spectrum reveals a negative c_1 and analysis gives a magnetic moment ratio of 3:1 between states 1 and 2. It appears that there is little if any intensity gained through the C_{3v} perturbation. Transition to the $^3\Pi_2$ component is observed in absorption for HI [14]. The mechanism conferring allowedness is associated with rotational-

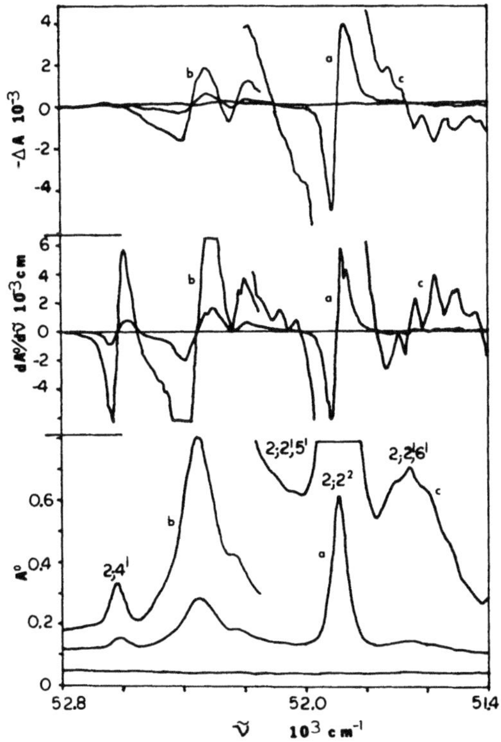

Fig 3. Absorption, $dA^0/d\tilde{\nu}$ and $-\Delta A$ spectra of gas-phase methyliodide. Sample pressures: a=0.6, b=3.0 and c=11.0 torr; magnet current = 44.2 amps. See Fig. 1 caption for other information.

electronic coupling. There is certainly no C_{3v} interaction in HI; however, the relative intensities of the transitions in HI and methyliodide are approximately the same.

While origin 3 is barely discernible in the absorption spectrum (Fig. 4), the MCD spectrum consists of a large c_0 term and no c_1 term. This origin is assigned as the transition to the $^3\Pi_{0+}$ component.

There are numerous vibrational bands joined to these origins. Those which involve excitations of the vibrations ν_1, ν_2 and ν_3 are Franck-Condon allowed. The vibrations ν_4, ν_5 and ν_6 are of e symmetry and single excitations of these are Franck-Condon forbidden. Approximate nuclear displacements associated with these

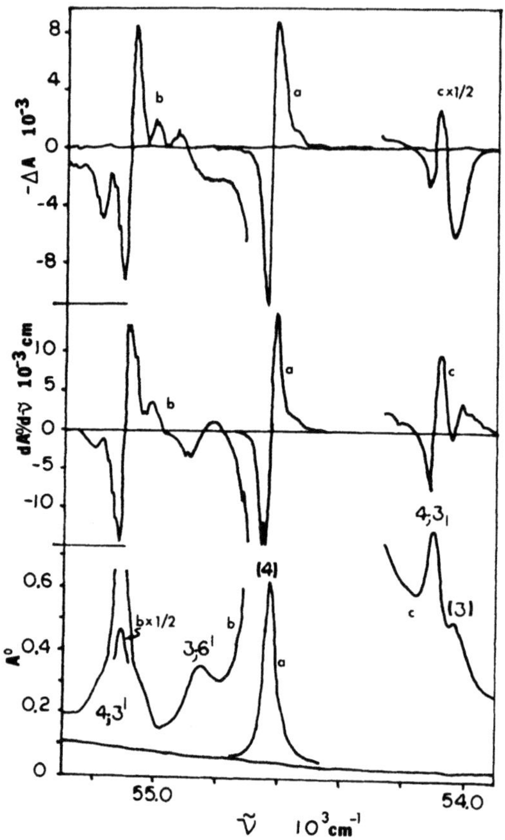

Fig. 4. Absorption, $dA^{O}/d\tilde{\nu}$ and $-\Delta A$ spectra of gas-phase methyliodide. Sample pressures: a=0.02, b=0.3 and c=15.0 torr; magnet current = 40.0 amps. See Fig. 1 caption for other information.

three doubly-degenerate vibrations are depicted in Fig. 7. Long series of Franck-Condon allowed components are joined to origins 2 and 4. The MCD spectra of these transitions reveal structure identical to their respective origins. This is expected since the angular momentum characteristics of these states are the same as that of their respective origins.

The Franck-Condon forbidden transitions demonstrate interesting points with respect to the coupling phenomena responsible for their observation in absorption. Single excitations in ν_4

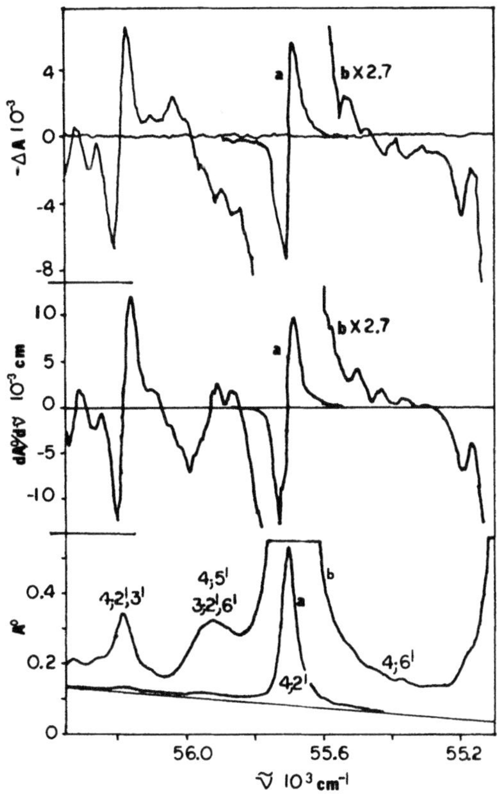

Fig. 5. Absorption, $dA^0/d\tilde{\nu}$ and $-\Delta A$ spectra of gas-phase methyliodide. Sample pressures: a=0.035 and b=0.65 torr; magnet current = 40.0 amps. See Fig. 1 caption for other information.

and ν_6 are joined to origin 1 ($1;6^1$ in Fig. 1, $1;2^1,6^1$ in Fig. 2 and $1;4^1$ in Fig. 3). Both vibrations are associated with negative c_1 values, the magnitudes of which are equal to that of origin 1. Only the $^1\Pi(\ell=0)$ contaminant appears to induce intensity; this may result from the lack of strongly allowed transitions to $^1\Sigma^+$ states in this region.

Joined to origins 2 and 4 are bands resulting from single excitations in each of the three Franck-Condon forbidden vibrations. In each series, ν_5 and ν_6 excitations result in positive c_1 values with magnitudes very nearly equal to that of the respective origins ($2;6^1$ and $2;5^1$ in Fig. 2 and $2;2^1,6^1$ and $2;2^1,5^1$ in

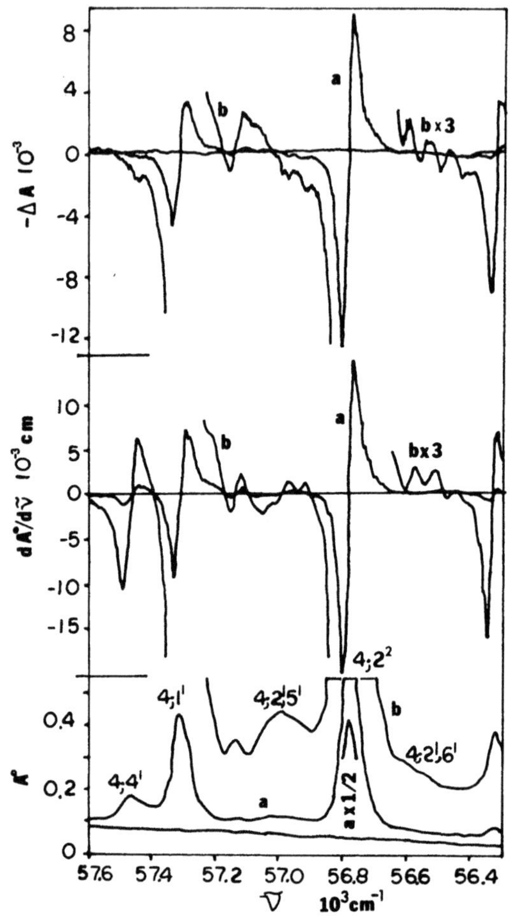

Fig. 6. Absorption, $dA^{0}/d\tilde{\nu}$ and $-\Delta A$ spectra of gas-phase methyliodide. Sample pressures: a=0.20 and b=0.65 torr; magnet current = 40.0 amps. See Fig. 1 caption for other information.

Fig. 3). The ratio of c_1 for 2 to that for $2;6^1$ and $2;2^1,6^1$ is $\sim$1:0.9. This may indicate slight intensity in the allowed nondegenerate component. The antiparallel combination of parent and contaminant, and the resulting positive c_1, were discussed in the previous section. The other vibration, ν_4, is also involved in vibronic coupling. The observed excitations ($2;4^1$ in Fig. 3 and $4;4^1$ in Fig. 6) reveal no c_1 terms but do yield large c_0 coefficients. This observation implies only Renner-Teller interaction

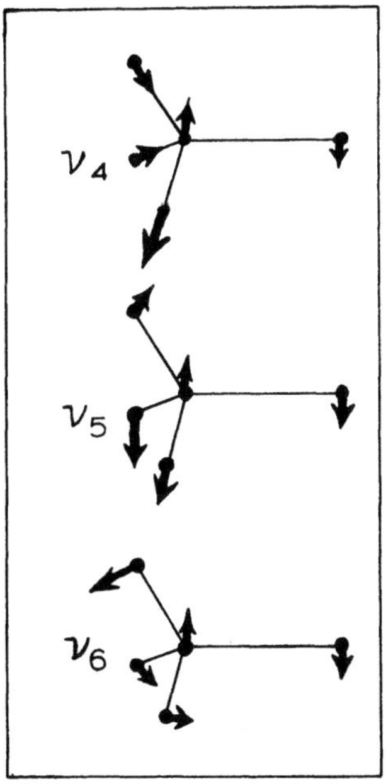

Fig. 7. Approximate nuclear motions associated with normal modes ν_4, ν_5 and ν_6 in methyliodide. Relative amplitudes are not to scale.

associated with this transition. The form of this vibration, shown in Fig. 7, indicates a deformation very much like that of the bending vibration in a linear triatomic, where Renner-Teller interaction is the only vibronic mechanism present.

Single excitations of ν_6 have been assigned joining to origin 3 ($3;6^1$ in Fig. 4 and $3;2^1,6^1$ in Fig. 5). Here, allowedness is gained by a mixing of the excited state with a $^1\Pi(\ell=0)$ state. While $\underset{\sim}{m}_k \neq 0$, the parent has $\underset{\sim}{\mu}_m = 0$ and c_1 should be quite small. The MCD spectra of these components demonstrate no derivative dependence, thus bearing out the $c_1 \simeq 0$ prediction.

From the MCD investigation of the Franck-Condon forbidden transitions joined to origins 1, 2 and 4, one may conclude that ν_4 components gain intensity through an interaction that is strictly Renner-Teller. The MCD of transitions associated with ν_5 and ν_6 excitations demonstrate both Renner-Teller (band system 1) and Jahn-Teller (band systems 2 and 4) interactions. The lack of evidence for Jahn-Teller interaction in band system 1, and Renner-Teller interaction in band systems 2 and 4, is probably a result of the absence of nearby, allowed $^1\Sigma^+$ states which are required to mix and induce intensity.

3. ELECTRIC FIELD-EFFECT SPECTROSCOPY

3.1 Theory

We now turn to a brief discussion of the theory underlying several properties measurable in an ELD or electrochromism experiment. Before any other property may be correctly evaluated by data analysis, the electronic symmetry of the excited state must be determined. This is done by examining the ratio of c_2'' to c_2' -- a positive ratio indicates a parallel transition dipole moment while a negative ratio indicates a perpendicular orientation. As discussed in Section 1.2, the orientation is taken with respect to the molecular figure axis of a molecule of C_{2v} or higher symmetry. There must also be a change in permanent dipole moment associated with the transition in question. Knowledge of the electronic symmetry is, of course, fundamentally important in the investigation of molecular states. Often, one must use oriented crystalline samples to determine such information, an impossible task in the case of most Rydberg transitions. Electric field-effect spectroscopy permits measurement of this property using gas-phase samples.

A second quantity determinable from c_2'' amd c_2' is the magnitude of the change in permanent electric-dipole moment. In addition, by examining the temperature dependence of c_1', the sign of the dipole moment change can be determined. Thus, the excited-state dipole can be evaluated.

We are currently developing a semiempirical model that should give insight into the properties associated with electric field investigations. It is being employed in the theoretical study of the hydrogen halides, HX. The hydrogen halide formed from the halogen having atomic number n-1 is treated as a perturbed rare-gas atom of atomic number n. The Schrödinger equation for this atom is approximated as a one-electron system:

$$\hat{H}^{\alpha}_{1e}\phi^{\alpha}_{k}(n) = E^{\alpha}_{k}\phi^{\alpha}_{k}(n) - E^{\alpha}_{k}\sum_{j=1}^{n-1} S^{\alpha,k}_{j}(n) + \sum_{j=1}^{n-1} H^{\alpha,k}_{j}(n) \tag{48}$$

where

$$S^{\alpha,k}_{j}(n) = \int\cdots\int \psi^{\alpha *}_{c}(i\neq n)\ \psi^{\alpha}_{c}(i\neq j)\ \phi^{\alpha}_{k}(j)d\tau_1\cdots d\tau_{n-1} \tag{49}$$

and

$$H_j^{\alpha,k}(n) = \int \cdots \int \psi_c^{\alpha *}(i \neq n) r_{jn}^{-1} \psi_c^{\alpha}(i \neq j) \phi_k^{\alpha}(j) d\tau_1 \cdots d\tau_{n-1} \qquad (50)$$

are the usual overlap and exchange functions, respectively, and where

$$\hat{H}_{1e}^{\alpha} = -\tfrac{1}{2}\nabla_n^2 - \frac{n}{r_n} + V_{eff}^{\alpha}(n) \ . \qquad (51)$$

The $V_{eff}^{\alpha}(n)$ is usually given as

$$V_{eff}^{\alpha}(n) = \sum_{j=1}^{n-1} \int \cdots \int \psi_c^{\alpha *}(i \neq n) r_{jn}^{-1} \psi_c^{\alpha}(i \neq n) d\tau_1 \cdots d\tau_{n-1} \ . \qquad (52)$$

In the above equations, α designates the αth state of the (n-1)-electron cation; k designates the Rydberg state and is referenced to the αth cation state; the indices in parenthesis denote coordinates of specified electrons; r_n is the distance between the nucleus and the nth electron; r_{jn} is the distance between electrons j and n; and ψ_c^{α} is the (n-1)-electron function of cation state α. The one-electron function ϕ_k^{α} is the "Rydberg" orbital. The state function is then the antisymmetrized product of ψ_c^{α} (i≠n) and ϕ_k^{α} (n). The energy E_k^{α} is equated with the negative of the experimentally determined term value T_k^{α} of the Rydberg transition, defined with respect to the αth cation state:

$$E_k^{\alpha} = -T_k^{\alpha} \ . \qquad (53)$$

Since the scope of this paper does not include complicated theoretical matters, no further discussion (core-orbital orthogonality, etc.) or justification of this form of approximation will be made.

We assume that the state functions of H-X can be expanded in terms of the state functions of the related, iso-valence rare gas. (This, of course, is not quite true because of function discontinuity at H.) Through first order, these functions are

$$\psi_{c,mol}^{\alpha} = \psi_c^{\alpha} + \psi_c^{\alpha'} \qquad (54)$$

and

$$\phi_{k,mol}^{\alpha} = \phi_k^{\alpha} + \phi_k^{\alpha'} \ . \qquad (55)$$

Neglecting the additional overlap and exchange terms introduced by the molecular symmetry, the one-electron molecular Hamiltonian becomes

$$\hat{H}^{\alpha}_{1e,\mathrm{mol}} = \hat{H}^{\alpha}_{1e} + \hat{H}^{\alpha'}_{1e} \tag{56}$$

where

$$\hat{H}^{\alpha'}_{1e} = [(1/r_n) - (1/r'_n)] + \sum_{j=1}^{n-1} \int\cdots\int r_{jn}^{-1} \Big[\psi_c^{\alpha *}(i\neq n)\psi_c^{\alpha'}(i\neq n) + \psi_c^{\alpha' *}(i\neq n)\psi_c^{\alpha}(i\neq n) + \psi_c^{\alpha' *}(i\neq n)\psi_c^{\alpha'}(i\neq n)\Big]\, d\tau_1\cdots d\tau_{n-1} \tag{57}$$

and where r'_n is the distance between electron n and the H nuc-nucleus. Evaluation of $\hat{H}^{\alpha'}_{1e}$ from atomic and molecular spectroscopy is one of the goals of the perturbation treatment. In order to accomplish this, $\hat{H}^{\alpha'}_{1e}$ is considered to be an expansion of electric multipoles, centered between H and X:

$$\hat{H}^{\alpha'}_{1e} \simeq \sum_{j=1} Q_j P_j(\cos\theta'_n)/R_n^{j+1} \tag{58}$$

where R_n and θ'_n are coordinates of electron n, transformed to the center of the bond. $P_j(\cos\theta'_n)$ is the jth-degree Legendre polynomial in $\cos\theta'_n$ and diatomic symmetry dictates use of only those multipoles having moments in the molecular axis. The form of Eq. (58) indicates point multipoles and the coefficients Q_j give the respective magnitudes and signs (j=1, dipole; j=2, quadrupole; etc.).

The basis set is the set of atomic functions; therefore, through first order

$$E^{\alpha}_{k,\mathrm{mol}} = E^{\alpha}_{k} + E^{\alpha'}_{k} \tag{59}$$

where

$$E^{\alpha'}_{k} = \int \phi_k^{\alpha *}(n)\hat{H}^{\alpha'}_{1e}\phi_k^{\alpha}(n)\,d\tau_n \ . \tag{60}$$

$E^{\alpha}_{k,\mathrm{mol}}$ is the negative of the molecular term value $T^{\alpha}_{k,\mathrm{mol}}$ involving the particular Rydberg transition and the proper ionization potential. Therefore,

$$E_k^{\alpha'} = T_k^{\alpha} - T_{k,mol}^{\alpha} \quad . \tag{61}$$

Combining atomic and molecular spectroscopic data in the context of Eqs. (58-61) provides approximations of the various Q_j values. Once these are evaluated, the first-order wavefunction corrections can be determined. The choice of zero-order wavefunctions is of primary concern. We have used orthogonalized STO's, the parameters of which are based on the atomic spectral data.

Knowledge of first-order wavefunction corrections permits calculation of important molecular properties such as excited-state permanent dipole moment and mean polarizability. The excited-state permanent electric-dipole moment is evaluated as the sum of the dipole term in the core (i.e., Q_1) and the orbital dipole moment. This latter quantity is evaluated from knowledge of the various orbital contributions combined with a positive point charge centered on X.

Our preliminary work with the first s Rydberg of HI yielded a core dipole moment of 2.60 D (negative end on I) and an orbital dipole moment of 5.84 D (negative end on H). Thus, the state dipole moment was calculated to be 3.24 D (negative end on H). The ground-state permanent moment is 0.42 D; therefore, the theoretical change is -3.66 D. This is probably quite a bit larger than the true value. The approximations made in the calculations included consideration of only the dipole term and only s and p_0 orbitals in the basis set. The wavefunction expansion was truncated after six s orbitals and six p_0 orbitals. The nature of a molecular Rydberg-state dipole moment is demonstrated, however. A large polarization of the core and the resultant polarization of the orbital is observed. Electric field data have not been measured for HI and, therefore, no comparison can yet be made.

3.2 Experiment

A number of reports on electric field-effect spectroscopy of molecular Rydberg transitions have appeared in recent years. Many of these are discussed elsewhere in this volume (see the paper by B. R. Russell). As an example, Fig. 8 depicts the electrochromism spectrum of the first allowed s Rydberg transition (with one quantum excitation in ν_2 of the excited electronic state) of methyliodide. The spectrum exhibits a striking resemblance to the second derivative of the absorption band, indicating a relatively large c_2 and, thus, a sizeable change in the permanent dipole moment. Data analysis results in a value $|\Delta\tilde{\mu}_e| = 1.04 \pm 0.05$ D. From the foregoing discussion of dipole moment theory, it is assumed that $\Delta\mu_e$ is negative. Using a value of 1.60 D for the ground-state dipole moment results in an experimental value of 0.56 D (negative end on I) for the excited state.

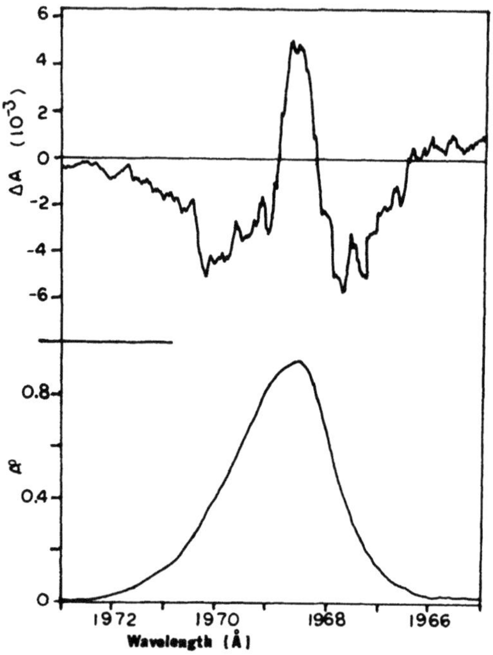

Fig. 8. Absorption and electrochromism spectra of gas-phase methyliodide. T = 300 K; sample pressure = 0.02 torr; and E = 40 kV/cm.

No other properties have been evaluated from these data. Improvements in instrumentation necessary to provide more precise data, more accurate data analysis and reliable temperature-dependence studies are underway.

4. PRESSURE-EFFECT SPECTROSCOPY

The theory of pressure effects on absorption spectra was introduced in Section 1.3. We have applied the technique outlined in that section to the first allowed s Rydberg transition of methyliodide using SF_6 as a perturber gas. Three of the spectra are given in Fig. 9. These data are qualitatively reliable; however, because of instrumentation problems the data cannot be aligned with the precision required to accurately evaluate molecular properties. The McPherson model 240 vacuum spectrophotometer used to measure these spectra is currently being modified to provide double-beam measurement and to allow computer control of wavelength scan and data acquisition.

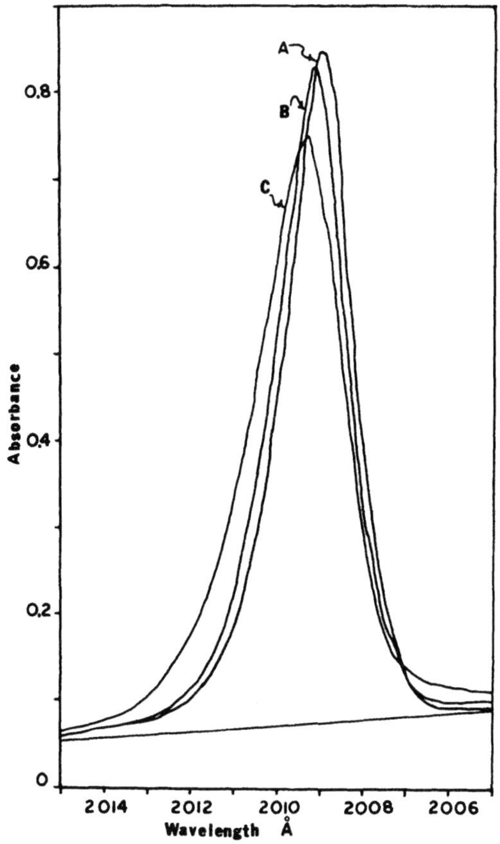

Fig 9. Absorption spectra of gas-phase methyl-iodide in the presence of SF_6 gas. SF_6 pressure: a=0, b=29 and c=93 psi; T = 300 K.

Fig. 10 contains plots of a_0 and a_1 as functions of SF_6 pressure. The non-zero slope of a_0 demonstrates the approximate nature of the $a_0 = 1$ assumption; thus, there is observable wavefunction perturbation. The a_1 plot appears linear, as was predicted by calculations using hypothetical ground-state and excited-state interaction parameters. The positive slope of a_1 indicates that $D_e' > D_e$ and that $r_e' < r_e$. Preliminary data analysis reveals both of these points to be the case. However, data uncertainty prevents reporting of D_e' and r_e' at this time.

Robin and Kuebler [8] reported an opposite effect in the sev-

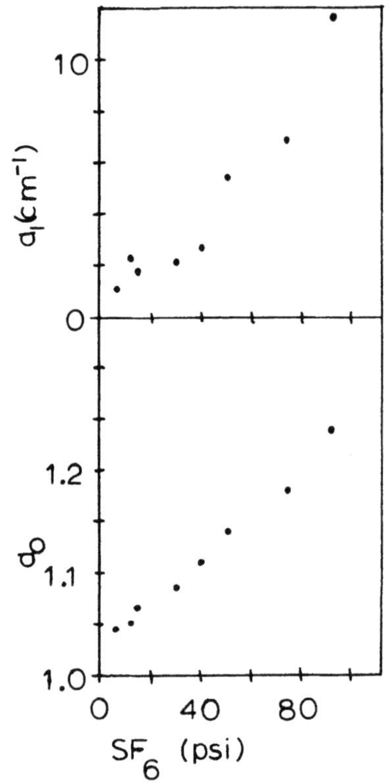

Fig. 10. The coefficients a_0 and a_1 evaluated from the pressure-effect spectra of methyliodide perturbed with SF_6 at 300 K, plotted as a function of SF_6 pressure.

eral systems they perturbed with inert gases and molecular nitrogen. One difference between their perturber gases and SF_6 is that SF_6 forms a stable gas-phase anion while the others do not. In other words, SF_6 has a positive electron affinity. This is an important fact with respect to the nature of the perturber gas. A species that forms stable anions should stabilize the energy of an electron in a Rydberg state. (In fact, the higher the state energy, the greater the expected stabilization.) A species that repels electrons (zero or negative electron affinity) should destabilize Rydberg electrons to a greater degree than it does valence electrons. This phenomenon is currently under investigation and emphasis is being placed on an electron-affinity model.

ACKNOWLEDGMENTS

The author wishes to acknowledge the support of the University of Montana Grants Administration, The Petroleum Research Fund of the American Chemical Society and The Research Corporation. Also, the patience and kindness of Professor B. R. Russell of North Texas State University and Professor S. P. McGlynn of Louisiana State University, in their willingness to share their time and their laboratories, is greatly appreciated.

REFERENCES

1. P. J. Stephens, Adv. Chem. Phys. 35, 197 (1976).
2. J. D. Scott, W. S. Felps, G. L. Findley and S. P. McGlynn, J. Chem. Phys. 68, 4678 (1978).
3. J. D. Scott, W. S. Felps and S. P. McGlynn, Nuc. Instrum. and Meth. 152, 231 (1978).
4. H. Labhart, Tetrahedron 19, 223 (1963).
5. W. Liptay, Z. Naturforsch. 20, 272 (1965).
6. J. D. Scott and B. R. Russell, J. Chem. Phys. 63, 3243 (1975).
7. H. Margenau and W. W. Watson, Rev. Mod. Phys. 8, 22 (1936).
8. M. B. Robin and N. A. Keubler, J. Mol. Spectrosc. 33, 274 (1970).
9. H. G. Drickamer, C. W. Franck and C. P. S. Lichter, Proc. Nat. Acad. Sci. USA 69, 933 (1972).
10. R. C. Tompkins, J. Chem. Phys. 72, 3449 (1980).
11. G. Herzberg, *Molecular Spectra and Molecular Structure. III. Electronic Spectra and Electronic Structure of Polyatomic Molecules* (Van Nostrand, Princeton, 1966) p. 576.
12. L. Pauling and E. B. Wilson, *Introduction to Quantum Mechanics* (McGraw-Hill, New York, 1935) pp. 105-111.
13. W. S. Felps, P. Hochmann, P. Brint and S. P. McGlynn, J. Mol. Spectrosc. 59, 355 (1976).
14. S. G. Tilford, M. L. Ginter and A. M. Bass, J. Mol. Spectrosc. 34, 327 (1970).

CIRCULAR DICHROISM AND MAGNETIC CIRCULAR DICHROISM STUDIES IN THE VACUUM ULTRAVIOLET

A. Gedanken

Department of Chemistry
Bar-Ilan University
Ramat Gan, Israel 52100

1. INTRODUCTION

The vacuum ultraviolet (VUV) absorption spectra of many molecular systems show broad and structureless continuous bands. Because of the lack of rotational and vibrational structure, it is often difficult, and in some cases impossible, to provide an unambiguous assignment for these absorption bands. An impetus exists, therefore, for the use of other techniques to help characterize these transitions. Circular dichroism (CD) and magnetic circular dichroism (MCD) are two techniques that are known to provide information which supplements absorption studies. Until recently, however, CD and MCD were limited to energy regions above the VUV because of a lack of polarizing elements which are transparent below 2000 Å. In the early seventies, when such elements became available, two instruments [1,2] were built at almost the same time and VUV-CD spectra were reported for many chromophores. Two reviews summarizing these VUV-CD studies have been published to date [3,4]. In the present paper, we review the CD and MCD measurements carried out at Bar-Ilan University.

1.1 Circular Dichroism

In natural CD, left and right circularly-polarized light is passed through an optically active molecule containing the chromophore of interest, and the difference in absorption of left and right circularly-polarized light is measured. In the work described below, the substituents which create the molecular asymmetry are an alkyl radical or a hydrogen atom, the excited electronic states of which are higher in energy than the electronic states of the chromophore and, therefore, perturb the spectros-

S. P. McGlynn et al. (eds.), Photophysics and Photochemistry in the Vacuum Ultraviolet, 765–781.

copy of the chromophore only slightly. The supplementary information that results from natural CD studies is as follows:

--- By its ability to measure both positive and negative signals, the CD technique enables one to uncover transitions that are not resolved in normal optical absorption measurements.

--- The measured quantity in CD is $\Delta\varepsilon$, defined as

$$\Delta\varepsilon = \varepsilon_L - \varepsilon_R$$

where ε is the molar extinction coefficient. $\Delta\varepsilon$ is related to quantum mechanical transition-moment matrix elements via

$$R \sim \int \frac{\Delta\varepsilon}{\nu}\, d\nu \sim \mathrm{Im}\left[\langle g|\underset{\sim}{\mu}_e|e\rangle\cdot\langle e|\underset{\sim}{\mu}_m|g\rangle\right] \quad (1)$$

where $|g\rangle$ and $|e\rangle$ are the ground and excited electronic states, respectively; and where $\underset{\sim}{\mu}_e$ and $\underset{\sim}{\mu}_m$ are the electric and magnetic dipole-moment operators, respectively.

For the absorption extinction coefficient, the analogous expression is

$$D \sim \int \frac{\varepsilon}{\nu}\, d\nu \sim \langle g|\underset{\sim}{\mu}_e|e\rangle\cdot\langle e|\underset{\sim}{\mu}_e|g\rangle \quad . \quad (2)$$

It can be shown that the anisotropy factor $g = \Delta\varepsilon/\varepsilon$ is proportional to $\langle g|\underset{\sim}{\mu}_m|e\rangle/\langle g|\underset{\sim}{\mu}_e|e\rangle$. Therefore, bands having a large g-value represent magnetic-dipole allowed and electric-dipole forbidden transitions, whereas a small g-value indicates the opposite case. CD measurements, as a result of the CD selection rules, can provide additional information for the assignment of magnetic-dipole allowed transitions.

--- The assignment of an absorption spectrum is always aided by the provision of quantum mechanical calculations of energies and oscillator strengths as a check on a proposed assignment. In the case of CD measurements, spectral assignments can be further substantiated since the sign of the rotational strength [cf. Eq. (1)] can be calculated theoretically.

1.2 Magnetic Circular Dichroism

In these measurements, circularly-polarized light is passed through a symmetric chromophore placed in a magnetic field, the direction of which is parallel to the propagation direction of the light. The measured quantity is $\Delta\varepsilon$, the differential absorption of left and right circularly-polarized light induced by the magnetic field [5]. The additional information obtainable from MCD studies is as follows:

--- $\Delta\varepsilon$ can be positive or negative and, therefore, overlapping electronic transitions can be resolved.

--- Two types of lineshapes are observed in MCD studies. The first type is a derivative lineshape, where the $\Delta\varepsilon = 0$ crossing point coincides with the absorption peak. The derivative lineshape is referred to as an *A*-term and is related to a degenerate excited state undergoing a first-order Zeeman effect [6]. One of the Zeeman components absorbs the left circularly-polarized light (lcp) while the other absorbs the right circularly-polarized light (rcp). Since the Zeeman splitting is much smaller than the width of the absorption band, and since the difference between lcp and rcp absorption is measured, the outcome is a derivative lineshape. Thus, the observation of a derivative lineshape leads to an immediate assignment of a degenerate excited state.

--- An absorption-type MCD lineshape is observed when two electronic states are mixed in the magnetic field. This lineshape, which is determined by the same parameters as those which determine the corresponding absorption bands, is known as a *B*-term. The two interacting states will show bands of opposite sign, with each band possessing equal intensity.

--- The analysis of an MCD spectrum provides an opportunity to evaluate relevant physical properties, properties that can be compared with *ab initio* quantum mechanical calculations, and to provide support for the characterization of excited electronic states. For example, the *A*-term is related to quantum mechanical matrix elements via

$$A(\mathrm{A} \rightarrow \mathrm{J}) = -\frac{1}{2} \sum_{\lambda,\lambda'} \mathrm{Im}\left[\langle \mathrm{A}|\underset{\sim}{\mu}_e|\mathrm{J}_\lambda\rangle \times \langle \mathrm{J}_{\lambda'}|\underset{\sim}{\mu}_e|\mathrm{A}\rangle \cdot \langle \mathrm{J}_\lambda|\underset{\sim}{\mu}_m|\mathrm{J}_{\lambda'}\rangle\right] \qquad (3)$$

where $|\mathrm{A}\rangle$ and $|\mathrm{J}\rangle$ are the ground (assumed to be nondegenerate) and excited states, respectively, and where $|\mathrm{J}_\lambda\rangle$, $|\mathrm{J}_{\lambda'}\rangle$ are the Zeeman components of the degenerate excited state. The dipole strength is expressed as before by

$$D(\mathrm{A} \rightarrow \mathrm{J}) = \sum_{\lambda} \langle \mathrm{A}|\underset{\sim}{\mu}_e|\mathrm{J}_\lambda\rangle \cdot \langle \mathrm{J}_\lambda|\underset{\sim}{\mu}_e|\mathrm{A}\rangle \; . \qquad (4)$$

The ratio *A*/*D* will therefore yield the expectation value of the magnetic moment in the degenerate state $|\mathrm{J}\rangle$. This ratio is in turn related to the ratio of the first moment of $\Delta\varepsilon$ to the zeroth moment of ε through [7]

$$A/D = -1.072 \times 10^4 \; \beta \frac{\langle\Delta\varepsilon\rangle_1}{\langle\varepsilon\rangle_0} \quad (5)$$

where β is the Bohr magneton. In Eq. (5), the nth moment of a quantity α, about the absorption frequency ν^0, is defined by

$$\langle\alpha\rangle_n = \int \frac{\alpha(\nu - \nu^0)^n}{\nu} d\nu \quad .$$

In addition, $\langle\Delta\varepsilon\rangle_1$ is defined per unit field strength.

The B-term, on the other hand, is related to quantum mechanical matrix elements via

$$B = \sum_\lambda \mathrm{Im}\left[\sum_{K_\kappa \neq J} \langle A|\underset{\sim}{\mu}_e|J_\lambda\rangle \times \langle K_\kappa|\underset{\sim}{\mu}_e|A\rangle \cdot \langle J_\lambda|\underset{\sim}{\mu}_m|K_\kappa\rangle / h\omega_{KJ}\right] \quad (6)$$

where the state $|K\rangle$, with components $|K_\kappa\rangle$, is mixed with state $|J\rangle$ in the magnetic field. (In Eq. (6) we have assumed that $|A\rangle$ and $|K\rangle$ are sufficiently separated in energy that we may neglect their field-induced mixing.) The B-term is expressed with respect to the zeroth moment of the MCD curve by

$$B/D = -1.072 \times 10^4 \; \beta \frac{\langle\Delta\varepsilon\rangle_0}{\langle\varepsilon\rangle_0} \quad (7)$$

where $\langle\Delta\varepsilon\rangle_0$ is defined per unit field strength. Theoretical calculations of the B-term are complicated since one has to sum the interaction of a specific excited state over a complete set of excited states. However, it is usually assumed that a limited set of excited states is dominant.

--- It has been recognized that MCD measurements can be fruitfully employed in the study of singlet-triplet transitions [8]. Since MCD is a sensitive modulation technique, the detection of singlet-triplet transitions by MCD can be performed with a short pathlength cell (e.g., 10 cm), whereas absorption measurements require a much longer cell (e.g., 10 m). The reason that singlet-triplet transitions give rise to an MCD signal lies in the nature of the magnetic moment operator $\underset{\sim}{\mu}_m$, which contains the total electronic spin angular momentum operator $\underset{\sim}{S}$ ($= \sum_i \underset{\sim}{s}_i$):

$$\underset{\sim}{\mu}_m = \frac{e}{2mc} \sum_i (\underset{\sim}{\ell}_i + \underset{\sim}{s}_i)$$

where ℓ_i is the orbital angular momentum of the ith electron.

2. INSTRUMENTATION

Eight VUV-CD instruments have been constructed since 1970; these are listed in a review article by W. C. Johnson [4]. The key element of a VUV-CD instrument is a quarter-wave plate. All of the above instruments use a stress-plate modulator made of CaF_2 (Morvue Electronic Systems, Inc. of Tigard, Oregon) and modulated at 50 kHz to convert linearly-polarized light into circularly-polarized light. The differences between the various existing instruments reside in the optical element employed to produce linearly-polarized light: This element is either a MgF_2 transmitting prism (Rochon or Wollaston), or a biotite plate from which the reflected light emerges as linearly-polarized light. Since MgF_2 shows an abrupt change in its birefringence around 1350 Å, the extension of VUV-CD measurements to higher energies requires the use of the biotite plate. The wavelength limit in this latter case is 1250 Å, which is imposed by the transmittance of the CaF_2 modulator.

The system operating in our laboratory has been described previously [9]. It employs a modified Hinteregger discharge lamp. (The main modification is that the hydrogen flows through the cathode along the capillary and is pumped out through the anode.) A 1 m normal-incidence McPherson 225 is used as the monochromator. This monochromator is equipped with a 1200 lines/mm grating, yielding a dispersion of 8 Å/mm. A MgF_2 Wollaston prism is placed inside the monochromator before the exit slit. Two linearly-polarized beams (parallel and perpendicular) are obtained, one of which is eventually discarded. The other linearly-polarized beam passes through the photoelastic modulator and is converted alternately to right and left circularly-polarized light at a frequency of 50 kHz. The modulated beam of light then passes through the sample and is detected by a solar blind photomultiplier tube. If the sample cell contains an optically active compound, the photomultiplier will detect an AC signal superimposed on a DC signal. To a good approximation, the CD is proportional to the intensity of the AC signal divided by the intensity of the DC signal. The DC signal is kept constant and the AC signal, detected by a lock-in amplifier, is directly proportional to the CD signal. When MCD is measured, the sample cell is placed in the bore of a superconducting magnet configured such that the circularly-polarized light propagates parallel to the magnetic field.

The detection limit of a CD instrument is determined by the number of photons N reaching the detector. And, since the noise in a CD spectrum is statistical, $\Delta\varepsilon/\varepsilon$ is proportional to $\sqrt{N}/N$. At present, the limit of sensitivity, with 2 mm slits and a time constant of 30 s, is about 10^{-5} in the 3300-1450 Å wavelength range.

3. RESULTS

3.1 Natural Circular Dichroism

3.1a CD of a Chiral Olefin: (R)-3-Methylcyclopentene (MCP)

We have measured the absorption and CD spectra of the chiral planar olefin (R)-3-Methylcyclopentene (MCP) [10]. This is a small olefin which possesses a rigid cyclic skeleton. The small size of the molecule enabled us to carry out *ab initio* SCF-CI calculations for the excited states. The assignments of the various observed transitions are based on the results of these calculations.

Fig. 1 shows the gas-phase absorption and CD spectra of MCP. The main features are as follows:

--- A negative CD band centered at around 205 nm exhibits two or three vibrational members with a spacing of $\sim$1200 cm^{-1}. These bands show the largest g-value ($\sim 4 \times 10^{-3}$) in MCP.

--- The strongest absorption band ($\varepsilon \simeq 8000$) peaks at $\sim$185 nm and a corresponding positive CD band is observed in the same wavelength region. The g-value calculated for this transition is 2×10^{-3}. The CD band is asymmetrically broadened toward shorter wavelengths.

--- A negative CD is observed at 161 nm, which has a corresponding absorption peak at 159 nm. The g-value of this transition is 2×10^{-3}.

--- A positive CD band which peaks at 149 nm is observed in the unstructured absorption at shorter wavelengths.

In Table 1 we summarize the assignment of the observed spectrum based on the *ab initio* calculations. The first band, which shows the largest g-value, is assigned as a $\pi \rightarrow R(3s)$ Rydberg transition; however, this assignment will not account for the large g-value and, therefore, we suggest the contribution of a $\sigma \rightarrow R(3s)$ transition. Moreover, this first band is an $A_{1g} \rightarrow B_{3g}$ transition (in the symmetric chromophore) and is therefore electric-dipole forbidden and magnetic-dipole allowed from the ground state, thus explaining the large g-value. The second band is assigned as $\pi \rightarrow \pi^*$, and the broad asymmetry is caused by a positive CD signal resulting from the third band, a $\pi \rightarrow R(3p_x)$ transition. The fourth band is assigned as $\pi \rightarrow R(3p_{yz})$, where the excited 3p molecular orbital lies in the plane of the double bond. Our calculated results enable us to account for the general pattern observed in the CD spectrum. In addition, the calculated rotational strengths predict the right sign but their magnitudes are system-

Table 1. Assignment of the Observed Spectrum of MCP

Band	Description	Energy (eV) obs.	Energy (eV) calc.	RS[a] obs.	RS[a] calc.[b]	f obs.	f calc.[c]
1		6.05		-16		0.03	
	$\pi \rightarrow R(3s)$		6.04		-2.03		0.014
	$\sigma \rightarrow R(3s)$				-6.01		0.006
2		6.7		+26		0.16	
	$\pi \rightarrow \pi^*$		7.4		+2.74		0.504
3		7.0					
	$\pi \rightarrow R(3p_x)$				+3.17		0.006
	$\sigma \rightarrow R(3p_x)$				+0.75		0.006
4		7.7		-16		0.09	
	$\pi \rightarrow R(3p_{yz})$				-1.36		0.005
	$\pi \rightarrow R(3p_{zy})$				-4.24		0.007

[a] RS ≡ rotational strength, in 10^{-40} cgs units.

[b] Average of $RS(\nabla)$ and $RS(r)$.

[c] $f = [f(r) \cdot f(\nabla)]^{1/2}$.

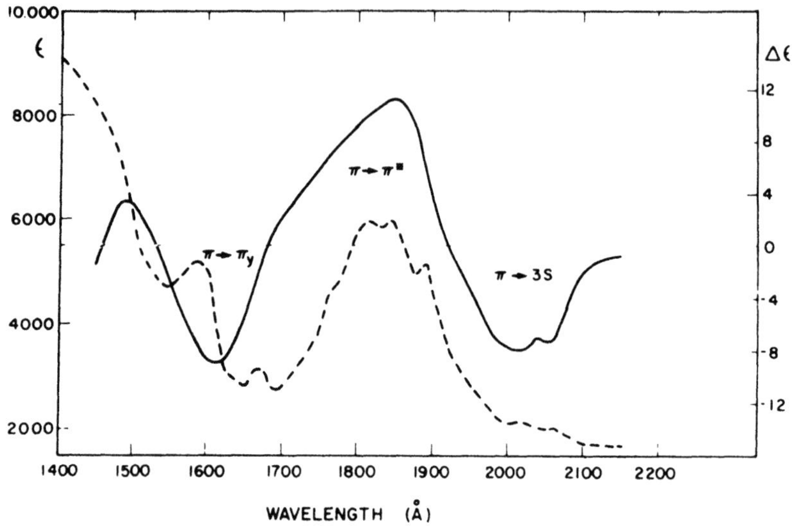

Fig. 1. Absorption (---) and CD (——) spectra of MCP. The resolution of the absorption spectrum is 8 Å, while that of the CD spectrum is 32 Å.

atically smaller.

3.1b CD of the Ethylene Oxide System: (-)-S-2-Methyloxirane (MO) and (-)-(S,S)-2,3-Dimethyloxirane (DMO)

The CD and absorption spectra of (-)-S-2-Methyloxirane (MO) and (-)-(S,S)-2,3-Dimethyloxirane (DMO) are presented in Figs. 2 and 3, respectiavely. *Ab initio* SCF-CI calculations were carried out on these molecules (Tables 2 and 3), and used in assigning the observed transitions [11].

The CD spectrum of MO is the simpler of the two and shows three bands with alternating signs, while the CD spectrum of DMO exhibits four transitions with alternating signs. The first positive band in MO and DMO is assigned as n(O) → R(3s) and the calculated sign is indeed positive. In MO we can assign the second negative band as n(O) → R(3p). However, as far as the third positive band in MO is concerned, it may be assigned either as one of the n(O) → R(3p) components or even as an n(O) → R(3d) transition.

Four transitions, which alternate in sign, are observed in the CD spectrum of DMO (maxima at 1755 Å, 1655 Å, 1620 Å and 1510 Å), one of which (the 1655 Å negative signal) does not have a cor-

Table 2. Summary of Calculated and Experimental Quantities for MO

	Calculated			Experimental		
Transition	ΔE^a	$f(\nabla)$	$RS(\nabla)^b$	ΔE^a	RS^b	f
n(O) → R(3s)	6.4	0.026	+4.92	7.12	11.8	0.025
n(O) → R(3p)	7.16	0.010	-1.30	7.75	-10.8	0.062
n(O) → R(3p)	7.34	0.026	+0.42			
n(O) → R(3p)	7.36	0.006	-0.02			

[a]Excitation energy from the ground state, in eV.
[b]RS ≡ rotational strength, in 10^{-40} cgs units.

Table 3. Summary of Calculated and Experimental Quantities for DMO

	Calculated			Experimental		
Transition	ΔE^a	$f(\nabla)$	$RS(\nabla)^b$	ΔE^a	RS^b	f
n(O) → R(3s)	6.36	0.0042	+3.44	7.06	+20	0.02
n(O) → R(3p)	6.64	0.0042	-7.34	7.49	-0.4	
n(O) → R(3p)	7.08	0.0024	+1.52	7.74 (both rows)	+14 (both rows)	0.05 (both rows)
n(O) → R(3p)	7.18	0.0002	+0.46			

[a]Excitation energy from the ground state, in eV.
[b]RS ≡ rotational strength, in 10^{-40} cgs units.

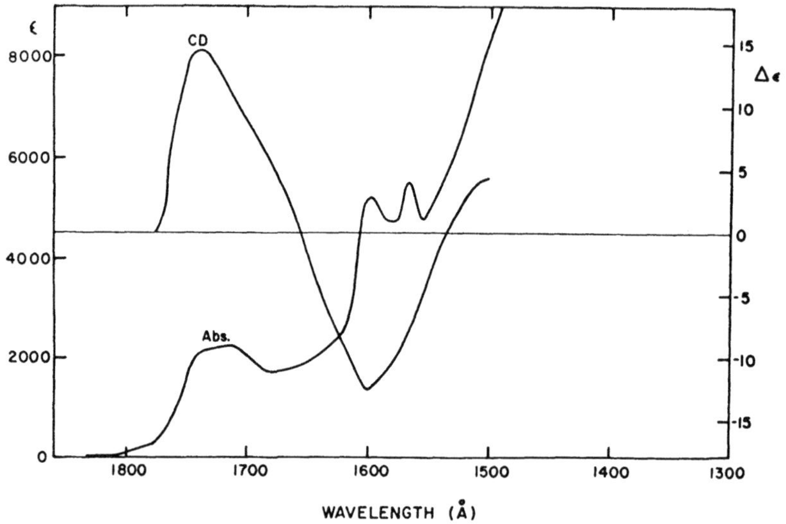

Fig. 2. Absorption and CD spectra of MO. The resolution of both spectra is 16 Å.

responding peak in the absorption spectrum. The bands at 1755 Å and 1620 Å are assigned as n(O) → R(3s) and n(O) → R(3p), respectively. This assignment is based on the calculated signs for these transitions and agrees well with the interpretation of the absorption spectrum of ethylene oxide [12]. Reasonable term values are obtained for the 3s and 3p Rydberg orbitals, which also supports this assignment. The negative CD band at 1510 Å cannot be unambiguously assigned on the basis of term values. An n(O) → R(3d), n(O) → R(4s) or even a valence transition corresponding to the 1440 Å transition in solid ethylene oxide are all possible assignments. The most interesting feature is the negative CD band at 1655 Å, which seems not to have any corresponding resolved absorption band in DMO. This band seems to relate to the 63,000 cm^{-1} shoulder in the absorption spectrum of solid ethylene oxide [12], which then leads to a valence-type n(O) → σ* assignment. However, our *ab initio* calculations suggest that another component of the n(O) → R(3p) Rydberg transition appears in this energy region and has a negative CD signal. The true location of the lowest-energy n(O) → σ* valence-type transition cannot be determined by our calculations for the following reasons:

--- The transition n(O) → Rydberg involves only a net 50% change in oxygen atom character. The valence-type n(O) → σ* transition, then, will probably have significant ionic character

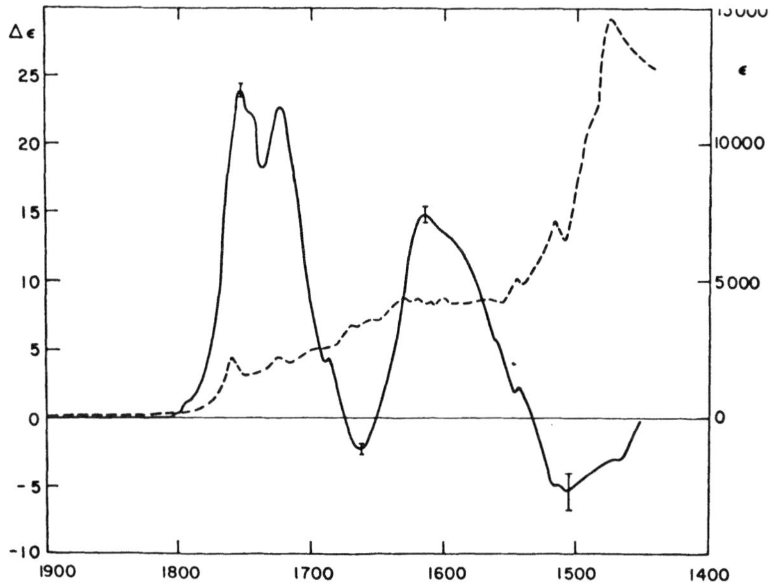

Fig. 3. Absorption (---) and CD (——) spectra of DMO. The resolution of both spectra is 16 Å.

which will tend to push it to higher calculated energies.

--- The SCF approximation seems to be incapable of projecting out the pure valence transitions when they lie among a series of Rydberg states, apparently for orthogonality reasons.

Thus, we are left with two proposed assignments for the transition at 1655 Å. CD experiments on solutions of DMO in perfluoro n-hexane are now in progress in order to determine the nature of this transition.

3.1c Excited Electronic States of the Ethylene Sulfide Chromophore: CD of (+)-(R,R)-2,3-Dimethylthiirane (DMT)

The absorption spectrum of ethylene sulfide reveals two broad and diffuse bands in the region 230-260 nm, and a sharp, structured spectrum below 210 nm. The two weak bands centered at 259 nm and 243 nm ($\varepsilon \simeq 30$) have been assigned as $n(S) \rightarrow \sigma^*$ transitions, the n orbital being the nonbonding 3p orbital on the sulfur atom [13]. The structured spectrum below 210 nm has been attributed to various Rydberg transitions [13].

We have measured the CD and absorption spectra of DMT down to 1500 Å, and carried out *ab initio* SCF-CI calculations for the energies, oscillator strengths and rotational strengths of the various excited states. The assignments of the absorption and CD spectra are based on the results of these calculations, which are given in Table 4.

Fig. 4 contains the absorption and CD spectra of DMT in the gas phase. Because of their low absorptivities, we could not detect the 250 nm bands even in a 10 cm cell. Three well-resolved transitions are observed in the CD and absorption spectra. Although small shifts in wavelength occur between related absorption and CD bands, they are assigned as corresponding to the same transitions.

The 2100-1800 Å region in ethylene sulfide and propylene sulfide clearly shows the appearance of two origins. These origins, which are assigned to 4s and 4p Rydberg transitions, are split by 4700 cm^{-1} and 4100 cm^{-1} in ethylene sulfide and propylene sulfide, respectively [14]. In DMT, however, we do not observe a clear splitting between the 4s and 4p Rydbergs. Nevertheless, we assign the band which covers the 2150-1900 Å region as being composed of two transitions, the 4s and 4p Rydbergs. The origins of these two transitions are assigned at 2103 Å and 1977 Å, respectively. The argument for these assignments is as follows:

--- The energy spacings between the first three bands are 1040 cm^{-1} and 1010 cm^{-1}, while this spacing is decreased between the third and fourth (2013 Å and 1977 Å) bands to 900 cm^{-1}. The change in the vibrational frequency supports the presence of more than one origin.

--- Robin [14] has shown that the term values for the 3s, 3p and 3d Rydberg orbitals, in many chromophores, decrease upon substitution and reach a limit of $\sim$22,000 cm^{-1} for 3s orbitals, $\sim$19,000 cm^{-1} for 3p orbitals and $\sim$12,000 cm^{-1} for 3d orbitals. These numbers also obtain for 4s, 4p and 4d orbitals of a molecule containing a sulfur atom. The ionization potential of DMT is not known; however, our assignment of the 4s and 4p Rydbergs would require a splitting of 3030 cm^{-1}. Even smaller splittings were observed in cyclic ethers.

--- The calculated CD signs for the 4s and 4p Rydbergs are negative, in agreement with the proposed assignment.

--- If we were to assign the shoulder at 1825 Å as the origin of the 4p Rydberg transition, it would require a 4s-4p splitting of 5100 cm^{-1}. This possibility is ruled out since the splitting decreases upon substitution and is 4700 cm^{-1} for ethylene sulfide and 4100 cm^{-1} for propylene sulfide. The absorption band

Table 4. Calculated and Experimental Excited-State Properties of DMT

	Calculated			Experimental		
Transition	ΔE^a	$f(\nabla)$	$RS(\nabla)^b$	ΔE^a	RS^b	f
$n(S) \to \sigma^*$	4.98	$2x10^{-4}$	-10.8	4.71 5.08	-1[c]	$5x10^{-5}$ [c]
$n(S) \to R(4s)$	5.39	0.053	-0.8	5.90	-8.1	0.053
$n(S) \to R(4p_x)$	5.90	0.004	-5.7	6.27		
$n(S) \to R(3d)$	6.46	0.003	+8.0	7.21	2.3	0.146

[a]Excitation energy from the ground state, in eV.
[b]RS ≡ rotational strength, in 10^{-40} cgs units.
[c]Ref. [12].

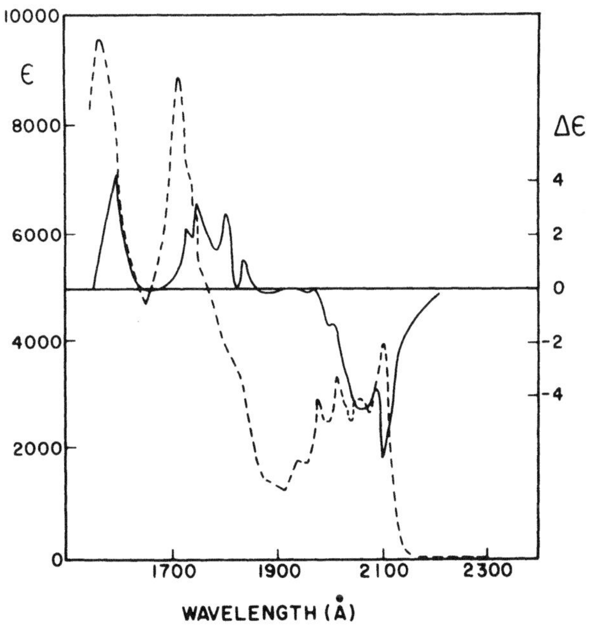

Fig. 4. Absorption (---) and CD (——) spectra of DMT. The resolution of both spectra is 16 Å.

at 1720 Å corresponds to the positive CD band centered at 1750 Å and is assigned as n(S) → R(4d). This assignment is based on the sign obtained for the 4d Rydberg transition in our calculations. It also agrees well with a previous interpretation attributing the 1740 Å band in ethylene sulfide to an $A_1 \rightarrow B_1$, n(S) → R(4d) transition [15].

3.2 Magnetic Circular Dichroism

3.2a MCD of 1,4-diazabicyclo[2,2,2]octane (DABCO)

Multiphoton (MPI) experiments with one [16] and two colors [17,18] were recently carried out on DABCO. These studies were able to identify two electronic transitions. The origin of the first transition is observed at 279.38 nm and is assigned as n → R(3s), while the second transition is located at 251.1 nm and is assigned as n → R(3p). In both transitions, n represents the symmetric combination of the two lone-pair, nonbonding orbitals on the two equivalent nitrogens. As far as the energetics of these transitions are concerned, the MPI results agree well with previous gas-phase absorption measurements reported by Halpern, *et al.* [18]. However, differences exist in the assignment of the B → X [n → R(3p)] transition. Whereas Halpern, *et al.* [18] and Hameda, *et al.* [19] prefer the $^1A_1' \rightarrow {}^1A_2''$ [n → $R(3p_z)$] assignment, Parker and Avouris [15b] conclude that it should be assigned as $^1A_1' \rightarrow {}^1E'$ [n → $R(3p_{xy})$]. MCD can provide an answer to the assignment problem since the appearance of a derivative lineshape (*A*-term) would permit an immediate assignment to a degenerate excited state. The absorption and MCD spectra of gas-phase DABCO are presented in Fig. 5. One sees that for each absorption line a pure derivative lineshape (*A*-term) is observed. Thus, in view of the possibilities discussed above for the assignment of the n → R(3p) transition, MCD indicates that the excited state is $^1E'$ and that the transition is n → $R(3p_{xy})$. Since the absorption and MCD were measured using the same cell and under the same conditions, the magnetic moment of the excited state was calculated from the ratio of the first moment of ΔA to the zeroth moment of A (the optical density), where ΔA is the difference in the optical density for left and right circularly-polarized light per gauss. The calculated ratio, according to Eq. (5), is related to A/D. In carrying out the moment analysis for this state, one can integrate over the envelope of all of the vibrational bands belonging to the n → R(3p) transition; alternatively, one can calculate $\langle \Delta A \rangle_1 / \langle A \rangle_0$ from the moments of the origin of the transition (which is very well-resolved) or from any other vibrational band in the progression. We have indeed evaluated the magnetic moment of the $^1E'$ state by calculating the moments of the absorption and MCD bands at 251.1 nm and 243.27 nm and obtained $A/D = -0.17\beta \pm 0.02$. These bands are well-resolved and show well-defined *A*-terms. As can be seen from Fig. 5, and from

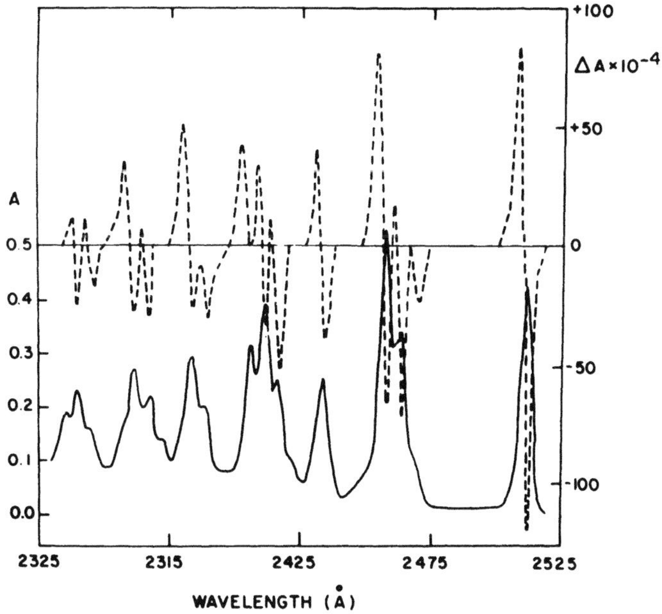

Fig. 5. Absorption (——) and MCD (---) spectra of gas-phase DABCO. The spectral resolution is 2.5 Å and ΔA is per 48.8 kgauss.

the results reported by Halpern, *et al.* [18], the n → R(3p) transition is superimposed on an underlying continuum. For the 243.27 nm transition (assigned as 0-0 + ν_1'), the underlying continuum was subtracted and *D* represents the net absorption of the n → R(3p) component.

The question that still remains is the location of the n → R($3p_z$) transition. This is an electric-dipole allowed transition and should be observed in the absorption spectrum. We have extended our measurements to the energy region of the hot bands built upon n → R($3p_{xy}$); the results are presented in Fig. 6. The strong *A*-term (going off scale) is that of the 0-0 band. At 0-0 - 430 cm^{-1}, a positive *B*-term is observed; this transition does not possess a corresponding absorption band. We tentatively assign this band as the n → R($3p_z$) 0-0 transition for the following reasons:

--- Since the excited state ($^1A_2''$) is non-degenerate, it should appear in the MCD spectrum as a *B*-term. If it were a hot band belonging to n → R($3p_{xy}$), it would have an *A*-term lineshape.

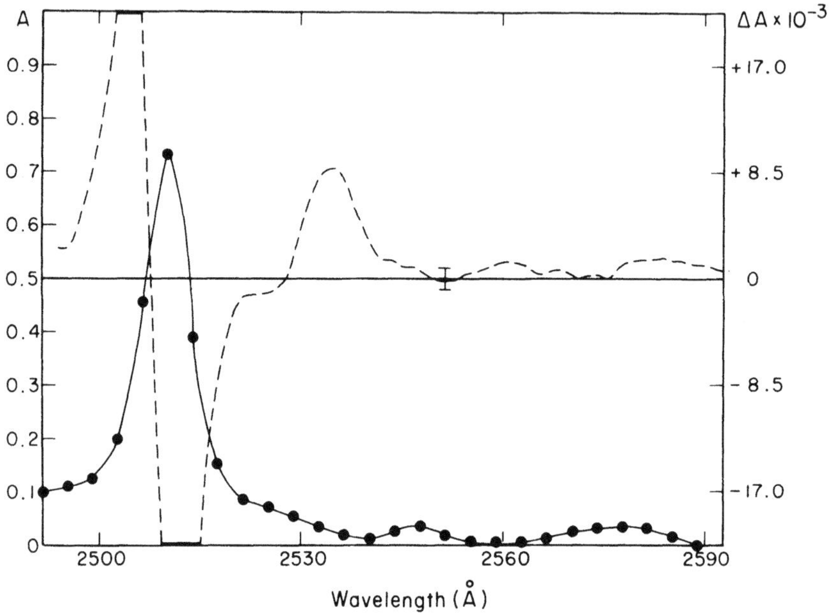

Fig. 6. Absorption (---) and MCD (——) spectra of gas-phase DABCO. The spectral resolution is 7 Å and ΔA is per 49 kgauss.

--- The fact that this transition is *not* observed in absorption and *is* observed in MCD is a result of the different matrix elements which determine the strength of this transition in absorption and MCD. For MCD, one can argue that, in Eq. (6), two factors are in favor of the observation of the n → R(3p_z) transition, the first one being the interaction between the two excited states coupled through the magnetic field. The interacting states are the 3p_{xy} Rydberg ($^1E'$) and the 3p_z Rydberg ($^1A_2'$); group theoretic multiplication yields $^1E' \otimes {}^1A_2'' = {}^1E''$, which transforms as R_x, R_y and, therefore, implies a *B*-term. The second factor is that the small energy denominator (430 cm^{-1}) enhances the "strength" of the *B*-term.

ACKNOWLEDGMENT

I would like to thank the Israeli National Council for Research and Development and the Israeli Commission for Basic Research for their financial help. I would also like to thank Profs. E. Gil-Av, V. Schurig, B. S. Green and R. Arad-Yellin for the donation of the optically active compounds which made this re search possible; Prof. Harold Basch and Dr. D. Cohen for carrying out the *ab initio* calculations; M. Levi for her technical help; and Dr. M. B. Robin for valuable discussions.

REFERENCES

1. O. Schnepp, S. Allen and E. F. Pearson, Rev. Sci. Instrum. 41, 1136 (1970).
2. W. C. Johnson, Jr., Rev. Sci. Instrum. 42, 1283 (1971).
3. E. S. Pysh, Ann. Rev. Biophys. Bioenerg. 5, 63 (1974).
4. W. C. Johnson, Jr., Ann. Rev. Phys. Chem. 29, 93 (1978).
5. MCD is related to Zeeman spectroscopy but provides no new information when the Zeeman components can be resolved. However, for broad bands, where conventional Zeeman effects are undetectable, MCD provides information that cannot be obtained from absorption spectroscopy.
6. P. J. Stephens, J. Chem. Phys. 52, 3489 (1970).
7. R. L. Mowery, P. N. Schatz and P. J. Stephens, J. Chem. Phys. 55, 224 (1971).
8. R. E. Linder, G. Barth, E. Bunnenberg, C. Djerassi, L. Seaman and A. Moscowitz, Chem. Phys. Lett. 39, 411 (1976).
9. A. Gedanken and M. Levi, Rev. Sci. Instrum. 48, 1661 (1977).
10. M. Levi, D. Cohen, V. Schurig, H. Basch and A. Gedanken, J. Am. Chem. Soc. 102, 6972 (1980).
11. D. Cohen, M. Levi, Harold Basch and A. Gedanken, J. Am. Chem. Soc., submitted.
12. H. Basch, M. B. Robin, N. A. Kuebler, C. Baker and D. W. Turner, J. Chem. Phys 51, 52 (1969).
13. G. L. Bendazzoli, G. Gottarelli and P. Palmieri, J. Am. Chem. Soc. 36, 11 (1974).
14. M. B. Robin, *Higher Excited States of Polyatomic Molecules*, Vol. 1 (Academic Press, New York, 1975).
15. D. Altenloh and B. R. Russell, Chem. Phys. Lett. 77, 217 (1981).
16. (a) D. H. Parker and Ph. Avouris, Chem. Phys. Lett. 53, 515 (1978); (b) D. H. Parker and Ph. Avouris, J. Chem Phys. 71, 1241 (1979).
17. D. H. Parker, R. B. Berenstein and D. A. Lichtin, J. Chem. Phys. 75, 2577 (1981); D. A. Lichtin, S. Detta-Ghosh, K. R. Newton and R. B. Berenstein, Chem. Phys. Lett. 75, 214 (1980); K. R. Newton, D. A. Lichtin and R. B. Berenstein J. Phys. Chem. 85, 15 (1981).
18. D. H. Parker and M. A. El-Sayed, Chem. Phys. 42, 379 (1979).
19. A. M. Halpern, J. L. Roebler and K. Weiss, J. Chem. Phys. 49, 1348 (1968).
20. Y. Hamada, A. Y. Hirakawa and M. Isuboi, J. Mol. Spectrosc. 47, 440 (1973).

ELECTRIC FIELD STUDIES IN THE VACUUM ULTRAVIOLET

Daniel D. Altenloh and B. R. Russell

Department of Chemistry
North Texas State University
Denton, Texas 76203, USA

1. INTRODUCTION

The application of intense electric fields induces a partial orientation of gaseous polar compounds. This orientation effect causes certain changes in the absorption band profile, changes which have been used to determine a number of excited-state parameters for several series of compounds [1-5]. These parameters include properties such as the change in dipole moment, change in polarizability and polarization of the transition moment. In general, the emphasis of these studies has been on characterizing absorption bands which occur in the lower-energy portion (200 to 170 nm) of the vacuum ultraviolet (VUV) spectral region. Many electronic transitions which occur in this spectral region have conflicting assignments as both intravalence and Rydberg. Excited-state properties obtained with the aid of electric field effects, however, can be used in the characterization of these absorptions and the clarification of their assignments.

The above excited-state parameters have been theoretically determined from field-effect expressions developed by Scott, *et al.* [1,2]. In the present paper, emphasis is placed on the experimental technique, and results which can be obtained from this technique, rather than on the theoretical development. (The theoretical development is covered by J. D. Scott in a separate paper in this volume.) There are two variants of the electric field technique -- electrochromism and electric linear dichroism (ELD). Electrochromism is defined as the differential photoabsorption by a molecule in a modulated (73 Hz) electric field, using either polarized or unpolarized radiation. ELD, on the other hand, is defined as the differential photoabsorption in a

S. P. McGlynn et al. (eds.), Photophysics and Photochemistry in the Vacuum Ultraviolet, 783–818.

static electric field, using linearly-polarized radiation which is modulated between 0° and 180° (at 50,000 Hz) with respect to the angle between the polarization vector and the electric field vector.

Lombardi [6,7] and others [8] previously developed a Stark-field method to determine excited-state dipole moments. This technique required that the absorption band contain resolvable rotational structure: The presence of the electric field resulted in a change in the rotational band envelope, which was then analyzed for the excited-state dipole moment. Electronic absorptions containing resolvable rotational bands are usually observed only for smaller molecules, which therefore restricts the use of the technique. The electric field method described here (i.e., electrochromism or ELD), however, is more general since it does not require resolvable rotational structure; nevertheless, much larger field strengths are needed. Recent studies on absorption bands in aniline have shown that results from the previous Stark-field method [6-8] are consistent with values obtained from the present electric field studies.

Electrochromism has provided detailed information about molecular Rydberg states and, in principle, can be applied to numerous types of absorption bands. However, Rydberg states were investigated initially since these absorptions are the most amenable to electric field techniques. From these studies, certain Rydberg-state properties, such as dipole moment and polarizability changes, have been reported [1-5]. These properties are important in determining the electronic structure of the excited state; moreover, experimental values for these properties provide parameters which can be used in comparisons with theory.

A brief description of the experimental apparatus used in our electric field studies is given in Section 2. A general discussion of the results obtainable from electric field techniques is presented in Section 3. Finally, the detailed discussion of experimental results is divided into two parts: The first part (Section 4) concentrates on excited-state parameters obtained for several compounds containing the chemically important carbonyl group, as well as those obtained for a number of sulfur-containing compounds; the second part (Section 5) concerns the effects which certain "solvent" gases have on the excited-state values obtained.

2. EXPERIMENTAL

The experimental apparatus is shown schematically in Fig. 1. Specific details have been given previously [1-5]. In general, the monochromatic radiation is collimated and passed between elec-

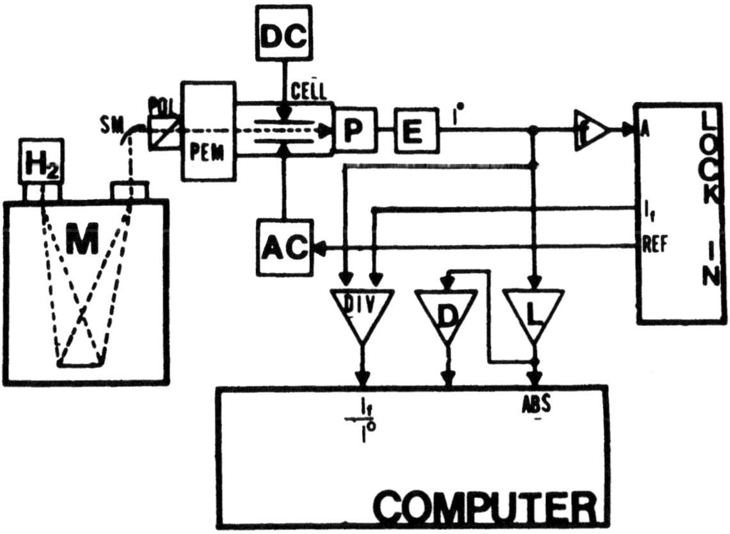

Fig. 1. Schematic of the electrochromism experimental configuration. Notation: H_2, hydrogen discharge lamp; M, monochromator; SM, spherical mirror used to collimate the monochromator output; POL, Wollaston prism; PEM, photoelastic modulator; P, photomultiplier detector (EMI 9635QB); E, electrometer; f, high-Q bandpass filter set to 73 Hz; AC/DC, electric field sources; L, logarithmic ratiometer; D, analog differentiator; DIV, analog divider. The computer is an Apple II+.

tric field plates, through a partially oriented gas in the sample cell. The intensity is detected by a photomultiplier tube and the resulting signals are acquired by a microprocessor.

This particular arrangement has been used to obtain data for four different types of spectra: unpolarized light electrochromism (UPE), zero-degree polarized electrochromism (ZPE), ninety-degree polarized electrochromism (NPE) and electric linear dichroism (ELD). By comparing the intensities of the electric field signal for these different experiments, one can determine the orientation of the transition-moment polarization vector. In addition, certain excited-state data can be determined, such as the dipole moment and polarizability, as discussed below.

Typical values for electric field intensity were $\sim 10^5$ V/cm. With this field intensity, a high dielectric gas (arc-suppressor gas) is necessary to prevent field breakdown due to corona discharge; sulfur hexafluoride (SF_6) or carbon tetrafluoride (CF_4) were used for this purpose. In order to maintain consistency between different data sets, electric field data were usually measured at an SF_6 pressure of 4.4 atm; this pressure is slightly above the minimum required for field stabilization.

In a preliminary study [9], the experimental values of excited-state dipole moments were found to be sensitive to the pressure of the arc-suppressor gas. This effect has been explored further using both SF_6 and CF_4; the results are presented in Section 5.

3. GENERAL RESULTS

A detailed theoretical description of the effects of external fields on excited-state molecular parameters has been given by Scott, *et al.* [1,2] and others [10,11], as well as elsewhere in this volume in the paper by J. D. Scott. The schematic form of the expression which describes changes in the spectral profile resulting from the presence of the orienting electric field, in terms of excited-state parameters, is:

$$S(\tilde{\nu}) = c_1\int A(\tilde{\nu})d\tilde{\nu} + c_2\int A'(\tilde{\nu})d\tilde{\nu} + c_3\int A''(\tilde{\nu})d\tilde{\nu} \qquad (1)$$

where $\tilde{\nu}$ is in cm^{-1}; $S(\tilde{\nu})$ is the electric field signal (defined below); $A(\tilde{\nu})$, $A'(\tilde{\nu})$, $A''(\tilde{\nu})$ are the absorption, and first and second derivatives of absorption with respect to wavenumber, respectively; and the coefficients c_1, c_2 and c_3 are specific functions of certain perturbation terms, and of several ground- and excited-state properties of the molecule. Qualitatively, the functional dependencies of these coefficients are:

--- $c_1 = f_1$(ground-state dipole moment, first- and second-order field-effect perturbation terms, polarization of the transition moment)

--- $c_2 = f_2$(ground-state dipole moment, dipole-moment change, polarizability change, field-effect perturbation terms, polarization of the transition moment)

--- $c_3 = f_3$(dipole-moment change, polarization of the transition moment)

Explicit expressions for these coefficients have been given elsewhere [1-3]; these expressions are also given in the paper by Scott in this volume.

The "electric field" signal $S(\tilde{\nu})$ is defined as the negative of the ratio of the light-intensity function in the presence of the field, $I^E(\tilde{\nu})$, to the light-intensity function in the absence of the field, $I^o(\tilde{\nu})$:

$$S(\tilde{\nu}) = - I^E(\tilde{\nu})/I^o(\tilde{\nu}) \quad . \tag{2}$$

This ratio is typically on the order of 10^{-3} - 10^{-4}, with a signal-to-noise ratio generally in the range 1:10 - 1:100. Thus, signal-enhancing techniques are necessary to separate (and amplify) the signal from the noise. By either oscillating the orienting field (electrochromism) or oscillating the polarization of the incident light with a photoelastic modulator (ELD), the signal can be separated from the noise using phase-sensitive detection.

Two methods have been developed to acquire these phase-sensitive signals [1-5, 9,10]: The original method used a lock-in amplifier; more recently, however, the use of a signal averager, such as a boxcar averager, has permitted considerable improvement in detection sensitivity and has enabled analysis of signal types which could not be investigated previously [12,13]. Thus, broader absorption bands, the analysis of which was either not feasible or very unreliable with the original method, can now be investigated. This second technique should be of particular use in investigating excited-state properties associated with broader absorptions, such as those typically resulting from intravalence transitions, since ELD signals are significantly reduced for broader absorption bands [13].

In general, electric field studies have permitted determination of the following parameters:

--- Polarization of the transition, either parallel or perpendicular to the principal axis of symmetry. (All compounds investigated thus far have been analyzed in terms of C_{2v} symmetry.)

--- Change in dipole moment between the ground and excited states. (This term has been found to be the largest and most reliable parameter and will be discussed in Section 3.2, below.)

--- Change in polarizability between the ground and excited states. (This term is less reliable due to certain approximations that are made to facilitate data analysis.)

--- Perturbation by nearby states. (The detailed expressions for the coefficients of Eq. (1) involve certain terms reflecting perturbations by nearby states. Only limited data have been obtained for these terms.)

--- Arc-suppressor gas effect. (The analysis of excited-state data indicates a dependence of the above parameters on the pressure of the arc-suppressor gas used to stabilize the intense electric fields. Section 5 is concerned with this effect, which is referred to as the "solvent" effect.)

3.1 Electric Field Dependence of $S(\tilde{\nu})$

The detailed theoretical expression of Eq. (1) was derived under the assumption that the electric field signal $S(\tilde{\nu})$ is *linearly* proportional to the *square* of the strength E of the orienting field, with higher-order terms being neglected. Thus, an experimental determination of the form of $S(\tilde{\nu})$ as a function of E^2 is of prime concern. This is shown in Fig. 2 for field strengths up to 250,000 V/cm (data are for acetone). From this figure we see that, indeed, $S(\tilde{\nu})$ depends linearly upon E^2, but that some degree of nonlinear behavior is indicated at higher field strengths. Since the degree of field-induced orientation depends upon the magnitude of the ground-state dipole moment as well as the strength of the applied field, it may become necessary to include higher-order terms when considering more intense fields. Further studies of the linearity factors are underway. In addition, low-

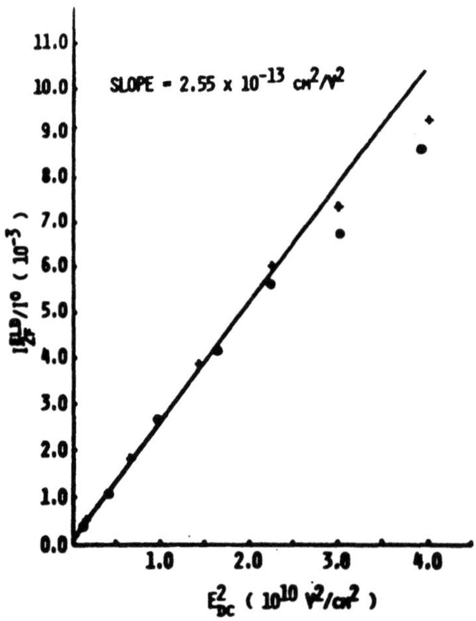

Fig. 2. Plot of the electric linear dichroism (ELD) signal maximum (acetone) as a function of the field strength squared.

er-energy Rydberg states have been found to be highly polarizable and the higher-energy members are expected to be even more polarizable. The effects of these highly polarizable states may require the inclusion of induced dipole-moment terms, and may become more important for the higher-energy Rydberg states, as the ionization potential is approached.

3.2 Dominating Terms

Certain prominant features of the electric field spectral data are evident from inspection. Provided that the electric field spectrum is not complicated by overlapping absorptions, it resembles the second derivative of the absorption spectrum. This is illustrated in Fig. 3: The absorption band for methyliodide at 201 nm, assigned as the 5p → 6s Rydberg transition, appears to be isolated and the electric field signal emphasizes the second-derivative dependence. For a specific polarization of the transition moment (parallel or perpendicular), the second-derivative

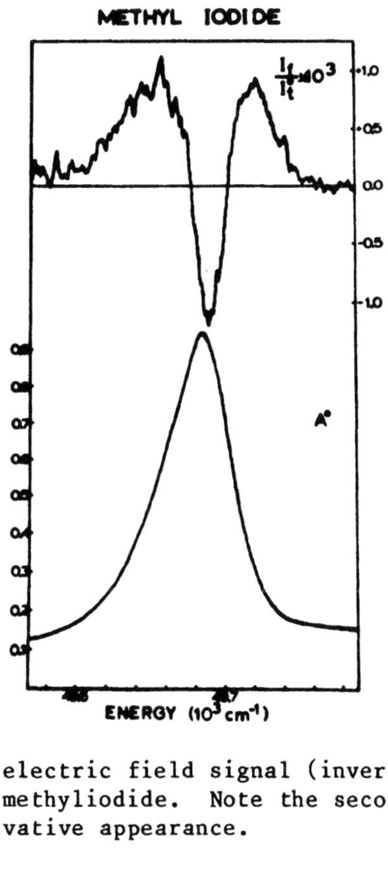

Fig. 3. The electric field signal (inverted) for methyliodide. Note the second derivative appearance.

coefficient (i.e., c_3), as determined by the fitting process, depends upon the change in dipole moment between the two states. The second-derivative term is the dominant and most reliable one; this is borne out by the finding that, for many compounds, the coefficient of the second-derivative term dominates the least-squares fitting. This dominating aspect of the second-derivative term points out the problem created by broad absorption bands: The second derivative of absorption with respect to wavelength results in values close to zero when the absorption band is broad. And, since intravalence transitions are typically characterized by broader absorptions, these transitions present difficulties for electric field studies.

Additional information can be extracted from the dominant term by the use of high pressures for studies of Rydberg bands. Robin has employed high pressures of inert gases to distinguish Rydberg absorptions from intravalence absorptions [14]. The electric field technique requires intense electric fields to partially orient the sample, and a high dielectric material to eliminate arcing between the orienting plates. In the Rydberg-state studies described below, the effect of the dielectric gas on the determination of the excited-state data is investigated. The results of these studies should help in understanding, on a more quantitative basis, the pressure broadening of Rydberg bands.

3.3 Polarization of the Transition Moment

One initial objective of these studies was to determine the polarization of the transition moment for a number of VUV absorptions. Specific interest has developed for certain absorptions that occur in the lower-energy portion (220 - 170 nm) of the VUV. Within this spectral region, multiple assignments have frequently been given to a specific absorption. Knowledge of the transition moment, however, can be used, from symmetry arguments, to substantiate assignments.

If a compound is of C_{2v} symmetry, the ground-state dipole-moment axis is the principal rotation axis. Because of the axial symmetry, the transition-moment polarization is either parallel or perpendicular to this axis. Inspection of the angular factors influencing this term in the electric field expression [i.e., Eq. (1)] indicates that the polarization of the transition can be determined by comparing the electric field signal, at one wavelength, for two different polarizations of the incident radiation. In practice, this technique requires scanning the absorption spectrum to locate a wavelength at which a strong electric field signal occurs; then, in comparing the relative signal by changing the polarization of the incident radiation, the parallel or perpendicular orientation of the transition moment can be obtained.

4. EXCITED-STATE PARAMETERS

4.1 Carbonyl Compounds

We begin our discussion of experimental results by considering carbonyl-substituted molecules. The compounds studied were formaldehyde, acetone, acetaldehyde, several straight-chain ketones, and the cyclic ketones from cyclobutanone to cycloheptanone. With a few exceptions, all of these molecules possess C_{2v} symmetry. From certain similarities among the spectra of the symmetric ketones (e.g., acetaldehyde and certain straight-chain ketones), these molecules can be viewed as having local C_{2v} symmetry about the carbonyl group.

The absorption band system investigated, for each of these compounds, was the second excited singlet, which occurs near 200 nm. These absorption bands have received numerous conflicting assignments, both intravalence, $\pi(2p) \rightarrow \pi^*$, and Rydberg, $\pi(2p) \rightarrow R(3s)$. In the case of acetone, a pressure study by Robin [14] revealed that there could be an underlying intravalence absorption in the same region. Nevertheless, the sharp absorption features appear to be Rydberg. These bands were investigated with electric field techniques in order to clarify the assignments.

One very important property in electric-field studies is the magnitude of the ground-state dipole moment. The ground-state dipole moment determines the degree of orientation according to a Boltzmann distribution. And, the subsequent electric field signal is proportional to the degree of orientation. Carbonyl compounds, therefore, will exhibit more intense electric field signals since these compounds generally possess large ground-state dipole moments ($\sim$3 D). A second important consideration for the electric field method is that the absorption bands be both well-resolved and intense. This factor affects the reliability of the electric field signal as described in the above discussion of the second-derivative term. In general, for all carbonyl compounds studied, the electric field data which were obtained were of sufficient reliability to ascertain characteristics indicative of the Rydberg assignment in every case. The specific data for each compound are discussed below. For ease of reference, the data for formaldehyde, acetone and acetaldehyde are given in Table 1; the data for three vibronic components of acetone are given in Table 2; the data for the straight-chain ketones are given in Table 3; and the data for the cyclic ketones are given in Table 4.

4.1a Straight-Chain Carbonyl Compounds

Formaldehyde:

This compound is the simplest of the carbonyl compounds studied. The electric field signal of the 174.9 nm absorption band is very weak in the UPE spectrum [3]. No signal could be resolved in the ZPE spectrum. This latter fact indicates that the transition moment is polarized perpendicular to the major axis, which supports an $A_1 \rightarrow B_2$ assignment for this absorption.

The dipole-moment change is ∿2.6 D, which implies a near-zero dipole moment for the excited state. In this excited state, the electron density is thought to be shifted from the oxygen atom toward the hydrogen atoms, with a localization of the positive charge in the center of the molecule. The polarizability change is ∿60 $Å^3$. The large dipole-moment change, and the correspondingly large polarizability of the excited state, support the Rydberg assignment.

Acetone:

ZPE, UPE and ELD electric field signals were obtained for the 195.4 nm band. Examples of the spectra are shown in Figs. 4-6. The UPE spectrum is more intense than the corresponding ZPE spectrum, thus indicating that the transition moment is polarized perpendicular to the major dipole axis. This supports an $A_1 \rightarrow B_2$ assignment for this absorption. The ELD signal is inverted from the other two signals, which further substantiates the perpendicular nature of the transition [3]. The dipole moment change is ∿3.1 D, resulting in a near-zero excited-state dipole moment. The corresponding polarizability of the excited state is ∿450 $Å^3$. These data, in conjunction with the polarization data, support the Rydberg assignment.

The absorption band for the second excited singlet of acetone contains three vibronic components, as shown in Fig. 7. The data for these components, given in Table 2, indicate that the dipole-moment change and the polarizability change are not significantly altered by the vibration.

Acetaldehyde:

ELD, ZPE and UPE spectra were obtained for the 181.8 nm absorption band of acetaldehyde [3]. This absorption band has been assigned as a Rydberg transition, n → R(3s) [A' → A'], based on the results of a pressure study [14]. The electric field signals obtained for acetaldehyde were the most intense of all the ketone compounds investigated.

The ZPE spectrum is more intense than the corresponding UPE spectrum. Therefore, assuming that the carbonyl group has local C_{2v} symmetry, the transition-moment polarization is near 0°. This result is valid provided that the molecule is not substantially altered in the excited state. If this is not the case, then

Table 1. Excited-State Parameters for the n → 3s Rydberg Transitions of Formaldehyde, Acetaldehyde and Acetone*

	Formaldehyde	Acetaldehyde	Acetone
$\Delta\mu$	2.66 ± 0.14	4.29 ± 0.29	3.08 ± 0.20
μ_{ex}	-(0.33 ± 0.16)	---	-(0.19 ± 0.23)
$\Delta\alpha$	-(60 ± 30)	---	-(440 ± 140)
α_{ex}	+(70 ± 30)	---	+(450 ± 140)
θ	$90^{\circ} \pm 10^{\circ}$	$(9^{\circ}$ or $171^{\circ}) \pm 5^{\circ}$	$90^{\circ} \pm 10^{\circ}$

*$\Delta\mu = \mu_g - \mu_{ex}$ = change in the dipole moment (Debye); μ_{ex} = excited-state dipole moment (Debye); α_{ex} = excited-state polarizability ($\mathring{A}^3$); θ = transition moment polarization

Table 2. Change in Dipole Moment for Three Vibrational Members of the n → 3s Rydberg Transition of Acetone*

Band (cm^{-1})	$\Delta\tilde{\nu}$ (cm^{-1})	$\Delta\mu$ (D)
51177	0	3.08
52493	1316	3.16
53619	2442	3.07

*See footnote to Table 1 for notation.

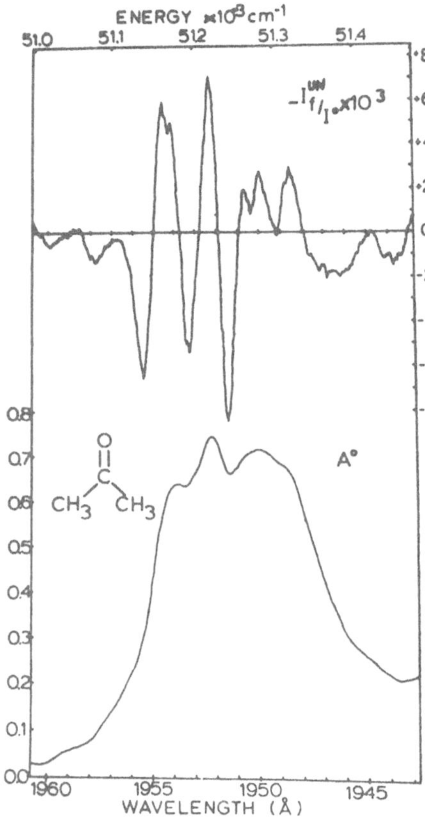

Fig. 4.

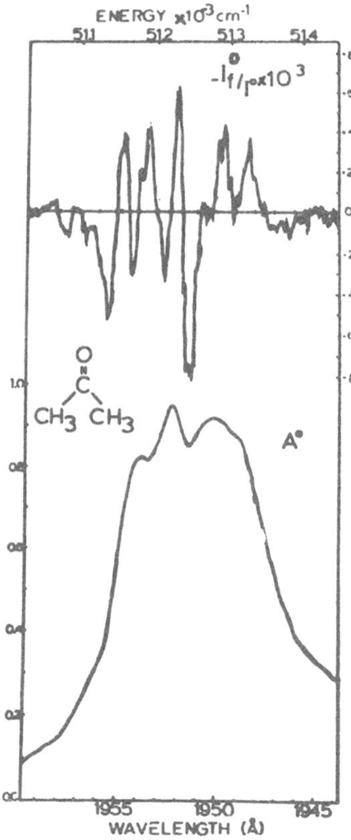

Fig. 5.

Fig. 4. Example of the unpolarized light electrochromism signal and the absorption band for the 195.4 nm C transition of acetone.

Fig. 5. Example of the zero-degree polarized light electrochromism signal and the absorption band for the 195.4 nm C transition of acetone.

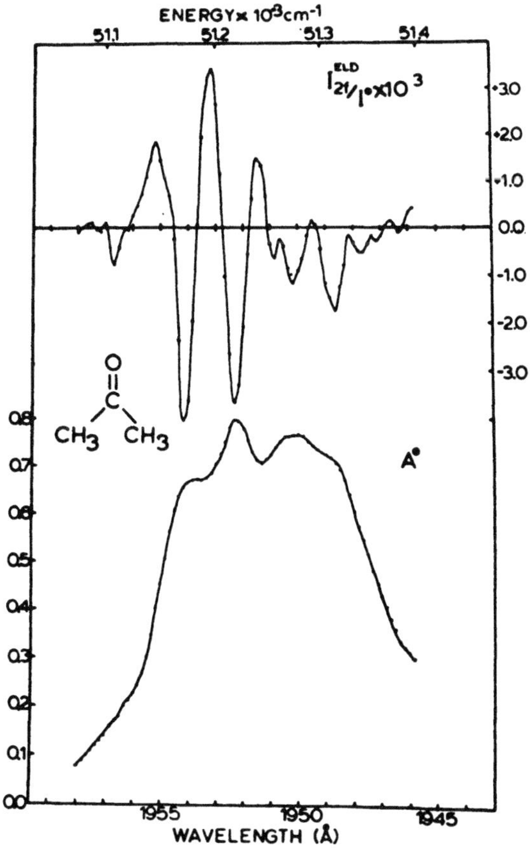

Fig. 6. Example of the electric linear dichroism signal and the absorption band for the 195.4 nm C transition of acetone.

there are no restrictions on the angular relations between the ground-state dipole moment and the transition moment. In a study of the 18 vibronic components associated with this transition, however, Lucazeau and Sandorfy [15] concluded that if the excited state is bent, it is not bent to any great extent.

The change in dipole moment is 4.29 D, with the electron density being transferred toward the carbon end of the molecule; the resulting dipole moment of the excited state is -1.6 D. (The negative sign denotes that the electron density is reversed in direction from that of the ground state.) The polarizability change and excited-state polarizability were not determined. The large value of the dipole-moment change, and the polarization of the transition moment, are consistent with the Rydberg assignment.

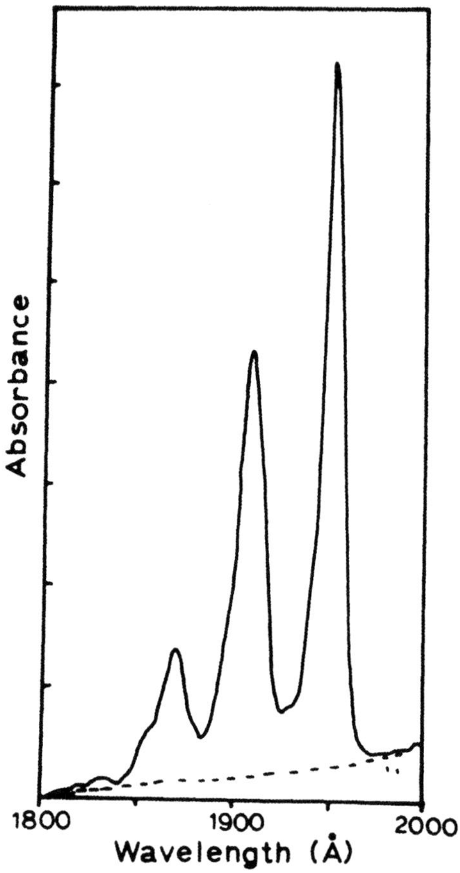

Fig. 7. Example of the absorption bands for the first three vibronic members of the 195.4 nm C transition of acetone.

Straight-chain ketones:

The electric field signals for straight-chain ketones were measured in order to assess the effect of increasing the number of carbon atoms on the change in dipole moment. The compounds investigated were, in addition to the above compounds, 2-butanone and 2-pentanone. The dipole-moment changes are listed in Table 3.

The effect of increasing the number of carbon atoms did not change the values, to any great extent, from the corresponding values obtained for the three compounds discussed above. The results indicate that the transition characteristics are indeed dom-

Table 3. Change in Dipole Moments for the $n \rightarrow 3s$ Rydberg Transition of Some Straight-Chain Carbonyl Compounds*

Compound	μ_g (D)	$\Delta\mu$ (D)
Acetaldehyde	2.69	4.29
Acetone	2.88	3.08
2-Butanone	2.78	3.60
2-Pentanone	2.72	3.76

*See footnote to Table 1 for notation.

inated by the carbonyl group, and that the local symmetry of the carbonyl group can be treated as "C_{2v}".

4.1b Cyclic Ketones

Electric field spectra were obtained for the 0-0 component of the second excited singlet absorption band of cyclobutanone (203.1 nm), cyclopentanone (199.8 nm), cyclohexanone (197.3 nm) and cycloheptanone (195.9 nm) [4]. The absorption spectra are shown in Fig. 8. These bands have received numerous assignments [14]; but, as in the case of the above ketones, the transition is thought to be the $\pi(2p) \rightarrow R(3s)$ $[A_1 \rightarrow B_2]$ Rydberg transition.

The results presented below support the Rydberg assignment in every case. The second emphasis of this study was on investigating the dependence of the excited-state data on ring size; the appropriate electric field data are presented in Table 4.

Cyclobutanone:

The UPE spectrum is more intense than the ZPE spectrum. This indicates that the transition moment is polarized perpendicular to the major axis, which supports an $A_1 \rightarrow B_2$ assignment.

Cyclobutanone exists in a "bent" configuration. However, if there is no significant geometry change in the excited state (e.g., to a planar state), the local symmetry of the carbonyl group can be considered to dominate the absorption characteristics. And, the vibrational structure for this absorption does not indicate that the excited state is planar.

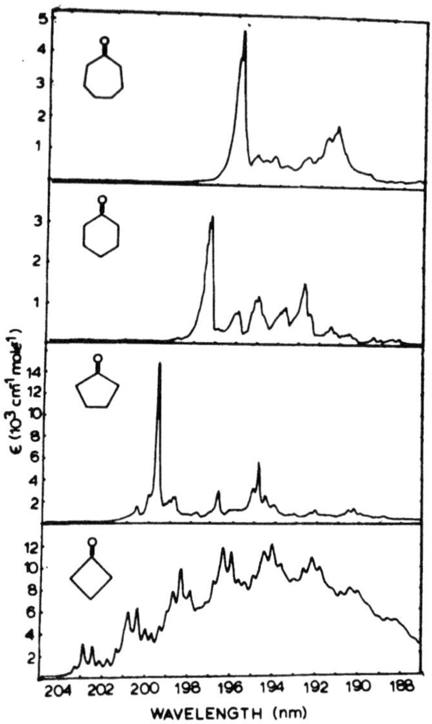

Fig. 8. Examples of the partial absorption spectra of several cyclic ketones. The first band, assigned as the n → 3s Rydberg transition, was investigated.

Table 4. Excited-State Parameters for the n → 3s Rydberg Transition of Some Cyclic Ketones*

	Cyclobutanone	Cyclopentanone	Cyclohexanone	Cycloheptanone
$\Delta\mu$	1.44	1.63	4.30	4.92
μ_{ex}	1.45	1.62	-1.22	-1.85
α_{ex}	210	300	700	1280
θ	77° (103°)	81° (98°)	76° (104°)	82° (98°)

*See footnote to Table 1 for notation.

The dipole-moment change is 1.44 D, resulting in an excited-state dipole moment of 1.45 D. The excited-state polarizability is ∿200 Å^3. The magnitudes associated with these properties support the Rydberg assignment.

Cyclopentanone:

The transition-moment polarization angle for the absorption band at 199.8 nm is 90°, thus supporting an $A_1 \rightarrow B_2$ assignment. The dipole-moment change and the excited-state dipole moment, evaluated from the experimental coefficients, are 1.63 D and 1.62 D, respectively. The excited-state polarizability is 300 Å^3.

Cyclohexanone and cycloheptanone:

The transition-moment polarization supports an $A_1 \rightarrow B_2$ assignment in both cases. The dipole-moment changes are 4.30 D and 4.92 D, respectively, for cyclohexanone and cycloheptanone. The values for the dipole-moment changes are greater than twice those determined for the other cyclic ketones. The ability of the larger ring systems to dissipate the charge is partially responsible, but other factors are thought to be important. In addition, the excited-state polarizabilities are 700 Å^3 and 1280 Å^3 for the two molecules, respectively.

In general, the dipole-moment change and the polarizability change for these transitions increase as the ring size increases. This is interpreted as resulting from electron density being more delocalized from the oxygen end of the molecule. In considering the data to be discussed for the sulfides below, this trend appears unique to the s Rydberg members; the p and d Rydberg members exhibit opposite trends.

The polarizabilities for these states follow a trend similar to that for the dipole-moment change, namely, the larger the number of atoms in the ring, the larger the polarizability. These data support the Rydberg nature of the excited-state MO. The ℓ-dependent differences, mentioned above, are thought to arise because of the penetration of the excited-state MO into the core orbitals. If the excited-state MO's for the s Rydberg states were centered about the positive charge, the dipole moment of the excited state would be zero. And, if we assume that the positive charge is localized on the oxygen atom, the large dipole-moment change, which results in a large shift of electron density into the ring, indicates that the center of the MO is displaced from the oxygen atom into the center of the molecule in the larger ring compounds. The electric field spectra of a number of sulfur-containing molecules were investigated to determine if this differential behavior for the s, p and d Rydberg absorption bands is unique to the ketone compounds, or if it is a characteristic of

Rydberg states in general. These results are reported in the next section.

4.2 Sulfide Compounds

The next group of compounds to be discussed consists of the cyclic sulfides and dimethylsulfide. In general, the ground-state dipole moments are 1.5-2.0 D. This results in correspondingly weaker electric field signals than those for the ketone compounds. However, these sulfide compounds possess several types (s, p and d) of intense Rydberg absorption bands, with well-resolved structure, in the near VUV. Thus, it is possible to investigate characteristic differences in the different types of Rydberg absorption bands, and to determine excited-state data which help in differentiating these absorption bands. For the cyclic sulfides, these data are given in Table 6 (s Rydberg), Table 7 (p Rydberg) and Table 8 (d Rydberg).

The primary reason for investigating these compounds was the wealth of Rydberg absorption bands available. The second reason was that the assignments have been well-established. It has been noted that if the absorption spectra are "lined up" with the ionization potentials, the energies of the Rydberg absorption bands will be aligned [16]. Thus, the emphasis in this study was directed toward the description of the various Rydberg states.

Dimethylsulfide and ethylenesulfide:

The absorption bands investigated were the 212.2 nm band of ethylenesulfide (also known as dimethylenesulfide or thiirane) and the 228.4 nm band of dimethylsulfide; the 192.2 nm band of ethylenesulfide and the 195.9 nm band of dimethylsulfide; and the 174.0 nm band of ethylenesulfide. These bands have been identified as the $n \rightarrow R(4s)$, $n \rightarrow R(4p)$ and $n \rightarrow R(3d)$ Rydberg transitions, respectively [14,17]. Dimethylsulfide and ethylenesulfide are the simplest of two classes of sulfur compounds, namely, the symmetric straight-chain alkyl sulfides and the cyclic sulfides, respectively. The basic difference between these two classes of compounds is the presence or absence of the ring. Assuming that the geometries remain unchanged upon excitation, variations in excited-state properties with respect to the structural restictions of the ring can be inferred from the electric field data. These data are given in Table 5.

The ground states for both dimethylsulfide and ethylenesulfide have been calculated to possess a highest occupied orbital of b_1 symmetry, localized in the lone-pair 3p orbital on the sulfur atom [16,18]. The $n \rightarrow R(4s)$ transitions (228.4 nm in dimethylsulfide and 212.2 nm in ethylenesulfide) are classified, by symmetry arguments, as $A_1 \rightarrow B_2$ transitions polarized perpendicular

Table 5. Excited-State Parameters for the First s, p and d Rydberg Transitions of Dimethylsulfide and Ethylenesulfide*

	Dimethylsulfide		Ethylenesulfide		
	s Rydberg	p Rydberg	s Rydberg	p Rydberg	d Rydberg
	228.4 nm	195.9 nm	212.2 nm	192.2 nm	174.0 nm
$\Delta\mu$	2.6	2.7	2.7	2.9	5.6
μ_{ex}	-1.2	-1.2	-0.9	-1.1	-3.8
α_{ex}	160	340	150	135	1625
θ	83	16	(90)	18	83

*See footnote to Table 1 for notation.

Table 6. Excited-State Parameters for the First s Rydberg Transition of Ethylenesulfide and Pentamethylenesulfide*

	Ethylenesulfide	Pentamethylenesulfide
	212.2 nm	223.7 nm
$\Delta\mu$	2.7	4.1
μ_{ex}	-1.1	-2.3
α_{ex}	(150)	(200)
θ	(90)	(90)

*See footnote to Table 1 for notation.

Table 7. Excited-State Parameters for the First p Rydberg Transition of Some Cyclic Sulfide Compounds.*

	Ethylenesulfide 192.2 nm	Trimethylenesulfide 202.7 nm	Tetramethylenesulfide 209.4 nm	Pentamethylenesulfide 208.4 nm	Pentamethylenesulfide 196.3 nm
$\Delta\mu$	2.9	2.8	1.9	(1.6)	(5.7)
μ_{ex}	-1.1	-1.1	-0.2	(0.2)	(-3.9)
α_{ex}	135	145	60	130	400
θ	0	0	0	0	0

*See footnote to Table 1 for notation.

Table 8. Excited-State Parameters for the First d Rydberg Transition of Some Cyclic Sulfide Compounds.*

	Ethylenesulfide 174.0 nm	Tetramethylenesulfide 186.9 nm	Pentamethylenesulfide 184.3 nm
$\Delta\mu$	5.6	4.2	3.1
μ_{ex}	-3.8	-2.5	-1.3
α_{ex}	1625	-410	(400)
θ	83	(90)	---

*See footnote to Table 1 for notation.

to the major axis. The polarization angles determined for these absorptions support this assignment.

Several possibilities exist for the n → R(4p) (195.9 nm in dimethylsulfide and 192.2 nm in ethylenesulfide) and the n → R(3d) (174 nm in ethylenesulfide) transitions. The p and d Rydbergs could result from either an $A_1 \rightarrow B_1$ transition or an $A_1 \rightarrow A_1$ transition. The experimental polarization angle supports the $A_1 \rightarrow A_1$ assignment for the p Rydberg and the $A_1 \rightarrow B_1$ assignment for the d Rydberg.

The dipole-moment changes associated with these transitions were found to be similar in magnitude and direction to those observed for the Rydberg transitions of the ketones [1-4]. The dipole-moment changes for the s and p Rydberg absorptions of dimethylsulfide and ethylenesulfide are ∿2.6-2.9 D; the value for the d Rydberg absorption in ethylenesulfide is 5.56 D, almost twice the value for the s and p Rydbergs.

The polarizabilities of the s Rydberg states of these two compounds are essentially the same (∿150 $Å^3$). The polarizability change for the p Rydberg of dimethylsulfide is greater than that for the corresponding p Rydberg of ethylenesulfide. This difference is attributed to the presence of the ring in ethylenesulfide and to the ability of the p Rydberg MO to penetrate the bonding orbitals of the core. The value for the d Rydberg state is considerably different from that for the s and p Rydbergs. Since this is the first d Rydberg considered, the significance of these findings will be discussed further after the data for the other cyclic members are presented.

Trimethylenesulfide:

Trimethylenesulfide (thietane) belongs to the C_s point group. However, for comparison with ethylenesulfide and the higher cyclic sulfides, we assume that the C_{2v} point group describes the local symmetry of the C-S-C unit. This local symmetry is suggested by the similarity of the spectrum of trimethylenesulfide to the spectra of the other sulfides, with respect to the Rydberg absorptions.

The n → R(4s) Rydberg transition at 226 nm contains numerous vibronic components. The structure of this band has been interpreted as resulting from an excitation to a planar state [19]. The electric field spectra of the origin and of several vibrational components were measured. The changes in excited-state properties, due to vibrational effects, were found to be less than experimental error. The change in the dipole moment is 0.8 D.

The 202.7 nm band of trimethylenesulfide has been assigned

as the n → R(4p) Rydberg transition [5,14]. In addition, this excited state has been shown to be nonplanar [19]. The UPE spectrum was found to be more intense than the NPE spectrum, which supports an assignment in which the transition moment is polarized parallel to the major dipole axis. Thus, an n → R(4p) [$A_1 \rightarrow A_1$] Rydberg assignment is indicated.

The dipole-moment change for the n → R(4p) transition is 2.8 ± 0.5 D. The polarizability change is -150 ± 300 Å^3, thus giving a value for the excited-state polarizability of ∿150 Å^3. From the value and sign of the polarizability change, the excited-state dipole moment is determined to be -1.1 ± 0.5 D. The negative sign is interpreted as arising from a shift of electron density from the sulfur end of the molecule to the carbon end, resulting in localization of the positive charge toward the sulfur atom. The dipole-moment change is in the same direction as, and similar in magnitude to, that determined for the ketone data presented.

Tetramethylenesulfide:

Since the absorption spectrum of tetramethylenesulfide (tetrahydrothiophene) is similar to those of the other cyclic sulfides, this compound will be discussed in terms of C_{2v} symmetry. The n → R(4s) absorption is broad and unstructured and, therefore, the electric field signal could not be studied. The two absorptions assigned as n → R(4p) and n → R(3d) were investigated. The n → R(4p) absorption occurs at 209.4 nm and examples of UPE and NPE spectra are given in Fig. 9. By comparison of the intensities for the two types of spectra, this absorption can be assigned as a parallel transition, thus supporting the n → R(4p) [$A_1 \rightarrow A_1$] Rydberg assignment. The polarization angle, as measured, is zero.

The dipole-moment change for the n → R(4p) transition is 1.9 ± 0.2 D. The polarizability change is -70 ± 150 Å^3, resulting in an excited-state polarizability of ∿70 ± 150 Å^3. The excited-state dipole moment is near zero.

The n → R(3d) absorption occurs at 186.9 nm. The transition moment is polarized perpendicular to the major dipole axis. Previous work on ethylenesulfide yielded the same symmetry assignment for the corresponding d Rydberg [17].

The dipole-moment change for the n → R(3d) transition is 4.2 ± 0.1 D, twice as large as that for the p Rydberg. The polarizability change is 420 ± 650 Å^3 and the excited-state dipole moment is -2.5 ± 0.1 D. This is the second d Rydberg that we have investigated by the electric field method: By comparing the ethylenesulfide and tetramethylenesulfide data, it is clear that the d Rydbergs are distinct from the other transitions, particularly with

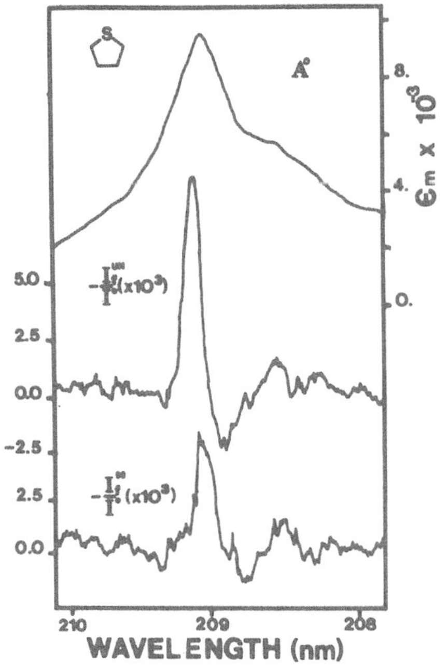

Fig. 9. Examples of the absorption, unpolarized light electrochromism ($-I_f^{UN}/I^o$) and the 90° polarized light electrochromism ($-I_f^{90}/I^o$) spectra for the 209.4 nm band of tetramethylenesulfide. The molar absorptivity ε_m (absorption spectrum) is in the units ℓ/mol·cm.

respect to excited-state polarizability.

Pentamethylenesulfide:

Pentamethylenesulfide exists in the chair conformation and belongs to the C_s point group. The general features of the spectrum are consistent with those for the other cyclic sulfides, with the first p Rydberg absorption being the most intense. The following discussion is based on the assumption that the molecule has local C_{2v} symmetry for the C-S-C bonds, and that this local symmetry for the sulfur atom dominates the absorption spectrum of the compound.

The bands investigated are labeled according to assignments given previously [20]. In general, these bands are broader than any of the others investigated; thus, the signal-to-noise ratio is considerably lower. A unique s series, two p series and several d series of Rydberg absorptions have been identified [20]. These assignments were made on the basis of term values, since no series leading to the ionization potential were found because of the complexity of the spectrum. The origin of the separate p and d series is thought to come from a mixing of the p-d orbitals as a result of the low symmetry of the molecule [20]. Although it is generally concluded that this mixing is not important in considering the ground-state MO's, such is not thought to be the case for the excited-state MO's [20-22].

In view of the above difficulties in interpreting the spectrum of pentamethylenesulfide, the electric field data were investigated: (i), to add data to the cyclic sulfide series; (ii), to provide semiquantitative results for the magnitudes of the important properties of the lower excited states; and (iii), to use these excited-state data in conjunction with the data for the other compounds to aid in the interpretation of the spectrum of pentamethylenesulfide. By considering the pentamethylenesulfide excited-state properties in conjunction with those of the cyclic sulfide series, certain generalities were noted which should prove useful in generating a spectral interpretation for this compound.

The absorption at 223.7 nm has been identified as the $n \rightarrow R(4s)$ Rydberg transition [20]. There is no discernable vibrational structure indicative of a geometry change. The dipole-moment change is 4.1 ± 0.2 D, resulting in an excited-state dipole moment of -2.3 ± 0.2 D. The negative sign is interpreted as arising from a shift of electron density into the ring, resulting in the sulfur atom becoming the positive end of the dipole. The value for the dipole-moment change is similar in magnitude to that determined for the d Rydberg states of the other cyclic sulfides. The polarization of the transiton is 90°. This, in conjunction with the fact that the polarizability of this excited state is similar to the $n \rightarrow R(4s)$ transition of ethylenesulfide, supports the $n \rightarrow R(4s)$ Rydberg assignment for pentamethylenesulfide.

There are two absorptions, labeled p and p', at 208.4 nm and 196.3 nm, respectively. The dipole-moment change for the $n \rightarrow R(4p)$ transition is 1.6 D while that for the $n \rightarrow R(4p')$ transition is 5.7 D. Because of the low signal-to-noise ratio for these bands, the uncertainties in the excited-state data are large. These high uncertainties preclude a determination of the polarization angle for these transitions. The dipole-moment change for the $n \rightarrow R(4p)$ transition (208.4 nm), however, is consistent with the $n \rightarrow R(4p)$ assignment in the other cyclic sulfides. The $n \rightarrow R(4p')$ absorption will be discussed below in conjunction with the d Rydberg transition.

The 184.3 nm absorption has been assigned as $n \rightarrow R(3d)$ [20].

The dipole-moment change is 3.1 ± 0.5 D, resulting in a -1.3 ± 0.5 D excited-state dipole moment. The dipole-moment change and the polarizability for the excited state are similar to the values determined for the five-member rings. These data suppport the d Rydberg assignment for this absorption.

Sandorfy, *et al.* state that the n → R(4p') absorption (196.3 nm) and the n → R(3d) absorption (184.3 nm) "may correspond to the first members of Rydberg series where the excited state is of mixed p-d character containing an increasing amount of d in this order" [20]. These p-d mixtures are possible from the low symmetry of this molecule. The excited-state data for the n → R(4p) absorption (208.4 nm), when viewed with respect to the other cyclic sulfides, indicate that this assignment is correct. The excited-state data for the n → R(3d) absorption (184.3 nm) also support the trend observed for the cyclic sulfides. It is clear from these data that the 196.3 nm band does not conform to either trend.

5. ARC-SUPPRESSOR "SOLVENT" EFFECTS

Our electric field studies require the use of relatively strong orienting fields (10^5 V/cm). And, at such high field strengths, arcing from corona formation within the sample cell can occur. At least 2.6 atm of sulfur hexafluoride (SF_6; a gaseous dielectric) were necessary to stabilize the field. The presence of this gaseous "arc suppressor" was found to alter the measured excited-state parameters [5,12].

In this section, we report results obtained for the variation in dipole-moment change, following excitation, with respect to the pressure of the arc-suppressor gas, for the n → R(3s) [A_1 → B_2] Rydberg transition of acetone (195.4 nm) and of cyclopentanone (199.8 nm), and for the π → π* intravalence transition of aniline. The polarizations of these transitions have been determined previously [2,3,6]. In addition, we report results concerning the corresponding changes in excited-state properties for the p and d Rydberg absorption bands of ethylenesulfide, using SF_6 and CF_4 as arc-suppressor gases. In general, the sample pressure was ∿1 torr and the arc-suppressor gas pressure was in the range 2.6 - 22 atm; thus, SF_6 is referred to as the "solvent" gas. The results of these studies, then, can be related to solvation effects.

Increasing the pressure of the arc-suppressor gas causes the absorption bands to broaden and shift to lower energy. An example of these effects, for the 199.8 nm band of cyclopentanone, is given in Fig. 10, along with an illustration of the subsequent effect on the electric field spectra. Robin has shown that a pres-

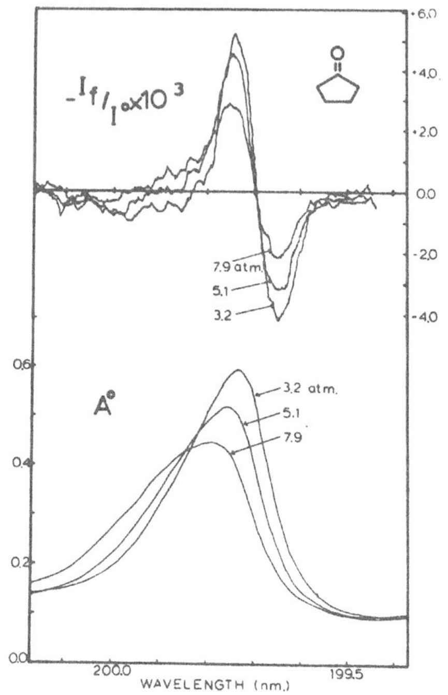

Fig. 10. Examples of the effect of increasing SF_6 pressure on the 199.8 nm absorption band, and the corresponding electric field signal, of cyclopentanone.

sure effect can be used to distinguish Rydberg absorptions from non-Rydberg absorptions [14]: Non-Rydberg absorptions show little or no pressure broadening and shifting. Since the $\pi \rightarrow \pi^*$ transition of aniline exhibits effects similar to those observed for the Rydberg absorptions in the ketones, one may surmise that this band possesses some Rydberg character.

The absolute values of the dipole-moment change $\Delta\mu$, at various pressures, for acetone, cyclopentanone and aniline were determined from the empirical coefficients. These values are plotted versus the molar density of SF_6 in Fig. 11. The molar density ($mmol/cm^3$) was determined from the pressure virial equation-of-state using the first three virial coefficients. From the plot of $|\Delta\mu|$ versus molar density, the "solvent-free" values of the dipole-moment change are 1.09 D for aniline, 1.65 D for cyclopentanone, and 2.94 D for acetone. (The error is estimated to be 10%.) Lombardi [6] has reported the dipole-moment change for the $\pi \rightarrow \pi^*$ transition of aniline, using an optical Stark effect on certain resolvable rotational lines of this transition. His

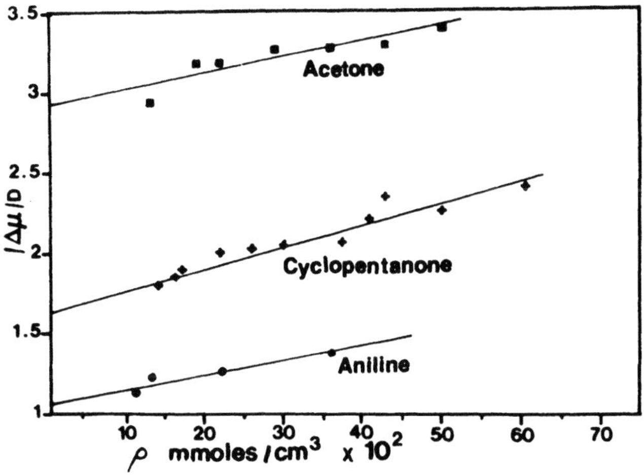

Fig. 11. Plots of the absolute values of the dipole-moment changes, versus pressure, for aniline, acetone and cyclopentanone.

value for the dipole-moment change is 0.95 D, which compares favorably with the 1.09 D value reported here.

The magnitude of the variation in the dipole-moment change is much larger than that predicted by the Lorentz or Onsager [10,23] correction for the local field. However, by considering the data in terms of a virial relation, the magnitude of the intermolecular forces for the excited state can be estimated.

The virial equation assumes that a measurable equilibrium property can be expanded in a power series with respect to molar density [23-26]:

$$Q = A_Q + B_Q/V + \ldots$$

where Q represents some measurable property; V is the molar volume (cm^3/mol); and A_Q, B_Q, etc. are the virial coefficients. The first virial coefficient (A_Q) represents the "perfect-gas" value for the measurable property, while the second virial coefficient (B_Q) represents the first-order correction due to two-particle interactions. For dielectric measurements, the previous equation becomes the Clausius-Mossotti (C-M) relation:

$$P_M = A_\varepsilon + B_\varepsilon \rho + \ldots$$

$$P_M = \frac{\varepsilon - 1}{\varepsilon + 2} V \; ; \; A_\varepsilon = \frac{4\pi N}{3}\left(\alpha + \frac{\mu^2}{3kT}\right).$$

In the above equations, ε is the dielectric constant; A_ε and B_ε are the first and second dielectric virial coefficients, respectively; P_M is the molar polarization; and V is the molar volume ($1/\rho$). A discussion of these coefficients has been given by Buckingham [24-26] and others [27-28].

The molar polarization contains a polarizability contribution P_α and a dipole-moment contribution P_μ (i.e., $P_M = P_\alpha + P_\mu$). Assuming that α and μ are independent of each other, each molar polarization can be expanded in a power series with respect to molar density:

$$P_M = (A_\alpha + A_\mu) + (B_\alpha + B_\mu)\rho + \ldots$$

with,

$$P_\alpha = A_\alpha + B_\alpha \rho + \ldots$$

$$P_\mu = A_\mu + B_\mu \rho + \ldots$$

where $A_\alpha = 4\pi N\alpha/3$ and $A_\mu = 4\pi N\mu^2/9kT$. For binary mixtures, each interaction will make a contribution to the total polarization and, therefore, to each virial coefficient (e.g., A_{11}, A_{22}; B_{11}, B_{12}, B_{22}; etc.). Each interaction must be included with an appropriate weight determined by the mole fraction [24-26]. Using the C-M relation, these contributions to the total molar polarization cannot be measured independently of the dielectric constant. Electric field studies, however, permit the determination of the change in both the dipole moment and the polarizability between the ground and excited states of a particular solute [1-5]. Thus, if the solute vapor pressure is held constant, variations in these properties as a function of arc-suppressor gas density indicate the presence of two, three, etc. particle interactions between the solute and the suppressor gas. Since the electrochromic effect is derived from the absorption band of the solute in a specific environment (and does not depend upon a bulk property such as the dielectric constant), the solute-solvent interaction term (B_{12}) can be evaluated directly from the variations in electric field data. The method used here has been reported previously [9].

The results for $B_{\mu,12}$, for the excited states investigated, are given in Table 9. From the data for the two ketones, the interaction of the 3s Rydberg state with the solvent environment is

Table 9. Second Dielectric Virial Coefficients: Aniline ($\pi \rightarrow \pi^*$), Cyclopentanone ($n \rightarrow 3s$) and Acetone ($n \rightarrow 3s$).

Compound	$B_{\mu,12}$ ($10^4 cm^6/mol^2$)
aniline ($\pi \rightarrow \pi^*$)	0.89
cyclopentanone ($n \rightarrow 3s$)	6.4
acetone ($n \rightarrow 3s$)	0.89

larger for the ring system. Further studies should clarify this trend for the ketones, as well as the significance of the virial coefficient for aniline.

The values of $B_{\mu,12}$ determined for aniline, cyclopentanone and acetone were on the order of 10^4. Typical values for this virial coefficient for a polar gas in its ground state in a nonpolar solvent are ∿10-100. Clearly, these data indicate that there is considerable dipolar interaction between the excited-state molecule and the solvent environment.

The polarizability data for aniline, cyclopentanone and acetone were not determined as a function of molar density and, therefore, the contributions to the molar polarization from the polarizability terms could not be evaluated. Our next investigation focused on the polarizability interactions between an excited-state molecule and the arc-suppressor "solvent", in order to determine the total molar polarization and to provide an estimate of the intermolecular forces responsible for the interaction. The absorption bands chosen for this study were the p and d Rydberg absorption bands of ethylenesulfide, which occur at 192.2 nm and 174 nm, respectively. The pressure effects were investigated using CF_4 and SF_6 as the nonpolar "solvent" gases. An example of the pressure effect on the d Rydberg absorption band, and on the corresponding electric field signal, for ethylenesulfide with SF_6 is given in Fig. 12. The electric field data (i.e., dipole-moment and polarizability changes) as a function of pressure for the p and d Rydberg absorption bands are given in Table 10. Finally, $\Delta\alpha$ versus molar density plots are provided in Figs. 13 and 14.

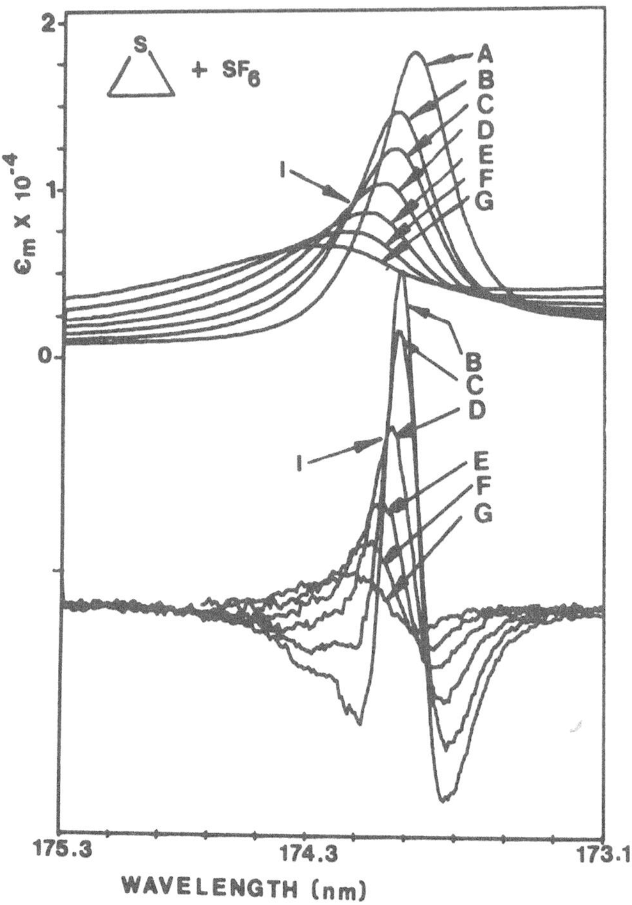

Fig. 12. Examples of the effect of increasing the SF_6 pressure on the absorption and electrochromism spectra of the 174 nm band of ethylenesulfide. The molar absorptivity ε_m is in the units of ℓ/mol·cm. The pressures (atm) used were: A, 0; B, 4.4; C, 7.8; D, 11.2; E, 14.6; F, 18.0; G, 21.4.

In contrast to the data for aniline, cyclopentanone and acetone, the dipole-moment change for ethylenesulfide showed little effect from increasing the pressure of the "solvent" gas. The polarizability change, however, was considerably affected. The dielectric virial coefficients determined from the $\Delta\alpha$ versus

Table 10. Excited-State Parameters for the 192.2 nm p Rydberg Transition and 174.0 nm d Rydberg Transition of Ethylenesulfide in the Presence of CF_4 and SF_6.*

	p Rydberg (192.2 nm)		d Rydberg (174.0 nm)	
Pressure (atm)	$\Delta\mu$	$\Delta\alpha$	$\Delta\mu$	$\Delta\alpha$
CF_4				
0.0	3.0 ± 0.1	-433 ± 120	5.1 ± 1.1	-1893 ± 600
10.5	3.0 ± 0.1	-135 ± 120	5.5 ± 1.1	-1370 ± 800
12.6	---	---	5.6 ± 1.0	-1020 ± 330
14.6	3.0 ± 0.1	-100 ± 120	---	---
18.0	3.0 ± 0.1	-90 ± 160	5.6 ± 0.8	-970 ± 400
21.4	2.9 ± 0.1	-135 ± 180	---	---
24.8	---	---	5.8 ± 1.0	-350 ± 400
SF_6				
0.0	2.9 ± 0.1	-210 ± 300	5.6 ± 0.7	-2150 ± 800
4.4	2.9 ± 0.1	-140 ± 300	5.7 ± 0.7	-1630 ± 900
7.8	3.0 ± 0.1	-140 ± 100	5.7 ± 0.5	-940 ± 630
11.2	2.9 ± 0.1	-80 ± 120	5.9 ± 1.0	-500 ± 400
14.6	2.9 ± 0.1	-70 ± 180	5.8 ± 1.0	-420 ± 400
18.0	2.9 ± 0.1	-120 ± 130	6.0 ± 1.2	-380 ± 280
21.4	3.0 ± 0.1	-190 ± 120	6.2 ± 1.3	-490 ± 400

*Standard deviations are given at the 95% confidence levels; see footnote to Table 1 for notation.

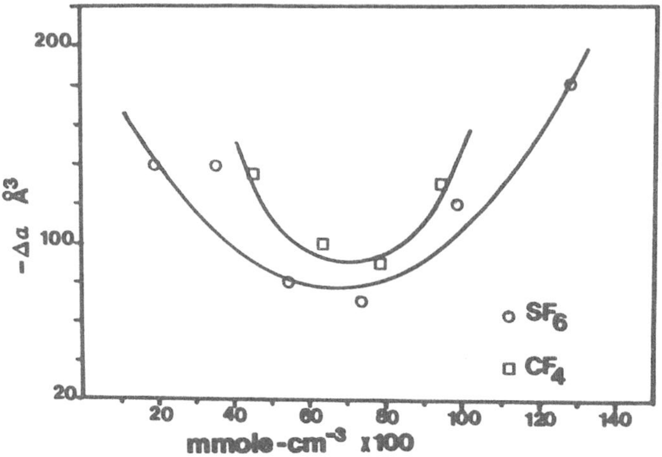

Fig. 13. Plot of $\Delta\alpha$ versus molar density of suppressor gas for the p Rydberg absorption band of ethylenesulfide.

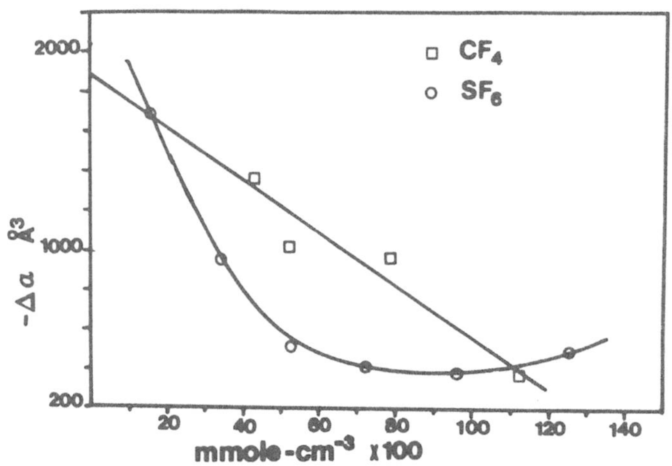

Fig. 14. Plot of $\Delta\alpha$ versus molar density of suppressor gas for the d Rydberg absorption band of ethylene sulfide.

molar density plots are given in Table 11.

The values for the polarizability contribution to the total molar polarizations are considerably greater than those for the dipole-moment contribution. In addition, the corresponding values for the ground state can be estimated to be 100-1000 cm^6/mol^2 [23-27]. Thus, the values listed in Table 11 can be used to approximate the apppropriate values for these excited states.

The magnitude of the excited-state intermolecular potential can be estimated from the coefficients of Table 11 once one has assumed an appropriate intermolecular potential function. The most realistic potential function of those typically used is the Lennard-Jones 6-12 potential:

$$\phi(r) = 4\varepsilon_{12}[(\sigma/r)^{12} - (\sigma/r)^{6}]$$

where ε_{12} is the depth of the intermolecular potential, and where σ is the collision radius. From these considerations, and from the theoretical forms of the virial coefficients [27-29], the depth of the intermolecular potential (i.e., ε_{12}) was determined to be ∿0.250 eV for the d Rydberg state.

We now turn to the spectrum of the 174 nm absorption band of ethylenesulfide, given in Fig. 12 for different pressures of the arc-suppressor gas SF_6. As the SF_6 pressure increases, the band maximum shifts to lower energy, with a concomitant broadening. In the process of broadening, the series of absorption curves cross at a common point. At this isosbestic point, the absorb-

Table 11. Virial Coefficients for p Rydberg (192.2 nm) and d Rydberg (174.0 nm) States of Ethylenesulfide.

Band + Gas	$A_{\alpha,12}$ (cm^3/mol)	$B_{\alpha,12}$ ($10^6 cm^6/mol^2$)	$C_{\alpha,12}$ ($10^9 cm^9/mol^3$)
p + CF_4	1187	2.45	1.74
p + SF_6	592	0.99	0.73
d + CF_4	5266	0.90	2.20
d + SF_6	6213	10.90	5.88

ance is constant at a specific wavelength for the series of different pressures of the arc-suppressor gas. A similar isosbestic point is seen in the electric field spectra.

The isosbestic point indicates an equilibrium which, in this case, is attributed to the formation of an aggregate between the solute (ethylenesulfide) and the "solvent" (SF_6). The strength of the binding (i.e., $\sim$0.250 eV) is thought to correspond to a van der Waals complex. When SF_6 is replaced by CF_4 or Ar, the energy shift of the band maximum, and the broadening of the absorption band, are to lower energies and also show isosbestic points.

The above broadening and shifting to lower energies is opposite to the trend reported by Robin [14] for studies of the effects of increased pressures of certain inert gases on Rydberg absorption bands. The energy shifts discussed here are thought to be related to the electron affinity and the polarizability of the solvent gas in comparison to the solute gas. Further studies of these pressure-dependent spectral variations are needed to clarify the specific nature of the phenomenon.

6. CONCLUSIONS

The preceeding description has illustrated the use of intense electric fields to obtain information about certain excited-state molecular parameters. This information has been available for only a few excited states of a limited number of compounds. The technique has been used primarily to investigate Rydberg states. The Rydberg states studied were those located in the lower-energy portion of the VUV. Studies of these absorptions are needed since they have received conflicting assignments. In addition to the polarization with respect to the polar axis of the molecule, the change in dipole moment and polarizability can be obtained from these experimental data. Studies of series of similar compounds have been made in order to determine the dependence of these parameters on the structural variations. For the lower-energy Rydberg states of the compounds investigated, the dipole-moment changes are large, with even larger changes in the polarizabilities. The variations in these parameters for the s, p and d Rydberg states are noticeably different.

The use of electric field techniques to investigate other states should be useful for probing variations in properties with respect to different types of changes within the molecules. One such variation that was discussed involved the changes in the excited-state dipole moments with respect to the pressure of suppressor/solvent gases. These solvation studies have provided insight into the interactions of certain highly polarizable states with SF_6 and CF_4.

REFERENCES

1. J. D. Scott and B. R. Russell, J. Chem. Phys. 63, 3243 (1975); J. D. Scott, W. S. Felps, G. L. Findley and S. P. McGlynn, J. Chem. Phys. 68, 4678 (1978).
2. J. D. Scott, B. R. Russell and G. C. Causley, Rev. Sci. Instrum. 48, 264 (1977); J. D. Scott, W. S. Felps and S. P. McGlynn, Nuc. Instrum. and Meth. 152, 231 (1978).
3. G. C. Causley and B. R. Russell, J. Am. Chem. Soc. 101, 5573 (1979).
4. G. C. Causley and B. R. Russell, J. Chem. Phys. 72, 2623 (1980).
5. Daniel D. Altenloh and B. R. Russel, J. Phys. Chem. 86, 1960 (1982).
6. J. R. Lombardi, J. Chem. Phys. 56, 2278 (1972).
7. W. Klemperer, D. E. Freeman and J. R. Lombardi, J. Chem. Phys. 46, 2746 (1967).
8. D. E. Freeman and W. Klemperer, J. Chem. Phys. 45, 52 (1966).
9. Daniel D. Altenloh, G. C. Causley and B. R. Russell, Chem. Phys. Lett. 89, 213 (1982).
10. W. Liptay, in: *Excited States*, Vol. I, E. C. Lim, ed. (Academic Press, New York, 1974), pp. 129-229.
11. H. Labhart, in: *Advances in Chemical Physics*, Vol. XIII, I. Prigogine, ed. (Interscience, New York, 1967), pp. 179-204.
12. Daniel D. Altenloh, Ph. D. Dissertation, (North Texas State University, Denton, Texas, 1982).
13. Daniel D. Altenloh, Linda R. Ashworth and B. R. Russell, J. Phys. Chem., in press.
14. M. B. Robin, *Higher Excited States of Polyatomic Molecules*, Vols. I and II (Academic Press, New York, 1974).
15. G. Lucazeau and C. Sandorfy, J. Mol. Spectros. 35, 214 (1970).
16. L. B. Clark and W. T. Simpson, J. Chem. Phys. 43, 3666 (1965).
17. Daniel D. Altenloh and B. R. Russell, Chem. Phys. Lett. 77, 217 (1981).
18. E. J. McAlduff and K. N. Houk, Can. J. Chem. 55, 318 (1977).
19. J. A. B. Whiteside and P. A. Warsop, J. Mol. Spectrosc. 29, 13 (1969).
20. A. A. Planckaert, J. Doucet and C. Sandorfy, J. Chem. Phys. 60, 4846 (1974).
21. S. P. McGlynn, S. D. Thompson, D. G. Carroll, F. Watson and M. O'Donnell, J. Chem. Phys. 45, 1367 (1966).

22. D. C. Frost, F. G. Herring, A. Katrib, C. A. McDowell and R. A. N. McLean, J. Phys. Chem. 76, 1030 (1972).
23. C. J. F. Bottcher, *Theory of Electric Polarization*, Vol. I (Elsevier, Amsterdam, 1974).
24. A. D. Buckingham and J. A. Pople, Trans. Faraday Soc. 51, 1029 (1955).
25. A. D. Buckingham and J. A. Pople, Trans. Faraday Soc 51, 1173, 1179 (1955).
26. A. D. Buckingham and J. A. Pople, Disc. Faraday Soc. 22, 17 (1956).
27. L. Jansen and P. Mazur, Physica 21, 193, 208 (1955).
28. R. D. Nelson and R. H. Cole, J. Chem. Phys. 54, 4033 (1971).
29. D. McQuarrie and H. Levine, Physica 31, 749 (1965).

VALENCE-SHELL AND RYDBERG TRANSITIONS IN LARGE MOLECULES

C. Sandorfy and L. S. Lussier

Département de Chimie
Université de Montréal
Montréal, Québec H3C 3V1, Canada

1. INTRODUCTION

The Rydberg (R) bands of most polyatomic molecules are located in the vacuum ultraviolet (VUV) spectral region, beyond 200 nm; in many cases, the lowest R bands are found between 200 and 120 nm (50,000-80,000 cm^{-1} or $\sim$6-10 eV). While a few molecular spectra in this energy region have been known for a long time, the systematic conquest of this part of the VUV began only in the 1940's, with the work by W. C. Price and his colleagues [1]. A large number of VUV spectra have been measured in the last twenty years. Robin [2] has provided a comprehensive summary of the state of the VUV art, up to 1975. Shorter reviews are also available [3,4].

The scope of the present article prohibits any recapping of the earlier reviews. Instead, our intent is to provide a succinct account of the remaining difficulties and uncertainties, particularly those pertaining to the question of Rydberg/valence mixing.

Section 2 presents a general account of the problems associated with distinguishing between Rydberg states and valence states in molecules; Section 3 briefly surveys the connections between VUV spectroscopy and photochemistry; and, finally, Section 4 discusses the importance of Rydberg states in the spectroscopy of transition metal complexes.

S. P. McGlynn et al. (eds.), Photophysics and Photochemistry in the Vacuum Ultraviolet, 819–840.

2. THE RYDBERG CONCEPT

For atoms, the definition of an *R* state is simple: If the principal quantum number increases during a transition to a given excited state, that transition, and the resultant excited state, are termed "Rydberg". Thus, the corresponding spectral band is a member of an *R* series and the size of the orbital of the excited electron is large in comparison to the size of the core.

For molecules, however, the definition of *R* states is problematic.

2.1 The Hydrogen Molecule

For simple molecules, one might initially presume that correlation diagrams would offer a basis for a definition. Consider the well-known correlation diagram for homonuclear diatomics, as given by Herzberg [5], and apply it to the hydrogen molecule. Using LCAO-MO language, from the 1s atomic orbitals of the two hydrogen atoms we can build two molecular orbitals: Omitting normalization factors these are, of course, σ_g1s ($1s_a + 1s_b$) and σ_u1s ($1s_a - 1s_b$). Between these two levels a transition is possible, namely $\sigma_u 1s \leftarrow \sigma_g 1s$. Since both the ground and excited states are built from atomically unexcited AO's, this transition is valence-shell (*VS*) and the molecular excited state is a *VS* state. However, the σ_u1s orbital correlates to 2pσ in the united-atom (UA) approximation, and to 2p at the UA limit, with an increase in principal quantum number from 1 to 2. Thus, if we wish to maintain the atomic *R*-state definition, we must call the σ_u1s state "Rydberg", at least for short internuclear separations. This example illustrates well the difficulty inherent in the molecular Rydberg concept: Orbitals and states that are *VS* at normal distances may become *R* at the UA limit, or approximately *R* at short molecular distances. Mulliken calls this process "Rydbergization" [6].

Now, if the above electronic transition were the only one possible, the H_2 spectrum would consist of only two bands, a singlet and a triplet. However, infinitely many other transitions are of course possible (see [5], p. 340). And, therefore, the higher-energy molecular orbitals, which will be truly Rydberg at any molecular distance, must be built from AO's which are atomically excited. The lowest energy one of these higher MO's is the σ_g2s ($2s_a + 2s_b$) orbital, which becomes 2sσ at short molecular distances and 2s at the UA limit. The energy of this orbital may be either higher or lower than that of the 2pσ, depending upon the internuclear separation. Thus, the following question immediately arises: Which comes first, the ($2s_a$ + $2s_b$) orbital, a bonding combination built from higher AO's, or the $(1s_a - 1s_b)$ orbital, an antibonding combination built from lower AO's? The answer follows from the correlation diagram: The ordering, in any given instance, depends upon the geometry of the diatomic molecule in question. Thus, the lowest-energy excited state may be Rydberg, valence-shell or of intermediate type.

2.2 Methane

We now turn our attention to methane, an example of a "simple" polyatomic molecule. Disregarding the carbon 1s $(1a_1)$ orbital, the lowest-energy MO is $2a_1$, which becomes 2s at the UA limit. This orbital has the LCAO form

$$2a_1 = 2s_C + \lambda(1s_{H_1} + 1s_{H_2} + 1s_{H_3} + 1s_{H_4}) \quad . \tag{1}$$

At higher energy, we next find two MO's ($3a_1$ and $1t_2$) which are energetically very close to one another at large C-H distances; however, at normal molecular distances, $1t_2$ is definitely the more stable of the two and is fully occupied in the ground state, while $3a_1$ is empty. ($1t_2$ and $3a_1$ become 2p and 3s, respectively, at the UA limit.) Thus, the lowest-frequency band in the VUV spectrum corresponds to the transition $3a_1 \leftarrow 1t_2$; and, since both $1t_2$ and $3a_1$ are built from atomically unexcited AO's, this can be considered a *VS* transition. At the UA limit, however, the transition becomes $3s \leftarrow 2p$, thus implying that the Rydbergization process is operative and that the upper orbital is best described as being of an intermediate, Rydberg/valence type.

The LCAO function corresponding to $3a_1$ is

$$3a_1 = 2s_C - \lambda'(1s_{H_1} + 1s_{H_2} + 1s_{H_3} + 1s_{H_4}) \quad . \tag{2}$$

This function possesses nodes between the carbon atom and each of the hydrogen atoms and, therefore, at short molecular distances it resembles the UA 3s orbital. We may, however, construct an alternative MO of the same symmetry by taking the positive combination of the carbon 3s orbital with the 1s orbitals of the hydrogens:

$$3'a_1 = 3s_C + \lambda''(1s_{H_1} + 1s_{H_2} + 1s_{H_3} + 1s_{H_4}) \quad . \tag{3}$$

Since this function is essentially $3s_C$, the MO is truly Rydberg; and, since it possesses two radial nodes, it will also become 3s at the UA limit.

The ambiguity implicit in Eqs. (2,3) raises a problem which has been the subject of much controversy (see also Section 2.4, below). One way out of this difficulty follows from an argument given by Schwarz [7]. While the conservation of orbital symmetry is mandatory in an approach to the UA limit, the conservation of the number of radial nodes may not be. Thus, in the isoelectronic series HCl, H_2S, PH_3, SiH_4, the lowest-energy unfilled orbital is *VS* for HCl but *R* for SiH_4, so that the 4s Rydberg orbital of HCl is correlated with the 5s orbital of SiH_4, the UA being Ar for both. Following a procedure inspired by Herzfeld [8], Schwarz showed that when we transform Ar into HCl "... by a repeated transfer of small amounts of positive charge from the heavy nucleus to the point where the proton is located in HCl, ... the molecular potential will change continuously, as will the orbital functions," but "... the nodes of the orbitals may change in number and topology" [7]. Thus, it appears to be possible that $3'a_1$ [Eq. (3)] correlates not to the 3s of the UA but to the 4s. Such a correlation would go some way towards alleviating the ambiguity discussed above. Moreover, it implies that the UA 4s may be of lower energy than the 3s, a result which of course appears strange. However, $3'a_1$ is an "all-plus" combination, while $3a_1$ is antibonding in the C-H bonds; thus, crossing between the resultant potential energy curves must be avoided, thereby giving rise to state mixing.

2.3 Ethylene

We take ethylene as our last example, but consider only the π-electron orbitals for the ground state. (This problem is similar to that for H_2, except that instead of two 1s AO's we now must consider two 2pπ AO's.) Under D_{2h} symmetry, the originating MO is b_{3u} and the LUMO is b_{2g}. (The z-axis is along the C-C bond and the x-axis is perpendicular to the molecular plane.) These are, of course, the well-known $(2p_{x_1} + 2p_{x_2})$ and $(2p_{x_1} - 2p_{x_2})$ MO's, respectively. The $b_{2g} \leftarrow b_{3u}$ transition leads to a *VS* state; however, b_{2g} "Rydbergizes" to a 3dπ UA AO, so again the upper orbital is of mixed Rydberg/valence character. (We have the vertical transition in mind and we disregard the fact that ethylene becomes perpendicular in the (π,π*) state.) This is, of course, the V ← N transition of ethylene.

The lowest energy *R* state of ethylene, however, is not b_{2g} but a_g; it is of 3s type and the corresponding spectral band is well known (see [9] and references therein). While in ethylene itself the *R*(3s) ← N and V ← N bands are in coincidence, in many ethylene derivatives the *R*(3s) ← N is clearly detached and is the singlet-singlet band of lowest energy [10,11]. According to Merer and Mulliken, this $3s(4a_g)$ orbital is "... the same MO as the first CH antibonding MO" [9]. However, the positive combination of the two $3s_C$ AO's also possesses the required characteristics; it is truly Rydberg and may be of lower energy than the $3s(4a_g)$.

From the above discussion, as well as from that of Sections

2.1 and 2.2, it follows that there is a difference between (i), *R* orbitals which are built from atomically unexcited AO's and which become Rydberg only at short molecular distances; and (ii), *R* orbitals which are built from atomically excited AO's and which are Rydberg at any distance. It would probably be better to call the former UA Rydberg, pre-Rydberg [12] or intermediate type orbitals, and to reserve the name Rydberg for the latter. (For further information on this problem, consult [6,13].)

More important than all of the above, however, is that Rydberg and valence-shell orbitals mix more often than not. This, we believe, is still not generally realized, but has far-reaching consequences. For example, Rydberg/valence mixing is almost certainly a major cause of the spectral broadening which occurs at higher frequencies, where Rydberg bands are numerous; it can also be a source of increased collisional broadening. (The sensitivity of *R* bands to intermolecular interactions has been stressed by Evans [14] and Robin [15]. If *VS* bands have partial *R* character, this mixing can serve to increase the sensitivity of these bands to the presence of neighboring molecular fields, which in turn may result in increased bandwidths.) Moreover, any comparison between observed spectra and the results of quantum chemical calculations may be misleading if no account is taken of the existence of *R* states and of Rydberg/valence mixing. And, finally, the photochemical consequences of Rydberg/valence mixing have only recently begun to be explored (see Section 3, below).

2.4 The Problem of Negative Combinations of Rydberg AO's

In 1970, Salahub and Sandorfy [16] proposed a quantum chemical schema in which Rydberg AO's were included in the basis. The results which were obtained [16] might appear naive today, especially after the spectacular developments which occurred during the subsequent decade, developments due mainly to Buenker, Peyerimhoff and their coworkers (see, for example, [17-20]). The earlier method [16] was a semi-empirical, Rydberg-CNDO scheme, which yielded many orbitals and many states. We may illustrate the difficulties associated with this method by an example touched upon in [16]. Consider the five carbon atoms of neopentane, $C(CH_3)_4$. If we wish to form MO's that can be built from, say, 3s AO's centered on the carbon atoms, we must construct linear combinations of all five $3s_C$ AO's. (Other σ-type AO's should also be included but we disregard this complication for the sake of simplicity.) These five $3s_C$ AO's yield, via the SCF procedure, five MO's of the type $\sum_{i=1}^{5} c_i 3s_{C_i}$. And, the question which immediately arises is: Are all five of these MO's physically acceptable? Can we build states from them and expect the related bands to appear in the spectrum? This question has been the subject of some controversy, which we summarize briefly below.

Sinanoglu [12] put it the following way. In conventional LCAO-MO-SCF calculations, in which only unexcited AO's are used, one obtains as many MO's as the number of AO's employed in the basis. This situation might change, however, if some of the AO-AO' overlap integrals approach unity ($S_{ab} \rightarrow 1$) and/or if the overlap matrix has zero determinant. The example of H_2 illustrates this nicely: If the internuclear separation $R \neq 0$ and $S_{ab} < 1$, then $(1s_a + 1s_b)$ and $(1s_a - 1s_b)$ are independent MO's; however, if $S_{ab} \rightarrow 1$ as $R \rightarrow 0$, then $(1s_a + 1s_b) \rightarrow 1s_a(=1s_b)$ and the MO becomes a one-center AO, essentially identical to either of the basis AO's. Furthermore, in the latter case $(1s_a - 1s_b) \rightarrow 0$ as $R \rightarrow 0$: That is, the negative combination disappears at zero internuclear separation!

A way around this impasse may seem to reside in the normalized orbital

$$1\sigma_u = [2(1 - S_{ab})]^{-\frac{1}{2}}(1s_a - 1s_b) \tag{4}$$

where $1\sigma_u \rightarrow 0/0$ as $R \rightarrow 0$; and where we take $2p\sigma_u$ as the UA limit. There are objections to this procedure, however, as Sinanoglu points out [12]: "Taking the limit is not an operation within the linear vector space on which the AO → MO, LCAO theory is based. [$2p\sigma_u$] cannot be expressed by any coefficients at $R \simeq 0$ in terms of the original AO's. One of the AO's has actually disappeared due to the linear dependence. Needing still two independent orbitals, one has simply invoked *from outside of the 2-dimensional linear vector space*, a new one, [$2p\sigma_u$]." The kernel of this argument is that it is not so easy to reach R=0 and, therefore, the UA concept stands in need of reassessment.

The situation is much worse for large Rydberg orbitals, where $S_{ab} \simeq 1$ even at normal molecular distances. As a result, Sinanoglu [12] recommends the elimination of all but the "all-plus" MO for *R* orbitals. This idea is a comforting one for spectroscopists, who normally presume that if all degeneracy is lifted there can be one ns, three np, five nd, etc. *R* series originating from any given ground-state orbital. This holds, of course, for atoms; and, if a molecule can have no states other than those which extrapolate to the UA limit, the assumption is logical for molecules as well.

Mulliken's argument [6] is slightly different from that given above, but arrives at the same conclusion. Since at R=0 the $1\sigma_u$ function goes over to $2p\sigma_u$, at intermediate distances this function can be written as the linear combination

$$1\sigma_u = A2p\sigma_u + B(1s_a - 1s_b) \qquad (5)$$

where $B \to 0$ and $A \to 1$ as $R \to 0$. Mulliken says [6]: "If one looks at the limiting form of the normalized $[1s_a - 1s_b]$ function at R=0, this becomes $1p\sigma$, i.e., $\cos\theta \; e^{-\alpha r}$, but that is unacceptable as an eigenfunction of the Schrödinger equation. In short, it then becomes redundant."

For methane, the $3a_1$ orbital becomes 3s at the UA limit (see Section 2.2, above); so, according to Mulliken [6], at intermediate distances this orbital can be written

$$3a_1 = c_1 3s + c_2(\lambda 2s_C - 1s_{H_1} - 1s_{H_2} - 1s_{H_3} - 1s_{H_4}) \quad . \qquad (6)$$

In the UA limit (i.e., neon), $c_1 = 1$ and $c_2 = 0$. This does not mean that there are two basis orbitals, a 3s orbital and a $(\lambda 2s_C - 1s_{H_1} - 1s_{H_2} - 1s_{H_3} - 1s_{H_4})$ orbital; rather, just as in the case of the $(1s_a - 1s_b)$ orbital of H_2, the second orbital of Eq. (6) - - which corresponds to Eq. (2), above - - becomes redundant at the UA limit and should be eliminated. (This is basically a continuity argument [6].) However, the 3s contribution to Eq. (6) is an orbital of the UA. Hence, there can be no objection to forming an orbital from $3s_C$ and the $1s_{H_i}$, namely $3s_C + \lambda'' (1s_{H_1} + 1s_{H_2} + 1s_{H_3} + 1s_{H_4})$; this is the first truly Rydberg orbital and is identical to our Eq. (3). Since this MO is essentially a carbon 3s orbital, it also becomes 3s at the UA limit. Must we eliminate this MO along with the second term of Eq. (6)? Certainly not, since Eq. (3) is an "all-plus" combination. (Eq. (3) is analogous to the $(2s_a + 2s_b)$ orbital of H_2, except that at the UA limit the latter becomes $2s\sigma$, not $2p\sigma$.) Nevertheless, both Eqs. (2) and (3) possess two radial nodes and seem to correlate to the UA 3s.

The above ambiguity serves to highlight the weakness of the procedure which underlies the construction of correlation diagrams. Both Mulliken and Sinanoglu appear to be right in suggesting that negative AO combinations should be eliminated for Rydberg MO's; however, when the negative combination is valence-shell and the "all-plus" combination is Rydberg, both should exist. This is the case for methane, and for similar molecules, and leads to what Robin has called "*VS*/Rydberg conjugates" [2]. In any particular instance, then, it devolves on calculations to decide which of the two members of a conjugate-pair is lower in energy. (According to Pauzat, Ridard and Levy [21], in methane the *R* orbital has a much lower energy than its *VS* counterpart.) The frustrating point is that it might never be possible to find both related bands in the spectrum: One of them may be at high frequencies, possibly beyond the ionization potential, and may be broad and lost in the background.

For methane, Eqs. (2) and (3) are of the same symmetry and can mix. Such mixing occurs generally for *R* and *VS* orbitals which are close enough in energy, since even if the orbitals are of different symmetry, they can mix vibronically. Thus, after eliminating negative AO combinations for Rydberg MO's, the theoretical treatment of electronic spectra, in LCAO-MO language at least, becomes largely a matter of mixing between *R* and *VS* orbitals. (For more information on this topic, refer to the review by Peyerimhoff [13] and numerous papers from the Buenker-Peyerimhoff group cited therein.)

The ambiguities which arise in the treatment of Rydberg levels are not a result of the molecular orbital approach *per se*, but of the use of LCAO functions and the concomitant reference to the united atom. An interesting attempt to avoid the "LCAO trap" has been made by Roberge and Salahub [22], who applied the Xα method to the ground state and a number of excited states of H_2S. In this method, the space around the molecule is divided into (i), spherical regions surrounding the atoms; (ii), an intersphere region; and (iii), the rest of space, which lies outside of a sphere surrounding the entire molecule. This division (i.e., the so-called "muffin-tin" approximation) suggests a natural measure of Rydberg character for a given orbital as being "the fractional charge for that orbital contained in the outer-sphere region" [22]. Moreover, this measure is independent of the choice of basis set. The solutions of the Xα equations are obtained through the use of Johnson's scattered-wave method [23], in which the radial functions are found independently for each level and are not constructed from a fixed set of basis functions. Roberge and Salahub [22] were able to match the best available configuration interaction calculations [24] as closely as 0.24 eV for the twenty-six lowest-energy excited states of H_2S. Most of these excited states possessed ∿90% *R* character; however, the lowest two, namely $6a_1 \leftarrow 2b_1$ and $3b_2 \leftarrow 2b_1$, possessed *R* characters of 83% and 47%, respectively, implying that the latter should be considered mainly *VS*. No *VS*/Rydberg conjugates were found.

Finally, the Xα method circumvents the uncertainties involved in the use of diffuse Rydberg AO's in an LCAO function. Significant advances can be expected as this method is applied to more complex molecules.

3. FROM SPECTRA TO PHOTOCHEMISTRY

After a period of intensive research on π-electron systems, and a more recent era of investigations on σ-electron systems and Rydberg transitions, we now have a basic understanding of the spectra of typical organic molecules, down to about 125 nm

(80,000 cm^{-1}). Photoelectron spectroscopy [25] and highly sophisticated quantum-chemical techniques [17-20] have been of indispensable help in achieving this understanding. The field has been reviewed several times [2-4]. Our purpose here is to provide a "nutshell" review.

Paraffin hydrocarbons are the only organic molecules which possess only σ-bonding electrons. These molecules have relatively high ionization potentials, which decrease from ∿13 eV to ∿10 eV in the series from methane to n-decane, a behavior not so different from that of the conjugated chains. All of the known excited states of n-paraffins are of intermediate (but closer to Rydberg) type. Except for methane, the n-paraffins have three closely-spaced originating orbitals, one of these being degenerate in ethane. This spacing is probably a major cause of the broadness of the spectral bands of these molecules since most of these bands are almost certainly composite. (Ethane is the only member of the series which exhibits vibrational fine-structure in its spectrum.) The transition leading to the lowest-energy excited states leaves the hole primarily in the C-H bonds, which is likely to be connected with the fact that the low-energy photolysis of n-paraffin yields mainly H_2 molecules. The many intersecting potential energy surfaces, at least a few of which are repulsive, makes the diffuse appearance of these spectra understandable.

Cyclopropane and cyclobutane have spectra about midway between σ- and π-type spectra. Transitions to *R* and *VS* excited states are intermingled among the lowest frequency bands.

Highly-branched paraffins exhibit spectra which are different from those discussed above. Contrary to the n-paraffins, the first band of the highly-branched paraffins is intense and is usually followed by a pronounced minimum. These spectra resemble that of methane, while the normal chain paraffins have spectra similar to that of ethane. (The difference is connected with the simple fact that under tetrahedral symmetry (methane) the first transition is allowed, while under D_{3d} symmetry (ethane) the center of symmetry makes it forbidden.) These two spectral types tend to persist in the more complicated molecules, with the result that the spectra of these species resemble that of either methane or ethane. Moreover, this distinction also exists in the analogous silane derivatives [26]. The lower-energy bands in n-paraffins undergo a regular bathochromic shift with increasing number of carbons; this shift comes to a halt for C_5 or C_6, at about 160 nm. However, highly-branched paraffins absorb at about 10 nm to longer wavelengths. Bands due to singlet-triplet transitions have not yet been identified: According to Buenker and Peyerimhoff [27], these bands must be very close to the corresponding singlets.

The spectra of perfluorocarbons are of C-C σ-type since the orbitals associated with either C-F bonds or F lone-pairs have high energies (beyond $\sim$15.5 eV) as a result of the high electron-attracting power of fluorine. There is some ambiguity with respect to the *R* or *VS* character of the bands in these spectra. Contrary to hydrocarbons, the bathochromic shift for the perfluorocarbons does not come to a halt at C_5 or C_6.

Saturated heteroatomic molecules, such as substituted halides, alcohols and amines, possess lone-pairs of electrons; in fact, the orbitals occupied by the lone-pair electrons are the originating orbitals for the photoelectron band of lowest energy, and for all of the known UV absorption bands. For halides (X=Cl, Br, I), the first excited state is *VS* (a C-X antibonding state); for alcohols, ethers and amines, however, it is of intermediate, Rydberg/valence type. The first *R*(3s) and *R*(3p) bands are usually well distinguished for these compounds. Alcohols present an interesting case, however, since the *R*(3p) manifold is largely split because of the nonlinearity of the C-O-H unit. Saturated molecules containing two or more heteroatoms have not been as extensively studied as have their π-electron counterparts.

In mono-olefins the (π,π*) band coincides with, or is preceded by, the 3s Rydberg band. In ethylene itself, the Rydberg band is superimposed on the low-frequency wing of the (π,π*) band, while in highly methylated or fluorinated olefins *R*(3s) becomes the first spectral band and is well separated from the (π,π*) band. Ethylene has served as a primary test-case for both theory and experiment since it is planar in its ground state, perpendicular in the singlet and triplet (π,π*) states, and has an angle of 25° between the two CH_2 planes in the 3s Rydberg state [9]. The protracted long-wavelength tail of the singlet (π,π*) band has been the topic of much discussion: McDiarmid [28] has shown that the longest-wavelength portion of it is due to traces of molecular oxygen; even so, the tail is quite long, thus demonstrating the great structural difference between the ground and excited states. The vibrational intervals in this tail correspond to the C-C stretching mode and to double quanta of the torsional mode.

Butadiene, and the further series of conjugated polyenes, are at present the target of intensive research. One reason for this interest is the relevance of these studies to the functioning of visual pigments and carotenoids. Butadiene can in principle exist in either the cis or trans configuration, with C_{2v} or

C_{2h} symmetry, respectively. Of course, the selection rules are more liberal for C_{2v}, which does not contain a center of symmetry, than for C_{2h}, which does. For this reason, two of the four (π,π^*) bands of trans-butadiene, which is the isomer encountered in practice, are forbidden by the Laporte rule. These bands can be observed, however, by two-photon spectroscopy, in which the selection rules permit $g \leftrightarrow g$ transitions. Furthermore, these two states can interact configurationally, at least in a mathematical sense, and this might push one of them, the A_g band, to lower frequencies. That an A_g^- band lies at energies lower than the allowed B_u band is an exciting possibility, one first put forward by Koutecky in 1967 [29]. For the last ten years, several attempts have been made to locate these bands in a number of conjugated polyenes [30-32]. The problem is rendered difficult, however, by the existence of the weak 3s Rydberg bands, which are also $g \leftrightarrow g$. Recently, McDiarmid and Doering [33] investigated the problem of the A_g bands of butadiene by electron-impact spectroscopy and found no such bands at frequencies lower than the B_u bands. As for the higher polyenes, Birge, *et al.* [34] found an A_g^- band in all-trans retinol by two-photon spectroscopy; its maximum is approximately 1600 cm^{-1} from the one-photon absorption maximum, to the long wavelength side.

The chromophore of visual pigments (rhodopsin) is the Schiff base of 11-cis retinal, which is covalently linked to the protein opsin on the nitrogen and is placed in the field of polar groups of the amino acid residues of the protein. Knowledge of the excited state in which the primary photochemical step of vision occurs is, of course, of utmost importance. There is general agreement at present that this excited state is the lowest-energy singlet (π,π^*) state [35-37]. If there were an A_g^- state at frequencies lower than the (π,π^*) state, all of our ideas about the mechanism of vision would have to be modified. Such, however, does not appear to be the case. Quantum chemical calculations show that the presence of a nitrogen atom in the polyene moves the A_g^- band to higher, not lower, frequencies; moreover, this shift is even more pronounced for a protonated Schiff base [38].

Finally, we must mention that theoreticians have been greatly concerned with butadiene for the past few years. Among the primary problems considered are the geometry changes which occur upon excitation, and the extent to which Rydberg/valence mixing is important in butadiene excited states [39].

The π-electron spectra of benzene and other aromatic compounds are among the most studied ones in molecular spectroscopy. In general, three out of the four possible (π,π^*) bands occur before the Rydberg bands are encountered. For benzene itself, the *R*(ns) bands (which are $g \leftrightarrow g$) have not been observed except, perhaps, for the *R*(3s). The *R*(np) series is known and a few (d or f) series with very low quantum defects are also known. All of these series converge to the first π ionization potential at 9.247 eV. A wealth of superexcited states are then found at higher energies.

Most organic molecules possess bonding σ, bonding π and lone-pair (n) electrons. There are often two originating orbitals, an n and a π, the relative ordering of which is different for different cases. Simple aldehydes and ketones exhibit only the well-known (n,π*) band in the near UV. This is followed, in the VUV, by three *R* bands: 3s, 3p and 3d. The (π,π*) band is thus at very high frequencies; in the case of formaldehyde, it is superexcited. Non-aromatic Schiff bases have spectra intermediate between those of olefins and carbonyl compounds. Because of the two relatively-close originating orbitals, and the many possibilities for crossing among potential energy surfaces, these spectra usually consist of broad bands; moreover, the *R* bands are difficult to locate. In short: Olefins have π spectra, aldehydes and ketones have lone-pair spectra, Schiff bases have π and lone-pair spectra [4].

In the spectra of amides the bands are, in order of increasing energy: (n,π*), *R*, (π,π*), *R*, (π,π*). This is typical of amides, acid halides, carboxylic acids, esters, etc. In these molecules there are three π levels instead of the two levels found in ketones and aldehydes; this puts the n level close to the second π level, pushes the (n,π*) band to higher frequencies and gives rise to close-lying n and π Rydberg bands.

In sum, if a molecule contains n and/or π electrons, transitions from these orbitals dominate even in the VUV, from 200 to 130 nm. Bands due to σ-electron transitions can only be observed when no other electrons are present: otherwise, these transitions readily move to higher energies. These σ electrons are strongly bound and do not easily come out into the limelight! Moreover, these σ excitations often cause dissociation. (Not always, however; remember the vibrational fine structure of ethane or cyclopropane.)

As a final point, we wish to say a few words about the photochemical consequences of the Rydberg/valence-shell distinction. The idea that primary photochemical steps should be different for photolytic reactions occurring via *R* states than for those occurring via *VS* excited states was put forward by Sandorfy in 1975 and subsequently elaborated in several papers [40-43]. Previously, in 1973, Kropp pointed out that in order to interpret the photochemical reactions of olefins, both *VS* and *R* excited states must be invoked [44,45].

Since most known photochemical reactions occur in the lowest-energy excited states, for organic molecules the *R*(3s) state is expected to be the most important Rydberg state photochemically. Let us first suppose that this is a *true* Rydberg state (i.e., one which remains Rydberg at any internuclear separation). Unlike the higher ns orbitals, 3s is not extremely large. Moreover, it possesses two radial nodes, one of which will lie, in molecules such as methane or ethane, between the carbon and the hydrogens. This is perhaps an important point: When referring to the bonding or antibonding character of orbital wavefunctions, one usually has the angular part of the function in mind. If, however, higher AO's (2s, 3s, 3p, ...) are involved, the radial nodes come into play; these nodes might cause antibonding character as well as the angular nodes. On the other hand, since ns orbitals are totally symmetric, the 3s orbital will provide some bonding between atoms which are not "chemically bound" in the usual sense. In other words, the 3s orbital will act as a weak one-electron bond.

In terms of Koopmans' theorem, then, it follows that if one excites an electron to a 3s orbital (or ionizes the molecule) from an orbital which is bonding, with respect to a given bond, that bond is likely to break in the excited state, the main reason being the loss of bonding character, and a lesser reason being that the 3s orbital contributes some antibonding character because of its radial nodes. (Rydberg orbitals are typically nonbonding rather than antibonding but the most important 3s orbitals are, to some extent, an exception to this.) Furthermore, the 3s orbital will bring together symmetrically-related atoms which are not chemically connected. Therefore, one may expect that the preferred primary process for photolysis from an *R*(3s) state will be molecular elimination rather than simple bond cleavage. An example of this is ethane, or any other normal paraffin, for which the most important primary process is elimination of H_2 molecules, *not* 2H [46,47].

If, on the other hand, the orbital of the excited electron is a *VS*-type orbital, which is typically antibonding, the expected primary process is simple bond cleavage. Typical examples are the alkyl chlorides, bromides, and iodides. The first excited state of these molecules is due to a (C-Cl)* ←n, etc. transition, where the (C-Cl)* orbital is antibonding; and, all of these molecules produce halogen atoms in the primary photolytic step.

There is little point in citing more examples. These ideas are possibly too simple to be applied to individual problems; we believe, however, that they contain significant elements of truth. The importance of radial nodes in producing antibonding character, and the ability of Rydberg orbitals to establish one-electron bonds, should not be dismissed lightly.

Such simple reasoning can at most apply to the vertical excited state. (Recall the use of static indices in the early days of quantum chemistry.) Naturally, in order to interpret the course of a photochemical reaction we must examine the potential energy surfaces and the evolution of these surfaces with time. Every reaction must be considered in its own right and under its own particular conditions. (In this regard, refer to the recent papers of Evleth and coworkers [48-50] and of Malrieu [51].) With respect to experimental photochemistry, results obtained by Kropp [44,45], von Sonntag and Schuchmann [52,53], Collin [54,55], Tschouikow-Roux [56], Wijnen [57] and others have shown that while these problems are complicated, the Rydberg/valence-shell distinction is indeed important for any understanding of photochemical reactions.

4. RYDBERG STATES OF TRANSITION METAL COMPLEXES

There is an immense literature on transition metal complexes (see, for example [58-61]). Our purpose here is to simply state the present position and then to see what one might expect to find in the VUV. We choose Cr(III) hexafluoroacetylacetonate (HFAA) as a typical example. HFAA complexes are sufficiently volatile to be studied in the gas phase; their photoelectron spectra have been determined by Evans, *et al.* [62] and by Furlani and Cauletti [63]. The gas-phase VUV spectra of HFAA complexes of Al(III), Sc(III), V(III), Cr(III), Fe(III), Mn(III) and Cu(III) have been measured (Lussier and Sandorfy, unpublished). From the point of view of the present discussion, all of these spectra yield similar results. The VUV absorption spectrum of Cr(III) $(HFAA)_3$ is shown in Fig. 1.

Under octahedral symmetry (O_h), the degeneracy of the five 3d atomic orbitals of Cr is partly lifted and two orbitals are obtained: The one of lower energy is triply-degenerate (t_{2g}), while the one of higher energy is doubly-degenerate (e_g). (These are also referred to as dε and dγ, respectively.) According to literature data (cf. [62]), the actual symmetry of tris-hexafluoroacetylacetonates is only D_3 in the crystal phase, and this is likely to be so in the gas phase as well. Under D_3 symmetry, t_{2g} becomes $a_1 + e_a$ while e_g remains degenerate and is called e_b:

Free Ion	O_h	D_3
$3d^5$	e_g	e_b
	t_{2g}	$a_1 + e_a$

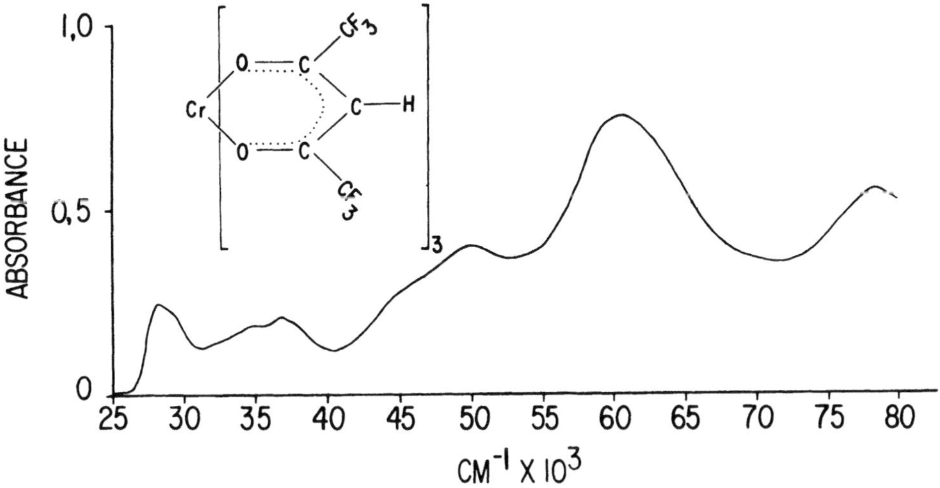

Fig. 1. VUV absorption spectrum of Cr(III) $HFAA_3$.

It is instructive, however, to consider both symmetries.

Cr(III) has three 3d electrons which half fill the t_{2g} orbital. From this configuration (d^3) of three equivalent d electrons, several states (all of which are gerade) result, namely, ^2P, $^2D(2)$, ^2F, ^2G, ^2H, ^4P and ^4F. (The angular momentum quantum number L may take any integer value from 1 to 5.) Among these many states, according to Hund's rules the ^4F term has the lowest energy. Under O_h symmetry, ^4F splits into three components, $^4A_{2g}$, $^4T_{2g}$ and $^4T_{1g}$, of which $^4A_{2g}$ is the lowest. This yields two transitions ($^4T_{2g} \leftarrow {}^4A_{2g}$ and $^4T_{1g} \leftarrow {}^4A_{2g}$, with the former being lower in energy than the latter) which correspond to the well-known characteristic visible bands of the Cr(III) complexes. These bands are very weak, being $g \leftrightarrow g$. Under D_3 symmetry, A_{2g} becomes A_2, T_{1g} becomes $A_2 + E$, and T_{2g} becomes $A_1 + E$. Then, from the A_2 ground state, transitions to A_1 and E will be allowed but transition to another A_2 state will be forbidden. (D_3, of course, does not possess a center of symmetry.) The bands will still be weak, however, since the departure from O_h symmetry is not very great.

The next higher state corresponds to the ^4P configuration of the free ion, which becomes T_{1g} under O_h symmetry and $A_2 + E$ under D_3 symmetry. This band is often covered by the much more intense charge-transfer bands, but for $Cr(III)(H_2O)_6$ it is found at about 38,000 cm^{-1}, 14,300 cm^{-1} higher than the second of the two ^4F bands. The bands corresponding to the doublet configurations are, of course, even weaker since they are not only symmetry but also spin forbidden. (For the heavy Cr(III) ion, however, this rule is only approximately valid.) Thus, the contribution of these bands to the overall appearance of the spectrum is expected to be negligible.

The HFAA ligand has a π orbital as its HOMO, π_3. The n_- and n_+ orbitals, which are combinations of the two oxygen lone-pair AO's, follow at higher energies. There are two empty π^* levels, π_4 and π_5. Under C_{2v} symmetry, the π_3 and n_- orbitals are, respectively, b_2 and b_1. With three ligands, under O_h symmetry, both π_3 and n_- are t_{1g}; and, under D_3 symmetry these orbitals transform as a_2 + e and are only slightly split. Mixing between the metal and ligand orbitals has been examined by previous authors [62] and we shall not discuss it here.

Evans, *et al.* [62] obtained 10.74 and 11.25 eV, respectively, for the π_3 and n_- Koopmans' ionization potentials of the free ligand. For Cr(III)$(HFAA)_3$, the first photoelectron band at 9.57 eV comes from the 3d ionization ($^4A_2$, top orbitals a_1e_a), followed by π_3 ionization at 10.18 eV and n_-/n_+ ionization at 11.10 and 11.61 eV. Thus, metal ionization precedes ligand ionization (approximate language) but the difference is only 0.61 eV.

The UV spectrum of HFAA has been published by Nakanishi, Morita and Nagakura [64]. They assigned the two intense bands at 37,400 and 63,700 cm^{-1} (with oscillator strengths of 0.20 and 0.30, respectively) to (π,π^*) transitions, but the second band is doubled and contains an (n,σ^*) band. A shoulder at about 57,300 cm^{-1} was attributed to a (π,σ^*) transition. The band at 70,400 cm^{-1} (f=0.15) was assigned to another (n,σ^*) transition.

We are now ready to have a look at the VUV spectrum of Cr(III)$(HFAA)_3$ (cf. Fig. 1). The two visible d-d bands, typical of Cr(III) complexes are not shown. Towards higher frequencies, at 29,200 cm^{-1} we find an intense band which is known to be a metal-to-ligand charge-transfer band, $\pi_4 \leftarrow 3d$ [65-68]. This band may split under D_3 symmetry, which would explain the observed shoulder.

At 35,000-36,750 cm^{-1} there are two intense, mutually overlapping bands. This is quite certainly the first ligand (π,π^*) band, $\pi_4 \leftarrow \pi_3$, and it is not dramatically displaced from its position in the free ligand. The second (π,π^*) band [or, rather, mixed (π,π^*) and (n,σ^*)] is also readily identified at 60,500 cm^{-1}. Both bands are evidently composite.

In the 44,000-46,000 cm^{-1} region there is a broad shoulder which can be assigned to a charge-transfer transition, $3d\gamma \leftarrow n$ (i.e., ligand to metal) [59,65,68]. The band at $\sim$50,000 cm^{-1} may contain still other charge-transfer bands. In fact, several other charge-transfer bands can be deduced from the orbital scheme of the complex [59]. These transitions, as well as those to σ^* orbitals, are often invoked in a general way. One wonders, however, if purely valence-shell considerations provide an adequate language to interpret the bands found at higher frequencies in the spectrum.

The bands shown in Fig. 1 are quite intense. Although we could not determine the molar absorption coefficients with precision, we did estimate their values from vapor/solution comparisons, at least in that part of the spectrum in which both vapor and solution spectra can be obtained. Thus, ε is of the order of 10,000 for the 44,000-46,000 cm^{-1} region; for the band at 50,000 cm^{-1} ε is about 20,000; and, for the broad band at 60,500 cm^{-1} it is about 40,000. Under these circumstances, the many forbidden d-d bands which were mentioned above cannot make any significant contribution.

Beyond the valence-shell bands discussed above, Rydberg bands must exist. The question is: May we expect any intense, allowed bands among them? To answer this question, we first look at the Rydberg states which are accessible by one-electron promotion from the d^3 ground state.

The d^2s configuration (first member: $3d^24s$)

The terms are:

$$^2S,\ ^2D,\ ^2G,\ ^4P,\ ^2P,\ ^4F,\ ^2F.$$

All of these terms are gerade. The lowest term is ^4F; under O_h symmetry ^4F splits into $^4A_{2g}$ + $^4T_{2g}$ + $^4T_{1g}$. All transitions from the $^4A_{2g}$ ground state to the ^4F components are forbidden by the Laporte rule; even if the actual symmetry is only D_3, we expect them to be very weak. The next state, ^4P, is $^4T_{1g}$ under O_h symmetry and, therefore, a transition to this state is also forbidden. Transitions to the doublet terms are even more strongly forbidden.

The d^2p configuration (first member: $3d^24p$)

This configuration gives a different picture. The terms are:

$$^2P,\ ^2F,\ ^2D,\ ^2P,\ ^2H,\ ^2G,\ ^2F,$$
$$^4D,\ ^4P,\ ^4S,\ ^2D,\ ^2P,\ ^2S,$$
$$^4G,\ ^4F,\ ^4D,\ ^2G,\ ^2F,\ ^2D.$$

All of these terms are ungerade. The lowest term is ^4G; under O_h symmetry ^4G splits into $^4A_{1u}$ + ^4E_u + $^4T_{1u}$ + $^4T_{2u}$. Transition from the $^4A_{2g}$ ground state to $^4T_{2u}$ is allowed, whereas the other components are symmetry forbidden. The ^4F and ^4D terms also yield allowed $^4T_{2u}$ components, but ^4P and ^4S do not: Under O_h symmetry ^4F becomes $^4A_{2u}$ + $^4T_{2u}$ + $^4T_{1u}$; ^4D becomes ^4E_u + $^4T_{2u}$; ^4P becomes $^4T_{1u}$; and ^4S becomes $^4A_{1u}$. Therefore, among the $3d^24p$ terms there are four allowed transitions (note that there are two ^4D terms) even under O_h symmetry. These transitions are both symmetry and spin allowed and are expected to be intense. They are also atomically allowed (i.e., $\Delta L = 0,\pm 1$).

The d^2d configuration (first member: $3d^24d$)

All terms arising from this configuration are $g \leftrightarrow g$ and, therefore, are forbidden.

The d^2f configuration (first member: $3d^24f$)

The terms are:

$$^2F,\ ^2H,\ ^2G,\ ^2F,\ ^2D,\ ^2P,\ ^2J,\ ^2I,\ ^2H,\ ^2G,\ ^2F,\ ^2D,\ ^2P,$$
$$^4G,\ ^4F,\ ^4D,\ ^2G,\ ^2F,\ ^2D,$$
$$^4I,\ ^4H,\ ^4G,\ ^4F,\ ^4D,\ ^4P,\ ^4S,$$
$$^2I,\ ^2H,\ ^2G,\ ^2F,\ ^2D,\ ^2P,\ ^2S.$$

All of these terms are ungerade. All of the quadruplet terms, except for ^4P and ^4S, give rise to allowed transitions, for a total of eight allowed nf series.

How can we locate these states? The Rydberg terms values for Cr(III) are not known. However, a look at the ionization potentials is of interest [69,70]:

Cr(III)	49.6 eV	(400,000 cm^{-1})
Cr(II)	30.95 eV	(249,700 cm^{-1})
Cr(I)	16.48 eV	(133,060 cm^{-1})
Cr	6.76 eV	(54,565 cm^{-1})

The ionization potential of the complex is 9.57 eV (77,190 cm^{-1}). Now, the electron configuration of these different ions of Cr, and the very high number of states that are issued from them, preclude any meaningful comparison. Nevertheless, it is tempting to guess from these data that the net positive charge around the central ion in the complex is much less than +1, perhaps +0.3 or so. Furthermore, IP = 9.57 eV is the same order of magnitude as those of organic molecules. The first s and p terms for most organic molecules are found at 27,000-22,000 cm^{-1} and 20,000-16,000 cm^{-1}, respectively. Given this, and the partial positive charge, one would expect strongly-penetrating Rydberg orbitals, with terms of the order of, perhaps, 35,000 cm^{-1} for the s and 30,000 cm^{-1} for the p. This would take us, for the first p-type Rydberg bands, to ~47,000 cm^{-1}. Then, four allowed p-type Rydberg terms might well be in the 42,000-52,000 cm^{-1} region where we actually see absorption which is neither ligand-ligand nor valence-shell d-d absorption. Instead of, or in addition to, invoking possible (but higher-energy) charge-transfer transitions, it is pleasant to propose that allowed Rydberg transitions contribute a significant part of the absorption intensity in this part of the spectrum.

There must also be ligand-to-Rydberg transitions. Taking π_3 as the originating orbital, transitions (O_h symmetry) to s-type Rydberg orbitals will be $a_{1g} \leftarrow t_{1g}$ (i.e., ${}^4T_{2g} \leftarrow {}^4A_{2g}$) which is forbidden; transitions to p-type Rydberg orbitals will be $t_{1u} \leftarrow t_{1g}$ (i.e., ${}^4A_{2u} + {}^4E_u + {}^4T_{1u} + {}^4T_{2u} \leftarrow {}^4A_{2g}$), with the ${}^4T_{2u}$ component being allowed. In this case, one might safely use the term values known from "organic" spectroscopy. These data, together with the second IP (10.18 eV = 82,100 cm^{-1}), predict p bands near 62,000-64,000 cm^{-1}. And, these bands may well contribute to the very broad (π,π^*) band which is found in this energy range. d-type Rydberg bands are again forbidden under O_h symmetry; however, both s and d might gain oscillator strength under D_3 symmetry.

For both metal-to-Rydberg and ligand-to-Rydberg transitions, the extent to which the Rydberg orbitals cover both the metal and the ligands seems to be a problem for the future. In addition, many of these transitions may possess some charge-transfer character.

To conclude, we wish to state only the following: Transition metal complexes must have allowed Rydberg bands; and, a search for these bands constitutes a worthwhile target for future research.

REFERENCES

1. W. C. Price, Advan. Spectrosc. 1, 56 (1959).
2. M. B. Robin, *Higher Excited States of Polyatomic Molecules*, Vols. I and II (Academic Press, New York, 1974, 1975).
3. C. Sandorfy, J. Mol. Struct. 19, 183 (1973).
4. C. Sandorfy, in: *Topics in Current Chemistry*, Vol. 86 (Springer-Verlag, Berlin, 1979), pp. 91-138.
5. G. Herzberg, *Molecular Spectra and Molecular Structure*. Vol. I. *Spectra of Diatomic Molecules* (Van Nostrand, Princeton, 1950).
6. R. S. Mulliken, Accts. Chem. Res. 9, 7 (1976).
7. W. H. E. Schwarz, Chem. Phys. 9, 157; 11, 217 (1975).

8. K. F. Herzfeld, Z. Naturforsch. 3a, 457 (1948).
9. A. J. Merer and R. S. Mulliken, Chem. Rev. 69, 639 (1969).
10. M. B. Robin, R. R. Hart and N. A. Kuebler, J. Chem. Phys. 44, 1803 (1966).
11. G. Belanger and C. Sandorfy, J. Chem. Phys. 55, 2055 (1971).
12. O. Sinanoglu, in: *Chemical Spectroscopy and Photochemistry in the Vacuum Ultraviolet*, C. Sandorfy, P. J. Ausloos and M. B. Robin, eds. (D. Reidel, Dordrecht, Holland, 1974), p. 337.
13. S. D. Peyerimhoff, Gazz. Chim. Ital. 108, 411 (1978).
14. D. F. Evans, Proc. Chem. Soc. London, 378 (1963).
15. M. B. Robin and N. A. Kuebler, J. Mol. Spectrosc. 33, 247 (1970).
16. D. R. Salahub and C. Sandorfy, Theoret. Chim. Acta 20, 227 (1971).
17. S. D. Peyerimhoff and R. J. Buenker, Adv. Quantum Chem. 9, 69 (1975).
18. R. J. Buenker and S. D. Peyerimhoff, Theoret. Chim. Acta 39, 217 (1975).
19. R. J. Buenker and S. D. Peyerimhoff, Chem. Rev. 74, 127 (1974).
20. R. J. Buenker and S. D. Peyerimhoff, in: *Excited States in Quantum Chemistry*, C. A. Nicolaides and D. R. Beck, eds. (D. Reidel, Dordrecht, Holland, 1979), pp. 45 79, 403.
21. F. Pauzat, J. Ridard and B. Levy, Mol. Phys. 23, 1163 (1972).
22. R. Roberge and D. R. Salahub, J. Chem. Phys. 70, 1177 (1979).
23. K. H. Johnson, Adv. Quantum Chem. 7, 143 (1973).
24. S. Shih, S. D. Peyerimhoff and R. J. Buenker, Chem. Phys. 17, 391 (1976).
25. D. W. Turner, C. Baker, A. D. Baker and C. R. Brundle, *Molecular Photoelectron Spectroscopy*, (Wiley-Interscience, New York, 1970).
26. R. Roberge, C. Sandorfy, J. I. Matthews and O. P. Strausz, J. Chem. Phys. 69, 5105 (1978).
27. R. J. Buenker and S. D. Peyerimhoff, in: *Chemical Spectroscopy and Photochemistry in the Vacuum Ultraviolet*, C. Sandorfy, P. J. Ausloos and M. B. Robin, eds., (D. Reidel, Dordrecht, Holland, 1974), p. 257.
28. R. McDiarmid, J. Chem. Phys. 55, 4669 (1971).
29. J. Koutecky, J. Chem. Phys. 47, 1501 (1967).
30. B. S. Hudson and B. E. Kohler, Chem. Phys. Lett. 14, 299 (1972).
31. K. Schulten and M. Karplus, Chem. Phys. Lett. 14, 305 (1972).
32. B. Hudson and B. E. Kohler, Ann. Rev. Phys. Chem. 25, 437 (1974).

33. R. McDiarmid and J. P. Doering, Chem. Phys. Lett. 88, 602 (1982).
34. R. R. Birge, J. A. Bennett, B. M. Pierce and T. M. Thomas, J. Am. Chem. Soc. 100, 1533 (1978).
35. B. Honig, Ann. Rev. Phys. Chem. 29, 31 (1978).
36. B. E. Kohler, Biophys. Struct. Mechanism 3, 101 (1977).
37. C. Sandorfy, J. Photochem. 17, 297 (1981).
38. J. -M. Leclercq, to be published.
39. R. J. Buenker, S. Shih and S. D. Peyerimhoff, Chem. Phys. Lett. 44, 385 (1976).
40. C. Sandorfy, Rev. React. Species Chem. React. 1, 159 (1975).
41. C. Sandorfy, Z. Phys. Chem. 101, 307 (1976).
42. C. Sandorfy, in: *Applications of MO Theory in Organic Chemistry*, I. G. Csizmadia, ed. (Elsevier, Amsterdam, 1977), p. 384.
43. C. Sandorfy, Int. J. Quantum Chem. 19, 1147 (1981).
44. P. J. Kropp, E. J. Reardon, Z. L. F. Gaibel, K. F. Williard and J. H. Hattaway, J. Am. Chem. Soc. 95, 7058 (1973).
45. T. R. Fields and P. J. Kropp, J. Am. Chem. Soc. 96, 7559 (1974).
46. H. Okabe and J. R. McNesby, J. Chem. Phys. 34, 668 (1961).
47. S. G. Lias, G. J. Collin, R. E. Rebbert and P. J. Ausloos, J. Chem. Phys. 52, 1841 (1970).
48. E. Kassab, J. T. Gleghorn and E. M. Evleth, Chem. Phys. Lett. 70, 151 (1980).
49. E. Kassab, E. M. Evleth, J. J. Dannenberg and J. C. Rayez, Chem. Phys. 52, 151 (1980).
50. E. M. Evleth and E. Kassab, J. Am. Chem. Soc., 100, 7859 (1978).
51. J. P. Malrieu, Theoret. Chim. Acta 59, 251 (1981).
52. C. von Sonntag and H. P. Schuchmann, *The Chemistry of Functional Groups* (Suppl. E), S. Patai, ed. (Wiley, New York, 1980).
53. C. von Sonntag and H. P. Schuchmann, in: *Advances in Photochemistry*, Vol. 10 (Wiley-interscience, New York, 1977).
54. G. J. Collin, H. Deslauriers and A. Wieckowski, J. Phys. Chem. 85, 944 (1981).
55. G. J. Collin, Rev. Chem. Int. 2, 377 (1979).
56. T. Yano and E. Tschouikow-Roux, J. Chem. Phys. 72, 3401 (1980).
57. R. Ausubel and M. H. J. Wijnen, J. Photochem. 4, 241 (1975).
58. H. L. Schlafer and G. Gliemann, *Basic Principles of Ligand Field Theory* (Wiley-Interscience, New York, 1969).
59. F. A. Cotton and G. Wilkinson, *Advanced Inorganic Chemistry*, 3rd Ed. (Wiley-Interscience, New York, 1972).

60. C. J. Ballhausen, *Introduction to Ligand Field Theory* (McGraw-Hill, New York, 1962).
61. C. K. Jorgensen, *Absorption Spectra and Chemical Bonding in Complexes* (Pergamon Press, Oxford, 1962).
62. S. Evans, A. Hamnett, A. F. Orchard and D. R. Lloyd, Disc. Faraday Soc. 54, 227 (1972).
63. C. Furlani and C. Cauletti, in: *Structure and Bonding*, Vol. 35 (Springer-Verlag, Berlin, 1978), p. 119.
64. H. Nakanishi, H. Morita and S. Nagakura, Bull. Chem. Soc. Japan 51, 1723 (1978).
65. R. H. Holm and F. A. Cotton, J. Am. Chem. Soc. 80, 5658 (1958).
66. D. W. Barnum, J. Inorg. Nucl. Chem. 21, 221 (1961).
67. D. W. Barnum, J. Inorg. Nucl. Chem. 22, 183 (1961).
68. I. Hanazaki, F. Hanazaki and S. Nagakura, J. Chem. Phys, 49, 276 (1968).
69. C. E. Moore, *Atomic Energy Levels*, Vols. 1 and 2 (Natl. Bur. Stand. (US) Spec. Pub. No. 35, 1971).
70. The atomic notation used here is that commonly found in the chemistry literature, namely, Cr(I), Cr(II),... correspond to Cr^{+1}, Cr^{+2},... . In the notation common to the physics literature, Cr(I), Cr(II),... correspond to Cr, Cr^{+1},... . This latter notation is adopted in the following paper and care should be taken to avoid confusion. The Editors.

MULTIPHOTON SPECTROSCOPY AND PHOTOCHEMISTRY*

M. B. Robin

Bell Laboratories
Murray Hill, New Jersey 07974, USA

1. INTRODUCTION

UV/VIS laser multiphoton ionization (MPI) of gas-phase polyatomic molecules is rapidly becoming a popular tool in the hands of the molecular spectroscopist. In this technique, the photocurrent flowing within a pulse-irradiated sample is monitored as the wavelength of the laser is scanned; photocurrent peaks arise whenever the absorption of an integral number of photons results in a resonance excitation in the target molecule [1]. A typical experimental setup for MPI spectroscopy is shown in Fig. 1. In the simplest experiment, 5 nsec pulses from a dye laser, pumped at $\sim$10 Hz, are focused with a 15 cm lens to a 100 μm spot between a pair of flat-plate electrodes within the sample cell. With 50 μJ pulses from the dye laser, peak powers of $\sim$100 MW/cm^2 are generated at the focus and, with a bias of 200-800 V on the plates, DC currents of $\sim 10^{-12}$A can be read with an electrometer or picoammeter. Sample gas is typically at 1-50 torr pressure in such static experiments.

How can a pulse of visible photons ionize a molecule possessing an ionization potential perhaps three times larger than the energy of one such photon? In general, at some unspecified frequency ν in the UV/VIS, the laser pulse will drive a molecule from the ground state Ψ_0 to an ionized state Ψ_f via a three-photon coherent absorption. This absorption proceeds through vir-

*This paper is a condensed version of A. Gedanken, M. B. Robin and N. A. Kuebler, "Nonlinear Photochemistry in Organic, Inorganic and Organometallic Systems," J. Phys. Chem. 86, 4096 (1982).

S. P. McGlynn et al. (eds.), Photophysics and Photochemistry in the Vacuum Ultraviolet, 841–853.

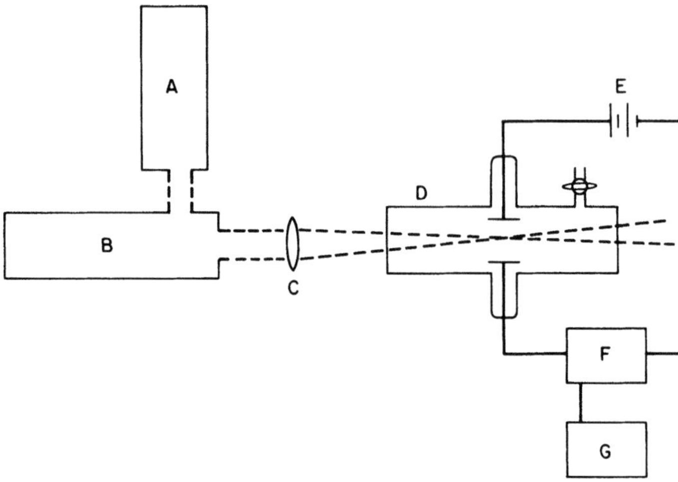

Fig. 1. Experimental setup for MPI spectroscopy. A, pump laser; B, dye laser; C, focusing lens; D, glass ionization cell with quartz windows; E, regulated voltage supply; F, electrometer; G, recorder.

tual levels Ψ_j and Ψ_k with a rate proportional to

$$I^3 \left| \frac{\sum\sum_{jk} \langle\Psi_0|er|\Psi_j\rangle\langle\Psi_j|er|\Psi_k\rangle\langle\Psi_k|er|\Psi_f\rangle}{(\Delta E_{0j}-h\nu)(\Delta E_{0k}-2h\nu)} \right|^2$$

where I is the instantaneous optical power, and where ΔE_{0j} is the energy separation between stationary levels Ψ_0 and Ψ_j. The factor given above is readily generalized to the case of n-photon absorption.

If the laser frequency is scanned, an m-photon resonance will eventually be achieved; ΔE_{0m} - $mh\nu$ then goes to zero and the overall ionization rate increases by several orders of magnitude. Thus, on scanning, an ionization spectrum results in which the peaks correspond to m-photon resonances in the absorption system. The I^n dependence of the ionization rate, with n>m and, typically, n=3 or 4, makes pulsed-laser ionization the only practical mode since, in this case, I is 10^{27} photons/cm^2 sec; whereas, for a CW source of the same bandwidth, I is only 10^{16} photons/cm^2 sec at best. The 10^{-7} duty cycle of the pulsed source is more than compensated for by the far larger value of I and the extremely nonlinear dependence of the ionization rate on I.

The MPI technique can be used to achieve several ends. In the hands of the molecular spectroscopist, this simple technique has revealed many two-photon excitations in high-symmetry molecules, excitations which are otherwise unobservable in one-photon absorption spectra. Moreover, the technique uncovers resonances in the VUV region (below 2000 Å) without the cumbersome VUV hardware normally required; and, the symmetry of the resonance state occasionally can be deduced from the ratio of ionization currents produced using linearly- and circularly-polarized light. The high sensitivity and specificity of MPI spectroscopy suggests its use as an analytical tool [2] and, indeed, single-molecule sensitivity has been claimed [3,4].

Another potentially interesting facet of MPI is multiphoton-driven photochemistry, a topic to which we now turn.

2. NONLINEAR PHOTOCHEMISTRY

In this paper, we describe the three most distinct types of nonlinear photochemical behavior and then proceed to illustrate these behavioral limits with MPI data from a number of organic, inorganic and organometallic compounds. The three general types of nonlinear photochemical behavior are shown schematically in Fig. 2 [5]. In this scheme, we postulate that the first excitation step is a two-photon coherent absorption from the ground state (Grnd) to the resonance state (Res), via a virtual level (Virt) in the polyatomic molecule MXYZ. While the lifetime in the virtual level ($\sim 10^{-15}$ sec) is too short to allow photochemical transformation, that in the resonance state is longer; and, if it is in the range 10^{-3} - 1 nsec (as compared with a laser pulse width of $\sim$5 nsec), the molecule may readily decompose or rearrange within the duration of the laser pulse. Thus, at the resonance state there is a branching into Class A and Class B systems (cf. Fig. 2), depending upon whether the molecule absorbs one or more photons and ionizes (Class A), or whether the molecule fragments/rearranges before another photon can be absorbed (Class B).

The n-photon absorption rate between any levels x and y is $I^n\sigma_{xy}$, where I is the optical power and σ_{xy} is the relevant cross-section. In Class A systems, the one-photon absorption rate $I\sigma_{RSE}$ from the resonance state to a superexcited state SE in the ionization continuum is much larger than the decomposition rate γ; hence, the molecule is preferentially excited to the superexcited level and then relaxes to the parent ion $MXYZ^+$. At low laser powers ($\sim 10^5$ W/cm^2), the species initially formed is often only $MXYZ^+$, which does not then further photofragment. However, in intense pulses with powers of $\sim 10^9$ W/cm^2, $MXYZ^+$ may be further excited and fragmented, eventually forming M^+, X^+, etc. The gen-

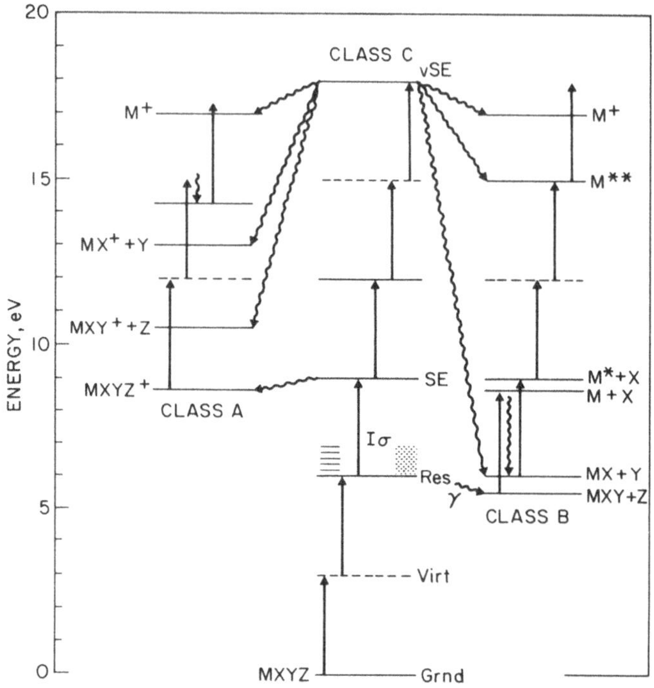

Fig. 2. The three paths of photochemical behavior under multiphoton excitation. Vibrational structure in the resonance state may be either discrete (left-hand side, Class A behavior), or continuous (right-hand side, Class B behavior).

eral course of Class A behavior is ionization followed by fragmentation (cf. Fig. 2).

In a Class B system, the photochemical relaxation rate γ in the resonance state is much larger than $I\sigma_{RSE}$ and, hence, the molecule fragments before another photon can be absorbed. If the molecular fragments are also multiphoton (or one-photon) resonant with the incident light, and are no more photostable than the parent system, these fragments will continue to absorb and to further fragment until only atoms remain. These atoms, as typified by M in Fig. 2, are then multiphoton ionized via atomic resonance states.

While most materials investigated to date may be categorized as either Class A or Class B, examples are also known of systems which are Class A in one spectral region and Class B in another;

and, if γ is approximately equal to $I\sigma_{RSE}$, mixed behavior can be expected as well. It may also happen that a parent molecule behaves as a Class B system in yielding a polyatomic neutral fragment, which then behaves as a Class A system. Class C behavior (cf. Fig. 2), although known in rare-gas atoms, has not been identified positively in a molecular system. Note, however, that even in sharp-line MPI spectra there is always observed an underlying ionization continuum which, in many cases, appears to be nonresonant and therefore corresponds to Class C. Characteristics of the various Classes are listed in Table 1.

In general, Class A molecules are multiply-bonded organics of low atomic weight atoms (e.g., acetaldehyde, benzene, butadiene, cyclopropane, etc.). Class B, on the other hand, encompasses inorganic and organometallic compounds having one or more heavy-metal atoms (e.g., $Cr(CO)_6$, $Mn_2(CO)_{10}$, ferrocene, $SnCl_4$, etc.). Clearly, multiphoton excitation of inorganic and organometallic systems serves as a clean source not only of gas-phase metallic ions, but of ground-and/or excited-state atoms as well.

3. TWO EXAMPLES

We now consider two examples, already in the literature, which illustrate many of the points listed in Table 1. In the region between 3600 and 3700 Å there is a sharp two-photon resonance in acetaldehyde [6-8] terminating at the (n,3s) Rydberg level; n $\rightarrow$ 3s is a well-known one-photon transition in this molecule. These simple observations (Characteristics 3, 6 and 10 of Table 1) firmly place acetaldehyde in Class A. In accord with this assessment, the mass spectrum of acetaldehyde ionized at low laser power shows a preponderance of CH_3CHO^+ parent ion; however, with increasing laser power, this gives way to CH_3CO^+, $CH_3{}^+$, $CH_2{}^+$, etc. (Characteristic 2). At low power (I), the ionization signal (S) obeys a power law of the form $S \propto I^n$, with n=3, and three photons are indeed required to ionize acetaldehyde at the n $\rightarrow$ 3s resonant wavelengths (Characteristic 5). Acetaldehyde is the prototypical Class A system.

The specifics of the MPI process in acetaldehyde are as follows: The first two photons are absorbed in a coherent excitation to (n,3s) [probably with (n,π^*) functioning as the virtual level] and a third photon then converts the (n,3s) resonance state into the parent ion. This is called a (2+1) process. At high optical powers, absorption of a fourth photon by the parent ion then leads to the photochemical production of the fragment ions CH_3CO^+ and HCO^+. The first of these fragments can absorb a fifth photon and dissociate to $CH_4{}^+$, $CH_3{}^+$ and $CH_2{}^+$. Among these, $CH_4{}^+$ can be further photolyzed to produce $CH_3{}^+$ and $CH_2{}^+$, while

Table 1. Characteristics of Molecular Systems Undergoing Multiphoton Ionization

Class A	Class B	Class C
1. System is ionized, then atomized	1. System is atomized, then ionized	1. System is atomized and ionized simultaneously
2. Produces $MXYZ^+$ at low power, giving way to MXY^+, MX^+, and finally M^+ at high power	2. Produces only M^+ at all powers	2. Produces constant ratios of various ions at all powers
3. MPI exhibits molecular resonances characteristic of target molecule	3. MPI exhibits atomic resonances independent of target molecule	3. Exhibits no resonances
4. MPI resonances originate in ground state only	4. MPI resonances originate in both ground and excited states	4. Exhibits no resonances
5. Low-power index = $IP/h\nu$	5. Low-power index >> $IP/h\nu$	5. Power index >> $IP/h\nu$ at all powers

Table 1. Continued

Class A	Class B	Class C
6. One-photon excitation to resonance state is spectrally discrete	6. One-photon excitation to resonance state is spectrally continuous	6. One-photon spectrum is irrelevant
7. Total energy of fragments lies above the resonance state	7. Total energy of fragments lies below the resonance state	7. Total energy of fragments lies below the final state
8. Ejected electron KE results from molecular ionization process; KE < $h\nu$	8. Ejected electron KE results from atomic ionization process; KE < $h\nu$	8. Ejected electron KE results from molecular ionization process; KE >> $h\nu$
9. At very high power, the course of ionization remains unchanged.	9. At very high power, Class B may become Class A	9. Products independent of power
10. Molecular resonance is always Rydberg	10. Molecular resonance is valence shell or, infrequently, Rydberg	10. Molecular resonance irrelevant

CH_2^+ can be photolyzed to yield CH^+. Each of the ionic species listed above is produced with its own characteristic power index. The total-ion power index, as conventionally determined using an ionization cell, is thus a complex average of these individual indices and has no meaning by itself, except at very low optical power where only the parent ion is formed.

When irradiated at 2485 Å, the acetaldehyde excitation is one-photon resonant with the smooth, continuous high-frequency wing of the $n \rightarrow \pi^*$ valence transition; and, by Characteristic 6, it follows that the nonlinear photochemistry in this region will be Class B, as observed [9].

The MPI spectrum and photochemistry of $Cr(CO)_6$ in the 4000-5000 Å region [10-12] stand in strong contrast to those of acetaldehyde discussed above. The one-photon spectrum of $Cr(CO)_6$ is broad and continuous at all wavelengths shorter than 3500 Å (Characteristic 6), whereas the MPI resonances in this molecule are atomically sharp and plentiful. These resonances are found in the MPI spectra at identical wavelengths regardless of which Cr compound is irradiated (Characteristic 3) and, moreover, the MPI-driven mass spectrum of $Cr(CO)_6$ reveals only a single ion, namely Cr^+ (Characteristic 2). This latter observation contrasts strongly with the results from conventional mass spectrometry, in which the mass spectrum is generated by electron impact, since in the conventional case over a dozen ions are observed, ranging from $Cr(CO)_6^+$ to Cr^+ [13]. Kinetic energy analysis of the MPI-ejected electrons shows that in $Cr(CO)_6$ these electrons originate in the ground and excited states of the Cr atom (Characteristic 8). Finally, at low power, a power index of 5-6 is observed, although only 3 photons are required to ionize $Cr(CO)_6$ itself (Characteristic 5). Clearly, $Cr(CO)_6$ exhibits Class B behavior in this spectral region.

To summarize the situation in $Cr(CO)_6$, the first two photons are coherently absorbed to excite the molecule in the broad-band region. This results in immediate dissociation to $Cr(CO)_5$, which then absorbs another two photons [probably incoherently, i.e., (1+1)] and dissociates to produce Cr(I) in its ground and excited states. These molecular absorptions are most likely valence-shell metal $\leftrightarrow$ ligand charge-transfer promotions. The Cr(I) atom then undergoes (2+1) excitations to form Cr^+. Although more than enough energy is supplied in this seven-photon excitation to form $Cr(CO)_6^+$, this ion is not observed.

4. OTHER SYSTEMS

The techniques and criteria described above for acetaldehyde and chromium hexacarbonyl have been applied to a large number of other organic, inorganic and organometallic systems. The results of some of these studies are summarized very briefly below.

4.1 Divalent Mercury

Divalent mercury systems are interesting because they show both Class A and Class B behavior. We have studied MPI processes in HgY_2, where Y = CH_3, Cℓ, Br and I. As already reported [14], the MPI spectrum of $Hg(CH_3)_2$ in the one-photon region (4000-3550 Å) displays very sharp two-photon molecular-resonance features, thus placing $Hg(CH_3)_2$ in Class A by Characteristic 3. In accord with this, the MPI mass spectrum of $Hg(CH_3)_2$ is rich in the parent ion at low power (Characteristic 2); the power index at low power is 3.0 (Characteristic 5); the one-photon absorption in the 2000-1800 Å region is sharp (Characteristic 6); and no Hg atomic lines are seen in the MPI spectrum (Characteristic 3). The $Hg(CH_3)_2$ two-photon excitation in question involves the promotion of a Hg-C σ-bonding electron into a spin-orbit component of the 6pπ Rydberg orbital (Characteristic 10). At low power, therefore $Hg(CH_3)_2$ exhibits the straightforward (2+1) molecular ionization of a Class A system.

By contrast, in the mercury halides the lowest one-photon excitation is to an A state, in which a halogen lone-pair electron is promoted into the Hg-X σ* MO. This excitation is valence shell (Characteristic 10) and is totally structureless in the one photon spectrum (Characteristic 6), thus leading one to expect Class B photochemical behavior. Indeed, the MPI spectra of the mercury halides are rich in Class B Hg-atom transitions (Characteristic 3). Note, however, that these transitions originate in atomic levels between 37,645 and 54,069 cm^{-1} above the $^1S_0$ ground state of Hg(I); moreover, three-photon resonances in atomic mercury vapor are observed in this same spectral region, resonances which originate in the $^1S_0$ ground state [14] but which are not observed when $HgBr_2$ serves as the source of Hg(I). Thus, the multiphoton dissociation of $HgBr_2$ in the 4300-3550 Å region yields only excited Hg(I) atoms. Many of the MPI lines originate in $^1P_1$, which is 54,069 cm^{-1} above the ground state [15]. Ionization of resonances originating in this state are then (1+1) processes.

Two-photon excitation of $HgBr_2$, using a fundamental wavelength in the vicinity of 4400 Å, should populate the A state; $\tilde{X} \rightarrow \tilde{A}$ is a transition with a continuous band shape. According to calculations by Wadt [16], the $\tilde{X} \rightarrow A$ promotion is immediately followed by dissociation into two Br atoms and the Hg atom in its ground state. Thus, Hg-atom MPI resonances resulting from A-band resonance and dissociation should originate in the $^1S_0$ ground state. This is contrary to our findings. Note, however, that Wadt's assignment of the A band ($^1\Sigma_g^+ \rightarrow {}^3\Pi_u$) is to that one state which is *one*-photon allowed, out of the five states predicted to lie in this region. Our experiment, on the other hand, involves a *two*-photon excitation which must be $^1\Sigma_g^+ \rightarrow {}^{1,3}\Pi_g$ (via an intermediate Π_u level); according to Wadt's calculation, these Π_g states relax to X atoms and the *stable* ground state of HgX. We presume, then, that HgBr($\tilde{X}$) is next promoted in a two-photon process to unbound levels which yield excited-state Hg atoms.

The strong contrast between the nonlinear photochemistries of $Hg(CH_3)_2$ and HgX_2 can be traced directly to the different orbital natures of the resonance states, and to the chemistry that then flows from these states. In the case of $Hg(CH_3)_2$, the resonance state is Rydberg and stable, whereas in HgX_2, it is valence shell and unstable (Characteristic 10).

4.2 Tetravalent Tin

The large differences in the photochemistries of $Hg(CH_3)_2$ and HgX_2 led us to investigate a similar pair, namely $Sn(CH_3)_4$ and $SnCl_4$. The first one-photon electronic transition in $SnCl_4$ [17] is found centered at 2000 Å (50,000 cm^{-1}), with a perfectly-smooth, symmetric vibronic profile. Compared to the $SnCl_4$ ionization potential (97,500 cm^{-1}), the frequency of the 50,000 cm^{-1} band is far too low for it to be a Rydberg excitation and, consequently, it has been assigned as valence shell [18]. It is most likely an A-band transition (chlorine 3p → σ*), similar to that in HgX_2 in the same region. According to Characteristics 6 and 10, the MPI of $SnCl_4$ should show Class B nonlinear photochemistry in the laser pulse.

The lowest-energy band in the $Sn(CH_3)_4$ electronic spectrum has been described as a continuum extending from 2100 to 2600 Å [19]. Nothing else is known of this molecule's electronic spectrum. It would seem that on the basis of Characteristic 6, $Sn(CH_3)_4$ is also Class B, in contrast to $Hg(CH_3)_2$.

The MPI resonances in $Sn(CH_3)_4$ consist of a very large number of lines, extending from 4100 Å down to 3587 Å. Many, but not all, of these lines are assignable as two-photon resonances in the tin atom, originating in the $^3P_0$ ground state and in the $^3P_1$, $^3P_2$ and $^1D_2$ excited states. A large number of other equally sharp and prominent lines are also observed but could not be assigned, under any assumptions which accord with the atomic levels listed in [15], as either Sn(I) or Sn(II) transitions. Although the two-photon region which is spanned by 4100-3587 Å in one-photon excitation is beyond the continuum absorption described for $Sn(CH_3)_4$, there is little doubt that we are exciting into a second, higher-energy dissociative continuum, and that this leads to Class B photochemistry.

We have measured the power exponent for ionization of $Sn(CH_3)_4$ at 3805 Å and find 5.5 ± 0.6 from the data at the lowest measurable powers. The difference between the measured power index of 5.5 and the anticipated value of 6.0 can be attributed to the action of the AC Stark effect (Characteristic 5). A power index of 5.5 is equivalent to 18 eV at 3805 Å, which clearly indicates that a large part of the overall photon energy (even at low optical powers) is being used to break bonds in molecular species, leading to ionic species.

The obvious production of Sn atoms in the laser photolysis of $Sn(CH_3)_4$ implies the presence of methyl-radical resonances, as well as Sn resonances, in the MPI spectrum of this compound. As described in another paper [20], such methyl-group two-photon resonances have been assigned to two broad features observed at 67,820 and 67,990 cm^{-1} in $Sn(CH_3)_4$.

MPI mass spectrometric measurements on $Sn(CH_3)_4$ are totally in line with the characteristics of Class B nonlinear photochemistry. Thus, we have excited several of the MPI resonance lines of $Sn(CH_3)_4$, within a TOF mass spectrometer, and have observed only Sn^+ at all optical powers, as expected for a Class B system (Characteristic 2). In contrast, the ions $Sn(CH_3)_{4-x}^+$, x=0,...,4, are found in the electron-impact mass spectrum of $Sn(CH_3)_4$ [21].

A brief study of $SnCl_4$ quickly confirms our expectation that this molecule will exhibit Class B photochemistry. This expectation is based upon (i), the continuous nature of the absorption bands (Characteristic 6); (ii), the valence nature of the resonance state (Characteristic 10); and (iii), the energy-level diagram, which looks much like that of $Sn(CH_3)_4$, with two fragmented configurations below the resonance state (Characteristic 7). The MPI spectrum of $SnCl_4$ shows all of the Sn(I) resonance lines observed for $Sn(CH_3)_4$ (Characteristic 3). There are no chlorine-atom resonances in this spectral region. The ions resulting from 3550 Å irradiation of $SnCl_4$, within the TOF mass spectrometer, proved to be 80% Sn^+, 19% $SnCl^+$ and 1% $SnCl_2^+$. In the electron-impact (31 eV) mass spectrum of $SnCl_4$ [22], $SnCl_3^+$ is the major ion, with all other ions occurring at about at about 1/5 of its intensity; however, with an electron-impact energy equal to the ionization threshold, $SnCl_4^+$ predominates.

4.3 Methyl Iodide

Methyl iodide is a hybrid system in the sense that it is clearly Class A in one spectral region, but Class B in another. This follows naturally from Characteristics 6 and 10 of Table 1: The one-photon spectrum of methyl iodide in the region 3600-2100 Å is continuous, corresponding to an $n \rightarrow \sigma^*$ A-band valence-shell excitation, whereas the second excitation beginning at 1950 Å is an n → 6s Rydberg transition which is quite sharply structured [23]. Thus, resonance in the first of these should give Class B behavior, while resonances in the second will behave as Class A.

Excitations which are one-photon resonant in the A-band of methyl iodide yield CH_3 and I fragments with near-unit quantum yield, with the I atoms in either the $^2P_{3/2}$ or $^2P_{1/2}$ state. Accordingly, when irradiated at wavelengths shorter than 3100 Å, MPI atomic resonances are observed to originate from these two states of the iodine atom, as expected for Class B behavior. Resonance features are also observed for the methyl radical [20]. On the other hand, pumping at 4000-3300 Å sidesteps the n → σ* resonance, but two-photon transitions excite the n → 6s and n → 6p molecular levels. Consequently, only the CH_3I molecular Rydberg resonances are observed [20,24]. The mass spectrum in this region shows a strong parent-ion peak (Characteristic 2), together with I^+ and CH_3^+ which must come from photodissociation of the parent ion, a process which is one-photon resonant in this wavelength region. Further, the thresholds for the production of CH_3^+ and I^+ from CH_3I^+ are exceeded by a one-photon absorption in the parent ion.

The Class A to Class B shift in nonlinear photochemistry displayed by methyl iodide is due to a change of resonance state on scanning through the UV/VIS region (i.e., from (5p,σ*) valence shell to (5p,6s) Rydberg). A somewhat similar change is observed in NO_2 [25], where a shift from Class A to Class B behavior is encountered upon exceeding the dissociation wavelength (4000 Å) of the valence excitation in the visible region.

ACKNOWLEDGMENTS

I wish to acknowledge helpful conversations with A. Gedanken, N. A. Kuebler, G. J. Fisanick, William Wadt and T. Bowmer.

REFERENCES

1. P. M. Johnson, Accts. Chem. Research 13, 20 (1980).
2. D. A. Lichtin, L. Zandee and R. B. Bernstein, in: *Lasers and Chemical Analysis*, G. M. Hieftje, F. E. Lytle and J. C. Travis, eds. (Humana Press, Clifton, N. J., 1980), p. 1.
3. C. Klimcak and J. Wessel, Appl. Phys. Lett. 37, 138 (1980).
4. V. S. Antonov, V. S. Letokhov and A. N. Shibanov, Optics Commun. 38, 182 (1981).
5. A diagram of this sort has also been presented by V. S. Antonov and V. S. Letokhov, Appl. Phys. 24, 89 (1981).

6. G. J. Fisanick, T. S. Eichelberger, IV, B. A. Heath and M. B. Robin, J. Chem. Phys. 72, 5571 (1980).
7. B. A. Heath, M. B. Robin, N. A. Kuebler, G. J. Fisanick and T. S. Eichelberger, IV, J. Chem. Phys. 72, 5565 (1980).
8. T. S. Eichelberger, IV and G. J. Fisanick, J. Chem. Phys. 74, 5962 (1981).
9. J. P. Riley and K. L. Kompa, Adv. Mass Spectrom. 8B, 1800 (1980).
10. D. P. Gerrity, L. J. Rothberg and V. Vaida, Chem. Phys. Lett. 74, 1 (1980).
11. G. J. Fisanick, A. Gedanken, T. S. Eichelberger, IV, N. A. Kuebler and M. B. Robin, J. Chem. Phys. 75, 5215 (1981).
12. M. A. Duncan, T. G. Dietz and R. E. Smalley, Chem. Phys. 44, 415 (1979).
13. G. A. Junk and H. J. Svec, Z. Naturforsch. 23b, 1 (1968).
14. A. Gedanken, M. B. Robin and N. A. Kuebler, Inorg. Chem. 20, 3340 (1981).
15. C. E. Moore, *Atomic Energy Levels*, Vol. III (National Bureau of Standards (US) Circular 467, 1958).
16. W. R. Wadt, J. Chem. Phys. 72, 2469 (1980).
17. W. R. Wadt, Appl. Phys. Lett. 34, 658 (1979).
18. M. B. Robin, *Higher Excited States of Polyatomic Molecules*, Vol. I (Academic Press, New York, 1974); Vol. II (Academic Press, New York, 1975).
19. H. W. Thompson and J. W. Linnett, Proc. Roy. Soc. London 156A, 108 (1936).
20. A. Gedanken and M. B. Robin, J. Chem. Phys. 76, 4798 (1982).
21. J. J. de Ridder and G. Dijkstra, Recueil 86, 737 (1967).
22. A. S. Buchanan, D. J. Knowles and D. L. Swingler, J. Phys. Chem. 73, 4394 (1969).
23. R. A. Boschi and D. R. Salahub, Mol. Phys. 24, 289 (1972).
24. D. H. Parker, R. Pandolfi, P. R. Stannard and M. A. El-Sayed, Chem. Phys. 45, 27 (1980).
25. M. B. Robin, Appl. Optics 19, 3941 (1980).

EXCESS ENERGY DEPENDENCE OF VIBRATIONAL RELAXATION AND PHOTO-PHYSICAL BRANCHING RATIOS IN ISOLATED AROMATIC MOLECULES: RELEVANCE TO VACUUM ULTRAVIOLET PHOTOCHEMISTRY

E. C. Lim

Department of Chemistry
Wayne State University
Detroit, Michigan 48202, USA

1. INTRODUCTION

During the last decade there has been a great deal of activity in the study of electronic relaxation in gaseous organic molecules under collision-free conditions. Early experimental investigations were primarily concerned with radiationless transitions in bulk gas at ambient and elevated temperatures [1]. Although these studies have yielded some useful information concerning vibronic level dependence of electronic relaxation processes, the interpretation of some of these results has been hampered by the fact that true photoselective excitation of well-defined initial quantum states cannot, in many cases, be accomplished because of the severe spectral congestion arising from thermal inhomogeneous broadening (vibrational sequence congestion and rotational broadening). Most of the recent experimental efforts, therefore, have been directed at the study of electronic relaxation in molecules which are cooled by supersonic expansion of seeded molecular beams, and which are excited by narrow-bandwidth dye lasers. Strong vibrational and rotational cooling (resulting from the supersonic expansion) significantly reduces thermal inhomogeneous broadening, thus providing greatly simplified and resolved spectra, the dynamical properties of which may be probed by laser-induced processes (e.g., fluorescence). Recent applications of this molecular beam technique have demonstrated the dramatic changes which can be brought about by internal cooling of molecules. Thus, nonradiative decay rates of the low-lying vibronic levels of the lowest-excited singlet states (S_1) of jet-cooled molecules have been shown to be significantly different from those of the corresponding static gases for several large aromat-

S. P. McGlynn et al. (eds.), Photophysics and Photochemistry in the Vacuum Ultraviolet, 855–873.

ic hydrocarbons [2]. Equally dramatic is the finding that a non-fluorescent electron donor-acceptor (EDA) complex can be made to fluoresce quite intensely when it is internally cooled by supersonic expansion [3]. For molecules with intermediate level structure, rotational cooling can effect changes from the intermediate case decay behavior to the small-molecule decay behavior [4]. These examples provide evidence of the type of new information about electronic relaxation which can be obtained when a true excitation of a well-defined quantum state of a large molecule is made, in the absence of thermal broadening effects. There is, however, one important conclusion from bulk-gas experiments which is unaffected by these new experimental developments: This conclusion is concerned with the variation in vibrational energy dependence exhibited by the different S_1 photophysical decay channels -- namely, internal conversion (IC) to the ground state (S_0) and intersystem crossing (ISC) to the triplet manifold (T_1). Since this topic is relevant to any analysis of the energy dependence of product channels in gas-phase photochemical reactions, especially in the vacuum ultraviolet, I wish to briefly describe the highlights of these bulk-gas investigations, which date back to 1974. I will also describe the effect that crossing the S_2 threshold has on the excess energy dependence of the S_1 nonradiative decay rate, an effect which can be related to the question of the efficiency of intramolecular vibrational relaxation. Although this paper focuses on aromatic hydrocarbons, the conclusions should also be applicable to other polyatomic molecules which undergo so-called "statistical" radiationless transitions (i.e., radiationless transitions which are characterized by irreversible, and exponential, decays).

2. SPECTRAL CHARACTERISTICS OF GAS-PHASE FLUORESCENCE

The fluorescence of organic molecules in the condensed phase or in the gas phase at high pressures, nearly always originates from the lowest-excited singlet state, regardless of the nature of the singlet state which is initially photoexcited. This finding, which is commonly referred to as Kasha's rule, is a familiar and important feature of molecular luminescence and demonstrates the efficiency of internal conversion (a radiationless transition between electronic states of the same multiplicity) [5]. A question of extraordinary importance is whether or not interaction with the surrounding medium, or collision with other molecules, is *essential* to the IC process. The answer to this question was contained, by and large, in an early experiment by Pringsheim [6], which showed that the fluorescence of anthracene vapor, under collision-free conditions, originates from S_1 even when the molecule is initially excited into a higher-energy singlet state. This important finding, which establishes $S_j(j>1) \rightarrow S_1$ IC as an *intra*molecular phenomenon, has been confirmed and extended to other molecules by a number of subsequent workers [7-13].

The above results are illustrated in Fig. 1, which shows the fluorescence spectra of dilute naphthalene vapor (0.07 torr), obtained by excitation into three different absorption bands. Upon excitation of higher-excited singlet states, the spectrum becomes diffuse but the spectral position of the emission is clearly that which follows excitation of S_1. The fluorescence is therefore $S_1 \rightarrow S_0$ in every case. The diffuseness and the slight red shift of the $S_1 \rightarrow S_0$ fluorescence which accompanies excitations of higher-energy singlet states can be rationalized in terms of the Franck-Condon (FC) principle and the assumption that the S_1 potential energy surface is shallower than the S_0 surface (see Fig. 2). The diffuseness arises from the FC-allowed radiative transition from high vibrational levels of S_1 to highly-excited and dense vibrational levels of S_0; the red shift, however, is a consequence of the decreased transition energy resulting from the disparity in the force constants of the two potential energy surfaces. Both the extent of the red shift and the diffuseness of the emission spectrum are expected (and found) to increase with increasing vibrational energy content of S_1. This behavior contrasts with that in the condensed phase, or in the gas phase under higher pressures, in which the spectral characteristics (band position and band shape) remain independent of excitation energy as a result of the rapid removal of the excess vibrational energy by collision with solvent (or other) molecules. Using the $S_2 \rightarrow S_0$ radiative decay rate, (based upon the integrated $S_0 \rightarrow S_2$ absorption coefficient and the upper limit of the quantum yield of the unobserved $S_2 \rightarrow S_0$ fluorescence), the rate of $S_2 \rightarrow S_1$ IC (competing with the $S_2 \rightarrow S_0$ radiative process) is estimated to be greater than $\sim 10^{11}$ sec^{-1}. This is at least four orders of magnitude greater than the hard-sphere collision rate at 0.07 torr and establishes $S_2 \rightarrow S_1$ internal conversion as a truly intramolecular phenomenon.

Since the fluorescence of dilute vapor originates from S_1 even under collision-free conditions, the excitation energy dependence of the emission reflects the excess energy dependence of radiative and nonradiative transitions from S_1. The magnitude of these decay rates can be determined from the lifetime ($\tau_F = 1/(k_r + k_{nr})$) and quantum yield ($\Phi_F = k_r/(k_r + k_{nr})$) of fluorescence through the expressions

$$k_r = \Phi_F/\tau_F \tag{1}$$

$$k_{nr} = (1 - \Phi_F)/\tau_F \quad . \tag{2}$$

Thus, by measuring the quantum yield and the lifetime at various

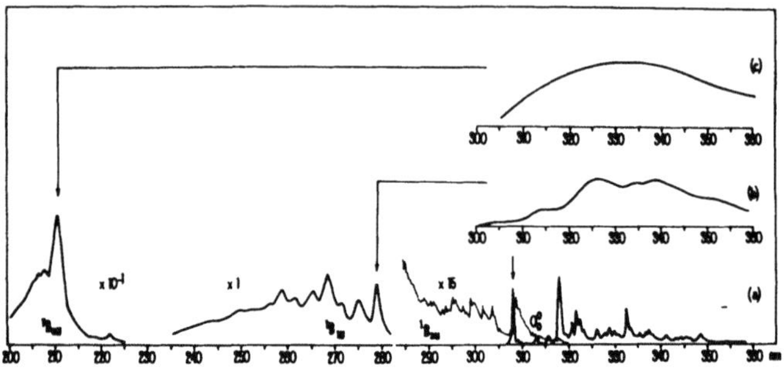

Fig. 1. Excitation energy dependence of fluorescence spectra in naphthalene vapor.

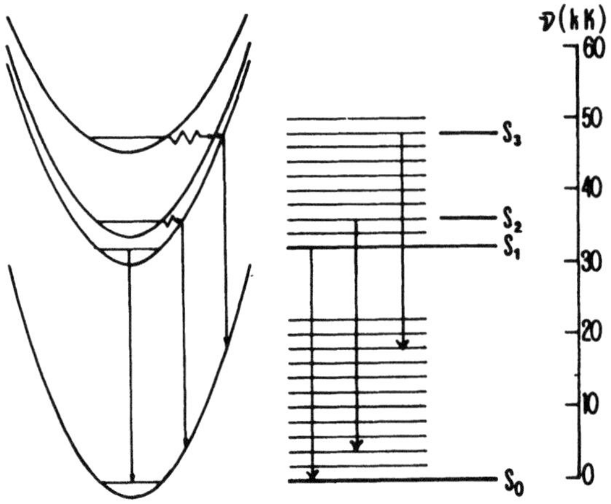

Fig. 2. Potential energy diagram explaining the excitation energy dependence of fluorescence spectra.

excitation wavelengths, it is possible to determine the manner in which the radiative and nonradiative decay rates vary with the S_1 vibrational energy content of the molecule.

3. EXCESS ENERGY DEPENDENCE OF RADIATIONLESS TRANSITIONS: THEORY

There are two major open channels for radiationless transitions from S_1: Internal conversion to S_0 and intersystem crossing (a radiationless transition between two electronic states of different multiplicities) to T_1. For most aromatic molecules in the condensed phase, or in the gas phase at high pressures, $S_1 \rightarrow T_1$ ISC is known to dominate over $S_1 \rightarrow S_0$ IC [5]. In order to understand this generalization, and in order to investigate the effect of vibrational excitation on radiationless processes, we consider a nonradiative transition between an initial electronic state $|\alpha\rangle$ (energy E_α) and a final state $|\beta\rangle$ (energy E_β). The rate expression for the nonradiative process may be written as [16]

$$k_{nr} = \frac{2\pi}{\hbar} |V^p_{\alpha\beta}|^2 \prod_j F_j \rho(\{\bar{n}_j\}) \tag{3}$$

where $V^p_{\alpha\beta}$ is the electronic matrix element involving the promoting mode p; F_j is the FC factor for mode j; and $\rho(\{\bar{n}_j\})$ is the "density-of-states weighted FC factor" for the set $\{\bar{n}_j\}$, where $\{\bar{n}_j\}$ represents the most probable distribution of the vibrational quanta in the final state. The function $\rho(\{\bar{n}_j\})$ accounts for the contributions of modes other than j. For a transition originating from the lowest vibrational level of mode j, the FC factor can be expressed as

$$F_j = |\langle\chi^\alpha_j(0)|\chi^\beta_j(\bar{n}_j)\rangle|^2 \tag{4}$$

where χ_j represents the vibrational wavefunction, and where the value of $\bar{n}_j$ is determined from the energy conservation condition [14]

$$E_\alpha - E_\beta - \hbar\omega_p = \sum_j \bar{n}_j \hbar\omega_j \ . \tag{5}$$

For an electronic energy gap $E_\alpha - E_\beta$ much larger than the vibrational spacing $\hbar\omega_p$, the FC factor decreases very rapidly with increasing values of $\bar{n}_j$; thus, Eq. (3) predicts a strong energy dependence of the radiationless decay rate on the energy gap (an energy gap law) [5]. The inefficiency of $S_1 \rightarrow S_0$ IC, relative to $S_j \rightarrow S_{j-1}$ ($j \geq 2$) IC, can be qualitatively understood in terms of a large $S_1 - S_0$ energy gap, as can the fact that $S_1 \rightarrow S_0$ IC appears to be important (compared to fluorescence or ISC) only in molecules with a low-lying S_1 (e.g., pentacene) [15].

For molecules which are vibrationally excited (m_j quanta, say) in the initial electronic state, the FC factor for radiationless transitions takes the form

$$F_j = |\langle\chi_j^\alpha(m_j)|\chi_j^\beta(\bar{k}_j)\rangle|^2 \tag{6}$$

where $\bar{k}_j$ is the new optimal distribution of the accepting mode j in the final electronic state. If the normal modes are parallel in the initial and final electronic states (i.e., if there is no frequency difference between the modes in these two states), $\bar{k}_j$ can be determined from the "retention of vibrational quanta" model [16], which assumes that the initially excited quanta m_j simply add to the final state distribution $\bar{n}_j$ for the vibrationless ($m_j = 0$) case to give the new distribution, $\bar{k}_j = \bar{n}_j + m_j$. Note that the energy conservation condition is satisfied in this approximation and that the necessity for finding a new optimal distribution for every given set of initial excitation conditions is obviated. Thus, the FC factor for radiationless transitions from the m_j vibrational level becomes

$$F_j = |\langle\chi_j^\alpha(m_j)|\chi_j^\beta(\bar{n}_j + m_j)\rangle|^2 \quad . \tag{7}$$

The excess vibrational energy dependence of the nonradiative decay rate is determined largely by the magnitude of this FC factor relative to the FC factor for the vibrationless case, since $\rho(\{\bar{n}_j\})$ in Eq. (3) varies only slowly with m_j and since the electronic matrix elements $V^p_{\alpha\beta}$ cancel in the calculation of relative rates. Thus [17],

$$\frac{k(\{m_j\})}{k(0)} = \prod_j \frac{|\langle\chi_j^\alpha(m_j)|\chi_j^\beta(\bar{n}_j + \bar{m}_j)\rangle|^2}{|\langle\chi_j^\alpha(0)|\chi_j^\beta(\bar{n}_j)\rangle|^2}$$

$$= \prod_j \frac{\Gamma(m_j + \bar{n}_j + 1)}{\Gamma(m_j + 1)\Gamma(\bar{n}_j + 1)} \left(1 - \frac{m_j}{\bar{n}_j + 1} g_j^2 + \ldots\right)^2 \tag{8}$$

where g_j represents a parameter which is related to the geometry change (along the vibrational coordinate j) accompanying the radiationless transition. Although the retention model is crude, its enormous simplicity permits qualitatively important conclusions to be drawn from "hand" calculation of the last term in Eq. (8). If $(m_j/\bar{n}_j + 1)g_j^2 \ll 1$, as would be the case for radiationless transitions between well-separated states of aromatic hydrocarbons, the excitation (m_j) dependence of the nonradiative decay rate can be written as

$$\frac{k(m_j)}{k(0)} \simeq \prod_j \frac{(\bar{n}_j + m_j)!}{\bar{n}_j!\, m_j!} \quad . \tag{9}$$

Since $\bar{n}_j$ is related to the electronic energy gap and vibrational frequency by the energy conservation condition [Eq. (5)], and since the ratio of the above factorials increases with increasing value of $\bar{n}_j$ or m_j, Eq. (9) leads to the following important qualitative predictions concerning the excess vibrational energy dependence of the radiationless decay rate:

Prediction 1: The initial excitation (m_j) of the accepting mode increases the nonradiative decay rate over that for the vibrationless case. This excitation dependence of the radiationless decay rate increases as the energy gap between initial and final electronic states becomes larger.

Prediction 2: The excitation energy dependence of the radiationless decay rate is larger in deuterated compounds than in the corresponding perprotonated compounds because of the reduced vibrational frequencies (larger $\bar{n}_j$) in the former. This isotope effect can be large for radiationless transitions involving large electronic energy gaps since the FC factors for such transitions are strongly influenced by high-frequency C-H stretching modes.

Prediction 1 implies that the $S_1 \rightarrow S_0$ IC rate should increase more rapidly with increasing vibrational excitation than the rate of $S_1 \rightarrow T_1$ ISC, owing to the much larger electronic energy gap of the former. Furthermore, the excitation dependence of the relative rates is highly dependent on both the frequency shift and the displacement in the equilibrium position of the accepting mode; that is, this dependence increases with an increase in the frequency shift. Since the frequency changes between an excited state and the ground state (as well as differences in equilibrium geometries and bond lengths) are in general much larger than those between two excited states, it follows that the $S_1 \rightarrow S_0$ rates should increase more rapidly than the $S_1 \rightarrow T_1$ rates, even if the electronic energy gaps were comparable for the two transitions. These conclusions suggest that $S_1 \rightarrow S_0$ IC may compete with, or even dominate, ISC for molecules with large excess vibrational energies in S_1. This prediction, which is schematically represented in Fig. 3, is in sharp contrast to the known photophysical behavior of aromatic molecules with low excess energies (e.g., molecules in the condensed phase), in which the FC vibrational overlap heavily favors $S_1 \rightarrow T_1$ ISC as the predominant

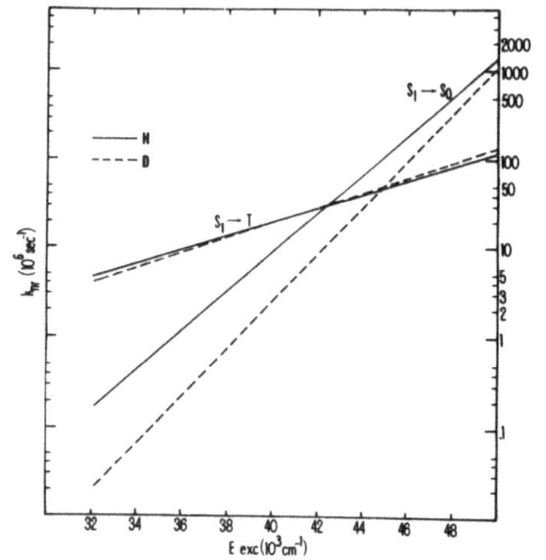

Fig. 3. Excitation energy dependence of radiationless transitions in perprotonated and perdeuterated naphthalenes.

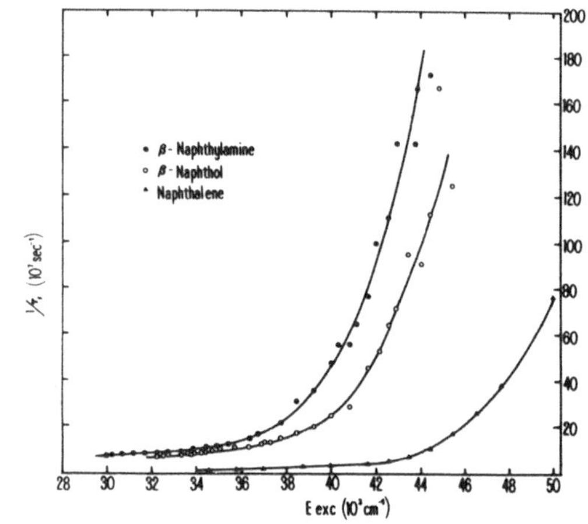

Fig. 4. Excitation energy dependence of total decay rate. See [19] for experimental details.

route of radiationless decay.

In the following section, we examine experimental results on radiationless transitions in a number of gaseous aromatic molecules to see how the findings can be rationalized in terms of the above theoretical ideas.

4. EXCESS ENERGY DEPENDENCE OF RADIATIONLESS TRANSITIONS: EXPERIMENT

One of the most striking features of excitation energy dependence of fluorescence lifetimes in isolated organic molecules is that the lifetime decreases first slowly, then rapidly, as the excess energy in S_1 is increased [19]. This is illustrated, in Fig. 4, by data for naphthalene and its derivatives. Since the quantum yield of fluorescence also shows a very similar dependence upon excitation energy, the variation of fluorescence lifetime must reflect the dependence of the nonradiative decay rate on excess vibrational energy. The slow initial increase in the radiationless decay decay rate, with increasing excess energy, clearly implies that the radiationless transition with relatively small energy gap dominates for low excess energies. This is in line with Prediction 1 of Section 3: For low excess energy, $S_1 \rightarrow T_1$ ISC is the predominant route of nonradiative decay because of more favorable FC factors.

The strong energy dependence of the nonradiative decay rate in the region of high excitation energy suggests that there is a new relaxation channel (not important in the low excess-energy regime) which becomes important for molecules with large excess energies. This channel is most likely IC to the ground state, which is predicted to become increasingly important with increasing S_1 excess vibrational energies. The large excess-energy dependence of the nonradiative decay rate is consistent with the large energy gap between the lowest excited state and the ground state. This interpretation is supported by the deuterium isotope effect on radiationless decay rate [19,20] (see Fig. 5), which shows that the differences in the decay rates of protonated and deuterated compounds are small for small excess energy and large for large excess energy. Interestingly, in the example shown in Fig. 5 (and in other examples), the onset of the large isotope effect practically coincides with the onset of the relatively strong excess-energy dependence of the nonradiative decay rate. The small isotope effect on the radiationless decay rate, when the excess energy is small, is not consistent with $S_1 \rightarrow S_0$ IC(the isotope effect on $S_1 \rightarrow S_0$ IC is expected to be large when excess energy is small); while a large isotope effect on the decay rate, when the excess energy is large, cannot be rationalized in terms of $S_1 \rightarrow T_1$ ISC (if ISC exhibits a small isotope effect at zero

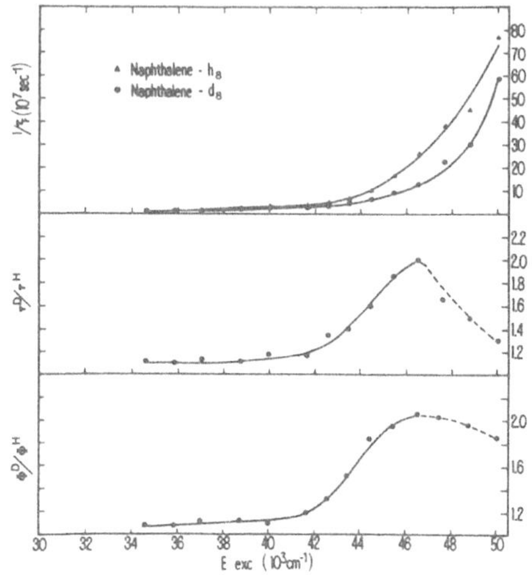

Fig. 5. Isotope dependence of fluorescence (lifetime and quantum yield) in naphthalene. See [19] for experimental details.

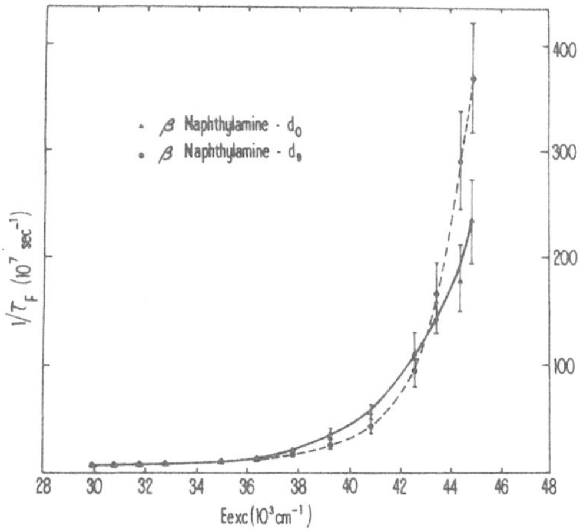

Fig. 6. Isotope dependence of fluorescence lifetime in β-naphthylamine. See [21] for experimental details.

excess energy, it should exhibit an inverse isotope effect at large excess energy).

The prediction of Section 3 (Prediction 2) that deuterated compounds should show a larger excitation dependence of radiationless decay rate than the corresponding protonated compounds is dramatically confirmed for β-naphthylamine (see Fig. 6) in which the discrepancy in the excitation dependence of the decay rate in isotopic compounds is so great that the ring deuterated (d_7) compound actually exhibits a greater decay rate than the protonated compound at relatively high excess energies [21].

If the rapid exponential increase in radiationless decay rate as a function of increasing vibrational excitation for large excess energies is a result of $S_1 \to S_0$ IC, as our arguments suggest, then the excess energy dependence of the decay rate should be related, in some fashion, to the $S_1 - S_0$ electronic energy gap according to Prediction 1 of Section 3. This is indeed confirmed [22] by the correlation given in Fig. 7, which shows a plot of the excess vibrational (E_{vib}) dependence of the radiationless decay rate, as measured by the quantity $\alpha = d\ln k/dE_{vib}$, against the $S_1 - S_0$ energy gap. This correlation, which applies to molecules which are initially excited into S_2 or S_3, would not have been expected if a nonradiative channel other than $S_1 \to S_0$ IC were to dictate the excess energy dependence of the radiationless decay rate.

An independent, and perhaps more conclusive, piece of evi-

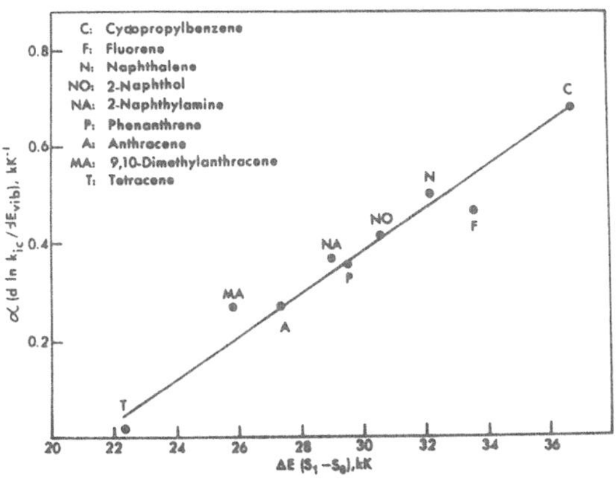

Fig. 7. Excess energy dependence of radiationless decay rate (in the high energy region) versus energy of the lowest-excited singlet state.

dence for the importance of internal conversion in molecules with large excess energies comes from the excitation energy dependence of triplet formation [23]. Fig. 8 shows the relative quantum yield of biacetyl phosphorescence, sensitized by energy transfer from naphthalene, as a function of excitation energy [23]. These experiments were carried out at room temperature, using a mixture of 0.07 torr of naphthalene with approximately 0.1 torr of biacetyl. Under these conditions the mixture displayed naphthalene fluorescence, the spectral shape and lifetimes of which were the same as those for naphthalene vapor (0.07 torr) without added biacetyl. (This indicates that the sensitized phosphorescence of biacetyl comes almost entirely from energy transfer from the triplet state of naphthalene.) The excitation energy dependence of sensitized phosphorescence shows that the quantum yield of the emission, as defined by the ratio of the number of quanta emitted by biacetyl to the number of quanta absorbed by naphthalene, decreases sharply when the excess vibrational energy in the naphthalene S_1 exceeds about 10,000 cm^{-1}. This result suggests that either the triplet yield of naphthalene vapor becomes very small for large excess energies, or the biacetyl triplet states which are formed by energy transfer from naphthalene do not emit (or emit with very low efficiencies) when their vibrational energy contents are large. The pressure effect on the excitation energy dependence of sensitized phosphorescence suggests, however, that

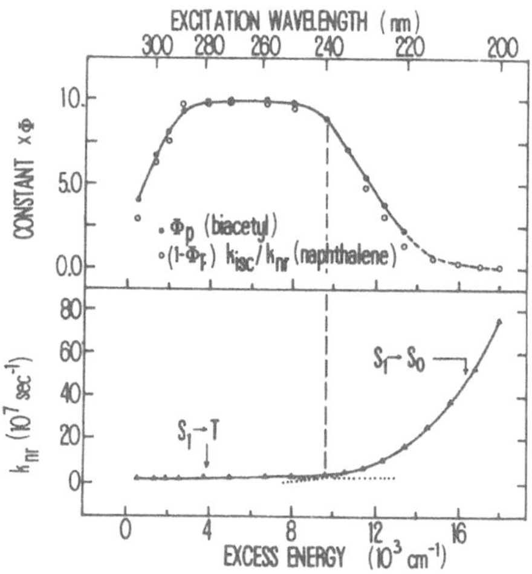

Fig. 8. Relative quantum yield of sensitized phosphorescence in biacetyl (top), and radiationless decay rate of naphthalene (bottom), as a function of excess vibrational energy. See [23] for experimental details.

the latter of these two possibilities is unlikely to be the cause of the drop in phosphorescence yield. This result, together with the photochemical stability of naphthalene and biacetyl under the above experimental conditions, leads to the conclusion that the quantum yield of naphthalene ISC is very small for large excess vibrational energies. Since the onset of the sharp drop in the triplet yield practically coincides with the onset of the sharp increases in the radiationless decay rate (see Fig. 8), the assumption that the rapid increase in the decay rate is due to the $S_1 \to S_0$ internal conversion appears to be confirmed. Interestingly, if it is assumed that the excitation energy dependence of the nonradiative decay rate (k_{nr}) in the region of low excess energies reflects the energy dependence of the intersystem crossing rate (k_{isc}), then the excitation dependence of the intersystem crossing yield (determined from $(1 - \Phi_F)k_{isc}/k_{nr}$) parallels that of the biacetyl experiments (cf. Fig. 8).

These results, and similar results on a number of other molecules, clearly support the predicted importance of $S_1 \to S_0$ internal conversion for molecules with large excess vibrational energies.

5. INTRAMOLECULAR VIBRATIONAL RELAXATION

The preceeding discussion described how a conclusion can be drawn about the nature of the dominant radiationless transition merely by assessing the very different excess energy dependences of the $S_1 \to S_0$ and $S_1 \to T_1$ radiationless processes. However, there is one additional important piece of information that can be obtained from the energy dependence of the nonradiative decay rate: This pertains to the efficiency of intramolecular vibrational relaxation, the importance of which -- for unimolecular reactions, including photochemistry -- is well-recognized. The occurrence of vibrational relaxation can be confirmed by the observation of fluorescence from vibrational levels other than the one initially prepared by optical excitation. In the first example of this kind (especially in a supersonic jet), Smalley, *et al.* [24] observed the spectral features of fluorescence resulting from the vibrational relaxation of a ring mode into alkyl chain modes for several long-chain alkyl benzenes. Less direct, but still useful, information concerning the efficiency of intramolecular vibrational redistribution can also be obtained from the excitation energy dependence of electronic relaxation [25]. One such approach is to study the effect of crossing the threshold into a new electronic manifold (S_2, say) on the radiationless decay rate of S_1. The vibrational level distribution prepared by direct optical excitation of S_1 will, in general, be different from that generated by $S_2 \to S_1$ IC, since the former is determined by FC factors for a vertical (isogeometric) transition while the latter is determined by generalized overlap integrals for a horizontal (isoenergetic) transition. Thus, if vibrational relaxation in S_1 (or in the triplet manifold) is incomplete during the S_1 lifetime, the change in the vibrational-level distribution following $S_2 \rightarrow S_1$ IC should lead to a change in the magnitude and excess energy dependence of the S_1 (or T_1) decay rate compared to what it would have been for optically prepared states of the same energy.

Experiments of the above type [19,21,26] have shown that the nonradiative decay rate of S_1 (or T_1), as a function of excess energy, exhibits discernible (or even dramatic) changes when the S_2 excitation threshold is crossed [19,21]. The results, which are illustrated in Figs. 9 and 10, suggest that $S_0 \rightarrow S_2$ excitation leads to changes in the vibrational distribution (relative to the distribution arising from $S_0 \rightarrow S_1$ excitation), changes which are substantially maintained during the lifetime of the emitting state. The data of Fig. 10 are especially interesting since the very different behavior of acetophenone (which has a triplet lifetime greater than one microsecond), when compared to that of 1-indanone (which has a triplet lifetime on the order of nanoseconds), may provide a clue to the timescale of intramolecular vibrational relaxation. Considering the moderate densities and relatively short lifetimes of S_1 vibronic levels at the electronic origin of S_2, the incomplete vibrational relaxation is not totally unexpected even for such large molecules possessing substantial excess energy. As the excess energy (and, therefore, the level density) increases, the rate of vibrational relaxation presumably increases, so that for large excess energies it may be possible for a very efficient intramolecular vibrational redistribution to take place prior to electronic relaxation of S_1. There are at least three observations which relate to the excess energy dependence of the radiationless decay rate and which are consistent with this possibility. These are: (i), the absence of a change in the energy dependence of the decay rate as the S_3 excitation threshold is crossed [19,21]; (ii), a linear relationship between the excess vibrational energy (E_{vib}) dependence of the $S_1 \rightarrow S_0$ IC rate (k_{ic}), as measured by $\alpha = d\ln k_{ic}/dE_{vib}$, and the S_1 - S_0 electronic energy gap of the molecule (the energy gap law) [22]; and (iii), the surprisingly good agreement between the estimated $S_1 \rightarrow S_0$ IC rates for naphthalene and anthracene in solution [27], and those obtained by extrapolation of the exponential energy dependence of the gas-phase IC rates to zero excess energy [20] (cf. Fig. 11). Observation (i) is clearly a characteristic of a molecule which loses memory of its mode of preparation, while Observation (iii) is consistent with the behavior of a molecule which acts as its own heat bath. Observation (ii) is in accord with the theoretical results of Fischer [28] and Lin [29], which show that the statistically-averaged nonradiative decay rate constant resulting from a random distribution of vibrational levels

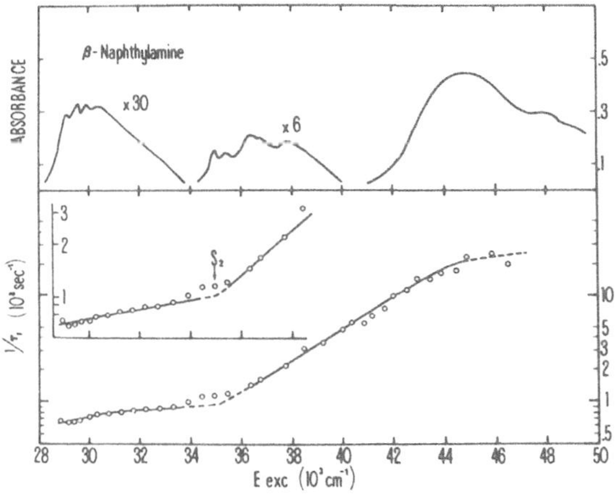

Fig. 9. Excitation energy dependence of fluorescence lifetime in β-naphthylamine. See [19] for experimental details.

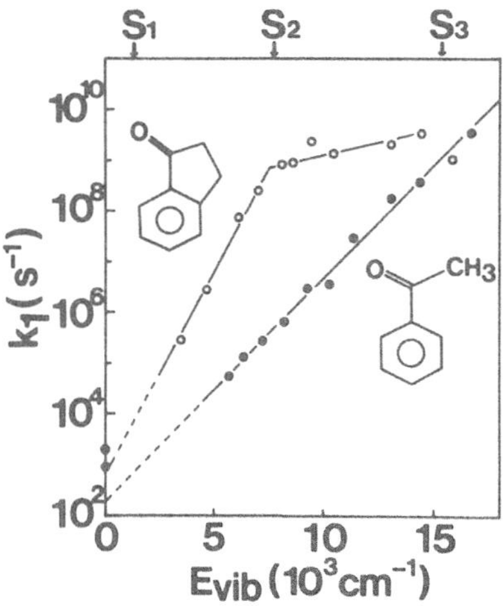

Fig. 10. Triplet state decay rate of phenyl ketones as a function of excess vibrational energy (in the triplet manifold). See [26] for experimental details.

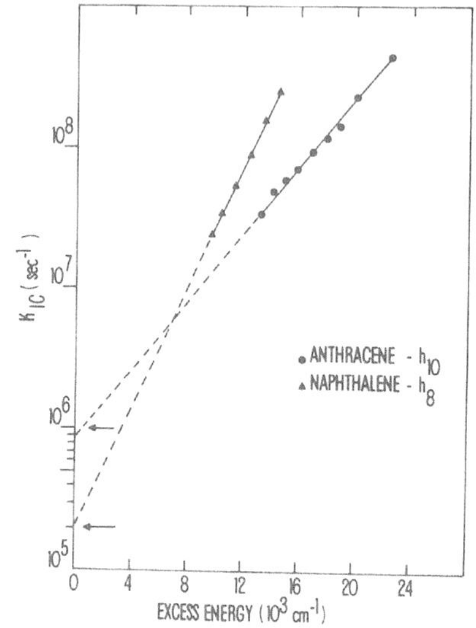

Fig. 11. Comparison between the estimated $S_1 \rightarrow S_0$ internal conversion rates in solution and that obtained by extrapolation of the gas-phase internal conversion rates to zero excess energy. See [20] for experimental details.

is a simple exponential function of excess vibrational energy, with the value of α depending upon the electronic energy gap in a roughly linear fashion. Observation (iii), if confirmed by further work, opens up the interesting possibility of determining the internal conversion rate in solution (or in gas phase under high pressures) from the energy dependence of the radiationless decay rate in dilute vapors. Such a possibility would be a valuable aid to the detailed theoretical understanding of $S_1 \rightarrow S_0$ internal conversion since direct measurement of internal conversion in solution is almost impossible for most molecules because of the small IC rate (relative to the decay rates of other photophysical processes).

6. PHOTOCHEMICAL IMPLICATIONS OF THE ENERGY DEPENDENCE OF ELECTRONIC RELAXATION PROCESSES

In Section 4 we concluded that, in aromatic molecules under collision-free conditions, the $S_1 \rightarrow S_0$ IC rate increases much faster with excitation energy than does the $S_1 \rightarrow T_1$ ISC rate. Because of the very different energy dependence of these radiationless decay rates, internal conversion to the ground state was proposed to be the dominant pathway of S_1 deactivation for large excess energy, even though this pathway is relatively unimportant for low excess energy (exceptions to this rule are molecules with a low-lying S_1, in which $S_1 \rightarrow S_0$ IC is important even for relatively low vibrational excitation [15]). The greater excess energy dependence of internal conversion can be attributed to both the larger $S_1 - S_0$ electronic energy gap and to the greater vibrational frequency changes and normal coordinate displacements between S_1 and S_0, relative to the corresponding $S_1 \rightarrow T_1$ ISC.

In aliphatic carbonyl compounds, the geometry and frequency changes and the electronic energy gap for $S_1(n\pi^*) \rightarrow S_0$ IC are especially large in comparison to those for $S_1(n\pi^*) \rightarrow T_1(n\pi^*)$ ISC; therefore, it is quite possible that internal conversion to the ground state is an important photophysical process even for molecules with modest vibrational excitation. In order to test this possibility, we carried out [30] a simple model calculation of the excess energy dependence of internal conversion for aliphatic ketones, with the carbonyl stretching vibration being considered as the only important optical accepting mode (an assumption which is probably justified by the fact that the carbonyl stretching mode is the highest frequency vibration which suffers the largest displacement and distortion in going from S_1 to S_0). The results indicate that the marked increase in the radiationless decay rate above a threshold excess energy in some aliphatic ketones [31] is consistent with the increasing importance of $S_1 \rightarrow S_0$ IC as a function of increasing vibrational excitation. This suggests the interesting possibility of controlling photochemical dissociation channels by simply altering the amount of excess vibrational energy in S_1. Majer, *et al.* [32], for example, find increasing production of chlorine atoms at shorter excitation wavelengths (within $S_0 \rightarrow S_1$ absorption) in the gas-phase photolysis of chloropentafluoroacetone. The C-Cl bond stretch in S_0 is approximately 75 kcal/mole and is unlikely to be significantly reduced in S_1, since the electronic excitation is primarily localized in the C=O bond. Thus, there is insufficient vibrational energy in S_1 to dissociate the C-Cl bond and it has been suggested [30] that the Cl atoms originate from dissociation of vibrationally-hot ground-state molecules resulting from $S_1 \rightarrow S_0$ IC. Likewise, Knecht [33] finds it necessary to invoke internal conversion and subsequent S_0 dissociation to explain the pressure dependence of product quantum yield in the photochemistry of hexafluoroacetone. It is perhaps even more relevant to mention that Lewis and Lee [34] find two photodecomposition channels (leading to different products) in perfluorocyclobutanone, a result which they interpret in terms of decomposition following $S_1 \rightarrow T_1$ and $S_1 \rightarrow S_0$ radiationless transitions, with the latter dominating at higher excitation energy.

A possible correlation between the energy dependence of unimolecular photochemistry and the energy dependence of electronic relaxation processes in polyatomic molecules is clearly an important problem deserving careful attention in the future.

ACKNOWLEDGMENT

The author wishes to thank his co-workers Sighart Fischer, Gregory Gillispie, Yoshinori Hirata, John Hsieh and Cheng-Schen Huang for their many contributions to the work described in this article. This research was supported by the U.S. Department of Energy.

REFERENCES

1. See, for recent reviews, (a) C. S. Parmenter, in: *MTP International Review of Science, Spectroscopy*, A. D. Buckingham and D. A. Ramsay, eds. (Butterworth, London, 1972), p. 297; (b) M. Stockburger, in: *Organic Molecular Photophysics*, Vol. II, J. B. Birks, ed. (Wiley-Interscience, London, 1974), p. 213; (c) K. F. Freed, in: *Topics in Applied Physics*, Vol. 15, F. K. Fong, ed. (Springer-Verlag, Heidelberg, 1976), p. 23; (d) P. Avouris, W. G. Gelbart and M. A. El-Sayed, Chem. Rev. 77, 793 (1977).
2. A. Amirav, U. Even and J. Jortner, J. Chem. Phys. 71, 2319 (1979); 75, 3770 (1981).
3. T. D. Russell and D. H. Levy, paper FB 7, Symposium on Molecular Spectroscopy, Columbus, Ohio, June 19, 1981; S. Okajima and E. C. Lim. J. Phys. Chem., in press.
4. H. Saigusa and E. C. Lim, J. Chem. Phys. 78, 91 (1983).
5. See, for example, G. W. Robinson, in: *Excited States*, Vol. I, E. C. Lim, ed. (Academic Press, N. Y. 1974), p. 1; B. R. Henry and W. Siebrand, in: *Organic Molecular Photophysics*, J. B. Birks, ed. (Wiley-Interscience, London, 1973), p. 153.
6. P. Pringsheim, Ann. Acad. Sci. Tech. Varsovie 5, 29 (1938).
7. B. Stevens and E. Hutton, Molec. Phys. 3, 71 (1960).
8. W. V. Gruzinskii and N. A. Borisevich, Opt. Spectrosc. 15, 246 (1963) [Opt. Spectrosk. 15, 457 (1963)].
9. M. Stockburger, Z. Physik. Chem. 35, 179 (1962).
10. R. Williams and G. J. Goldsmith, J. Chem. Phys. 39, 2008 (1963).
11. R. J. Watts and S. J. Strickler, J. Chem. Phys. 44, 2423 (1966).
12. J. M. Blondeau and M. Stockburger, Chem. Phys. Lett. 8, 436 (1971).
13. J. O. Uy and E. C. Lim, Chem. Phys. Lett. 7, 306 (1970).

14. Note that the effective energy gap is smaller than the actual gap by the frequency of the promoting mode. This arises from the selection rule on $V^p_{\alpha\beta}$, which requires that the radiationless transition from the lowest vibrational component of $|\alpha\rangle$ must gain one quantum of mode p in going to $|\beta\rangle$.

15. B. Soep, A. Kellmann, M. Martin and L. Lindquist, Chem. Phys Lett. 13, 241 (1972).
16. S. F. Fischer, A. L. Stanford and E. C. Lim, J. Chem. Phys. 61, 582 (1974).

17. A more general expression, which includes the effects of frequency change ($\Delta\omega_j$) and of initial excitation of the promoting mode, can be found in [16,18].

18. S. F. Fischer and E. C. Lim, Chem. Phys. Lett. 26, 312 (1974).
19. J. C. Hsieh, C. -S. Huang and E. C. Lim, J. Chem. Phys. 60, 4345 (1974).
20. C. -S. Huang, J. C. Hsieh and E. C. Lim, Chem. Phys. Lett. 28, 130 (1974).
21. C. -S. Huang and E. C. Lim, Chem. Phys. Lett. 37, 349 (1976).
22. C. -S. Huang and E. C. Lim, J. Chem. Phys. 62, 3826 (1975).
23. J. C. Hsieh and E. C. Lim, J. Chem. Phys. 61, 736 (1974).
24. J. B. Hopkins, D. E. Powers and R. E. Smalley, J. Chem. Phys. 73, 683 (1980), and references therein.
25. See, for example, the paper by Sidney Leach in this volume.
26. Y. Hirata and E. C. Lim, Chem. Phys. Lett. 71, 167 (1980).
27. W. Siebrand and D. F. Williams, J. Chem. Phys. 49, 1860 (1968).
28. S. Fischer, Chem. Phys. Lett. 4, 333 (1969); S. Fischer and E. W. Schlag, Chem. Phys. Lett. 4, 393 (1969).
29. S. H. Lin, J. Chem. Phys. 58, 5760 (1973).
30. G. Gillispie and E. C. Lim, Chem. Phys. Lett. 34, 513 (1975).
31. See, for example, J. Metcalfe and D. Phillips, J. Chem. Soc. Farady III 72, 1574 (1976); P. Avouris, W. D. Hopewell and M. A. El-Sayed, J. Chem. Phys. 66, 1376 (1977).
32. J. R. Majer, C. Olavesen and J. C. Robb, J. Chem. Soc. (A), 893 (1969).
33. D. A. Knecht, Chem. Phys. Lett. 33, 325 (1975).
34. R. S. Lewis and E. K. C. Lee, J. Phys. Chem. 79, 187 (1975).

THEORETICAL STUDIES OF THE ELECTRONIC STRUCTURE AND SPECTRA OF NH_3^+

Angelo R. Rossi

Department of Chemistry
University of Connecticut
Storrs, Connecticut 06268, USA

Phaedon Avouris

IBM Thomas J. Watson Research Center
Yorktown Heights, New York 10598, USA

1. INTRODUCTION

Molecular radical cations, and open-shell molecules in general, are currently attracting a great deal of interest because of an increased awareness of the importance of these molecules (i), in studies of the upper atmosphere, stellar/interstellar media, electrical discharges and flames; (ii), in laser and radiation chemistry; and (iii), in analyses of more conventional chemical systems such as solution-phase oxidation-reduction reactions.

The advent of the laser has vastly enhanced our ability to determine properties of molecular ions. Recently developed techniques for studying molecular ions include laser ionization mass spectroscopy [1], coupled visible and UV multiphoton ionization with consequent fragmentation of the parent ion [2], and emission and laser-induced fluorescence techniques in the gas phase [3] and in cryogenic matrices [4]. These experimental techniques have produced a wealth of information on molecular radical cations and have posed many interesting questions to be explained by theoretical calculations.

The present work is a theoretical study of the electronic and geometric structure of NH_3^+ in its ground [5a] and lower excited states [5b]. We pay particular attention to the interpretation of the two lowest-energy photoelectron bands of NH_3.

S. P. McGlynn et al. (eds.), Photophysics and Photochemistry in the Vacuum Ultraviolet, 875–893.

1.1 The $^2A_1$ (NH_3^+) ← $^1A_1$ (NH_3) Photoelectron Spectrum

The photoelectron spectrum of NH_3 has been obtained by several groups of investigators [6-9]. The lowest ionization potential involves the ionization of the lone pair; it represents the transition (... $1e^4$ $3a_1{}^1$) $^2A_1$ ($NH_3{}^+$) ← (... $1e^4$ $3a_1{}^2$) $^1A_1$ (NH_3, C_{3v}) leading to a planar ground state of the ion. The band structure appears to be characterized by a simple extended progression in the ν_2 bending vibration. In early low-resolution experiments [7], the progression reaches a maximum at $v_2 = 5$. Arguments analogous to those related to the $\tilde{A} \leftarrow \tilde{X}$ transition of NH_3 [5a] led to the expectation that the photoionization spectrum should exhibit a progression in the ν_1 stretching mode as well. A subsequent high-resolution photoelectron spectrum [9], shown in Fig. 1 (top), revealed evidence of ν_1 excitation. Rosenstock and Botter [10, 11] and Harshbarger [12] investigated theoretically the Franck-Condon (FC) activity of ν_1 in the photoelectron spectrum. Botter and Rosenstock [11] found that the FC activity of the ν_1 mode could be quenched provided that the N-H bond length *decreases* by ∿0.06-0.07 Å upon ionization. Alternatively, however, they found [10] that the best argeement with the experimental FC structure reported by Branton, *et al.* [7] was encountered when there was no change in the N-H bond length, or a slight *compression* of it, in the excited state. Under these conditions, FC promotion of the stretching frequency must occur, so that (ν_1, ν_2) combination bands are responsible for some of the intensity at higher energies. On the other hand, Harshbarger [12] found good agreement with experiment [7] if the N-H bond length is taken to *increase* by 0.06 Å upon ionization. Again, combination bands are present. In either case, the lack of any structure assignable to ν_1 promotion had to be rationalized by taking $\nu_1 \simeq 3\nu_2$.

In the present paper we report (Section 2) two-dimensional Franck-Condon factor calculations employing a uniquely determined force field for the ground state of NH_3. The effect of the vibronic linewidth is included. The theoretical FC spectrum is compared with the experimental photoelectron spectrum [9] and conclusions are reached regarding the geometry of the $^2A_1$ ion state.

1.2 The $^2E(NH_3{}^+)$ ← $^1A_1(NH_3)$ Photoelectron Spectrum

After the $^2A_1$ ← $^1A_1$ transition, the next higher-energy photoelectron band of NH_3 corresponds to $(...1e^3 3a_1{}^2)^2E$ ← $(...1e^4 3a_1{}^2)$ $^1A_1$. Photoelectron spectra in this region have been recorded by several groups [9]. Fig. 1 contains the spectrum obtained by Rabalais, *et al.* [9]. Since the 1e orbital has strong bonding character, the ^2E ionic state may be expected to exhibit strong Jahn-Teller (JT) distortions. The experimental spectrum of Fig.

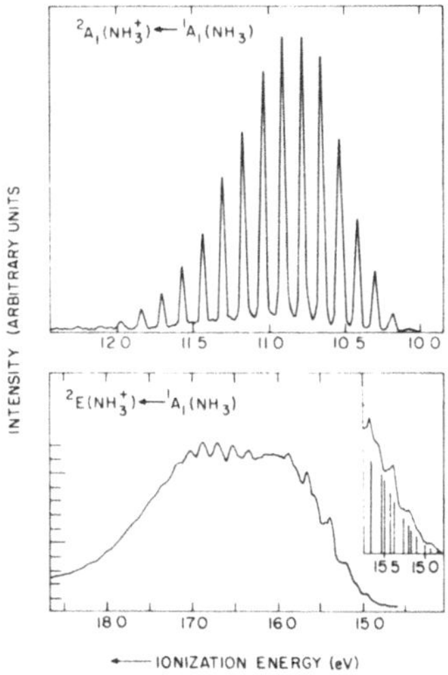

Fig. 1. Top: The first photoelectron band of NH_3. Adapted from [9].
Bottom: The second photoelectron band of NH_3. Adapted from [9].

1 (bottom), however, is very broad (∿2 eV) and possesses a plateau of ∿1 eV, with complicated vibronic structure making its analysis particularly difficult. Symmetry allows all four modes of NH_3 to be excited upon photoionization from the molecular orbital. Only the two e vibrational modes (ν_3 and ν_4) can lead to Jahn-Teller distortions; however, the two totally-symmetric modes (ν_1 and ν_2) are also excited and contribute to the complicated appearance and width of the spectrum.

Rabalais, *et al.* [9] attempted to describe the Jahn-Teller distortion of $^2E(NH_3^+)$ by assuming that only one of the two degenerate modes, namely ν_4, is important. The JT stabilization energy thus obtained was ∿0.5 eV. Köppel, *et al.* [13], on the other hand, have argued in favor of an E X e Jahn-Teller effect (i.e., the interaction of both e modes with the ^2E ionic state). An *ab initio* calculation of the linear coupling constants between the electronic and nuclear motions yields [13] a JT stabilization energy of 0.88 eV by coupling through ν_4, and 0.15 eV by coupling through ν_3. The complete stabilization energy calculated by Köppel, *et al.* [13], therefore, is twice that suggested by Rabalais, *et al.* [9].

In Section 3 we present *ab initio* calculations on the low-lying excited states of NH_3^+, including the Jahn-Teller active ^2E state. Moreover, we calculate the potential energy minima of these states, including distortions along all e and a_1 vibrational modes.

2. FRANCK-CONDON ANALYSIS OF THE $^2A_1(NH_3^+) \leftarrow {}^1A_1(NH_3)$ SPECTRUM

2.1 Computational Procedure

The band shape of an electronic transition, within the Born-Oppenheimer approximation and with neglect of the nuclear-coordinate dependence of the electronic transition moment, may be written as

$$W(\nu,T) = \frac{1}{2\pi} \sum_i B_i(T) \sum_N \frac{\Gamma_{iv}|\langle v|i\rangle|^2}{(E_{iv}-h\nu)^2 + (\Gamma_{iv}/2)^2} \tag{1}$$

where $|\langle v|i\rangle|^2$ is the FC factor which couples vibrational level $|i\rangle$ of the ground electronic state with vibrational level $|v\rangle$ of the excited electronic state; Γ_{iv} is the associated linewidth; E_{iv} is the transition energy; and B_i is the normalized Boltzmann factor for the ith level of the ground electronic state. In applying Eq. (1) to the ammonia problem, the 3N-6 dimensional nuclear coordinate space of $|i\rangle$ and $|v\rangle$ can be restricted to only two dimensions, consisting of the totally symmetric N-H stretching motion (ν_1 mode) and the N-H bending motion (ν_2 mode). Furthermore, we can take $B_i = 0$ for $i \neq 0$ and $B_0 = 1$ at room temperature. The linewidth approximation $\Gamma_{v_i} = \Gamma_{v_1,v_2;0,0} = \Gamma_{v_1}$ introduces the simplifying assumption (cf. Section 1.1) that the linewidth does not vary with the quantum number of the bending mode, but may do so with that of the stretching mode. Thus, for calculating band shapes for the photoelectron spectrum of NH_3, when discrete rotational structure is absent, Eq. (1) reduces to

$$W(\nu) = \sum_{v_1} W_{v_1}(\nu) \tag{2}$$

where

$$W_{v_1} = \frac{\Gamma_{v_1}}{2\pi} \sum_{v_2} \frac{|\langle v_1,v_2|0,0\rangle|^2}{(E_{v_1v_2;00}-h\nu) + (\Gamma_{v_1}/2)^2} \quad . \tag{3}$$

The calculation of the band profile involves enumeration of all two-dimensional FC factors $|\langle v_1,v_2|0,0\rangle|^2$, which are obtained using the harmonic oscillator approximation in the Sharp-Rosenstock [14] formalism. The use of this formalism requires knowledge of geometry differences and force fields for the two

Table 1. Parameters for NH_3 and NH_3^+ Used in the Two-Dimensional Franck-Condon Calculations

	$\tilde{X}$ ($^1A_1$)	Ion $\tilde{X}$ ($^2A_1$)
ν_1, cm^{-1}	3506[a]	2740[c]
ν_2, cm^{-1}	1021[a]	950[c]
r_e(NH), Å	1.012[b]	----[d]
θ_e(HNH), deg	106.7[b]	120[c]

[a]Ref. [15].
[b]Ref. [16].
[c]Ref. [9].
[d]Determined by the calculation. See text for a discussion.

electronic states involved in the transition. The parameters used for the NH_3 and NH_3^+ calculations are collected in Table 1.

2.2 Results and Discussion

The calculations of the structure of the photoelectron spectrum rest on the results of the most recent, high-resolution experiments of Rabalais, *et al.* [9]. The N-H distance in the $^2A_1$ ion state is not known and, therefore, the two-dimensional FC calculations were performed as a function of the N-H bond length of a planar NH_3^+. The best agreement with experiment was found when the N-H bond length in NH_3^+ exceeded that in NH_3 by 0.06 Å, and when $\Gamma_{v_1} = \Gamma_0 = 230$ cm^{-1}. The calculated spectrum is shown in Fig. 2. Under these conditions, the peak-to-peak calculated intensities agree with experiment to within 10% or better. The continuum is also found to be significantly reduced, which is in accord with the most recent photoionization studies [9].

As mentioned in Section 1.1, Rosenstock and Botter [10,11] found that compression of the N-H bond by 0.06 Å eliminates FC activity of the stretching frequency in the NH_3^+ spectrum and removes the requirement that $\nu_1 \simeq 3\nu_2$. To test this case, v_1 of NH_3^+ was taken equal to that in the neutral. However, the experimentally observed FC structure was not well-reproduced: A compression of the N-H bond by 0.06 Å upon ionization gives very

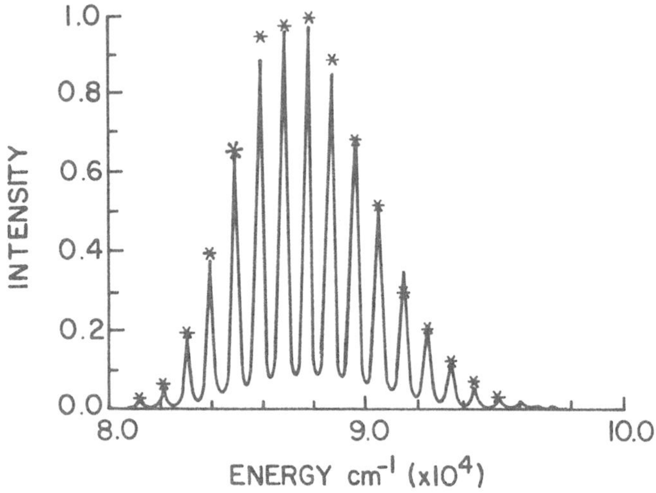

Fig. 2. Theoretical photoelectron spectrum of $^2A_1(NH_3^+) \leftarrow {}^1A_1(NH_3)$. Calculated under the following conditions: Both ν_1 and ν_2 modes are excited; $\nu_1(NH_3^+) = 2740$ cm^{-1}; $\nu_2(NH_3^+) = 950$ cm^{-1}; $\Gamma_0 = \Gamma_{v1} = 230$ cm^{-1}; and, the N-H bond length in the ion was taken to be longer than that in the neutral by 0.06 Å.

poor agreement with observation. Furthermore, contrary to the results of Rosenstock and Botter [10], we find that agreement is also not good when there is no change (or a slight compression) in the N-H bond length. (It is important to note, however, that the present work uses the more recent high-resolution spectrum [9] for comparing calculations with experiment.) That the N-H bond is longer in the ion than in NH_3 is now supported by *ab initio* quantum-chemical calculations [5b, 17]. Harshbarger [12] also reached the conclusion that the N-H bond length in NH_3^+ is 1.07 Å but he fitted the earlier low-resolution spectrum of Branton, *et al.* [7] and used a different harmonic force field for ammonia than the one used here. Thus, congruence between the present and earlier [12] calculations must be coincidental. Domcke, *et al.* [18] obtained good agreement between the observed spectrum of Rabalais, *et al.* [9] with a calculational method which assumed the same force field in the ground state of the neutral and the ground state of the ion. However, the N-H bond length change induced by the ionization was not computed [18].

The present FC calculations suggest that the planar ground state of NH_3^+ has an N-H bond longer than the neutral by $\sim$0.06 Å. We also find that at least two quanta of ν_1 must be excited in order to obtain good agreement with experiment. This suggests a stable NH_3^+, and is in agreement with the studies of Wight, *et al.* [19] which show no other photofragment besides NH_3^+ when NH_3 is ionized with $h\nu < 15$ eV. The stability of NH_3^+ justifies using $\Gamma_{v1} = \Gamma_0$.

3. *AB INITIO* CALCULATIONS ON LOW-LYING STATES OF NH_3^+

The ground electronic state of ammonia is usually taken to have a pyramidal (C_{3v}) geometry with a $1a_1^2 2a_1^2 1e^4 3a_1^2$ orbital configuration. The ground state of NH_3^+ is planar (D_{3h}) at equilibrium and is formed by removing a lone-pair electron ($3a_1$ in C_{3v} or $1a_2''$ in D_{3h}) to yield the ground state electronic configuration $1a_1'^2 2a_1'^2 1e'^4 1a_2''^1$; $\tilde{X}\ ^2A_2''$ (D_{3h}). A schematic representation of the NH_3 valence orbitals is given in Fig. 3. Several types of excitations in NH_3^+ can lead to different kinds of excited states. In particular, we are interested in: (i), excitations from the half-filled orbital into unfilled orbitals; and (ii), transitions from doubly-occupied orbitals to the half-occupied orbital. This latter type of transition is formally equivalent to higher ionization events in a photoelectron spectrum.

In the calculations that follow, we focus on determining the equilibrium geometries of the ground and excited states of NH_3^+,

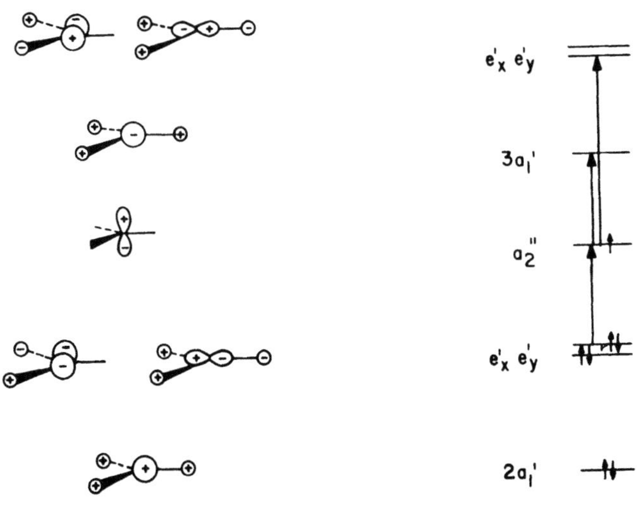

Fig. 3. Valence orbitals of the planar NH_3^+ ion.

as well as on obtaining energy differences between these states. The structural parameters which are varied in these calculations are the N-H bond distances (r_i) and <HNH bond angles (θ_i), as shown in Fig. 4.

3.1 Calculational Details

Ab initio calculations were performed with the GAUSSIAN 80 [20] and GAMESS [21] series of programs on an IBM 370/3033 computer. Molecular geometries were optimized under the appropriate symmetry constraints: [D_{3h}: $^2A_2''(NH_3^+)$, $^2A_1'(NH_3^+)$]; [C_{3v}: $^2E(NH_3^+)$, $^1A_1(NH_3)$]; [C_s: $^2A''(NH_3^+)$, $^2A'(NH_3^+)$]. Geometry optimizations were carried out using Hartree-Fock wavefunctions (restricted (RHF) [22] for closed-shell singlet states, spin-unrestricted (UHF) [23] and spin-restricted (ROHF)[24] for open-shell doublet states.) Correlation-energy calculations were performed at the level of third-order Moller-Plesset perturbation theory [25] (MP3 for closed-shell, UMP3 for open-shell) and at the level of multiconfigurational Hartree-Fock (MCSCF) theory. Geometry optimizations at the Hartree-Fock level (RHF, UHF, ROHF), as well as those at the MCSCF level, utilized a gradient method in which the first derivative of the energy, with respect to nuclear coordinates, was calculated analytically at each point and the result used to determine the direction of steepest descent [26]; geometry optimizations at the correlated level (MP3, UMP3) utilized a Fletcher-Powell optimization procedure with finite difference gradients. The MCSCF calculations were performed by distributing seven electrons from all of the filled orbitals (except for the

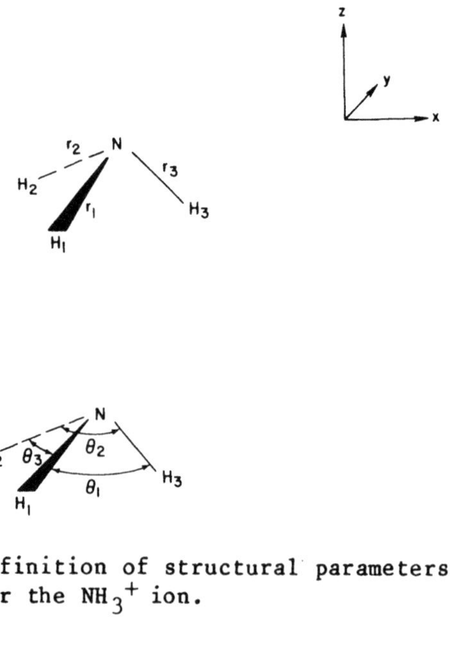

Fig. 4. Definition of structural parameters for the NH_3^+ ion.

core $1a_1'$ orbital) into a space comprised of seven orbitals, consisting of the filled ($2a_1'$, $1e'$, $1a_2''$) and unfilled ($3a_1'$ and $2e'$) valence orbitals, constituting a complete active space (CAS) calculation with about 400 configurations. The basis used in all calculations was the 6-31G** [27] polarization set, which is split-valence and which includes a set of d-polarization functions for the nitrogen and p-polarization functions for the hydrogens.

3.2 Results and Discussion

3.2a The Planar Nondegenerate $^2A_2''$ and $^2A_1'$ States of NH_3^+

Both the ground $^2A_2''$ state and the highly-excited $^2A_1'$ state are planar (D_{3h} symmetry) and only the N-H bond must be varied to obtain the calculated equilibrium geometry. Energies and bond lengths obtained from the different levels of calculation are summarized in Table 2 for the $^2A_2''$ and $^2A_1'$ states. The results presented in Table 2 provide a basis for comparing geometry trends in the several different calculational methods. As expected, the UMP3 level (which includes electron correlation) tends to give a longer N-H bond distance for the $^2A_2''$ state than do the HF methods. Both the UHF and ROHF methods produce comparable geometries for the $^2A_2''$ ground state of NH_3^+. (Comparing results given in Table 2 with those for the ground state of NH_3, given in Table 3, we see that there is a small (~0.01 Å) increase in the N-H bond

Table 2. Energies (au), N-H Bond Distances (Å) and $\langle S^2 \rangle$ Expectation Values for Planar D_{3h} States of NH_3^+

State		Level of Calculation		
		UHF	UMP3	ROHF
$^2A_2''$	Energy	-55.88489	-56.04536	-55.88002
	r_e(NH)	1.0119	1.0211	1.0106
	$\langle S^2 \rangle$	0.760	0.760	0.750
$^2A_1'$	Energy	-55.48905	-55.66187	-55.48380
	r_e(NH)	1.2250	1.2353	1.2070
	$\langle S^2 \rangle$	0.783	0.789	0.750

Table 3. Energy and Optimized Structural Parameters for NH_3 Ground State.

Parameter	Level of Calculation		Experiment [16]
	RHF	MP3	
Energy, au	-56.19554	-56.39599	---
r_e(NH), Å	1.001	1.012	1.012
θ_e(NHN), deg	107.6	106.1	106.7

length following removal of an electron from an a_1 orbital, when electron correlation is included. This result is in qualitative agreement with our FC calculations [5a].) However, differences between the UHF and ROHF methods begin to appear in the excited $^2A_1{}'$ state, where the ROHF value is shorter by about 0.02 Å. While this difference is not large, it does point to the fact that the UHF method incorporates a degree of electron correlation and may yield more reasonable results for those states with longer bond distances. It is also important to note that $\langle S^2 \rangle > 0.75$ for the UHF calculations, thus implying some contamination of the wavefunction by higher spin states which could lengthen the N-H bond distance.

The optimized structural parameters for neutral NH_3 (C_{3v} geometry) are given in Table 3 for the RHF and MP3 calculations. The vertical and adiabatic ionization potentials which follow from the calculations summarized in Tables 2 and 3 are shown in Table 4. This latter table reveals an interesting trend in cor-

Table 4. Comparison of Vertical and Adiabatic Ionization Potentials (eV) for Different Levels of Approximation.

Ionization Potential	Koopmans' Theorem	SCF	UMP3	Experiment [9]
Vertical	11.4	9.2	10.4	10.85
Adiabatic	--	8.5	9.5	10.073

relation effects for states of NH_3^+ and neutral NH_3: Namely, correlation seems to be more important for neutral NH_3 than for NH_3^+ since the increase in both the vertical and adiabatic ionization potentials results from a correlation-induced decrease in energy of about 1 eV more for NH_3 than for NH_3^+.

Several parameters for the $^2A_2''$ ground state of NH_3^+ computed in the present work are compared, in Table 5, with the results of other calculations. The calculated N-H bond distance is similar in every case. On the other hand, the total energy (SCF) is lowest for [28] since, in that work, a triple-zeta plus polarization basis set was used (leading to a better description of core electrons). The a_2'' bending vibration calculated from the present work is in good agreement with experiment; however, the a_1' stretching vibration is not. This latter result corresponds to previous findings [29,30] that calculated vibrational frequencies can deviate substantially from experimental values when electron correlation effects are not included.

3.2b The Degenerate ^2E State of NH_3^+

The ^2E excited state of NH_3^+ is pyramidal and can undergo a Jahn-Teller distortion. The valence orbitals on the nitrogen, and the three hydrogen s orbitals, form the basis for an a_1 and an e representation in C_{3v}:

$$\begin{aligned} \phi_{3a_1} &= (2s + 2p_z)_N - (s_1 + s_2 + s_3)_H \\ \phi_{e_x} &= (2p_x)_N + (s_2 - s_1)_H \\ \phi_{e_y} &= (2p_y)_N + (2s_3 - s_1 - s_2)_H \quad . \end{aligned} \qquad (4)$$

From the above orbitals, and from those of Fig. 3, the following symmetry-adapted doublet spin configurations for NH_3^+ can be constructed:

$$\begin{aligned} \chi_1(^2A_2'') &= |\cdots (e_x')^2(e_y')^2(1a_2'')^1\rangle \quad &\text{(Planar)} \\ \chi_2(^2E_x) &= |\cdots (e_x)^1(e_y)^2(3a_1)^2\rangle \quad &\text{(Pyramidal)} \\ \chi_2(^2E_y) &= |\cdots (e_x)^2(e_y)^1(3a_1)^2\rangle \quad &\text{(Pyramidal)} \quad . \end{aligned} \qquad (5)$$

The Jahn-Teller theorem guarantees the existence of a non-totally symmetric vibration along which the molecule can distort to lift the degeneracy. The symmetry vibrations for NH_3^+ in C_{3v} symmetry

Table 5. Comparison of Properties of the NH_3^+ $^2A_2''$ Ground State with Experiment and with Other Calculations

Parameter	This Work	[28]	[27]	[17]	Experiment [9]
Energy, au	-55.88489	-55.8913	-55.87323	-55.80167	---
r_e(NH), Å	1.012	1.011	1.012	1.010	---
$\nu_1(a_1')$, cm^{-1}	3558	3508	---	---	2742
$\nu_2(a_2'')$, cm^{-1}	910	814	---	---	895
ν_3(e, stretch), cm^{-1}	3763	---	---	---	---
ν_4(e, bend), cm^{-1}	1656	---	---	---	---

are (see Fig. 4 for a definition of parameters):

$$\left.\begin{aligned} S_1 &= 1/\sqrt{3}\ (r_1 + r_2 + r_3) \\ S_2 &= r_e/\sqrt{3}\ (\theta_1 + \theta_2 + \theta_3) \end{aligned}\right\} \quad (a_1 \text{ modes})$$

$$\left.\begin{aligned} S_{3a} &= 1/\sqrt{6}\ (2r_3 - r_2 - r_1) \\ S_{3b} &= 1/\sqrt{2}\ (r_2 - r_1) \end{aligned}\right\} \quad (e \text{ mode}) \qquad (6)$$

$$\left.\begin{aligned} S_{4a} &= r_e/\sqrt{6}\ (2\theta_3 - \theta_2 - \theta_1) \\ S_{4b} &= r_e/\sqrt{2}\ (\theta_2 - \theta_1) \ . \end{aligned}\right\} \quad (e \text{ mode})$$

The totally symmetric vibrations S_1 and S_2 cannot lift the ^2E degeneracy; however, both sets of degenerate vibrations S_3 and S_4 are adequate to distort the $NH_3{}^+$ ion. In order to determine how the Jahn-Teller effect distorts a degenerate species, we assume initially that only the degenerate stretching vibration (S_3) is active, so that the following definitions can be made [31]:

$$S_{3a} = q_y = \Delta\sin\theta$$

$$S_{3b} = q_x = \Delta\cos\theta \qquad (7)$$

$$\Delta = (q_x{}^2 + q_y{}^2)^{1/2} \ .$$

If the Hamiltonian is then expanded in terms of q_x and q_y about the ^2E minimum in C_{3v}, we find

$$H = H_0 + q_x\left(\frac{\partial H}{\partial q_x}\right)_0 + q_y\left(\frac{\partial H}{\partial q_y}\right)_0 + \ldots \ . \qquad (8)$$

A first-order Jahn-Teller distortion produces an energy which is linear in Δ and independent of θ. That is,

$$E = E_0 \pm \lambda\Delta \qquad (9)$$

where $\lambda = \langle\chi_2(^2E_y)|\left(\frac{\partial H}{\partial q_y}\right)_0|\chi_2(^2E_y)\rangle$. And, the corresponding wavefunctions are

$$\chi_+ = \cos\tfrac{1}{2}(\theta-90^\circ)\chi_2(^2E_x) - \sin\tfrac{1}{2}(\theta-90^\circ)\chi_2(^2E_y)$$

$$\chi_- = \sin\tfrac{1}{2}(\theta-90^\circ)\chi_2(^2E_x) - \cos\tfrac{1}{2}(\theta-90^\circ)\chi_2(^2E_y) \quad . \tag{10}$$

When the term linear in Δ is added to the energy E_0, the energy surfaces can be described as two conical sheets (an upper and a lower sheet), as shown in Fig. 5., which touch at a point where there is no distortion ($\Delta = 0$). If the wavefunctions in Eq. (10) are then used as a basis for adding second-order effects, Davidson, *et al.* [31] have shown that the form of the energy expression is

$$E = E_0 \pm \lambda\Delta \pm (\mu \pm \tfrac{1}{2}\nu \sin 3\theta)\, \Delta^2 \tag{11}$$

where the constants μ and ν represent sums which couple excited states to the degenerate zeroth-order state.

The important point to note is that second order effects introduce a θ dependence in the energy. Thus, from Eq. (11) there are two sets of minima, one set at $\theta = 30^\circ, 150^\circ, 270^\circ$ and the

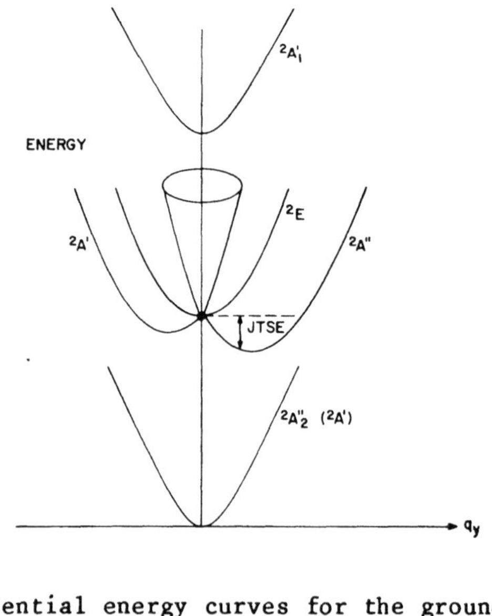

Fig. 5. Potential energy curves for the ground $^2A_2''$ and excited ^2E, $^2A_1'$ states of NH_3^+ as a function of the distortion coordinate q_y, leading to C_s symmetry.

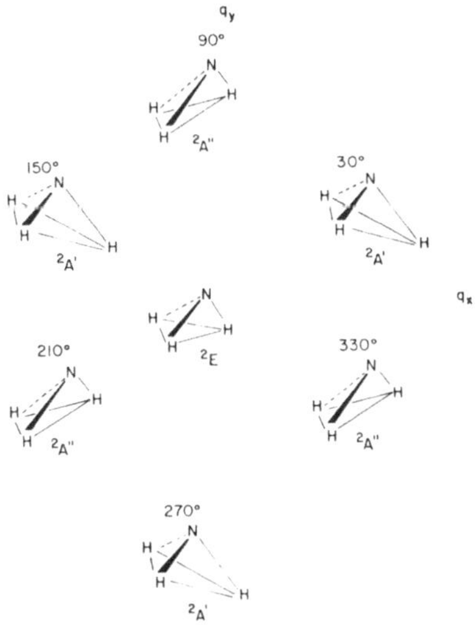

Fig. 6. The geometry of the $^2A'$ and $^2A''$ extrema of NH_3^+ as a function of the pseudorotation angle θ (see Eq.(7) of text).

other set at $\theta = 90^\circ, 210^\circ, 330^\circ$. These equivalent sets are given in Fig. 6, where the three-fold periodicity of each set is shown for fixed values of Δ. Another important point, as shown in Fig. 6, is that the two minima which occur for $\theta = 90^\circ$ and $\theta = 270^\circ$ are both of C_s symmetry, with distortion along the q_y vibration. This allows subsequent calculations for relevant distorted geometries to be carried out in C_s. Of the two sets of angles shown in Fig. 6, that set which corresponds to true minima, and that set which corresponds to saddle points, must be determined by calculation. In Table 6, we summarize the geometries which were obtained when the optimizations were carried out with C_s symmetry for the two distorted Jahn-Teller components ($^2A'$ and $^2A''$) and with C_{3v} symmetry for the ^2E state. There is close agreement between the UHF and ROHF results, with a bond length deviation of no more than 0.002 Å and a bond angle deviation of $\sim 0.05^\circ$. However, when those methods (UMP3 and MCSCF) which incorporate some degree of electron correlation are compared, the differences between bond lengths and bond angles are much larger. (The UMP3 results actually agree quite closely with the UHF and ROHF results.) Only the MCSCF method could be used to determine the energy and geometry of the $^2A'$ component since the distorted ground state has the same $^2A'$ symmetry; thus, the excited $^2A'$ component was obtained as a second root in the MCSCF calculation.

Table 6. Geometries of the ^2E State and Its Distorted Jahn-Teller Components.

State	Geometrical Parameters[a]	Method of Calculation			
		UHF	ROHF	UMP3	MCSCF
^2E	$r_1 = r_2 = r_3$	1.0929	1.0935	1.1092	1.1300
	$\theta_1 = \theta_2 = \theta_3$	94.5	93.6	96.4	89.9
$^2A''$	r_3	1.0220	1.0209	1.0367	1.0512
	$r_1 = r_2$	1.1361	1.1347	1.1537	1.1819
	$\theta_1 = \theta_2$	105.6	105.7	103.7	101.2
	θ_3	53.0	53.5	52.7	50.6
$^2A'$	r_1	---	---	---	1.2484
	$r_1 = r_2$	---	---	---	1.0642
	$\theta_1 = \theta_2$	---	---	---	83.5
	θ_3	---	---	---	109.8

[a]Values of r_i are in angstroms; values of θ_i are in degrees. See Fig. 4 for a definition of r_i and θ_i.

The total energies of the ^2E state and its distorted $^2A'$ and $^2A''$ components, along with the Jahn-Teller stabilization energy (JTSE) for each component, are given in Table 7. The JTSE is the difference in energy between the undistorted ^2E state and the lower distorted ($^2A'$ or $^2A''$) component. The first thing to note in Table 7 is that the $^2A''$ component is $\sim$0.5 eV more stable than the $^2A'$ component. Also, from Table 7 we see that the JTSE (relative to the $^2A''$ component) is predicted to be greater than 1 eV by all of the calculational methods. These calculations indicate that the JTSE is 1.2-1.4 eV, which should be contrasted with the value of $\sim$0.5 eV obtained by Rabalais, *et al.* [9] when only the

Table 7. Total Energies (au) for the ^2E State and Its Distorted Components and the Jahn-Teller Stabilization Energy [JTSE] (eV)

State	Calculational Method			
	UHF	ROHF	UMP3	MCSCF
^2E	-55.65264	-55.64598	-55.82230	-55.72234
$^2A'$	---	---	---	-55.75028
$^2A''$	-55.69977	-55.69656	-55.86916	-55.76745
JTSE[a]	1.28	1.38	1.28	1.23

[a] The JTSE is the difference in energy between the undistorted ^2E state and the lowest distorted component ($^2A''$).

stretching mode was assumed to be active in the distortion. Since the calculations presented here involve optimization over all bond lengths and bond angles, we conclude that distortions over both of these sets of variables contributes substantially to the JT-induced energy lowering. This result also agrees with the findings of Köppel, *et al.* [13].

ACKNOWLEDGMENT

This work was supported in part by a generous grant from the Research Corporation.

REFERENCES

1. D. H. Parker, R. B. Bernstein and D. A. Lichten, J. Chem. Phys. 75, 2577 (1981).
2. (a) U. Boesl, H. J. Neusser and E. W. Schlag, J. Chem. Phys. 72, 4327 (1980); (b) R. S. Pandolfi, D. A. Gobeli and M. A. El-Sayed, J. Phys. Chem. 85, 1779 (1981); (c) K. R. Newton, D. Lichten and R. B. Bernstein, J. Phys. Chem. 85, 15 (1981).
3. (a) T. A. Miller and V. E. Bondybey, J. Chim. Phys. 77, 695 (1980) and references cited therein; (b) J. P. Maier, Angew. Chem. 20, 638 (1981) and references cited therein.

4. L. Andrews, in: *Molecular Ions: Geometric and Electronic Structure*, NATO-ASI, J. Berkowitz, ed. (Plenum Press, New York, 1981).
5. (a) Ph. Avouris, A. R. Rossi and A. C. Albrect, J. Chem. Phys. 74, 5516 (1981); (b) A. R. Rossi and Ph. Avouris, to be published.
6. M. I. Al-Joboury and D. W. Turner, J. Chem. Soc. 4434 (1964).
7. G. R. Branton, D. C. Frost, F. G. Herring, C. A. McDowell and I. A. Stenhouse, Chem. Phys. Lett. 3, 581 (1969).
8. M. J. Weiss and G. M. Lawrence, J. Chem. Phys. 53, 214 (1970).
9. J. W. Rabalais, L. Karlson, L. O. Werme, T. Bergmark and K. Siegbahn, J. Chem. Phys. 58, 3370 (1973).
10. H. M. Rosenstock and R. Botter, in: *Recent Developments in Mass Spectrometry*, K. Ogata and T. Hayakawa, eds. (University of Tokyo, 1970), p. 797.
11. R. Botter and H. M. Rosenstock, Adv. Mass Spectrom. 4, 479 (1968).
12. W. R. Harshbarger, J. Chem. Phys. 56, 177 (1972).
13. H. Koppel, L. S. Cederbaum, W. Domcke and W. Von Niessen, Mol. Phys. 35, 1283 (1978).
14. T. E. Sharp and H. M. Rosenstock, J. Chem. Phys. 41, 3453 (1964).
15. W. S. Benedict and E. K. Plyer, Can. J. Phys. 35, 1235 (1957).
16. Y. Morino, K. Kichitsu and S. Yamamoto, Spectrochim. Act. 24, Part A, 335 (1968).
17. W. A. Lathan, W. J. Hehre, L. A. Curtiss and J. A. Pople, J. Am. Chem. Soc. 93, 5377 (1971).
18. W. Domcke, L. S. Cederbaum, H. Koppel and W. Von Niessen, Mol. Phys., 34, 1759 (1977).
19. G. R. Wight, M. J. Van der Wiel and C. E. Brion, J. Phys. B10, 1863 (1977).
20. (a) J. S. Binkley, R. A. Whiteside, R. Krishnan, R. Seeger, D. J. DeFrees, H. B. Schlegel, S. Topiol, L. R. Kahn and J. A. Pople, QCPE 10, 416 (1981); (b) A more complete description of the work in this section can be found in A. R. Rossi and P. Avouris, J. Chem. Phys. 78, 0000 (1983).
21. M. Dupius, D. Spangler and J. J. Wendoloski, QCPE 1, 91 (1981).
22. C. C. J. Roothaan, Rev. Mod. Phys. 23, 69 (1951).
23. J. A. Pople and R. K. Nesbet, J. Chem. Phys. 22, 571 (1954).
24. E. R. Davidson, Chem. Phys. Lett. 21, 565 (1973).
25. J. A. Pople, J. S. Binkley and R. Seeger, Int. J. Quantum Chem. 9, 229 (1975).
26. J. A. Pople, R. Krishnan, H. B. Schlegel and J. S. Binkley, Int. J. Quantum Chem. S13, 225 (1979).
27. P. C. Hariharan and J. A. Pople, Mol. Phys. 27, 209 (1974).
28. H. Ågren, J. Müller and J. Nordgren, J. Chem. Phys. 72, 4078 (1980).

29. R. F. Hout, Jr., B. A. Levi and W. J. Hehre, J. Comp. Chem. 3, 234 (1982).
30. Y. Yamaguchi and H. F. Schaefer, III, J. Chem. Phys. 73, 2310 (1980).
31. The formalism for the Jahn-Teller distortion used here is due to Borden and Davidson: (a) For reviews see W. T. Borden and E. R. Davidson, Ann. Rev. Phys. Chem. 30, 125 (1979); Acc. Chem. Res. 14, 69 (1981); (b) R. L. Martin and E. R. Davidson, Mol. Phys. 35, 1713 (1978); (c) E. R. Davidson and W. T. Borden, J. Chem. Phys. 67, 2191 (1977).

THEORETICAL CORRELATIONS OF ORGANIC PHOTOCHEMICAL REACTIONS IN THE VUV

Angelo R. Rossi

Department of Chemistry
University of Connecticut
Storrs, Connecticut 06268, USA

R. Srinivasan

IBM Thomas J. Watson Research Center
Yorktown Heights, New York 10598, USA

1. INTRODUCTION

One of the most widely-held and important generalizations in solution-phase photochemistry is Kasha's Rule [1], which states that emission occurs from vibrationally relaxed molecules in their lowest excited singlet (S_1) or triplet (T_1) state, independently of the state which was intitially excited. The conventional picture is that excitation to vibrationally excited levels of the lowest electronic state, or to higher electronic states, results in rapid radiationless decay to an equilibrium distribution of vibronic states of S_1 or T_1 before emission or photochemical reactions occur. However, it is now generally accepted that photochemical reactions of organic molecules are not limited to those occurring from the lowest excited state of the relevant manifolds [2]. A molecule containing excess electronic and vibrational energy can undergo processes such as nonradiative decay, radiative decay and photochemistry. A comparison of spectroscopic and photochemical rate constants is given in Table 1: Rate constants for nonradiative decay have values of $\sim 10^{12}$ - 10^{14} s^{-1}; however, the rate constant for a unimolecular reaction, assuming $E_a \ll E^*$ (cf. Table 1), can reach a limiting value of $\sim 10^{12}$ s^{-1}. Moreover, fluorescence rate constants are of the order of 10^9 s^{-1}. Thus, there is a basis for supposing that photochemical reactions in solution can compete effectively with nonradiative decay and

S. P. McGlynn et al. (eds.), Photophysics and Photochemistry in the Vacuum Ultraviolet, 895–911.

Table 1. Comparison of Spectroscopic and Reaction Rate Constants†

Nonradiative Decay in Solutions:	$k_{ic} \approx 10^{12}\ s^{-1} - 10^{14}\ s^{-1}$
Radiative Decay in Solutions:	$k_f \simeq 10^{9}\ s^{-1}$
Rate Constant for a Unimolecular Reaction (RRK):	$k \simeq A \left(\frac{E^*-E_a}{E^*}\right) s^{-1} \simeq A \approx 10^{12}\ s^{-1}$

†E^* ≡ excitation energy; E_a ≡ activation energy; A ≡ Arrhenius pre-exponential factor.

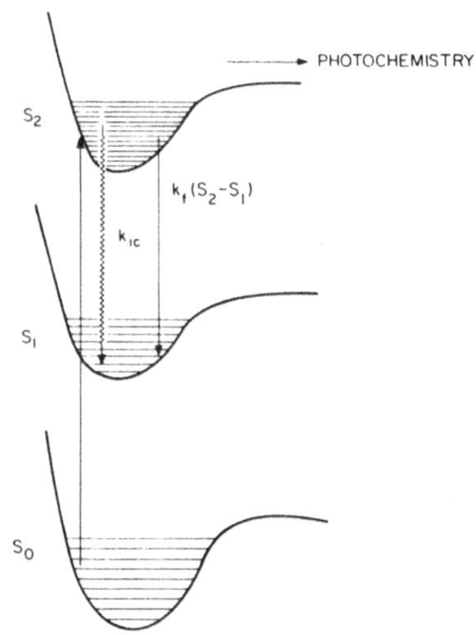

Fig. 1. State diagram indicating competing rates of radiative decay, nonradiative decay and photochemistry.

fluorescent decay. This type of competition between photochemical reactions and fluorescent or nonradiative decay processes is shown schematically in Fig. 1.

There has been relatively little attention given to higher excited states of dienes (either conjugated or isolated). The principal problem has been the limited experimental accessibility of even the *lowest* excited state of an isolated olefin, and the consequent impossibility of irradiating any other excited state with photons of sufficient energy. We have developed techniques [3] for the study of vacuum ultraviolet photochemistry of organic molecules in the condensed phase and have used these techniques in the photochemistry of cyclic systems containing chromophores. Some of the characteristics of VUV photochemistry (185 nm) are:

(i) Absorptions are usually allowed and tend to be intense.
(ii) Quantum yields for decomposition are high (0.1-1.0).
(iii) Only singlet states are involved -- triplet states rarely play a role (population of triplets will be small since $k_{ic} > k_p$).
(iv) Bare chromophores are affected, with no involvement of substituents.
(v) Density of states is large (multiple valence and Rydberg states).
(vi) Radiative lifetime is very short (1.0×10^{-9} s).

The progressive increase in the density of states as one probes deeper into the VUV, and the decreasing lifetime of such states with respect to decomposition (the principal photodecay process), makes the region from 230 to 185 nm very attractive for the study of wavelength and temperature effects on photochemistry. The object of this paper is to compare calculations performed on 1,4-cyclohexadiene and methylene cyclopropane with experimental results on these molecules at 6.7 eV (185 nm).

2. 1,4-CYCLOHEXADIENE

The reversible system shown in Eq. (1),

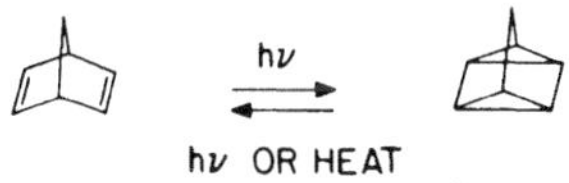

(1)

1 *2*

which involves the conversion of norbornadiene (*1*) to quadricyclane (*2*), has been observed on direct irradiation [4] and triplet sensitization with a variety of compounds [5]. Other work has focused on (i), the use of photosensitization [6,7] and photolysis of charge-transfer complexes [8] of quadricyclane to bring about the reverse reaction; and (ii), the application of this system for the storage of solar energy.

A related system is provided by 1,4-cyclohexadiene (*3*):

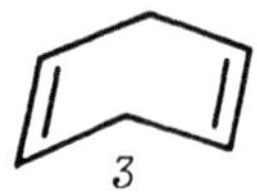

3

The properties of *3*, and its photochemistry in the VUV, are of interest from two points of view: (i), While the interaction between the double bonds in the ground state of *3* has been examined theoretically [9], such an analysis has not been extended to the excited electronic states in any detail. Moreover, experimental results [10,11] identifying the positions of the excited states have been a subject of some controversy [11]. (ii), The excited states and photochemistry of *1* have been extensively studied [4,11,12]. Thus, one anticipates that information concerning reaction pathways and excited-state energies of *1* may be applied correlatively to *3*.

2.1 Theory and Discussion

The purpose of the present work is to extend the earlier theoretical analysis [9] of 1,4-cyclohexadiene by providing a detailed description of the excited states of this molecule. In Fig. 2, the relative energy (GAUSSIAN-70 [13], STO-3G basis [14])

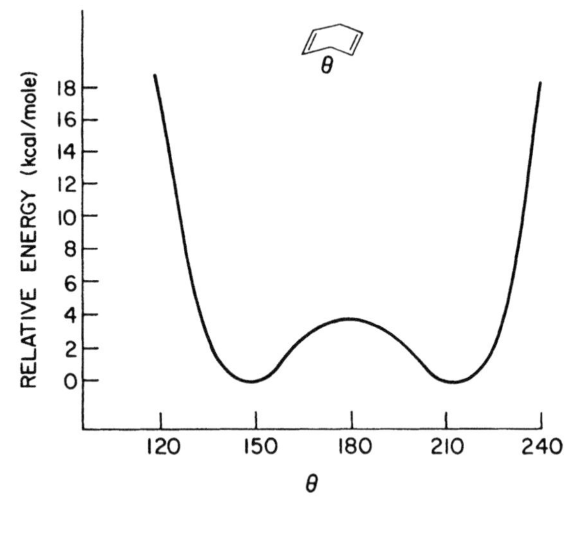

Fig. 2. Calculated energy of ground-state 1,4-cyclohexadiene as a function of dihedral angle θ.

of ground-state 1,4-cyclohexadiene is shown as a function of the dihedral angle θ. Although experimentally θ = 160° [15], the computed minimum energy is at θ ≃ 150°. For θ > 130°, however, the potential energy surface is soft, since only 4 kcal/mole are required to achieve a planar geometry. (This barrier height represents an upper limit since an idealized geometry was assumed and geometry optimization was not performed.)

An orbital interaction diagram, similar to that presented by Hoffmann [9], showing the interaction of the methylene groups with the π orbitals in 1,4-cyclohexadiene is presented in Fig. 3. One of the salient features of this diagram is the manner in which the hyperconjugative interaction of the intervening methylene groups serves to destabilize the $\pi_1 + \pi_2$ combination, while the $\pi_1 - \pi_2$ combination remains unaffected. However, the influence of the hyperconjugative interaction of the methylene groups is a sensitive function of dihedral angle, as shown in Fig. 4. The behavior of $\pi_1 + \pi_2$ and $\pi_1 - \pi_2$ as a function of θ indicates that the $\pi_1 + \pi_2$ combination crosses the $\pi_1 - \pi_2$ combination at θ ≃ 140°. There is also a crossing of the unfilled virtual $\pi_1^* + \pi_2^*$ and $\pi_1^* - \pi_2^*$ molecular orbitals. It is the relationship between the final interacted orbitals for 1,4-cyclohexadiene, and the possible excited states derived from these orbitals, which is the concern of the present analysis.

The combinations of ethylenic π orbitals are labelled in

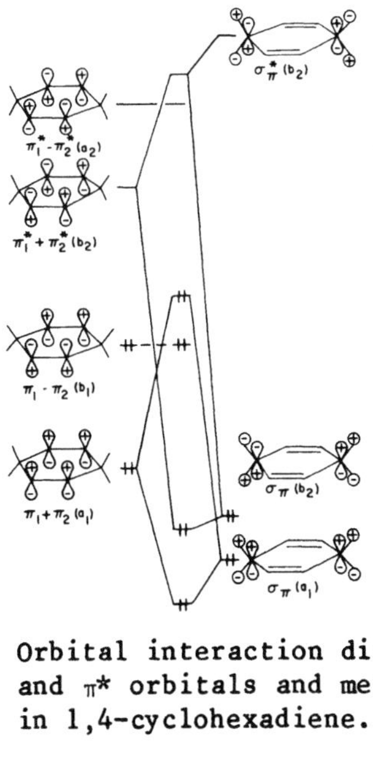

Fig. 3. Orbital interaction diagram for π and π* orbitals and methylene groups in 1,4-cyclohexadiene.

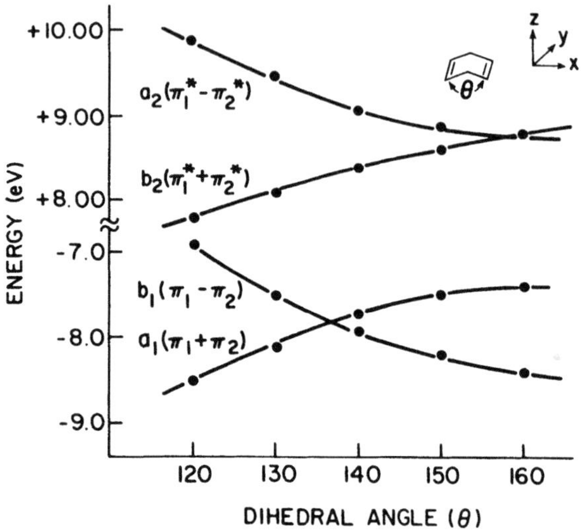

Fig. 4. Energies of the valence π and π* orbitals of 1,4-cyclohexadiene as a function of dihedral angle θ.

Fig. 4 according to C_{2v} symmetry (with respect to the coordinate system shown in the figure) as $b_1(\pi_1 - \pi_2)$, $a_1(\pi_1 + \pi_2)$, $a_2(\pi_1^* - \pi_2^*)$ and $b_2(\pi_1^* + \pi_2^*)$. All of the singlet/triplet excited states resulting from electron promotion within this set of orbitals belong to either the A_2 or the B_2 irreducible representation. Of these, the four lowest-energy singlet states and the two lowest-energy triplet states are shown in Fig. 5 as a function of dihedral angle θ. (The calculations were performed using the INDO/S program of Zerner [16]. Configuration interaction for each symmetry type involved all single excitations from the ten highest occupied orbitals to the ten lowest unoccupied orbitals.) A more complete description of these excited states, for $\theta = 160^{\circ}$ and $\theta = 120^{\circ}$, is given in Table 2. Several important conclusions follow from these calculations. *First*, the lowest-energy $^3A_2$ ($\theta = 160^{\circ}$) state is a mixture of two configurations, whereas the lowest-energy corresponding $^1A_2$ ($\theta = 160^{\circ}$) state is predominantly one configuration. Moreover, the higher-excited singlet states incorporate increasing amounts of σ* character (from C=C), as evidenced by the 6.4 eV $^1B_2$ state in which there is a 66% contribution arising from electron promotion from the $a_1(\pi_1 + \pi_2)$ orbital to a $\sigma_1^* + \sigma_2^*$ orbital. The trend of increasing σ* (C=C) character of the higher excited states is maintained even when θ is decreased, thus implying that this trend may be of importance in molecules, such as norbornadiene, which possess small dihedral angles. *Second*, the results reported in Table 2 are in satisfac-

Table 2. Description and Composition of the Three Lowest Excited State Singlets and Two Lowest Triplets for 1,4-Cyclohexadiene

State	Excitation Energy (eV)	Percent Composition				
		$a_1^1 a_2^1$	$a_1^1 b_2^1$	$b_1^1 a_2^1$	$b_1^1 b_2^1$	$a_1^1 (\sigma_1^* + \sigma_2^*)^1$
		Dihedral Angle θ = 160°				
$^3A_2$	3.7	64	--	--	34	--
$^3B_2$	3.8	--	56	41	--	--
$^1A_2$	5.9	92	--	--	--	--
$^1B_2$	6.3	--	59	16	--	20
$^1B_2$	6.4	--	20	--	--	66
		Dihedral Angle θ = 120°				
$^3A_2$	2.6	9	--	--	89	--
$^3B_2$	3.3	--	25	71	--	--
$^1A_2$	4.0	--	--	--	95	--
$^1B_2$	5.4	--	--	--	--	93
$^1B_2$	5.4	--	19	80	--	--

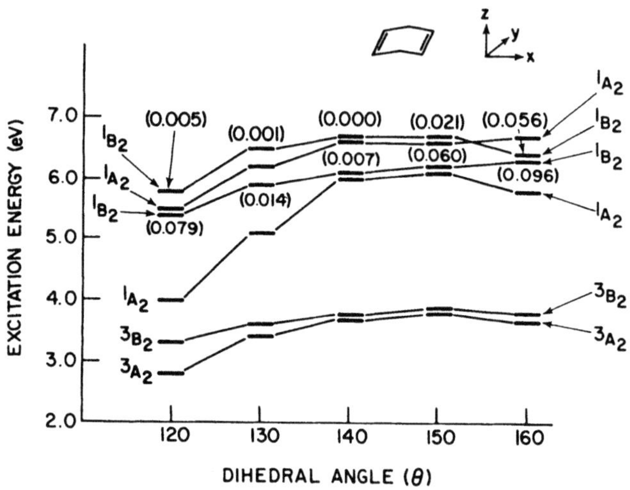

Fig. 5. Energies (eV) of the singlet and triplet states of 1,4-cyclohexadiene relative to the ground state ($^1A_1$). Numbers in parentheses indicate oscillator strengths for allowed transitions.

tory agreement with previous results on the electron-impact spectrum of 1,4-cyclohexadiene [11], in which there is a low-energy spectral feature, extending from 3.4 to 5.4 eV, attributable to the lowest-energy triplet. Other high-lying singlet states are found in the range 6.14-7.95 eV. The lower-lying singlet state could be of Rydberg character and, since Rydberg functions are not included in the basis, care should be used when comparing the present results with experimental values. *Third*, the $\theta = 120^\circ$ results provide a basis for understanding the photochemical reactions of norbornadiene, which has a smaller dihedral angle than 1,4-cyclohexadiene. The percent composition of the $b_1{}^1b_2{}^1$ configuration is very large for both the $^{3,1}A_2$ states. Although direct optical excitation to either of these states is forbidden, the excited-state molecule can readily relax to the $^1A_2$ state to form quadricyclane. Quadricyclane formation from norbornadiene is an extremely efficient reaction and accounts for over 95% of the norbornadiene that disappears in the 250-200 nm wavelength region [4]. In addition, the $^3A_2$ state is calculated to have ∿90% of the $b_1{}^1b_2{}^1$ configuration, which can also lead to quadricyclane by triplet sensitization.

The primary photochemical reactions that are observed to take place on the irradiation of 1,4-cyclohexadiene at 185 nm are shown in Eqs. (2), (3) and (4).

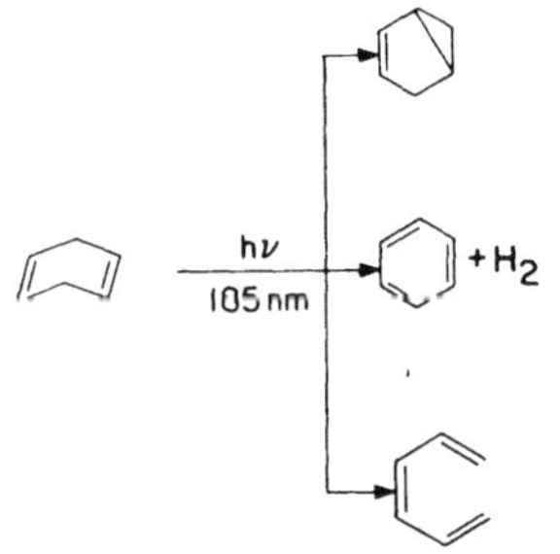

(2)

(3)

(4)

The identities of those excited states of 1,4-cyclohexadiene that take part in these reactions are best discussed in terms of the $\theta = 160^\circ$ data of Table 2. Starting at low energies, the lowest triplet state ($^3A_2$) can be identified as that state which is potentially capable of leading to (2 + 2) addition. That this reaction is *not* observed is not surprising since $^3A_2$ has only 34% of the $b_1{}^1b_2{}^1$ configuration required for this reaction. This can be compared to the 89% $b_1{}^1b_2{}^1$ contribution that is found for $^3A_2$ in the $\theta = 120^\circ$ conformation, which is the geometry of norbornadiene. The only other triplet state that is available is $^3B_2$ and this must be identified (by a process of elimination) with the di-π-methane rearrangement, Eq. (2) [17].

The singlet states of 1,4-cyclohexadiene (Table 2, $\theta = 160^\circ$) are surprisingly closely spaced. The highest of them ($^1B_2$) is readily accessible at $h\nu = 6.7$ eV. In view of the large contribution to this state from the $a_1{}^1\ (\sigma_1{}^* + \sigma_2{}^*)^1$ configuration, it is tempting to attribute all three of the observed hydrogen migration reactions to $^1B_2$. The representation of the $\sigma_1{}^* + \sigma_2{}^*$ orbital given in *4*

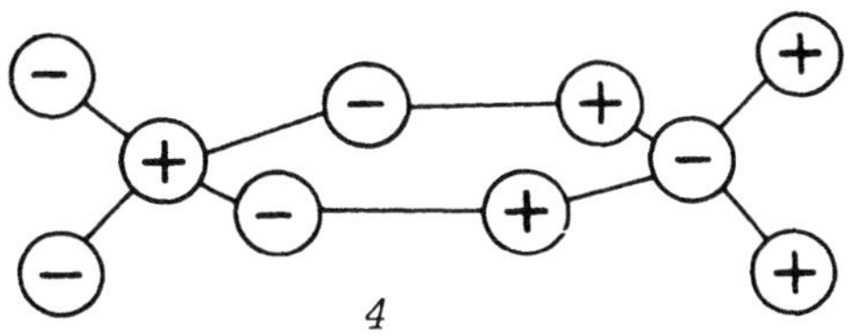

4

shows that there is considerable C-H antibonding character in this orbital; thus, excitation of an electron to $\sigma_1{}^* + \sigma_2{}^*$ could promote facile cleavage of the C-H bond.

3. METHYLENE CYCLOPROPANE

Even though the thermal rearrangement of methylene cyclopropane *5* , shown in Eq. (5),

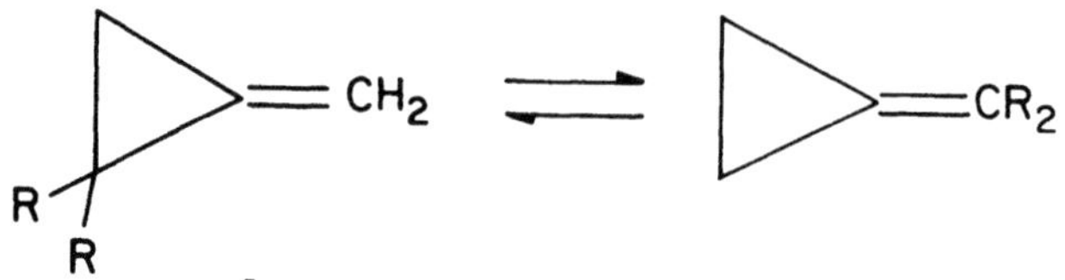

(5)

5

is a well-studied reaction [18], there remain many unanswered questions concerning the reaction mechanism. This isomerization of methylene cyclopropane is an allowed *thermal* ($\sigma_S{}^2 + \pi_A{}^2$ or $\sigma_A{}^2 + \pi_S{}^2$) process [19]; however, it is an allowed *photochemical* ($\sigma_A{}^2 + \pi_A{}^2$ or $\sigma_S{}^2 + \pi_S{}^2$) reaction as well [19]. Moreover, the two reaction pathways have different stereochemical consequences.

When a solution containing substituted methylene cyclopropane is irradiated at 185 nm (6.7 eV), all possible methylene cyclopropane rearrangement products are obtained in comparable yields. This result is in striking contrast to the thermal arrangement reaction, which leads mostly to one rearranged product [20]. The other important photochemical reaction at this wavelength leads to an olefin and an acetylene, as shown in Eq. (6),

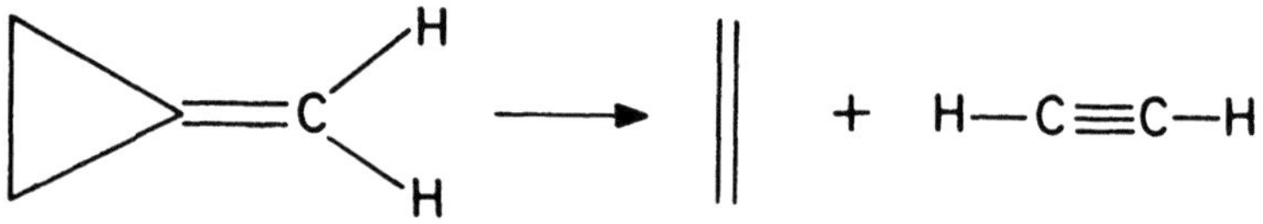

(6)

with retention of the olefin stereochemistry. A likely intermediate in the rearrangement of methylene cyclopropane (MCP) is the trimethylenemethane (TMM) diradical, which could be responsible for the geometric isomerization of the starting material.

3.1 Theory and Discussion

The valence π orbitals for the planar (D_{3h}) trimethylenemethane radical are shown in Fig. 6. Those configurations which produce electronic states important in the thermally accessible reactions of trimethylenemethane are given at the top of Table 3 [21]. The lowest-energy electronic configuration (neglecting core electrons) is $(\ldots 1a_2''^2\ 1e''^2)$, which gives rise to a $^3A_2'$ ground state, a $1^1E'$ first excited state and a $^1A_1'$ second excited state [21]. The $1^1E'$ state undergoes Jahn-Teller distortion to planar $^1A_1$ and $^1B_2$ (or an orthogonal (twisted) $^1B_1$) components [22]. The large differences in the yields of stereochemical products formed by the thermal and photochemical processes [20] suggest that an excited state of TMM other than those discussed above may be involved.

To begin with, it is important to look at the excited states of the starting material, methylene cyclopropane. The three

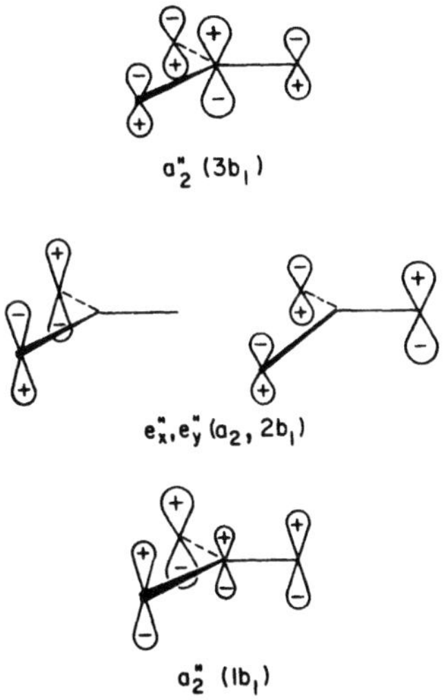

Fig. 6. The valence π orbitals of the planar (D_{3h}) trimethylenemethane radical. Symmetry designations for the C_{2v} point group are given in parentheses.

Table 3. Relevant Configurations for Electronic States of the Trimethylenemethane Radical for Both Thermal and Photochemical Reactions

	Thermal			
$2a_2''$	—	—	—	—
e_x'', e_y''	↑ ↑	↑ ↓	↑↓ —	— ↑↓
$1a_2''$	↑↓	↑↓	↑↓	↑↓
	Photochemical			
$2a_2''$	↓	↓	—	—
e_x'', e_y''	↑ —	— ↑	↑↓ ↓	↓ ↑↓
$1a_2''$	↑↓	↑↓	↑	↑

Table 4. Excitation Energies for Low-Lying States of Methylene Cyclopropane

Transition	Excitation Energy (eV)	Oscillator Strength
$\sigma_A \rightarrow \pi^*$	6.5	0.000
$\pi \rightarrow \sigma_S^*$	6.7	0.003
$\pi \rightarrow \pi^*$	6.9	0.644

lowest-energy excited states of methylene cyclopropane (calculated by the INDO [16,23] method) are presented in Table 4, and the orbitals of methylene cyclopropane are shown on the left hand side of Fig. 7. All excitation energies given in Table 4 are calculated to be near 6.7 eV. The $\pi \rightarrow \sigma_S^*$ transition, which has a small oscillator strength, could be responsible for the olefin

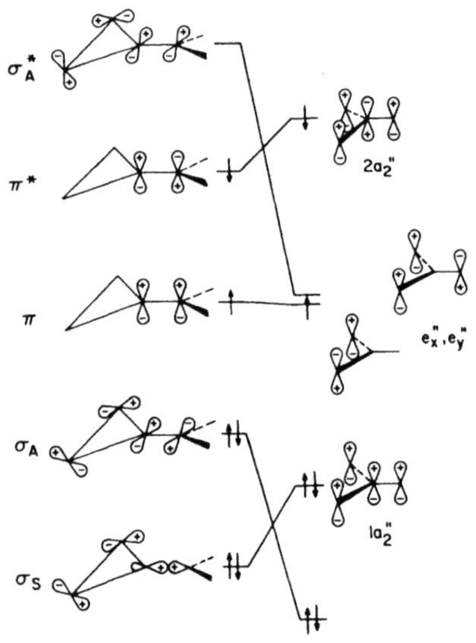

Fig. 7. Correlation diagram showing the disrotatory photochemical opening of methylene cyclopropane to form trimethylenemethane.

plus acetylene formation [Eq. (6)]. The σ_S* orbital, shown in *6*,

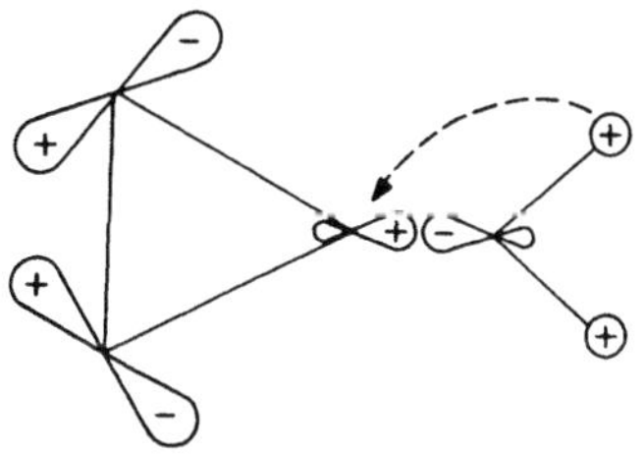

6

is beautifully set up to promote two-bond cleavage, as well as hydrogen transfer, to yield an acetylene and an olefin. The $\sigma_A \rightarrow \pi^*$ transition has no oscillator strength and does not appear to provide a low-energy pathway for ring opening and subsequent closure to an isomerized product. However, the $\pi \rightarrow \pi^*$ transition is calculated to be the most intense absorption; and, through a *disrotatory* [19] ring opening, as shown in Fig. 7, this transition may provide an energetically accessible excited-state TMM intermediate. The *conrotatory* [19] opening, given in Fig. 8, also results from a $\pi \rightarrow \pi^*$ transition and may provide yet another pathway leading to a TMM excited state. Table 3 (bottom) shows those

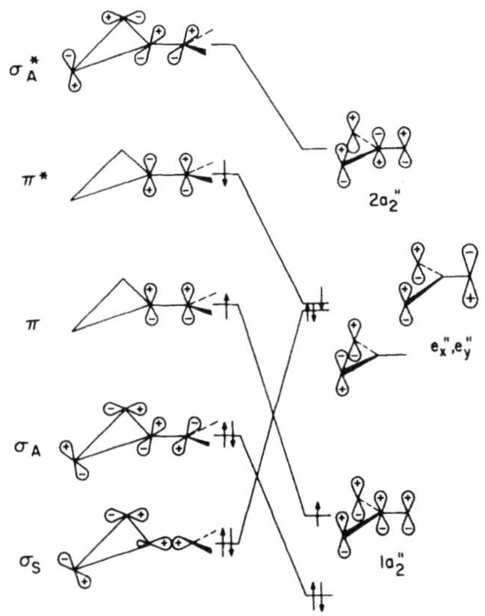

Fig. 8. Correlation diagram showing the conrotatory photochemical opening of methylene cyclopropane to form trimethylenemethane.

Table 5. Excitation Energies for Higher Excited States of Trimethylenemethane (TMM) in Planar Undistorted (D_{3h}) and Distorted (C_{2v}) Geometries

Configuration[a] of Valence π Orbitals	State (energy)		
	D_{3h}[b] (eV)	C_{2v}[c] (eV)	C_{2v}[d] (eV)
$1b_1^{2}(1a_2^{2} + 2b_1^{2})$	$^1A_1'$ (4.0)	$^1A_1$ (4.1)	$^1A_1$ (4.5)
$1b_1^{2}2b_1^{1}3b_1^{1}$	$2^1E'$ (6.4)	$^1A_1$ (6.3)	$^1A_1$ (7.6)
$1b_1^{2}1a_2^{1}3b_1^{1}$		$^1B_2$ (8.3)	$^1B_2$ (6.2)
$1b_1^{1}1a_2^{2}2b_1^{1}$	$3^1E'$ (7.3)	$^1A_1$ (9.5)	$^1A_1$ (8.5)
$1b_1^{1}1a_2^{1}2b_1^{2}$		$^1B_2$ (8.1)	$^1B_2$ (8.3)

[a]Symmetry designations are with respect to C_{2v}.

[b]C-C bonds = 1.430 Å, C-H bonds = 1.10 Å, all bond angles = 120° [22a]. Energies are relative to the lowest $1^1E'$ state.

[c]Unique C-C bond = 1.477 Å, symmetry-related C-C bonds = 1.34 Å, C-H bonds = 1.10 Å, all bond angles = 120° [25]. Energies are relative to the lowest $^1B_2$ component.

[d]Unique C-C bond = 1.380 Å, symmetry-related C-C bonds = 1.509 Å, C-H bonds = 1.10 Å, all bond angles = 120° [25]. Energies are relative to the lowest $^1A_1$ component.

configurations which arise from either a conrotatory or disrotatory photochemical opening of methylene cyclopropane to form trimethylenemethane. The excited states arising from these configurations are $2^1E'$ and $3^1E'$; however, it is more likely that the excited trimethylenemethane distorts to lower symmetry. Another possibility is that the distorted components of the higher-energy $^1E'$ states mix with lower-energy states of the same symmetry. The question then comes down to whether or not there are any trimethylenemethane excited states which are energetically accessible in the 6.7 eV range.

INDO [16,24] calculations were performed on the excited states of trimethylenemethane and the results are presented in Table 5. Neither the $2^1E'$ and $3^1E'$ states, nor the distorted C_{2v} components, are energetically low enough for the 6.7 eV excitation, since very accurate calculations [22d] indicate that the lowest-excited singlet state of trimethylenemethane ($1^1E'$) is at least 1 eV above the methylene cyclopropane ground state. However, the $^1A_1'$ state at 4.0 eV in D_{3h} becomes $^1A_1$ in C_{2v} and can mix with a $^1A_1$ state of higher energy. Yarkony and Schaeffer [22a] calculate a 4.7 eV difference in energy between the lowest planar singlet state of trimethylenemethane and the $^1A_1'$ state, a result which is in reasonable agreement with the value of 4.0 eV given in Table 5, especially considering the qualitative nature of the present calculations. Since the $^1A_1'(^1A_1)$ state may be accessible in the 6.7 eV range, as Table 5 demonstrates, a reasonable hypothesis is that the ring opening of methylene cyclopropane initially takes place along a higher $^1A_1$ potential energy surface, mixes with the lower $^1A_1'(^1A_1)$ surface, and then proceeds to form products. Further calculations along these lines are in progress.

ACKNOWLEDGMENT

This work was supported in part by the National Science Foundation (Grant No. CHE-8015355 to Dr. Gary Epling, Department of Chemistry, University of Connecticut).

REFERENCES

1. M. Kasha, Discuss. Faraday Soc. 9, 14 (1950).
2. (a) For a review of reactions occurring from upper excited states, see N. J. Turro, V. Ramamurthy, W. Cherry and W. Farneth, Chem. Rev. 78, 125 (1978); (b) P. de Mayo, Afinidad 34, 32 (1977).

3. (a) R. Srinivasan, J. Studebaker and K. H. Brown, Tetra. Lett. 1955 (1979); (b) R. Srinivasan and J. A. Ors, J. Am. Chem. Soc. 101, 3411 (1979); (c) R. Srinivasan and J. A. Ors, J. Am. Chem. Soc. 100, 7089 (1978).
4. W. G. Dauben and R. L. Cargill, Tetrahedron 15, 197 (1961).
5. G. S. Hammond, N. J. Turro and A. Fischer, J. Am. Chem. Soc. 83, 4674 (1961).
6. S. Murov and G. S. Hammond, J. Phys. Chem. 72, 3797 (1968).
7. (a) S. B. S. Solomon, C. Steel and A. Weller, Chem. Commun. 927 (1969); (b) G. N. Taylor, Chem. Phys. Lett. 10, 355 (1971).
8. G. Jones, S. -H. Chiang, W. G. Becker and D. P. Greenberg, J. Chem. Soc., Chem. Commun. 681 (1980).
9. R. Hoffmann, Acc. Chem. Res. 4, 1 (1971).
10. (a) M. B. Robin, *Higher Excited States of Polyatomic Molecules*, Vol. II (Academic Press, New York, 1975), p. 170ff; (b) R. P. Frueholz, W. M. Wicker, O. A. Mosher and A. J. Kuppermann, J. Chem. Phys. 70, 1986 (1979).
11. R. Srinivasan, L. S. White, A. R. Rossi and G. A. Epling, J. Am. Chem. Soc. 103, 7299 (1981).
12. [10], p. 176 ff.
13. W. J. Hehre, W. A. Lathan, R. Ditchfield, M. Newton and J. A. Pople, QCPE 10, 236 (1973).
14. W. J. Hehre, R. F. Stewart and J. A. Pople, J. Chem. Phys. 51, 2657 (1969).
15. H. Oberhammer and S. H. Bauer, J. Am. Chem. Soc. 91, 10 (1969).
16. J. Ridley and M. Zerner, Theoret. Chim. Acta (Berl.) 32, 111 (1973).
17. S. S. Hixson, P. S. Mariano and H. E. Zimmerman, Chem. Rev. 75, 531 (1973).
18. For recent reviews on the thermal rearrangement of MCP see: (a) P. Dowd and M. Chow, Tetrahedron 38, 799 (1982); (b) J. E. Baldwin and G. E. C. Chang, Tetrahedron 38, 825 (1982); (c) J. J. Gajewski, C. W. Benner, B. N. Stakly, R. E. Hall and R. I. Sato, Tetrahedron 38, 853 (1982); (d) J. J. Gajewski, *Hydrocarbon Thermal Isomerization* (Academic Press, New York, 1981), pp.51-61.
19. R. B. Woodward and R. Hoffmann, *The Conservation of Orbital Symmetry* (Verlag Chemie, Berlin, 1970).
20. T. Baum, R. Srinivasan and A. R. Rossi, J. Am. Chem. Soc., to be submitted.
21. For excellent reviews on diradicals see: (a) W. T. Borden and E. R. Davidson, Ann. Rev. Phys. Chem. 30, 125 (1979); (b) W. T. Borden and E. R. Davidson, Acc. Chem. Res. 14, 69 (1981).

22. (a) D. R. Yarkony and H. F. Schaefer, III, J. Am. Chem. Soc. 96, 3754 (1974); (b) E. R. Davidson and W. T. Borden, J. Am. Chem. Soc. 99, 2053 (1977); (c) D. A. Dixon, T. H. Dunning, Jr., R. A. Eades and D. A. Kleier, J. Am. Chem. Soc. 103, 2878 (1981); (d) D. Feller, K. Tanaka, E. R. Davidson and W. T. Borden, J. Am. Chem. Soc. 104, 967 (1982).

23. The calculations were performed on an idealized MCP with bond lengths C-C = 1.52 Å, C=C = 1.34 Å, C-H = 1.10 Å and with <HCH = 120°. The cyclopropane fragment was taken as an equilateral triangle.

24. The calculations on trimethylenemethane include all single and double excitations within the π valence orbital manifold.

25. S. B. Auster, R. M. Pitzer and M. S. Platz, J. Am. Chem. Soc. 104, 3812 (1982).

PHOTOCHEMISTRY OF SATURATED ALCOHOLS AND OPEN-CHAIN ETHERS AT 185 nm IN THE LIQUID PHASE

Clemens von Sonntag

Max-Planck-Institut für Strahlenchemie
Stiftstrasse 34-36
D-4330 Mülheim a.d. Ruhr, West Germany

1. INTRODUCTION

The photochemistry of saturated alcohols and ethers has been reviewed in some detail [1,2]. In the present paper, salient points of these photolyses will be given again, but some emphasis will also be placed upon experimental results that have not yet found a proper interpretation. The reactions to be discussed are presented in Schemes 1-5.

2. DISCUSSION

Saturated alcohols and ethers start to absorb around 200 nm, the maximum of the first absorption band being at $\sim$185 nm. The low-pressure Hg arc emits two major lines, one at 254 nm and one at 185 nm. The 254 nm line, the one commonly used in organic photochemistry, is not absorbed by these systems to any significant extent and has been shown to be photochemically inactive, at least as compared to the 185 nm line. At the latter wavelength, the quantum yield of product formation approaches unity.

Both in saturated alcohols and in ethers, the first absorption band has been attributed to an n $\rightarrow$ Rydberg transition [3,4]. In primary and secondary alcohols [5-12], the major processes are (i), homolytic scission of the O-H bond (e.g., Reaction 1); and (ii), elimination of molecular hydrogen with concomitant formation of the corresponding carbonyl compound (e.g., Reaction 2). Other reactions, such as scission of a C-H, C-C or C-O bond, play a minor role (cf. Scheme 2).

S. P. McGlynn et al. (eds.), Photophysics and Photochemistry in the Vacuum Ultraviolet, 913–925.

Scheme 1. 185 nm photolysis of liquid methanol: Primary processes and their quantum yields, and subsequent free radical processes [5].

$$CH_3OH \xrightarrow{h\nu} \begin{cases} \xrightarrow{\phi = 0.69} CH_3O^{\bullet} + H^{\bullet} & (1) \\ \xrightarrow{\phi = 0.06} CH_2O + H_2 & (2) \\ \xrightarrow{\phi = 0.05} CH_3^{\bullet} + {}^{\bullet}OH & (3) \\ \xrightarrow{\phi = 0.08} {}^{\bullet}CH_2OH + H^{\bullet} & (4) \\ \xrightarrow{\phi = 0.12} \text{deactivation} & (5) \end{cases}$$

$$CH_3O^{\bullet} + CH_3OH \longrightarrow CH_3OH + {}^{\bullet}CH_2OH \quad (6)$$

$$H^{\bullet} + CH_3OH \longrightarrow H_2 + {}^{\bullet}CH_2OH \quad (7)$$

$${}^{\bullet}CH_3 + CH_3OH \longrightarrow CH_4 + {}^{\bullet}CH_2OH \quad (8)$$

$${}^{\bullet}OH + CH_3OH \longrightarrow H_2O + {}^{\bullet}CH_2OH \quad (9)$$

$$2\ {}^{\bullet}CH_2OH \longrightarrow \begin{cases} HOCH_2 - CH_2OH & (10) \\ CH_2O + CH_3OH & (11) \end{cases}$$

$${}^{\bullet}CH_3 + {}^{\bullet}CH_2OH \longrightarrow \begin{cases} CH_3 - CH_2OH & (12) \\ CH_4 + CH_2O & (13) \end{cases}$$

$$2\ {}^{\bullet}CH_3 \longrightarrow CH_3 - CH_3 \quad (14)$$

A long-standing open question concerns why, upon n → Rydberg excitation, there is such a difference between the ease of scission of the O-H bond (e.g., Reaction 1) as compared to that of the C-O bond (e.g., Reaction 3). This is especially puzzling since, in saturated ethers, the C-O bond is cleaved preferentially over other processes (see below). Calculations have shown [13] that in the excited state both bonds are dissociative. Thus, there is no strong barrier to impede C-O bond scission. One may conclude, therefore, that the difference in the ease of the two reactions must arise mainly from a mass effect: The H atom, with its low mass, leaves the excited molecule more rapidly than does the heavier OH group. As a consequence of this, deuteration at oxygen should reduce the ease of Reactions 1 and 2 and, hence, favor the other competing reactions. This has been studied in some detail for methanol. In Scheme 1, the complete set of methanol reactions, both primary and secondary, are given. In Table 1, the quantum yields of those key products that can be measured with some accuracy are shown for both CH_3OH and CH_3OD.

The radicals $CH_3O^\bullet$, $H^\bullet$ and $^\bullet OH$ react rapidly with methanol (Reactions 6, 7 and 9) and do not undergo combination and disproportionation reactions with the other radicals present. The methyl radical reacts by H abstraction (Reaction 8); in addition, it reacts with other radicals present but the combination reactions (Reactions 12 and 14) are certainly favored over the disproportionation Reaction 13. This is also true for the hydroxymethyl radical (Reactions 10 and 11). For this reason, the formaldehyde yield is an approximate measure of Reaction 2.

Key products are H_2, CH_4 and C_2H_6, the former representing Reactions 1 and 2 and the latter two representing Reaction 3. Isotope labeling has shown that Reaction 4 is not too important and will not drastically change the general picture. The values given in Table 1 indicate that, upon O-deuteration, $\phi(1+2)$ decreases whereas $\phi(3)$ increases. The decrease in $\phi(1+2)$ is not fully compensated by the increase in $\phi(3)$. This is to be expected since, even with CH_3OH, the quantum yield of the sum of all product-forming processes is only $\sim$0.85. Deuteration at oxygen slows down the rate of Reaction 1; this leaves more time for the other reactions to occur, including deactivation (Reaction 5).

The nature of the deactivation process is not known. First, one might consider a cage recombination of two radicals formed in one of the primary processes. However, considering the $\sim$200 kJ/mol of excess energy remaining in the two particles after Reaction 1 has occurred, it is unlikely that the H atom could be retained in the cage. Although there might be some back reaction of the methyl and OH radicals formed in Reaction 3, this cannot fully account for all of the deactivation observed. Hydrogen bonding might play a role; however, this is far from being under-

Table 1. Quantum Yields of Some Key Products in the 185 nm Photolysis of Liquid Methanol and Methanol-OD.

	MeOH		MeOD
	Ref. [5]	[13]	[13]
H_2	0.83	0.79	absent
HD			0.71
D_2			absent
CH_4	0.05	0.035	0.050
C_2H_6	0.002	0.0052	0.0074

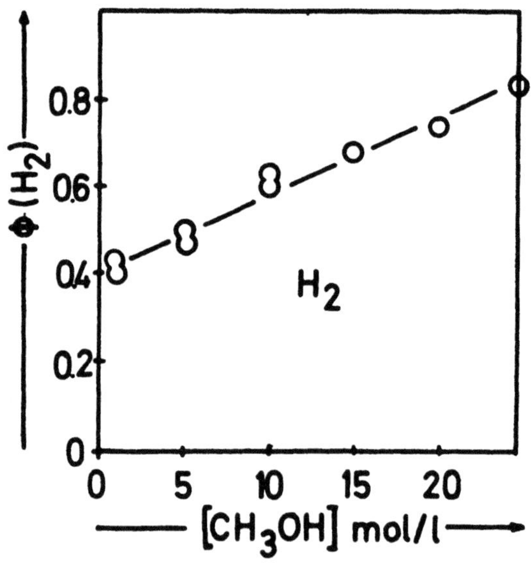

Fig. 1. 185 nm photolysis of methanol/water mixtures: Quantum yields of hydrogen formation.

Scheme 2. 185 nm photolysis of liquid isopropanol: Primary processes and their quantum yields [7].

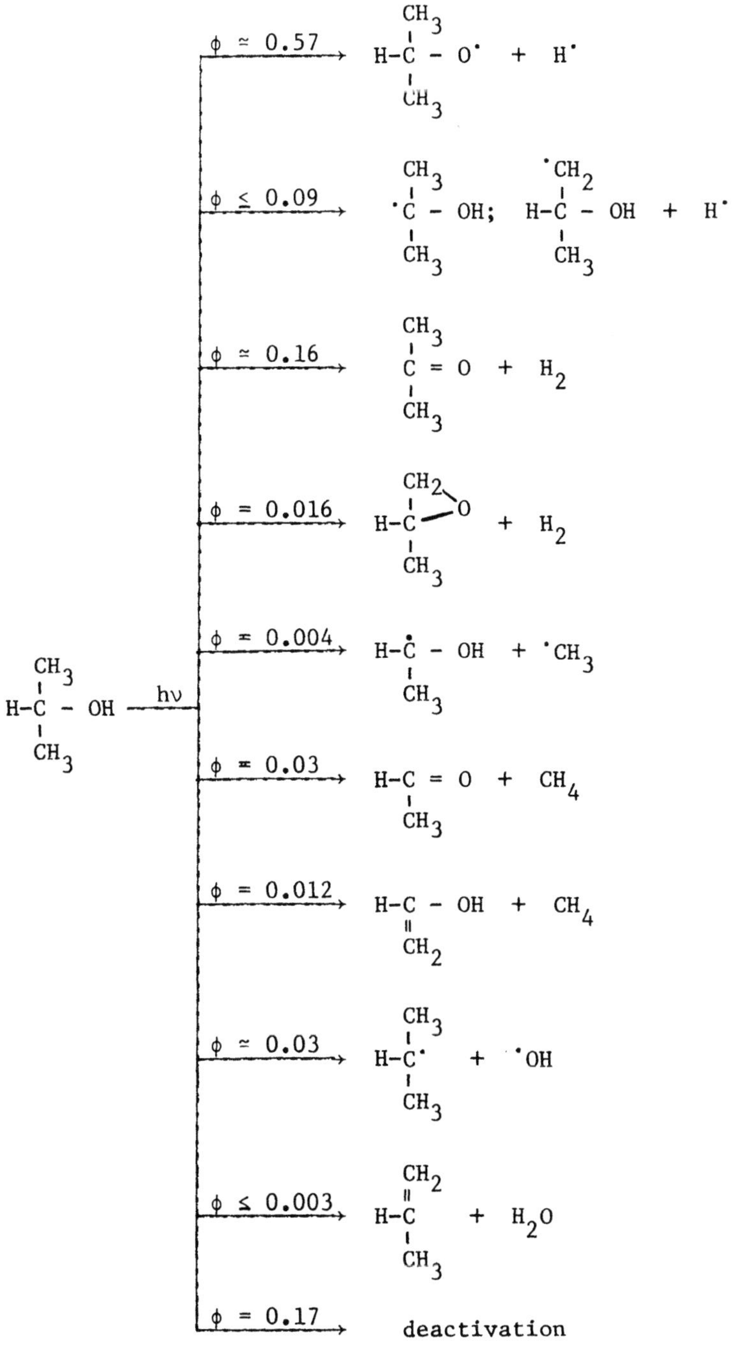

(15)
(16)
(17)
(18)
(19)
(20)
(21)
(22)
(23)
(24)

stood. Experimentally, one observes that, upon dilution with water (which does not absorb down to low alcohol concentrations), the quantum yields of some products drop considerably (cf. Fig. 1). Such an effect has not only been observed with alcohols [5-7, 14] but also with ethers [15]; and, since hydrogen bonding would affect the lone-pair at oxygen, it is not unlikely that this bonding could interfere with the excited state which, being of the n → Rydberg type, involves the lone-pair at oxygen.

In Scheme 2, the reaction pathways of excited liquid isopropanol are given. In this molecule, the major processes (Reactions 15 and 17) are the same as those discussed above. There is, however, another interesting route to molecular H_2, namely, Reaction 18. If this process occurs from a very short-lived excited state, the ground-state configuration must have an important effect on the quantum yield. In the ground state, alcohol molecules arrange in such a way as to minimize steric interference of neighboring groups with the oxygen-bound H atom. In Scheme 3, Newman projections of the most favored configurations are given, together with the quantum yields of epoxide formation. In n-propanol, the hydrogens needed for epoxide formation do not interact with the H atom at oxygen; however, such an interaction does occur in isopropanol, sec-butanol (ϕ(epoxide) = 0.04) [16] and, especially, t-butanol.

In the 185 nm photolysis of isopropanol, methane formation [8] is largely due to molecular processes (Reactions 20 and 21). The reaction involving the elimination of the oxygen-bound H atom (Reaction 20) is favored over that involving an H atom of the other methyl group (Reaction 21). The methyl radical (from Reac-

Scheme 3. Newman projections of some alcohols and quantum yields of epoxide formation upon irradiation at 185 nm.

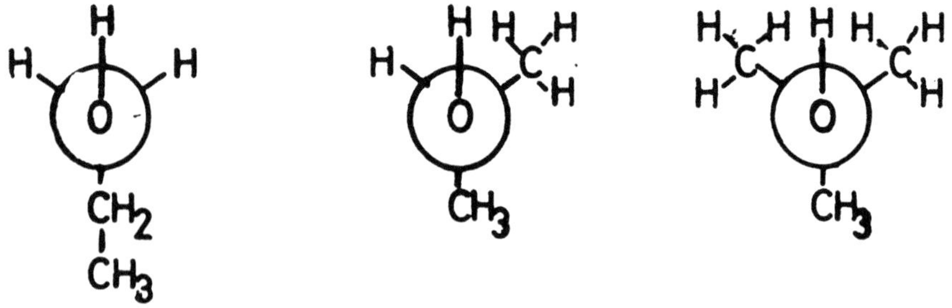

n-Propanol ϕ(epoxide) = nil

Isopropanol ϕ(epoxide) = 0.016

t-Butanol ϕ(epoxide) = 0.08

Scheme 4. 185 nm photolysis of liquid t-butanol: primary processes and their quantum yields [19].

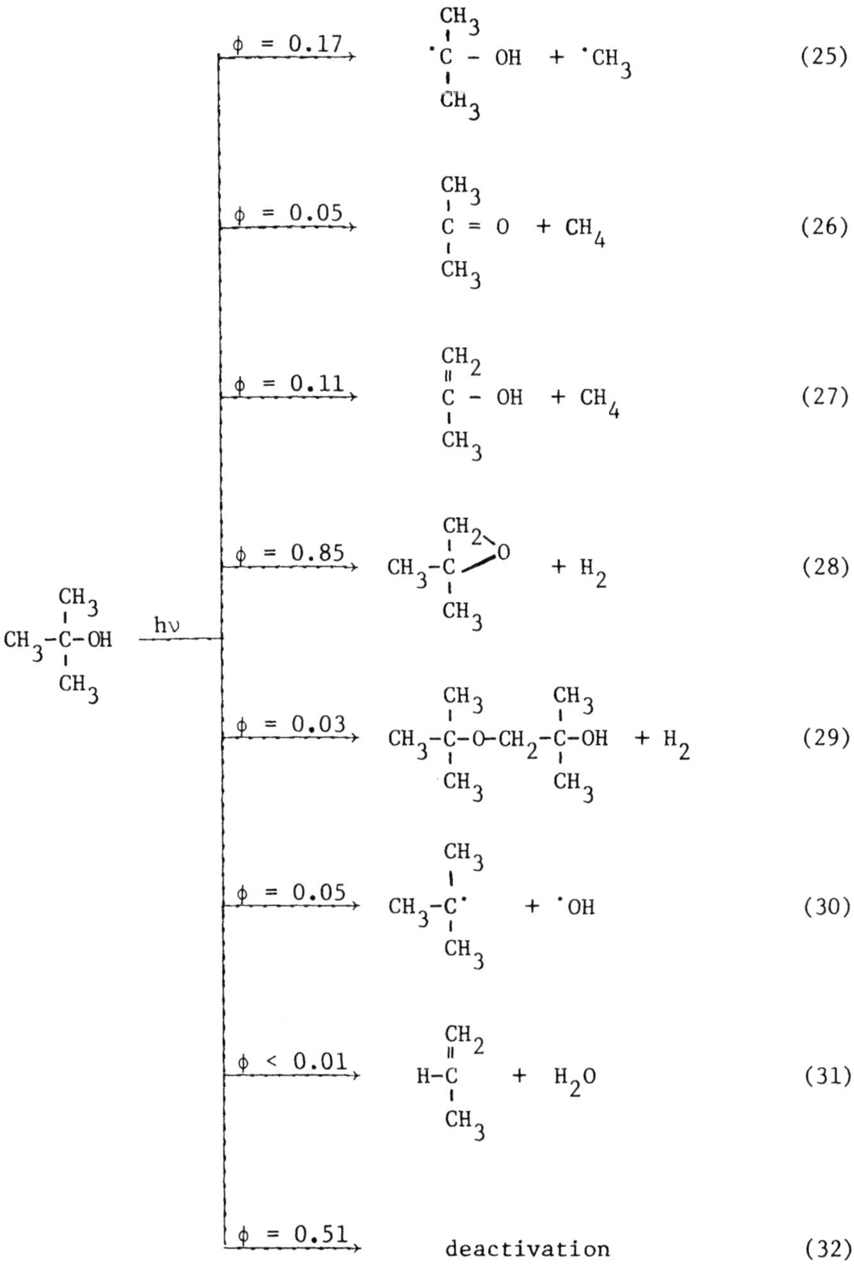

tion 19) plays a minor role. This is quite in contrast to the γ-radiolysis of isopropanol [17,18]. There, methyl radicals (some of them hot) are mainly responsible for the formation of methane. And, as in the methanol system, C-O bond scission (Reactions 22 and 23) is of little importance. With respect to the two molecular processes (Reactions 20 and 21), it must be emphasized that, in these studies, true molecular processes and cage disproportionation reactions cannot be distinguished. This always holds if larger radicals (or diradicals) [1,2] might be involved, but probably does not have to be considered in H_2 elimination reactions (e.g., Reactions 17 and 18).

The photolysis of neat t-butanol [19], the primary processes and quantum yields of which are given in Scheme 4, is very different from that of the primary and secondary alcohols. The major reaction of primary and secondary alcohols, namely, the homolytic O-H bond scission, no longer occurs. Although molecular hydrogen is a significant product, its formation can be fully accounted for by two processes: (i), epoxide formation (Reaction 28), which has been discussed above; and (ii), formation of the ether t-butoxy-2-methyl-2-propanol (Reaction 29). It has been shown that the latter product must be formed in the excited state and requires a second t-butanol molecule as next neighbor [20]. By far the most important processes are C-C bond fragmentations (Reactions 25-27). The quantum yield of the homolytic process (Reaction 25) has possibly been underestimated because a recombination of the two radicals has not been considered.

The sum of the molecular processes (Reactions 30 and 31) are again of minor importance. One should note that deactivation processes (Reaction 32) play a much larger role in the t-butanol system than in primary and secondary alcohols. This could be due partly to a considerable back reaction of Reaction 25 within the solvent cage. We feel, however, that this cannot be the only reason. When t-butanol is photolyzed in aliphatic hydrocarbon solution [14,21,22], $\phi(H_2)$ increases but $\phi(CH_4)$ decreases strongly upon dilution; and, in cyclohexane solutions containing 0.05 M t-butanol, $\phi(H_2)$ reaches a value of 0.6, whereby Reaction 33

$$(CH_3)_3C\text{-}O\text{-}H \xrightarrow{h\nu} (CH_3)_3C\text{-}O^{\bullet} + H^{\bullet} \qquad (33)$$

has a quantum yield as high as 0.54. C-C bond scission in this case becomes practically negligible [22]. It is, of course,

Table 2. 185 nm Photolysis of Liquid Methyl-n-Propylether: Products and Their Quantum Yields [24].

Product	Quantum Yield
Hydrogen	0.019
Methane	0.29
Ethylene	0.001
Ethane	0.083
Propene	0.020
Cyclopropane	0.0007
Propane	0.046
n-Butane	0.029
n-Pentane	0.0005
n-Hexane	0.003
Formaldehyde	0.004
Methanol	0.16
Propionaldehyde	0.13
n-Propanol	0.70
Dimethylether	(0.002)
Methyl-vinylether	$\leq$ 0.01
Ethyl-methylether	(0.002)
Allyl-methylether	0.004
cis-Methyl-n-propenylether	0.048
trans-Methyl-n-propenylether	0.056
Ethyl-n-propylether	0.080
Methyl-2-butylether	0.20
Methyl-2-methylpropylether	0.015
Methyl-n-butylether	0.002
Di-n-propylether	0.002
Methyl-3-pentylether	0.0035
n-Butyl-n-propylether	0.029
3-Hexyl-methylether	0.046
3,4-Dimethoxyhexane	0.093
1-Methoxy-3-n-propoxypropane	0.03
1-n-Propoxy-2-methoxybutane	0.077
1,2-Di-n-propoxyethane	0.014

Scheme 5. 185 nm photolysis of liquid methyl-n-propylether: primary processes and their quantum yields [24].

$$CH_3-O-CH_2-CH_2-CH_3 \xrightarrow{h\nu}$$

$$\xrightarrow{\phi\ =0.7} {}^{\bullet}CH_3 + {}^{\bullet}O-CH_2-CH_2-CH_3 \qquad (34)$$

$$\xrightarrow{\phi \simeq 0.13} CH_4 + CHO-CH_2-CH_3 \qquad (35)$$

$$\xrightarrow{\phi \simeq 0.15} CH_3O^{\bullet} + {}^{\bullet}CH_2-CH_2-CH_3 \qquad (36)$$

$$\xrightarrow{\phi \leq 0.01} CH_2O + CH_3-CH_3-CH_3 \qquad (37)$$

$$\xrightarrow{\phi \leq 0.02} CH_3OH + CH_2=CH-CH_3 \qquad (38)$$

$$\xrightarrow{\phi \simeq 0.02} \begin{cases} H_2 + CH_3O-\dot{C}H_2=CH-CH_3 & (39) \\ H^{\bullet} + R^{\bullet} & (40) \end{cases}$$

$$\xrightarrow{\phi \leq 0.01} H_2 + CH_3O-CH_2-CH=CH_2 \qquad (41)$$

$$\xrightarrow{\phi \leq 0.01} CH_5O-CH_2^{\bullet} + {}^{\bullet}CH_2-CH_3 \qquad (42)$$

$$\xrightarrow{\phi \leq 0.01} CH_3-O-CH_3 + CH_2=CH_2^{\bullet} \qquad (43)$$

$$\xrightarrow{\phi \leq 0.01} CH_3-O-CH_2^{\bullet} + {}^{\bullet}CH_3 \qquad (44)$$

$$\xrightarrow{\phi \leq 0.01} CH_3-O-CH=CH_2 + CH_4 \qquad (45)$$

$$\xrightarrow{\phi \simeq\ 0} \text{deactivation} \qquad (46)$$

tempting to correlate this behavior with the different photolytic properties of monomers and H-bonded oligomers. No satisfactory correlation, however, has been reached [22].

In the series of those open-chain aliphatic ethers that have been studied in some detail (diethyl [23], methyl-n-propyl [24], methyl-t-butyl [25], and di-t-butyl [26] ether), homolytic scission of the C-O bond always predominates, except in the case of di-t-butylether (see below). We discuss n-propylether as a typical example. The results of the exhaustive product study (cf. Table 2) boil down to a very simple reaction scheme (Scheme 5). Practically only Reactions 34-36 are of any importance. The scission of the methyl-O bond (Reaction 34) is much favored over the scission of the propyl-O bond (Reaction 36). This effect becomes even more noticable if the molecular processes (Reactions 35, 37 and 38) are also considered. The preferential scission of the C-O bond to the smaller alkyl group appears to be quite general and one might speculate that, as in the case of the preferred O-H splitting in alcohols, this could be due to a mass effect. Elimination of molecular hydrogen, as shown by Reaction 39, is not very prominent in methyl-n-propylether, but might attain considerable quantum yields in, for example, diethylether.

The di-t-butylether is already strained in its ground state. There, it readily decomposes into t-butanol and isobutene. This is also the major process in the excited state (Reaction 47, $\phi \simeq 0.73$)

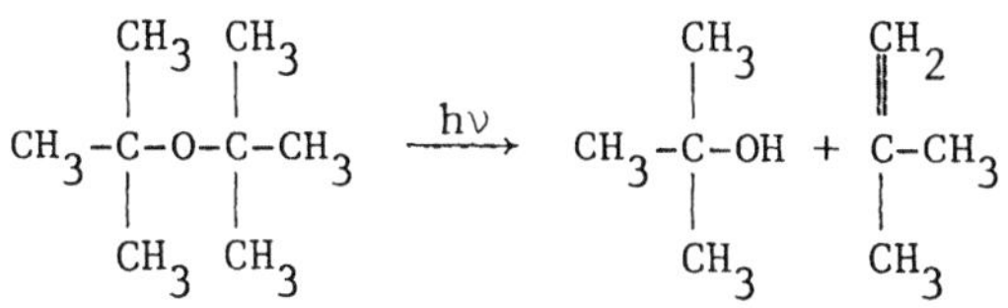

(47)

and, therefore, one always must consider cage disproportionation as an alternative to a true molecular process.

In the two ethers discussed here, the sum of the quantum yields of product-forming reactions is close to unity and little or no deactivation (and cage recombination) occurs. This is, however, not always so. In diethylether and methyl-t-butylether, the sum of the product-forming processes reaches quantum yields between only 0.6 and 0.7.

As mentioned above, quantum yields are drastically reduced in aqueous solutions. The photolysis of dimethylether [15] serves as an example. In cyclohexane solutions, the quantum yield for homolytic C-O bond scission in dimethylether has a lower limit of 0.5, whereas in aqueous solutions this value is only 0.09. The aqueous medium may also have an effect on the subsequent free-radical reactions. Indeed, in the case of dimethylether, the fast rearrangement of the methoxy radical into the hydroxymethyl radical (Reaction 48)

$$CH_3O^{\bullet} \longrightarrow {}^{\bullet}CH_2OH \tag{48}$$

a reaction that seems to be restricted to aqueous solutions, governs the product distribution. In the photolysis of these aqueous solutions, an intermediate is formed that can react with N_2O and the proton. The nature of this intermediate is discussed in the following paper.

REFERENCES

1. C. von Sonntag and H. -P. Schuchmann, Advan. Photochem. 10 59 (1977).
2. C. von Sonntag and H. -P. Schuchmann, in: *The Chemistry of Functional Groups* (Suppl. E), S. Patai, ed. (Wiley, New York, 1980), pp. 903-922.
3. M. B. Robin, *Higher Excited States of Polyatomic Molecules*, Vol. I (Academic Press, New York, 1974), p. 254.
4. C. Sandorfy, Top. Curr. Chem. 86, 92 (1979).
5. C. von Sonntag, Tetrahedron 25, 5853 (1969).
6. C. von Sonntag, Z. Physik. Chem. N.F. 69, 292 (1970).
7. C. von Sonntag, Z. Naturforsch. 27b, 41 (1972).
8. C. von Sonntag, Int. J. Radiat. Phys. Chem. 1, 33 (1969).
9. C. von Sonntag, Tetrahedron 24, 117 (1968).
10. C. von Sonntag and D. Schulte-Frohlinde, Z. Physik. Chem. N.F. 55, 329 (1967).
11. H. J. van der Linde and C. von Sonntag, Photochem. Photobiol. 13, 147 (1971).
12. R. Ford, H. -P. Schuchmann and C. von Sonntag, J. Chem. Soc. Perkin Trans. II 1975, 1338.
13. R. J. Buenker, G. Olbrich, H. -P. Schuchmann, B. Schurmann and C. von Sonntag, to be published.
14. D. Schulte-Frohlinde, D. Sanger and C. von Sonntag, Z. Naturforsch. 27b, 205 (1972).
15. H. -P. Schuchmann and C. von Sonntag, J. Photochem. 16, 289 (1981).
16. C. von Sonntag and D. Sanger, Tetrahedron Lett. 1968, 4515.
17. C. von Sonntag and W. Bruning, Int. J. Radiat. Phys. Chem. 1, 25 (1969).
18. C. von Sonntag, Z. Naturforsch. 25b, 654 (1970).
19. D. Sanger and C. von Sonntag, Tetrahedron 26, 5489 (1970).

20. D. Sanger and C. von Sonntag, Z. Naturforsch. 25b, 1491 (1970).
21. H. -P. Schuchmann, C. von Sonntag and D. Schulte-Frohlinde, J. Photochem. 3, 267 (1974/75).
22. H. -P. Schuchmann, C. von Sonntag and D. Schulte-Frohlinde, J. Photochem. 4, 63 (1975).
23. C. von Sonntag, H. -P. Schuchmann and G. Schomburg, Tetrahedron 28, 4333 (1972).
24. H. -P. Schuchmann and C. von Sonntag, Z. Naturforsch. 30b, 399 (1975).
25. H. -P. Schuchmann and C. von Sonntag, Tetrahedron 29, 1811 (1973).
26. H. -P. Schuchmann and C. von Sonntag, Tetrahedron 29, 3351 (1973).

PHOTOLYSIS OF CYCLIC ETHERS AND ACETALS AT 185 nm IN THE LIQUID PHASE

Clemens von Sonntag

Max-Planck-Institut für Strahlenchemie
Stiftstrasse 34-36
D-4330 Mülheim a.d. Ruhr, West Germany

1. DISCUSSION

As in the case of open-chain ethers discussed in the preceding paper, the first absorption band of cyclic ethers reached by the 185 nm light emitted by the low-pressure Hg arc must be attributed to an n $\rightarrow$ Rydberg transition [1,2]. In open-chain ethers, the major process following excitation at 185 nm is the scission of a C-O bond (Reaction 1) [3,4]:

$$R\text{-}O\text{-}R' \xrightarrow{h\nu} R\text{-}O^{\bullet} + R'^{\bullet} \qquad (1)$$

In cyclic ethers, such a reaction will lead to a diradical (Reaction 2):

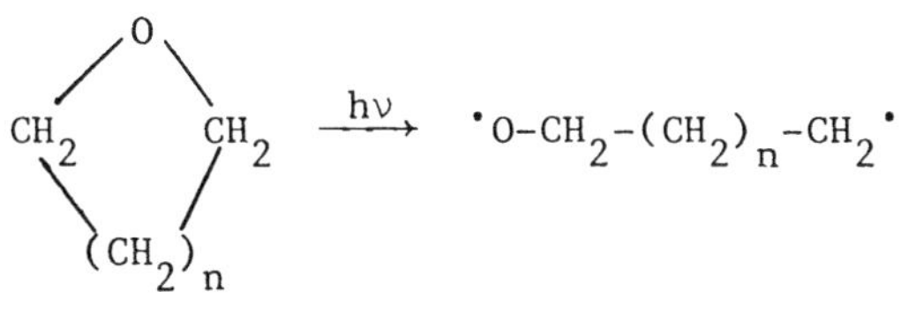

(2)

The intermediacy of a diradical may be evidenced only in systems that undergo a detectable change within the short time before the diradical recloses. In the series of tetrahydrofurans, cis- and trans-2,5-dimethyltetrahydrofuran illustrate this effect (cf. Scheme 1) [5].

The diradical reached upon ring-opening (Reaction 4) in the excited state may invert at the carbon which carries the free

S. P. McGlynn et al. (eds.), Photophysics and Photochemistry in the Vacuum Ultraviolet, 927–943.

Scheme 1. 185 nm photolysis of liquid cis-2,5-dimethyl-tetrahydrofuran: Reaction scheme.

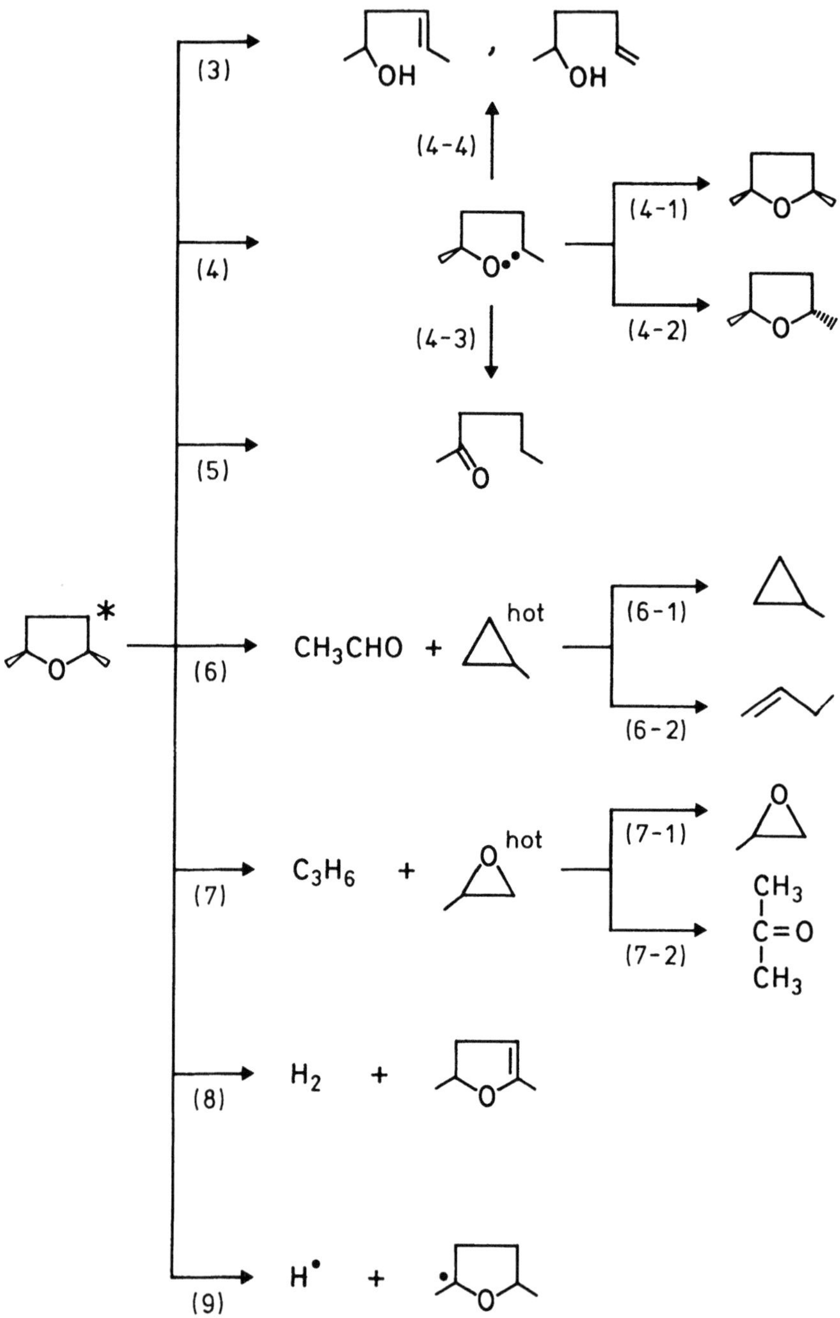

Table 1. Photolysis of Some Neat Liquid Tetrahydrofurans (THF): Quantum Yields Grouped by Reaction Type.

Compound	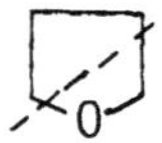	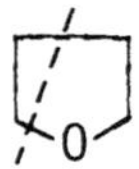	Dehydrogenation	Ring-opened isomers	cis-trans Rearrangement
THF	0.20	0.02	0.07	0.08	
2-Me THF	0.24	0.03	0.17	0.07	
cis-2,5-Me_2 THF	0.22	0.02	0.27	0.09	0.22
trans-2,5-Me_2 THF	0.08	0.05	0.29	0.08	0.2
2,2,5,5,-Me_4 THF	0.39	0.05	0.00	0.27	

Scheme 2. Product-forming reactions of the oxylalkyl diradical generated in the 185 nm photolysis of liquid oxepane.

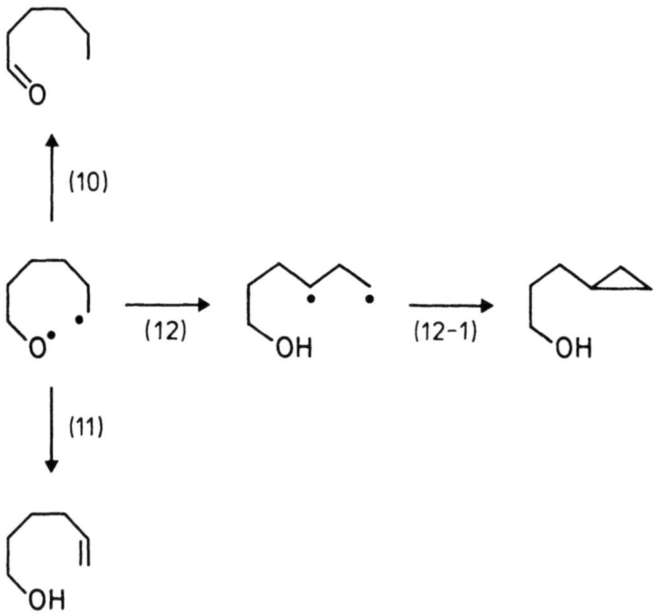

spin and, upon reclosure, the other isomer may be formed (Reaction 4-2). Indeed, it has been found that the cis-compound yields the trans-compound with a quantum yield of as much as 0.2 (cf. Table 1). The same quantum yield has been found for the inverse case. It thus appears that ring-opening followed by reclosure may be a major process in all of these systems, although such a process might not be observable by product analysis. In competition with reclosure, the diradical might undergo a disproportionation reaction (e.g., Reactions 4-4 and 4-5), or the oxyl end might abstract internally an H atom if a six-membered transition state favors this latter process (e.g., Reaction 12). These three processes are shown for the oxepane system [6] in Scheme 2.

Although oxyl radicals are known to fragment by β-cleavage (e.g., Reaction 13)

$$CH_2\text{-}CH_2\text{-}CH_2\text{-}O^{\bullet} \rightarrow CH_3\text{-}CH_2^{\bullet} + CH_2O \qquad (13)$$

this reaction appears not to occur in neat ethers because ethers are very good H-donors and because Reaction 14

$$PrO^{\bullet} + CH_3\text{-}CH_2\text{-}CH_2\text{-}O\text{-}CH_3 \longrightarrow PrOH + CH_3\text{-}CH_2\text{-}\dot{C}H\text{-}O\text{-}CH_3 \quad (14)$$

is considerably faster, as has been shown in the 185 nm photolysis of methyl-n-propylether [7], where the propoxyl radical is the most important primary radical. One therefore would expect that diradicals of the type $^{\bullet}O\text{-}CH_2\text{-}(CH_2)_n\text{-}CH_2^{\bullet}$ will also undergo β-fragmentation, except perhaps for the four-membered diradical which we will not discuss here.

In the series of cyclic ethers, however, fragmentation products are found, in some cases with very high quantum yields (cf. Scheme 3 and Table 1). There are marked differences in the quantum yield of methylcyclopropane formation depending upon whether the cis- or the trans-2,5-dimethyltetrahydrofuran is the educt. This would not be expected if the (common) diradical was the precursor. Considerable differences are also observed with the six-membered cyclic ethers 1,4-dioxane [9], 1,3-dioxane [10] and tetrahydropyran [8]. We therefore conclude that these fragmentation products are formed by true molecular processes. Such a process could possibly be considered as a redistribution of the molecular orbitals on the way from the excited state to the ground state. We note that for these fragmentations only a minor change in the arrangements of the nuclei would be necessary. These reactions should be governed by symmetry rules and it has been suggested [11] that the big difference in ϕ(methylcyclopropane) between cis- and trans-2,5-dimethyltetrahydrofuran might be a result of this.

Educts and fragment products are, for our purposes, practically isoenergetic. The excitation energy, therefore, fully remains in the products (i.e., they are hot). At low pressures in the gas phase, hot cyclopropane rearranges into propene [5,12,13]. This rearrangement can only be quenched by very high pressures (cf. Fig. 1). At 150 atm of N_2, the propene/(cyclopropane + propene) ratio drops to 6.5%. An even lower value (3.6%) has been found in the liquid phase where thermal quenching is even more effective [5]. It is known that hot oxiranes undergo a considerably faster H-1,2-shift than do the corresponding cyclopropanes [14]. It is, therefore, not surprising that the hot oxirane which is formed in Reaction 7 (Scheme 1) should rearrange (Reaction 7-2) to a larger extent (33%) than does hot cyclopropane (3.6%; Reaction 6-2). Similar results have been obtained with cyclic acetals (see below).

Another true molecular process is the elimination of molecular hydrogen. In the series of tetrahydrofurans (cf. Table 1), one observes that the quantum yield for this product (cf. Reaction 8) is strongly enhanced by a methyl substituent in the α-position. An α-H, however, is also required since 2,2,5,5-tetra-

Scheme 3. 185 nm photolysis of liquid cyclic ethers: Major fragmentation processes and their quantum yields.

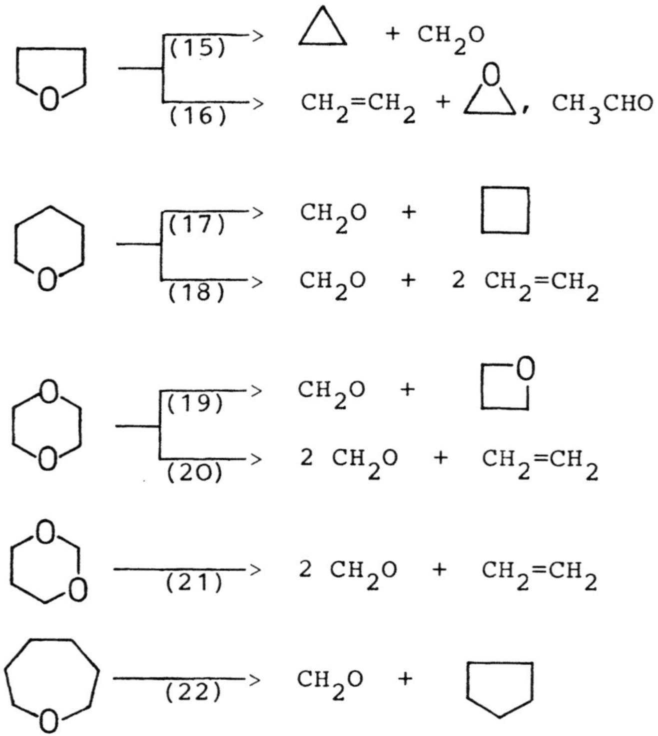

Reaction	Quantum Yield	References
(15)	0.19	[5]
(16)	0.025	
(17)	0.002	[8]
(18)	0.005	
(19)	0.04	[9]
(20)	0.15	
(21)	0.04	[10]
(22)	nil	[6]

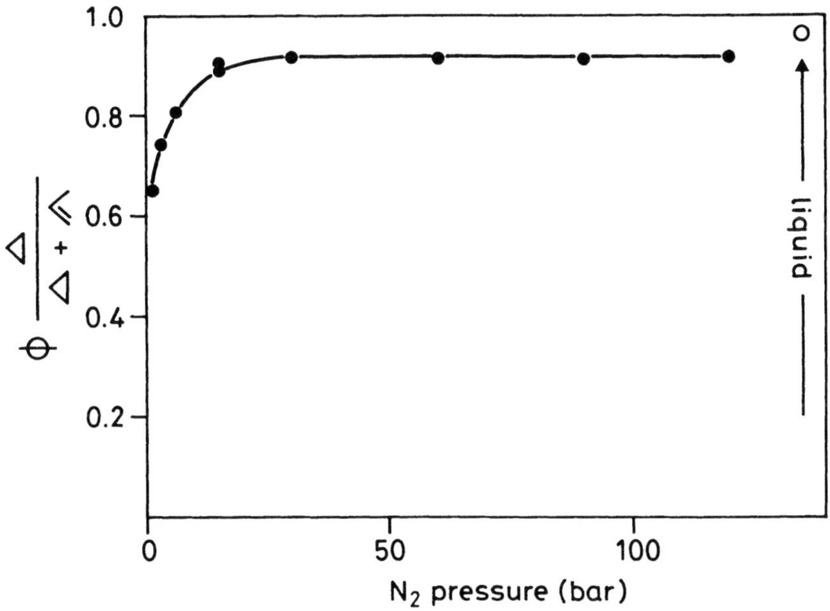

Fig. 1. 185 nm photolysis of tetrahydrofuran: Pressure dependence of cyclopropane and propene formation.

methyltetrahydrofuran no longer yields H_2.

In addition to the above reactions, which can all be rationalized on the basis of our present knowledge, there is one type of process which is not yet understood. In the tetrahydropyran system [8], the major product is a dimer:

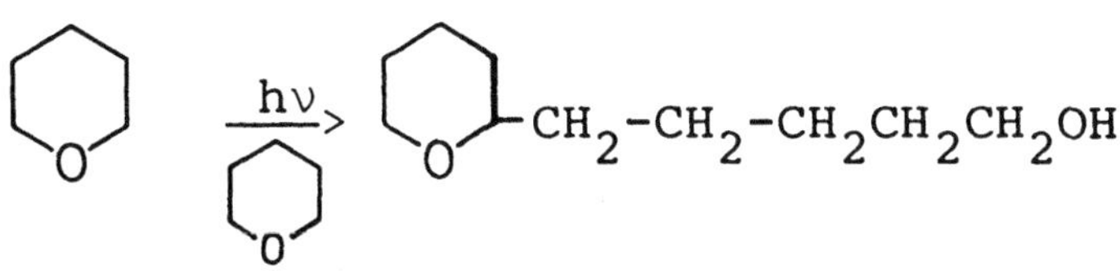

(23)

The quantum yield of Reaction 23 is as high as 0.25. In the tetrahydrofuran [5] and oxepane [6] systems, the corresponding products are also formed, albeit with lower quantum yields (Φ = 0.03 and 0.04, respectively). We judge that these products cannot be interpreted as being formed from the diradical, which abstracts an H atom from the substrate and then reacts in the cage by forming the product in question. It has been suggested instead that an eximer breaks down to yield the product. Excited aggregates appear to be common in the ether series. 1,4-dioxane is a case in point and some aspects of its photophysics and photochemistry will be dealt with next.

Liquid 1,4-dioxane fluoresces, while the gas-phase system does not [15,16]. The maximum of the emission is strongly shifted by nonpolar solvents (to the blue) and by water (to the red). Water also quenches the emission, as does nitrous oxide [9,17]. It has been suggested that this emission is due to an exciplex; in neat dioxane it has a lifetime of ∿2 ns [16]. The system, however, is not yet fully understood. Nitrous oxide quenches the emission more effectively than it quenches the formation of dioxane products. Concomitantly, N_2 is formed. In saturated solution, the quantum yield for N_2 formation is as high as 0.6. It has been suggested that in neat 1,4-dioxane the quenching occurs by energy transfer (Scheme 4, Reaction 24). The excited N_2O decomposes into N_2 and an O atom (Reaction 25), the latter of which then inserts into 1,4-dioxane yielding 2-hydroxy-1,4-dioxane as the major product (Reaction 26, cf. Table 2). In N_2O-saturated aqueous solutions, this product is no longer formed. The major 1,4-dioxane derivative is now the dehydrodimer and the quantum yield of N_2 formation approaches unity. There is a strong competition between the N_2O and the proton for the precursor, as shown in Fig. 2. These results can be accounted for by at least two mechanisms, both of which are given in Scheme 5. In Mechanism A, the excited 1,4-dioxane (or 1,4-dioxane/water complex) ejects an electron (Reaction 27). This process is just possible energetically. N_2O and the proton can then compete for the solvated electron (Reactions 28 and 29). The subsequent processes are represented by Reactions 33-36. In the alternative Mechanism B, a free solvated electron is not formed but, in the excited state, an electron is transferred either to N_2O (Reaction 30) or to the proton (Reaction 32).

In principle, one should be able to distinguish experimentally between the above two mechanisms. N_2O and protons compete for two species which possess different lifetimes. The excited 1,4-dioxane (1,4-dioxane/water complex) should live for a few nanoseconds (several tens of nanoseconds at the most). In aqueous solutions, albeit free of any impurities that can act as scavengers, the solvated electron will live for up to several milliseconds. A reduction of the N_2O concentration will have a strong effect on (N_2) if the short-lived excited state is the precursor (Mechanism B), and none or almost none if the solvated electron is the intermediate (Mechanism A). Relevant experiments are not so easy to perform as one might initially presume. Preliminary, and by no means confirmed, results [10] indicate that the intermediate species is shorter-lived than the solvated electron generated by γ-radiolysis (in the aqueous 1,4-dioxane system containing N_2O), but also that the lifetime of the intermediate must ex-

Scheme 4. Reaction of excited 1,4-dioxane with N_2O in neat 1,4-dioxane.

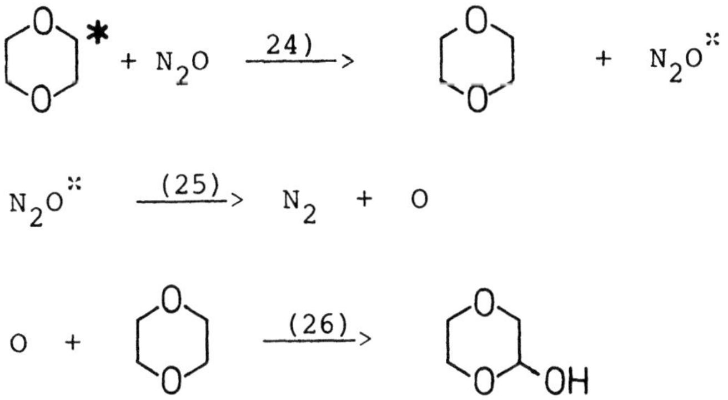

Table 2. Quantum Yields of Products from the 185 nm Photolysis of Dry Liquid 1,4-Dioxane in the Absence and Presence of N_2O (N_2O at Atmospheric Pressure).

Product	Neat	N_2O-Saturated
hydrogen	0.04	0.01
carbon monoxide	0.01	not determined
ethylene	0.15	0.10
formaldehyde	0.23	0.14
acetaldehyde	0.03	0.02
oxetane	0.04	0.03
2-hydroxyethyl vinyl ether	0.17	0.10
bidioxanyl (meso + dl)	0.01	0.02
N_2	absent	0.59
hydroxydioxane	absent	0.4

Scheme 5. Reaction of excited 1,4-dioxane with N_2O in aqueous solutions.

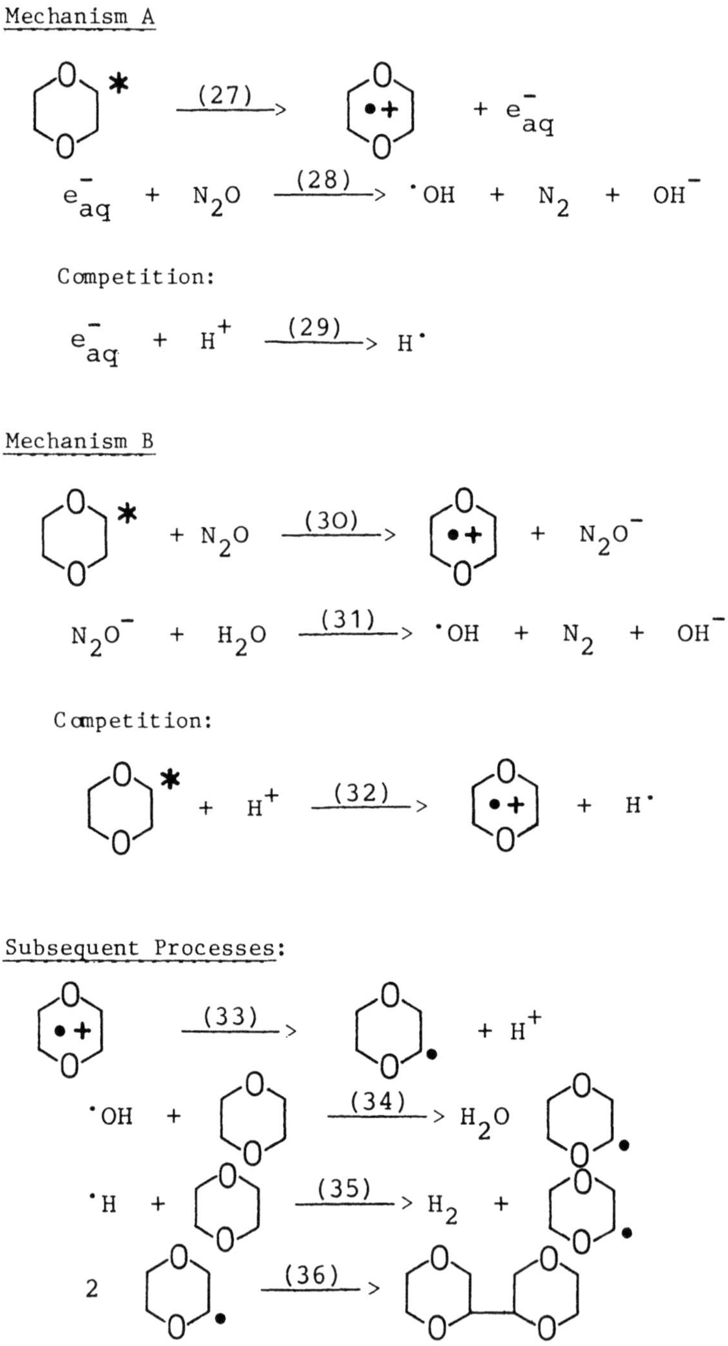

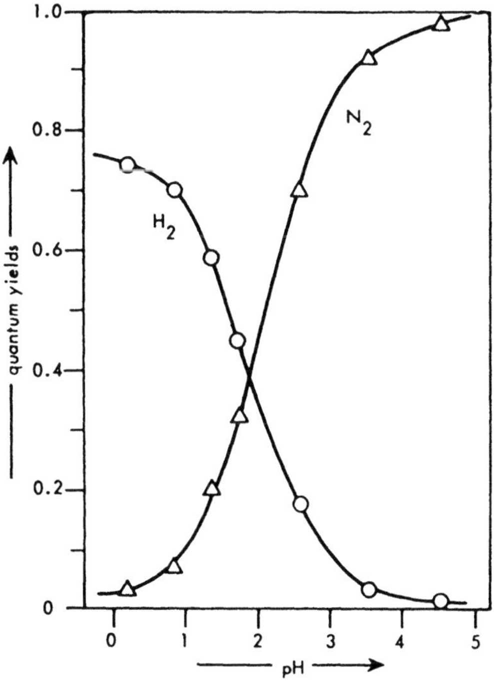

Fig. 2. pH dependence of the quantum yields of H_2 (o) and N_2 (Δ) formation in N_2O-saturated solutions of p-dioxane in water (20 vol %).

ceed 10 ns considerably. In this context, one might consider results obtained on dimethylether [18]. In aqueous N_2O-containing solutions of dimethylether, N_2 is also liberated, but with a very low quantum yield. Protons also compete for this reaction. Owing to a higher ionization potential of dimethylether as compared to that of 1,4-dioxane, the formation of a free solvated electron (cf. Mechanism A) is no longer possible. The lower $\phi(N_2)$ in the dimethylether system could be due either to (i), a shorter lifetime of the excited dimethylether (dimethylether/water complex) -- that is, only a small surviving fraction could be scavenged at the N_2O saturation level; or (ii), to the intermediate species being long-lived but formed only with a low probability.

Open-chain acetals [19-21] behave in a fashion very similar to the open-chain ethers in that C-O bonds are broken preferentially. If the system is substituted with a t-butyl group, as in pivalaldehyde dimethylacetal [21], there is not much discrimination between the two possible routes (Reactions 37 + 38 and 39 + 40 of Scheme 6).

Scheme 6. Primary processes and their quantum yields in the 185 nm photolysis of liquid formaldehyde dimethylacetal.

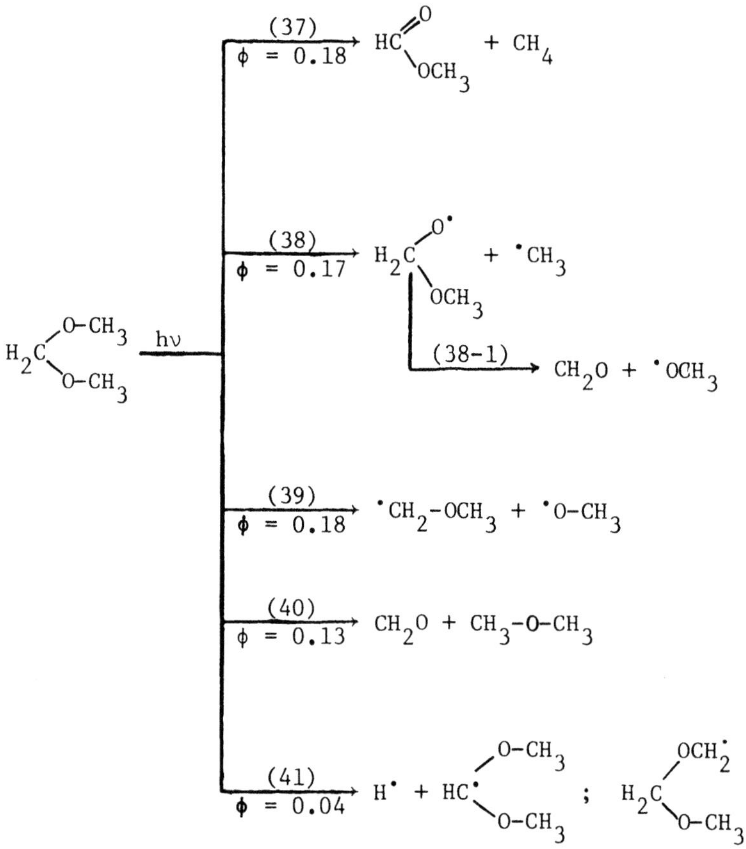

Scheme 7. Primary processes in the 185 nm photolysis of liquid 1,3-dioxolan (R = H) and the 2,2-dimethyl-substituted compound.

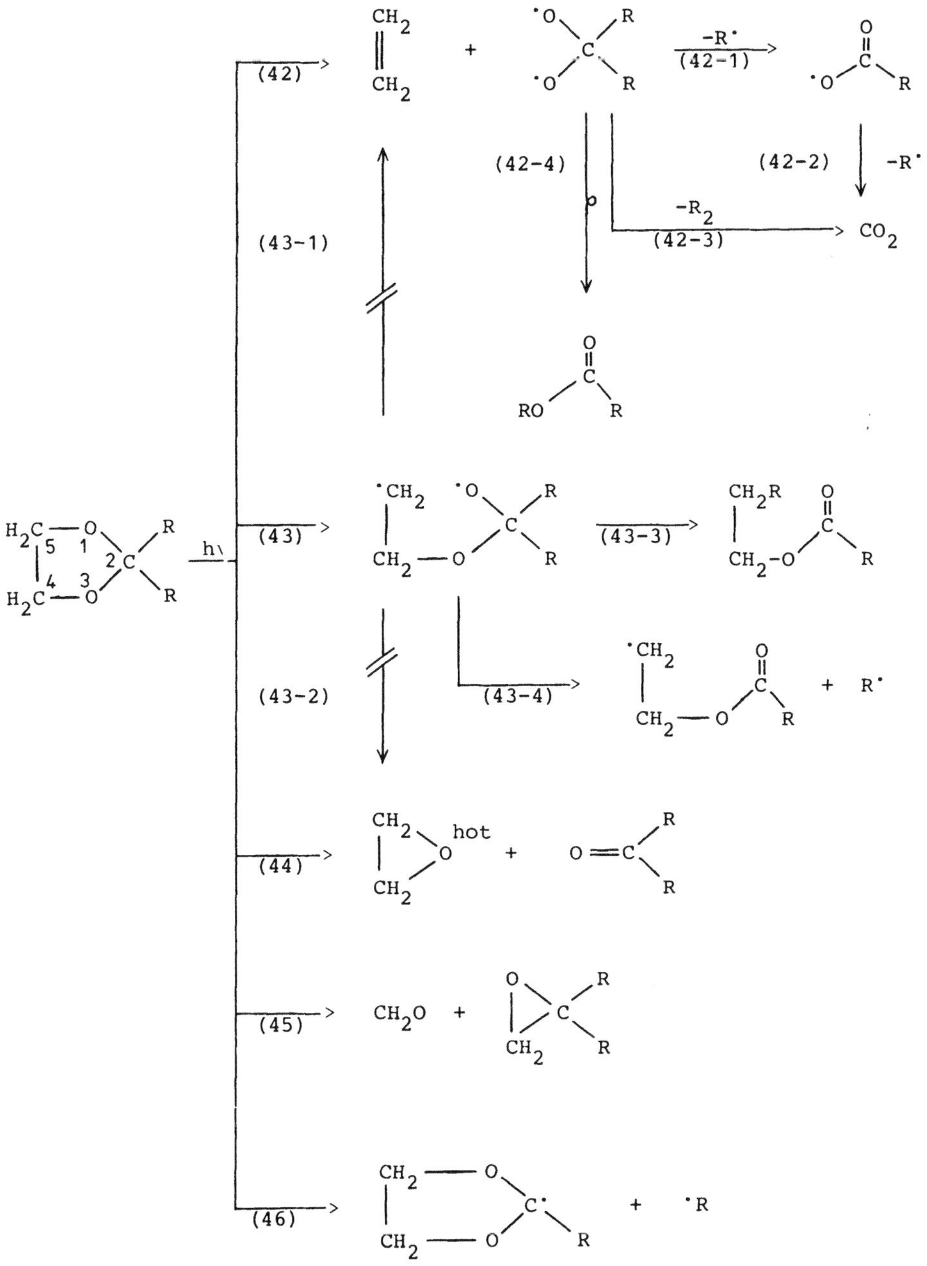

Table 3. UV-Photolysis of Cyclohexanodioxolans. Quantum Yields of Diastereoisomerization.

Educt	$\phi(A)$	$\phi(B)$	$\phi(C)$
A 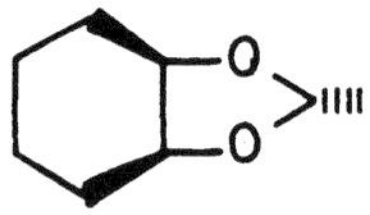	—	0.02_1	0.02_1
B 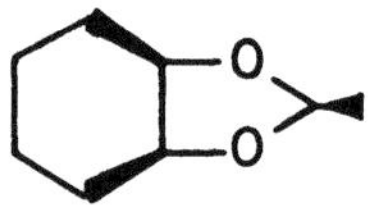	0.01_6	—	0.01_6
C 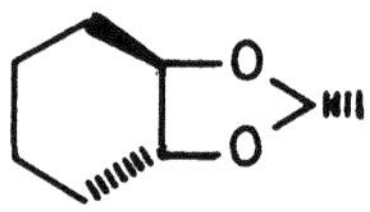	0.01_5	0.01_0	—

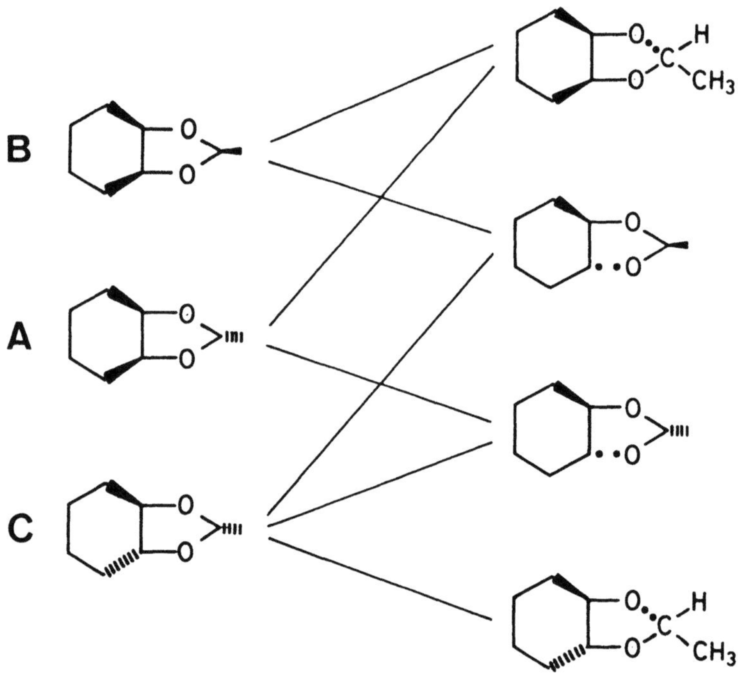

Scheme 8. Relationship between diastereoisomeric cyclohexanodioxolans and the corresponding oxyl-alkyl 1,5-diradicals.

In the series of cyclic acetals, only one type of scission, namely that of the O(1)-C(5) bonds, has been recognized by product analysis [22] (see Scheme 7) and one must ask the following question: Do the cyclic acetals differ in this respect from their open-chain analogs, or is this mode of reaction not recognized because the diradical formed by the scission of the O(1)-C(2) bond recloses before it can undergo other reactions? To answer this question adequately, substituted dioxolans have been synthesized and quantum yields of isomerization determined [23] (cf. Table 3). From Scheme 8, one sees that the isomerizations shown in Table 3 cannot be accounted for if only the O(1)-C(5) bond is broken, but that the experimental results can be reconciled if both O(1)-C(5) and O(1)-C(0) bond scissions occur. This result eliminates the apparent discrepancies between open-chain and cyclic acetals.

Another question is whether the olefin and the oxirane are formed from the diradical intermediate (Reactions 43-1 and 43-2), or whether they originate from true molecular processes (Reactions 42 and 44). This question can be answered by photolyzing cis- and trans-2,2,4,5-tetramethyldioxolan [23]. The quantum yields presented in Table 4 show that, in the olefin product,

Table 4. Quantum Yields of Some Products in the 185 nm Photolysis of Liquid cis- and trans-2,2,4,5-Tetramethyldioxolans.

	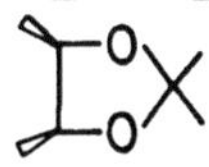		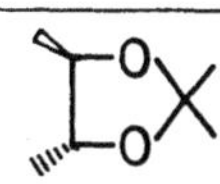	
	Φ	%	Φ	%
	0.07	97	0.005	7
	0.002	3	0.065	93
	0.04	57	0.01	-
	0.01	14	0.03	-
$CH_3-\overset{O}{\overset{\Vert}{C}}-CH_2-CH_3$	0.02	29	n.d.	-

n. d. = not determined

there is a very high retention of configuration. This strongly argues against a diradical intermediate as the major precursor and is an indication that the olefin is mainly formed in a true molecular process. The oxiranes results are not that clearcut. In this case there is also considerable retention of configuration but, in addition to the other isomer, high yields of the rearranged (H-1,2-shift) product have been observed (cf. Table 4). As in the case of the cyclic ethers, one must postulate that if the oxirane is formed in the excited state as a molecular product (Reaction 44), it must be hot. The ease with which a thermally-hot oxirane rearranges [14], especially undergoing an H-1,2-shift could explain the results. Although thermalization in the liquid will be very rapid, only part of the intermediate will become thermalized before undergoing the H-1,2-shift. In this fraction, however, the expected isomer predominates. This leaves little room for a contribution from the diradical (Reaction 43-2).

REFERENCES

1. M. B. Robin, *Higher Excited States of Polyatomic Molecules*, Vol. I (Academic Press, New York, 1974), p. 254.
2. C. Sandorfy, Top. Curr. Chem. 86, 91 (1979).
3. C. von Sonntag and H. -P. Schuchmann, Advan. Photochem. 10, 59 (1977).
4. C. von Sonntag and H. -P. Schuchmann, in: *The Chemistry of Functional Groups* (Supplement E), S. Patai, ed. (Wiley, New York, 1980), pp. 903-922.
5. N. Kizilkilic, H. -P. Schuchmann and C. von Sonntag, Can. J. Chem. 58, 2819 (1980).
6. H. -P. Schuchmann and C. von Sonntag, J. Photochem. 13, 347 (1980).
7. H. -P. Schuchmann and C. von Sonntag, Z. Naturforsch. 30b, 399 (1975).
8. H. -P. Schuchmann, P. Naderwitz and C. von Sonntag, Z. Naturforsch. 33b, 942 (1978).
9. H. -P. Schuchmann, H. Bandmann and C. von Sonntag, Z. Naturforsch. 34b, 327 (1979).
10. H. -P. Schuchmann and C. von Sonntag, unpublished results.
11. E. A. Halevi, private communication.
12. B. C. Roquitte, J. Phys. Chem. 70, 1334 (1966).
13. B. C. Roquitte, J. Am. Chem. Soc. 91, 7664 (1969).
14. M. C. Flowers and R. M. Parker, J. Chem. Soc. (B) 1971, 1980.
15. F. Hirayama, C. W. Lawson and S. Lipsky, J. Phys. Chem. 74, 2411 (1970).
16. A. M. Halpern and W. A. Ware, J. Phys. Chem. 74, 2413 (1970).
17. C. von Sonntag and H. Bandmann, J. Phys. Chem. 78, 2181 (1974).

18. H. -P. Schuchmann and C. von Sonntag, J. Photochem. 16, 289 (1981).
19. H. -P. Schuchmann and C. von Sonntag, J. Chem. Soc. Perkins Trans. II 1976, 1408.
20. H. -P. Schuchmann and C. von Sonntag, Z. Naturforsch. 32b, 205 (1977).
21. P. Naderwitz, H. -P. Schuchmann and C. von Sonntag, Z. Naturforsch. 32b, 209 (1977).
22. E. Cetinkaya, H. -P. Schuchmann and C. von Sonntag, J. Chem. Soc. Perkins Trans. II 1978, 985.
23. H. -P. Schuchmann and C. von Sonntag, to be published.

LIST OF PARTICIPANTS

Dr. Daniel Altenloh
Department of Chemistry
North Texas State University
Denton, TX 76203
U.S.A.

Dr. Michael N. R. Ashfold
School of Chemistry
University of Bristol
Bristol BS8 1TS
ENGLAND

Prof. H. H. Brongersma
Department of Physics
Eindhoven University of Technology
5600 MD Eindhoven
THE NETHERLANDS

Mr. Dennis Burnes
Department of Chemistry
University of Montana
Missoula, MT 59812
U.S.A.

Mr. Michael Cavagnero
Department of Physics
University of Chicago
Chicago, IL 60637
U.S.A.

Dr. Wendell Chen
Center for Applied Science
University of Texas
Arlington, TX 76019
U.S.A.

Dr. Robert Compton
Oak Ridge National Laboratory
P.O. Box X
Oak Ridge, TN 37830
U.S.A.

Dr. H. M. Crosswhite
Argonne National Laboratory
9700 South Cass Avenue
Argonne, IL 60439
U.S.A.

Mr. J. A. Dagata
Department of Chemistry
Louisiana State University
Baton Rouge, LA 70803
U.S.A.

Dr. Joseph L. Dehmer
Argonne National Laboratory
9700 South Cass Avenue
Argonne, IL 60439
U.S.A.

Dr. Patricia M. Dehmer
Argonne National Laboratory
9700 South Cass Avenue
Argonne, IL 60439
U.S.A.

Dipl. Chem. Ourania Dimopoulou
Institut für Physikalische Chemie
Freie Universität Berlin
D-1000 Berlin 33
WEST GERMANY

Dr. Gerald Dujardin
Laboratoire de Photophysique Moleculaire
Universite de Paris-Sud
F-91405 Orsay
FRANCE

Dr. Walter S. Felps
Department of Chemistry
Louisiana State University
Baton Rouge, LA 70803
U.S.A.

Dr. Ann M. Findley
Department of Chemistry
New York University
New York, NY 10003
U.S.A.

Prof. Gary L. Findley
Department of Chemistry
New York University
New York, NY 10003
U.S.A.

Dr. Johann-Heinrich Fock
Institut für Experimentalphysik
Universität Hamburg
D-2000 Hamburg 50
WEST GERMANY

Prof. Jean-Claude Gay
Laboratoire de Spectroscopie Hertzienne
Ecole Normale Superieure
F-75230 Paris
FRANCE

Prof. Aharon Gedanken
Department of Chemistry
Bar-Ilan University
Ramat-Gan
ISRAEL

Prof. Chris H. Greene
Department of Physics
Louisiana State University
Baton Rouge, LA 70803
U.S.A.

Ms. Joyce A. Guest
Department of Chemistry
Stanford University
Stanford, CA 94306
U.S.A.

Prof. Petr Hochmann
College of Science and Math
University of Texas
San Antonio, TX 78230
U.S.A.

Dr. Richard Holroyd
Brookhaven National Laboratory
Upton, NY 11973
U.S.A.

Dr. Keh-Ning Huang
Argonne National Laboratory
9700 South Cass Avenue
Argonne, IL 60439
U.S.A.

Dr. Russell H. Huebner
Argonne National Laboratory
9700 South Cass Avenue
Argonne, IL 60439
U.S.A.

Prof. Yuksel Inel
Department of Chemistry
Bogazici University
Bebek P.K. 2, Istanbul
TURKEY

Prof. Joshua Jortner
Department of Chemistry
University of Tel-Aviv
Tel-Aviv 69978
ISRAEL

Prof. Darrell L. Judge
Space Sciences Center
University of Southern California
Los Angeles, CA 90089
U.S.A.

Dr. Cornelius E. Klots
Oak Ridge National Laboratory
P.O. Box X
Oak Ridge, TN 37830
U.S.A.

Dr. E. E. Koch
Hamburger Synchrotron-
strahlungslabor
DESY
2 Hamburg 52
WEST GERMANY

Ms. Rengin Konuk
Department of Chemistry
Louisiana State University
Baton Rouge, LA 70803
U.S.A.

Dr. G. A. Kourouklis
Oak Ridge National Laboratory
P.O. Box X
Oak Ridge, TN 37830
U.S.A.

Dr. Devendra Kumar
Department of Chemistry
Louisiana State University
Baton Rouge, LA 70803
U.S.A.

Dr. K. T. Lu
Argonne National Laboratory
9700 South Cass Avenue
Argonne, IL 60439
U.S.A.

Prof. S. Leach
Laboratoire de Photophysique
Moleculaire
Universite de Paris-Sud
F-91405 Orsay
FRANCE

Dr. Bruno Leclerc
Department of Physics
Universite Laval
Quebec G1K 7P4
CANADA

Dr. Bee T. Lim
Department of Chemistry
Wayne State University
Detroit, MI 48202
U.S.A.

Prof. Edward C. Lim
Department of Chemistry
Wayne State University
Detroit, MI 48202
U.S.A.

Prof. Einar Lindholm
Department of Physics
Royal Institute of Technology
S-100 44 Stockholm
SWEDEN

Prof. Sean P. McGlynn
Department of Chemistry
Louisiana State University
Baton Rouge, LA 70803
U.S.A.

Dr. John C. Miller
Oak Ridge National Laboratory
P.O. Box X
Oak Ridge, TN 37830
U.S.A.

Prof. Cleanthes A. Nicolaides
Theoretical and Physical
Chemistry Institute
Hellenic Research Foundation
Athens 501
GREECE

Prof. Robert F. O'Connell
Department of Physics
Louisiana State University
Baton Rouge, LA 70803
U.S.A.

Mr. Patrick F. O'Mahoney
Department of Physics
University of Chicago
Chicago, IL 60637
U.S.A.

Dr. Ronald S. Pandolfi
Department of Chemistry
University of California
Los Angeles, CA 90024
U.S.A.

Dr. Klaus Rademann
Institut für Physikalische Chemie
Freie Universität Berlin
D-1000 Berlin 33
WEST GERMANY

Mr. Maurice Raoult
Laboratoire de Photophysique Moleculaire
Universite de Paris-Sud
F-91405 Orsay
FRANCE

Prof. A. R. P. Rau
Department of Physics
Louisiana State University
Baton Rouge, LA 70803
U.S.A.

Dr. M. B. Robin
Bell Laboratories
600 Mountain Avenue
Murray Hill, NJ 07974
U.S.A.

Dr. Marko Robnik
Institut für Astrophysik
Universität Bonn
D-5300 Bonn
WEST GERMANY

Prof. A. R. Rossi
Department of Chemistry
University of Connecticut
Storrs, CT 06268
U.S.A.

Ms. Pascale Roy
Department of Physics
Universite Laval
Quebec G1K 7P4
CANADA

Prof. B. R. Russell
Department of Chemistry
North Texas State University
Denton, TX 76203
U.S.A.

Prof. C. Sandorfy
Departement de Chemie
Universite de Montreal
Montreal, Quebec H3C 3V1
CANADA

Prof. J. D. Scott
Department of Chemistry
University of Montana
Missoula, MT 59812
U.S.A.

Ms. Sue Scott
Department of Chemistry
Louisiana State University
Baton Rouge, LA 70803
U.S.A.

Prof. Patricia Ann Snyder
Department of Chemistry
Florida Atlantic University
Boca Raton, FL 33431
U.S.A.

Prof. Mark G. Steinmetz
Department of Chemistry
Marquette University
Milwaukee, WI 53233
U.S.A.

Dr. Frank S. Tomkins
Argonne National Laboratory
9700 South Cass Avenue
Argonne, IL 60439
U.S.A.

Dr. Gabriel Vazquez
Lehrstuhl für Theoretische Chemie
Universität Bonn
D-5300 Bonn 1
WEST GERMANY

Prof. Clemens von Sonntag
Max-Planck-Institut für
Strahlenchemie
D-4330 Muelheim/Ruhr
WEST GERMANY

Prof. S. Wallace
Department of Chemistry
University of Toronto
Toronto, Ontario M5S 1A1
CANADA

Dr. Ken-Guang Wang
Argonne National Laboratory
9700 South Cass Avenue
Argonne, IL 60439
U.S.A.

SUBJECT INDEX

Adiabatic invariant, 631,633, 662,666-672,680,684,689, 691,700-701
Anti-Stokes Raman effect, 105-106,121
Area-preserving mapping, 591-597,603-604
Arnold's cat, 597-598
Arnold diffusion, 591
Arnold web, 591
Associative ionization, 473, 476,495-498
Auger electrons, 343,376
Autodetachment, 167,266-267, 273-279,288
Autoionization, 10,12,14,36, 136,139,141,166,202,210, 214,217-218,220,225-234, 237,239,245-246,254,256, 272,278,297-308,314-318, 330,343,366,373,375,377ff, 470-471,476,495,497,500, 515-516,544,702
  and MQDT, 382-390,395-397

Beam foil spectroscopy, 273
Bernoulli system, 579
Beutler-Fano autoionization spectrum, 10,12,14,60,88, 218,234,304-305,318,377, 383-384,483,511,544
Birkhoff-Gustavson normal form, 604,606,623-625
Bohr-Sommerfeld quantization conditions, 608,631,687-689
Born-Oppenheimer approximation, 6,18,169,185,249-254,324, 377,383,686,689,878

Caustic, 609-610
Centrifugal barrier, 7,13, 34-35,343-356,362
Channel, 14-16,195ff,217ff, 245ff,341ff
  closed, 200,214,218,223,225, 231-233,245,251
  dissociation, 397,425, 430-431
  eigenchannel, 197,211-214, 220,222-225,232-233,235, 237,239,248-251,255, 355-360
  fragmentation, 196,198,211, 245-246
  interaction, 14-16,218,233, 341,346,355,377
  ionization, see open channel
  mixing angle, 218,235
  mixing coefficient, 224, 233-235
  nonresonant, 362,365
  open, 200-201,211-214,218, 220,222-225,231-233,235, 239,248,251,253,342,353, 362,366-367,370,375,378, 383,386,389,395,397
Chaos, 579ff
  classical, 662-663,666,677, 690
  transition to, in quantum mechasnical systems, 718

Charge-exchange reactions, 37-38,261,264,282-283
Charge-transfer processes, 142, 515-516,833-835,837
Charge-transfer-to-solvent (CTTS), 36
Charmonium, 721-722,726-727
Circular dichroism (CD), 765-766,769-778
Clebsch-Gordan coefficients, 248,638,682
Coherent radiation,
harmonic generation of, 105-106
higher-order nonlinear production of, 105-106,112
Collisional detachment, 278
Collisional dissociation, 286, 476
Collisional ionization, 142, 145,147,262-263,477
Collisional processes, electron and molecule,
direct excitation, 164-166
dissociative attachment, 164, 166,168,175-177,180,258
elastic scattering, 164,173
inelastic scattering, 173,181
radiative attachment, 164, 166,262
resonance excitation, 164-166
resonant elastic scattering, 164
superexcited state, 164
Condensed media effcts on Rydberg states, 8,33,36-37
Condensed media spectroscopy, 59,64-67
Conduction band, 36-37,79-81, 525,529,536
Constant final state spectroscopy, 74-75
Constant initial state spectroscopy, 74
Core effects, 8,14,16-22
spin-orbit coupling, 8,16-20, 29,210,228,234,236,274,415, 447,468,482,493,527-528, 535-536,608,708,748,750
Core effects,
vibronic and spin-vibronic coupling, 7,18-22,397, 740-748,754
Core hole, 72
Correlation algorithms,
Rydberg term values and ionization energies, 8,22-28,62-64
Crystal-field mixing, 75, 519-520,558

Davydov splitting, 518-520
Deep-level spectroscopy, 43
De-excitation processes, 38, 45,67,77,88,91,164-166, 297ff,377,409,428,441, 515-517,564,566,855ff
dissociative, 165
ionic, 321,323
nondissociative, 165
nonionic, 305-309
nonradiative, 45,67,301, 321-336,441,515-517,564, 566,855ff,895-897, see also, Internal conversion and Intersystem crossing
radiative, 45,67,164-165, 329,428,564,566,857,859, 869,895-897
rigid gas-to-solid shift, 77
superexcited state, 297-321
vibrational energy effects, 859ff
Diamagnetism, 29,32,221, 235-236,631ff,707-713, 722-727
inter-ℓ mixing regime, 654ff
inter-n mixing regime, 654, 671,679-682,695
Landau regime, 654ff
strong-field mixing regime, 654ff
Dielectic polarization, 76
Dissociative attachment, 164, 166,168,175-177,180,258

Dissociative ionization, 302, 318-321
Dissociative recombination, 253,257
Dopant molecules, 36-37

Einstein-Brillouin-Keller (EBK) quantization, 687,689
Effective mass approximation, 525
Electric field effects, 8,27, 32-33,119,145,221,235-236, 732-733,756-760,783ff,851
semi-empirical model, 756-759
Electric linear dichroism (ELD), 33,729,732,756, 783-788,792,795
Electrochromism, 729,732,756, 759,783-787,792,794, 804-805,810,831
Electron affinity, adiabatic, 262-264,288
Electron at surface of liquid helium, 722,725-726
Electron attachment, 264ff
unimolecular, 265-270
Electron beam experiments, 288
Electron donor acceptor (EDA) complex, 856
Electron-electron scattering, 74
Electron-hole recombination, 515,551
Electron-impact excitation, 163ff,289,300,343,475
Electron-impact-induced light emission (EIILE) spectroscopy, 163ff
Electron-phonon coupling, 67, 559,566
Electron-phonon scattering, 74
Electronic relaxation processes, see De-excitation processes
Electron transmission spectroscopy, 173-174
Energy-transfer processes in solids, 67,75-79,91
Ergodic motion, 579ff
Excimer, 106-109,467,500
Exciplex, 106-107,934
Excited-state dynamics in rare-gas solids, 551
Exciton, 43,67,81,515ff,631, 686
and impurity states in rare-gas fluids, 540-545
bulk, 531-532,535
electronic-vibrational, 519-523
energy transfer in rare-gas solids, 567-575
free, 567-570
Frenkel, 67,515-524,527, 533,540
in rare-gas solids, 525-539
intermediate, 524
longitudinal, 519
-phonon states, 516
progression, 67
self-trapping in rare-gas solids, 552-558,573
transport, 567
transverse, 519
transverse-longitudinal splitting, 533,535
Wannier, 36,515-516,524ff
Exciton-phonon coupling, 67
Exciton-vibron coupling, 518-520
Extended x-ray absorption fine-structure (XAFS), 352

Fast electron differential scattering, 301
Fermi energy, 74
Fermi golden rule expression, 329
Field-induced mixing, 631ff, 708ff,
crossed electric and magnetic fields, 708
ℓ-mixing, 708,710,713,718
m-mixing, 708,710-713
n-mixing, 710,713-715,718
motional Stark, 713
parity, 707-708

Field ionization, 1,32,38
Fine-structure splitting, 196, 289,
Fine-structure threshold, 482-486,510
Flowing afterglow technique, 263
Fluorescence excitation spectra, 315
Fock hypersphere, 666,684
Fock representation, 646-649
Forward-scattering electron energy loss spectrum, 302
Franck-Condon effects, 45,107, 134,139,166-167,169,265, 286,288,318,330,360-367, 370,372-374,389,394,411, 424,428,438-439,452, 454-455,470,475-476,487, 496-497,500,504,506,518, 742-755,859-861,863,867, 876,878-881
Frenkel excitons, 67,515-524, 527,533,540
Frequency conversion, 133-134, 147
Frequency mixing,
difference, 110,112,116
sum, 110,112,115-118,126

Gauge, 643,645-646,651-653
Landau, 645
symmetric, 645-646, 651-652
Glauber operator, 649

Hartree-Fock wavefunctions,
multiconfigurational (MCSCF), 882-884,889
restricted (RHF), 882-884
Herzberg-Teller expansion, 18
Homoclinic point, 597-598,602
Host-guest interaction, 36
Hyperconjugative interaction, 899
Hyperfine structure, 713

Impurity states in solids, 534-550
electronic relaxation of, 564-566
lattice relaxation of, 558-564
Insulators, 515ff
large gap, 65
molecule-doped, 36
van der Waals, 76
Interband transitions in condensed systems, 515-516
Intermediate compound state, 298
Intermediate coupling, 17-20, 212
Internal conversion, 856ff
Intersystem crossing, 856ff
Intramolecular vibrational relaxation, 856,867-868
Intravalence state, 411,443
Ion cyclotron resonance (ICR) mass spectrometry, 171,173, 175,178,268,286
Ion fluorescence decay, 324-326,332-334
Ion pair production, 261, 279-280,306-307
Ion yield spectroscopy, 314

jj-coupling, 196,202,211-212, 223
Jahn-Teller effects, 740,744, 746,748,755,876-878,885, 887-891,904
Jost function, 193
Jost matrix, 200-201

K-entropy, 602
K-impact, 176-177,179
K-matrix, 212,220,223,235, 246-248,255,257,354-355
K-system, 602,604,606,623
KAM (Kolmogorov-Arnold-Moser) theorem, 579,587-591,593, 595,626-627
Kasha's rule, 856,895

Kerr effect,
electro-optical, 732
quadratic, 119
Koopmans theorem, 267,480,831, 884

(Λ,S)-coupling, 17,239,415,445
LS-coupling, 197-198,202,211, 223,233
LS→jj recoupling transformation, 198-199,212
ℓ-mixing, 236,356
Landau resonance region, 638-649,652-653
Landé g factor, 221
Larmor frequency, 656
Laser, 105ff,133ff
excimer-pumped dye, 105ff, 133ff,416
free-electron, 42
negative-ion gas phase, 269,279
VUV, 106ff
XUV, 105,119-121
Laser-induced fluorescence (LIF) detection, 121-127, 411,424-425,457-459
Lattice relaxation, 555,558-564
Levi-Civita regularization procedure, 665
Liouville measure, 599
Liouville theorem, 601
Lorentz force, 655
Lu-Fano plot, 14-16, 210,213,218, 224,226-234,236-241
Luminescence spectroscopy, 42-44,67-72,91,559
Lyapunov exponent, 601

Magnetic circular dichroism (MCD), 29-31,729,731-732, 736-756,765-768,778-780
absorption-type lineshape (*B* term), 767-768,779-780
derivative-type lineshape (*A* term), 767-768,778-779
Magnetic circular dichroism,
perturbation theory model, 736-748
spectral interpretation, 748-756
state mixing, 747-748
vibronic coupling, 740-748, 754-755
Magnetic field effects, 8, 27-32,631ff,707ff,729, 731-732,736-756,765ff
linear, 710
quadratic, 710-713
Magnetic resonance techniques, 275
Maslov index, 609-611,687
Maslov quantization conditions, 609,611
Mass spectrometry, 43-45, 134-141,153,171ff,266ff, 317,321,324,410ff,467ff, 845,848,851
conventional, 175,848,851
ion cyclotron resonance, 171,173,175,178,268,286
multiphoton ionization, 134-135,137,845,848,851
Penning source, 268
photoionization, 317,321, 467,470-473,493,511
tandem, 269,274-275
time-of-flight, 135-136, 141,266,271,324,410, 416-419,422,458
using synchrotron radiation, 43-45
Matrix isolation spectroscopy, 45,71
Microcanonical distribution, 597-598
Mixing system, 598-599,601, 603,606,617
Molecular beams, 691-694,855
effusive, 371,495-495
supersonic, 286-288,290, 371,410-411,416,457, 473-477,487,493-496
Molecular clusters, 91, 134-135,279-280,286, 474ff,550

Molecular crystals, 41,44,58, 67,73,75,79
Molecular thin films, 82,89, 180-183
Moller-Plesser perturbation theory, 882-884,889-891
Moser twist theorem, 594
Muffin-tin approximation, 826
Multichannel quantum defect theory (MQDT), 8,10-18,209ff, 217-218,222,234ff,245ff, 301,318,342,382-390, 395-397,467,511,702
Multiphoton ionization (MPI), 106,119,133ff,311,396, 413,459-462,778,841ff
  and third-harmonic generation, 145ff
  nonresonant, 136,143-144, 148-152
  -photoelectron spectroscopy (MPI-PES), 134-145
  resonantly-enhanced (REMPI), 134-145,149
  sensitivity and specificity, 843

Negative-ion states, 164ff, 195-197,202,211,261ff
  decay of, 164,166
  dipole-bound, 285
  dissociation limit of, 166
  energy, 164-166,175-179
  lifetimes of, 275
  long-lived, 274-279
  metastable, 276
  multiply-charged, 270-271
  of molecular clusters, 286
  parent, 166-168,285
  photoabsorption by, 271-274
  production of, 279-281
  vibrational energy, 174
Ninety-degree polarized electrochromism (NPE), 785,804
Nonadiabatic effects, 250,253, 255,257,366,377,383
Nonlinear optical phenomena, 105ff
Nonlinear photochemistry, 843ff
Nonlinear spectroscopy, 133ff
Nonradiative decay processes, see De-excitation processes

(Ω,ω)-coupling, 17,445
Orbital types,
  intermediate, 821-823
  pre-Rydberg, 823
  Rydberg, 820-826,831,836
  valence-shell, 820-821,826, 831

Paramagnetism, 632,652-656, 700-701,708,710
Partial wave formalism, 194
Penning-type ionization, 567
Period-doubling bifurcations, 603-604
Phase amplitude method, 12
Phase shift, 192,203,208, 211-212,214,220,225,227, 249,251,256,349
Phonon broadening, 544,548
Photodesorption, 44-45
Photodetachment, 195-199,202, 211,261-262,272,276,278, 286
Photodissociation dynamics, 256-257,261,280-281,286, 409ff,550
Photoelectron emission, 279
Photoelectron spectroscopy, 41-45,62,72-90,134-145, 262,267,270,314ff,360, 362,392,397,483,876-881
  and multiphoton ionization, 134-145,149
  difference, 83,86
  energy distribution curve, 73,76-77,82,85-90
  using synchrotron radiation, 41-45,62,72-90
Photoemission spectroscopy, 42,44,67,72-90,536-539, 548,564-572
Photofragment spectroscopy, 410ff

Photoionization, 236,251-257, 282,297ff,341ff,467ff
asymmetry parameter, 364ff
branching ratios, 341ff
cross-section, 236,253, 297-311,318,320,341ff, 467ff
mass spectrometry, 317,321, 467,470-473,493,511
polarization, 369-370
resonant, 342,391
triply-differential measurements, 342,367-372,382, 386,388,390-397
Photon excitation sources, 310-313
Poincaré-Birkhoff theorem, 592, 595
Poincaré index, 596
Poincaré surface-of-section, see Surface-of-section
Polarized photofluorescence excitation spectroscopy, 431-438,446-447
Positive-ion states, 166,172, 178
Positronium, 652,727
Potential,
barrier, 725
centrifugal, 247
confining, 727
Coulomb, 202-208,223, 246-247
Coulomb + Stark, 235
diamagnetic, 236
double-well, 34-35,71,685,724
Henon-Heiles, 604,606
image, 725
ion-atom Morse-Morse spline van der Waals, 497
ion-induced dipole, 492-493
long-range, 193,202-210
non-Coulombic, 217,222-223
one-dimensional well, 192-193,721,724-725
short-range, 193,206-208, 217,222-223
Stark, 235
spherically symmetric, 193-195,249
Predissociation, 14,136, 157-158,250,253,256,299, 306-307,315,320,323,333, 411,428,435,438-439,441, 443,447,455,459,462,471, 476,497,500
Pressure-effect spectroscopy, 729,734-736,760-762
Pseudopotential theory, 530-531

Quantum defect, 14,63,203, 210-211,213,217ff, 249-250,252-253,255,314, 382,531,685,695,708,710, 713
Quantum defect plot, see Lu-Fano plot
Quantum defect theory, 3, 8-18,191ff,217ff,245ff, 301,318,341-342,382-390, 395-397,467,511,530-533, 544,573,702
multichannel, 8,10-18, 209ff,217-218,222,234ff, 245ff,301,318,342, 382-390,395-397,467,511, 702
single-channel, 8-13, 202-208
Quasi-Landau resonance region, 710,718-719,722,724-725
Quasi-Landau spectrum, 722

Radiative attachment, 164,166, 262
Radiative decay processes, see De-excitation processes
Rare-gas dimers, 467ff
Rare-gas solids, 515ff
Reflectance spectroscopy, 43-45,57-67,535
Relaxation processes, see De-excitation processes
Renner-Teller effects, 415, 740-745,747-748

Resonance,
  autoionization, see autoionization
  Feshbach, 63,168-169,265,290
  Hill's (classical), 654, 663-664
  intermediate, 396
  Landau, 638ff
  predissociation, see predissociation
  quasi-Landau, 710,718-719, 722,724-725
  shape, 81,168,258,261,265, 267,270,272,342-377
  sub-Landau, 715-719
  window, 272
Resonance scattering technique, 262
Resonance state, 841ff
  absorption to, 841-843
  and nonlinear photochemistry, 843
  atomic, 844,846
  molecular, 846-847
  symmetry of, 843
Rotational-electronic coupling, 750-751
Rydberg state, 1ff,819ff
  atomic, 823,826
  definition of, 3-8,820
  Hamiltonian, 4-8
  molecular, 821ff
  series interaction 14
Rydberg/valence mixing, 8,38, 819ff
"Rydbergization" process, 820, 822
Runge-Lenz vector, 587,634-637, 653,666,668,670,677,689

S-matrix, 355
Scattering length, 191,194-198, 202
Semiclassical quantization, 608,612-616,631,633,646, 686-691,696
Shape resonance, 81,168,258, 261,265,267,270,272, 342-377
  discrete, 351
  non-Franck-Condon effects, 360-367,370,372-374
Sinai billiard, 599-600, 619-620,622
Single-channel quantum defect theory, 8-13,202-208
Solid-state continuum model, 555-556
Solid-state spectroscopy, 59, 64-67
Spin-orbit coupling, 8,16-20, 29,210,228,234,236,274, 415,447,468,482,493, 527-528,535-536,608,708, 748,750
Spin-spin coupling, 274
Spin-vibronic coupling, 7,18, 20-22
Stark effects, see Electric field effects
Stationary afterglow technique, 286
Stochastic transition, 601-607,623
Stochasticity criterion, 603
Strong-field mixing regime, 654ff,721,727
Superexcited state, 164,297ff
Surface-of-section (SOS), 591-595,602-603,660-663
Symmetry breaking effects, 542
Synchrotron radiation sources, 41ff,105,121,146,310-315, 342-343,366-368,374,422, 429,459,560
  undulators, 42-43,50-53,72, 92,311
  wigglers, 42-43,50-54,72, 92,311

Third harmonic generation, 106,110,112-117,133, 145ff,463
  and multiphoton ionization, 145ff

Time-of-flight mass spectrometry, 135-136,141,266,271, 324,410,416-419,422,458
Trapped electron spectroscopy, 164,171-173,182
Torus, 582ff
invariant, 582-587,590,602, 604-605,608,687,689
irrational, 614,617
KAM invariant, 595
nonresonant, 585
quantization of, 609,687
rational, 588
resonant, 585,590,592-595, 614

United atom approximation, 3, 351,820-826
Unpolarized light electrochromism (UPE), 785,792,794,797, 804-805

Valence band, 79,81,83-84,525, 529
Valence-band hole,
polarization energy, 76
screened, 76
Valence-shell state, 820ff
Valence-shell/Rydberg conjugates, 825-826
van der Waals clusters, see Molecular clusters
van der Waals dimers, see Rare-gas dimers
Vibronic coupling, 7,18,20,397, 740-748,754

Wannier excitons, 36,515-516, 524ff
Window resonance, 272

X-ray diffraction, soft, 42
X-ray emission spectroscopy, 72

Zeeman effects, see Magnetic field effect
Zero-degree polarized electrochromism (ZPE), 785,792, 794,797

GPSR Compliance

The European Union's (EU) General Product Safety Regulation (GPSR) is a set of rules that requires consumer products to be safe and our obligations to ensure this.

If you have any concerns about our products, you can contact us on ProductSafety@springernature.com

In case Publisher is established outside the EU, the EU authorized representative is:

Springer Nature Customer Service Center GmbH
Europaplatz 3
69115 Heidelberg, Germany

Batch number: 10607216

Printed by Printforce, the Netherlands